Handbook of Numerical Analysis
Volume 25

Numerical Analysis Meets Machine Learning

Handbook of Numerical Analysis
Volume 25

Numerical Analysis Meets Machine Learning

Edited by

Siddhartha Mishra
Department of Mathematics
ETH Zürich
Zürich, Switzerland

Alex Townsend
Department of Mathematics
Cornell University
Ithaca, NY, United States

ELSEVIER
N·H
North-Holland
An imprint of Elsevier

North-Holland is an imprint of Elsevier
Radarweg 29, PO Box 211, 1000 AE Amsterdam, Netherlands
125 London Wall, London EC2Y 5AS, United Kingdom

ISBN: 978-0-443-23984-7
ISSN: 1570-8659

For information on all North-Holland publications
visit our website at https://www.elsevier.com/books-and-journals

Publisher: Zoe Kruze
Acquisitions Editor: Sam Mahfoudh
Editorial Project Manager: Naiza Ermin Mendoza
Production Project Manager: James Selvam
Cover Designer: Gopalakrishnan Venkatraman

Typeset by VTeX

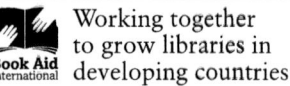

Contents

Contributors

Jonas Actor, Center for Computing Research, Sandia National Laboratories, Albuquerque, NM, United States

Ben Adcock, Simon Fraser University, Burnaby, BC, Canada

Jules Berman, Courant Institute of Mathematical Sciences, New York University, New York, NY, United States

Holger Boche, Technical University of Munich, Munich, Germany
Ruhr University Bochum, Bochum, Germany
Munich Center for Quantum Science and Technology, Munich, Germany
Munich Quantum Valley, Munich, Germany
BMBF Research Hub 6G-life, Munich, Germany

David M. Bortz, Department of Applied Mathematics, University of Colorado, Boulder, CO, United States

Nicolas Boullé, Department of Applied Mathematics and Theoretical Physics, University of Cambridge, Cambridge, United Kingdom

Simone Brugiapaglia, Concordia University, Montreal, QC, Canada

Matthew J. Colbrook, Department of Applied Mathematics and Theoretical Physics, University of Cambridge, Cambridge, United Kingdom

Nick Dexter, Florida State University, Tallahassee, FL, United States

Leon Gerard, Faculty of Mathematics, University of Vienna, Vienna, Austria

Philipp Grohs, Faculty of Mathematics, University of Vienna, Vienna, Austria
Johann Radon Institute for Computational and Applied Mathematics, Austrian Academy of Sciences, Linz, Austria

Shuai Jiang, Center for Computing Research, Sandia National Laboratories, Albuquerque, NM, United States

George Em Karniadakis, Division of Applied Mathematics, Brown University, Providence, RI, United States

Nikola B. Kovachki, NVIDIA, Santa Clara, CA, United States

Gitta Kutyniok, Ludwig-Maximilians-Universität München, Munich, Germany
Munich Center for Machine Learning, Munich, Germany
University of Tromsø, Tromsø, Norway
German Aerospace Center, Oberpfaffenhofen, Germany

Samuel Lanthaler, Computing and Mathematical Sciences, California Institute of Technology, Pasadena, CA, United States

Yunseok Lee, Ludwig-Maximilians-Universität München, Munich, Germany

Tao Luo, School of Mathematical Sciences, Institute of Natural Sciences, MOE-LSC, and Qing Yuan Research Institute, Shanghai Jiao Tong University, Shanghai, PR China

Daniel A. Messenger, Department of Applied Mathematics, University of Colorado, Boulder, CO, United States

Sebastian Moraga, Simon Fraser University, Burnaby, BC, Canada

Benjamin Peherstorfer, Courant Institute of Mathematical Sciences, New York University, New York, NY, United States

Scott Roberts, Engineering Sciences Center, Sandia National Laboratories, Albuquerque, NM, United States

Michael Scherbela, Faculty of Mathematics, University of Vienna, Vienna, Austria

Paul Schwerdtner, Courant Institute of Mathematical Sciences, New York University, New York, NY, United States

Yeonjong Shin, Department of Mathematics, North Carolina State University, Raleigh, NC, United States

Andrew M. Stuart, Computing and Mathematical Sciences, California Institute of Technology, Pasadena, CA, United States

Alex Townsend, Department of Mathematics, Cornell University, Ithaca, NY, United States

April Tran, Department of Applied Mathematics, University of Colorado, Boulder, CO, United States

Nathaniel Trask, Department of Mechanical Engineering and Applied Mechanics, University of Pennsylvania, Philadelphia, PA, United States

Haizhao Yang, Department of Mathematics and Department of Computer Science, University of Maryland, College Park, MD, United States

Zhongqiang Zhang, Department of Mathematical Sciences, Worcester Polytechnic Institute, Worcester, MA, United States

Preface

This volume of the Handbook is dedicated to the connections between numerical analysis and machine learning, where the rigor of computational algorithms meets the predictive power of data-driven techniques. Numerical analysis has long served as the backbone of computational methodologies, addressing problems that span science, engineering, and mathematics. It is only natural that the well-established tools from computational mathematics and partial differential equations (PDEs) have much to offer in the rapidly evolving domain of machine learning. To realize the full potential of data-driven techniques and their limitations, we must deeply understand their theoretical foundations. In this Handbook, we recognize important experts in this area and survey some influential contributions.

The past several years have seen a remarkable increase in the number of computational mathematicians working on the theoretical underpinning of machine learning. There are tools from high-dimensional approximation theory that help to explain the approximation power of neural networks. Monte Carlo simulations have become one of the go-to methods for probabilistic learning and inference in machine learning models. The wavelet and Fourier transform are essential in feature extraction, time series analysis, signal processing, and image recognition. Optimization is fundamental to training neural networks with all their complications with large, high-dimensional, non-convex loss functions. Optimal transport is now a popular computational framework used to compare and align probability distributions in data assimilation.

There is more than a symbiotic relationship between the mathematical foundations laid by numerical analysis and the applications being explored by machine learning. For example, in PDE learning, one often finds that a state-of-the-art machine learning technique looks eerily close to a hyper-adaptive variant of a well-known numerical PDE solver. Some computational mathematicians worry that the problem-agnostic approach in machine learning threatens their livelihoods. However, machine learning is challenging our understanding and opens up many new research directions. In applied mathematics, discerning which problem features necessitate direct exploitation by an efficient solver is crucial, as is identifying those that are red herrings. Machine learning shows us that specific details, previously deemed essential, are distractions. The onus is now on numerical analysts to explain why efficient solvers can be constructed without these superfluous details.

The machine-learning hype will probably be over in a few decades. The long-lasting legacy of machine learning on applied mathematics will likely be the technical contributions drawn from numerical analysts. With this in mind, we hope the insights presented in this Handbook will remain relevant and valuable for many years.

The chapters in this Handbook serve as both a primer and an in-depth treatise on the indispensable connections between numerical analysis and machine learning. To whet your appetite, we provide a brief description of the contents of each chapter:

Chapter 1. This chapter addresses the significant strides made in learning approximations to smooth target functions from finite pointwise samples, which are pivotal in computational science and engineering. Highlighting advancements from sparse polynomial approximation to deep neural networks, The authors explain the challenges and breakthroughs in learning infinite-dimensional, Banach-valued holomorphic functions to establish a foundation for neural network approximation theory.

Chapter 2. In data-driven modeling, weak form-based methodologies have recently gained traction in scientific machine learning. These methods stand out for their computational efficiency and robustness to noise, and they are applicable in domains like equation discovery and reduced order modeling. The chapter surveys how to appropriately select test functions and autotune hyperparameters with benchmarks from ordinary, partial, and stochastic differential equations.

Chapter 3. Operator learning seeks to unravel the characteristics of dynamical systems or PDEs through data. This chapter provides a comprehensive guide and motivates many of the neural network architectures and numerical strategies currently used in operator learning. The authors seek to provide practical advice on the management of training data and the intricacies of training neural operators.

Chapter 4. Dynamic Mode Decomposition is a powerful tool for deconstructing nonlinear systems into discernible modes. The chapter emphasizes the connections between DMD and Koopman operators, facilitating a linear perspective on nonlinear dynamics and enhancing the understanding of spectral computations within DMD methodologies. The chapter offers an excellent survey of a vast, popular, and ever-expanding literature.

Chapter 5. Chapter 5 discusses the numerical solutions of the electronic Schrödinger equation, underscoring its significance in predicting molecular properties. The chapter covers the evolution of numerical schemes, the advent of deep learning techniques, and the implications for drug and material discovery, offering a mathematician's perspective on overcoming the equation's many challenges.

Chapter 6. Physics-informed neural networks (PINNs) are discussed in Chapter 6 with their application in solving forward and inverse problems. The chapter discusses the neural networks' role as trainable parameters and the unified error estimation approach, enhancing the understanding of PINNs in computational science. Additionally, it examines the integration of physical laws into machine learning, demonstrating how PINNs can bridge the gap between theoretical physics and practical computation.

Chapter 7. Chapter 7 spotlights the optimization challenges inherent in AI and Data Science, focusing on computational constraints and their implications for machine learning, operational research, and logistics. The chapter critically analyzes computability and optimization theory, highlighting the nuanced issues faced in digital computation.

Chapter 8. Neural Galerkin schemes are an exciting mix of classical numerical methods and modern neural network architectures, offering a dynamic approach to solving time-dependent PDEs. In Neural Galerkin schemes, the traditional Galerkin method's basis functions are replaced with neural network approximations, allowing for flexible and adaptive solution spaces. The chapter discusses the implementation of these schemes, highlighting the integration of neural networks in approximating solution manifolds and using variational principles to guide training. The advantages of Neural Galerkin methods, such as their ability to handle complex geometries and non-linearities inherent in many PDEs, are examined alongside potential challenges and limitations. The chapter showcases the efficacy and versatility of Neural Galerkin schemes in capturing the nuanced behaviors of dynamical systems.

Chapter 9. In operator learning, a significant challenge is efficiently approximating complex mappings between function spaces, often associated with nonlinear PDEs. The chapter explains how to use deep learning to approximate these operators, thereby enabling efficient computation and analysis of systems governed by PDEs. There is a balance between presenting the theoretical underpinnings of neural operator models and discussing their practical design, training, and application in solving high-dimensional problems. Special attention is given to how numerical analysis enhances the accuracy and computational feasibility of these neural operators, focusing on their application in fluid dynamics, material science, and other engineering disciplines.

Chapter 10. This chapter introduces a novel approach combining domain decomposition with machine learning to solve PDEs when the underlying physical laws are unknown or complex. It surveys how data-driven models can be integrated into a domain decomposition framework to solve large-scale problems. The method's efficacy in handling multi-scale and multi-physics problems offers insights into its potential to revolutionize computational science, particularly in areas where traditional modeling and simulation techniques fall short.

Chapter 11. Concluding the Handbook, the final chapter delves into the least-squares formulation for solving PDEs using neural networks. It outlines how this approach leverages the optimization capabilities of machine learning to identify solutions to PDEs. The chapter also discusses the convergence, stability, and error analysis of the least-squares method, bridging the gap between numerical PDE-solving techniques and machine learning-driven optimization, and underscores the potential of machine learning and numerical analysis to advance computational methodologies in various scientific disciplines.

Siddharta Mishra and Alex Townsend

Chapter 1

Learning smooth functions in high dimensions

From sparse polynomials to deep neural networks

Ben Adcock[a,*], Simone Brugiapaglia[b], Nick Dexter[c], and Sebastian Moraga[a]

[a]*Simon Fraser University, Burnaby, BC, Canada,* [b]*Concordia University, Montreal, QC, Canada,*
[c]*Florida State University, Tallahassee, FL, United States*
**Corresponding author: e-mail address: ben_adcock@sfu.ca*

Contents

Handbook of Numerical Analysis, Volume 25, ISSN 1570-8659, https://doi.org/10.1016/bs.hna.2024.05.001

Abstract

Learning approximations to smooth target functions of many variables from finite collections of pointwise samples is an important task in scientific computing and its many applications in computational science and engineering. Despite well over half a century of research on high-dimensional approximation, this remains a challenging problem. Yet, significant advances have been made in the last decade towards efficient methods for doing this, commencing with so-called *sparse polynomial approximation* methods and continuing most recently with methods based on *Deep Neural Networks (DNNs)* and *Deep Learning (DL)*. In tandem, there have been substantial advances in the relevant approximation theory and analysis of these techniques. In this work, we survey this recent progress. We describe the contemporary motivations for this problem, which stem from parametric models and computational uncertainty quantification; the relevant function classes, namely, classes of infinite-dimensional, Banach-valued, holomorphic functions; fundamental limits of learnability from finite data for these classes; and finally, sparse polynomial and DNN methods for efficiently learning such functions from finite data. In the case of the latter, there is currently a significant gap between the approximation theory of DNNs and the practical performance of DL. Aiming to narrow this gap, we develop the topic of *practical existence theory*, which asserts the existence of dimension-independent DNN architectures and DL training strategies that achieve provably near-optimal generalization errors in terms of the amount of training data.

Keywords

High-dimensional approximation, Holomorphy, Scarce data, Polynomial approximation, Deep neural networks, Deep learning, Practical existence theory

MSC Codes

65D40, 41A10, 41A63, 41A25, 68T07

1 Introduction

This article reviews the problem of learning infinite-dimensional functions from finite data. In this section, with begin by describing motivations for this problem and its challenges (§1.1). We then present an overview of the remainder of the article (§1.2). Finally, we discuss relevant literature (§1.3).

1.1 Motivations and challenges

Approximating functions of many variables is a classical topic in approximation theory and numerical analysis, which has been studied intensively for more than seventy years. The contemporary resurgence in this problem stems from applications to parametric models, parametric *Differential Equations (DEs)* and computational *Uncertainty Quantification (UQ)*. In such models, the target function represents some quantity-of-interest of a given physical system and its variables the parameters of the system. Complex physical models involve many parameters, which naturally leads to functions of many variables. It is also increasingly common to consider countably infinite parameterizations, a standard example being the use of a Karhunen–Loève expansion to represent a random field in, for example, parametric models of porous media (see, e.g., Le Maître and Knio, 2010, §2.1, or Sullivan, 2015, §11.1). In this case, the target function depends on countably many variables.

Therefore, the first challenge in approximating such functions is that the dimension is high or countably infinite. The second is that the function is usually expensive to evaluate. Typically, each evaluation involves either an expensive numerical simulation of the DE model or a costly physical experiment. Hence, one strives to learn an accurate approximation from a limited set of samples. This approximation – or *surrogate model* as it is often termed (see Smith, 2013, Chpt. 13, or Sullivan, 2015, Chpt. 13) – can then be used in place of the true function for various downstream tasks, such as parameter optimization, inverse parametric problems or uncertainty quantification. Nonetheless, since the target function is expensive to evaluate, the amount of data available to learn this approximation is usually highly limited. Put another way, the learning problem is highly *data-starved.*

A final challenge is that the target function is often not scalar-valued. In the case of parametric DEs, the output of the target function is the solution of the DE with the given input parameters. Hence this function takes values in an infinite-dimensional Banach or Hilbert space. This adds complications, both theoretically and practically, stemming from the need to discretize this output space in order to perform computations.

1.2 Overview

We now give a detailed overview of this article.

Problem statement and notation (§2)

We consider an unknown target function f depending on countably many variables that range between finite maximum and minimum values and take values in a Banach space $(\mathcal{V}, \|\cdot\|_{\mathcal{V}})$. After normalizing, we may assume that

$$f : \mathcal{U} \to \mathcal{V}, \quad \text{where } \mathcal{U} = [-1, 1]^{\mathbb{N}}.$$

We equip $\mathcal{U}$ with the uniform probability measure ϱ and consider training data

$$\{(\boldsymbol{y}_i, f(\boldsymbol{y}_i) + e_i)\}_{i=1}^m \subseteq \mathcal{U} \times \mathcal{V}, \tag{1.1}$$

where $\boldsymbol{y}_1, \ldots, \boldsymbol{y}_m \sim_{\text{i.i.d.}} \varrho$ and $e_i \in \mathcal{V}$ represents measurement error. The objective is to learn an approximation $\hat{f}$ to f from the data (2.2).

Holomorphic functions of infinitely many variables (§3)

An important property of many parametric model problems is that the parametric map f is a smooth function of its variables. There is a large body of literature, which we review further in §3, that establishes that solution maps for various parametric DEs are *holomorphic* functions – in other words, they admit holomorphic extensions to certain complex regions $\mathcal{U} \subset \mathcal{O} \subseteq \mathbb{C}^{\mathbb{N}}$.

In one dimension, it is common to consider complex regions $[-1, 1] \subset \mathcal{O} \subset \mathbb{C}$ defined by *Bernstein ellipses* (see, e.g., Trefethen, 2013). The convergence of the s-term polynomial expansion is then *exponential* in s, with the rate being characterized by the largest Bernstein ellipse to which the function can be extended (see, e.g., Trefethen, 2013, Chpt. 8). For functions of $d \geq 1$ variables, it is natural to look at regions $[-1, 1]^d \subset \mathcal{O} \subset \mathbb{C}^d$ defined by *Bernstein polyellipses*, i.e., tensor-products of one-dimensional Bernstein ellipses. One can then show exponential convergence in $s^{1/d}$ of a suitable s-term polynomial expansion, with rate depending once more on the largest polyellipse to which f admits an extension (see, e.g., Adcock et al., 2022, §3.5-3.6).

$(\boldsymbol{b}, \varepsilon)$-holomorphic functions. The situation changes in infinite dimensions. Holomorphy in an arbitrary Bernstein polyellipse is no longer sufficient to guarantee convergence of a (in fact, any) s-term polynomial expansion. At the very least, one needs *anisotropy*, namely, increasing regularity with the variables y_i as $i \to \infty$. Moreover, for parametric DE applications, one needs to consider not just a single Bernstein ellipse, but complex regions $\mathcal{R}(\boldsymbol{b}, \varepsilon) \subset \mathbb{C}^{\mathbb{N}}$ defined by certain *unions* of Bernstein polyellipses. Here $\boldsymbol{b} = (b_j)_{j \in \mathbb{N}} \in [0, \infty)^{\mathbb{N}}$ with $\boldsymbol{b} \in \ell^1(\mathbb{N})$ is a sequence and $\varepsilon > 0$ is a scalar that parametrize the corresponding region. In Definition 3.3 we formalize the corresponding class of so-called *$(\boldsymbol{b}, \varepsilon)$-holomorphic functions* (see, e.g., Chkifa et al., 2015b; Schwab and Zech, 2019). This class was first introduced in the context of parametric DEs (Cohen and DeVore, 2015), and has since become a standard setting in which to consider the approximation of holomorphic functions in infinite dimensions (Adcock et al., 2022, Chpt. 3). Throughout the majority of this work, we set $\varepsilon = 1$; a property which can always be guaranteed by rescaling $\boldsymbol{b}$. For convenience, we also define the set

$$\mathcal{H}(\boldsymbol{b}) = \left\{ f : \mathcal{U} \to \mathcal{V} \ (\boldsymbol{b}, 1)\text{-holomorphic} : \sup_{\boldsymbol{z} \in \mathcal{R}(\boldsymbol{b}, 1)} \|f(\boldsymbol{z})\|_{\mathcal{V}} \leq 1 \right\}. \tag{1.2}$$

This is the main class of functions considered in the remainder of this work. In §3 we explain the relevance of this class to parametric DEs in more detail.

Unknown anisotropy. Having done so, we then explain how the sequence $\boldsymbol{b}$ controls the *anisotropy* of $f \in \mathcal{H}(\boldsymbol{b})$. Larger b_i implies that f is less smooth with respect to its ith variable y_i, while smaller b_i implies more smoothness. In some specific situations, one may have knowledge of a suitable $\boldsymbol{b}$ – information which could then be used when designing a learning algorithm. However, in practice, it is more often the case that $\boldsymbol{b}$ is unknown. The focus of this work is the more realistic *unknown anisotropy* setting. In short, the learning method must be independent of $\boldsymbol{b}$, with the assumption $f \in \mathcal{H}(\boldsymbol{b})$ being used only to derive bounds for the resulting generalization error.

Orthogonal polynomials and best s-term polynomial approximation (§4)

As previously observed, holomorphy is intimately related to polynomial approximation. In §4 we introduce multivariate orthogonal polynomials, orthogonal polynomial expansions of infinite-dimensional, Banach-valued functions and the concept of *best s-term polynomial approximation.* As we argue, best s-term polynomial approximation is an important theoretical benchmark against which to compare methods for learning holomorphic functions from data.

A signature result we recap in this section is the following.

Informal Result 1 (Algebraic convergence of the best s-term approximation, Theorem 4.1). *The best s-term polynomial approximation of $f \in \mathcal{H}(\boldsymbol{b})$ converges algebraically fast in s. Specifically, if $\boldsymbol{b} \in \ell^p(\mathbb{N})$ for some $0 < p < 1$, then the L^2_ϱ-norm error decreases like $\mathcal{O}(s^{1/2-1/p})$ as $s \to \infty$.*

This result implies that best s-term polynomial approximation in $\mathcal{H}(\boldsymbol{b})$ is free from the *curse of dimensionality*, whereas the aforementioned exponential rates witnessed in $d < \infty$ dimensions (which are exponential in $s^{1/d}$) quickly succumb to this curse. Although the focus in this work is infinite-dimensional functions, in §4 we demonstrate that such algebraic rates also typically describe the true convergence behavior seen for functions of finitely many variables, whenever d is large enough (e.g., $d \geq 16$).

Limits of learnability from data (§5)

We next turn our attention to learning holomorphic functions from the data (1.1). In §5, we study lower bounds using concepts from *information-based complexity*, in particular, *adaptive m-widths* (Definition 5.1). In order to consider the unknown anisotropy setting, we introduce the function classes

$$\mathcal{H}(p) = \bigcup \left\{ \mathcal{H}(\boldsymbol{b}) : \boldsymbol{b} \in \ell^p(\mathbb{N}),\ \boldsymbol{b} \in [0,\infty)^{\mathbb{N}},\ \|\boldsymbol{b}\|_p \leq 1 \right\}, \quad 0 < p < 1,$$

where $\mathcal{H}(\boldsymbol{b})$ is as in (1.2), and

$$\mathcal{H}(p,\mathsf{M}) = \bigcup \left\{ \mathcal{H}(\boldsymbol{b}) : \boldsymbol{b} \in \ell^p_{\mathsf{M}}(\mathbb{N}),\ \boldsymbol{b} \in [0,\infty)^{\mathbb{N}},\ \|\boldsymbol{b}\|_{p,\mathsf{M}} \leq 1 \right\}, \quad 0 < p < 1.$$

Here $\ell^p_{\mathsf{M}}(\mathbb{N})$ is the *monotone* ℓ^p-space (see §2). We then have the following.

Informal Result 2 (Limits of learnability, Theorem 5.2). *It is impossible to learn functions from $\mathcal{H}(p)$. Specifically, there does not exist a method for learning functions from m (adaptive) linear samples for which the L^2_ϱ-norm error decreases as $m \to \infty$ uniformly for functions in $\mathcal{H}(p)$. Moreover, when restricting to the space $\mathcal{H}(p, \mathsf{M})$, or even $\mathcal{H}(\boldsymbol{b})$, the error cannot decrease faster that $c \cdot m^{1/2-1/p}$ for some universal constant $c > 0$, even if the method is allowed to depend on the anisotropy parameter $\boldsymbol{b} \in \ell^p_{\mathsf{M}}(\mathbb{N})$.*

This theorem illustrates a fundamental gap between approximation theory and learning from data. The best s-term approximation converges like $s^{1/2-1/p}$ for any $f \in \mathcal{H}(p)$, yet no method can learn such functions from data. Note that this result holds not just for i.i.d. pointwise samples, as in (1.1), but *any* linear measurements, which may also be *adaptive*.

This gap is narrowed by restricting to the class $\mathcal{H}(p, \mathsf{M})$, wherein a function's variables are "on average" ordered in terms of importance. Notably, this lower bound also implies that there is no benefit to knowing $\boldsymbol{b}$, at least in terms of sample complexity. The same lower bound $c \cdot m^{1/2-1/p}$ holds, even if the method is allowed to depend on $\boldsymbol{b}$.

With this in mind, the remainder of this work is devoted to describing learning methods that achieve (close to) the rate $m^{1/2-1/p}$ uniformly for functions in $\mathcal{H}(p, \mathsf{M})$ for *any* $0 < p < 1$. Specifically, we focus on sparse polynomial and *Deep Neural Network (DNN)* methods.

Learning sparse polynomial approximations from data (§6)

Inspired by the benchmark results on best s-term approximation, we first consider methods that learn polynomial approximations. This approach is heavily based on techniques from *compressed sensing* (Foucart and Rauhut, 2013; Vidyasagar, 2019). Superficially, this seems straightforward. Fast decay of the best s-term polynomial approximation means that the polynomial coefficients are approximately sparse. Thus, a natural first idea would be to formulate an ℓ^1-minimization problem for the coefficients and leverage standard compressed sensing theory to derive recovery guarantees.

Unfortunately, this approach fails to deliver optimal rates. The specific reason is that there are s-term polynomials that require approximately $m = \mathcal{O}(s^2)$ i.i.d. samples to be stably recovered. The higher level reason is that the assumed *low-complexity* model (functions in $\mathcal{H}(p)$ have approximately sparse coefficients) is simply too crude. To lower the sample complexity requirement, we refine the model. We do this via *weighted sparsity* (Rauhut and Ward, 2016), which encodes the additional information that coefficients are both approximately sparse and *decaying*.

Informal Result 3 (Near-optimal learning via polynomials, Theorem 6.8). *There is a method (based on weighted ℓ^1-minimization) for learning functions* *from the data* (1.1) *that achieves an L^2_ϱ-norm error that decays like*

$$\mathcal{O}((m/\log^4(m))^{1/2-1/p}), \quad m \to \infty, \tag{1.3}$$

with high probability, for functions in $\mathcal{H}(p, \mathsf{M})$ and any $0 < p < 1$.

Comparing with the lower bound of $m^{1/2-1/p}$, we see that this procedure is optimal up to the logarithmic term. Note that in the full theorem, we also account for other errors in learning process: namely, *measurement error* (i.e., the effect of the terms e_i in (1.1)), *physical discretization error* (i.e., the effect of discretizing the space $\mathcal{V}$ in order to perform computations) and *optimization error* (i.e., inexact solution of the optimization problem).

DNN existence theory (§7)

As we discuss further in §1.3, the last five years have seen a growing interest in the application of *Deep Learning (DL)* to challenging parametric model problems. DL has the potential to achieve significant performance gains. Yet, it is poorly understood from a theoretical perspective, especially in terms of the sample complexity, i.e., the amount of training data need to learn good DNN approximations for specific classes of functions.

We commence §7 by briefly reviewing the approximation theory of DNNs, which has been an area of significant research in the last few years. Broadly, this area aims to establish *existence theorems*. Building on the classical *universal approximation theory* of shallow *Neural Networks (NNs)*, these results assert the existence of DNNs of a given complexity – in terms of the width and depth, number of nonzero weights and biases, or other pertinent metrics – that approximate functions in a given class to within a prescribed accuracy.

Such results are typically proved through *emulation*. One first shows that DNNs can emulate a standard approximation scheme (e.g., polynomials, piecewise polynomials, splines, wavelets, and so forth), either exactly or up to some error that can be made arbitrarily small. Then one leverages known bounds for the standard scheme to derive the corresponding existence theorem.

Having briefly reviewed this literature, we then describe an exemplar existence theorem for holomorphic functions, which is obtained by emulating a certain s-term polynomial approximation.

Informal Result 4 (Existence theorem for DNNs, Theorem 7.2). *Let $(\mathcal{V}, \|\cdot\|_{\mathcal{V}}) = (\mathbb{R}, |\cdot|)$. Then for any $\boldsymbol{b} \in [0,\infty)^{\mathbb{N}}$ with $\boldsymbol{b} \in \ell^p(\mathbb{N})$ for some $0 < p < 1$ and $s \in \mathbb{N}$, there is a family $\mathcal{N}$ of DNNs of width $\mathcal{O}(s^2)$ and depth $\mathcal{O}(\log(s))$ with the following property. For any $f \in \mathcal{H}(\boldsymbol{b})$ there is an element of $\mathcal{N}$ that achieves an L^2_ϱ-norm error of $\mathcal{O}(s^{1/2-1/p})$.*

Practical existence theory: near-optimal DL (§8)

While important, existence theorems such as this say little about the performance of DL, i.e., computing a DNN from training data by minimizing a loss function. As noted, analysis of this practical scenario is currently lacking, certainly in terms of sample complexity. It has also been well documented that there is a gap between existence theory and the practical performance of DNNs when trained from finite numbers of samples (Abedeljawad and Grohs, 2023; Adcock and Dexter, 2021; Grohs and Voigtlaender, 2023).

In this section, we describe a theoretical framework, termed *practical existence theory* (Adcock et al., 2021, 2023a; Franco and Brugiapaglia, 2024), that seeks to narrow this gap. Its aim is to show that there not only exists a family DNNs, but also a training strategy akin to standard practice – i.e., minimizing a loss function – that attains the near-optimal rates.

Informal Result 5 (Practical existence theorem for DNNs, Theorem 8.1). *There is a class of DNNs $\mathcal{N}$ (of width and depth depending on m), a regularized ℓ^2-loss function and a choice regularization parameter (depending on m only) such that any DNN obtained by solving the resulting training problem (minimizing the loss function over $\mathcal{N}$) achieves the error decay rate* (1.3).

Like existence theorems, this result is proved by emulating the sparse polynomial method with DNNs. The class $\mathcal{N}$ is 'handcrafted' in the sense that the weights and biases in the hidden layers are fixed, and chosen specifically to emulate certain polynomials. Only the weights in the final layer are trained.

Epilogue: the benefits of practical existence theory and the gap between theory and practice (§9)

Practical existence theory does not explain the performance of standard DL strategies based on *fully trained models* (i.e., those where all layers of a DNN are trained). Nonetheless, it provides a number of key insights and benefits, which we now list (see §9 for details).

1. It narrows the gap between theory and practice, by showing the existence of good training procedures, which are near-optimal for certain function classes.
2. It thereby emphasizes the potential to achieve even better performance in practice with suitably chosen architectures and training strategies.
3. It also stresses the sample complexity aspect of DL for high-dimensional approximation. As argued, this is vital to parametric modelling applications, where data scarcity is of primary concern.
4. The architectures in Theorem 8.1 are much wider than they are deep. Smoother activation functions also lead to narrower and shallow DNNs. Both insights agree with the empirical performance of fully trained models.
5. Once trained, the DNN in Theorem 8.1 can be sparsified via *pruning*. This lends credence to the idea that sparse DNNs can perform well in practice.
6. The methodology of practical existence theory is flexible, and can be applied to other problems, such as reduced order models based on deep autoencoders (Franco and Brugiapaglia, 2024) and Physics-Informed Neural Networks (PINNs) for PDEs (Brugiapaglia et al., 2024).

While narrowed, the gap between theory and practice still persists. With an eye towards future research, we end §9 with some further discussion. We discuss how current techniques for analyzing fully trained models may at best achieve a rate of $\mathcal{O}(m^{-1/2})$, regardless of p, which is strictly slower than those asserted by practical existence theorems. However, the far greater expressivity of DNNs – in particular, their ability to approximate continuous and discontinuous functions alike – sets them apart from more standard scientific computing tools such as polynomials. While the latter are near-optimal for holomorphic functions, the flexibility of former is a significant potential advantage.

1.3 Further literature

See Ghanem et al. (2017); Le Maître and Knio (2010); Smith (2013); Sullivan (2015) for general introductions to computational UQ and Adcock et al. (2022); Cohen and DeVore (2015) for more on parametric models and parametric DEs. Note that in the context of UQ, orthogonal polynomial expansions are often termed polynomial *chaos* expansions (see, e.g., Ghanem and Spanos, 2003). Best s-term polynomial approximation is a type of nonlinear approximation (DeVore, 1998). It was developed in a series of works in the context of parametric DEs (Beck et al., 2014, 2012; Bieri et al., 2010; Bonito et al., 2021; Chkifa et al., 2015b; Cohen et al., 2010, 2011; Hansen and Schwab, 2013a; Todor and Schwab, 2007; Tran et al., 2017), focusing on Taylor, Legendre or Chebyshev polynomial expansions in infinite dimensions. See, e.g., Cohen and DeVore (2015) and Adcock et al. (2022, Chpt. 3) for in-depth reviews. See §3.2 for a discussion holomorphic regularity of solutions of various parametric DEs, including the $(\boldsymbol{b}, \varepsilon)$-holomorphic class and relevant references. Our results on limits of learnability are from Adcock et al. (2024b) and, as noted, use ideas from information-based complexity. For overviews of this topic, see Novak (1988); Novak and Woźniakowski (2008, 2010); Traub et al. (1988).

In tandem with best s-term polynomial approximation theory, there has been a focus on learning polynomial approximations from data. Some early approaches in the context of parametric DEs included (adaptive) interpolation using sparse grids (Bäck et al., 2011; Babuška et al., 2007; Chkifa et al., 2014; Ganapathysubramanian and Zabaras, 2007; Gunzburger et al., 2014b; Ma and Zabaras, 2009; Mathelin et al., 2005; Nobile et al., 2008a,b; Xiu and Hesthaven, 2005). Another significant line of investigation has been on least-squares methods (Chkifa et al., 2015a; Cohen et al., 2013; Migliorati, 2013; Migliorati et al., 2014; Berveiller et al., 2006; Migliorati et al., 2013). See Guo et al. (2020); Hadigol and Doostan (2018) and Adcock et al. (2022, Chpt. 5) for reviews. While simpler than compressed sensing methods, these methods require *a priori* knowledge of a good multiindex set in which to construct the polynomial approximation, and are therefore best suited to the simpler *known anisotropy* setting. *Adaptive* least-squares methods (Migliorati, 2015, 2019) can address unknown anisotropy, but they currently lack theoretical guarantees. Compressed sensing for learning polynomial approximations began with the works of Blatman and Sudret (2011); Doostan and Owhadi (2011); Mathelin and Gallivan (2012); Rauhut and Ward (2012); Yan et al. (2012); Yang and Karniadakis (2013), and has seen much subsequent development. See Hampton and Doostan (2017) and Adcock et al. (2022, Chpt. 7) and references therein. The focus in this work on *weighted* sparsity originated in Rauhut and Ward (2016). See also Adcock (2017); Adcock et al. (2019); Peng et al. (2014) and references therein. The results described in this work are based on Adcock et al. (2024a).

Early works on the application of DL to parametric DEs include Adcock et al. (2021); Cyr et al. (2020); Dal Santo et al. (2020); Geist et al. (2021); Khoo et al. (2021); Laakmann and Petersen (2021). See also Becker et al. (2023); Cicci et al. (2022); Heiß et al. (2021, 2023); Khara et al. (2021); Lei et al. (2022); Scarabosio (2022) and references therein for more recent developments. Closely related to this topic is that of *operator learning* with DNNs – an area of significant current interest (Bhattacharya et al., 2021; Boullé and Townsend, 2023; Kovachki et al., 2023, 2024; Li et al., 2021; L. Lu et al., 2021). One can view holomorphic, Banach-valued functions as an example of holomorphic operators with specific parameterizations of the input space (Herrmann et al., 2022; Schwab et al., 2023). These operators constitute an important example in the field of operator learning for which there are strong guarantees on both parametric complexity (i.e., the number of DNN parameters required) and sample complexity. The relevant theory for these examples is based on the theory of holomorphic, Banach-valued functions described in this work. See Kovachki et al. (2024, §5.2) or Lanthaler (2023, §3.4) for further discussion.

See Cybenko (1989); Hornik et al. (1989) and Pinkus (1999) for the classical *universal approximation theory* of (shallow) NNs. The modern study of existence theory was initiated in Yarotsky (2017), and has since seen a wealth of developments. We review this literature in §7. The polynomial emulation results used in this work are based on Daws and C. Webster (2019); De Ryck et al. (2021); Opschoor et al. (2022).

The theory-to-practice gap in DL was studied empirically (Adcock and Dexter, 2021) and theoretically in Abedeljawad and Grohs (2023); Grohs and Voigtlaender (2023). Practical existence theorems were first established in Adcock et al. (2021); Adcock and Dexter (2021). Our results are based on Adcock et al. (2023a). This approach is not unique to high-dimensional regression problems. Similar ideas have been used to assert that it is possible to compute stable, accurate and efficient DNNs for inverse problems in imaging (Colbrook et al., 2022; Neyra-Nesterenko and Adcock, 2023). Here, one starts with a standard regularization problem, such as TV minimization, then exploits the fact that n steps of a first-order optimization method for solving this problem can be reinterpreted as a DNN with $O(n)$ layers – a process known as *unrolling* (Monga et al., 2021) (see also Adcock and Hansen, 2021, Chpts. 19-21). Interestingly, this also leads to principled ways to design DNN architectures for DL in inverse problems (Monga et al., 2021), although the trained networks may not be robust (Antun et al., 2023, 2020).

2 Problem statement and notation

In this section, we formalize the problem studied in this work and establish important notation.

Throughout this work, $\mathcal{U} = [-1, 1]^{\mathbb{N}}$ and $(\mathcal{V}, \|\cdot\|_{\mathcal{V}})$ is a Banach space over $\mathbb{R}$. We write $\boldsymbol{y} = (y_i)_{i \in \mathbb{N}}$ for the independent variable in $\mathcal{U}$. We equip $\mathcal{U}$ with the uniform probability measure ϱ and, for $1 \leq p \leq \infty$, write $L^p_\varrho(\mathcal{U}; \mathcal{V})$ for the Lebesgue–Bochner space of (equivalence classes of) strongly ϱ-measurable functions $f : \mathcal{U} \to \mathcal{V}$ for which $\|f\|_{L^p_\varrho(\mathcal{U};\mathcal{V})} < \infty$, where

$$\|f\|_{L^p_\varrho(\mathcal{U};\mathcal{V})} := \begin{cases} \left(\int_{\mathcal{U}} \|f(\boldsymbol{y})\|^p_{\mathcal{V}} \, \mathrm{d}\varrho(\boldsymbol{y})\right)^{1/p} & 1 \leq p < \infty, \\ \operatorname{ess\,sup}_{\boldsymbol{y} \in \mathcal{U}} \|f(\boldsymbol{y})\|_{\mathcal{V}} & p = \infty. \end{cases} \tag{2.1}$$

When $\mathcal{V} = (\mathbb{R}, |\cdot|)$, we just write $L^p_\varrho(\mathcal{U})$ for the corresponding Lebesgue space of real-valued functions.

Let $f \in L^2_\varrho(\mathcal{U}; \mathcal{V})$ be the unknown target function and consider sample points $\boldsymbol{y}_1, \ldots, \boldsymbol{y}_m \sim_{\text{i.i.d.}} \varrho$. Then we consider training data

$$\{(\boldsymbol{y}_i, f(\boldsymbol{y}_i) + e_i)\}_{i=1}^m \subseteq \mathcal{U} \times \mathcal{V}, \tag{2.2}$$

where $e_i \in \mathcal{V}$ represents measurement error. In this work, we assume the noise is adversarial and of small norm, i.e., $\sum_{i=1}^m \|e_i\|^2_{\mathcal{V}} \ll 1$. Statistical models can also be considered – see, e.g., Migliorati et al. (2015). The problem is then to learn an approximation to f based on the data (2.2).

We require some notation for sequences indexed via possibly multiindices. Let $d = \mathbb{N} \cup \{\infty\}$. We write $\boldsymbol{\nu} = (\nu_k)_{k=1}^d$ for an arbitrary multiindex in $\mathbb{N}_0^d$. Let $\Lambda \subseteq \mathbb{N}_0^d$ be a finite or countable set of multiindices. For $0 < p \leq \infty$, we write $\ell^p(\Lambda; \mathcal{V})$ for the space of all $\mathcal{V}$-valued sequences $\boldsymbol{c} = (c_{\boldsymbol{\nu}})_{\boldsymbol{\nu} \in \Lambda}$ for which $\|\boldsymbol{c}\|_{p;\mathcal{V}} < \infty$, where

$$\|\boldsymbol{c}\|_{p;\mathcal{V}} = \begin{cases} \left(\sum_{\boldsymbol{\nu} \in \Lambda} \|c_{\boldsymbol{\nu}}\|^p_{\mathcal{V}}\right)^{\frac{1}{p}} & 0 < p < \infty, \\ \sup_{\boldsymbol{\nu} \in \Lambda} \|c_{\boldsymbol{\nu}}\|_{\mathcal{V}} & p = \infty. \end{cases}$$

When $(\mathcal{V}, \|\cdot\|_{\mathcal{V}}) = (\mathbb{R}, |\cdot|)$, we just write $\ell^p(\Lambda)$ and $\|\cdot\|_p$. Given a sequence $\boldsymbol{c} = (c_{\boldsymbol{\nu}})_{\boldsymbol{\nu} \in \Lambda}$, we define its support

$$\operatorname{supp}(\boldsymbol{c}) = \{\boldsymbol{\nu} \in \Lambda : c_{\boldsymbol{\nu}} \neq 0\} \subseteq \Lambda.$$

Now let $\Lambda = \mathbb{N}$ and $0 < p \leq \infty$. Given a real-valued sequence $z = (z_i)_{i \in \mathbb{N}} \in \ell^\infty(\mathbb{N})$, we define its minimal monotone majorant as

$$\tilde{z} = (\tilde{z}_i)_{i \in \mathbb{N}}, \quad \text{where } \tilde{z}_i = \sup_{j \geq i} |z_j|, \ \forall i \in \mathbb{N} \tag{2.3}$$

and the *monotone ℓ^p-space* $\ell^p_{\mathsf{M}}(\mathbb{N})$ as

$$\ell^p_{\mathsf{M}}(\mathbb{N}) = \{z \in \ell^\infty(\mathbb{N}) : \|z\|_{p,\mathsf{M}} := \|\tilde{z}\|_p < \infty\}. \tag{2.4}$$

Let $d \in \mathbb{N} \cup \{\infty\}$. We write $\boldsymbol{e}_j$, $j = 1, \ldots, d$, for the canonical basis vectors in $\mathbb{R}^d$. Given multiindices $\boldsymbol{\nu} = (\nu_k)_{k=1}^d$ and $\boldsymbol{\mu} = (\mu_k)_{k=1}^d$, the inequality $\boldsymbol{\nu} \geq \boldsymbol{\mu}$ is interpreted componentwise, i.e., $\nu_k \geq \mu_k$, $\forall k$. We define $\boldsymbol{\nu} > \boldsymbol{\mu}$ analogously. We write $\mathbf{0}$ for the multiindex of zeros and $\mathbf{1}$ for the multiindex of ones.

Remark 2.1 (Other measures and domains). We work with the uniform probability measure on $\mathcal{U} = [-1, 1]^{\mathbb{N}}$ for convenience. This means that we construct polynomial approximations using the corresponding multivariate Legendre polynomials. It is possible to work more generally, for instance by considering ultraspherical measures and the resulting Jacobi polynomials. See Adcock et al. (2024b). The cases where $\mathcal{U} = \mathbb{R}^{\mathbb{N}}$ with the Gaussian measure (corresponding to Hermite polynomials) or $\mathcal{U} = [0, \infty)^{\mathbb{N}}$ with the gamma measure (corresponding to Laguerre) have also been studied, although the theory is less complete. Results on best s-term polynomial approximation are known in this case, as are DNN existence theorems (Dũng et al., 2023; Schwab and Zech, 2023). However, results akin to those we present in this work on learning sparse polynomial or DNN approximations from data are lacking.

Remark 2.2 (Error metric). Throughout this work, we measure the error of the various approximations in the $L^2_\varrho(\mathcal{U}; \mathcal{V})$-norm. The majority of the results we present extend to the stronger $L^\infty_\varrho(\mathcal{U}; \mathcal{V})$-norm. In particular, the various upper bounds that establish algebraic rates with index $1/2 - 1/p$ can be modified to show algebraic rates with index $1 - 1/p$ (see, e.g., Adcock et al., 2024a). For succinctness, we do not present this modification.

3 Holomorphic functions of infinitely many variables

In this section, we first formally define the classes of holomorphic functions considered in this work and then discuss their relevance to parametric DEs.

3.1 $(\boldsymbol{b}, \varepsilon)$-holomorphic functions

We commence with the definition of holomorphy.

Definition 3.1 (Holomorphy). Let $\mathcal{O} \subseteq \mathbb{C}^{\mathbb{N}}$ be an open set and $\mathcal{V}$ be a Banach space. A function $f : \mathcal{O} \to \mathcal{V}$ is *holomorphic in $\mathcal{O}$* if it is holomorphic with respect to each variable in $\mathcal{O}$. That is to say, for any $z \in \mathcal{O}$ and any $j \in \mathbb{N}$, the following limit exists in $\mathcal{V}$:

$$\lim_{\substack{h \in \mathbb{C} \\ h \to 0}} \frac{f(z + h\boldsymbol{e}_j) - f(z)}{h} \in \mathcal{V}.$$

Definition 3.2 (Holomorphic extension). Let $\mathcal{U} \subset \mathcal{O} \subseteq \mathbb{C}^{\mathbb{N}}$ be an open set. A function $f : \mathcal{U} \to \mathcal{V}$ has a *holomorphic extension to* $\mathcal{O}$ (or simply, *is holomorphic in* $\mathcal{O}$) if there is a $\tilde{f} : \mathcal{O} \to \mathcal{V}$ that is holomorphic in $\mathcal{O}$ for which $\tilde{f}|_{\mathcal{U}} = f$. In this case, we also define $\|f\|_{L^\infty(\mathcal{O};\mathcal{V})} := \|\tilde{f}\|_{L^\infty(\mathcal{O};\mathcal{V})} = \sup_{z \in \mathcal{O}} \|\tilde{f}(z)\|_{\mathcal{V}}$ or, when $\mathcal{V} = \mathbb{C}$, simply $\|f\|_{L^\infty(\mathcal{O})}$. If $\mathcal{O}$ is a closed set, then we say that f is holomorphic in $\mathcal{O}$ if it has a holomorphic extension to some open neighborhood of $\mathcal{O}$.

In this work, we consider functions that possess holomorphic extensions to (unions of) tensor-products of Bernstein ellipses. The *Bernstein ellipse* with parameter $\rho > 1$ is the set

$$\mathcal{E}(\rho) = \left\{ (z + z^{-1})/2 : z \in \mathbb{C},\ 1 \le |z| \le \rho \right\} \subset \mathbb{C}.$$

See, e.g., Trefethen (2013, Chpt. 8). This is an axis-aligned ellipse with foci at ± 1 and with major and minor semiaxis lengths given by $(\rho + 1/\rho)/2$ and $(\rho - 1/\rho)/2$, respectively. By convention, we let $\mathcal{E}(\rho) = [-1, 1]$ when $\rho = 1$. Next, we define the *Bernstein polyellipse* with parameter $\boldsymbol{\rho} = (\rho_i)_{i \in \mathbb{N}} > \boldsymbol{1}$ by

$$\mathcal{E}(\boldsymbol{\rho}) = \mathcal{E}(\rho_1) \times \mathcal{E}(\rho_2) \times \cdots \subset \mathbb{C}^{\mathbb{N}}.$$

Definition 3.3 (($\boldsymbol{b}, \varepsilon$)-holomorphic functions). Let $\boldsymbol{b} = (b_j)_{j \in \mathbb{N}} \in [0, \infty)^{\mathbb{N}}$ with $\boldsymbol{b} \in \ell^1(\mathbb{N})$ and $\varepsilon > 0$. A function $f : \mathcal{U} \to \mathcal{V}$ is $(\boldsymbol{b}, \varepsilon)$*-holomorphic* if it has a holomorphic extension to the region

$$\mathcal{R}(\boldsymbol{b}, \varepsilon) = \bigcup \left\{ \mathcal{E}(\boldsymbol{\rho}) : \boldsymbol{\rho} \ge \boldsymbol{1},\ \sum_{j=1}^{\infty} \left(\frac{\rho_j + \rho_j^{-1}}{2} - 1 \right) b_j \le \varepsilon \right\} \subset \mathbb{C}^{\mathbb{N}}. \quad (3.1)$$

See, e.g., Chkifa et al. (2015b); Schwab and Zech (2019). We discuss the motivations for this definition next. Beforehand, it is worth noting that the parameter ε is technically redundant, since we can always rescale $\boldsymbol{b}$. In the remainder of this work, we assume that $\varepsilon = 1$. For later use, we now introduce the class of $(\boldsymbol{b}, 1)$-holomorphic functions with norm at most one, i.e.,

$$\mathcal{H}(\boldsymbol{b}) = \left\{ f : \mathcal{U} \to \mathcal{V}\ (\boldsymbol{b}, 1)\text{-holomorphic} : \|f\|_{L^\infty(\mathcal{R}(\boldsymbol{b},1);\mathcal{V})} \le 1 \right\}.$$

Remark 3.4 (Functions of finitely many variables). In finite dimensions (in particular, $d = 1$), one normally considers functions that are holomorphic in a single Bernstein polyellipse. For such functions, it is well known that one can find polynomial approximations that converge exponentially fast. As we see later, in infinite dimensions we use the assumption of holomorphy in the region (3.1) in order to obtain algebraic rates of convergence.

It is, however, worth noting that finite-dimensional holomorphy in a Bernstein polyellipse implies $(\boldsymbol{b}, \varepsilon)$-holomorphy. Let $f : [-1, 1]^d \to \mathbb{R}$ be a function of finitely many variables that is holomorphic in $\mathcal{E}(\bar{\rho}_1) \times \cdots \times \mathcal{E}(\bar{\rho}_d) \subset \mathbb{C}^d$. We can extend f to a function $\bar{f} : \mathcal{U} \to \mathbb{C}$ in the standard way as $\bar{f}(y_1, y_2, \ldots) = f(y_1, \ldots, y_d)$, $\forall \boldsymbol{y} = (y_i)_{i \in \mathbb{N}} \in \mathcal{U}$. Now define

$$b_i = \varepsilon((\bar{\rho}_i + \bar{\rho}_i^{-1})/2 - 1)^{-1}, \ i \in [d], \qquad b_i = 0, \ i \in \mathbb{N} \backslash [d].$$

Then $\bar{f}$ is $(\boldsymbol{b}, \varepsilon)$-holomorphic. Therefore, all the results that follow on approximation of $(\boldsymbol{b}, \varepsilon)$-holomorphic functions also apply *mutatis mutandis* to functions of d variables that are holomorphic in a single Bernstein ellipse.

3.2 Holomorphy and parametric DEs

As mentioned, $(\boldsymbol{b}, \varepsilon)$-holomorphic functions were first developed in the context of parametric DEs. There is now a wealth of literature that establishes that solution maps of many parametric DEs posses this regularity.

The classical example of a parametric PDEs is the parametric stationary diffusion equation with parametrized diffusion coefficient

$$-\nabla_{\boldsymbol{x}} \cdot (a(\boldsymbol{x}, \boldsymbol{y}) \nabla_{\boldsymbol{x}} u(\boldsymbol{x}, \boldsymbol{y})) = F(\boldsymbol{x}), \ \boldsymbol{x} \in \Omega, \qquad u(\boldsymbol{x}, \boldsymbol{y}) = 0, \ \boldsymbol{x} \in \partial\Omega. \quad (3.2)$$

Here $\boldsymbol{x} \in \Omega$ is the spatial variable, $\Omega \subset \mathbb{R}^k$ is the domain, which is assumed to have Lipschitz boundary, and $\nabla_{\boldsymbol{x}}$ is the gradient with respect to $\boldsymbol{x}$. The function F is the forcing term, and is assumed to be nonparametric. We assume that (3.2) satisfies a *uniform ellipticity condition*

$$\operatorname*{ess\,inf}_{\boldsymbol{x} \in \Omega} a(\boldsymbol{x}, \boldsymbol{y}) \geq r, \quad \forall \boldsymbol{y} \in \mathcal{U}, \quad (3.3)$$

for some $r > 0$. Let $\mathcal{V} = H_0^1(\Omega)$ be the standard Sobolev space of functions with weak first-order derivatives in $L^2(\Omega)$ and traces vanishing on $\partial\Omega$. Then the problem (3.2) has a well-defined *parametric solution map*

$$u : \mathcal{U} \to \mathcal{V}, \ \boldsymbol{y} \mapsto u(\cdot, \boldsymbol{y}). \quad (3.4)$$

Now suppose that the diffusion coefficient has the affine parametrization

$$a(\boldsymbol{x}, \boldsymbol{y}) = a_0(\boldsymbol{x}) + \sum_{j=1}^{\infty} y_j \psi_j(\boldsymbol{x}), \quad (3.5)$$

where the functions $a_0, \psi_1, \psi_2, \ldots \in L^\infty(\Omega)$. Notice that uniform ellipticity (3.3) for this problem is equivalent to the condition

$$\sum_{j=1}^{\infty} |\psi_j(\boldsymbol{x})| \leq a_0(\boldsymbol{x}) - r, \quad \forall \boldsymbol{x} \in \Omega.$$

Under this assumption, one can show (see, e.g., Adcock et al., 2022, Prop. 4.9) that the solution map (3.4) is well defined and $(\boldsymbol{b}, \varepsilon)$-holomorphic for $0 < \varepsilon < r$ with

$$\boldsymbol{b} = (b_i)_{i=1}^{\infty}, \quad \text{with } b_i = \|\psi_i\|_{L^\infty(\Omega)}, \ \forall i \in \mathbb{N}. \tag{3.6}$$

This example was first considered in Bieri et al. (2010); Cohen et al. (2010, 2011). Subsequently, these results were generalized to many other classes of parametric DEs and PDEs. One such extension (see, e.g., Chkifa et al., 2015b) considers the problem (3.2) with various nonaffine parametric diffusion coefficients, such as

$$a(\boldsymbol{x}, \boldsymbol{y}) = a_0(\boldsymbol{x}) + \left(\sum_{j=1}^{\infty} y_j \psi_j(\boldsymbol{x}) \right)^2 \quad \text{or} \quad a(\boldsymbol{x}, \boldsymbol{y}) = \exp\left(\sum_{j=1}^{\infty} y_j \psi_j(\boldsymbol{x}) \right).$$

This is part of a general framework for showing holomorphy for parametric weak problems in Hilbert spaces. See Chkifa et al. (2015b, Thm. 4.1), Cohen and DeVore (2015, Cor. 2.4) or Adcock et al. (2022, §4.3.1). See also Kunoth and Schwab (2013); Rauhut and Schwab (2017). Another related general framework considers parametric implicit operator equations (see Chkifa et al., 2015b, Thm. 4.3, Cohen and DeVore, 2015, Thm. 2.5, or Adcock et al., 2022, §4.3.1).

These frameworks can be used to establish holomorphy results for various problems beyond (3.2). This includes parabolic problems (see Chkifa et al., 2015b, §5.1, or Cohen and DeVore, 2015, §2.2) and various types of nonlinear, elliptic PDEs (see Chkifa et al., 2015b, §5.2, or Cohen and DeVore, 2015, §2.3). Another prominent example includes PDEs such as (3.2) over parametrized domains $\Omega = \Omega_{\boldsymbol{y}}$. Holomorphy of the solution map is often referred to as *shape holomorphy* in this context. See Chkifa et al. (2015b, §5.3), Cohen and DeVore (2015, §2.2), Castrillón-Candás et al. (2016); Cohen et al. (2018) and references therein. Other examples include parametric hyperbolic problems (Hoang and Schwab, 2012) and certain classes of parametric control problems (Kunoth and Schwab, 2013).

See Cohen and DeVore (2015) or Adcock et al. (2022, Chpt. 4) for further reviews. Note that the above problems all involve PDEs. Classes of parametric ODEs also admit holomorphic regularity. See Adcock et al. (2022, §4.1) and Hansen and Schwab (2013b). Recent work in this direction has also shown such regularity for parametric ODEs arising from diffusion on graphs (Ajavon, 2024).

3.3 Known and unknown anisotropy

Functions that are $(\boldsymbol{b}, \varepsilon)$-holomorphic are *anisotropic*: they depend more smoothly on some variables rather than others. This behavior is dictated by the parameter $\boldsymbol{b}$, with a larger (smaller) value of an entry b_i corresponding to less (more) smoothness with respect to the variable y_i. This can be seen by noting that the condition

$$\sum_{i=1}^{\infty} \left(\frac{\rho_i + \rho_i^{-1}}{2} - 1 \right) b_i \leq 1,$$

in (3.1) holds only for smaller (larger) values of ρ_i when b_i is large (small), meaning that f only admits an extension in the y_i variable to a relatively small (large) Bernstein ellipse.

In practice, *a priori* analysis of a given problem may establish that the target function is $(\boldsymbol{b}, \varepsilon)$-holomorphic for some $\boldsymbol{b}$. However, an optimal value for $\boldsymbol{b}$ may be unknown. This situation arises in parametric DEs. For problems such as the stationary diffusion equation (3.2) with affine diffusion, one can find a sufficient value (3.6) of $\boldsymbol{b}$ for which the parametric solution map is $(\boldsymbol{b}, \varepsilon)$-holomorphic, although this value may not be sharp. In general, since $\mathcal{R}(\boldsymbol{b}, \varepsilon) \subseteq \mathcal{R}(\boldsymbol{b}', \varepsilon)$ whenever $\boldsymbol{b}' \leq \boldsymbol{b}$, it is difficult to know whether a value of $\boldsymbol{b}$ obtained from some analysis is optimal. Further, for more complicated problems, such as some of those described above, one can derive theoretical guarantees that assert holomorphy, but without an estimate for the region itself (Cohen and DeVore, 2015, Rem. 2.6).

For this reason, the primary focus of this work is the *unknown anisotropy* setting. We may assume the target function f is $(\boldsymbol{b}, \varepsilon)$-holomorphic for some $\boldsymbol{b}$, but we do not have access to $\boldsymbol{b}$ itself. In particular, this means that we cannot use $\boldsymbol{b}$ to design a method for learning f from data: the holomorphy assumption can only be used to provide bounds for the resulting approximation error.

Remark 3.5. To emphasize this consideration further, consider the function

$$f(y_1, y_2) = \sin(1000 y_2)/(1.1 - y_1).$$

This function is entire with respect to the variable y_2, but has a pole at $y_1 = 1.1$. By Remark 3.4, it is $(\boldsymbol{b}, 1)$-holomorphic for $\boldsymbol{b} = (b_1, b_2, \ldots) = (10, 0, 0, \ldots)$. When building a polynomial approximation to f, knowledge of $\boldsymbol{b}$ may lead one to include only low-degree terms in y_2 and more higher-degree terms in y_1. Yet this is completely the opposite of what one should do in practice. The function $\sin(1000 y_2)$, while entire, is highly oscillatory, and can only be resolved by using high-degree polynomials. On the other hand, the function

$$g(y_1, y_2) = 1/(1.1 - y_1)$$

is also $(\boldsymbol{b}, 1)$-holomorphic with the same $\boldsymbol{b}$. Yet, it requires no nonconstant terms in y_2 in order to approximate it accurately with a polynomial.

3.4 ℓ^p-summability and the $\mathcal{H}(p)$ and $\mathcal{H}(p, \mathsf{M})$ classes

As we will see in §4, it is generally impossible to approximate a $(\boldsymbol{b}, \varepsilon)$-holomorphic function without a further assumption on the sequence $\boldsymbol{b}$. In particular, the terms of $\boldsymbol{b}$ need to decay sufficiently fast. Henceforth, we will assume that $\boldsymbol{b}$ is not just in $\ell^1(\mathbb{N})$, but that, for some $0 < p < 1$, either

$$\boldsymbol{b} \in \ell^p(\mathbb{N}) \text{ or } \boldsymbol{b} \in \ell^p_{\mathsf{M}}(\mathbb{N}).$$

Here we recall that $\ell^p_{\mathsf{M}}(\mathbb{N})$ is the monotone ℓ^p-space (2.4). Since we consider the unknown anisotropy setting, we also define the function classes

$$\mathcal{H}(p) = \bigcup \left\{ \mathcal{H}(\boldsymbol{b}) : \boldsymbol{b} \in \ell^p(\mathbb{N}),\ \boldsymbol{b} \in [0, \infty)^{\mathbb{N}},\ \|\boldsymbol{b}\|_p \leq 1 \right\} \tag{3.7}$$

and

$$\mathcal{H}(p, \mathsf{M}) = \bigcup \left\{ \mathcal{H}(\boldsymbol{b}) : \boldsymbol{b} \in \ell^p_{\mathsf{M}}(\mathbb{N}),\ \boldsymbol{b} \in [0, \infty)^{\mathbb{N}},\ \|\boldsymbol{b}\|_{p,\mathsf{M}} \leq 1 \right\}. \tag{3.8}$$

In what follows we will ask for a method to provide a uniform error bound (depending on m, the amount of data, and p) for *any* function in these classes.

4 Best s-term polynomial approximation

In this section, we make a first foray into polynomial approximation of holomorphic functions. We introduce orthogonal polynomials and orthogonal polynomial expansions in $L^2_\varrho(\mathcal{U}; \mathcal{V})$. We then introduce best s-term polynomial approximation and provide a key theorem on algebraic convergence for it. See, e.g., Adcock et al. (2022, Chpts. 2-3) or Chkifa et al. (2015b); Cohen and DeVore (2015), for further information on this material.

4.1 Orthogonal polynomials

Let $P_0, P_1, \ldots$ denote the classical one-dimensional Legendre polynomials with the normalization

$$\|P_\nu\|_{L^\infty([-1,1])} = P_\nu(1) = 1. \tag{4.1}$$

We consider the orthonormalized version of these polynomials. Since

$$\int_{-1}^{1} |P_\nu(y)|^2 \, \mathrm{d}y = \frac{2}{2\nu + 1}, \quad \forall \nu \in \mathbb{N}_0,$$

we define these as

$$\psi_\nu(y) = \sqrt{2\nu + 1} P_\nu(y), \quad \forall \nu \in \mathbb{N}_0. \tag{4.2}$$

The set $\{\psi_\nu\}_{\nu \in \mathbb{N}_0} \subset L^2_\rho([-1, 1])$ forms an orthonormal basis, where ρ is the one-dimensional uniform measure.

We construct an orthonormal basis of $L^2_\varrho(\mathcal{U})$ by tensorization. Let

$$\mathcal{F} = \left\{ \boldsymbol{\nu} = (\nu_k)_{k=1}^{\infty} \in \mathbb{N}_0^{\mathbb{N}} : |\{k : \nu_k \neq 0\}| < \infty \right\}$$

be the set of multiindices with finitely many nonzero terms and set

$$\Psi_{\boldsymbol{\nu}}(\boldsymbol{y}) = \prod_{i \in \mathbb{N}} \psi_{\nu_i}(y_i), \quad \forall \boldsymbol{y} \in \mathcal{U}, \ \boldsymbol{\nu} \in \mathcal{F}.$$

Note that $\psi_0 = 1$ by construction. Therefore, this is equivalent to a product over finitely many terms, i.e.,

$$\Psi_{\boldsymbol{\nu}}(\boldsymbol{y}) = \prod_{i : \nu_i \neq 0} \psi_{\nu_i}(y_i). \tag{4.3}$$

Given these functions, it can now be shown that the set

$$\{\Psi_{\boldsymbol{\nu}}\}_{\boldsymbol{\nu} \in \mathcal{F}} \subset L^2_\varrho(\mathcal{U}) \tag{4.4}$$

constitutes an orthonormal basis for $L^2_\varrho(\mathcal{U})$ (Cohen and DeVore, 2015, §3).

Using (4.1)–(4.3), we also deduce that

$$\|\Psi_{\boldsymbol{\nu}}\|_{L^\infty_\varrho(\mathcal{U};\mathcal{V})} = |\Psi_{\boldsymbol{\nu}}(\boldsymbol{1})| = \prod_{k \in \mathbb{N}} \sqrt{2\nu_k + 1} =: u_{\boldsymbol{\nu}}. \tag{4.5}$$

The values $u_{\boldsymbol{\nu}}$ will be of use later. For convenience, we also define

$$\boldsymbol{u} = (u_{\boldsymbol{\nu}})_{\boldsymbol{\nu} \in \mathcal{F}}. \tag{4.6}$$

4.2 Orthogonal polynomial expansions

Let $f \in L^2_\varrho(\mathcal{U};\mathcal{V})$. Then we have the convergent expansion (in $L^2_\varrho(\mathcal{U};\mathcal{V})$)

$$f = \sum_{\boldsymbol{\nu} \in \mathcal{F}} c_{\boldsymbol{\nu}} \Psi_{\boldsymbol{\nu}}, \quad \text{where } c_{\boldsymbol{\nu}} = \int_{\mathcal{U}} f(\boldsymbol{y}) \Psi_{\boldsymbol{\nu}}(\boldsymbol{y}) \, \mathrm{d}\varrho(\boldsymbol{y}) \in \mathcal{V}. \tag{4.7}$$

Note that the *coefficients* $c_{\boldsymbol{\nu}}$ are elements of $\mathcal{V}$ and defined by Bochner integrals. For convenience, we denote the infinite sequence of coefficients of f by

$$\boldsymbol{c} = (c_{\boldsymbol{\nu}})_{\boldsymbol{\nu} \in \mathcal{F}}. \tag{4.8}$$

Parseval's identity gives that $\boldsymbol{c} \in \ell^2(\mathcal{F};\mathcal{V})$ with $\|\boldsymbol{c}\|_{2;\mathcal{V}} = \|f\|_{L^2_\varrho(\mathcal{U};\mathcal{V})}$.

4.3 Best s-term polynomial approximation

Let $s \in \mathbb{N}$. An *s-term polynomial approximation* to f is an approximation

$$f \approx f_S = \sum_{\boldsymbol{\nu} \in S} c_{\boldsymbol{\nu}} \Psi_{\boldsymbol{\nu}} \tag{4.9}$$

for some multiindex set $S \subset \mathcal{F}$, $|S| = s$. This raises the question: which index set S should one choose? One answer is to select a set S which provides the best approximation, an approach known as *best s-term approximation* and itself a type of nonlinear approximation (DeVore, 1998). Formally, a *best s-term approximation* f_s of f (with respect to the L^2_ϱ-norm) is defined as

$$f_s = f_{S^*}, \qquad \text{where } S^* \in \operatorname{argmin}\{\|f - f_S\|_{L^2_\varrho(\mathcal{U};\mathcal{V})} : S \subset \mathcal{F},\ |S| = s\}. \tag{4.10}$$

This approximation can also be characterized in terms of the coefficients of f. Due to Parseval's identity, the error of the approximation f_S is precisely

$$\|f - f_S\|^2_{L^2_\varrho(\mathcal{U};\mathcal{V})} = \sum_{\boldsymbol{\nu} \in \mathcal{F} \setminus S} \|c_{\boldsymbol{\nu}}\|^2_{\mathcal{V}}. \tag{4.11}$$

Therefore, any set S^* that yields a best s-term approximation consists of multiindices corresponding to the largest s coefficient norms $\|c_{\boldsymbol{\nu}}\|_{\mathcal{V}}$. Specifically,

$$S^* = \{\boldsymbol{\nu}_1, \boldsymbol{\nu}_2, \ldots, \boldsymbol{\nu}_s\},$$

where $\boldsymbol{\nu}_1, \boldsymbol{\nu}_2, \ldots$ is an ordering of the multiindex set $\mathcal{F}$ such that $\|c_{\boldsymbol{\nu}_1}\|_{\mathcal{V}} \geq \|c_{\boldsymbol{\nu}_2}\|_{\mathcal{V}} \geq \cdots$. It follows immediately from this and (4.11) that the error

$$\|f - f_s\|^2_{L^2_\varrho(\mathcal{U};\mathcal{V})} = \sum_{i > s} \|c_{\boldsymbol{\nu}_i}\|^2_{\mathcal{V}} \tag{4.12}$$

is precisely the $\ell^2(\mathcal{F};\mathcal{V})$-norm of the sequence of coefficients (4.8), excluding those coefficients with indices in S^*. Note that S^* (and therefore f_s) is generally nonunique. However, this fact causes no difficulties in what follows.

4.4 Rates of best s-term polynomial approximation

The best s-term approximation describes the best possible approximation obtainable with an s-term polynomial expansion. It is therefore important to provide bounds for this *benchmark* approximation in the case of holomorphic functions. Besides providing insight into limits of approximation, these bounds are also useful when we come to learn sparse polynomial approximations from data.

As we discussed above, Parseval's identity relates questions of best s-term approximation to f (in the $L^2_\varrho(\mathcal{U};\mathcal{V})$ norm) to approximation of its coefficients

$\boldsymbol{c}$ (in the $\ell^2(\mathcal{F};\mathcal{V})$-norm). This motivates one to study best s-term approximation of sequences, rather than functions. To this end, we now introduce some additional notation. Let $\Lambda \subseteq \mathcal{F}$, $0 < p \leq \infty$, $\boldsymbol{c} \in \ell^p(\Lambda;\mathcal{V})$ and $s \in \mathbb{N}_0$ with $s \leq |\Lambda|$. The *ℓ^p-norm best s-term approximation error* of the sequence $\boldsymbol{c}$ is defined as

$$\sigma_s(\boldsymbol{c})_{p;\mathcal{V}} = \min\left\{\|\boldsymbol{c} - \boldsymbol{z}\|_{p;\mathcal{V}} : \boldsymbol{z} \in \ell^p(\Lambda;\mathcal{V}),\ |\mathrm{supp}(\boldsymbol{z})| \leq s\right\}. \tag{4.13}$$

Note that when $p = 2$ we have $\sigma_s(\boldsymbol{c})_{2;\mathcal{V}} = \|f - f_s\|_{L^2_\varrho(\mathcal{U};\mathcal{V})}$, where f_s is a best s-term approximation to f (4.10).

We now state a well known result regarding the best s-term approximation (of coefficients) of holomorphic functions (see, e.g., Adcock et al., 2022, Thm. 3.28, or Cohen and DeVore, 2015, §3.2).

Theorem 4.1 (Algebraic convergence of the best s-term approximation). *Let $\boldsymbol{b} \in [0,\infty)^{\mathbb{N}}$ be such that $\boldsymbol{b} \in \ell^p(\mathbb{N})$ for some $0 < p < 1$. Then for any $s \in \mathbb{N}$ and $p \leq q \leq 2$, there exists a set $S \subset \mathcal{F}$ with $|S| \leq s$ such that*

$$\sigma_s(\boldsymbol{c})_{p;\mathcal{V}} \leq \|\boldsymbol{c} - \boldsymbol{c}_S\|_{q;\mathcal{V}} \leq C(\boldsymbol{b}, p) \cdot s^{\frac{1}{q} - \frac{1}{p}},$$

for all $f \in \mathcal{H}(\boldsymbol{b})$ with coefficients $\boldsymbol{c}$ as in (4.8).

Remark 4.2 (Sharpness of the algebraic rate). Theorem 4.1 provides an upper bound for the best s-term approximation error. However, it is also possible to provide a lower bound. As shown in Adcock and Monte (2023, Thm. 5.6), the are choices of $\boldsymbol{b} \in \ell^p(\mathbb{N})$ and functions $f \in \mathcal{H}(\boldsymbol{b})$ for which

$$\limsup_{s\to\infty}\left\{s^{\frac{1}{r}-\frac{1}{2}} \cdot \|f - f_s\|_{L^2_\varrho(\mathcal{U};\mathcal{V})}\right\} = +\infty$$

for any $0 < r < p$. Thus, the algebraic exponent $1/2 - 1/p$ is sharp.

4.5 How high is high dimensional?

As in Remark 3.4, let $f : [-1,1]^d \to \mathbb{R}$ be holomorphic in a single Bernstein ellipse $\mathcal{E}(\bar{\boldsymbol{\rho}}) = \mathcal{E}(\bar{\rho}_1) \times \cdots \times \mathcal{E}(\bar{\rho}_d) \subset \mathbb{C}^d$. Then its best s-term polynomial approximation converges with exponential rate in $s^{1/d}$: specifically

$$\|f - f_s\|_{L^2_\varrho(\mathcal{U};\mathcal{V})} \leq \|f\|_{L^\infty(\mathcal{R};\mathcal{V})} C(\epsilon)\sqrt{s}\exp\left(-\left(\frac{s d! \prod_{j=1}^d \log(\bar{\rho}_j)}{1+\epsilon}\right)^{1/d}\right),$$

for any $0 < \epsilon < 1$ and all $s \geq \bar{s}$, where $\bar{s}$ is a constant depending on d, ϵ and $\bar{\boldsymbol{\rho}}$ only. See Tran et al. (2017) or Adcock et al. (2022, Thm. 3.21). In low dimensions, this rate accurately predicts the convergence of the best s-term approximation. However, for larger d this rate is generally not witnessed for

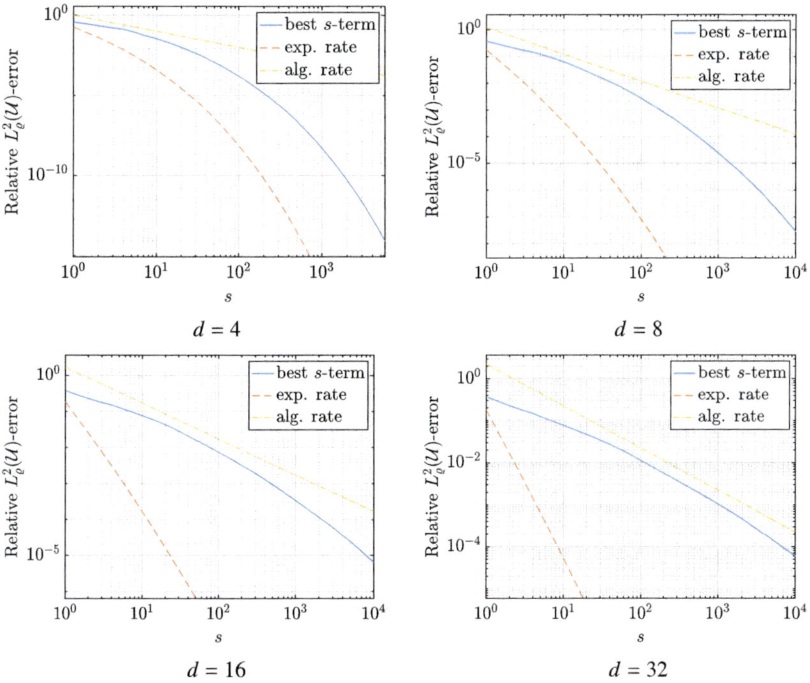

FIGURE 1 Best s-term approximation error in the $L^2_\varrho(\mathcal{U})$-norm for (4.14) with $\delta_i = i^{3/2}$. This figure also shows the exponential rate "exp. rate", defined as $C_{\text{exp}} \cdot \exp\left(-\left(sd!\prod_{i=1}^d \log(\rho_i)\right)^{1/d}\right)$, where ρ_i is such that $(\rho_i + 1/\rho_i)/2 = 1 + \delta_i$, and the algebraic rate "alg. rate", defined as $C_{\text{alg}} \cdot s^{-1}$. The constants C_{exp} and C_{alg} are chosen empirically to aid visualization.

any finite s, due to the exceedingly slow growth of the term $s^{1/d}$. For example, $s^{1/d} \leq 3.2$ in $d = 8$ dimensions for all $1 \leq s \leq 10{,}000$ (the range of s that is quite reasonable in practice).

For larger d, the algebraic rates of Theorem 4.1 better describe the convergence behavior. An example of this effect is shown in Fig. 1 for

$$f(\boldsymbol{y}) = \prod_{i=1}^{d} \frac{(2\delta_i + \delta_i^2)^{1/2}}{y_i + 1 + \delta_i}, \qquad \forall \boldsymbol{y} = (y_i)_{i=1}^d \in [-1, 1]^d. \tag{4.14}$$

This function is holomorphic in $\mathcal{E}(\boldsymbol{\rho})$ for any ρ_i satisfying $(\rho_i + 1/\rho_i)/2 < 1 + \delta_i$. Thus, by Remark 3.4, it is also $(\boldsymbol{b}, \varepsilon)$-holomorphic for any $b_i > \varepsilon/\delta_i$. In this figure, we consider $\delta_i = i^{3/2}$, meaning that $\boldsymbol{b}$ can be chosen so that $\boldsymbol{b} \in \ell^p(\mathbb{N})$ for any $2/3 < p < 1$. Thus, Theorem 4.1 predicts dimension-independent algebraic convergence with order that is arbitrarily close to $1/2 - 1/(2/3) = -1$. As we see, in $d = 4$ dimensions the error follows the exponential rate. However, once $d = 16$ or $d = 32$, the algebraic rate better predicts the true error.

This discussion and example motivates the study of algebraic rates. Simply put, they typically better describe the convergence behavior, even in finite dimensions as long as d is not too small (e.g., $d \geq 16$).

5 Limits of learnability from data

We now consider learning $(\boldsymbol{b}, \varepsilon)$-holomorphic functions from data. In this section, we present lower bounds for the amount of data that is necessary to learn such functions. We do this by appealing to techniques from information-based complexity (Novak, 1988; Novak and Woźniakowski, 2008, 2010; Traub et al., 1988) – in particular, *(adaptive) m-widths.*

Since our present focus is on lower bounds, in this section we do not assume that the training data takes the form of pointwise evaluations of the target function, as in (2.2). In fact, we can allow for arbitrary (adaptive) linear sampling operators. We will, for simplicity, consider only scalar-valued function approximation, i.e., $(\mathcal{V}, \|\cdot\|_{\mathcal{V}}) = (\mathbb{R}, |\cdot|)$. See Remark 5.3 for discussion.

5.1 Adaptive m-widths

Let $C(\mathcal{U})$ be the Banach space of continuous functions $f : \mathcal{U} \to \mathbb{R}$. We define an *adaptive sampling operator* as a map of the form

$$\mathcal{S} : C(\mathcal{U}) \to \mathbb{R}^m, \qquad \mathcal{S}(f) = \begin{bmatrix} S_1(f) \\ S_2(f; S_1(f)) \\ \vdots \\ S_m(f; S_1(f), \ldots, S_{m-1}(f)) \end{bmatrix},$$

where $S_1 : C(\mathcal{U}) \to \mathbb{R}$ is a bounded linear functional and, for $i = 2, \ldots, m$, $S_i : C(\mathcal{U}) \times \mathbb{R}^{i-1} \to \mathbb{R}$ is bounded and linear in its first component.

This definition includes standard bounded linear operators as a special case. However, it also allows for situations where the ith sample is selected adaptively based on the existing measurements. In machine learning, this is commonly referred to as *active learning* (Settles, 2012). In information-based complexity, it is commonly termed *adaptive information* (Novak and Woźniakowski, 2008, Sec. 4.1.1). As noted, our primary concern in this work is where each S_i is a pointwise evaluation operator (so-called *standard information* (Novak and Woźniakowski, 2008, Sec. 4.1.1)). In this case, $\mathcal{S}(f) = (f(\boldsymbol{y}_i))_{i=1}^m \in \mathbb{R}^m$, where $\boldsymbol{y}_i \in \mathcal{U}$ is the ith sample point, which is potentially chosen adaptively based on the previous measurements $f(\boldsymbol{y}_1), \ldots, f(\boldsymbol{y}_{i-1})$.

Definition 5.1 (Adaptive m-width). The *(adaptive) m-width* of a subset $\mathcal{K} \subseteq C(\mathcal{U})$ is

$$\Theta_m(\mathcal{K}) = \inf_{\mathcal{S}, \mathcal{R}} \sup_{f \in \mathcal{K}} \|f - \mathcal{R}(\mathcal{S}(f))\|_{L^2_{\varrho}(\mathcal{U})}, \tag{5.1}$$

where the infimum is taken over all adaptive sampling operators $\mathcal{S} : C(\mathcal{U}) \to \mathbb{R}^m$ and reconstruction maps $\mathcal{R} : \mathbb{R}^m \to C(\mathcal{U})$.

The adaptive m-width is related to the concept of *information complexity* (Novak and Woźniakowski, 2008, §4.1.4). It measures how well one can approximate functions from $\mathcal{K}$ using an arbitrary (adaptive) linear sampling operator $\mathcal{S}$ and an arbitrary (potentially nonlinear) reconstruction map $\mathcal{R}$. Due to the inner supremum, the approximation is measured in a worst-case (or uniform) sense; in other words, the sampling-recovery pair $(\mathcal{S}, \mathcal{R})$ is required to provide a guaranteed bound simultaneously for all functions in the class.

5.2 Lower bounds for adaptive m-widths

We now present the main results of this section. Following §3.4, we let

$$\theta_m(p) = \Theta_m(\mathcal{H}(p)), \quad \theta_m(p, \mathsf{M}) = \Theta_m(\mathcal{H}(p, \mathsf{M})).$$

We also consider slightly weaker notions, which are defined as follows.

$$\overline{\theta_m}(p) = \sup\left\{\Theta_m(\mathcal{H}(\boldsymbol{b})) : \boldsymbol{b} \in \ell^p(\mathbb{N}),\ \boldsymbol{b} \in [0,\infty)^{\mathbb{N}},\ \|\boldsymbol{b}\|_p \leq 1\right\},$$
$$\overline{\theta_m}(p, \mathsf{M}) = \sup\left\{\Theta_m(\mathcal{H}(\boldsymbol{b})) : \boldsymbol{b} \in \ell^p_{\mathsf{M}}(\mathbb{N}),\ \boldsymbol{b} \in [0,\infty)^{\mathbb{N}},\ \|\boldsymbol{b}\|_{p,\mathsf{M}} \leq 1\right\}.$$

The widths $\theta_m(p)$ and $\theta_m(p, \mathsf{M})$ pertain to the unknown anisotropy setting. A sampling-recovery pair $(\mathcal{S}, \mathcal{R})$ does not have access to the anisotropy parameter $\boldsymbol{b}$, and must work uniformly for all holomorphic functions with ℓ^p or ℓ^p_{M}-summable $\boldsymbol{b}$ (of norm at most one). Conversely, $\overline{\theta_m}(p)$ and $\overline{\theta_m}(p, \mathsf{M})$ are weaker and pertain to the known anisotropy setting, since they allow the sampling-recovery pair $(\mathcal{S}, \mathcal{R})$ to depend on $\boldsymbol{b}$. In particular, we have

$$\theta_m(p) \geq \overline{\theta_m}(p) \text{ and } \theta_m(p, \mathsf{M}) \geq \overline{\theta_m}(p, \mathsf{M}).$$

Theorem 5.2 (Limits of learnability). *Let $m \geq 1$ and $0 < p < 1$. Then*

- *the m-widths $\overline{\theta_m}(p)$ and $\overline{\theta_m}(p, \mathsf{M})$ satisfy*

$$\overline{\theta_m}(p) \geq \overline{\theta_m}(p, \mathsf{M}) \geq c \cdot 2^{-\frac{1}{p}} \cdot m^{\frac{1}{2}-\frac{1}{p}}, \tag{5.2}$$

- *the m-width $\theta_m(p)$ satisfies*

$$\theta_m(p) \geq c \cdot 2^{\frac{1}{2}-\frac{2}{p}}, \tag{5.3}$$

- *and the m-width $\theta_m(p, \mathsf{M})$ satisfies*

$$\theta_m(p, \mathsf{M}) \geq \overline{\theta_m}(p, \mathsf{M}) \geq c \cdot 2^{-\frac{1}{p}} \cdot m^{\frac{1}{2}-\frac{1}{p}}. \tag{5.4}$$

Here $c > 0$ is a universal constant.

See Adcock et al. (2023c, Thms. 4.4 & 4.5). This theorem has several consequences. First, (5.3) shows that it is impossible to uniformly learn functions from the class $\mathcal{H}(p)$. In other words, mere ℓ^p-summability of the anisotropy parameter $\boldsymbol{b}$ is insufficient. By contrast, (5.4) shows that it may be possible to learn functions in $\mathcal{H}(p, \mathsf{M})$, where $\boldsymbol{b}$ is now ℓ^p_{M}-summable, but only at a rate that is algebraically decaying in m. We will see in the next section that these rates can in fact be attained (up to log terms) by practical methods. Finally, (5.2) asserts that knowledge of the anisotropy parameter $\boldsymbol{b}$ does not help. The lower bound for the nonuniform width $\overline{\theta_m}(p, \mathsf{M})$ is the same as the lower bound for the uniform width $\theta_m(p, \mathsf{M})$. Therefore, in terms of sample complexity, knowledge of $\boldsymbol{b}$ conveys no benefit – we may as well deal exclusively with the unknown anisotropy setting.

It is worth relating this result back to Theorem 4.1. Theorem 5.2 shows that we can at best achieve a rate of $m^{1/2-1/p}$ when learning functions in $\mathcal{H}(\boldsymbol{b})$ from data, whenever $\boldsymbol{b} \in \ell^p_{\mathsf{M}}(\mathbb{N})$. Theorem 4.1 (with $p = 2$) shows that the best s-term polynomial approximation can achieve a rate of $s^{1/2-1/p}$ subject to the weaker assumption $\boldsymbol{b} \in \ell^p(\mathbb{N})$. Hence, the discrepancy between $\ell^p(\mathbb{N})$ (where learning is impossible) and $\ell^p_{\mathsf{M}}(\mathbb{N})$ (where learning is possible) must stem from the limited amount of data, not intrinsic properties of the function class itself.

Remark 5.3 (Extensions). For simplicity, we have restricted this discussion to scalar-valued functions with ϱ being the uniform measure. The latter assumption can be easily relaxed to an arbitrary tensor-product probability measure. One may also consider Banach-valued functions, although some care is needed when defining an adaptive sampling operator in this case. See Adcock et al. (2023c) for details.

5.3 Towards methods

In summary, we have now seen that functions in $\mathcal{H}(p, \mathsf{M})$ can be learned from training sets of size m with rates that are at best $O(m^{1/2-1/p})$, regardless of the training data and learning method. In the remainder of this review, we focus on describing methods that achieve (close to) these rates when the training data consists of i.i.d. pointwise samples, as in (2.2). We consider, firstly, polynomial-based methods (§6) and, secondly, methods based on DNNs (§7-8).

6 Learning sparse polynomial approximations from data

We know from Theorem 4.1 that the best s-term approximation converges at the desired rate $s^{1/2-1/p}$. Therefore, our goal is to design methods that can compute best (or quasibest) s-term approximations using roughly s samples (up to constant and log factors). We shall do this using tools from compressed sensing (Foucart and Rauhut, 2013; Vidyasagar, 2019), which recast the problem of learning sparse polynomial approximations as recovering approximately sparse vectors. Unfortunately, as we describe in this section, this task is easily laid out, but not so easily completed.

6.1 Setup

Throughout this section, we assume that $\mathcal{V}$ is a Hilbert space (see §6.7 for some discussion on the Banach case). Moreover, in a practical scenario, we often cannot work directly with the space $\mathcal{V}$, since it may be infinite dimensional. Therefore, we now consider a finite-dimensional discretization $\mathcal{V}_h \subseteq \mathcal{V}$, where $h > 0$ denotes a discretization parameter. For parametric PDEs, this is typically the mesh size when a Finite Element Method (FEM) is used. We also assume that the training data (2.2) belongs to $\mathcal{U} \times \mathcal{V}_h$, rather than $\mathcal{U} \times \mathcal{V}$, i.e.,

$$\{(\boldsymbol{y}_i, f(\boldsymbol{y}_i) + e_i)\}_{i=1}^m \subseteq \mathcal{U} \times \mathcal{V}_h. \tag{6.1}$$

This encapsulates the notion that f is evaluated using some numerical simulation (e.g., a FEM) that outputs values in $\mathcal{V}_h$. Notice that the error term e_i incorporates any error incurred by this computation. For convenience, we also define the orthogonal projection

$$\mathcal{P}_h : \mathcal{V} \to \mathcal{V}_h.$$

Moreover, if $f \in L^2_\varrho(\mathcal{U}; \mathcal{V})$, then we let $\mathcal{P}_h f \in L^2_\varrho(\mathcal{U}; \mathcal{V}_h)$ be the function defined almost everywhere as $(\mathcal{P}_h f)(\boldsymbol{y}) = \mathcal{P}_h(f(\boldsymbol{y}))$, $\boldsymbol{y} \in \mathcal{U}$.

Now consider a function $f \in L^2_\varrho(\mathcal{U}; \mathcal{V})$ with expansion (4.7) and coefficients (4.8). Standard compressed sensing involves the recovery of finite vectors. Therefore, we first need to truncate the expansion (4.7). Let $\Lambda \subset \mathcal{F}$ be finite with $|\Lambda| = N$ and an enumeration $\Lambda = \{\boldsymbol{\nu}_1, \ldots, \boldsymbol{\nu}_N\}$. Then

$$f(\boldsymbol{y}_i) + e_i = f_\Lambda(\boldsymbol{y}_i) + (f - f_\Lambda)(\boldsymbol{y}_i) + e_i = \sum_{\boldsymbol{\nu} \in \Lambda} c_{\boldsymbol{\nu}} \Psi_{\boldsymbol{\nu}}(\boldsymbol{y}_i) + (f - f_\Lambda)(\boldsymbol{y}_i) + e_i,$$

where $f_\Lambda = \sum_{\boldsymbol{\nu} \in \Lambda} c_{\boldsymbol{\nu}} \Psi_{\boldsymbol{\nu}}$. Now let $\boldsymbol{c}_\Lambda = (c_{\boldsymbol{\nu}_i})_{i=1}^N \in \mathcal{V}^N$. Then we have

$$\boldsymbol{A}\boldsymbol{c}_\Lambda + \boldsymbol{e} + \boldsymbol{e}' = \boldsymbol{f}, \tag{6.2}$$

where

$$\boldsymbol{f} = \frac{1}{\sqrt{m}} \left(f(\boldsymbol{y}_i) + e_i\right)_{i=1}^m \quad \boldsymbol{A} = \frac{1}{\sqrt{m}} \left(\Psi_{\boldsymbol{\nu}_j}(\boldsymbol{y}_i)/\sqrt{m}\right)_{i,j=1}^{m,N} \in \mathbb{R}^{m \times N}, \tag{6.3}$$

and

$$\boldsymbol{e} = \frac{1}{\sqrt{m}} (e_i)_{i=1}^m, \quad \boldsymbol{e}' = \frac{1}{\sqrt{m}} \left((f - f_\Lambda)(\boldsymbol{y}_i)\right)_{i=1}^m \in \mathbb{R}^m.$$

The vector $\boldsymbol{c}_\Lambda$ is approximately s-sparse. Therefore, we have recast the problem into a standard compressed sensing form: namely, the recovery of an approximately sparse vector from noisy linear measurements (6.2).

Of course, this problem is not quite standard, since the vector $\boldsymbol{c}_\Lambda$ has entries taking values in $\mathcal{V}$, rather than $\mathbb{R}$ or $\mathbb{C}$. Fortunately, this challenge can be dealt with by 'lifting' the appropriate theoretical tools from $\mathbb{R}$ or $\mathbb{C}$ to Hilbert spaces (see Remark 6.1 below). A more delicate challenge is the following.

Challenge 1. How should the truncation set Λ be chosen.

In general, f has an infinite expansion. Since we recover only those coefficients with indices in Λ (i.e., the vector $\boldsymbol{c}_\Lambda$), the error in recovering f will always involve the expansion tail $f - f_\Lambda$. On the one hand, Λ should not be too large, since, as we see later in (6.21), $\log(N)$ will enter into the sample complexity bound. Moreover, larger N will also increase the computational cost. On the other hand, we need $f - f_\Lambda$ to be sufficiently small so as to obtain optimal learning rates. To do this, we need to ensure no large coefficients of f lie outside Λ. And here is where the problem lies. The result on best s-term approximation, Theorem 4.1, which has served as our rationale up to now for using compressed sensing, asserts approximate sparsity of the coefficients, but gives no guarantees whatsoever on where the largest s coefficients should lie within the set $\mathcal{F}$.

6.2 Sampling discretizations for multivariate polynomials

We will return to Challenge 1 in a moment. But, first, we also need to introduce a second challenge. Let us imagine some oracle gives us a 'good' set $S \subset \mathcal{F}$ of size $|S| = s$. For example, this could even be a set $S = S^*$ corresponding to the best s-term approximation (4.10). Then one would naturally learn an approximation to f via the empirical least-squares fit

$$\hat{f} \in \underset{p \in \mathbb{P}_{S;\mathcal{V}}}{\operatorname{argmin}} \frac{1}{m} \sum_{i=1}^{m} \| f(\boldsymbol{y}_i) + e_i - p(\boldsymbol{y}_i) \|_{\mathcal{V}}^2, \tag{6.4}$$

where $\mathbb{P}_{S;\mathcal{V}}$ is the s-dimensional subspace

$$\mathbb{P}_{S;\mathcal{V}} = \left\{ \sum_{\boldsymbol{\nu} \in S} c_{\boldsymbol{\nu}} \Psi_{\boldsymbol{\nu}} : c_{\boldsymbol{\nu}} \in \mathcal{V} \right\} \subset L^2_{\varrho}(\mathcal{U}; \mathcal{V})$$

of Hilbert-valued polynomials with nonzero coefficients in S.

The behavior of the estimator (6.4) is intimately related to the existence of a *sampling discretization* of the $L^2_\rho(\mathcal{U})$-norm for the scalar-valued analogue of this space, i.e., $\mathbb{P}_S = \mathbb{P}_{S;\mathbb{R}} = \operatorname{span}\{\Psi_{\boldsymbol{\nu}} : \boldsymbol{\nu} \in S\}$. A sampling discretization (Kashin et al., 2022) (also known as a *Marcinkiewicz–Zygmund inequality* (Temlyakov, 2018)) is an inequality of the form

$$\alpha \|p\|^2_{L^2_\varrho(\mathcal{U})} \le \frac{1}{m} \sum_{i=1}^{m} |p(\boldsymbol{y}_i)|^2 \le \beta \|p\|^2_{L^2_\varrho(\mathcal{U})}, \quad \forall p \in \mathbb{P}_S, \tag{6.5}$$

for constants $0 < \alpha \leq \beta < \infty$. It is known that (6.5) ensures both accuracy of the estimator (6.4) in the scalar-valued case and robustness to measurement error (see, e.g., Adcock et al., 2022, Thm. 5.3).

Remark 6.1 (Lifting to Hilbert spaces). This is an instance of the aforementioned 'lifting' concept. The sampling discretization (6.5) is formulated for the space $\mathbb{P}_S = \mathbb{P}_{S;\mathbb{R}}$. It is a short argument (see, e.g., Adcock et al., 2024a, Lem. 7.5) to show that it is, in fact, equivalent to a sampling discretization for the space $\mathbb{P}_{S;\mathcal{V}}$, i.e.,

$$\alpha \|p\|^2_{L^2_\varrho(\mathcal{U};\mathcal{V})} \leq \frac{1}{m} \sum_{i=1}^{m} \|p(\boldsymbol{y}_i)\|^2_{\mathcal{V}} \leq \beta \|p\|^2_{L^2_\varrho(\mathcal{U};\mathcal{V})}, \quad \forall p \in \mathbb{P}_{S;\mathcal{V}}.$$

In turn, this implies accuracy and robustness of the estimator (6.4) in the Hilbert-valued case.

Sufficient conditions for sampling discretizations in linear subspaces with i.i.d. samples can be derived using standard matrix concentration inequalities. See, e.g., Cohen et al. (2013) or Adcock et al. (2022, Chpt. 5). These conditions involve the quantity

$$\kappa(\mathbb{P}_S) = \|\mathcal{K}(\mathbb{P}_S)\|_{L^\infty_\rho(\mathcal{U})}, \tag{6.6}$$

where $\mathcal{K}(\mathbb{P}_S)$ is the (reciprocal) *Christoffel function* of the subspace $\mathbb{P}_S$:

$$\mathcal{K}(\mathbb{P}_S)(\boldsymbol{y}) = \sum_{\boldsymbol{\nu} \in S} |\Psi_{\boldsymbol{\nu}}(\boldsymbol{y})|^2, \quad \forall \boldsymbol{y} \in \mathcal{U}. \tag{6.7}$$

Specifically, one can show that if

$$m \geq c \cdot \kappa(\mathbb{P}_S) \cdot \log(2s/\epsilon), \tag{6.8}$$

where $s = |S|$ and $c > 0$ is a universal constant, then (6.5) holds with constants $\beta \leq 2$ and $\alpha \geq 1/2$ (these values are arbitrary) with probability at least $1 - \epsilon$ on the draw of the sample points $\boldsymbol{y}_1, \ldots, \boldsymbol{y}_m \sim_{\text{i.i.d.}} \varrho$ (see, e.g., Adcock et al., 2022, Thm. 5.12).

Therefore, the *sample complexity* of the 'oracle' estimator (6.4) is governed by the maximal behavior of the Christoffel function (6.6). Note that $\kappa(\mathbb{P}_S) \geq s$ for any set S with $|S| = s$ (Adcock et al., 2022, §5.3). To use (6.8) to achieve the optimal rates described in Theorem 5.2 for this oracle estimator, we require that $\kappa(\mathbb{P}_S) \lesssim s$ as well. Unfortunately, this is not the case: $\kappa(\mathbb{P}_S)$ can be arbitrarily large in comparison to s. To see why, we recall (4.1). This implies that

$$\kappa(\mathbb{P}_S) = \mathcal{K}(P_S)(\mathbf{1}) = \sum_{\boldsymbol{\nu} \in S} u^2_{\boldsymbol{\nu}}, \tag{6.9}$$

where the $u_{\boldsymbol{\nu}}$ are as in (4.5). Since $u_{\boldsymbol{\nu}} \to \infty$ as $\boldsymbol{\nu} \to \infty$, we deduce the claim.

Challenge 2. Even if it is known, the index set S of the best s-term approximation may lead to a sample complexity estimate (6.8) that is arbitrarily large.

Similar to Challenge 1, this difficulty arises because Theorem 4.1 gives no guarantees about the set $S = S^*$ which attains the best s-term approximation. In particular, it says nothing about the term $\kappa(\mathbb{P}_{S^*})$. The paths to resolving each challenge are therefore similar. We need to show that *near-best* s-term approximations can be obtained using suitably *structured* index sets.

Remark 6.2 (Necessity of sampling discretizations). The sampling discretization (6.5) is sufficient condition for robustness of the estimator $\hat{f}$. However, the lower inequality is also essentially necessary. Indeed, let $\mathcal{S} : C(\mathcal{U}) \to \mathbb{R}^m$, $f \mapsto (f(\mathbf{y}_i))_{i=1}^m/\sqrt{m}$ and $\mathcal{R} : \mathbb{R}^m \to C(\mathcal{U})$ be any reconstruction map that is δ-*accurate* over $\mathbb{P}_S$, i.e.,

$$\|p - \mathcal{R}(\mathcal{S}(p))\|_{L^2_\varrho(\mathcal{U})} \leq \delta \|p\|_{L^2_\varrho(\mathcal{U})}, \quad \forall p \in \mathbb{P}_s, \tag{6.10}$$

for some $\delta > 0$. Then it is a short argument to show that the ϵ-Lipschitz constant

$$L_\epsilon = \sup_{f \in C(\mathcal{U})} \sup_{0<\|e\|_2 \leq \epsilon} \|\mathcal{R}(\mathcal{S}(f) + e) - \mathcal{R}(\mathcal{S}(f))\|_{L^2_\varrho(\mathcal{U})}/\|e\|_2$$

satisfies $L_\epsilon \geq (1 - \delta)/\sqrt{\alpha}$. Thus, when α is small, a reconstruction map cannot be simultaneously accurate (in the sense of (6.10)) and stable.

6.3 Resolving Challenge 1: lower and anchored sets

In order to address Challenge 1, we now introduce the following concept.

Definition 6.3 (Lower and anchored sets). A multiindex set $S \subseteq \mathcal{F}$ is *lower* if the following holds for every $\boldsymbol{\nu}, \boldsymbol{\mu} \in S$:

$$(\boldsymbol{\nu} \in S \text{ and } \boldsymbol{\mu} \leq \boldsymbol{\nu}) \Rightarrow \boldsymbol{\mu} \in \Lambda.$$

A multiindex set $S \subseteq \mathcal{F}$ is *anchored* if it is lower and if the following holds for every $j \in \mathbb{N}$:

$$\boldsymbol{e}_j \in S \Rightarrow \{\boldsymbol{e}_1, \boldsymbol{e}_2, \dots, \boldsymbol{e}_j\} \subseteq S.$$

Lower sets are classical objects in multivariate approximation theory. Anchored sets were introduced in the context of infinite-dimensional approximations. See, e.g., Adcock et al. (2022, §2.3.3 & 2.5.3) and references therein.

We now state a result that shows that the $s^{1/q-1/p}$ rate asserted in Theorem 4.1 can also be obtained using a lower or (under an additional assumption) anchored set. See, e.g., Adcock et al. (2022, Thm. 3.33) or Cohen and DeVore (2015, §3.8).

Theorem 6.4 (Algebraic convergence in lower or anchored sets). *Let $\boldsymbol{b} \in [0,\infty)^{\mathbb{N}}$ be such that $\boldsymbol{b} \in \ell^p(\mathbb{N})$ for some $0 < p < 1$. Then for any $s \in \mathbb{N}$ and $p \le q \le 2$, there exists an lower set $S \subset \mathcal{F}$ with $|S| \le s$ such that*

$$\|\boldsymbol{c} - \boldsymbol{c}_S\|_{q;\mathcal{V}} \le C(\boldsymbol{b}, p) \cdot s^{\frac{1}{q} - \frac{1}{p}}$$

for all $f \in \mathcal{H}(\boldsymbol{b})$ with coefficients $\boldsymbol{c}$ as in (4.8). If $\boldsymbol{b} \in \ell^p_{\mathsf{M}}(\mathbb{N})$ then S can also be chosen as an anchored set.

In particular, when $q = 2$, this theorem and Parseval's identity imply the existence of a lower or anchored set S with $|S| \le s$ such that

$$\|f - f_S\|_{L^2_\varrho(\mathcal{U};\mathcal{V})} \le C(\boldsymbol{b}, p) \cdot s^{\frac{1}{2} - \frac{1}{p}}, \quad \forall f \in \mathcal{H}(\boldsymbol{b}). \tag{6.11}$$

Now recall that our aim is to learn functions in the space $\mathcal{H}(p, \mathsf{M})$ defined by (3.8). Any $f \in \mathcal{H}(p, \mathsf{M})$ satisfies $f \in \mathcal{H}(\boldsymbol{b})$ for some $\boldsymbol{b} \in \ell^p_{\mathsf{M}}(\mathbb{N})$. Therefore, for any such function, we know there is an anchored set S, $|S| \le s$, that achieves (6.11). How does this allow us to overcome Challenge 1? The reason is because anchored sets of a fixed size lie within a *finite* subset of $\mathcal{F}$. Indeed, one can show (see, e.g., Cohen et al., 2017, or Adcock et al., 2022, Prop. 2.18) that $S \subset \Lambda^{\mathsf{HCI}}_s$ for all $S \subset \mathcal{F}$ with $|S| \le s$, where Λ^{HCI}_s is the finite set

$$\Lambda^{\mathsf{HCI}}_s = \left\{ \boldsymbol{\nu} = (\nu_k)_{k=1}^{\infty} \in \mathcal{F} : \prod_{k=1}^{s-1} (\nu_k + 1) \le s,\ \nu_k = 0,\ \forall k \ge s \right\}. \tag{6.12}$$

This set is in fact isomorphic to the $(s-1)$-dimensional *hyperbolic cross* set of order $s-1$, a very well-known object in high-dimensional approximation (Dũng et al., 2018). With this knowledge, we set $\Lambda = \Lambda^{\mathsf{HCI}}_s$ and then apply Theorem 6.4:

$$\|f - f_\Lambda\|_{L^2_\varrho(\mathcal{U};\mathcal{V})} = \|\boldsymbol{c} - \boldsymbol{c}_\Lambda\|_{2;\mathcal{V}} \le \|\boldsymbol{c} - \boldsymbol{c}_S\|_{2;\mathcal{V}} \le C(\boldsymbol{b}, p) \cdot s^{\frac{1}{2} - \frac{1}{p}},$$

for all $f \in \mathcal{H}(\boldsymbol{b})$ and $\boldsymbol{b} \in \ell^p_{\mathsf{M}}(\mathbb{N})$. This resolves Challenge 1.

Remark 6.5. Theorem 5.2 states that no method can learn functions in $\mathcal{H}(p)$ from finite data. Theorem 6.4 says that we can always find a lower set that yields the desired rate (6.11) when $\boldsymbol{b} \in \ell^p(\mathbb{N})$. However, the union of all lower sets in infinite dimensions is not a finite set. This is precisely what prohibits one from learning functions in $\mathcal{H}(p)$ using compressed sensing (which, of course, must not be possible in view of Theorem 5.2): namely, there is no way to construct a finite set Λ that ensures a uniformly small truncation error $f - f_\Lambda$.

6.4 Resolving Challenge 2: weighted k-term approximation

We now return to Challenge 2. In the previous subsection, we identified anchored sets as structured sets which achieve near-best s-term approximation rates. Unfortunately, while anchored sets do ameliorate the sample complexity issue, they do not fully resolve it. On the one hand, it is possible to show that $\kappa(\mathbb{P}_S) \le s^2$ whenever S is anchored (see, e.g., Adcock et al., 2022, Prop. 5.17), where $\kappa(\mathbb{P}_S)$ is as in (6.6). Unfortunately, this bound is sharp. For example, the anchored set

$$S = \{j\boldsymbol{e}_1 : j = 0, \dots, s-1\}, \quad \text{where } \boldsymbol{e}_1 = (1, 0, 0, \dots),$$

satisfies $\kappa(\mathbb{P}_S) = s^2$. This follows from (6.9) and the fact that $u_{j\boldsymbol{e}_1} = u_j = \sqrt{2j+1}$. Therefore, the sample complexity bound (6.8) for learning with anchored sets is generically log-quadratic in s.

To overcome Challenge 2 we need a different concept, *weighted k-term approximation*. Motivated by (6.9), we now define the *weighted cardinality* of set $S \subset \mathcal{F}$ with respect to *weights* $\boldsymbol{u}$ as

$$|S|_{\boldsymbol{u}} := \sum_{\boldsymbol{\nu} \in S} u_{\boldsymbol{\nu}}^2.$$

Thus we may reinterpret (6.8) as follows: the sample complexity is determined not by the cardinality of S, but by its weighted cardinality.

Fortunately, as we next show, there exist sets of a given weighted cardinality that achieve near-optimal approximation rates. For this, we need some additional notation. Given $1 \le p \le 2$, we define the weighted $\ell^p_{\boldsymbol{u}}(\Lambda; \mathcal{V})$ space as the space of $\mathcal{V}$-valued sequences $\boldsymbol{c} = (c_{\boldsymbol{\nu}})_{\boldsymbol{\nu} \in \Lambda}$ for which $\|\boldsymbol{c}\|_{p,\boldsymbol{u};\mathcal{V}} < \infty$, where

$$\|\boldsymbol{c}\|_{p,\boldsymbol{u};\mathcal{V}} = \left(\sum_{\boldsymbol{\nu} \in \Lambda} u_{\boldsymbol{\nu}}^{2-p} \|c_{\boldsymbol{\nu}}\|_{\mathcal{V}}^p\right)^{\frac{1}{p}}.$$

Notice that $\|\cdot\|_{2,\boldsymbol{u};\mathcal{V}}$ coincides with the unweighted norm $\|\cdot\|_{2;\mathcal{V}}$.

Theorem 6.6 (Algebraic convergence of the weighted best k-term approximation). *Let $\boldsymbol{b} \in [0,\infty)^{\mathbb{N}}$ be such that $\boldsymbol{b} \in \ell^p(\mathbb{N})$ for some $0 < p < 1$. Then for any $k > 0$ and $p \le q \le 2$ there exists a set $S \subset \mathcal{F}$ with $|S|_{\boldsymbol{u}} \le k$ such that*

$$\|\boldsymbol{c} - \boldsymbol{c}_S\|_{q,\boldsymbol{u};\mathcal{V}} \le C(\boldsymbol{b}, p) \cdot k^{\frac{1}{q} - \frac{1}{p}}$$

for all $f \in \mathcal{H}(\boldsymbol{b})$ with coefficients $\boldsymbol{c}$ as in (4.8).

This theorem shows that we can construct near-best approximations using set of weighted cardinality at most k. In particular, when $q = 2$, this theorem and Parseval's identity give that there exists a set S with $|S|_{\boldsymbol{u}} \le k$ such that

$$\|f - f_S\|_{L^2_\varrho(\mathcal{U};\mathcal{V})} \le C(\boldsymbol{b}, p) \cdot k^{\frac{1}{2} - \frac{1}{p}}, \quad \forall f \in \mathcal{H}(\boldsymbol{b}).$$

Note that we do not require $\boldsymbol{b} \in \ell^p_{\mathsf{M}}(\mathbb{N})$ for this result. This additional regularity is only needed to resolve Challenge 1.

6.5 Weighted ℓ^1-minimization

Challenge 2 is now resolved, at least insofar as the 'oracle' least-squares estimator goes. Of course, in practice we do not have access to such an oracle. However, Theorem 6.6 implies that the sequence $\boldsymbol{c}$, and therefore the finite vector $\boldsymbol{c}_\Lambda$, is approximately *weighted sparse*. In other words, it is well approximated by $\boldsymbol{c}_S$ for some index set $S \subseteq \Lambda$ of weighted cardinality $|S|_{\boldsymbol{u}} \leq k$.

Weighted sparsity is a well-understood concept in compressed sensing (see Adcock, 2017; Rauhut and Ward, 2016 and references therein). Much like how sparsity can be promoted by solving a minimization problem involving the ℓ^1-norm, weighted sparsity can be solving a minimization problem involving the weighted $\ell^1_{\boldsymbol{u}}$-norm. A large weight penalizes the corresponding coefficient, much like how a large weight increases the weighted sparsity of any vector that has a nonzero coefficient at the corresponding index. There are a number of ways to formulate the weighted ℓ^1-minimization problem, but, following Adcock et al. (2019, 2024a), we will consider the Hilbert-valued, *weighted square-root LASSO* program

$$\min_{z \in \mathcal{V}_h^N} \mathcal{G}(z), \quad \text{where } \mathcal{G}(z) := \lambda \|z\|_{1,\boldsymbol{u};\mathcal{V}} + \|\boldsymbol{A}z - \boldsymbol{f}\|_{2;\mathcal{V}}. \tag{6.13}$$

Here $\|z\|_{1,\boldsymbol{u};\mathcal{V}} = \sum_{\boldsymbol{\nu} \in \Lambda} u_{\boldsymbol{\nu}} \|z_{\boldsymbol{\nu}}\|_{\mathcal{V}}$ is the $\ell^1_{\boldsymbol{u};\mathcal{V}}$-norm and $\lambda > 0$ is a parameter. Notice that this problem is posed over $\mathcal{V}_h^N$, which means it can be numerically solved. See Remark 6.7 below. In order to account for inexact solution of (6.13), given $\gamma \geq 0$ we say that $\hat{\boldsymbol{c}} = (\hat{c}_{\boldsymbol{\nu}})_{\boldsymbol{\nu} \in \Lambda}$ is a γ*-minimizer* of (6.13) if

$$\mathcal{G}(\hat{\boldsymbol{c}}) \leq \min_{z \in \mathcal{V}_h^N} \mathcal{G}(z) + \gamma.$$

For such $\hat{\boldsymbol{c}}$, we define the corresponding sparse polynomial approximation

$$\hat{f} = \sum_{\boldsymbol{\nu} \in \Lambda} \hat{c}_{\boldsymbol{\nu}} \Psi_{\boldsymbol{\nu}}. \tag{6.14}$$

Remark 6.7 (Numerical solution of (6.13)). In practice, (6.13) is solved by first introducing a basis $\{\varphi_i\}_{i=1}^K$ for the space $\mathcal{V}_h$, where $K = \dim(\mathcal{V}_h)$ (in particular, $K = 1$ in the scalar-valued case $\mathcal{V} = (\mathbb{R}, |\cdot|)$). Instead of searching for a $\mathcal{V}_h$-valued vector of coefficients $\hat{\boldsymbol{c}} \in \mathcal{V}_h^N$, one now searches for an equivalent matrix of coefficients $\widehat{\boldsymbol{C}} \in \mathbb{R}^{N \times K}$. It is a short exercise to show that (6.13) is equivalent to the problem

$$\min_{\boldsymbol{Z} \in \mathbb{R}^{N \times K}} \lambda \|\boldsymbol{Z}\|_{2,1,\boldsymbol{u}} + \|(\boldsymbol{A}\boldsymbol{Z} - \boldsymbol{F})\boldsymbol{G}^{1/2}\|_F, \tag{6.15}$$

where $\|\cdot\|_F$ is the Frobenius norm, $\boldsymbol{G} = \left(\langle \varphi_i, \varphi_j \rangle\right)_{i,j=1}^{K} \in \mathbb{R}^{K \times K}$ and

$$\|\boldsymbol{Z}\|_{2,1,\boldsymbol{u}} = \sum_{i=1}^{N} u_{\boldsymbol{\nu}_i} \sqrt{\sum_{j=1}^{K} |z_{ij}|^2}.$$

As discussed in Adcock et al. (2024a), the convex optimization problem (6.15) can be solved efficiently using Chambolle & Pock's primal-dual iteration (Chambolle and Pock, 2011, 2016) in combination with a restart scheme (Adcock et al., 2023b; Roulet and d'Aspremont, 2020). Specifically, one can obtain a γ-minimizer in at most $\log(1/\gamma)$ iterations, where the cost-per-iteration is bounded by

$$c \cdot (m \cdot N \cdot K + (m + N) \cdot (F(\boldsymbol{G}) + K)). \tag{6.16}$$

Here $F(\boldsymbol{G}) \leq K^2$ is the cost of performing the matrix-vector multiplication $\boldsymbol{x} \mapsto \boldsymbol{G}\boldsymbol{x}$. See Adcock et al. (2024a, Thm. 3.9 & Lem. 4.3).

6.6 Theoretical guarantee

We are now ready to present a theoretical guarantee for this estimator. For convenience, we define

$$L = L(m, \epsilon) = \log(m) \cdot (\log^3(m) + \log(\epsilon^{-1})). \tag{6.17}$$

Theorem 6.8 (Near-optimal learning via polynomials). *Let $m \geq 3$, $0 < \epsilon < 1$ and $n = \lceil m/L \rceil$, where $L = L(m, \epsilon)$ is as in* (6.17). *Let $0 < p < 1$, $\boldsymbol{b} \in \ell^p_{\mathsf{M}}(\mathbb{N})$, $f \in \mathcal{H}(\boldsymbol{b})$, $\boldsymbol{y}_1, \ldots, \boldsymbol{y}_m \sim_{\text{i.i.d.}} \varrho$ and consider the training data* (6.1). *Then with probability at least $1 - \epsilon$ the approximation $\hat{f}$ defined by* (6.14) *for any γ-minimizer $\hat{\boldsymbol{c}}$ of* (6.13), *$\gamma \geq 0$, with $\Lambda = \Lambda_n^{\mathsf{HCI}}$ satisfies*

$$\|f - \hat{f}\|_{L^2_\varrho(\mathcal{U};\mathcal{V})} \leq c \cdot \zeta, \tag{6.18}$$

where $c \geq 1$ is a universal constant and

$$\zeta := C \cdot (m/L)^{1/2-1/p} + \|\boldsymbol{e}\|_{2;\mathcal{V}}/\sqrt{m} + \|f - \mathcal{P}_h(f)\|_{L^\infty(\mathcal{U};\mathcal{V})} + \gamma. \tag{6.19}$$

See Adcock et al. (2024a, Thm. 3.7). The above theorem is slightly more general than Adcock et al. (2024a, Thm. 3.7) since it does not require $\boldsymbol{b}$ to be monotonically decreasing. This generalization follows from techniques given in the proof of Theorem 7.2 in Adcock and Monte (2023).

This result demonstrates that functions in $\mathcal{H}(p, \mathsf{M})$ can be learned from i.i.d. samples with a rate $O((m/\log^4(m))^{1/2-1/p})$ which, in view of Theorem 5.2, is near-optimal. Moreover, the estimator is robust to other sources of error in the problem. Indeed, the term (6.19) is a linear combination of the following.

(a) An *approximation error* $C \cdot (m/L)^{1/2-1p}$, as discussed.
(b) A *measurement error* $\|\boldsymbol{e}\|_{2;\mathcal{V}}/\sqrt{m}$, which accounts for any errors in computing the sample values $f(\boldsymbol{y}_i)$.
(c) A *physical discretization error* $\|f - \mathcal{P}_h(f)\|_{L^\infty(\mathcal{U};\mathcal{V})}$, which accounts for the error induced when working over the finite-dimensional space $\mathcal{V}_h$ instead of $\mathcal{V}$ and depends on the orthogonal projection $\mathcal{P}_h(f)$.
(d) An *optimization error* γ, which depends on the optimality gap of the computed solution $\hat{\boldsymbol{c}}$. See Remark 6.7 for further discussion on this term.

The proof of this theorem relies heavily on tools from compressed sensing. To relate it back to the discussion in §6.2, we remark that a key step involves establishing a sampling discretization of the form

$$\alpha\|p\|^2_{L^2_\varrho(\mathcal{U})} \leq \frac{1}{m}\sum_{i=1}^{m}|p(\boldsymbol{y}_i)|^2 \leq \beta\|p\|^2_{L^2_\varrho(\mathcal{U})}, \quad \forall p \in \mathbb{P}_S,\ S \subseteq \Lambda,\ |S|_{\boldsymbol{u}} \leq k. \tag{6.20}$$

This is stronger than (6.5), since it is required to hold simultaneously for all sets $S \subseteq \Lambda$ of weighted cardinality at most k. It is an example of a *universal sampling discretization* (Dai and Temlyakov, 2023). The proof first shows that it is equivalent to a certain *weighted Restricted Isometry Property (wRIP)*, then uses known results for the wRIP (see Brugiapaglia et al., 2021, as well as Rauhut and Ward, 2016). As proved therein, a sufficient condition for (6.20) to hold with probability at least $1-\epsilon$ and constants $\alpha \geq 1-\delta$ and $\beta \leq 1+\delta$ for some $0 < \delta < 1$ is

$$m \geq c \cdot \delta^{-2} \cdot k \cdot (\log(eN) \cdot \log^2(k/\delta) + \log(2/\epsilon)). \tag{6.21}$$

Crucially, up to the log terms, there is now a linear scaling between m and k. This is what leads to the near-optimal approximation error term.

Remark 6.9 (Sparse polynomial approximations). Generally, the vector $\hat{\boldsymbol{c}}$ computed by solving (6.13) will not be sparse. Therefore, $\hat{f}$ will not be a *sparse polynomial* approximation per se. Fortunately, one can always postprocess the solution to obtain such an approximation. This was shown in Adcock and Monte (2023, Thm. 7.2) in the scalar-valued case $\mathcal{V} = (\mathbb{R}, |\cdot|)$, but the technique extends straightforwardly to the Hilbert-valued case. In the case of Theorem 6.8, given a γ-minimizer $\hat{\boldsymbol{c}} = (\hat{c}_{\boldsymbol{\nu}})_{\boldsymbol{\nu}\in\Lambda} \in \mathcal{V}_h^N$ one first computes the index set $S \subseteq \Lambda$ of the largest $n = \lceil m/L \rceil$ entries of the vector $(\|\hat{c}_{\boldsymbol{\nu}}\|_{\mathcal{V}})_{\boldsymbol{\nu}\in\Lambda}$, and then replaces $\hat{f}$ with the n-sparse polynomial approximation $\check{f} = \sum_{\boldsymbol{\nu}\in S} \hat{c}_{\boldsymbol{\nu}} \Psi_{\boldsymbol{\nu}}$. As shown in Adcock and Monte (2023, Thm. 7.2), $\check{f}$ attains the same error bounds up to possible changes in the constants. The computational cost of this postprocessing step, roughly $O(N\log(N) + N \cdot F(\boldsymbol{G}))$ operations, is generally negligible in comparison to the cost of computing the initial γ-minimizer $\hat{\boldsymbol{c}}$.

6.7 Discussion and extensions

We conclude this section a short discussion. First, Theorem 6.8 asserts that i.i.d. pointwise samples constitute *near-optimal information*. This is an interesting facet of infinite-dimensional holomorphic function approximation, which contrasts starkly with the finite-, and specifically, low-dimensional case, where samples drawn i.i.d. from the uniform measure are distinctly suboptimal. See Adcock and Monte (2023) for a detailed analysis of this phenomenon.

On the other hand, Theorem 6.8 assumes that $\mathcal{V}$ is a Hilbert space, whereas the lower bounds in Theorem 5.2 allow for Banach spaces. Theorem 6.8 can be extended to Banach spaces, but with the suboptimal rate $(m/L)^{1/2(1/2-1/p)}$ (Adcock et al., 2023a, Thm. 4.1). It is currently unknown whether this rate can be improved.

Moreover, the computational cost of this procedure can be prohibitive. As discussed in Adcock et al. (2024a), combining (6.16) with a standard estimate for $N = |\Lambda_n^{\mathsf{HCI}}|$ leads to a per-iteration costs that scales like $m^{3+\log_2(m)}$ – i.e., subexponential, but superpolynomial in m. The reason for this is the need to construct the matrix $\boldsymbol{A}$ corresponding to all polynomials in the large index set Λ_n^{HCI}. Whether functions in $\mathcal{H}(p, \mathsf{M})$ can be learned to algebraic accuracy in m with polynomial computational cost in m is currently an open problem. Sublinear time algorithms (Choi et al., 2021a,b) may yield a solution to this problem.

7 DNN existence theory

We now turn our attention to DNNs. In this section, we discuss *existence theory*. Theorems of this ilk describe the *expressivity* of NNs: namely, their ability to approximation functions from specific classes to a desired accuracy. We commence with a review, before showing how to establish an existence theorem for $(\boldsymbol{b}, \varepsilon)$-holomorphic functions via the technique of polynomial emulation.

7.1 Review

Arguably, the first results on existence theory are the various *universal approximation theorems* for NNs. See Cybenko (1989); Hornik et al. (1989) and, in particular, Pinkus (1999). These show that shallow NNs with one hidden layer can approximate any continuous function.

Unfortunately, these classical results only apply to shallow networks, and do not always give quantitative bounds on the complexity (i.e., the width) of the corresponding networks (some notable exceptions to this include Mhaskar, 1996, and references therein). These issues have been investigated intensively over the last five years. Some of the first results were obtained in Yarotsky (2017). Here, explicit width and depth bounds were derived for DNNs with the *Rectified Linear Unit (ReLU)* activation function for approximating Lipschitz continuous functions in the L^{∞}-norm over compact sets. This inspired many subsequent works. Other activations have been studied, including *hyperbolic tangents (tanh)*, *Rectified Quadratic Units (ReQU)* and *Rectified Polynomial Units (RePU)*, and various others. And quantitative bounds have been shown for various different function classes. A partial list includes: functions in Sobolev spaces with ReLU (Yarotsky, 2017; Gühring et al., 2020), RePU (Li et al., 2020), tanh (De Ryck et al., 2021) or rational (Boullé et al., 2020) activations; piecewise smooth functions with ReLU (Petersen and Voigtlaender, 2018); C^k and Hölder smooth functions with ReLU (J. Lu et al., 2021; Schmidt-Hieber, 2020) or general (Ohn and Kim, 2019) activations; uniformly continuous functions with ReLU activations (Yarotsky, 2018); functions in Besov spaces with ReLU (Opschoor et al., 2020; Suzuki, 2019); spaces of mixed smoothness with ReLU (Blanchard and Bennouna, 2020; Montanelli and Du, 2019; Dũng and Nguyen, 2021; Suzuki, 2019), RePU (Li et al., 2020) or smooth activations (Blanchard and Bennouna, 2020); Gevrey functions with ReLU (Opschoor et al., 2020), RePU (Opschoor et al., 2022) or tanh (De Ryck et al., 2021) activations; bandlimited functions with ReLU activations (Montanelli et al., 2021); functions in Barron spaces (E et al., 2021) with ReLU and non-ReLU activations; compositional functions (Poggio et al., 2017; Liang and Srikant, 2017; Schmidt-Hieber, 2020); smooth functions on manifolds (Chen et al., 2022; Shaham et al., 2018); finite-dimensional, analytic functions with smooth (Mhaskar, 1996), ReLU (E and Wang, 2018) or RePU (Opschoor et al., 2022) activation functions; and, most relevantly to this work, infinite-dimensional holomorphic functions with ReLU (Schwab and Zech, 2019, 2023; Dũng et al., 2023), tanh (De Ryck et al., 2021) or RePU (Schwab and Zech, 2023) activations. See also DeVore et al. (2021); Elbrächter et al. (2021) for reviews.

These existence theorems are based on emulating suitable classical approximation schemes with DNNs. Emulation results include (localized) Taylor polynomials (De Ryck et al., 2021; Gühring and Raslan, 2021; Liang and Srikant, 2017; Li et al., 2020; J. Lu et al., 2021; Yarotsky, 2017; Schwab and Zech, 2019), orthogonal polynomials (Adcock et al., 2023a; Daws and C. Webster, 2019; Mhaskar, 1996; Opschoor et al., 2022; Tang et al., 2019; Dũng et al., 2023; Opschoor and Schwab, 2023; Schwab and Zech, 2023), rational functions (Boullé et al., 2020; Telgarsky, 2017), wavelets (Shaham et al., 2018) and general affine systems (Bölcskei et al., 2019), B-splines (Mhaskar, 1993), free-knot splines (Opschoor et al., 2020), finite elements (Longo et al., 2023; Opschoor et al., 2020; Opschoor and Schwab, 2023) and sparse grids (Blanchard and Bennouna, 2020; Montanelli and Du, 2019; Dũng and Nguyen, 2021; Suzuki, 2019).

To highlight some of these themes, we next describe an existence theorem for $(\boldsymbol{b}, \varepsilon)$-holomorphic functions. This is achieved by first emulating Legendre polynomials with DNNs. For ease of presentation, we consider tanh DNNs. But, as we discuss, other activation functions can also be used.

7.2 Neural network architectures

In this and subsequent sections, we consider standard feedforward DNN architectures of the form

$$\Phi : \mathbb{R}^n \to \mathbb{R}^k, \; z \mapsto \Phi(z) = \mathcal{A}_{D+1}(\sigma(\mathcal{A}_D(\sigma(\cdots\sigma(\mathcal{A}_0(z))\cdots)))), \tag{7.1}$$

where $\mathcal{A}_l : \mathbb{R}^{N_l} \to \mathbb{R}^{N_{l+1}}$, $l = 0, \ldots, D+1$ are affine maps and σ is the activation function, which we assume acts componentwise. The values $\{N_l\}_{l=1}^{D+1}$ are the widths of the hidden layers, and for convenience, we write $N_0 = n$ and $N_{D+2} = k$. Given such a DNN Φ, we write

$$\text{width}(\Phi) = \max\{N_1, \ldots, N_{D+1}\}, \qquad \text{depth}(\Phi) = D.$$

We denote a class of DNNs of the form (7.1) with a fixed architecture (i.e., fixed activation function, depth and widths) as $\mathcal{N}$, and define

$$\text{width}(\mathcal{N}) = \max\{N_1, \ldots, N_{D+1}\}, \qquad \text{depth}(\mathcal{N}) = D.$$

Finally, since the DNNs (7.1) take a finite input, yet the functions considered in this work take inputs in $\mathcal{U} \subseteq \mathbb{R}^{\mathbb{N}}$, we also need to introduce a restriction operator. Let $\Theta \subset \mathbb{N}$ with $|\Theta| = n$. Then we define the *variable restriction operator* as

$$\mathcal{T}_\Theta : \mathbb{R}^{\mathbb{N}} \to \mathbb{R}^n, \; \boldsymbol{y} = (y_i)_{i\in\mathbb{N}} \to (y_i)_{i\in\Theta}. \tag{7.2}$$

If $\Theta = \{1, \ldots, n\}$ then we simply write $\mathcal{T}_\Theta = \mathcal{T}_n$.

7.3 Emulating polynomials with DNNs: typical result

We now present a result on emulating a finite set of multivariate Legendre polynomials with a tanh DNN.

Theorem 7.1 (Emulating multivariate Legendre polynomials). *Let $\Lambda \subset \mathcal{F}$ be a finite multiindex set and $\Theta \subset \mathbb{N}$, $|\Theta| = n$, be such that* $\text{supp}(\boldsymbol{\nu}) \subseteq \Theta$, $\forall \boldsymbol{\nu} \in \Lambda$. *Then for every $0 < \delta < 1$ there exists a tanh DNN $\Phi_{\Lambda,\delta} : \mathbb{R}^n \to \mathbb{R}^{|\Lambda|}$, such that, if $\Phi_{\Lambda,\delta}(z) = (\Phi_{\boldsymbol{\nu},\delta}(z))_{\boldsymbol{\nu}\in\Lambda}$, $z = (z_j)_{j\in\Theta} \in \mathbb{R}^n$ and $\mathcal{T}_\Theta$ is as in (7.2), then*

$$\|\Psi_{\boldsymbol{\nu}} - \Phi_{\boldsymbol{\nu},\delta} \circ \mathcal{T}_\Theta\|_{L^\infty(\mathcal{U})} \le \delta, \quad \forall \boldsymbol{\nu} \in \Lambda,$$

where $\Psi_{\boldsymbol{\nu}}$ is the corresponding multivariate Legendre polynomial. The width and depth of this network satisfy

$$\text{width}(\Phi_{\Lambda,\delta}) \le c_1 \cdot |\Lambda| \cdot m(\Lambda), \quad \text{depth}(\Phi_{\Lambda,\delta}) \le c_2 \cdot \log_2(m(\Lambda)),$$

for universal constants $c_1, c_2 > 0$, where $m(\Lambda) = \max_{\boldsymbol{\nu}\in\Lambda} \|\boldsymbol{\nu}\|_1$.

This result was shown in Adcock et al. (2023a, Thm. 7.4), and is based on techniques of Daws and C. Webster (2019); De Ryck et al. (2021); Opschoor et al. (2022). To avoid unnecessary complications, we have stated it for tanh DNNs. It also applies without changes to DNNs with the sigmoid activation, since this is obtained from tanh via shifting and scaling. The result (Adcock et al., 2023a, Thm. 7.4) also considers the ReLU and RePU activations. As discussed therein, the proof readily extends to more general classes of activation functions. See also De Ryck et al. (2021, Rem. 3.9) and Opschoor et al. (2022, §2.4) for related discussion.

7.4 Elements of the proof of Theorem 7.1

The proof of Theorem 7.1 involves three key ingredients:

(i) The map $(x, y) \mapsto xy$ can be approximated by a shallow tanh NN.
(ii) Using (i), the multiplication of d numbers $(x_1, \dots, x_d) \mapsto x_1 \cdots x_d$ can be approximated by a tanh DNN of depth $c\lceil \log_2(d) \rceil$ for some constant $c > 0$.
(iii) By the fundamental theorem of algebra, the multivariate Legendre polynomial $\Psi_{\boldsymbol{\nu}}$ can be expressed as a product of $\|\boldsymbol{\nu}\|_1$ terms.

The constructions that lead to the proofs of (i) and (ii) have become standard. In the case of tanh DNNs, they can be found in De Ryck et al. (2021, Lem. 3.8). For (i), the basic idea is to use the relation $xy = ((x + y)^2 - (x - y)^2)/4$ in combination with a tanh DNN for approximating the map $x \mapsto x^2$. The latter is constructed via finite differences (see De Ryck et al., 2021, Lem. 3.1, and Gühring and Raslan, 2021). Having shown (i), the next ingredient (ii) follows from arguments given (Schwab and Zech, 2019, Prop. 3.3), which formulates a DNN for multiplying d numbers as a binary tree of depth $\lceil \log_2(d) \rceil$. The final ingredient (iii) follows Daws and C. Webster (2019) and involves writing

$$\Psi_{\boldsymbol{\nu}}(\boldsymbol{y}) = \prod_{i \in \operatorname{supp}(\boldsymbol{\nu})} \prod_{j=1}^{\nu_i} a_{ij}(y_i - r_{ij}), \quad \forall \boldsymbol{y} \in \mathcal{U}, \ \boldsymbol{\nu} \in \mathcal{F}, \tag{7.3}$$

for scalars $a_{ij}, r_{ij} \in \mathbb{R}$. We deduce that $\Psi_{\boldsymbol{\nu}}(\boldsymbol{y})$ can be approximated by the composition of the affine map $\boldsymbol{y} \mapsto (a_{ij}(y_i - r_{ij}))_{ij}$ and the tanh DNN that approximately computes the product in (7.3).

The precise bounds on the architecture given in Theorem 7.1 are now evident. First, each polynomial $\Psi_{\boldsymbol{\nu}}$ involves a product of $\|\boldsymbol{\nu}\|_1 \le m(\Lambda)$ numbers. The depth bound now follows immediately. The width bound follows from the fact that the DNNs emulating each polynomial are stacked vertically to form the overall network $\Phi_{\Lambda,\delta}$ which simultaneously emulates all $|\Lambda|$ polynomials.

It is also evident that this proof can be adapted to any other activation function for which the multiplication map in (i) can be approximated with a DNN of a quantifiable width and depth. ReLU DNNs were also considered in Adcock et al. (2023a, Thm. 7.4), with ingredients (i) and (ii) being based on Opschoor et al. (2022, Prop. 2.6). The resulting depth and width bounds are worse than the tanh case. Conversely, a RePU network of depth one and constant width can *exactly* represent the multiplication of two numbers (see, e.g., Li et al., 2020, Lem. 2.1). Therefore, a RePU network with the same width and depth bounds as in Theorem 7.1 can *exactly* emulate the Legendre polynomials, as opposed to approximately in the tanh case. The same holds for DNNs based on *rational* activation functions (Boullé et al., 2020).

Finally, we note that the proof can be easily adapted to other polynomial systems, such as Chebyshev or, more generally, Jacobi polynomials. In the RePU case, as before, this emulation is exact.

7.5 Existence theorem for $(\boldsymbol{b}, \varepsilon)$-holomorphic functions

We now present an existence theorem for $(\boldsymbol{b}, \varepsilon)$-holomorphic functions.

Theorem 7.2 (Existence theorem for DNNs). *Let $\mathcal{V} = (\mathbb{R}, |\cdot|)$ and $\boldsymbol{b} \in [0, \infty)^{\mathbb{N}}$ be such that $\boldsymbol{b} \in \ell^p(\mathbb{N})$ for some $0 < p < 1$. Then, for every $s \in \mathbb{N}$ there exists a class of tanh DNNs $\Phi : \mathbb{R}^s \to \mathbb{R}$ with*

$$\mathrm{width}(\mathcal{N}) \leq c_1 \cdot s^2, \quad \mathrm{depth}(\mathcal{N}) \leq c_2 \cdot \log(s),$$

for universal constants $c_1, c_2 > 0$ such that the following holds. For every $f \in \mathcal{H}(\boldsymbol{b})$, there is a $\Phi \in \mathcal{N}$ such that

$$\|f - \Phi \circ \mathcal{T}_s\|_{L^2_{\varrho}(\mathcal{U})} \leq C(\boldsymbol{b}, p) \cdot s^{\frac{1}{2} - \frac{1}{p}}.$$

This result follows immediately from Theorems 6.4 and 7.1. The former asserts the existence of a lower set S of size $|S| \leq s$ and depending on $\boldsymbol{b}$ and p only which attains the desired rate of $s^{1/2-1/p}$. Then the latter asserts the existence of the DNN $\Phi_{S,\delta}$, with $\delta = s^{-1/p}$, of the desired size. Note that $m(S) \leq s$ for any lower set. Indeed, if $\boldsymbol{\nu} \in S$ then so does the set $\{\boldsymbol{\mu} \in \mathcal{F} : \boldsymbol{\mu} \leq \boldsymbol{\nu}\}$. Therefore

$$\|\boldsymbol{\nu}\|_1 \leq \prod_{i \in \mathbb{N}} (\nu_i + 1) = |\{\boldsymbol{\mu} \in \mathcal{F} : \boldsymbol{\mu} \leq \boldsymbol{\nu}\}| \leq |S| \leq s.$$

Theorem 7.2 only deals with scalar-valued functions. We will tackle Hilbert-valued functions in the next section when we consider learning via DNNs.

8 Practical existence theory: near-optimal DL

While existence theory provides crucial insight into the *expressivity* of DNNs, it says nothing about whether networks with similar approximation guarantees can be trained in practice and, in particular, how much training data suffices to do so. To narrow this gap between theory and practice, in this section we develop the topic of *practical existence theory*. The goal of this endeavor is to show that there exists both an architecture *and* a training strategy that is similar to what is used in practice (i.e., minimizing a loss function) from which one provably learns near-optimal DNN approximations from the training data (6.1).

8.1 Setup

As in §6, we assume that $\mathcal{V}$ is a Hilbert space with a finite-dimensional discretization $\mathcal{V}_h$. Following in Remark 6.7, we let $\{\varphi_i\}_{i=1}^K$ be a basis for $\mathcal{V}_h$. If $f : \mathcal{U} \to \mathcal{V}$ is the function to recover, then we write

$$f(\boldsymbol{y}) \approx \sum_{i=1}^{K} d_i(\boldsymbol{y})\varphi_i,$$

where the coefficients $d_i : \mathcal{U} \to \mathbb{R}$ are scalar-valued functions. We now seek to approximate these functions using a DNN with K neurons on the output layer, with the ith neuron corresponding to the approximation of the function d_i. Let $\mathcal{N}$ be a class of DNNs of the form $\Phi : \mathbb{R}^n \to \mathbb{R}^K$ for some $n \in \mathbb{N}$. Then our aim is to use the training data (6.1) to compute a suitable $\Phi \in \mathcal{N}$ that yields an approximation $f_{\hat{\Phi}} \approx f$ defined by

$$f_{\Phi}(\boldsymbol{y}) := \sum_{i=1}^{K} (\Phi \circ \mathcal{T}_n(\boldsymbol{y}))_i \varphi_i, \quad \forall \boldsymbol{y} \in \mathcal{U}. \tag{8.1}$$

8.2 Practical existence theorem

The first practical existence theorems were shown in Adcock and Dexter (2021) (scalar-valued case) and Adcock et al. (2021) (Hilbert-valued case) for holomorphic function approximation in finite dimensions with ReLU DNNs. This was extended in Adcock et al. (2023a) to infinite-dimensional, holomorphic functions with ReLU, RePU or tanh activation functions. The following result is based on Adcock et al. (2023a, Thm. 4.4).

Theorem 8.1 (Practical existence theorem for DNNs). *There are universal constants $c_1, c_2, c_3 \geq 1$ such that the following holds. Let $m \geq 3$, $0 < \epsilon < 1$ and $n = \lceil m/L \rceil$, where $L = L(m, \epsilon)$ is as in* (6.17). *There exists*

1. *a class $\mathcal{N}$ of DNNs $\Phi : \mathbb{R}^n \to \mathbb{R}^K$ with tanh activation function satisfying*

$$\mathrm{width}(\mathcal{N}) \leq c_1 \cdot m^{3+\log_2(m)}, \quad \mathrm{depth}(\mathcal{N}) \leq c_2 \cdot \log(m);$$

2. *a regularization function $\mathcal{J} : \mathcal{N} \to [0, \infty)$ equivalent to a certain norm of the trainable parameters;*
3. *and a choice of regularization parameter λ involving only m and ϵ;*

such that following holds for every $0 < p < 1$ and $\boldsymbol{b} \in \ell^p_{\mathsf{M}}(\mathbb{N})$. Let $f \in \mathcal{H}(\boldsymbol{b})$, $\boldsymbol{y}_1, \ldots, \boldsymbol{y}_m \sim_{\text{i.i.d.}} \varrho$ and consider the training data (2.2). *Then, with probability* *at least* $1-\epsilon$, *every* γ*-minimizer* $\hat{\Phi}$, $\gamma \geq 0$, *of the training problem*

$$\min_{\Phi \in \mathcal{N}} \mathcal{G}(\Phi), \quad \text{where } \mathcal{G}(\Phi) = \sqrt{\frac{1}{m}\sum_{i=1}^{m} \| f_\Phi(\boldsymbol{y}_i) - d_i \|_{\mathcal{V}}^2} + \lambda \mathcal{J}(\Phi), \tag{8.2}$$

where f_Φ *is as in* (8.1)*, satisfies, with* ζ *as in* (6.19),

$$\| f - f_{\hat{\Phi}} \|_{L^2_\varrho(\mathcal{U};\mathcal{V})} \leq c_3 \cdot \zeta. \tag{8.3}$$

In this theorem, as before, we term $\hat{\Phi}$ a *γ-minimizer* of (8.2) if $\mathcal{G}(\hat{\Phi}) \leq \min_{\Phi \in \mathcal{N}} \mathcal{G}(\Phi) + \gamma$. Comparing with Theorem 6.8, we conclude that there is a DNN architecture and training procedure such that the resulting learned approximations achieve the same error bounds (up to possible constants) as sparse polynomial approximation based on weighted ℓ^1-minimization. In particular, this procedure also achieves the near-optimal approximation rate $(m/L)^{\frac{1}{2}-\frac{1}{p}}$.

We term Theorem 8.1 a *practical existence theorem.* It not only asserts the existence of a DNN with a given architecture that achieves some desired rate of approximation, but demonstrates how to construct it from training data and gives generalization bounds that are explicit in the amount of training data m. Moreover, the training procedure is similar to standard DL procedures, in that it involves minimizing a (regularized) least-squares loss function.

While Theorem 8.1 only considers tanh DNNs, it can be readily adapted to other activations (Adcock et al., 2023a). As discussed next, all one requires to do this are variants of Theorem 7.1 for other activations, a topic we discussed previously in §7.4.

8.3 The mechanism of practical existence theorems

We now give some insight into the proof of Theorem 8.1, since this provides a general recipe for establishing practical existence theorems.

The overall mechanism involves not just emulating polynomials with DNNs, but also emulating the weighted ℓ^1-minimization problem of §6.1 as a DNN training problem. More precisely, following §6.1 we proceed as follows.

(i) Approximate the Legendre polynomials $\{\Psi_{\boldsymbol{\nu}}\}_{\boldsymbol{\nu} \in \Lambda}$ using DNNs.
(ii) Replace the polynomials in the matrix (6.3) by these DNNs, leading to a matrix $\boldsymbol{A}' \approx \boldsymbol{A}$. Then replace $\boldsymbol{A}$ by $\boldsymbol{A}'$ in (6.15).
(iii) Re-cast (6.13) with $\boldsymbol{A}'$ as a training problem of the form (8.2), where the unknowns $\boldsymbol{Z} \in \mathbb{R}^{N \times K}$ correspond to the trainable parameters of the DNNs.
(iv) Emulate the proof steps of Theorem 6.8, showing, as needed, that each step remains valid for the perturbed matrix $\boldsymbol{A}'$.

Step (i) is accomplished straightforwardly by using the emulation result, Theorem 7.1. Since Λ is taken to be $\Lambda = \Lambda_n^{\mathsf{HCI}}$, we have $\operatorname{supp}(\boldsymbol{\nu}) \subseteq \{1, \ldots, n\}$

for $\boldsymbol{\nu} \in \Lambda$. Hence we set $\Theta = \{1, \ldots, n\}$ and consider a suitable parameter δ that is chosen later in the proof to balance the ensuing error terms. Let $\Phi_{\Lambda,\delta} : \mathbb{R}^n \to \mathbb{R}^{|\Lambda|}$ be the resulting DNN.

Step (ii) warrants no further discussion. Now consider Step (iii). Let $N = |\Lambda|$ as before, and define the class of DNNs

$$\mathcal{N} = \left\{ \Phi = \boldsymbol{Z}^\top \Phi_{\Lambda,\delta} : \boldsymbol{Z} \in \mathbb{R}^{N \times K} \right\}.$$

This is a class of DNNs, where only the weight matrix on the output layer is trainable. The remaining layers are nontrainable, and are handcrafted to emulate the Legendre polynomials $\{\Psi_{\boldsymbol{\nu}}\}_{\boldsymbol{\nu} \in \Lambda}$.

Now let $z = (z_{\boldsymbol{\nu}_j})_{j=1}^N \in \mathcal{V}_h^N$. Then we can associate z with its matrix of coefficients $\boldsymbol{Z} = (Z_{ij}) \in \mathbb{R}_{i,j=1}^{N \times K}$ via the relation

$$z_{\boldsymbol{\nu}_i} = \sum_{j=1}^{K} Z_{ij} \varphi_j, \quad \forall i = 1, \ldots, N, \tag{8.4}$$

and consequently with the DNN $\Phi = \boldsymbol{Z}^\top \Phi_{\Lambda,\delta} \in \mathcal{N}$. Using this, it is a short argument to show that

$$f_\Phi(\boldsymbol{y}) = \sum_{\boldsymbol{\nu} \in \Lambda} z_{\boldsymbol{\nu}} \Phi_{\boldsymbol{\nu},\delta} \circ \mathcal{T}_\Theta(\boldsymbol{y}).$$

Therefore,

$$\|\boldsymbol{A}' z - \boldsymbol{f}\|_{2;\mathcal{V}}^2 = \frac{1}{m} \sum_{i=1}^{m} \left\| \sum_{\boldsymbol{\nu} \in \Lambda} z_{\boldsymbol{\nu}} \Phi_{\boldsymbol{\nu},\delta}(\boldsymbol{y}_i) - d_i \right\|_{\mathcal{V}}^2 = \frac{1}{m} \sum_{i=1}^{m} \| f_\Phi(\boldsymbol{y}_i) - d_i \|_{\mathcal{V}}^2.$$

Now let $\mathcal{J} : \mathcal{N} \to [0, \infty)$ be the regularization functional defined by

$$\mathcal{J}(\Phi) = \sum_{j=1}^{N} u_{\boldsymbol{\nu}_j} \left\| \sum_{k=1}^{K} Z_{jk} \varphi_k \right\|_{\mathcal{V}} = \sum_{j=1}^{N} u_{\boldsymbol{\nu}_j} \|z_{\boldsymbol{\nu}_j}\|_{\mathcal{V}} = \|z\|_{1,\boldsymbol{u};\mathcal{V}},$$

for $\Phi = \boldsymbol{Z}^\top \Phi_{\Lambda,\delta} \in \mathcal{N}$, where $z = (z_{\boldsymbol{\nu}_i})_{i=1}^N$ is as defined by (8.4). We readily see that $\mathcal{J}$ is a norm over the trainable parameters. Using this and the previous expression we see that (6.13) with the matrix $\boldsymbol{A}'$ in place of $\boldsymbol{A}$ can be re-cast as a training problem of the form (8.2).

Finally, consider step (iv). This step is facilitated by the perturbation bound $\|\boldsymbol{A} - \boldsymbol{A}'\|_2 \leq \sqrt{N}\delta$, which in turn follows from a short argument via the Cauchy-Schwarz inequality:

$$\|(A - A')z\|_2^2 = \frac{1}{m}\sum_{i=1}^{m}\left|\sum_{\nu \in \Lambda}(\Psi_\nu(y_i) - \Phi_{\nu,\delta}(y_i))z_\nu\right|^2 \leq \left(\sum_{\nu \in \Lambda}\delta|z_\nu|\right)^2$$
$$\leq \delta^2 N\|z\|_2^2.$$

The remainder of this step involves a careful modification of the proof of Theorem 6.8 to take into account this perturbation. Theorem 6.8 involves compressed sensing techniques, and relies on first asserting that the matrix A has a certain *weighted robust Null Space Property (rNSP)* (Rauhut and Ward, 2016) (see also Adcock et al., 2022, Chpt. 6). This is a weighted variant of the classical (unweighted) rNSP, and like the latter, is a slightly weaker condition than the better known (weighted) RIP (Foucart and Rauhut, 2013). Crucially, it can be shown that the (weighted) rNSP is preserved under sufficiently small perturbations (Adcock and Hansen, 2021, Lem. 8.5).

We note in passing that the width and depth bounds in Theorem 8.1 follow quite directly from Theorem 7.1. Recall that $\Lambda = \Lambda_n^{\mathrm{HCI}}$ is as in (6.12). A standard estimate (see Kühn et al., 2015, Thm. 4.9) gives that

$$|\Lambda| \leq e n^{2+\log_2(n)}, \quad \forall n \in \mathbb{N}. \tag{8.5}$$

Moreover, by definition, any $\nu \in \Lambda$ satisfies $\|\nu\|_1 \leq \prod_{k=1}^{n-1}(\nu_k + 1) \leq n$. Hence

$$m(\Lambda) \leq n. \tag{8.6}$$

The desired width and depth bounds now follow immediately from Theorem 7.1 and these estimates, along with the (somewhat loose) bound $n \leq m$.

9 Epilogue

This work has been about the approximation of smooth, infinite-dimensional functions from limited data. We close with a discussion on the benefits and consequences of practical existence theory and the gap between the *handcrafted* models on which it is based and the *fully trained* models used in practice.

9.1 Scientific computing and data scarcity

As discussed in §1.1, parametric DE problems are often *data scarce*. This is the case for many problems in scientific computing in which machine learning and, specifically, DL, is currently being applied. It also stands in stark contrast to more classical DL applications such as image classification, where datasets usually contain tens of millions of images or more. Therefore, understanding the sample complexity is crucial. Practical existence theorems show that DNNs can be learned from data in a sample-efficient manner.

9.2 Potential benefits to DNNs over sparse polynomials

Having said this, we emphasize that practical existence theorems are, at least in this work, intended primarily as *theoretical* contributions. Since the strategy in Theorem 8.1 involves emulating the sparse polynomial approximation scheme constructed in §6, there is no benefit to implementing it over the latter.

However, in related work in inverse problems in imaging (recall the discussion in §1.3), unrolling is both used to establish practical existence theorems *and* as a principled way to design DNN architectures which can then be trained as part of a DL strategy (Monga et al., 2021), potentially using the theoretical weights and biases as initialization. It remains to see whether similar ideas could be effective in the parametric DE setting. Some initial work in this direction can be found in Daws and C.G. Webster (2019).

In addition, existence theory establishes that DNNs have the capacity to approximate broad classes of functions efficiently. This is *not* the case for polynomials, which fail dramatically on, for instance, discontinuous functions. This work has focused on classes of holomorphic functions, where polynomials are well suited (in fact, near optimal). DNN-based schemes have the potential to succeed on quite different function classes, something that distinguishes them from traditional methods of scientific computing. We remark in passing that discontinuous or sharp transitions arise frequently in parametric model problems, and are difficult to treat with standard methods (Elman and Miller, 2012; Gunzburger et al., 2014a,b; Jakeman et al., 2011; Ma and Zabaras, 2009; Zhang et al., 2016).

9.3 Theorem 8.1 does not eliminate the theory-practice gap

DNNs in Theorem 8.1 are handcrafted to emulate Legendre polynomials, with only the final layer being trained. Standard DL methods use fully trained models, where all layers are trained. Theorem 8.1 says nothing directly about this practice. However, it does lead to some insights, as we now discuss.

9.4 Practical insights

Width and depth bounds

The aim of practical existence theory is to express the error in terms of the sample complexity m. This differs from standard existence theory in which express the error is expressed in terms of the complexity of the network, e.g., its width and depth, or its *size* (number of nonzero weights and biases). Nonetheless, it is worth discussing the network complexity in Theorem 8.1.

This theorem describes architectures that are much wider than they are deep. In fact, the depth grows very slowly with the number of samples m, like $\log(m)$. This broadly agrees with empirical insights from the application of DL in scientific computing, where it is often observed that relatively shallow networks perform well. See De Ryck et al. (2021) and references therein, as well as Adcock et al. (2021); Adcock and Dexter (2021).

However, while the depth bound in Theorem 8.1 is somewhat reasonable, the width bound of $\mathcal{O}(m^{3+\log_2(m)})$ grows extremely rapidly with m (albeit subexponentially). As discussed in §8.3, the large estimate for the width arises, in great part, from having to emulate all polynomials in the hyperbolic cross Λ_n^{HCI}, whose size, as shown in (8.5), behaves like $\mathcal{O}(n^{2+\log_2(n)})$. This has nothing to do with DNN approximation itself. It stems from the sparse polynomial approximation scheme and, as discussed in §6.3, the need to build a finite search set Λ outside of which the error is ensured to be small. It is unclear whether the near-optimal approximation rates of Theorem 6.8 can be obtained with a smaller search space without further assumptions on the functions being approximated. This is an interesting prospect for future work.

Post-training pruning and sparsification

On the other hand, it is always possible to sparsify the DNN $\hat{\Phi}$ learned in Theorem 8.1 after training – a process known as *pruning* (see Frankle and Carbin, 2019, and references therein). This exploits the fact that the networks in $\mathcal{N}$, while very wide, are sparsely connected. This is done in much the same way as the in polynomial case (Remark 6.9). If $\hat{\boldsymbol{C}} \in \mathbb{R}^{N \times K}$ are the weights of $\hat{\Phi}$, then one first forms $\hat{\boldsymbol{c}} = (\hat{c}_{\boldsymbol{\nu}})_{\boldsymbol{\nu}\in\Lambda} \in \mathcal{V}_h^N$ using (8.4), then computes the index set $S \subseteq \Lambda$ of the largest n entries of $(\|\hat{c}_{\boldsymbol{\nu}}\|_{\mathcal{V}})_{\boldsymbol{\nu}\in\Lambda}$, and finally replaces $\hat{\Phi}$ with $\check{\Phi} = \boldsymbol{Z}_S^\top \Phi_{S,\delta}$, where $\boldsymbol{Z}_S \in \mathbb{R}^{n \times K}$ is formed from the rows of $\boldsymbol{Z}$ with indices in S. By Theorem 7.1 and the bound $m(S) \leq m(\Lambda) \leq n$ (recall (8.6)), we have $\text{width}(\check{\Phi}) \leq c_1 m^2$ and $\text{depth}(\check{\Phi}) \leq c_2 \log(m)$. In particular, $\check{\Phi}$ is significantly narrower than $\hat{\Phi}$.

This suggests that methods to promote sparsity in training (see Hoefler et al., 2021, and references therein) may be beneficial in practice when training fully-connected models for scientific computing problems. This requires further investigation.

9.5 Eliminating the gap: beating the Monte Carlo rate is key

Despite these insights, the gap between theory and practice persists. It is worth noting that standard approaches to estimating the generalization error for fully-trained models based on statistical learning theory and the *bias-variance decomposition* (see, e.g., Beck et al., 2022; Chen et al., 2022; Ohn and Kim, 2019; Schmidt-Hieber, 2020; Suzuki, 2019 and references therein) are not immediately applicable to this setting. These approaches use estimates for the covering number or Rademacher complexity of the relevant DNN classes. Unfortunately, they typically lead to rates that decay at best like $m^{-1/2}$. These rates are strictly slower than near-optimal rates $m^{1/2-1/p}$, up to log factors, asserted in Theorem 8.1. A major theme in parametric DEs and computational UQ is building methods that beat the *Monte Carlo* rate $m^{-1/2}$ (Adcock et al., 2022, Chpt. 1).

Whether these approaches could be meaningfully combined with practical existence theorems is an interesting question for future work.

9.6 Conclusion

To summarize, practical existence theory is a promising way to study DNN approximation which has the potential to give new insights into the promises and challenges of data scarce applications in computational science and engineering. In addition to those outlined above, several interesting future avenues include the design of improved training methodologies and novel architectures and activation functions. Studying these areas is key for pushing the boundaries of what DNNs can achieve in scientific computing, particularly in the face of scarce data and complex computational tasks. These efforts may not only narrow the theory-to-practice gap but also unlock new DL approaches, making it more efficient, accurate, and applicable across a broader spectrum of scientific challenges and guiding the way towards more sophisticated and capable DNN models.

References

Abedeljawad, A., Grohs, P., 2023. Sampling complexity of deep approximation spaces. arXiv:2312.1337.

Adcock, B., 2017. Infinite-dimensional ℓ^1 minimization and function approximation from pointwise data. Constr. Approx. 45 (3), 343–390.

Adcock, B., Bao, A., Brugiapaglia, S., 2019. Correcting for unknown errors in sparse high-dimensional function approximation. Numer. Math. 142 (3), 667–711.

Adcock, B., Brugiapaglia, S., 2023. Monte Carlo is a good sampling strategy for polynomial approximation in high dimensions. arXiv:2208.09045.

Adcock, B., Brugiapaglia, S., Dexter, N., Moraga, S., 2021. Deep neural networks are effective at learning high-dimensional Hilbert-valued functions from limited data. In: Bruna, J., Hesthaven, J.S., Zdeborová, L. (Eds.), Proceedings of the Second Annual Conference on Mathematical and Scientific Machine Learning. In: Proc. Mach. Learn. Res. (PMLR), vol. 145. PMLR, pp. 1–36.

Adcock, B., Brugiapaglia, S., Dexter, N., Moraga, S., 2023a. Near-optimal learning of Banach-valued, high-dimensional functions via deep neural networks. arXiv:2211.12633.

Adcock, B., Brugiapaglia, S., Dexter, N., Moraga, S., 2024a. On Efficient Algorithms for Computing Near-Best Polynomial Approximations to High-Dimensional, Hilbert-Valued Functions from Limited Samples. Mem. Eur. Math. Soc. https://dx.doi.org/10.4171/MEMS/13.

Adcock, B., Brugiapaglia, S., Webster, C.G., 2022. Sparse Polynomial Approximation of High-Dimensional Functions. Comput. Sci. Eng. Society for Industrial and Applied Mathematics, Philadelphia, PA.

Adcock, B., Colbrook, M.J., Neyra-Nesterenko, M., 2023b. Restarts subject to approximate sharpness: a parameter-free and optimal scheme for first-order methods. arXiv:2301.02268.

Adcock, B., Dexter, N., 2021. The gap between theory and practice in function approximation with deep neural networks. SIAM J. Math. Data Sci. 3 (2), 624–655.

Adcock, B., Dexter, N., Moraga, S., 2023c. Optimal approximation of infinite-dimensional holomorphic functions II: recovery from i.i.d. pointwise samples. arXiv:2310.16940.

Adcock, B., Dexter, N., Moraga, S., 2024b. Optimal approximation of infinite-dimensional holomorphic functions. Calcolo 61 (12).

Adcock, B., Hansen, A.C., 2021. Compressive Imaging: Structure, Sampling, Learning. Cambridge University Press, Cambridge, UK.

Ajavon, K., 2024. Surrogate models for diffusion on graphs: a high-dimensional polynomial approach. Master's thesis. Concordia University.

Antun, V., Gottschling, N.M., Hansen, A.C., Adcock, B., 2023. Am (A)I hallucinating? Non-robustness, hallucinations and unpredictable performance of AI for MR image reconstruction. Preprint.

Antun, V., Renna, F., Poon, C., Adcock, B., Hansen, A.C., 2020. On instabilities of deep learning in image reconstruction and the potential costs of AI. Proc. Natl. Acad. Sci. USA 117 (48), 30088–30095.

Babuška, I.M., Nobile, F., Tempone, R., 2007. A stochastic collocation method for elliptic partial differential equations with random input data. SIAM J. Numer. Anal. 43 (3), 1005–1034.

Bäck, J., Nobile, F., Tamellini, L., Tempone, R., 2011. Stochastic spectral Galerkin and collocation methods for PDEs with random coefficients: a numerical comparison. In: Hesthaven, Jan S., Rønquist, Einar M. (Eds.), Spectral and High Order Methods for Partial Differential Equations. In: Lect. Notes Comput. Sci. Eng., vol. 76. Springer, Berlin, Heidelberg, Germany, pp. 43–62.

Beck, C., Jentzen, A., Kuckuck, B., 2022. Full error analysis for the training of deep neural networks. Infin. Dimens. Anal. Quantum Probab. Relat. Top. 25 (2), 2150020.

Beck, J., Nobile, F., Tamellini, L., Tempone, R., 2014. Convergence of quasi-optimal stochastic Galerkin methods for a class of PDEs with random coefficients. Comput. Math. Appl. 67 (4), 732–751.

Beck, J., Tempone, R., Nobile, F., Tamellini, L., 2012. On the optimal polynomial approximation of stochastic PDEs by Galerkin and collocation methods. Math. Models Methods Appl. Sci. 22 (9), 1250023.

Becker, S., Jentzan, A., Müller, M.S., von Wurstemberger, P., 2023. Learning the random variables in Monte Carlo simulations with stochastic gradient descent: machine learning for parametric PDEs and financial derivative pricing. Math. Finance 34 (1), 90–150.

Berveiller, M., Sudret, B., Lemaire, M., 2006. Stochastic finite element: a non intrusive approach by regression. Eur. J. Comput. Mech. 15 (1–3), 81–92.

Bhattacharya, K., Hosseini, N., Kovachki, B., Stuart, A., 2021. Model reduction and neural networks for parametric PDEs. J. Comput. Math. 7, 121–157.

Bieri, M., Andreev, R., Schwab, C., 2010. Sparse tensor discretization of elliptic SPDEs. SIAM J. Sci. Comput. 31 (6), 4281–4304.

Blanchard, M., Bennouna, M.A., 2020. The representation power of neural networks: breaking the curse of dimensionality. arXiv:2012.05451.

Blatman, G., Sudret, B., 2011. Adaptive sparse polynomial chaos expansion based on least angle regression. J. Comput. Phys. 230, 2345–2367.

Bölcskei, H., Grohs, P., Kutyniok, G., Petersen, P., 2019. Optimal approximation with sparsely connected deep neural networks. SIAM J. Math. Data Sci. 1 (1), 8–45.

Bonito, A., DeVore, R., Guignard, D., Jantsch, P., Petrova, G., 2021. Polynomial approximation of anisotropic analytic functions of several variables. Constr. Approx. 53, 319–348.

Boullé, N., Nakatsukasa, Y., Townsend, A., 2020. Rational neural networks. In: Advances in Neural Information Processing Systems, pp. 14243–14253.

Boullé, N., Townsend, A., 2023. A mathematical guide to operator learning. arXiv:2312.14688.

Brugiapaglia, S., Dexter, N., Karam, S., Wang, W., 2024. Physics-informed deep learning and compressive collocation for high-dimensional diffusion-reaction equations: practical existence theory and numerics. Preprint.

Brugiapaglia, S., Dirksen, S., Jung, H.C., Rauhut, H., 2021. Sparse recovery in bounded Riesz systems with applications to numerical methods for PDEs. Appl. Comput. Harmon. Anal. 53, 231–269.

Castrillón-Candás, J.E., Nobile, F., Tempone, R., 2016. Analytic regularity and collocation approximation for elliptic PDEs with random domain deformations. Comput. Math. Appl. 71 (6), 1173–1197.

Chambolle, A., Pock, T., 2011. A first-order primal-dual algorithm for convex problems with applications to imaging. J. Math. Imaging Vis. 40 (1), 120–145.

Chambolle, A., Pock, T., 2016. On the ergodic convergence rates of a first-order primal-dual algorithm. Math. Program. 159 (1–2), 253–287.
Chen, M., Jiang, H., Liao, W., Zhao, T., 2022. Nonparametric regression on low-dimensional manifolds using deep ReLU networks: function approximation and statistical recovery. Inf. Inference 11 (4), 1203–1253.
Chkifa, A., Cohen, A., Migliorati, G., Nobile, F., Tempone, R., 2015a. Discrete least squares polynomial approximation with random evaluations - application to parametric and stochastic elliptic PDEs. ESAIM: Math. Model. Numer. Anal. 49 (3), 815–837.
Chkifa, A., Cohen, A., Schwab, C., 2014. High-dimensional adaptive sparse polynomial interpolation and applications to parametric PDEs. Found. Comput. Math. 14 (4), 601–633.
Chkifa, A., Cohen, A., Schwab, C., 2015b. Breaking the curse of dimensionality in sparse polynomial approximation of parametric PDEs. J. Math. Pures Appl. 103 (2), 400–428.
Choi, B., Iwen, M.A., Krahmer, F., 2021a. Sparse harmonic transforms: a new class of sublinear-time algorithms for learning functions of many variables. Found. Comput. Math. 21 (2), 275–329.
Choi, B., Iwen, M.A., Volkmer, T., 2021b. Sparse harmonic transforms II: best s-term approximation guarantees for bounded orthonormal product bases in sublinear-time. Numer. Math. 148 (2), 293–362.
Cicci, L., Fresca, S., Manzoni, A., 2022. Deep-HyROMnet: a deep learning-based operator approximation for hyper-reduction of nonlinear parametrized PDEs. J. Sci. Comput. 93, 57.
Cohen, A., Davenport, M.A., Leviatan, D., 2013. On the stability and accuracy of least squares approximations. Found. Comput. Math. 13, 819–834.
Cohen, A., DeVore, R.A., 2015. Approximation of high-dimensional parametric PDEs. Acta Numer. 24, 1–159.
Cohen, A., DeVore, R.A., Schwab, C., 2010. Convergence rates of best N-term Galerkin approximations for a class of elliptic sPDEs. Found. Comput. Math. 10, 615–646.
Cohen, A., DeVore, R.A., Schwab, C., 2011. Analytic regularity and polynomial approximation of parametric and stochastic elliptic PDE's. Anal. Appl. (Singap.) 9 (1), 11–47.
Cohen, A., Migliorati, G., Nobile, F., 2017. Discrete least-squares approximations over optimized downward closed polynomial spaces in arbitrary dimension. Constr. Approx. 45, 497–519.
Cohen, A., Schwab, C., Zech, J., 2018. Shape holomorphy of the stationary Navier–Stokes equations. SIAM J. Math. Anal. 50 (2), 1720–1752.
Colbrook, M.J., Antun, V., Hansen, A.C., 2022. The difficulty of computing stable and accurate neural networks: on the barriers of deep learning and Smale's 18th problem. Proc. Natl. Acad. Sci. USA 119 (12), e2107151119.
Cybenko, G., 1989. Approximation by superpositions of a sigmoidal function. Math. Control Signals Syst. 2 (4), 303–314.
Cyr, E.C., Gulian, M.A., Patel, R.G., Perego, M., Trask, N.A., 2020. Robust training and initialization of deep neural networks: an adaptive basis viewpoint. In: Lu, Jianfeng, Ward, Rachel (Eds.), Proceedings of the First Mathematical and Scientific Machine Learning Conference. In: Proceedings of Machine Learning Research, vol. 107. Princeton University/PMLR, Princeton, NJ, USA, pp. 512–536.
Dai, F., Temlyakov, V., 2023. Universal sampling discretization. Constr. Approx. 58, 589–613.
Dal Santo, N., Deparis, S., Pegolotti, L., 2020. Data driven approximation of parametrized PDEs by reduced basis and neural networks. J. Comput. Phys. 416, 109550.
Daws, J., Webster, C., 2019. Analysis of deep neural networks with quasi-optimal polynomial approximation rates. arXiv:1912.02302.
Daws, J., Webster, C.G., 2019. A polynomial-based approach for architectural design and learning with deep neural networks. arXiv:1905.10457.
De Ryck, T., Lanthaler, S., Mishra, S., 2021. On the approximation of functions by tanh neural networks. Neural Netw. 143, 732–750.
DeVore, R., Hanin, B., Petrova, G., 2021. Neural network approximation. Acta Numer. 30, 327–444.
DeVore, R.A., 1998. Nonlinear approximation. Acta Numer. 7, 51–150.

Doostan, A., Owhadi, H., 2011. A non-adapted sparse approximation of PDEs with stochastic inputs. J. Comput. Phys. 230 (8), 3015–3034.

Dũng, D., Nguyen, V.K., 2021. Deep ReLU neural networks in high-dimensional approximation. Neural Netw. 142, 619–635.

Dũng, D., Nguyen, V.K., Pham, D.T., 2023. Deep ReLU neural network approximation in Bochner spaces and applications to parametric PDEs. J. Complex. 79, 101779.

Dũng, D., Temlyakov, V., Ullrich, T., 2018. Hyperbolic Cross Approximation. Adv. Courses Math. CRM Barcelona. Birkhäuser, Basel, Switzerland.

E, W., Ma, C., Wu, L., 2021. The Barron space and the flow-induced function spaces for neural network models. Constr. Approx. 55, 369–406.

E, W., Wang, Q., 2018. Exponential convergence of the deep neural network approximation for analytic functions. Sci. China Math. 61 (10), 1733–1740.

Elbrächter, D., Perekrestenko, D., Grohs, P., Bölcskei, H., 2021. Deep neural network approximation theory. IEEE Trans. Inf. Theory 67 (6), 2581–2623.

Elman, H.C., Miller, C.W., 2012. Stochastic collocation with kernel density estimation. Comput. Methods Appl. Mech. Eng. 245–246, 36–46.

Foucart, S., Rauhut, H., 2013. A Mathematical Introduction to Compressive Sensing. Appl. Numer. Harmon. Anal. Birkhäuser, New York, NY.

Franco, N.R., Brugiapaglia, S., 2024. A practical existence theorem for reduced order models based on convolutional autoencoders. arXiv:2402.00435.

Frankle, J., Carbin, M., 2019. The lottery ticket hypothesis: finding sparse, trainable neural networks. In: ICLR.

Ganapathysubramanian, B., Zabaras, N., 2007. Sparse grid collocation schemes for stochastic natural convection problems. J. Comput. Phys. 225 (1), 652–685.

Geist, M., Petersen, P., Raslan, M., Schneider, R., Kutyniok, G., 2021. Numerical solution of the parametric diffusion equation by deep neural networks. J. Sci. Comput. 88 (22).

Ghanem, R.G., Spanos, P.D., 2003. Stochastic Finite Elements: A Spectral Approach, revised edition. Dover Publications, Inc., Mineola, NY.

Ghanem, Roger, Higdon, David, Owhadi, Houman (Eds.), 2017. Handbook of Uncertainty Quantification. Springer, Switzerland.

Grohs, P., Voigtlaender, F., 2023. Proof of the theory-to-practice gap in deep learning via sampling complexity bounds for neural network approximation spaces. Found. Comput. Math. https://doi.org/10.1007/s10208-023-09607-w.

Gühring, I., Kutyniok, G., Petersen, P., 2020. Error bounds for approximations with deep ReLU neural networks in $W^{s,p}$ norms. Anal. Appl. (Singap.) 18 (05), 803–859.

Gühring, I., Raslan, M., 2021. Approximation rates for neural networks with encodable weights in smoothness spaces. Neural Netw. 134, 107–130.

Gunzburger, M., Webster, C.G., Zhang, G., 2014a. An adaptive wavelet stochastic collocation method for irregular solutions of partial differential equations with random input data. In: Garcke, Jochen, Pflüger, Dirk (Eds.), Sparse Grids and Applications – Munich 2012. In: Lect. Notes Comput. Sci. Eng., vol. 97. Springer, Cham, Switzerland, pp. 137–170.

Gunzburger, M., Webster, C.G., Zhang, G., 2014b. Stochastic finite element methods for partial differential equations with random input data. Acta Numer. 23, 521–650.

Guo, L., Narayan, A., Zhou, T., 2020. Constructing least-squares polynomial approximations. SIAM Rev. 62 (2), 483–508.

Hadigol, M., Doostan, A., 2018. Least squares polynomial chaos expansion: a review of sampling strategies. Comput. Methods Appl. Mech. Eng. 332, 382–407.

Hampton, J., Doostan, A., 2017. Compressive sampling methods for sparse polynomial chaos expansions. In: Ghanem, Roger, Higdon, David, Owhadi, Houman (Eds.), Handbook of Uncertainty Quantification. Springer, Cham, Switzerland, pp. 827–855.

Hansen, M., Schwab, C., 2013a. Analytic regularity and nonlinear approximation of a class of parametric semilinear elliptic PDEs. Math. Nachr. 286 (8–9), 832–860.

Hansen, M., Schwab, C., 2013b. Sparse adaptive approximation of high dimensional parametric initial value problems. Vietnam J. Math. 41 (2), 181–215.

Heiß, C., Gühring, I., Eigel, M., 2021. A neural multilevel method for high-dimensional parametric PDEs. In: Advances in Neural Information Processing Systems.

Heiß, C., Gühring, I., Eigel, M., 2023. Multilevel CNNs for parametric PDEs. J. Mach. Learn. Res. 24, 1–42.

Herrmann, L., Schwab, C., Zech, J., 2022. Neural and spectral operator surrogates: unified construction and expression rate bounds. arXiv:2207.04950v1.

Hoang, V.H., Schwab, C., 2012. Regularity and generalized polynomial chaos approximation of parametric and random second-order hyperbolic partial differential equations. Anal. Appl. (Singap.) 10 (3), 295–326.

Hoefler, T., Alistarh, D., Ben-Nun, T., Dryden, N., Peste, A., 2021. Sparsity in deep learning: pruning and growth for efficient inference and training in neural networks. J. Mach. Learn. Res. 23, 1–124.

Hornik, K., Stinchcombe, M., White, H., 1989. Multilayer feedforward networks are universal approximators. Neural Netw. 2 (5), 359–366.

Jakeman, J.D., Archibald, R., Xiu, D., 2011. Characterization of discontinuities in high-dimensional stochastic problems on adaptive sparse grids. J. Comput. Phys. 230 (10), 3977–3997.

Kashin, B., Kosov, E., Limonova, I., Temlyakov, V., 2022. Sampling discretization and related problems. J. Complex. 71, 101653.

Khara, B., Balu, A., Joshi, A., Sarkar, S., Hegde, C., Krishnamurthy, A., Ganapathysubramanian, B., 2021. NeuFENet: neural finite element solutions with theoretical bounds for parametric PDEs. arXiv:2110.01601.

Khoo, Y., Lu, J., Ying, L., 2021. Solving parametric PDE problems with artificial neural networks. Eur. J. Appl. Math. 32 (3), 421–435.

Kovachki, N., Li, Z., Liu, B., Azizzadnesheli, K., Bhattacharya, K., Stuart, A., Anandkumar, A., 2023. Neural operator: learning maps between function spaces with applications to PDEs. J. Mach. Learn. Res. 24, 1–97.

Kovachki, N.B., Lanthaler, S., Stuart, A.M., 2024. Operator learning: algorithms and analysis. arXiv:2402.15715.

Kühn, T., Sickel, W., Ullrich, T., 2015. Approximation of mixed order Sobolev functions on the d-torus: asymptotics, preasymptotics, and d-dependence. Constr. Approx. 42, 353–398.

Kunoth, A., Schwab, C., 2013. Analytic regularity and GPC approximation for control problems constrained by linear parametric elliptic and parabolic PDEs. SIAM J. Control Optim. 51 (3), 2442–2471.

Laakmann, F., Petersen, P., 2021. Efficient approximation of solutions of parametric linear transport equations by ReLU DNNs. Adv. Comput. Math. 47 (11).

Lanthaler, S., 2023. Operator learning with PCA-Net: upper and lower complexity bounds. arXiv: 2303.16317.

Le Maître, O., Knio, O.M., 2010. Spectral Methods for Uncertainty Quantification: With Applications to Computational Fluid Dynamics. Sci. Comput. Springer, Dordrecht, Netherlands.

Lei, Z., Shi, L., Zeng, C., 2022. Solving parametric partial differential equations with deep rectified quadratic unit neural networks. J. Sci. Comput. 93, 80.

Li, B., Tang, S., Yu, H., 2020. Better approximations of high dimensional smooth functions by deep neural networks with rectified power units. Commun. Comput. Phys. 27, 379–411.

Li, Z., Kovachki, N., Azizzadenesheli, K., Liu, B., Bhattacharya, K., Stuart, A., Anandkumar, A., 2021. Fourier neural operator for parametric partial differential equations. In: ICLR.

Liang, S., Srikant, R., 2017. Why deep neural networks for function approximation? In: ICLR.

Longo, M., Opschoor, J.A.A., Disch, N., Schwab, C., Zech, J., 2023. De Rham compatible deep neural network FEM. Neural Netw. 165, 721–739.

Lu, J., Shen, Z., Yang, H., Zhang, S., 2021. Deep network approximation for smooth functions. SIAM J. Math. Anal. 53 (5), 5465–5506.

Lu, L., Jin, P., Pang, Z., Zhang, G., Karniadakis, G.E., 2021. Learning nonlinear operators via DeepONet based on the universal approximation theorem of operators. Nat. Mach. Intell. 3, 218–229.

Ma, X., Zabaras, N., 2009. An adaptive hierarchical sparse grid collocation algorithm for the solution of stochastic differential equations. J. Comput. Phys. 228 (8), 3084–3113.

Mathelin, L., Gallivan, K.A., 2012. A compressed sensing approach for partial differential equations with random input data. Commun. Comput. Phys. 12 (4), 919–954.

Mathelin, L., Hussaini, M.Y., Zang, T.A., 2005. Stochastic approaches to uncertainty quantification in CFD simulations. Numer. Algorithms 38 (1–3), 209–236.

Mhaskar, H., 1993. Approximation properties of a multilayered feedforward artificial neural network. Adv. Comput. Math. 1, 61–80.

Mhaskar, H., 1996. Neural networks for optimal approximation of smooth and analytic functions. Neural Comput. 8 (1), 164–177.

Migliorati, G., 2013. Polynomial approximation by means of the random discrete L^2 projection and application to inverse problems for PDEs with stochastic data. PhD thesis. Politecnico di Milano.

Migliorati, G., 2015. Adaptive polynomial approximation by means of random discrete least squares. In: Abdulle, Assyr, Deparis, Simone, Kressner, Daniel, Nobile, Fabio, Picasso, Marco (Eds.), Numerical Mathematics and Advanced Applications – ENUMATH 2013. Springer, Cham, Switzerland, pp. 547–554.

Migliorati, G., 2019. Adaptive approximation by optimal weighted least squares methods. SIAM J. Numer. Anal. 57 (5), 2217–2245.

Migliorati, G., Nobile, F., Tempone, R., 2015. Convergence estimates in probability and in expectation for discrete least squares with noisy evaluations at random points. J. Multivar. Anal. 142, 167–182.

Migliorati, G., Nobile, F., von Schwerin, E., Tempone, R., 2013. Approximation of quantities of interest in stochastic PDEs by the random discrete L^2 projection on polynomial spaces. SIAM J. Sci. Comput. 35 (3), A1440–A1460.

Migliorati, G., Nobile, F., von Schwerin, E., Tempone, R., 2014. Analysis of the discrete L^2 projection on polynomial spaces with random evaluations. Found. Comput. Math. 14, 419–456.

Monga, V., Li, Y., Eldar, Y.C., 2021. Algorithm unrolling: interpretable, efficient deep learning for signal and image processing. IEEE Signal Process. Mag. 38 (2), 18–44.

Montanelli, H., Du, Q., 2019. New error bounds for deep ReLU networks using sparse grids. SIAM J. Math. Data Sci. 1 (1), 78–92.

Montanelli, H., Yang, H., Du, Q., 2021. Deep ReLU networks overcome the curse of dimensionality for bandlimited functions. J. Comput. Math. 39 (6), 801–815.

Neyra-Nesterenko, M., Adcock, B., 2023. NESTANets: stable, accurate and efficient neural networks for analysis-sparse inverse problems. Sampl. Theory Signal Process. Data Anal. 21 (4).

Nobile, F., Tempone, R., Webster, C.G., 2008a. An anisotropic sparse grid stochastic collocation method for partial differential equations with random input data. SIAM J. Numer. Anal. 46 (5), 2411–2442.

Nobile, F., Tempone, R., Webster, C.G., 2008b. A sparse grid stochastic collocation method for partial differential equations with random input data. SIAM J. Numer. Anal. 46 (5), 2309–2345.

Novak, E., 1988. Deterministic and Stochastic Error Bounds in Numerical Analysis. Number 1. Springer Berlin, Heidelberg.

Novak, E., Woźniakowski, H., 2008. Tractability of Multivariate Problems, Volume I: Linear Information, vol. 6. European Math. Soc. Publ. House, Zürich.

Novak, E., Woźniakowski, H., 2010. Tractability of Multivariate Problems, Volume II: Standard Information for Functionals, vol. 12. European Math. Soc., Zürich.

Ohn, I., Kim, Y., 2019. Smooth function approximation by deep neural networks with general activation functions. Entropy 21 (7), 627.

Opschoor, J.A.A., Petersen, P.C., Schwab, C., 2020. Deep ReLU networks and high-order finite element methods. Anal. Appl. (Singap.) 18 (5), 715–770.

Opschoor, J.A.A., Schwab, C., 2023. Deep ReLU networks and high-order finite element methods II: Chebyshev emulation. arXiv:2310.07261.

Opschoor, J.A.A., Schwab, Ch., Zech, J., 2022. Exponential ReLU DNN expression of holomorphic maps in high dimension. Constr. Approx. 55, 537–582.

Peng, J., Hampton, J., Doostan, A., 2014. A weighted ℓ_1-minimization approach for sparse polynomial chaos expansions. J. Comput. Phys. 267, 92–111.

Petersen, P., Voigtlaender, F., 2018. Optimal approximation of piecewise smooth functions using deep ReLU neural networks. Neural Netw. 108, 296–330.

Pinkus, A., 1999. Approximation theory of the MLP model in neural networks. Acta Numer. 8, 143–195.

Poggio, T., Mhaskar, H., Rosasco, L., Miranda, B., Liao, Q., 2017. Why and when can deep-but not shallow-networks avoid the curse of dimensionality: a review. Int. J. Autom. Comput. 14, 503–519.

Rauhut, H., Schwab, C., 2017. Compressive sensing Petrov-Galerkin approximation of high-dimensional parametric operator equations. Math. Comput. 86, 661–700.

Rauhut, H., Ward, R., 2012. Sparse Legendre expansions via ℓ_1-minimization. J. Approx. Theory 164 (5), 517–533.

Rauhut, H., Ward, R., 2016. Interpolation via weighted ℓ^1 minimization. Appl. Comput. Harmon. Anal. 40 (2), 321–351.

Roulet, V., d'Aspremont, A., 2020. Sharpness, restart, and acceleration. SIAM J. Optim. 30 (1), 262–289.

Scarabosio, L., 2022. Deep neural network surrogates for nonsmooth quantities of interest in shape uncertainty quantification. SIAM/ASA J. Uncertain. Quantificat. 10 (3), 975–1011.

Schmidt-Hieber, J., 2020. Nonparametric regression using deep neural networks with ReLU activation function. Ann. Stat. 48 (4), 1875–1897.

Schwab, C., Stein, A., Zech, J., 2023. Deep operator network approximation rates for Lipschitz operators. arXiv:2307.09835.

Schwab, C., Zech, J., 2019. Deep learning in high dimension: neural network expression rates for generalized polynomial chaos expansions in UQ. Anal. Appl. (Singap.) 17 (1), 19–55.

Schwab, C., Zech, J., 2023. Deep learning in high dimension: neural network expression rates for analytic functions in $L^2(\mathbb{R}^d, \gamma_d)$. SIAM/ASA J. Uncertain. Quantificat. 11 (1), 199–234.

Settles, B., 2012. Active Learning. Synthesis Lectures on Artificial Intelligence and Machine Learning. Springer, Cham, Switzerland.

Shaham, U., Cloninger, A., Coifman, R.R., 2018. Provable approximation properties for deep neural networks. Appl. Comput. Harmon. Anal. 44, 537–557.

Smith, R.C., 2013. Uncertainty Quantification: Theory, Implementation, and Applications. Comput. Sci. Eng. Society for Industrial and Applied Mathematics, Philadelphia, PA.

Sullivan, T.J., 2015. Introduction to Uncertainty Quantification. Texts Appl. Math., vol. 63. Springer, Cham, Switzerland.

Suzuki, T., 2019. Adaptivity of deep ReLU network for learning in Besov and mixed smooth Besov spaces: optimal rate and curse of dimensionality. In: ICLR.

Tang, S., Li, B., Haijun, Y., 2019. ChebNet: efficient and stable constructions of deep neural networks with rectified power units via Chebyshev approximation. arXiv:1911.05467.

Telgarsky, M., 2017. Neural networks and rational functions. In: ICML.

Temlyakov, V.N., 2018. The Marcinkiewicz-type discretization theorems. Constr. Approx. 48 (2), 337–369.

Todor, R.A., Schwab, C., 2007. Convergence rates for sparse chaos approximations of elliptic problems with stochastic coefficients. IMA J. Numer. Anal. 27 (2), 232–261.

Tran, H., Webster, C.G., Zhang, G., 2017. Analysis of quasi-optimal polynomial approximations for parameterized PDEs with deterministic and stochastic coefficients. Numer. Math. 137 (2), 451–493.

Traub, J.F., Woźniakowski, H., Wasilkowski, G.W., 1988. Information-Based Complexity. Elsevier Science and Technology Books.

Trefethen, L.N., 2013. Approximation Theory and Approximation Practice. Society for Industrial and Applied Mathematics, Philadelphia, PA.

Vidyasagar, M., 2019. An Introduction to Compressed Sensing. Comput. Sci. Eng. Society for Industrial and Applied Mathematics, Philadelphia, PA.

Xiu, D., Hesthaven, J.S., 2005. High-order collocation methods for differential equations with random inputs. SIAM J. Sci. Comput. 27 (3), 1118–1139.

Yan, L., Guo, L., Xiu, D., 2012. Stochastic collocation algorithms using ℓ_1-minimization. Int. J. Uncertain. Quantificat. 2 (3), 279–293.

Yang, X., Karniadakis, G.E., 2013. Reweighted ℓ_1 minimization method for stochastic elliptic differential equations. J. Comput. Phys. 248, 87–108.

Yarotsky, D., 2017. Error bounds for approximations with deep ReLU networks. Neural Netw. 94, 103–114.

Yarotsky, D., 2018. Optimal approximation of continuous functions by very deep ReLU networks. In: Bubeck, Sébastien, Perchet, Vianney, Rigollet, Philippe (Eds.), Proceedings of the 31st Conference on Learning Theory. In: Proceedings of Machine Learning Research, vol. 75. PMLR, pp. 639–649.

Zhang, G., Webster, C.G., Gunzburger, M., Burkardt, J., 2016. Hyperspherical sparse approximation techniques for high-dimensional discontinuity detection. SIAM Rev. 58 (3), 517–551.

Chapter 2

Weak form-based data-driven modeling

Computationally efficient and noise robust equation learning and parameter inference

David M. Bortz*, Daniel A. Messenger, and April Tran
Department of Applied Mathematics, University of Colorado, Boulder, CO, United States
**Corresponding author: e-mail address: david.bortz@colorado.edu*

Contents

Abstract

Recent developments in the field of data-driven modeling have led to the advancement of weak form-based methodologies in scientific machine learning. This class of methods offers several compelling advantages, including high computational efficiency and high noise robustness. In this chapter, we present an overview of the weak form approach as well as discuss several categories of applications including equation discovery, constitutive parameter inference, and reduced order modeling. In particular, we illustrate the performance on several benchmark examples for ordinary, partial, and stochastic differential equations.

Handbook of Numerical Analysis, Volume 25, ISSN 1570-8659, https://doi.org/10.1016/bs.hna.2024.05.002

Keywords
Equation learning, Data-driven modeling, Weak form, WSINDy, WENDy, WLaSDI

MSC Codes
37M10, 62-07, 62FXX, 62JXX, 65L09, 65M32, 65R99, 92-08

1 Introduction

The *weak form* of a differential equation is created by multiplying both sides by a smooth function ϕ and integrating over a domain of interest. Historically, this idea originated from the fact that frequently physical conservation laws can intuitively be cast as integral equations. Moreover, simple integral (i.e., $\phi \equiv 1$) and variational formulations consistently allow for easier analysis and simulation of a broad class of models (including those with shocks and other solution discontinuities). In the twentieth century, Sobolev was the first to suggest that ϕ be (what Friedrichs later named a *mollifier*) a compactly supported C^∞ function which integrates to one, while Schwartz rigorously recasts the classical notion of a function acting on a point to one acting on a measurement structure or *test function* (ϕ) (Schwartz, 1950). Modern computational approaches based on the weak form (such as the finite element method (FEM) for solving an equation) originated with the work of many researchers including Argyris, Courant, Friedrichs, Galerkin, Hrennikoff, Oganesya, and others (see Liu et al., 2022, for an overview of the history). Lax and Milgram then built on these efforts to establish a theoretical foundation by proving the existence of weak solutions (in a Hilbert Space) to certain classes of PDEs (Lax and Milgram, 1955).

While it is clear that the weak form is used widely in the analysis and simulation of differential equations, recent advances suggest that there is an incredible potential for novel methods applied to statistical and machine learning problems. In particular, weak form versions of sparse regression-based equation learning, direct parameters estimation, and reduced order modeling can offer advantages in terms of noise robustness and computational speedups over both conventional methods and black-box neural network approaches.

The overall goal of this handbook chapter is to both introduce the basics of weak form system identification as well as communicate the breadth of possible applications. Accordingly, in §2, we provide an overview of the Weak form Sparse Identification of Nonlinear Dynamics (WSINDy) method (§2.2) in the context of equation discovery for ODEs (§2.3) and PDEs (§2.4), summarizing the contributions from Messenger and Bortz (2021a,b). Motivated by the fact that weak form-based learning is demonstrably more robust to noise than strong form-based approaches, in §3 we present a novel theoretical explanation for the improved performance (under certain assumptions).

In §4 we explore how a direct parameter estimation method substantially benefits from casting the equation in the weak form. We build on our Weak form Estimation of Nonlinear Dynamics (WENDy) applied to ODEs (Bortz et al., 2023) and we extend the method to PDEs and SDEs. Lastly, in §5 we demonstrate how WENDy can complement existing SVD- and autoencoder-based reduced order modeling (ROM) techniques by substantially enhancing the discovery of the latent space dynamics from noisy data.

2 Weak form-based equation discovery

In this section, we discuss a class of methods that use sparse regression to learn governing equations directly from data. The idea is to consider a large library of potential functional forms for terms on the right side of an evolution equation. As described below, the library and the measured data are combined to create a sparse regression problem, the answer to which reveals the subset of terms on the right side which best relate the data to its time derivative.

We will begin by describing the first widely successful equation discovery method and then discuss how to cast this problem as one in which the equation is learned in the weak form. We will then present examples using conventional ODEs and PDEs. However, we also note that the examples presented here are only a subset of models and applications to which we have applied the weak form. We have considered integro-differential equation models (Messenger et al., 2022b), online learning (Messenger et al., 2022a), and coarse-graining of stochastic differential equations (Messenger and Bortz, 2022b; Messenger et al., 2023).

2.1 The sparse identification of nonlinear dynamics (SINDy) method for learning governing equations

To begin, let us consider an evolution equation system

$$\frac{d}{dt}\mathbf{u}(t) = \mathbf{F}(\mathbf{u}(t)), \quad \mathbf{u}(0) = \mathbf{u}_0 \in \mathbb{R}^d, \quad 0 \le t \le T, \tag{1}$$

with data $\mathbf{U} \in \mathbb{R}^{M \times d}$ observed at M timepoints $\mathbf{t} = (t_1, \ldots, t_M)^T$ by

$$\mathbf{U}_{md} = \mathbf{u}_i(t_m) + \epsilon_{mi}, \qquad m \in [M],\ i \in [d].$$

Note that we use the bracket notation $[M] := \{1, \ldots, M\}$ and that the variable $\epsilon \in \mathbb{R}^{M \times d}$ is a matrix of i.i.d. measurement noise. The goal of the equation learning problem is to discover the dynamics (1) from the measurements $\mathbf{U}$.

In 2016, Brunton, Proctor, and Kutz published an article in the Proceedings of the National Academy (Brunton et al., 2016) introducing the Sparse Identification of Nonlinear Dynamics (SINDy) algorithm. This algorithm has been successful used in learning parsimonius nonlinear dynamics when (among other considerations) noise is small. This framework assumes that the ith element of the function $\mathbf{F} : \mathbb{R}^d \to \mathbb{R}^d$ in (1) is

$$\mathbf{F}_i(\mathbf{u}(t)) = \sum_{j=1}^{J} \mathbf{w}^{\star}_{ji}\, f_j(\mathbf{u}(t)) \tag{2}$$

for some known family of functions $(f_j)_{j\in[J]}$ and a coefficient weight matrix $\mathbf{w}^\star \in \mathbb{R}^{J\times d}$ which is mostly zeros, i.e., sparse.

To learn the model, first a data matrix is built $\Theta(\mathbf{U}) \in \mathbb{R}^{M\times J}$ by substituting the data directly into the equation

$$[\Theta(\mathbf{U})]_{(m,j)} = f_j(\mathbf{U}_m), \qquad \mathbf{U}_m := (\mathbf{U}_{m1}, \ldots, \mathbf{U}_{md}),$$

so that the candidate functions are directly evaluated at the measured data. Model discovery is thus recast as solving for a sparse $\mathbf{w}^\star$ by minimizing (subject to a sparsity penalty) the L^2 norm of the residual

$$\left\|\dot{\mathbf{U}} - \Theta(\mathbf{U})\,\widehat{\mathbf{w}}\right\|_2, \tag{3}$$

where $\dot{\mathbf{U}}$ is the numerical time derivative of the data $\mathbf{U}$. In SINDy, sequential-thresholding least squares on the parameters $\mathbf{w}$ are then used to arrive at a sparse solution.

2.2 Weak form SINDy (WSINDy)

While SINDy has been used successfully in a wide range of areas, it is well known that using a simple finite difference approximation for the derivatives will lead to poor performance in the presence of even modest amounts of noise. There have been several efforts to address the problem using methods such as total-variation regularized derivatives (Brunton et al., 2016), linear multistep derivative approximations (Chen, 2023), and ensembling (Fasel et al., 2022) (among others), each with varying degrees of success.

Weak-form extensions to SINDy have been independently discovered by several groups (Gurevich et al., 2019; Messenger and Bortz, 2021a,b; Pantazis and Tsamardinos, 2019; Wang et al., 2019) over the past few years. Although our group was not the first to propose a weak-form methodology, we have investigated its use for equation learning in a wide range of model structures and applications including: ODEs (Messenger and Bortz, 2021b), PDEs (Messenger and Bortz, 2021a), interacting particle systems of the first (Messenger and Bortz, 2022b) and second (Messenger et al., 2022b) order, and online streaming (Messenger and Bortz, 2022b). As mentioned below, we have also studied the theoretical convergence properties for WSINDy in the continuum data limit (Messenger and Bortz, 2022a). This work led to specification of a broad class of models for which the asymptotic limit of continuum data can overcome any noise level to produce both an accurately learned equation and a correct parameter estimate (see Messenger and Bortz, 2022a, for more information).

To describe WSINDy, we begin by recalling that for any smooth test function $\phi : \mathbb{R} \to \mathbb{R}$ and interval $(a, b) \subset [0, T]$, Eq. (1) admits the weak formulation

$$\phi(b)\mathbf{u}(b) - \phi(a)\mathbf{u}(a) - \int_a^b \dot{\phi}(t)\,\mathbf{u}(t)\,dt = \int_a^b \phi(t)\,\mathbf{F}(\mathbf{u}(t))\,dt, \tag{4}$$
$$0 \le a < b \le T.$$

With $\phi = 1$, we arrive at the integral equation of the dynamics explored in Schaeffer and McCalla (2017). If we instead take ϕ to be nonconstant and compactly supported in (a, b), we arrive at

$$-\int_a^b \dot{\phi}(t)\,\mathbf{u}(t)\,dt = \int_a^b \phi(t)\,\mathbf{F}(\mathbf{u}(t))\,dt. \tag{5}$$

Assuming a representation is of the form (2), we then define the generalized residual $\mathcal{R}(\mathbf{w}; \phi)$ for a given test function ϕ and a set of candidate terms $(f_j)_{j\in[J]}$ and with data $\mathbf{U}$ as follows:

$$\mathcal{R}(\mathbf{w}; \phi) := \int_a^b \left(\dot{\phi}(t)\,\mathbf{U}(t) + \phi(t) \left(\sum_{j=1}^{J} \mathbf{w}_j\, f_j(\mathbf{U}(t)) \right) \right) dt. \tag{6}$$

Clearly, with $\mathbf{w} = \mathbf{w}^\star$ and $\mathbf{U} = \mathbf{u}(t)$ we have $\mathcal{R}(\mathbf{w}; \phi) = 0$ for all ϕ compactly-supported in (a, b); however, $\mathbf{U}$ is a finite dimensional vector of data, hence the integral in (6) must be approximated numerically. Measurement noise then presents a further barrier to accurate identification of $\mathbf{w}^\star$.

In Algorithm 1, we now state the Weak SINDy method in full generality. We propose a generalized least squares approach with approximate covariance matrix Σ. Below we derive a particular choice of Σ which utilizes the action of the test functions $(\phi_k)_{k\in[K]}$ on the data $\mathbf{y}$. Sequential-thresholding on the weight coefficients $\mathbf{w}$ with thresholding parameter λ is used to enforce sparsity, where $\lambda \le \min_{\mathbf{w}^\star \neq 0} |\mathbf{w}^\star|$ is necessary for recovery. Lastly, an ℓ_2-regularization term with coefficient γ is included for problems involving rank deficiency.

WSINDy has proven to be successful in discovering governing equations for a wide class of differential equations. In this section, we will illustrate the application to ODEs, PDEs, and the mean field limit equation of a specific class of SDEs. We note that the ODE example in Section 2.3 is a straightforward application of the formulae in Algorithm 1. However, the later examples for PDEs (Section 2.4) need more exposition and notation to be fully described.

2.3 WSINDy for ordinary differential equations

Recall the canonical Lorenz system with a set of parameters known to induce solutions with chaotic dynamics

$$\begin{aligned}\dot{x} &= 10(y - x)\\ \dot{y} &= x(28 - z) - y\\ \dot{z} &= xy - \tfrac{8}{3}z.\end{aligned}$$

We solve this ODE using a variable order, variable stepsize solver based on numerical differentiation formulas of orders 1 to 5, as implemented in Matlab®'s ode15s with absolute and relative tolerances of 10^{-12}.

Algorithm 1: Weak form Sparse Identification of Nonlinear Dynamics (WSINDy)

input : Data $\{\mathbf{t}, \mathbf{U}\}$, Candidate functions $(f_j)_{j\in[J]}$, Test Functions $(\phi_k)_{k\in[K]}$, Variance Σ, Regularization (λ, γ)
output: Parameter Estimate $\widehat{\mathbf{w}}$

```
// Construct matrix of trial gridfunctions
```
$\Theta(\mathbf{U}) \leftarrow \left[f_1(\mathbf{U}) \mid \ldots \mid f_J(\mathbf{U})\right]$

```
// Construct integration matrices
```
$[\mathbf{\Phi}]_{(k,m)} \leftarrow \Delta t \phi_k(t_m)$
$[\dot{\mathbf{\Phi}}]_{(k,m)} \leftarrow \Delta t \dot{\phi}_k(t_m)$

```
// Compute Gram matrix and right side vector
```
$[\mathbf{G}]_{(k,j)} = \langle \phi_k, f_j(\mathbf{U})\rangle$
$[\mathbf{b}]_{(k,i)} = -\langle \phi_k', \mathbf{U}_i\rangle$

```
// Solve the following generalized least squares problem
   with ℓ2-regularization using
// sequential thresholding with parameter λ to enforce
   sparsity.
```
$\widehat{\mathbf{w}} = \operatorname{argmin}_{\mathbf{w}} \left\{(\mathbf{G}\mathbf{w} - \mathbf{b})^T \Sigma^{-1}(\mathbf{G}\mathbf{w} - \mathbf{b}) + \gamma^2 \|\mathbf{w}\|_2^2\right\}$

To apply WSINDy to the ODE equation recovery problem, a test function must be specified. In our initial efforts to learn ODEs (Messenger and Bortz, 2021b), we empirically discovered that a smooth piecewise polynomial provides excellent results

$$\phi(t, r, p) := \left(1 - \left(\frac{t}{r}\right)^2\right)^p.$$

In Messenger and Bortz (2021b), we proposed a strategy for choosing the compact support radius r and the power p. While we have discovered more accurate and sophisticated strategies for tuning these parameters (see Section 2.4), there are several general principles. First, in general, the power p will need to be much larger than the value needed to perform integration by parts. Second, the radius depends upon the dynamics and typically larger radii work better. Notably, in the absence of noise and using a large radius and a highly smooth test function, WSINDy recovers the parameters to machine precision-level accuracy.

Depicted in Fig. 1 are the result of using WSINDy to discover the governing Lorenz equations from data with 10% additive Gaussian noise. In the upper left subfigure, the (high numerical accuracy) solution is in black and the sampled data are in red. In the upper right subfigure, the black is the same as in the upper

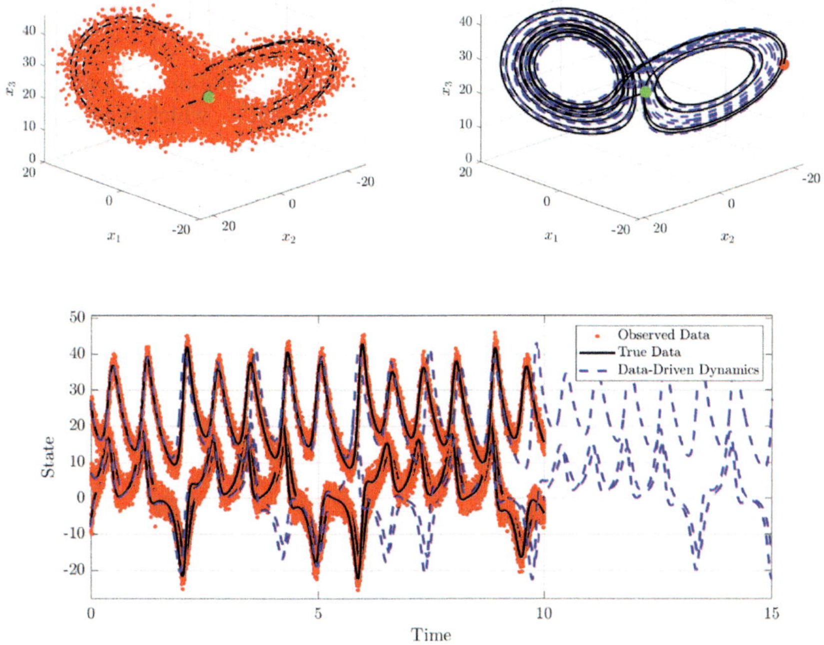

FIGURE 1 Lorenz system with $(x_0, y_0, z_0) = (-8,\ 7,\ 27)^T$. All correct terms were identified with an error in the weights of $E_2(\widehat{w}t) = 0.0084$ and trajectory error $\mathcal{E}(\widehat{w}) = 0.56$. The large trajectory error is expected due to the chaotic nature of the solution. Using data up until $t = 1.5$ (first 1500 timepoints) the trajectory error is 0.027. Figure from Messenger and Bortz (2021b).

left and the blue dotted curve is a simulation using the discovered parameters. As is clear in the bottom subfigure, the simulated solution is close to the highly accurate one, but given that the system is chaotic, it is notable that the discovered solution provides a nearly perfect match until about 4 time units.

For more ODE examples including Lotka-Volterra, Van der Pol, Duffing, etc., we direct the interested reader to Messenger and Bortz (2021b).

2.4 WSINDy for partial differential equations

We are now ready to discuss the formulation of WSINDy for PDEs. We assume that the set of multiindices $(\alpha^s)_{s\in[S]}$ together with α^0 enumerates the set of possible true differential operators that govern the evolution of u and that $(g_s)_{s\in[S]} \subset \text{span}(f_j)_{j\in[J]}$ where the family of functions $(f_j)_{j\in[J]}$ (referred to as the *trial functions*) is known beforehand. This enables us to write down a general class of PDEs as

$$\mathscr{D}^{\alpha^0} u = \sum_{s=1}^{S} \sum_{j=1}^{J} \mathbf{w}^{\star}_{(s-1)J+j} \mathscr{D}^{\alpha^s} f_j(u), \tag{7}$$

so that discovery of the correct PDE is reduced to a finite-dimensional problem of recovering the true vector of coefficients $\mathbf{w}^{\star} \in \mathbb{R}^{SJ}$, which is assumed to be sparse. We emphasize that a wide variety of PDEs can be written in the form (7) including inviscid Burgers, Korteweg-de Vries, Kuramoto-Sivashinsky, nonlinear Schrödinger's, Sine-Gordon, a class of reaction-diffusion systems, and Navier-Stokes.

To convert the PDE into its weak form, we multiply Eq. (7) by a smooth *test function* $\psi(x, t)$, compactly-supported in $\Omega \times (0, T)$, and integrate over the spacetime domain,

$$\left\langle \psi,\ \mathscr{D}^{\alpha^0} u \right\rangle = \sum_{s=1}^{S} \sum_{j=1}^{J} \mathbf{w}^{\star}_{(s-1)J+j} \left\langle \psi,\ \mathscr{D}^{\alpha^s} f_j(u) \right\rangle,$$

where the L^2-inner product is defined $\langle \psi, f \rangle := \int_0^T \int_\Omega \psi(x,t) f(x,t)\, dx dt$. Using the compact support of ψ and Fubini's theorem, we then integrate by parts as many times as necessary to arrive at the following weak form of the dynamics:

$$\left\langle (-1)^{|\alpha^0|} \mathscr{D}^{\alpha^0} \psi,\ u \right\rangle = \sum_{s=1}^{S} \sum_{j=1}^{J} \mathbf{w}^{\star}_{(s-1)J+j} \left\langle (-1)^{|\alpha^s|} \mathscr{D}^{\alpha^s} \psi,\ f_j(u) \right\rangle, \tag{8}$$

where $|\alpha^s| := \sum_{d=1}^{D+1} \alpha^s_d$ is the order of the multiindex.[1] Using an ensemble of test functions $(\psi_k)_{k \in [K]}$, we then discretize the integrals in (8) with $f_j(u)$ replaced by $f_j(\mathbf{U})$ (i.e. evaluated at the observed data $\mathbf{U}$) to arrive at the linear least squares problem

$$\min_{w} \|\mathbf{b} - \mathbf{G}\mathbf{w}\|_2^2$$

defined by

$$\begin{cases} \mathbf{b}_k = \left\langle (-1)^{|\alpha^0|} \mathscr{D}^{\alpha^0} \psi_k,\ \mathbf{U} \right\rangle, \\ \mathbf{G}_{k,(s-1)J+j} = \left\langle (-1)^{|\alpha^s|} \mathscr{D}^{\alpha^s} \psi_k,\ f_j(\mathbf{U}) \right\rangle, \end{cases} \tag{9}$$

where $\mathbf{b} \in \mathbb{R}^K$, $\mathbf{G} \in \mathbb{R}^{K \times SJ}$ and $\mathbf{w} \in \mathbb{R}^{SJ}$ are using the inner product both in the sense of a continuous and exact integral in (8) and a numerical approximation in (9) which depends on a chosen quadrature rule.[2]

1 For example, with $\mathscr{D}^{\alpha^s} = \frac{\partial^{2+1}}{\partial x^2 \partial y}$, integration by parts occurs twice with respect to the x-coordinate and once with respect to y, so that $|\alpha^s| = 3$ and $(-1)^{|\alpha^s|} = -1$.

2 In all cases in this chapter, we use the trapezoidal rule, see Messenger and Bortz (2021b) for a discussion.

For the WSINDy algorithm, the key pieces of the algorithm are (i) the choice of reference test function ψ, (ii) the method of a sparsification, (iii) the method of regularization, (iv) selection of convolution query points $\{(\mathbf{x}_k, t_k)\}_{k\in K}$, and (v) the model library. Full guidance for choices for each of these hyperparameters (in the case of PDEs) is provided in Messenger and Bortz (2021a).

Here we will simply note that the choice of test function is central to the performance and it is still an open question as to the optimal functions for different scenarios. For the illustrating examples below, we use a piecewise polynomial

$$\phi(v) = \begin{cases} C(v-a)^p(b-v)^q & a < v < b, \\ 0 & \text{otherwise,} \end{cases} \tag{10}$$

where $p, q \geq 1$ and v is a (time or space) variable. The normalization

$$C = \frac{1}{p^p q^q}\left(\frac{p+q}{b-a}\right)^{p+q}$$

ensures that $\|\phi\|_\infty = 1$. For ease of computation, multiplicative test functions are used, e.g.,

$$\psi(t, \mathbf{x}) = \phi_0(t)\prod_{i=1}^{D}\phi_i(x_i)$$

for a D-dimensional space. Examples of these test functions in one time and one space dimension are depicted in Fig. 2. Lastly, we note that the test functions defined here are functions of both time and space (instead of just of time as in Section 2.3).

Table 1 lists the PDEs used to demonstrate the performance of WSINDy. We numerically solved[3] these equations and then created artificial data by adding i.i.d. Gaussian noise with variance σ^2 to each data point. The value of σ is constructed to depend on the root mean squares (RMS) of the data, i.e., $\sigma := \sigma_{NR}\|\mathbf{U}^\star\|_{RMS}$ and we refer to σ_{NR} as the *noise ratio*.

There are standard metrics for measuring the accuracy of parameter estimates, such as the relative ℓ^2 error

$$E_2(\widehat{w}) := \frac{\|\widehat{w} - \mathbf{w}^\star\|_{RMS}}{\|\mathbf{w}^\star\|_{RMS}}, \tag{11}$$

which we will employ. For the equation discovery problem, we will use the *true positivity ratio* introduced in Lagergren et al. (2020) and defined by

$$\text{TPR}(\widehat{w}) = \frac{\text{TP}}{\text{TP} + \text{FN} + \text{FP}}, \tag{12}$$

[3] Details on the numerical methods and boundary conditions used to simulate each PDE can be found in Messenger and Bortz (2021a).

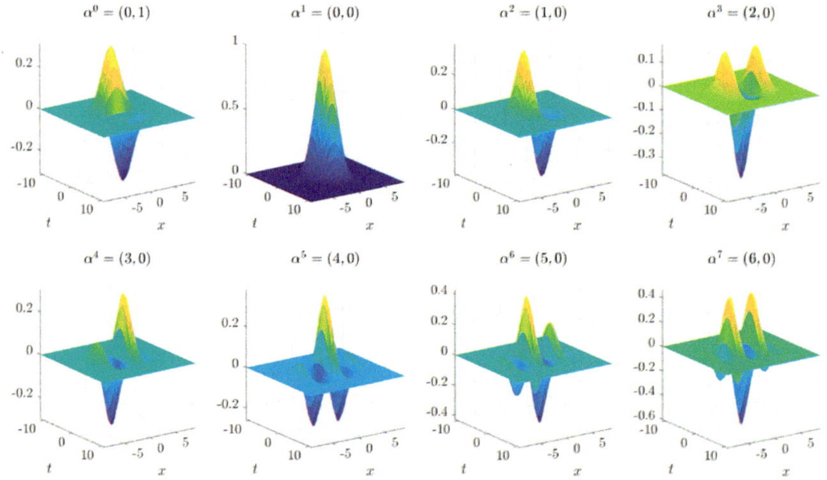

FIGURE 2 Plots of reference test function ψ and partial derivatives $\mathscr{D}^{\alpha^s}\psi$ used for identification of the Kuramoto-Sivashinsky equation. The upper left plot shows $\partial_t \psi$, the bottom right shows $\partial_x^6 \psi$. See Tables 1–2 for more details. Reproduced from Messenger and Bortz (2021a).

TABLE 1 PDEs used in numerical experiments, written in the form identified by WSINDy. Domain specification and boundary conditions are given in Messenger and Bortz (2021a).

Inviscid Burgers (IB)	$\partial_t u = -\frac{1}{2}\partial_x(u^2)$
Korteweg-de Vries (KdV)	$\partial_t u = -\frac{1}{2}\partial_x(u^2) - \partial_{xxx} u$
Kuramoto-Sivashinsky (KS)	$\partial_t u = -\frac{1}{2}\partial_x(u^2) - \partial_{xx} u - \partial_{xxxx} u$
Nonlinear Schrödinger (NLS)	$\begin{cases} \partial_t u = \frac{1}{2}\partial_{xx} v + u^2 v + v^3 \\ \partial_t v = -\frac{1}{2}\partial_{xx} u - u v^2 - u^3 \end{cases}$
Anisotropic Porous Medium (PM)	$\partial_t u = (0.3)\partial_{xx}(u^2) - (0.8)\partial_{xy}(u^2) + \partial_{yy}(u^2)$
Sine-Gordon (SG)	$\partial_{tt} u = \partial_{xx} u + \partial_{yy} u - \sin(u)$
Reaction-Diffusion (RD)	$\begin{cases} \partial_t u = \frac{1}{10}\partial_{xx} u + \frac{1}{10}\partial_{yy} u - u v^2 - u^3 + v^3 + u^2 v + u \\ \partial_t v = \frac{1}{10}\partial_{xx} v + \frac{1}{10}\partial_{yy} v + v - u v^2 - u^3 - v^3 - u^2 v \end{cases}$
2D Navier-Stokes (NS)	$\partial_t \omega = -\partial_x(\omega u) - \partial_y(\omega v) + \frac{1}{100}\partial_{xx}\omega + \frac{1}{100}\partial_{yy}\omega$

where TP is the number of correctly identified nonzero coefficients, FN is the number of coefficients falsely identified as zero, and FP is the number of coefficients falsely identified as nonzero. Identification of the true model results in a TPR of 1, while identification of half of the correct nonzero terms and no falsely identified nonzero terms results in TPR of 0.5.

TABLE 2 Computational efficiency of WSINDy for learning the listed PDEs. The $\tilde{G}$ column reports the size of the matrix used with in the sparse regression where the tilde denotes that the data has been scaled to improve the computation stability. The next column indicates the condition number after the rescaling. The last column shows the start-to-finish walltime with all computations in serial measured on a laptop with an 8-core Intel i7-2670QM CPU. Notably, none of the computations take more than 75 seconds. Moreover, these walltimes are independent of the noise since the same algorithm is being used regardless of the σ_{NR} level.

PDE	$\tilde{G}$	$\kappa(\tilde{G})$	Walltime (sec)
IB	784×43	1.4×10^6	0.12
KdV	1443×43	3.2×10^6	0.39
KS	1806×43	3.7×10^3	0.24
NLS	1804×190	1.2×10^5	2.5
NS	3872×50	8.2×10^2	12
PM	4608×65	2.4×10^4	16
SG	$13{,}000 \times 73$	1.3×10^4	29
RD	$11{,}638 \times 181$	4.5×10^3	75

For each system in Table 1 and a range of noise levels $\sigma_{NR} \in [0, 1]$ we ran[4] WSINDy with 200 realizations of noise and average the results.

Fig. 3 depicts the TPR for all the PDEs investigated. Notably, for inviscid Burgers, Korteweg-de Vries, Kuramoto-Sivashinsky and Sine-Gordon, the average TPR stays above 0.95 even for $\sigma_{NR} = 1$. The average TPR for the nonlinear Schrödinger and porous medium equations stays above 0.95 until 50% noise, after which identification of the correct monomial nonlinearity is not as reliable. For NLS, this is a drastic improvement over previous studies (Rudy et al., 2017), especially considering the large library of 190 terms used.

Fig. 4 depicts the E_2 error in the recovered coefficients for $\sigma_{NR} \in [0, 1]$. Intriguingly, for a noise level of 10%, the relative error is typically 10% or less in almost all cases. This is surprising as in many cases, conventional methods (depending on a comparison of data with numerical solutions) yield relative errors on the order of or higher than the σ_{NR} from the data (see §4.1.1 and Bortz et al., 2023).

[4] Computations were run on a 16-core Intel Xeon 5218 CPU node in the Colorado Research Computing condo system.

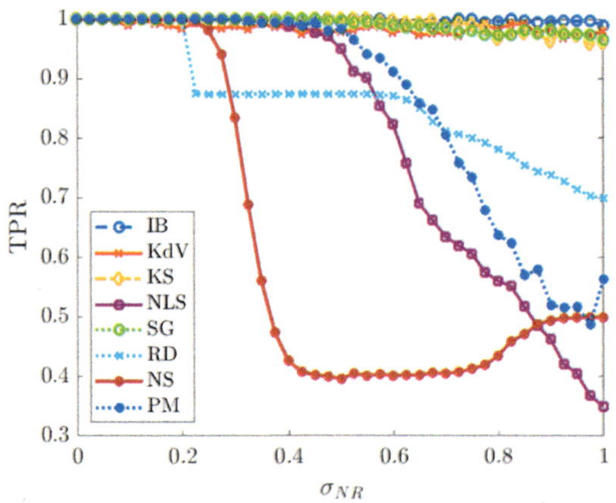

FIGURE 3 Average TPR (true positivity ratio, defined in (12)) for each of the PDEs in Table 1 computed from 200 instantiations of noise for each noise level σ_{NR}. Figure from Messenger and Bortz (2021a).

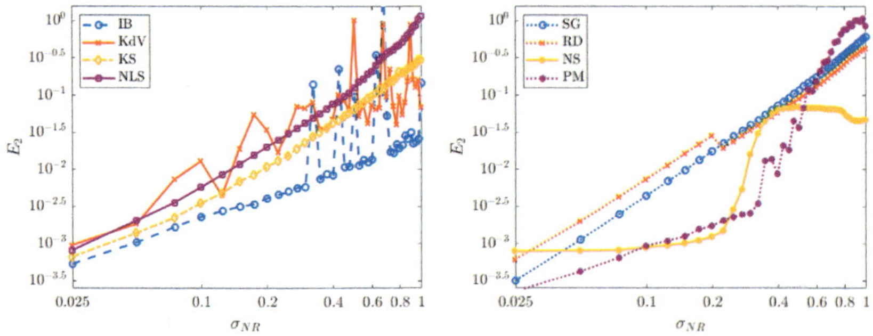

FIGURE 4 Coefficient errors E_2 for each of the seven models Table 1. The plot on the left contains PDEs in one spatial dimension, and the plot on the right contains PDEs in two dimensions. Figure from Messenger and Bortz (2021a).

3 Theoretical results

In the application of WSINDy to several classes of differential equations, the general pattern has emerged that the system identification using the weak form offers substantial advantages over using the strong form. However, there is no rigorous explanation for this behavior and thus a theoretical analysis of WSINDy is an ongoing project. Here we report briefly on recent theoretical results presented in Messenger and Bortz (2022a) concerning the performance of WSINDy in the limit of continuum data. Theorem 1 below (Messenger and Bortz, 2022a, Theorem 4.2) demonstrates that within a broad class of models, the WSINDy estimate converges in probability to the correct model, provided that the noise level σ is below a critical noise threshold σ_c (see Assumptions A for a full list of assumptions). This provides the following:

(I) An explanation for the empirically observed robustness of weak-form equation learning methods
(II) Quantification of the effects of nonsmoothness (e.g. weak solutions) on the rate of convergence
(III) Specification of a class of models for which WSINDy converges for any noise level (i.e. $\sigma_c = \infty$)

In addition, in Messenger and Bortz (2022a, Theorem 4.3) it is also proved that suitably denoising the data (e.g. with a simple moving-average filter) results in unconditional convergence of WSINDy over the class of models with locally-Lipschitz nonlinearities, extending the class specified in (III) above.

3.1 Assumptions

Theorem 1 concerns the performance of WSINDy in the limit of *continuum data*. Using the notation from Section 2.4, this limit is defined by a sequence of noisy samples $\{\mathbf{U}^{(n)}\}_{n=1}^{\infty}$ of the solution u observed on successively finer computational grids $\{(\mathbf{X}^{(n)}, \mathbf{t}^{(n)})\}_{n=1}^{\infty} \subset \Omega \times (0, T)$, each of which is equally spaced with resolution $\Delta x^{(n)}$ in all spatial coordinates and $\Delta t^{(n)}$ in time. For each $\mathbf{U}^{(n)}$, a weak-form linear system $(\mathbf{G}^{(n)}, \mathbf{b}^{(n)})$ is constructed according to (9), and WSINDy is employed to produce an estimate $\widehat{\mathbf{w}}^{(n)}$ of the true weight vector $\mathbf{w}^{\star}$. We now describe the core assumptions used in the proof of Theorem 1 are below.

We define the following admissible solution spaces for the underlying solution u. For an open, bounded domain $D \subset \mathbb{R}^{d+1}$, codomain $D' \subset \mathbb{R}^N$, $p \in [1, \infty]$, and $k > 0$, define the function spaces

$$\mathcal{H}^{k,p}(D, D') := \left\{ f \in L^p(D, D') \ : \ \exists \text{ disjoint, open } (D_i)_{i=1}^{\ell} \right. \\ \left. \text{s.t. } \overline{D} = \bigcup_{i=1}^{\ell} \overline{D}_i\,, \ f\big|_{D_i} \in H^k(D_i, D'),\ \partial D_i \in C^{0,1} \right\}, \tag{13}$$

where $H^k(D, D')$ is the space of functions from D to D' with weak derivatives up to order k in $L^2(D, D')$. In what follows, we take $D' = \mathbb{R}^N$ and suppress explicit reference to the codomain. The spaces $\mathcal{H}^{k,p}(D) \subset L^p(D)$ are similar to the broken Sobolev spaces used in the analysis of discontinuous Galerkin methods (see e.g. Houston et al., 2002). With $u \in \mathcal{H}^{k,\infty}(\Omega \times (0, T))$ such that $k > (d + 1)/2$, we have that pointwise evaluations of u are well-defined (apart from a set of measure zero, e.g. when considering solutions with shocks) by the Sobolev embedding theorem.

In addition, define the *discretization level* m_n by[5]

$$m_n := \#\{\mathrm{supp}\,(\psi) \cap (\mathbf{X}^{(n)}, \mathbf{t}^{(n)})\}. \tag{14}$$

[5] $\#\{\cdot\}$ indicates the set cardinality.

In words, m_n is the number of points that the reference test function ψ is supported on within the grid $(\mathbf{X}^{(n)}, \mathbf{t}^{(n)})$. In Messenger and Bortz (2022a, Theorem 3.1), it is proved that $(\mathbf{G}^{(n)}, \mathbf{b}^{(n)})$ concentrates around its mean at a rate $O(\exp(-cm_n^{2/p}))$, where c is a universal constant and $p_{\max}$ is the maximum polynomial degree in the model library.

Assumptions A. Let $p \geq 1$ be fixed.

(A.1) (Regularity of u) $u \in \mathcal{H}^{k,\infty}(\Omega \times (0, T))$ for some $k > (d+1)/2$ is a weak solution to (7) with coefficients $\mathbf{w} = \mathbf{w}^\star$.

(A.2) (Noise distribution) Measurement noise $\epsilon = \mathbf{U}^{(n)} - u(\mathbf{X}^{(n)}, \mathbf{t}^{(n)})$ is i.i.d. according to a symmetric and sub-Gaussian probability distribution ρ.[6] We refer to the standard deviation $\sigma = \sqrt{\mathbb{V}[\rho]}$ as the *noise level.*

(A.3) (Computational grids) Each grid $(\mathbf{X}^{(n)}, \mathbf{t}^{(n)})$ has uniform spacing $\Delta x^{(n)}$ in all spatial dimensions and $\Delta t^{(n)}$ in time, and the collection of grids is dense in $\Omega \times (0, T)$, or $\overline{\bigcup_{n=1}^\infty (\mathbf{X}^{(n)}, \mathbf{t}^{(n)})} = \Omega \times (0, T)$.

(A.4) (Model library) The family of functions $\mathcal{F} = (f_j(u))_{j\in[J]}$ consists of $P^{(p_{\max})}$, the set of monomials of total degree[7] at most $p_{\max}$ on $\mathbb{R}^N$, as well as $F^{\boldsymbol{\omega}} = \{\exp(i\omega^T u)\}_{\omega\in\boldsymbol{\omega}}$, a finite collection of Fourier modes on $\mathbb{R}^N$ (i.e. $\boldsymbol{\omega} \subset \mathbb{R}^N$ is a finite set). Furthermore we assume[8] $\widehat{\rho}(\omega) \neq 0$ for $\omega \in \boldsymbol{\omega}$.

(A.5) (Reference test function and Query points) We assume that $\psi \in C^{|\boldsymbol{\alpha}|}(\Omega \times (0, T))$ with compact support in $\Omega \times (0, T)$, and that for all $(\mathbf{x}_k, t_k) \in Q$,

$$\operatorname{supp}(\psi(\mathbf{x}_k - \cdot, t_k - \cdot)) \subset \Omega \times (0, T) \tag{15}$$

(A.6) (Conditioning of $(\mathbf{G}^\star, \mathbf{b}^\star)$) The noise-free continuum matrix $\mathbf{G}^\star$[9] has full column rank. Moreover, the true dynamics have a stable representation in weak-form quantified by

$$\mu^\star := \min_{S \subsetneq S^\star} \frac{\left\|\mathbf{P}^\perp_{\mathbf{G}^\star_{S^\star\setminus S}} \mathbf{b}^\star\right\|}{\|\mathbf{b}^\star\|} - \frac{|S|+1}{\mathfrak{J}} > 0 \tag{16}$$

where $\mathfrak{J}$ is number of columns in $\mathbf{G}^\star$, $S^\star$ is the support of the true weight vector $\mathbf{w}^\star$, and $\mathbf{P}^\perp_{\mathbf{G}^\star_{S^\star\setminus S}}$ denotes the projection onto space orthogonal to the columns $\mathbf{G}^\star_{S^\star\setminus S}$ lying in the set $S^\star \setminus S$. In words, $\mathbf{b}^\star$ cannot be approximated arbitrarily well from a strict subset of $\mathbf{G}^\star_{S^\star}$.

Under these assumptions, we have the following.

---

[6] That is, ρ satisfies $\|\rho\|_{SG} := \inf\{\lambda > 0 : \mathbb{E}_{\epsilon\sim\rho}\left[\exp(\epsilon^2/\lambda^2)\right] \leq 2\} < \infty$. This includes e.g. Gaussian and uniform white noise, see Vershynin (2018) for more details.

[7] For a monomial $p(x_1, \ldots, x_N) = \prod_{i=1}^N x_i^{q_i}$, the total degree is defined by $\sum_{i=1}^N q_i$.

[8] $\widehat{\rho}(\omega) = \int_{\mathbb{R}^N} e^{i\omega\cdot y}\rho(y)dy$ is the Fourier transform of ρ.

[9] Entries $\mathbf{G}^\star_{k,(s-1)J+j}$ and $\mathbf{b}^\star_k$ are given by the right and left sides of (9).

Theorem 1. *Provided Assumptions A hold, there exists a critical noise level $\sigma_c > 0$ and a stability tolerance θ_* such that for all $\sigma < \sigma_c$, $\theta < \theta_*$, and sufficiently large n, it holds that*

$$supp\left(\widehat{\mathbf{w}}^{(n)}\right) = supp\left(\mathbf{w}^\star\right) \qquad and \qquad \left\|\widehat{\mathbf{w}}^{(n)} - \mathbf{w}^\star\right\|_\infty < C'(\theta + \sigma^2) \quad (17)$$

with probability exceeding $1 - 4K(\mathfrak{J}+1)\exp\left(-\frac{c}{2}(m_n\theta)^{2/p_{\max}}\right)$, *where* $\widehat{\mathbf{w}}^{(n)} = \text{MSTLS}^{(1)}(\mathbf{G}^{(n)}, \mathbf{b}^{(n)})$ *and* $c, C' > 0$ *are independent of n.*

The proof of Theorem 1 is contained in Messenger and Bortz (2022a, Theorem 4.2).[10]

4 Weak form-based parameter estimation

The previous sections describe how WSINDy can be used to discover governing equations. However, in the case where there is high confidence in the model equation itself, one can also perform parameter estimation using the weak form of the model. Versions of this idea have existed since the mid 1950's (Shinbrot, 1954), but the proposed test functions either did not have enough smoothness or were spectrally mismatched with the data and did not yield highly accurate estimates. Accordingly, a natural research direction is to build on the success of WSINDy and combine it with the theoretical results in §3 as well as modern statistical regression to create a novel parameter estimation method. In Bortz et al. (2023), we introduced an improved weak-form parameter estimation algorithm WENDy (Weak-form Estimation of Nonlinear Dynamics) which works for the class of differential equations with right sides that are linearly separable. In §4.1, we will summarize the results for ODEs and then discuss how to apply WENDy to PDEs §4.2 and SDEs §4.3.

In this section we assume that the model is known and will attempt to solve this parameter estimation problem

$$\widehat{\mathbf{w}} := \arg\min_{\mathbf{w}\in\mathbb{R}^J} \|u(\mathbf{t};\mathbf{w}) - \mathbf{U}\|_2^2, \tag{18}$$

where the data $\mathbf{U} \in \mathbb{R}^{(M+1)\times d}$ is sampled at $M+1$ timepoints $t := \{t_i\}_{i=0}^M$, and the function $u : \mathbb{R} \to \mathbb{R}^d$ is a solution to a differential equation model

$$\begin{aligned} \dot{u} &= \sum_{j=1}^{J} w_j f_j(u), \\ u(t_0) &= u_0 \in \mathbb{R}^d. \end{aligned} \tag{19}$$

[10] Also note that $\text{MSTLS}^{(1)}$ refers to the Modified Sequential Thresholding Least Squares algorithm with a single thresholding step per $\lambda \in \boldsymbol{\lambda}$, or $\ell = 0$ (see Messenger and Bortz, 2021a).

4.1 Ordinary differential equations

We begin by considering a d-dimensional matrix form of (19), i.e., an ordinary differential equation system model

$$\dot{u} = \Theta(u)W \tag{20}$$

with row vector of the d solution states

$$u(t;W) := \left[\; u_1(t;W) \;\middle|\; u_2(t;W) \;\middle|\; \cdots \;\middle|\; u_d(t;W)\right],$$

row vector of J features (i.e., right side terms where $f_j : \mathbb{R}^d \to \mathbb{R}$ is C_c^2) such that $\Theta(u) := \left[\; f_1(u) \;\middle|\; f_2(u) \;\middle|\; \cdots \;\middle|\; f_J(u)\right]$, and the matrix of unknown parameters $W \in \mathbb{R}^{J\times d}$. The matrix version of these terms evaluated at the timepoints is thus

$$\mathbf{t} := \begin{bmatrix} t_0 \\ \vdots \\ t_M \end{bmatrix}, \qquad \mathbf{u} := \begin{bmatrix} u_1(t_0) & \cdots & u_d(t_0) \\ \vdots & \ddots & \vdots \\ u_1(t_M) & \cdots & u_d(t_M) \end{bmatrix},$$

$$\Theta(\mathbf{u}) := \begin{bmatrix} f_1(u(t_0)) & \cdots & f_J(u(t_0)) \\ \vdots & \ddots & \vdots \\ f_1(u(t_M)) & \cdots & f_J(u(t_M)) \end{bmatrix}.$$

Multiplication by a set of compactly supported test functions $\{\phi_{m_k}\}$ (centered at timepoints $\{t_{m_k}\}$, a subset of the entries of $\mathbf{t}$) using trapezoidal quadrature yields

$$-\dot{\Phi}_k\mathbf{u} \approx \Phi_k\Theta(\mathbf{u})W, \tag{21}$$

where

$$\Phi_k := \left[\; \phi_k(t_0) \;\middle|\; \cdots \;\middle|\; \phi_k(t_M) \;\right], \qquad \dot{\Phi}_k := \left[\; \dot{\phi}_k(t_0) \;\middle|\; \cdots \;\middle|\; \dot{\phi}_k(t_M) \;\right].$$

The core idea of the weak-form-based direct parameter estimation is to identify W as a least squares solution to

$$\min_W \|\text{vec}(\mathbf{G}W - \mathbf{B})\|_2^2 \tag{22}$$

where "vec" vectorizes a matrix,

$$\mathbf{G} := \Phi\Theta(\mathbf{U}) \in \mathbb{R}^{K\times J},$$
$$\mathbf{B} := -\dot{\Phi}\mathbf{U} \in \mathbb{R}^{K\times d},$$

where $\mathbf{U} \in \mathbb{R}^{(M+1)\times d}$ represents the data, and the integration matrices are

$$\Phi = \begin{bmatrix} \Phi_1 \\ \vdots \\ \Phi_K \end{bmatrix} \in \mathbb{R}^{K\times(M+1)} \quad \text{and} \quad \dot{\Phi} = \begin{bmatrix} \dot{\Phi}_1 \\ \vdots \\ \dot{\Phi}_K \end{bmatrix} \in \mathbb{R}^{K\times(M+1)}.$$

4.1.1 WENDy using iterative reweighting

We note that the posed regression problem does not fit within the framework of ordinary least squares, and is actually an Errors-In-Variables problem. We will also derive a linearization that reveals a covariance structure which depends on the jacobian of the right side as well as the true parameters. First, we denote the vector of true (but unknown) parameter values used in all state variable equations as $\mathbf{w}^\star$ and let $u^\star := u(t; \mathbf{w}^\star)$ and $\Theta^\star := \Theta(u^\star)$. The system measurements are assumed to be noisy so that at each timepoint t all states are observed with additive noise

$$U(t) = u^\star(t) + \varepsilon(t) \tag{23}$$

where each element of $\varepsilon(t)$ is i.i.d. $\mathcal{N}(0, \sigma^2)$. Lastly, we note that there are d variables, J feature terms, and $M+1$ timepoints. In what follows, we present the expansion using Kronecker products (denoted as $\otimes$).

We begin by considering the sampled data $\mathbf{U} := \mathbf{u}^\star + \boldsymbol{\varepsilon} \in \mathbb{R}^{(M+1)\times d}$ and vector of parameters to be identified $\mathbf{w} \in \mathbb{R}^{Jd}$. We use bolded variables to represent evaluation at the timegrid $\mathbf{t}$, and use superscript $\star$ notation to denote quantities based on true (noise-free) parameter or states. We now consider the residual

$$\mathbf{r}(\mathbf{U}, \mathbf{w}) := \mathbf{G}\mathbf{w} - \mathbf{b}, \tag{24}$$

where we redefine

$$\begin{aligned} \mathbf{G} &:= [\mathbb{I}_d \otimes (\Phi\Theta(\mathbf{U}))], \\ \mathbf{b} &:= -\text{vec}(\dot{\Phi}\mathbf{U}). \end{aligned}$$

We decompose and linearize $\mathbf{r}$ such that

$$\mathbf{r}(\mathbf{U}, \mathbf{w}) \approx \mathbf{G}^\star\mathbf{W} - \mathbf{b}^\star + \mathbf{L}_\mathbf{w}\text{vec}(\boldsymbol{\varepsilon}), \tag{25}$$

where

$$\begin{aligned} \mathbf{G}^\star &:= [\mathbb{I}_d \otimes (\Phi\Theta(\mathbf{u}^\star))], \\ \mathbf{b}^\star &:= -\text{vec}(\dot{\Phi}\mathbf{u}^\star), \\ \mathbf{L}_\mathbf{w} &:= [\text{mat}(\mathbf{w})^T \otimes \Phi]\nabla\Theta\mathbf{K} + [\mathbb{I}_d \otimes \dot{\Phi}], \end{aligned}$$

where "mat" is the matricization operation and $\mathbf{K}$ is the commutation matrix such that $\mathbf{K}\text{vec}(\boldsymbol{\varepsilon}) = \text{vec}(\boldsymbol{\varepsilon}^T)$. The matrix $\nabla\Theta$ contains derivatives of the features

$$\nabla\Theta := \left[\begin{array}{ccc} \nabla f_1(\mathbf{U}_0) & & \\ & \ddots & \\ & & \nabla f_1(\mathbf{U}_M) \\ \hline & \vdots & \\ \hline \nabla f_J(\mathbf{U}_0) & & \\ & \ddots & \\ & & \nabla f_J(\mathbf{U}_M) \end{array}\right], \tag{26}$$

where

$$\nabla f_j(\mathbf{U}_m) = \left[\ \frac{\partial}{\partial u_1} f_j(\mathbf{U}_m) \ \middle|\ \cdots \ \middle|\ \frac{\partial}{\partial u_d} f_j(\mathbf{U}_m)\ \right],$$

and $\mathbf{U}_m \in \mathbb{R}^{1\times d}$ is the row vector of data at t_m.

If all elements of $\boldsymbol{\varepsilon}$ are i.i.d. Gaussian, i.e., $\mathcal{N}(0, \sigma^2)$ then to first order

$$\mathbf{r}(\mathbf{U}, \mathbf{w}) - (\mathbf{r}_0 + \mathbf{e}_{\text{int}}) \sim \mathcal{N}(\mathbf{0}, \sigma^2 \mathbf{L}_{\mathbf{w}}(\mathbf{L}_{\mathbf{w}})^T). \tag{27}$$

We note that in (27), the covariance is dependent upon the parameter vector $\mathbf{w}$. In the statistical inference literature, the Iteratively Reweighted Least Squares (IRLS) (Jorgensen, 2012) method provide a good strategy to account for a parameter-dependent covariance by iterating between solving for $\mathbf{w}$ and updating the covariance matrix $\mathbf{C}$. In Algorithm 2 we present the WENDy method, updating $\mathbf{C}^{(n)}$ (at the n-th iteration step) in lines 7-8 and then the new parameters $\mathbf{w}^{(n+1)}$ are computed in line 9 by weighted least squares.

Depicted in Fig. 5 are the results of comparing WENDy and FSNLS for the FitzHugh-Nagumo equation on a scatterplot of walltime (in seconds) vs. the relative accuracy in the estimated parameters.

$$\begin{aligned} \dot{u}_1 &= 3u_1 - 3u_1^3 + 3u_2 \\ \dot{u}_2 &= -\,1/3u_1 + 17/150 + 1/15u_2 \end{aligned}$$

It is clear in this figure that WENDy is on average both more accurate and faster. In Bortz et al. (2023), WENDy is applied to several other benchmark ODE parameter estimation problems including Logistic growth, Lotka-Volterra, Hindmarsh-Rose, and a Protein Transduction system. In almost all cases, WENDy offered more accurate parameter estimates. And, in *all* cases, WENDy was faster, sometimes by several orders of magnitude.

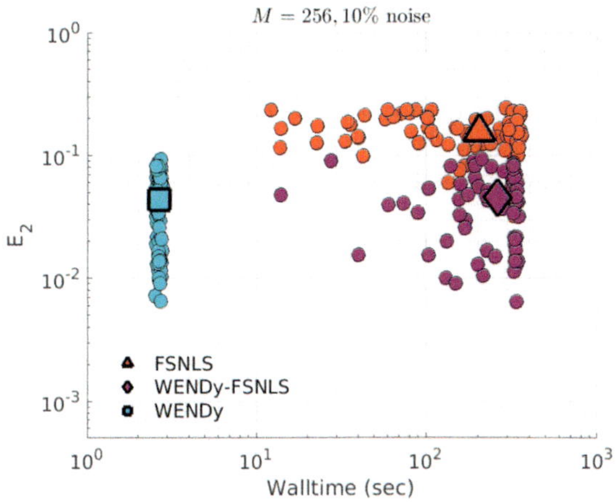

FIGURE 5 Comparison between Forward Solver-based Nonlinear Least Squares (FSNLS), Weak form Estimation of Nonlinear Dynamics (WENDy), and using the results of WENDy as an initial estimate for FSNLS (WENDy-FSNLS) for the FitzHugh-Nagumo model. Both variables are observed at 256 timepoints and with 10% additive Gaussian noise level. Figure from Bortz et al. (2023).

4.2 Partial differential equations

Extending WENDy to PDEs is straightforward. Consider the weak form (9) written as a convolution between the test function and the equation terms, and assume that the ith solution component obeys

$$\left(\mathscr{D}^{\alpha^0}\psi\right) * u_i(Q) - \sum_{j=1}^{J} \mathbf{W}_{ji}\left(\mathscr{D}^{\alpha^j}\psi\right) * f_j(u)(Q) = 0.$$

We can write this in discrete form as

$$\Phi^{\alpha_0}\mathbf{U}_i^\star - \sum_{j=1}^{J} \Phi^{\alpha_j} f_j(\mathbf{U}^\star)\mathbf{W}_i = \mathbf{b}_i - \mathbf{G}\mathbf{W}_i \approx 0$$

where Φ^{α^j} are associated discrete matrices that enact the convolution with $\mathscr{D}^{\alpha^j}\psi$ over spacetime, evaluated at the query points Q, and $\mathbf{W}_i$ is the ith column of $\mathbf{W}$. Performing similar analysis to Section 4.1.1, we arrive at the linear transformation

$$\mathbf{L_W} := \left[\mathbf{W}_1^T \otimes \Phi^{\alpha_1} \quad \mathbf{W}_2^T \otimes \Phi^{\alpha_2} \quad \dots \quad \mathbf{W}_J^T \otimes \Phi^{\alpha_J}\right] \nabla\Theta\mathbf{K} + \left[\mathbb{I}_d \otimes \Phi^{\alpha^0}\right].$$

Here $\mathbf{W}_j^T$ is the jth column of $\mathbf{W}$ and $\nabla\Theta$ is as in (26) suitably reindexed to account for the vectorization of multidimensional arrays. Appealing to sparse matrix constructions, the results can similarly accelerate and improve on estimates obtained using forward simulation-based nonlinear least squares, as

Algorithm 2: WENDy

input : Data $\{\mathbf{U}\}$, Feature Map $\{\Theta, \nabla\Theta\}$, Test Function Matrices $\{\Phi, \dot{\Phi}\}$, Stopping Criteria $\{SC\}$, Covariance Relaxation Parameter $\{\alpha\}$, Variance Filter $\{\mathbf{f}\}$

output: Parameter Estimate $\{\widehat{\mathbf{w}}, \widehat{\mathbf{C}}, \widehat{\sigma}, \mathbf{S}, \texttt{stdx}\}$

```
// Compute weak-form linear system
```
$\mathbf{G} \leftarrow [\mathbb{I}_d \otimes (\Phi\Theta(\mathbf{U}))]$

$\mathbf{b} \leftarrow -\text{vec}(\dot{\Phi}\mathbf{U})$

```
// Solve Ordinary Least Squares Problem
```
$\mathbf{w}^{(0)} \leftarrow (\mathbf{G}^T\mathbf{G})^{-1}\mathbf{G}^T\mathbf{b}$

```
// Solve Iteratively Reweighted Least Squares Problem
```
$n \leftarrow 0$

`check` $\leftarrow$ true

while `check` is true **do**

  $\mathbf{L}^{(n)} \leftarrow [\text{mat}(\mathbf{w}^{(n)})^T \otimes \Phi]\nabla\Theta(\mathbf{U})\mathbf{K} + [\mathbb{I}_d \otimes \dot{\Phi}]$

  $\mathbf{C}^{(n)} = (1-\alpha)\mathbf{L}^{(n)}(\mathbf{L}^{(n)})^T + \alpha\mathbf{I}$

  $\mathbf{w}^{(n+1)} \leftarrow (\mathbf{G}^T(\mathbf{C}^{(n)})^{-1}\mathbf{G})^{-1}\mathbf{G}^T(\mathbf{C}^{(n)})^{-1}\mathbf{b}$

  `check` $\leftarrow \text{SC}(\mathbf{w}^{(n+1)}, \mathbf{w}^{(n)})$

  $n \leftarrow n+1$

end

```
// Return estimate and standard statistical quantities
```
$\widehat{\mathbf{w}} \leftarrow \mathbf{w}^{(n)}$

$\widehat{\mathbf{C}} \leftarrow \mathbf{C}^{(n)}$

$\widehat{\sigma} \leftarrow (Md)^{-1/2}\|\mathbf{f} * \mathbf{U}\|_F$

$\mathbf{S} \leftarrow \widehat{\sigma}^2((\mathbf{G}^T\mathbf{G})^{-1}\mathbf{G}^T)\,\widehat{\mathbf{C}}\,(\mathbf{G}(\mathbf{G}^T\mathbf{G})^{-1}))$

`stdx` $\leftarrow \sqrt{\texttt{diag}(\mathbf{S})}$

evidenced in Fig. 6 for the Kuramoto-Sivashinsky PDE. All circles and squares in the figure represent different realizations of the noisy data. While there is certainly some variability, on average WENDy is two orders of magnitude faster and an order of magnitude more accurate than the conventional FSNLS method.

4.3 Stochastic differential equations

Weak-form estimation can also be easily extended to stochastic differential equations. The focus is to identify the drift f and diffusivity σ of an Itô SDE

$$dX_t = f(X_t, t)dt + \sigma(X_t, t)dB_t \tag{28}$$

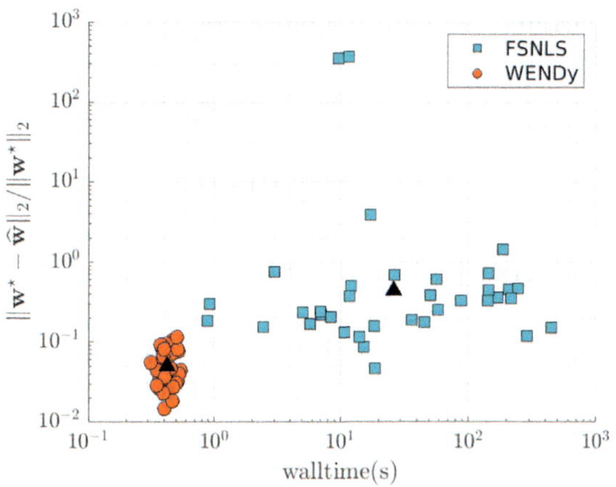

FIGURE 6 Coefficient error vs. walltime for WENDy applied to the Kuramoto-Sivashinsky equation, as described in Section 4.2, vs. forward solver-based nonlinear least-squares (FSNLS). The underlying data has 128 × 64 points in space and time and 20% added Gaussian white noise.

given discrete-time observations of its solution $X_t \in \mathbb{R}^d$. In this setting the input data is a collection of L discrete-time realizations $\mathbf{Y} = \{\mathbf{Y}^{(\ell)}\}_{\ell=1}^{L}$, where each realization $\mathbf{Y}^{(\ell)} = (Y_{t_0}^{(\ell)}, \ldots, Y_{t_{M+1}}^{(\ell)})$ occurs over the time grid $\mathbf{t} = (t_0, \ldots, t_{M+1})$, and at each time $t_i \in \mathbf{t}$ the observations are given by $Y_{t_i}^{(\ell)} = X_{t_i}^{(\ell)} + \epsilon$. Here ϵ represents possible measurement noise and $X_t^{(\ell)}$, $\ell = 1, \ldots, L$ are solutions to (28) with initial conditions $x_0^{(\ell)}$ each drawn independently from the distribution ρ_0.

Weak formulation in time

Itô calculus can be employed to formulate the weak-form discovery problem. Let $\psi(x,t) : \mathbb{R}^d \times \mathbb{R}_+ \to \mathbb{R}$ be a C^2 function compactly-supported in the time interval $(0, T)$ for all $x \in \mathbb{R}^d$. Itô's formula applied (28) then gives

$$d\left(\psi(X_t,t)\right) = \left(\partial_t \psi(X_t,t) + \nabla\psi(X_t,t)\cdot f(X_t,t) + \frac{1}{2}\mathrm{H}\psi(X_t,t) : (\sigma\sigma^T)\right)dt \\ + \nabla\psi(X_t,t)\cdot\sigma(X_t,t)dB_t,$$

where H denotes the Hessian and $A : B = \mathrm{vec}(A)\cdot\mathrm{vec}(B)$. Integrating in time and using compact support, we get

$$-\int_0^T \partial_t\psi(X_t,t)\,dt = \int_0^T \left(\nabla\psi(X_t,t)\cdot f(X_t,t) + \frac{1}{2}\mathrm{H}\psi(X_t,t) : (\sigma\sigma^T)\right)dt \\ + \int_0^T \nabla\psi(X_t,t)\cdot\sigma(X_t,t)dB_t. \tag{29}$$

Estimation algorithm: test and trial functions

As in the PDE case, the set of test functions $\Psi = \{\psi_1, \ldots, \psi_K\}$ can be chosen in a flexible and efficient manner by letting each ψ_k be separable, of the form

$$\psi_k(x,t) = \theta_k(t)\phi_k(x).$$

This leads to a very general scheme, with the only requirements being that $\theta_k \in C^1$, compactly support in $[0, T]$, and $\phi_k \in C^2$.

We search for linear representations of the drift f and squared diffusivity $\sigma\sigma^T$. That is, we define a drift basis $\mathbb{F} = (f_1, \ldots, f_{J_f})$ and a basis of upper-triangular matrices $\mathbb{S} = (\Sigma_1, \ldots, \Sigma_{J_\sigma})$, and search for linear representations

$$\widehat{f} = \sum_{j=1}^{J_f} w_j^{\mathbb{F}} f_j, \qquad \widehat{\Sigma} = \sum_{j=1}^{J_\sigma} w_j^{\mathbb{S}} \Sigma_j.$$

Our diffusivity estimator is then

$$\widehat{\sigma\sigma^T} = \widehat{\Sigma} + \widehat{\Sigma}^T - \mathrm{diag}(\widehat{\Sigma}).$$

Note that for two symmetric matrices A, B, letting $U(A)$ and $U(B)$ denote their upper-triangular parts, we have

$$\begin{aligned} A : B &= \sum_{i=1}^{n}\sum_{j=1}^{n} A_{ij}B_{ij} = 2\sum_{i=1}^{n}\sum_{j=i}^{n} A_{ij}B_{ij} - \mathrm{diag}(A) : \mathrm{diag}(B) \\ &=: (U(A) : U(B)) \end{aligned}$$

which defines an inner product $(\cdot : \cdot)$ on upper triangular matrices, so the representation of the squared diffusivity $\widehat{\sigma\sigma^T}$ above in terms of $\mathbb{S}$ is valid.

Quadrature

Integrals in (28) are discretized using the trapezoidal rule, which achieves an optimal minimax rate for integrands with the same regularity as Brownian motion (Diaconis, 1988), which is shared by solutions to Itô SDEs.[11] For each observed realization $\mathbf{Y}^{(\ell)}$, we define the matrices $(\mathbf{G}^{\mathbb{F}}(\mathbf{Y}^{(\ell)}), \mathbf{G}^{\mathbb{S}}(\mathbf{Y}^{(\ell)}))$ by

$$\mathbf{G}_{kj}^{\mathbb{F}}(\mathbf{Y}^{(\ell)}) = \sum_{i=1}^{M} \nabla\psi_k(\mathbf{Y}_{t_i}^{(\ell)}, t_i) \cdot f_j(\mathbf{Y}_{t_i}^{(\ell)}, t_i)\left(\frac{\Delta t_i + \Delta t_{i-1}}{2}\right)$$

$$\mathbf{G}_{kj}^{\mathbb{S}}(\mathbf{Y}^{(\ell)}) = \sum_{i=1}^{M} \left(\mathrm{H}\psi_k(\mathbf{Y}_{t_i}^{(\ell)}, t_i) : \Sigma_j(\mathbf{Y}_{t_i}^{(\ell)}, t_i)\right)\left(\frac{\Delta t_i + \Delta t_{i-1}}{2}\right),$$

11 Strictly speaking, if the diffusivity is nonconstant, then regularity can be slightly worse than this, in which case there may be a better quadrature. It may even be the case that Riemann sums have lower variance, this needs to be fully explored.

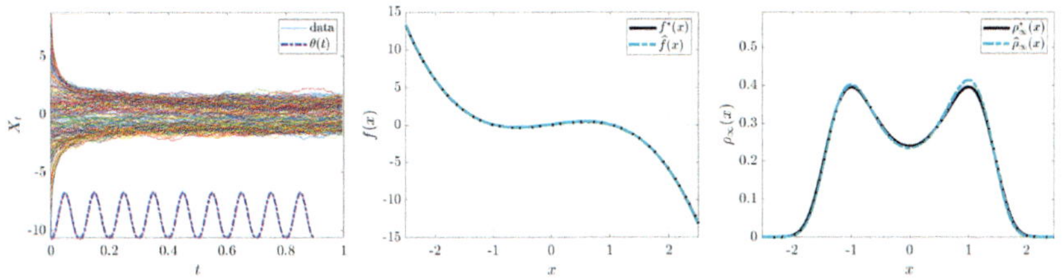

FIGURE 7 WENDy applied to the stochastic double-well potential dynamics (30). Left: data realizations and temporal testfunctions ϕ_k. Middle: comparison between learned ($\hat{f}$) and true ($f^\star$) drift functions. Right: comparison between learned ($\hat{\rho}_\infty$) and true ($\rho^\star_\infty$) stationary measure.

where timesteps are given by $\Delta t_i = t_{i+1} - t_i$. We then form the concatenated matrix

$$\mathbf{G}(\mathbf{Y}^{(\ell)}) = \left[\mathbf{G}^{\mathbb{F}}(\mathbf{Y}^{(\ell)}) \quad \mathbf{G}^{\mathbb{S}}(\mathbf{Y}^{(\ell)})\right].$$

The response vector $\mathbf{b}(\mathbf{Y}^{(\ell)})$ is defined by

$$\mathbf{b}_k(\mathbf{Y}^{(\ell)}) = -\sum_{i=1}^{M} \partial_t \psi_k(\mathbf{Y}^{(\ell)}_{t_i}, t_i) \left(\frac{\Delta t_i + \Delta t_{i-1}}{2}\right).$$

It then holds that

$$\mathbf{b}(\mathbf{Y}^{(\ell)}) \approx \mathbf{G}(\mathbf{Y}^{(\ell)})\mathbf{w}$$

where $\mathbf{w} = ((\mathbf{w}^{\mathbb{F}})^T, (\mathbf{w}^{\mathbb{S}})^T)^T$. To combine the L observed realizations $\mathbf{Y}^{(\ell)}$, we average the L linear systems together, arriving at a small linear system $(\overline{\mathbf{G}}, \overline{\mathbf{b}})$, which reduces variance at the Monte-Carlo rate in L.

Optimization

The residuals

$$\mathbf{r}^{(\ell)}(\mathbf{w}) = \mathbf{b}(\mathbf{Y}^{(\ell)}) - \mathbf{G}(\mathbf{Y}^{(\ell)})\mathbf{w}$$

are amenable to the same analysis employed in WENDy in the context of ODEs and PDEs, leading to an iteratively reweighted least-squares approach for the weights $\widehat{\mathbf{w}}$. Specifically, the right-most term in (29) is an Itô integral and constitutes the residual for each test function. Its covariance against other residuals can be computed using Itô calculus and properties of local martingales. However, as a first pass ordinary least squares can be used, which can be seen to perform well applied to the double-well potential dynamics (30) (see Fig. 7).

Test problem: double-well potential

Consider $X_t \in \mathbb{R}$ satisfying (28)

$$f(x,t) = x - x^3, \qquad \sigma(x,t) = 1. \tag{30}$$

The dynamics, drift and stationary measure are pictured in Fig. 7.

TABLE 3 Comparison between state-of-the-art deep learning-based estimation and WENDy.

Parameter	λ_0	λ_1	λ_2	λ_3	μ	runtime
True value	0	1	0	−1	1	—
PINN (Chen et al., 2021)	0.0051	0.8422	−0.0071	−0.8994	1.0347	minutes-hours
WENDy	0.0064	1.0145	0.00045	−1.0071	0.958	2.5 sec

An existing PINN-based approach by Chen et al. (2021) for learning f and σ in (30) used $\mathbf{Y}$ containing 40,000 datapoints $Y_{t_i}^{(\ell)}$. Runtime information is not available in Chen et al. (2021), however can reasonably be estimated from the method. The authors used an architecture of 4 hidden layers each with 20 neurons and performed optimization via Adam with learning rate 10^{-4} and 200,000 iterations. The high iteration count and slow learning rate, combined with optimization over an 80-dimensional space, is guaranteed to far exceed the $O(1)$ seconds required for WENDy, and so is simply listed as "minutes-hours" in Table 3.

To compare with WENDy, we use the same amount of data spread over $L = 200$ realizations $\mathbf{Y}^{(\ell)}$ each containing $M + 2 = 200$ timepoints $t_0 = 0, \ldots, 1 = t_{M+1}$. We use the same cubic trial drift basis, $\mathbb{F} = (1, x, x^2, x^3)$, as in Chen et al. (2021). For test functions we use $\theta_k(t) = \theta(t - t_k)$ and $\phi_k(x) = \phi(x - x_k)$, for query points $Q = \{(x_k, t_k)\}_{k=1}^K$, with base functions $\theta(t) = (1 - (t/r_t)^2)^3$ and $\psi = \exp(-x^2/(2r_x^2))$. We use ~ 1000 equally-spaced query points.[12] We set $(r_t, r_x) = (5\Delta t, \mu/3)$ so that ϕ and ψ are reasonably localized with respect to $\mathbf{Y}$, however, the radii (r_t, r_x) can easily be informed from the dynamics.

The results are in Table 3. Estimated parameters are significantly more accurate than those in Chen et al. (2021) (taken from Table4: Appendix B). Moreover, WENDy takes approximately 2.5 seconds to run.

As well as parameter estimates, it is instructive to measure the point-wise drift error between the learned drift $\hat{f}$ and f, as well as the resulting stationary measure $\hat{\rho}_\infty$ to the true stationary measure ρ_∞, which can be computed using $\hat{f}$ and $\hat{\mu}$:

$$\hat{\rho}_\infty(x) = Z^{-1} \exp\left(-\hat{\mu}^{-1} \int^x \hat{f}(v)dv\right)$$

where Z is computed such that $\|\hat{\rho}_\infty\|_1 = 1$ using suitable quadrature, and $\hat{f}$ can be integrated exactly, being polynomial. For the results in Fig. 7, we have

$$\frac{\|\hat{f} - f\|_2}{\|f\|_2} = 0.00552, \qquad \|\hat{\rho}_\infty - \rho_\infty\|_2/\|\rho_\infty\| = 0.0274,$$

[12] Note that local, smooth, decaying functions are advantageous, even though not required for ψ. It can be shown that $\psi_k(x) = x^2$ will amplify the stochastic noise in X_t, decreasing quadrature accuracy.

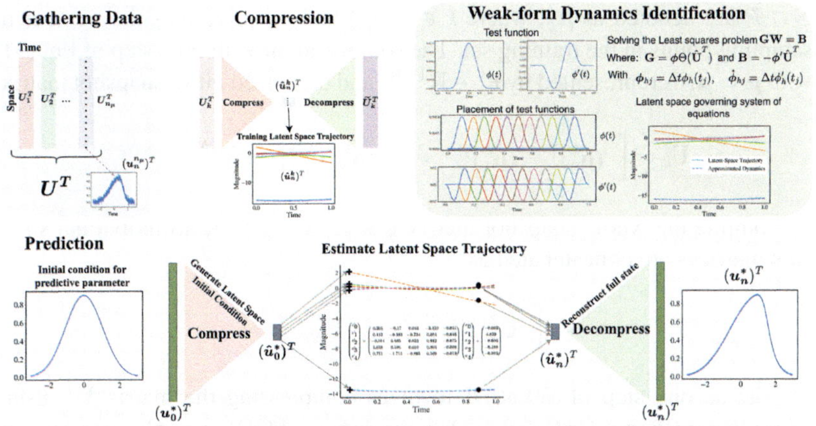

FIGURE 8 Diagram illustrating the WLaSDI algorithm's application to 1D Burgers' simulations, featuring four key steps: data gathering, compression, weak form dynamics identification, and prediction. Diagram from Tran et al. (2024).

providing a highly accurate read on the short-time as well as long-time underlying dynamics.

5 Weak form-based reduced order modeling

Reduced Order Models (ROMs) can be used to accelerate simulations while maintaining high accuracy. The Weak Form Latent Space Dynamics Identification (WLaSDI) employs projection-based reduced order modeling (pROM) in conjunction with weak form parameter estimation. As depicted in Fig. 8, WLaSDI first compresses data, then projects onto the test functions and learns the local latent space models. The variance reduction of the weak form offers robust and precise latent space recovery, hence allowing for a fast and accurate simulation.

To illustrate WLaSDI, consider an N_s-dimensional full-order model, characterized by:

$$\frac{d\mathbf{u}}{dt} = \mathbf{h}(\mathbf{u}, t), \quad \mathbf{u}(0; \boldsymbol{\mu}) = \mathbf{u}_0(\boldsymbol{\mu}) \tag{31}$$

where $t \in [0, T]$. The solution, denoted as $\mathbf{u}(t; \boldsymbol{\mu})$, maps from $[0, T] \times \mathcal{P}$ to $\mathbb{R}^{1\times N_s}$. The initial condition is given by $\mathbf{u}_0$, where $\boldsymbol{\mu} \in \mathcal{P}$ represents a parameter affecting only the initial conditions. We assume a uniform time step, $\Delta t \in \mathbb{R}$. Throughout this chapter, we use $\mathbf{u}_n := \mathbf{u}(t_n; \boldsymbol{\mu})$.

5.1 WLaSDI algorithm

To begin, WLaSDI draws samples from the parameter space $\mathcal{P}$ and collects together snapshots of artificial data. The sampling points within a training set

$\mathcal{S} \subset \mathcal{P}$ are denoted as $\boldsymbol{\mu}_k$, where $k \in \mathbb{N}(n_\mu)$ and n_μ represents the number of sampling points in the training set. The solution at the n-th time step of Eq. (31) with $\boldsymbol{\mu} = \boldsymbol{\mu}_k$ is represented by $\mathbf{u}_n^k \in \mathbb{R}^{1\times N_s}$ and organized into a snapshot matrix:

$$\mathbf{U}_k = \begin{bmatrix} (\mathbf{u}_0^k)^T & (\mathbf{u}_1^k)^T & \cdots & (\mathbf{u}_{N_t}^k)^T \end{bmatrix}^T \in \mathbb{R}^{(N_t+1)\times N_s}$$

To compile the whole snapshot matrix $\mathbf{U} \in \mathbb{R}^{(N_t+1)n_\mu \times N_s}$, all individual snapshot matrices are concatenated as

$$\begin{bmatrix} \mathbf{U}_0^T & \mathbf{U}_1^T & \cdots & \mathbf{U}_{N_t}^T \end{bmatrix}^T$$

The second step of WLaSDI involves compressing the matrix $\mathbf{U}^T$ using either linear (Proper Orthogonal Decomposition - POD) or nonlinear (Autoencoder) techniques. The linear compression method is referred to as WLaSDI-LS (Linear Subspace), and the nonlinear method is denoted as WLaSDI-NM (Nonlinear Manifold).

- **Proper Orthogonal Decomposition**
  POD generates a spatial basis $\widehat{\boldsymbol{\Psi}}$ that compactly represents $\mathbf{U}^T$, minimizing the projection error:

$$\widehat{\boldsymbol{\Psi}} := argmin_{\boldsymbol{\Psi}\in\mathbb{R}^{N_s\times n_s}, \boldsymbol{\Psi}^T\boldsymbol{\Psi}=\mathbf{I}} \left\| \mathbf{U}^T - \boldsymbol{\Psi}\boldsymbol{\Psi}^T\mathbf{U}^T \right\|_F^2$$

  We set $\widehat{\boldsymbol{\Psi}} = \begin{bmatrix} \mathbf{v}_1 & \mathbf{v}_2 & \cdots & \mathbf{v}_{n_s} \end{bmatrix}$ for $n_s < n_\mu(N_t+1)$. We choose n_s to be the dimension of the latent space. The vector $\mathbf{v}_k$ is the k-th column of the left singular matrix $\mathbf{V}$ from the Singular Value Decomposition (SVD), $\mathbf{U}^T = \mathbf{V}\boldsymbol{\Sigma}\mathbf{W}$. Projecting the snapshot matrix $\mathbf{U}^T$ onto the subspace spanned by the column vectors of $\widehat{\boldsymbol{\Psi}}^T$ results in a reduced snapshot matrix $\widehat{\mathbf{U}}^T \in \mathbb{R}^{n_s\times(N_t+1)n_\mu}$, i.e.,

$$\widehat{\mathbf{U}}^T := \widehat{\boldsymbol{\Psi}}^T\mathbf{U}^T$$

- **Autoencoder**
  Auto-encoders function as a nonlinear counterpart to POD. Two neural networks undergo training: one for the encoder $\mathcal{G}_{en} : \mathbb{R}^{N_s} \to \mathbb{R}^{n_s}$ and another for the decoder: $\mathcal{G}_{de} : \mathbb{R}^{n_s} \to \mathbb{R}^{N_s}$. The objective is to minimize the mean square error:

$$MSE(\mathbf{U^T} - \mathcal{G}_{de}(\mathcal{G}_{en}(\mathbf{U^T}))$$

  Analogous to POD, we obtain reduced space data through $\widehat{\mathbf{U}}^T := \mathcal{G}_{en}(\mathbf{U}^T) \in \mathbb{R}^{n_s\times(N_t+1)n_\mu}$

With the data projected onto the latent space, we can then use WENDy to construct a surrogate ODE model which is substantially faster to simulate than a

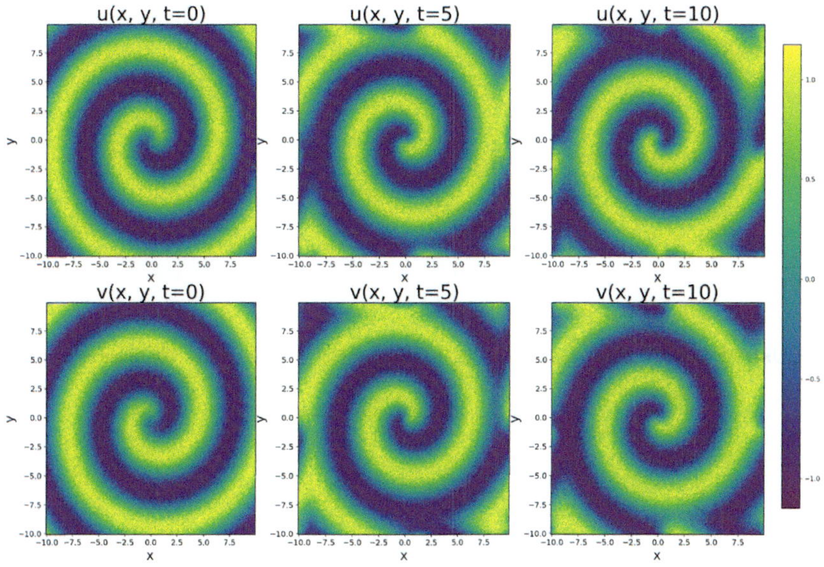

FIGURE 9 Solution to the reaction-diffusion system with 10% noise.

5.2 WLaSDI example

We illustrate WLaSDI using the following reaction-diffusion system:

$$\begin{cases} \partial_t u = \frac{1}{10}\partial_{xx}u + \frac{1}{10}\partial_{yy}u - uv^2 - u^3 + v^3 + u^2v + u \\ \partial_t v = \frac{1}{10}\partial_{xx}v + \frac{1}{10}\partial_{yy}v + v - uv^2 - u^3 - v^3 - u^2v \end{cases} \tag{32}$$

The simulation of system (32) takes place across a periodic domain (x, y) within the range of $[-10, 10] \times [-10, 10]$, and the temporal domain spans $[0, 10]$. The simulation employs Fourier spectral differentiation in spatial dimensions and Python's scipy integration. The computational domain is characterized by dimensions $nx = ny = 128$ and $nt = 201$. The initial condition for the system takes the form of a spiral, expressed parametrically as follows:

$$\begin{cases} u(x, y, 0; a, b) = \tanh(a\sqrt{x^2 + y^2})\cos(\theta - b\sqrt{x^2 + y^2}) \\ v(x, y, 0; a, b) = \tanh(a\sqrt{x^2 + y^2})\sin(\theta - b\sqrt{x^2 + y^2}) \end{cases}$$

This results in unstable spirals that break apart over time. An example of the solution is provided with parameters $a = b = 1$ and 10% added white noise as shown in Fig. 9.

The parameter space is defined by $a = [0.9, 1.0.1.1]$ and $b = 0.9, 1.0, 1.1$ The testing parameter $\mu^* = (0.95, 1.05)$. In the compression step, we employ Proper Orthogonal Decomposition (POD) on the noisy data, selecting a latent

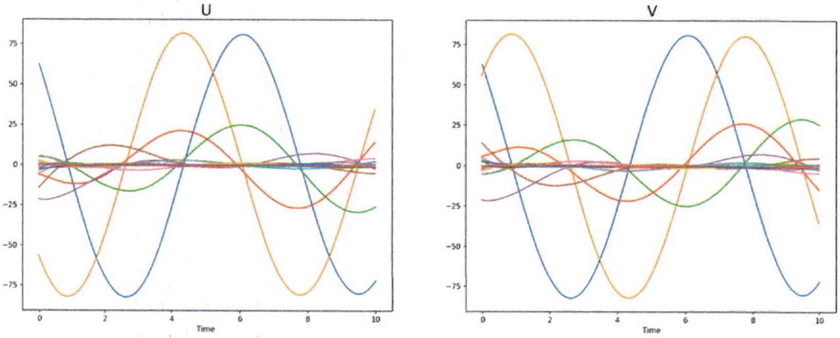

FIGURE 10 Latent space trajectories of the reaction-diffusion system with 10% noise using POD.

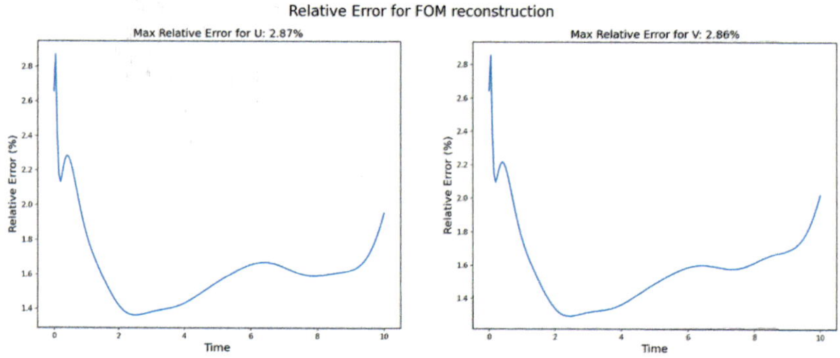

(a) Relative error of the system simulation with 10% noise

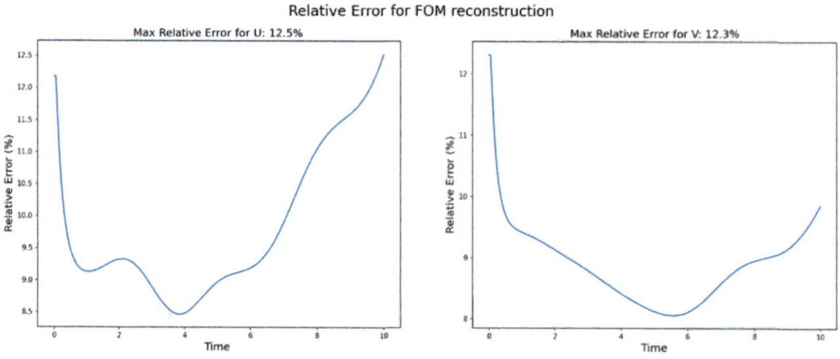

(b) Relative error of the system simulation with 100% noise

FIGURE 11 Relative reconstruction error in simulations.

dimension of 15. This process yields the latent space trajectories for both u and v, as illustrated in Fig. 10.

We calculate the relative error of the simulation results for both u and v. Notably, with a 10% noise level, we attain a remarkable relative error below

3%. Furthermore, when subject to 100% noise, the relative error remains below 13% as illustrated in Fig. 11. Particularly, with WLaSDI, the simulation is completed about 250 times faster compared to numerically solving the equation. Additional details regarding WLaSDI, along with a comprehensive comparison between the weak and strong form when applying to projection-based reduced order modeling, can be found in Tran et al. (2024).

6 Conclusions

Weak form methods are a novel class of scientific machine learning algorithms. They have found success in a wide range of areas including sparse regression-based automated discovery of governing equations, direct parameter estimation, and reduced order models. In all cases the use of the weak form allows improves robustness of the method to noisy state measurements.

Acknowledgments

This work was supported in part by NSF grant MCB-2054085, DOE grant DE-SC0023346, and NIH grant R35GM149335.

References

Bortz, David M., Messenger, Daniel A., Dukic, Vanja, 2023. Direct estimation of parameters in ODE models using WENDy: weak-form estimation of nonlinear dynamics. Bull. Math. Biol. 85 (110).

Brunton, Steven L., Proctor, Joshua L., Kutz, J. Nathan, 2016. Discovering governing equations from data by sparse identification of nonlinear dynamical systems. Proc. Natl. Acad. Sci. 113 (15), 3932–3937.

Chen, Hao, 2023. Data-driven sparse identification of nonlinear dynamical systems using linear multistep methods. Calcolo 60 (1), 11.

Chen, Xiaoli, Yang, Liu, Duan, Jinqiao, Karniadakis, George Em, 2021. Solving inverse stochastic problems from discrete particle observations using the Fokker–Planck equation and physics-informed neural networks. SIAM J. Sci. Comput. 43 (3), B811–B830.

Diaconis, Persi, 1988. Bayesian numerical analysis. In: Gupta, S.S., Berger, James O. (Eds.), Statistical Decision Theory and Related Topics IV, vol. 1, pp. 163–175.

Fasel, U., Kutz, J.N., Brunton, B.W., Brunton, S.L., 2022. Ensemble-SINDy: robust sparse model discovery in the low-data, high-noise limit, with active learning and control. Proc. R. Soc., Math. Phys. Eng. Sci. 478 (2260), 20210904.

Gurevich, Daniel R., Reinbold, Patrick A.K., Grigoriev, Roman O., 2019. Robust and optimal sparse regression for nonlinear pde models. Chaos 29 (10), 103113.

Houston, Paul, Schwab, Christoph, Süli, Endre, 2002. Discontinuous hp-finite element methods for advection-diffusion-reaction problems. SIAM J. Numer. Anal. 39 (6), 2133–2163.

Jorgensen, Murray, 2012. Iteratively reweighted least squares. In: El-Shaarawi, Abdel H., Piegorsch, Walter W. (Eds.), Encyclopedia of Environmetrics, first edition. Wiley.

Lagergren, John H., Nardini, John T., Lavigne, G. Michael, Rutter, Erica M., Flores, Kevin B., 2020. Learning partial differential equations for biological transport models from noisy spatio-temporal data. Proc. R. Soc. A 476 (2234), 20190800.

Lax, P.D., Milgram, A.N., 1955. IX. Parabolic Equations. Annals of Mathematical Studies, vol. 33. Princeton University Press, pp. 167–190.

Liu, Wing Kam, Li, Shaofan, Park, Harold S., 2022. Eighty years of the finite element method: birth, evolution, and future. Arch. Comput. Methods Eng. 29 (6), 4431–4453.
Messenger, Daniel A., Bortz, David M., 2021a. Weak SINDy for partial differential equations. J. Comput. Phys. 443, 110525.
Messenger, Daniel A., Bortz, David M., 2021b. Weak SINDy: Galerkin-based data-driven model selection. Multiscale Model. Simul. 19 (3), 1474–1497.
Messenger, Daniel A., Bortz, David M., 2022a. Asymptotic consistency of the wsindy algorithm in the limit of continuum data. arXiv preprint. arXiv:2211.16000.
Messenger, Daniel A., Bortz, David M., 2022b. Learning mean-field equations from particle data using wsindy. Phys. D: Nonlinear Phenom. 439, 1474–1497, 133406.
Messenger, Daniel A., Burby, Joshua W., Bortz, David M., 2023. Coarse-Graining Hamiltonian Systems Using WSINDy. arXiv preprint. arXiv:2310.05879.
Messenger, Daniel A., Dall'Anese, Emiliano, Bortz, David M., 2022a. Online weak-form sparse identification of partial differential equations. In: Proceedings of The Third Mathematical and Scientific Machine Learning Conference, vol. 190, pp. 241–256.
Messenger, Daniel A., Wheeler, Graycen E., Liu, Xuedong, Bortz, David M., 2022b. Learning anisotropic interaction rules from individual trajectories in a heterogeneous cellular population. J. R. Soc. Interface 19 (195), 20220412.
Pantazis, Yannis, Tsamardinos, Ioannis, 2019. A unified approach for sparse dynamical system inference from temporal measurements. Bioinformatics 35 (18), 3387–3396.
Rudy, Samuel H., Brunton, Steven L., Proctor, Joshua L., Kutz, J. Nathan, 2017. Data-driven discovery of partial differential equations. Sci. Adv. 3 (4), e1602614.
Schaeffer, Hayden, McCalla, Scott G., 2017. Sparse model selection via integral terms. Phys. Rev. E 96 (2), 023302.
Schwartz, Laurent, 1950. Théorie Des Distributions, volume I. Hermann et Cie, Paris, France.
Shinbrot, Marvin, 1954. On the analysis of linear and nonlinear dynamical systems for transient-response data. Technical Report NACA TN 3288. Ames Aeronautical Laboratory, Moffett Field, CA.
Tran, April, He, Xiaolong, Messenger, Daniel A., Choi, Youngsoo, Bortz, David M., 2024. Weak-form latent space dynamics identification. Comput. Methods Appl. Mech. Eng. 356, 116998.
Vershynin, Roman, 2018. High-Dimensional Probability: An Introduction with Applications in Data Science, vol. 47. Cambridge University Press.
Wang, Zhenlin, Huan, Xun, Garikipati, Krishna, 2019. Variational system identification of the partial differential equations governing the physics of pattern-formation: inference under varying fidelity and noise. Comput. Methods Appl. Mech. Eng. 356, 44–74.

Chapter 3

A mathematical guide to operator learning

Nicolas Boullé[a,*] and Alex Townsend[b]
[a]*Department of Applied Mathematics and Theoretical Physics, University of Cambridge, Cambridge, United Kingdom,* [b]*Department of Mathematics, Cornell University, Ithaca, NY, United States*
Corresponding author: e-mail address: nb690@cam.ac.uk

Contents

Abstract

Operator learning aims to discover properties of an underlying dynamical system or partial differential equation (PDE) from data. Here, we present a step-by-step guide to operator learning. We explain the types of problems and PDEs amenable to operator learning, discuss various neural network architectures, and explain how to employ numerical PDE solvers effectively. We also give advice on how to create and manage training data and conduct optimization. We offer intuition behind the various neural network architectures employed in operator learning by motivating them from the point-of-view of numerical linear algebra.

Handbook of Numerical Analysis, Volume 25, ISSN 1570-8659, https://doi.org/10.1016/bs.hna.2024.05.003

Keywords
Scientific machine learning, Deep learning, Operator learning, Partial differential equations

MSC Codes
47-02, 47A58, 47F99, 65-02, 65F55, 65J10

1 Introduction

The recent successes of deep learning (LeCun et al., 2015) in computer vision (Krizhevsky et al., 2012), language model (Brown et al., 2020), and biology (Jumper et al., 2021) have caused a surge of interest in applying these techniques to scientific problems. The field of scientific machine learning (SciML) (Karniadakis et al., 2021), which combines the approximation power of machine learning (ML) methodologies and observational data with traditional modeling techniques based on partial differential equations (PDEs), sets out to use ML tools for accelerating scientific discovery.

SciML techniques can roughly be categorized into three main areas: (1) PDE solvers, (2) PDE discovery, and (3) operator learning (see Fig. 1). First, PDE solvers, such as physics-informed neural networks (PINNs) (Raissi et al., 2019; Lu et al., 2021b; Cuomo et al., 2022; Wang et al., 2023), the deep Galerkin method (Sirignano and Spiliopoulos, 2018), and the deep Ritz method (E and Yu, 2018), consist of approximating the solution a known PDE by a neural network by minimizing the solution's residual. At the same time, PDE discovery aims to identify the coefficients of a PDE from data, such as the SINDy approach (Brunton et al., 2016; Champion et al., 2019), which relies on sparsity-promoting algorithms to determine coefficients of dynamical systems. There are also symbolic regression techniques, such as AI Feynman introduced by Udrescu and Tegmark (2020); Udrescu et al. (2020) and genetic algorithms (Schmidt and Lipson, 2009; Searson et al., 2010), that discover physics equations from experimental data.

Here, we focus on the third main area of SciML, called operator learning (Lu et al., 2021a; Kovachki et al., 2023). Operator learning aims to discover or approximate an unknown operator $\mathcal{A}$, which often takes the form of the solution operator associated with a differential equation. In mathematical terms, the problem can be defined as follows. Given pairs of data (f, u), where $f \in \mathcal{U}$ and $u \in \mathcal{V}$ are from function spaces on a d-dimensional spatial domain $\Omega \subset \mathbb{R}^d$, and a (potentially nonlinear) operator $\mathcal{A} : \mathcal{U} \to \mathcal{V}$ such that $\mathcal{A}(f) = u$, the objective is to find an approximation of $\mathcal{A}$, denoted as $\hat{\mathcal{A}}$, such that for any new data $f' \in \mathcal{U}$, we have $\hat{\mathcal{A}}(f') \approx \mathcal{A}(f')$. In other words, the approximation should be accurate for both the training and unseen data, thus demonstrating good generalization.

This problem is typically approached by representing $\hat{\mathcal{A}}$ as a neural operator, which is a generalization of neural networks as the inputs and outputs

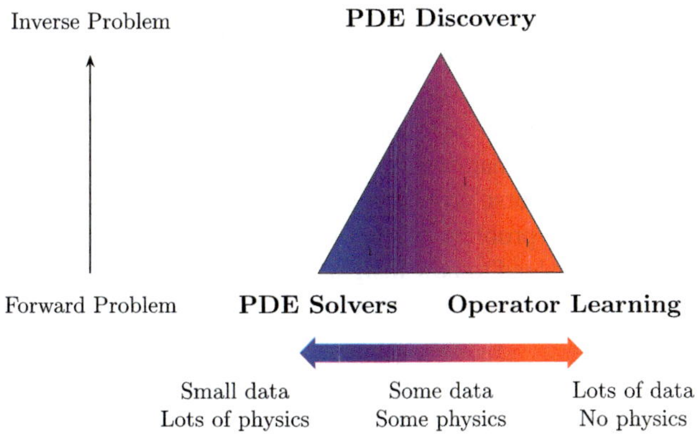

FIGURE 1 Illustrating the role of operator learning in SciML. Operator learning aims to discover or approximate an unknown operator $\mathcal{A}$, which often corresponds to the solution operator of an unknown PDE. In contrast, PDE discovery aims to discover coefficients of the PDE itself, while PDE solvers aim to solve a known PDE using ML techniques.

are functions, not vectors. After discretizing the functions at sensor points $x_1, \ldots, x_m \in \Omega$, one then parametrizes the neural operator with a set of parameters $\theta \in \mathbb{R}^N$, which could represent the weights and biases of the underlying neural network. Then, one typically formulates an optimization problem to find the best parameters:

$$\min_{\theta \in \mathbb{R}^N} \sum_{(f,u)\in \text{data}} L(\hat{\mathcal{A}}(f;\theta), u), \tag{1}$$

where L is a loss function that measures the discrepancy between $\hat{A}(f;\theta)$ and u, and the sum is over all available training data pairs (f, u). The challenges of operator learning often arise from selecting an appropriate neural operator architecture for $\hat{A}$, the computational complexities of solving the optimization problem, and the ability to generalize to new data.

A typical application of operator learning arises when learning the solution operator associated with a PDE, which maps a forcing function f to a solution u. One can informally think of it as the (right) inverse of a differential operator. One of the simplest examples is the solution operator associated with Poisson's equation with zero Dirichlet conditions:

$$-\nabla^2 u = f, \quad x \in \Omega \subset \mathbb{R}^d, \quad u|_{\partial\Omega} = 0, \tag{2}$$

where $\partial\Omega$ means the boundary of Ω. In this case, the solution operator, $\mathcal{A}$, can be expressed as an integral operator:

$$\mathcal{A}(f) = \int_\Omega G(\cdot, y) f(y)\, \mathrm{d}y = u,$$

where G is the Green's function associated with Eq. (2) (Evans, 2010, Chapt. 2). A neural operator is then trained to approximate the action of $\mathcal{A}$ using training data pairs $(f_1, u_1), \ldots, (f_M, u_M)$.

In general, recovering the solution operator is challenging, as it is often nonlinear and high-dimensional, and the available data may be scarce or noisy. Nevertheless, unlike inverse problems, which aim to recover source terms from solutions, the forward problem is usually well-posed. As we shall see, learning solution operators lead to new insights or applications that can complement inverse problem techniques, as described in two surveys (Stuart, 2010) and (Arridge et al., 2019).

1.1 What is a neural operator?

Neural operators (Kovachki et al., 2023; Lu et al., 2021a) are analogues of neural networks with infinite-dimensional inputs. Neural operators were introduced to generalize standard deep learning techniques to learn mappings between function spaces instead of between discrete vector spaces $\mathbb{R}^{d_1}$ to $\mathbb{R}^{d_L}$, where d_1 is the input dimension of a neural network and d_L is the output dimension. In its most traditional formulation, a fully connected neural network can be written as a succession of affine transformations and nonlinear activation functions as

$$\mathcal{N}(x) = \sigma(A_L(\cdots \sigma(A_1 x + b_1) \cdots) + b_L),$$

where $L \geq 1$ is the number of layers, A_i are the weight matrices, b_i are the bias vectors, and $\sigma : \mathbb{R} \to \mathbb{R}$ is the activation function, often chosen to be the ReLU function $\sigma(x) = \max\{x, 0\}$. Neural operators generalize this architecture, where the input and output of the neural network are functions instead of vectors. Hence, the input of a neural operator is a function $f : \Omega \to \mathbb{R}^{d_1}$, where $\Omega \subset \mathbb{R}^d$ is the domain of the function, and the output is a function $u : \Omega \to \mathbb{R}^{d_L}$. The neural operator is then defined as a composition of integral operators and nonlinear functions, which results in the following recursive definition at layer i:

$$u_{i+1}(x) = \sigma\left(\int_{\Omega_i} K^{(i)}(x, y) u_i(y)\, dy + b_i(x)\right), \quad x \in \Omega_{i+1}, \tag{3}$$

where $\Omega_i \subset \mathbb{R}^{d_i}$ is a compact domain, b_i is a bias function, and $K^{(i)}$ is the kernel. The kernels and biases are then parameterized and trained similarly to standard neural networks. However, approximating the kernels or evaluating the integral operators could be computationally expensive. Hence, several neural operator architectures have been proposed to overcome these challenges, such as DeepONets (Lu et al., 2021a) and Fourier neural operators (Li et al., 2021a).

1.2 Where is operator learning relevant?

Operator learning has been successfully applied to many PDEs from different fields, including fluid dynamics with simulations of fluid flow turbulence in the

Navier–Stokes equations at high Reynolds number (Li et al., 2023b; Peng et al., 2022), continuum mechanics (You et al., 2022), astrophysics (Mao et al., 2023), quantum mechanics with the Schrödinger equation (Li et al., 2021a), and weather forecasting (Kurth et al., 2023; Lam et al., 2023). The following four types of applications might directly benefit from operator learning.

Speeding up numerical PDE solvers

First, one can use operator learning to build reduced-order models of complex systems that are computationally challenging to simulate with traditional numerical PDE solvers. For example, this situation arises in fluid dynamics applications such as modeling turbulent flows, which require a very fine discretization or the simulation of high dimensional PDEs. Moreover, specific problems in engineering require the evaluation of the solution operator many times, such as in the design of aircraft or wind turbines. In these cases, a fast but less accurate solver provided by operator learning may be used for forecasting or optimization. This is one of the main motivations behind Fourier neural operators in Li et al. (2021a). There are also applications of operator learning (Zheng et al., 2023) to speed up the sampling process in diffusion models or score-based generative models (Sohl-Dickstein et al., 2015; Ho et al., 2020; Song et al., 2021), which require solving complex differential equations. However, one must be careful when comparing performance against classical numerical PDE solvers, mainly due to the significant training time required by operator learning.

Parameter optimization

In our experience, the computational efficiency of operator learning is mainly seen in downstream applications such as parameter optimization. Once the solution operator has been approximated, it can be exploited in an inverse problem framework to recover unknown parameters of the PDE, which may be computationally challenging to perform with existing numerical PDE solvers. Additionally, neural operators do not rely on a fixed discretization as they are mesh-free and parameterized by a neural network that can be evaluated at any point. This property makes them suitable for solving PDEs on irregular domains or transferring the model to other spatial resolutions (Kovachki et al., 2023).

Benchmarking new techniques

Operator learning may also be used to benchmark and develop new deep learning models. As an example, one can design specific neural network architectures to preserve quantities of interest in PDEs, such as symmetries (Olver, 1993a), conservation laws (Evans, 2010, Sec. 3.4), and discretization independence. This could lead to efficient architectures that are more interpretable and generalize better to unseen data, and exploit geometric priors within datasets (Bronstein et al., 2021). Moreover, the vast literature on PDEs and numerical solvers can be leveraged to create datasets and assess the performance of these models in various settings without requiring significant computational resources for training.

Discovering unknown physics

Last but not least, operator learning is helpful for the discovery of new physics (Lu et al., 2021a). Indeed, the solution operator of a PDE is often unknown, or one may only have access to a few data points without any prior knowledge of the underlying PDE. In this case, operator learning can be used to discover the PDE or a mathematical model to perform predictions. This can lead to new insights into the system's behavior, such as finding conservation laws, symmetries, shock locations, and singularities (Boullé et al., 2022a). However, the complex nature of neural networks makes them challenging to interpret or explain, and there are many future directions for making SciML more interpretative.

1.3 Organization of the paper

This paper is organized as follows. We begin in Section 2 by exploring the connections between numerical linear algebra and operator learning. Then, we review the main neural network architectures used to approximate operators in Section 3. In Section 4, we focus on the data acquisition process, a crucial step in operator learning. We discuss the choice of the distribution of source terms used to probe the system, the numerical PDE solver, and the number of training data. Along the way, we analyze the optimization pipeline, including the possible choices of loss functions, optimization algorithms, and assessment of the results. Finally, in Section 5, we conclude with a discussion of the remaining challenges in the field that include the development of open-source software and datasets, the theoretical understanding of the optimization procedure, and the discovery of physical properties in operator learning, such as symmetries and conservation laws.

2 From numerical linear algebra to operator learning

There is a strong connection between operator learning and the recovery of structured matrices from matrix-vector products. Suppose one aims to approximate the solution operator associated with a linear PDE using a single layer neural operator in the form of Eq. (3) without the nonlinear activation function. Then, after discretizing the integral operator using a quadrature rule, it can be written as a matrix-vector product, where the integral kernel $K : \Omega \times \Omega \to \mathbb{R}$ is approximated by a matrix $A \in \mathbb{R}^{N \times N}$. Moreover, the structure of the matrix is inherited from the properties of the Green's function (see Table 1). The matrix's underlying structure—whether it is low-rank, circulant, banded, or hierarchical low-rank—plays a crucial role in determining the efficiency and approach of the recovery process. This section describes the matrix recovery problem as a helpful way to gain intuition about operator learning and the design of neural operator architectures (see Section 3). Another motivation for recovering structured solution operators is to ensure that the neural operators are fast to evaluate, which is essential in applications involving parameter optimization and benchmarking (Kovachki et al., 2023).

TABLE 1 Solution operators associated with linear PDEs can often be represented as integral operators with a kernel called a Green's function. The properties of a linear PDE induce different structures on the Green's function, such as translation-invariant or off-diagonal low-rank. When these integral operators are discretized, one forms a matrix-vector product, and hence the matrix recovery problem can be viewed as a discrete analogue of operator learning. The dash in the first column and row means no PDE has a solution operator with a globally smooth kernel.

Property of the PDE	Solution operator's kernel	Matrix structure
—	Globally smooth	Low rank
Periodic BCs & const. coeffs	Convolution kernel	Circulant
Localized behavior	Off-diagonal decay	Banded
Elliptic / Parabolic	Off-diagonal low rank	Hierarchical

Consider an unknown matrix $A \in \mathbb{R}^{N\times N}$ with a known structure such as rank-k or banded. We assume that the matrix A is a black box and cannot be seen, but that one can probe A for information via matrix-vector products, i.e., the maps $x \mapsto Ax$ and $x \mapsto A^\top x$, with $A^\top$ representing the transpose of A. The matrix recovery problem is the task of approximating the matrix A using as few queries to $x \mapsto Ax$ and $x \mapsto A^\top x$ as possible. Every matrix with N columns can be deduced in a maximum of N matrix-vector product queries, as Ae_j for $1 \leq j \leq N$ returns the jth column, where e_j denotes the jth standard basis vector. However, if the matrix A has a specific structure such as low-rank, circulant, banded, or hierarchical low-rank, it is often possible to recover A using far fewer queries. This section describes how to recover structured matrices efficiently using matrix-vector products. We prefer doing matrix recovery with Gaussian random vectors because the infinite-dimensional analogue of these vectors are random functions drawn from a Gaussian process, which is a widespread choice of training input data in operator learning.

2.1 Low rank matrix recovery

Let $A \in \mathbb{R}^{N\times N}$ be a rank-k matrix, then it can be expressed for some $C \in \mathbb{R}^{N\times k}$ and $R \in \mathbb{R}^{k\times N}$ as

$$A = CR.$$

Halikias and Townsend (2023) showed that at least $2k$ queries are required to capture the k-dimensional row and column spaces and deduce A. Halko et al. (2011); Martinsson and Tropp (2020) introduced the randomized singular value decomposition (SVD) as a method to recover a rank-k matrix with probability one in $2k$ matrix-vector products with random Gaussian vectors. The randomized SVD can be expressed as a recovery algorithm in Algorithm 1.

Algorithm 1 Randomized singular value decomposition.

1: Draw a random matrix $X \in \mathbb{R}^{N\times k}$ with i.i.d. standard Gaussian entries.
2: Perform k queries with A: $Y = AX$.
3: Compute the QR factorization of $Y = QR$.
4: Perform k queries with $A^\top$: $Z = A^\top Q$.
5: Return $A = QZ^\top$.

A randomized algorithm is crucial for low-rank matrix recovery to prevent input vectors from lying within the $N - k$ dimensional nullspace of A. Hence, recovering any low-rank matrix with a deterministic algorithm using fixed input vectors is impossible. Then, for the rank-k matrix recovery problem, Algorithm 1 recovers A with probability one. A small oversampling parameter $p \geq 1$ is used for numerical stability, such as $p = 5$. This means that $X \in \mathbb{R}^{N\times(k+p)}$, preventing the chance that a random Gaussian vector might be highly aligned with the nullspace of A.

A convenient feature of Algorithm 1 is that it also works for matrices with numerical rank[1] k, provided that one uses a random matrix X with $k + p$ columns. In particular, a simplified statement of Halko et al. (2011, Thm. 10.7) shows that the randomized SVD recovers a near-best low-rank matrix in the sense that

$$\mathbb{P}\left[\|A - QZ^\top\|_\mathrm{F} \leq \left(1 + 15\sqrt{k+5}\right) \min_{\mathrm{rank}(A_k)\leq k} \|A - A_k\|_\mathrm{F}\right] \geq 0.999,$$

where $\|\cdot\|_\mathrm{F}$ is the Frobenius norm of A. While other random embeddings can be used to probe A, Gaussian random vectors give the cleanest probability bounds (Martinsson and Tropp, 2020). Moreover, ensuring that the entries in each column of X have some correlation and come from a multivariable Gaussian distribution allows for the infinite-dimensional extension of the randomized SVD and its application to recover Hilbert–Schmidt operators (Boullé and Townsend, 2022, 2023). This analysis allows one to adapt Algorithm 1 to recover solution operators with low-rank kernels.

Low-rank matrix recovery is one of the most straightforward settings to motivate DeepONet (see Section 3.1). One of the core features of DeepONet is to use the trunk net to represent the action of a solution operator on a set of basis functions generated by the so-called branch net. Whereas in low-rank matrix recovery, we often randomly draw the columns of X as input vectors, DeepONet is trained with these functions. However, like DeepONet, low-rank matrix

[1] For a fixed $0 < \epsilon < 1$, we say that a matrix A has numerical rank k if $\sigma_{k+1}(A) < \epsilon\sigma_1(A)$ and $\sigma_k(A) \geq \epsilon\sigma_1(A)$, where $\sigma_1(A) \geq \sigma_2(A) \geq \ldots \geq \sigma_N(A) \geq 0$ are the singular values of A.

recovery is constructing an accurate approximant whose action is on these vectors. Many operators between function spaces can often be represented to high accuracy with DeepONet in the same way that the kernels of solution operators associated with linear PDEs often have algebraically fast decaying singular values.

2.2 Circulant matrix recovery

The FNO (Li et al., 2021a) structure is closely related to circulant matrix recovery. Consider an $N \times N$ circulant matrix C_c, which is parameterized a vector $c \in \mathbb{R}^N$ as follows:

$$C_c = \begin{bmatrix} c_0 & c_{N-1} & \cdots & c_2 & c_1 \\ c_1 & c_0 & c_{N-1} & \ddots & c_2 \\ \vdots & c_1 & c_0 & \ddots & \vdots \\ c_{N-2} & \ddots & \ddots & \ddots & c_{N-1} \\ c_{N-1} & c_{N-2} & \cdots & c_1 & c_0 \end{bmatrix}.$$

To recover C_c with a random Gaussian vector g, we recall that C_c can be interpreted as a multiplication operator in the Fourier basis. By associativity of multiplication, we have

$$C_c g = C_g c.$$

If we perform the matrix-vector product query $y = C_c g$, we can find the vector c by solving the linear system $C_g c = y$. Since c completely defines C_c, we have recovered the circulant matrix. Moreover, the linear system $C_g c = y$ can be solved efficiently using the fast Fourier transform (FFT) in $\mathcal{O}(N \log N)$ operations. A convenient feature of circulant matrices is that given a new vector $x \in \mathbb{R}^N$, one can compute $C_c x$ in $\mathcal{O}(N \log N)$ operations using the FFT.

Circulant matrix recovery motivates Fourier neural operators. Hence, FNOs leverage the fast Fourier transform to efficiently parameterize the kernel of a solution operator, essentially capturing the operator in a spectral sense. Similarly, circulant matrices are diagonalized by the discrete Fourier transform matrix. The infinite-dimensional analogue of a circulant matrix is a solution operator with a periodic and translation invariant kernel, and this is the class of solution operators for which the FNO assumptions are fully justified. FNOs are extremely fast to evaluate because of their structure, making them popular for parameter optimization and favorable for benchmarking against reduced-order models.

2.3 Banded matrix recovery

We now consider a banded matrix $A \in \mathbb{R}^{N \times N}$ with a fixed bandwidth w, i.e.,

$$A_{ij} = 0, \quad \text{if } |i - j| > w.$$

The matrix A can be recovered with $w+2$ matrix-vector products, but not fewer, using the $w+2$ columns of the following matrix as input vectors:

$$\begin{bmatrix} I_{w+2} & \cdots & I_{w+2} \end{bmatrix}^\top,$$

where I_{w+2} is the $(w+2) \times (w+2)$ identity matrix. Since every wth column has disjoint support, these input vectors recover the columns of A. Of course, this also means that a A can be recovered with $w+2$ Gaussian random vectors.

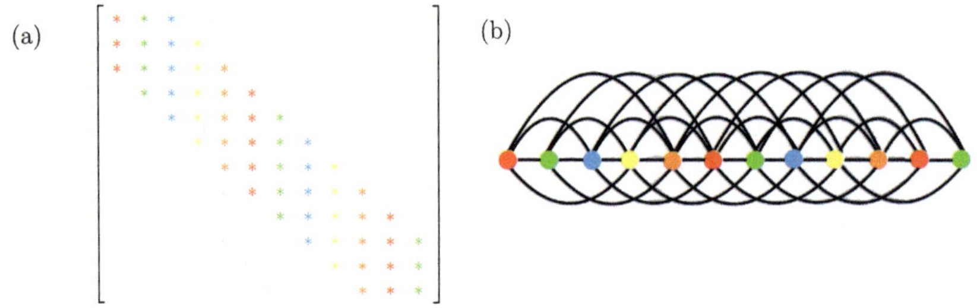

FIGURE 2 (a) A generic 12×12 banded matrix with bandwidth 2, with a maximum of 5 diagonals, and the corresponding graph (b). Here, each vertex is a column of the banded matrix, and two vertices are connected if their corresponding columns do not have disjoint support. The coloring number of 5 determines the minimum number of matrix-vector products needed to recover the structure. Generally, an $N \times N$ banded matrix with bandwidth w can be recovered in $2w+1$ matrix-vector products.

There is a way to understand how many queries one needs as a graph-coloring problem. Consider the graph of size N, corresponding to an $N \times N$ banded matrix with bandwidth w, where two vertices are connected if their corresponding columns do not have disjoint support (see Fig. 2). Then, the minimum number of matrix-vector product queries needed to recover A is the graph coloring number of this graph.[2] One can see why this is the case because all the columns with the same color can be deduced simultaneously with a single matrix-vector product as they must have disjoint support.

Banded matrix recovery motivates Graph neural operators (GNOs), which we will describe later in Section 3.4, as both techniques exploit localized structures within data. GNOs use the idea that relationships in nature are local and can be represented as graphs with no faraway connections. By only allowing local connections, GNOs can efficiently represent solution operators corresponding to local solution operators, mirroring the way banded matrices

2 Recall that the coloring number of a graph is the minimum number of colors required to color the vertices so that no two vertices connected by an edge are identically colored.

are concentrated on the diagonal. Likewise, with a strong locality, GNOs are relatively fast to evaluate, making them useful for parameter optimization and benchmarking. However, they may underperform if the bandwidth increases or the solution operator is not local.

2.4 Hierarchical low rank matrix recovery

An $N \times N$ rank-k hierarchical off-diagonal low rank (HODLR) matrix, denoted as $H_{N,k}$, is a structure that frequently appears in the context of discretized solution operators associated with elliptic and parabolic PDEs (Hackbusch et al., 2004). To understand its recursive structure, we assume N to be a power of 2 and illustrate the structure in Fig. 3(a).

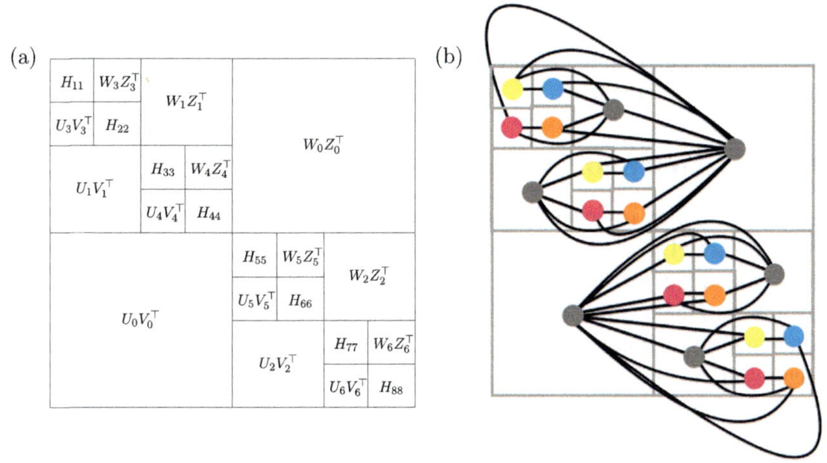

FIGURE 3 (a) A HODLR matrix $H_{N,k}$ after three levels of partitioning. Since $H_{N,k}$ is a rank-k HODLR matrix, U_i, V_i, W_i, and Z_i have at most k columns. The matrices A_{ii} are themselves rank-k HODLR matrices of size $N/8 \times N/8$ and can be further partitioned. (b) Graph corresponding to a hierarchical low-rank matrix with three levels. Here, each vertex is a low-rank block of the matrix, where two vertices are connected if their low-rank blocks occupy the same row. At each level, the number of required matrix-vector input probes to recover that level is proportional to the coloring number of the graph when restricted to submatrices of the same size. In this case, the submatrices that are identically colored can be recovered simultaneously.

Since a rank-k matrix requires $2k$ matrix-vector product queries to be recovered (see Section 2.1), a naive approach to deducing $H_{N,k}$ is to use $2k$ independent queries on each submatrix. However, one can show that some of the submatrices of $H_{N,k}$ can be recovered concurrently using the same queries. We use a graph coloring approach (Levitt and Martinsson, 2022b) to determine which submatrices can be recovered concurrently. This time, we consider the graph where each vertex is a low-rank submatrix of $H_{N,k}$ and connect two vertices if their corresponding low-rank submatrices occupy the same column as in Fig. 3(b). The low-rank submatrices that are identically colored at each level can be recovered concurrently in only $2k$ queries. Hence, it can be shown that an $N \times N$ hierarchical rank-k matrix can be recovered in fewer than $10k\lceil \log_2(N) \rceil$ matrix-vector products (Halikias and Townsend, 2023). The precise coloring of the graph in Fig. 3(b) can also be used to derive a particular algorithm for hierarchical matrix recovery known as peeling (Levitt and Martinsson, 2022a,b; Lin et al., 2011; Martinsson, 2011). These peeling algorithms have been recently generalized to the infinite-dimensional setting by Boullé et al. (2023).

HODLR recovery can be seen as the simplest version of a multipole graph neural operator (MGNO) (see Section 3.5) as both emphasize the importance of capturing operators at multiple scales. MGNOs are based on a hierarchical graph with interactions at different scales or levels (see Section 3.4). By incorporating local (near-field) and global (far-field) interactions, MGNOs can effectively learn complex patterns. MGNOs are often great at representing solution operators due to their multiscale nature. The price to pay is that the final neural operator can be computationally expensive to evaluate, and it is a complicated structure to implement.

3 Neural operator architectures

In this section, we review the main neural operator architectures used in the literature, namely DeepONets (Lu et al., 2021a), Fourier neural operators (Li et al., 2021a), and Deep Green networks (Gin et al., 2021; Boullé et al., 2022a). We also refer to the recent survey by Goswami et al. (2023) for a review of the different neural operator architectures and their applications. Each of these architectures employs different discretization and approximation techniques to make the neural operator more efficient and scalable by enforcing certain structures on the kernel such as low-rank, periodicity, translation invariance, or hierarchical low-rank structure (see Table 2).

TABLE 2 Summary table of neural operator architectures, describing the property assumption on the operator along with the discretization of the integral kernels.

Neural operators	Property of the operator	Kernel parameterization
DeepONet	Low-rank	Branch and trunk networks
FNO	Translation-invariant	Fourier coefficients
GreenLearning	Linear	Rational neural network
DeepGreen	Semilinear	Kernel matrix
Graph neural operator	Diagonally dominant	Message passing network
Multipole GNO	Off-diagonal low rank	Neural network

Most neural operator architectures also come with theoretical guarantees on their approximation power. These theoretical results essentially consist of universal approximation properties for neural operators (Chen and Chen, 1995;

Kovachki et al., 2023; Lu et al., 2021a), in a similar manner as neural networks (DeVore, 1998), and quantitative error bounds based on approximation theory to estimate the size, i.e., the number of trainable parameters, of a neural operator needed to approximate a given operator between Banach spaces to within a prescribed accuracy (Lanthaler et al., 2022; Yarotsky, 2017).

3.1 Deep operator networks

Deep Operator Networks (DeepONets) are a promising model for learning nonlinear operators and capturing the inherent relationships between input and output functions (Lu et al., 2021a). They extend the capabilities of traditional deep learning techniques by leveraging the expressive power of neural networks to approximate operators in differential equations, integral equations, or more broadly, any functional maps from one function space to another. A key theoretical motivation for DeepONet is the universal operator approximation theorem (Chen and Chen, 1995; Lu et al., 2021a). This result can be seen as an infinite dimensional analogue of the universal approximation operator for neural networks (Cybenko, 1989; Hornik, 1991), which guarantee that a sufficiently wide neural network can approximate any continuous function to any accuracy. Since the introduction of DeepONets by Lu et al. (2021a), several research works focused on deriving error bounds for the approximation of nonlinear operators by DeepONets in various settings, such as learning the solution operator associated with Burger's equation or the advection-diffusion equation (Deng et al., 2022), and the approximation of nonlinear parabolic PDEs (De Ryck and Mishra, 2022; Lanthaler et al., 2022).

A DeepONet is a two-part deep learning network consisting of a branch network and a trunk network. The branch net encodes the operator's input functions f into compact, fixed-size latent vectors $b_1(f(x_1), \dots, f(x_m)), \dots, b_p(f(x_1), \dots, f(x_m))$, where $\{x_i\}_{i=1}^m$ are the sensor points at which the input functions are evaluated. The trunk net decodes these latent vectors to produce the final output function at the location $y \in \Omega$ as

$$\mathcal{N}(f)(y) = \sum_{k=1}^{p} b_k(f(x_1), \dots, f(x_m)) t_k(y).$$

A schematic of a deep operator network is given in Fig. 4. The defining feature of DeepONets is their ability to handle functional input and output, thus enabling them to learn a wide array of mathematical operators effectively. It's worth mentioning that the branch network and the trunk network can have distinct neural network architectures tailored for different purposes, such as performing a feature expansion on the input of the trunk network as $y \to \begin{pmatrix} y & \cos(\pi y) & \sin(\pi y) & \dots \end{pmatrix}$ to take into account any potential oscillatory patterns in the data (Di Leoni et al., 2023). Moreover, while the interplay of the branch and trunk networks is crucial, the output of a DeepONet does not

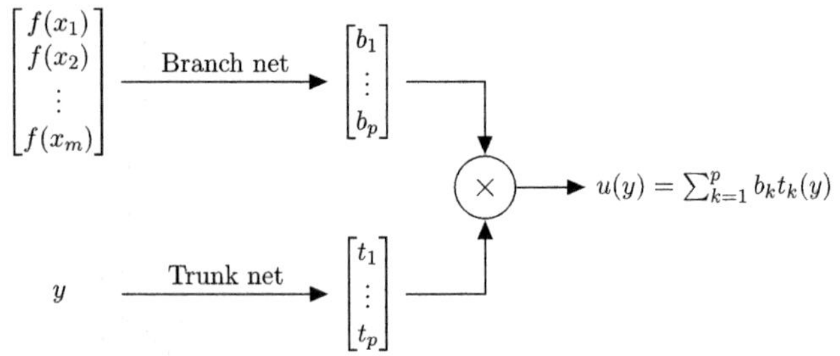

FIGURE 4 Schematic diagram of a deep operator network (DeepONet). A DeepONet parametrizes a neural operator using a branch network and a truncation (trunk) network. The branch network encodes the input function f as a vector of p features, which is then multiplied by the trunk network to yield a rank-p representation of the solution u.

necessarily depend on the specific input points but rather on the global property of the entire input function, which makes it suitable for learning operator maps.

One reason behind the performance of DeepONet might be its connection with the low-rank approximation of operators and the SVD (see Section 2.1). Hence, one can view the trunk network as learning a basis of functions $\{t_k\}_{k=1}^{p}$ that are used to approximate the operator, while the branch network expresses the output function in this basis by learning the coefficients $\{b_k\}_{k=1}^{p}$. Moreover, the branch network can be seen as a feature extractor, which encodes the input function into a compact representation, thus reducing the problem's dimensionality to p, where p is the number of branch networks. Additionally, several architectures, namely the POD-DeepONet (Lu et al., 2022) and SVD-DeepONet (Venturi and Casey, 2023), have been proposed to strengthen the connections between DeepONet and the SVD of the operator and increase its interpretability.

A desirable property for a neural operator architecture is to be discretization invariant in the sense that the model can act on any discretization of the source term and be evaluated at any point of the domain (Kovachki et al., 2023). This property is crucial for the generalization of the model to unseen data and the transferability of the model to other spatial resolutions. While DeepONets can be evaluated at any location of the output domain, DeepONets are not discretization invariant in their original formulation by Lu et al. (2021a) as the branch network is evaluated at specific points of the input domain (see Fig. 4). However, this can be resolved using a low-rank neural operator (Kovachki et al., 2023), sampling the input functions at local spatial averages (Lanthaler et al., 2022), or employing a principal component analysis (PCA) alternative of the branch network (de Hoop et al., 2022).

The training of DeepONets is performed using a supervised learning process. It involves minimizing the mean-squared error between the predicted output $\mathcal{N}(f)(y)$ and the actual output of u the operator on the training functions at random locations $\{y_j\}_{j=1}^n$, i.e.,

$$\min_{\theta\in\mathbb{R}^N} \frac{1}{|\text{data}|} \sum_{(f,u)\in\text{data}} \frac{1}{n}\sum_{j=1}^{n} |\mathcal{N}(f)(y_j) - u(y_j)|^2. \tag{4}$$

The term inside the first sum approximates the integral of the mean-squared error, $|\mathcal{N}(f) - u|^2$, over the domain Ω using Monte-Carlo integration. The optimization is typically done via backpropagation and gradient descent algorithms, which are the same as in traditional neural networks. Importantly, DeepONets allow for different choices of loss functions, depending on the problem. For example, mean squared error is commonly used for regression tasks, but other loss functions might be defined to act as a regularizer and incorporate prior physical knowledge of the problem (Goswami et al., 2022; Wang et al., 2021b). The selection of an appropriate loss function is a crucial step in defining the learning process of these networks and has a substantial impact on their performance (see Section 4.2.1).

DeepONet has been successfully applied and adapted to a wide range of problems, including predicting cracks in fracture mechanics using a variational formulation of the governing equations (Goswami et al., 2022), simulating the New York-New England power grid behavior with a probabilistic and Bayesian framework to quantify the uncertainty of the trajectories (Moya et al., 2023), as well as predicting linear instabilities in high-speed compressible flows with boundary layers (Di Leoni et al., 2023).

3.2 Fourier neural operators

Fourier neural operators (FNOs) (Li et al., 2021a; Kovachki et al., 2023) are a class of neural operators motivated by Fourier spectral methods. FNOs have found their niche in dealing with high-dimensional PDEs, which are notoriously difficult to solve using traditional numerical methods due to the curse of dimensionality. They've demonstrated significant success in learning and predicting solutions to various PDEs, particularly those with periodic boundary conditions or those that can be transformed into the spectral domain via Fourier transform. This capability renders FNOs an invaluable tool in areas where PDEs play a central role, such as fluid dynamics, quantum mechanics, and electromagnetism.

The main idea behind FNOs is to choose the kernels $K^{(i)}$ in Eq. (3) as translation-invariant kernels satisfying $K^{(i)}(x, y) = k^{(i)}(x - y)$ (provided the input and output domains are torus) such that the integration of the kernel can be performed efficiently as a convolution using the Fast Fourier Transform (FFT) (Cooley and Tukey, 1965), i.e., multiplication in the feature space of Fourier coefficients. Hence, the integral operation in Eq. (3) can be performed as

$$\int_{\Omega_i} k^{(i)}(x-y)u_i \,\mathrm{d}y = \mathcal{F}^{-1}(\mathcal{F}(k^{(i)})\mathcal{F}(u_i))(x) = \mathcal{F}^{-1}(\mathcal{R}\cdot\mathcal{F}(u_i))(x), \quad x \in \Omega_i,$$

where $\mathcal{F}$ denotes the Fourier transform and $\mathcal{F}^{-1}$ its inverse. The kernel $K^{(i)}$ is parametrized by a periodic function $k^{(i)}$, which is discretized by a (trainable) weight vector of Fourier coefficients $\mathcal{R}$, and truncated to a finite number of Fourier modes. Then, if the input domain is discretized uniformly with m sensor points, and the vector $\mathcal{R}$ contains at most $k_{\max} \leq m$ modes, the convolution can be performed in quasilinear complexity in $\mathcal{O}(m \log m)$ operations via the FFT. This is a significant improvement over the $\mathcal{O}(m^2)$ operations required to evaluate the integral in Eq. (3) using a quadrature rule. In practice, one can restrict the number of Fourier modes to $k_{\max} \ll m$ without significantly affecting the accuracy of the approximation whenever the input and output functions are smooth so that their representation in the Fourier basis enjoy rapid decay of the coefficients, thus further reducing the computational and training complexity of the neural operator.

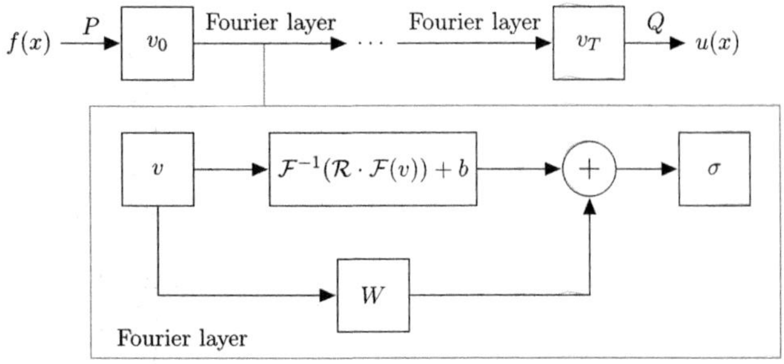

FIGURE 5 Schematic diagram of a Fourier neural operator (FNO). The networks P and Q, respectively, lift the input function f to a higher dimensional space and project the output of the last Fourier layer to the output dimension. An FNO mainly consists of a succession of Fourier layers, which perform the integral operations in neural operators as a convolution in the Fourier domain and component-wise composition with an activation function σ.

We display a diagram of the architecture of an FNO in Fig. 5. The input function f is first lifted to a higher dimensional space by a neural network P. Then, a succession of Fourier layers is applied to the lifted function, which is parametrized by a vector of Fourier coefficients $\mathcal{R}_i$, a bias vector b_i, and a weight matrix W. Then, the output of the FNO at the ith layer is given by

$$v_i = \sigma(W_i v_{i-1} + \mathcal{F}^{-1}(\mathcal{R}_i \cdot \mathcal{F}(u_{i-1})) + b_i),$$

where $\sigma : \mathbb{R} \to \mathbb{R}$ is the activation function whose action is defined component-wise, often chosen to be the ReLU function. The weight matrix W_i and bias vector b_i perform a linear transformation of the input function v_i. After the last Fourier layer, the output of the FNO is obtained by applying a final neural network Q on the output of the last Fourier layer to project it to the output dimension.

The training of FNOs, like DeepONets, is carried out via a supervised learning process. It typically involves minimizing a loss function that measures the discrepancy between the predicted and the true output of the operator on the input functions with respect to the trainable parameters of the neural network as in Eq. (4). Here, one needs to perform backpropagation through the Fourier layers, which is enabled by the implementation of fast GPU differentiable FFTs (Mathieu et al., 2014) in deep learning frameworks such as PyTorch (Paszke et al., 2019) and TensorFlow (Abadi et al., 2016).

While FNOs have been proposed initially to alleviate the computational expense of performing integral operations in neural operators by leveraging the FFT, they have a distinctive advantage in learning operators where computations in the spectral domain are more efficient or desirable. This arises naturally when the target operator, along with the input and output functions, are smooth so that their representation as Fourier coefficients decay exponentially fast, yielding an efficient truncation. Hence, by selecting the architecture of the FNO appropriately, such as the number of Fourier modes $k_{\max}$, or the initialization of the Fourier coefficients, one can obtain a neural operator that preserves specific smoothness properties. However, when the input or output training data is not smooth, FNO might suffer from Runge's phenomenon near discontinuities (de Hoop et al., 2022).

One main limitation of the FNO architectures is that the FFT should be performed on a uniform grid and rectangular domains, which is not always the case in practice. This can be overcome by applying embedding techniques to transform the input functions to a uniform grid and extend them to simple geometry, using a Fourier analytic continuation technique (Bruno et al., 2007). Recently, several works have been proposed to extend the FNO architecture to more general domains, such as using a zero padding, linear interpolation (Lu et al., 2022), or encoding the geometry to a regular latent space with a neural network (Li et al., 2022, 2023a). However, this might lead to a loss of accuracy and additional computational cost. Moreover, the FFT is only efficient for approximating translation invariant kernels, which do not occur when learning solution operators of PDEs with nonconstant coefficients.

Other related architectures aim to approximate neural operators directly in the feature space, such as spectral neural operators (SNO) (Fanaskov and Oseledets, 2022), which are based on spectral methods and employ a simple feed-forward neural network to map the input function, represented as a vector of Fourier or Chebyshev coefficients to an output vector of coefficients. Finally, Raonic et al. (2023) introduce convolutional neural operators (CNOs) to alleviate the aliasing phenomenon of convolutional neural networks (CNNs) by learning the mapping between bandlimited functions. Contrary to FNOs, CNOs parameterize the integral kernel on a $k \times k$ grid and perform the convolution in the physical space as

$$\int_{\Omega} k(x-y) f(y)\,\mathrm{d}y = \sum_{i,j=1}^{k} k_{ij} f(x - z_i j), \quad x \in \Omega,$$

where z_{ij} are the grid points.

Similarly to DeepONets, Fourier neural operators are universal approximators, in the sense that they are dense in the space of continuous operators (Bhattacharya et al., 2021; Kovachki et al., 2023, 2021). However, even while being universal approximators, FNOs could, in theory, require a huge number of parameters to approximate a given operator to a prescribed accuracy $\epsilon > 0$. As an example, Kovachki et al. (2021) showed that the size of the FNO must grow exponentially fast as ϵ decreases to approximate any operator between rough functions whose Fourier coefficients decay only at a logarithmic rate. Fortunately, these pessimistic lower bounds are not observed in practice when learning solution operators associated with PDEs. Indeed, in this context, one can exploit PDE regularity theory and Sobolev embeddings to derive quantitative bounds on the size of FNOs for approximating solution operators that only grow sublinearly with the error. Here, we refer to the analysis of Darcy flow and the two-dimensional Navier–Stokes equations by Kovachki et al. (2021).

3.3 Deep Green networks

Deep Green networks (DGN) employ a different approach compared to DeepONets and FNOs to approximate solution operators of PDEs (Gin et al., 2021; Boullé et al., 2022a). Instead of enforcing certain properties on the integral kernel in Eq. (3), such as being low-rank (DeepONet) or translation-invariant (FNO), DGN learns the kernel directly in the physical space. Hence, assume that the underlying differential operator is a linear boundary value problem of the form

$$\mathcal{L}u = f \text{ in } \Omega, \quad \text{and} \quad u = 0 \text{ on } \partial\Omega. \tag{5}$$

Under suitable regularity assumptions on operator $\mathcal{L}$ (e.g., uniform ellipticity or parabolicity), the solution operator $\mathcal{A}$ can be expressed as an integral operator with a Green kernel $G : \Omega \times \Omega \to \mathbb{R} \cup \{\infty\}$ as (Evans, 2010; Boullé et al., 2022b; Boullé and Townsend, 2023)

$$\mathcal{A}(f)(x) = u(x) = \int_\Omega G(x, y) f(y)\, \mathrm{d}y, \quad x \in \Omega.$$

Boullé et al. (2022a) introduced GreenLearning networks (GL) to learn the Green kernel G directly from data. The main idea behind GL is to parameterize the kernel G as a neural network $\mathcal{N}$ and minimize the following relative mean-squared loss function to recover an approximant to G:

$$\min_{\theta \in \mathbb{R}^N} \frac{1}{|\text{data}|} \sum_{(f,u) \in \text{data}} \frac{1}{\|u\|^2_{L^2(\Omega)}} \int_\Omega \left(u(x) - \int_\Omega \mathcal{N}(x, y) f(y)\, \mathrm{d}y \right)^2 \mathrm{d}x. \tag{6}$$

Once trained, the network $\mathcal{N}$ can be evaluated at any point in the domain, similarly to FNO and DON. A key advantage of this method is that it provides a more interpretable model, as the kernel can be visualized and analyzed to recover properties of the underlying differential operators (Boullé et al., 2022a). However, this comes at the cost of higher computational complexity, as the integral operation in Eq. (6) must be computed accurately using a quadrature rule and typically requires $\mathcal{O}(m^2)$ operations, as opposed to the $\mathcal{O}(m \log m)$ operations required by FNOs, where m is the spatial discretization of the domain Ω.

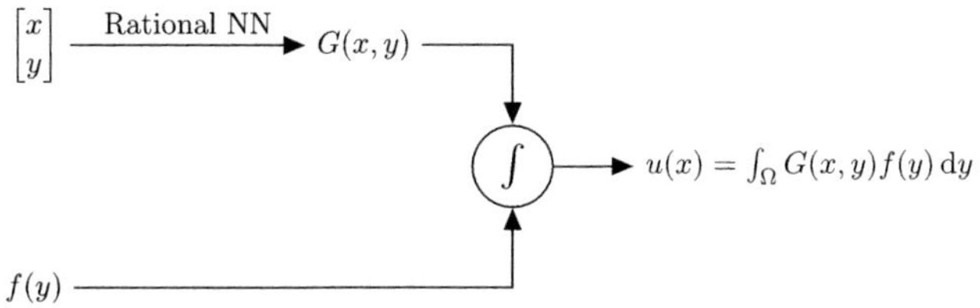

FIGURE 6 Schematic diagram of a GreenLearning network (GL), which approximates the integral kernel (Green's function) associated with linear PDEs using rational neural networks.

As Green's functions may be unbounded or singular, Boullé et al. (2022a) propose to use a rational neural network (Boullé et al., 2020) to approximate the Green kernel. A rational neural network is a neural network whose activation functions are rational functions, defined as the ratio of two polynomials whose coefficients are learned during the training phase of the network. This choice of architecture is motivated by the fact that rational networks have higher approximation power than the standard ReLU networks (Boullé et al., 2020), in the sense that they require exponentially fewer layers to approximate continuous functions within a given accuracy and may take arbitrary large values, which is a desirable property for approximating Green kernels. A schematic diagram of a GL architecture is available in Fig. 6.

When the underlying differential operator is nonlinear, the solution operator $\mathcal{A}$ cannot be written as an integral operator with a Green's function. In this case, Gin et al. (2021) propose to learn the solution operator $\mathcal{A}$ using a dual autoencoder architecture Deep Green network (DGN), which is a neural operator architecture that learns an invertible coordinate transform map that linearizes the nonlinear boundary value problem. The resulting linear operator is approximated by a matrix, which represents a discretized Green's function but could also be represented by a neural network if combined with the GL technique. This approach has been successfully applied to learn the solution operator of the nonlinear cubic Helmholtz equation non-Sturm–Liouville equation and discover an underlying Green's function (Gin et al., 2021).

Other deep learning-based approaches (Lin et al., 2023; Peng et al., 2023; Sun et al., 2023) have since been introduced to recover Green's functions using deep learning, but they rely on a PINN technique in the sense that they require the knowledge of the underlying PDE operator. Finally, Stepaniants (2023) proposes to learn the Green kernel associated with linear partial differential operators using a reproducible kernel Hilbert space (RKHS) framework, which leads to a convex loss function.

3.4 Graph neural operators

As described in Section 3.3, solution operators associated with linear, elliptic, or parabolic PDEs of the form $\mathcal{L}u = f$ can be written as an integral operator with a Green kernel G (Evans, 2010, Sec. 2.2.4). For simplicity, we consider Green kernels associated with uniformly elliptic operators in divergence form defined on a bounded domain Ω in spatial dimension $d \geq 3$:

$$\mathcal{L}u = -\operatorname{div}(A(x)\nabla u) = f, \quad \text{on } \Omega \subset \mathbb{R}^d, \tag{7}$$

where $A(x)$ is a bounded coefficient matrix satisfying the uniform ellipticity condition $A(x)\xi \cdot \xi \geq \lambda|\xi|^2$ for all $x \in \Omega$ and $\xi \in \mathbb{R}^d$, for some $\lambda > 0$. In this section, we present a neural operator architecture that takes advantage of the local structure of the Green kernel associated with Eq. (7), inferred by PDE regularity theory.

This architecture is called graph neural operator (GNO) (Li et al., 2020a) and is inspired by graph neural network (GNN) models (Scarselli et al., 2008; Wu et al., 2020; Zhou et al., 2020). It focuses on capturing the Green kernel's short-range interactions to reduce the integral operation's computational complexity in Eq. (3). The main idea behind GNO is to perform the integral operation in Eq. (5) locally on a small ball of radius r, $B(x, r)$, around x for each $x \in \Omega$ as follows:

$$\mathcal{A}(f)(x) = u(x) \approx \int_{B(x,r)} G(x, y) f(y) \, dy, \quad x \in \Omega. \tag{8}$$

Here, Li et al. (2020a) propose to discretize the domain Ω using a graph, whose nodes represent discretized spatial locations, and use a message passing network architecture (Gilmer et al., 2017) to perform an average aggregation of the nodes as in Eq. (8). The approach introduced by Li et al. (2020a) aims to approximate the restriction G_r to the Green's function G on a band of radius r along the diagonal of the domain $\Omega \times \Omega$, defined as

$$G_r(x, y) = \begin{cases} G(x, y), & \text{if } |x - y| \leq r, \\ 0, & \text{otherwise}, \end{cases}$$

where $|\cdot|$ is the Euclidean distance in $\mathbb{R}^d$.

This neural architecture is justified by the following pointwise bound satisfied by the Green's function and proven by Grüter and Widman (1982, Thm. 1.1):

$$|G(x, y)| \leq C(d, A)|x - y|^{2-d}, \quad x, y \in \Omega, \tag{9}$$

where Ω is a compact domain in $\mathbb{R}^d$ for $d \geq 3$, and C is a constant depending only on d and the coefficient matrix $A(x)$. Similar bounds have been derived in spatial dimension $d = 2$ by Cho et al. (2012); Dong and Kim (2009) and for Green's functions associated with time-dependent, parabolic, PDEs (Hofmann and Kim, 2004; Cho et al., 2008). Then, integrating Eq. (9) over the domain $\Gamma_r := \{(x, y) \in \Omega \times \Omega \colon |x - y| > r\}$ yields a bound on the approximation error between G and G_r that decays algebraically fast as r increases:

$$\begin{aligned}\|G - G_r\|_{L^2(\Omega\times\Omega)} &= \left(\int_{\Gamma_r} |G(x, y)|^2 \,\mathrm{d}x\,\mathrm{d}y\right)^{1/2} \\ &\leq C(d, A)\left(\int_{\Gamma_r} r^{4-2d} \,\mathrm{d}x\,\mathrm{d}y\right)^{1/2} \\ &\leq |\Omega| C(d, A) r^{2-d}.\end{aligned}$$

This implies that the Green's function can be well approximated by a bandlimited kernel G_r and that the approximation error bound improves in high dimensions. To illustrate this, we plot in Fig. 7 the Green's function associated with the one-dimensional Poisson equation on $\Omega = [0, 1]$ with homogeneous Dirichlet boundary conditions, along with the error between the Green's function G and its truncation G_r along a bandwidth of radius r along the diagonal of the domain.

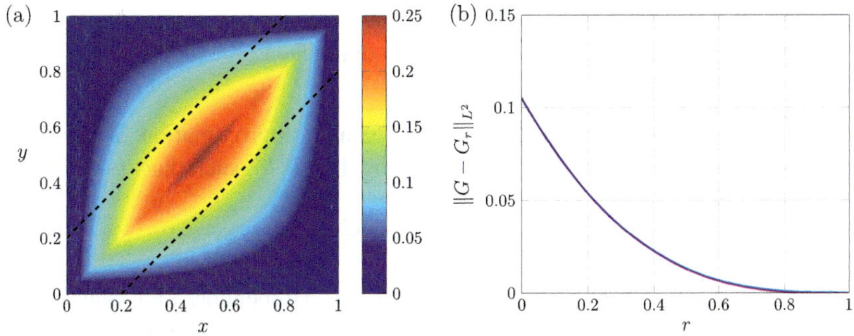

FIGURE 7 (a) Green's function associated with the one-dimensional Poisson equation on $\Omega = [0, 1]$ with homogeneous Dirichlet boundary conditions. The dashed lines highlight a band of radius $r = \sqrt{2}/10$ around the diagonal. (b) L^2-norm of the error between the Green's function G and its truncation G_r along a bandwidth of radius r along the diagonal of the domain.

3.5 Multipole graph neural operators

Multipole graph neural operator (MGNO) has been introduced by Li et al. (2020b) and is a class of multiscale networks that extends the graph neural operator architecture described in Section 3.4 to capture the long-range interactions of the Green kernel. The main idea behind MGNO is to decompose the Green kernel G into a sum of low-rank kernels as $G = K_1 + \cdots + K_L$, which approximates the short and wide-range interactions in the PDEs. This architecture is motivated by the same reasons that led to the development of hierarchical low-rank matrices (see Section 2.4), such as the fast multipole method (Greengard and Rokhlin, 1997; Ying et al., 2004). It allows for the evaluation of the integral operation in Eq. (3) in linear complexity.

MGNO is based on low-rank approximations of kernels, similar to DeepONets or low-rank neural operators (see Section 3.1 and Li et al., 2020a), but is more flexible than vanilla DeepONets since it does not require the underlying kernels to be low-rank. Hence, if we consider a Green's function G associated with a uniformly elliptic PDE in the form of Eq. (7), then Weyl's law (Weyl, 1911; Canzani, 2013; Minakshisundaram and Pleijel, 1949) states that the eigenvalues of the solution operator associated with Eq. (7) decay at an algebraic rate of $\lambda_n \sim cn^{-2/d}$ for a constant $c > 0$. This implies that the approximation error between the solution operator and its best rank-k approximant decays only algebraically with k. Moreover, the decay rate deteriorates in high dimensions. In particular, the length p of the feature vector in DeepONets must be significantly large to approximate the solution operator to a prescribed accuracy.

However, Bebendorf and Hackbusch (2003, Thm. 2.8) showed that the Green's function G associated with Eq. (7) can be well approximated by a low-rank kernel when restricted to separated subdomains $D_X \times D_Y$ of $\Omega \times \Omega$, satisfying the strong admissibility condition: $\text{dist}(D_X, D_Y) < \text{diam}(D_Y)$. Here, the distance and diameter in $\mathbb{R}^d$ are defined as

$$\text{dist}(D_X, D_Y) = \inf_{x \in D_X, y \in D_Y} |x - y|, \qquad \text{diam}(D_Y) = \sup_{y_1, y_2 \in D_Y} |y_1 - y_2|.$$

Then, for any $\epsilon \in (0, 1)$, there exists a separable approximation of the form $G_k(x, y) = \sum_{i=1}^{k} u_i(x) v_i(y)$, with $k = \mathcal{O}(\log(1/\epsilon)^{d+1})$, such that

$$\|G - G_k\|_{L^2(D_X \times D_Y)} \leq \epsilon \|G\|_{L^2(D_X \times \hat{D}_Y)},$$

where $\hat{D}_Y$ is a domain slightly larger than D_Y (Bebendorf and Hackbusch, 2003, Thm. 2.8). This property has been exploited by Boullé and Townsend (2023); Boullé et al. (2023, 2022b) to derive sample complexity bounds for learning Green's functions associated with elliptic and parabolic PDEs. It motivates the decomposition of the Green kernel into a sum of low-rank kernels $G = K_1 + \ldots + K_L$ in MGNO architectures. Indeed, one can exploit the low-rank structure of Green's functions on well-separated domains to perform a hierarchical decomposition of the domain $\Omega \times \Omega$ into a tree of subdomains satisfying the admissibility condition. In Fig. 8, we illustrate the decomposition of the Green's function associated with the 1D Poisson equation on $\Omega = [0, 1]$ with homogeneous Dirichlet boundary conditions into a hierarchy of $L = 3$ levels of different range of interactions. The first level captures the long-range interactions, while the last level captures the short-range interactions. Then, the integral

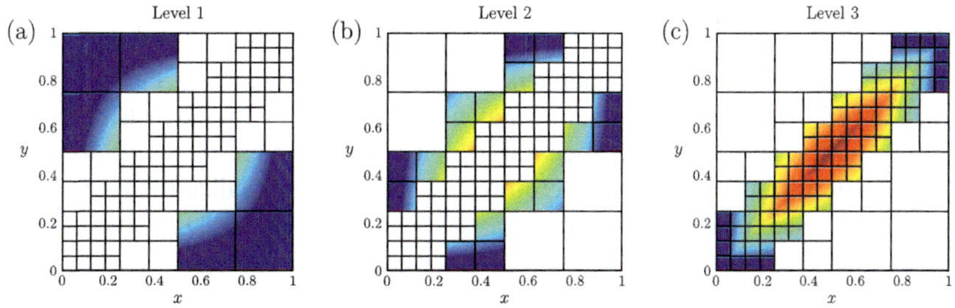

FIGURE 8 Decomposition of the Green's function associated with the 1D Poisson equation, displayed in Fig. 7(a), into a hierarchy of kernels capturing different ranges of interactions: from long-range (a) to short-range (c) interactions.

operation in Eq. (3) can be performed by aggregating the contributions of the subdomains in the tree, starting from the leaves and moving up to the root. This allows for the evaluation of the integral operation in Eq. (3) in linear complexity in the number of subdomains. The key advantage is that the approximation error on each subdomain decays exponentially fast as the rank of the approximating kernel increases.

One alternative approach to MGNO is to encode the different scales of the solution operators using a wavelet basis. This class of operator learning techniques (Feliu-Faba et al., 2020; Gupta et al., 2021; Tripura and Chakraborty, 2022) is based on the wavelet transform and aims to learn the solution operator kernel at multiple scale resolutions. One advantage over MGNO is that it does not require building a hierarchy of meshes, which could be computationally challenging in high dimensions or for complex domain geometries.

Finally, motivated by the success of the self-attention mechanism in transformers architectures for natural language processing (Vaswani et al., 2017) and image recognition (Dosovitskiy et al., 2020), several architectures have been proposed to learn global correlations in solution operators of PDEs. In particular, Cao (2021) introduced an architecture based on the self-attention mechanism for operator learning and observed higher performance on benchmark problems when compared against the Fourier Neural Operator. More recently, Kissas et al. (2022) propose a Kernel-Coupled Attention mechanism to learn correlations between the entries of a vector feature representation of the output functions. In contrast, Hao et al. (2023a) designed a general neural operator transformer (GNOT) that allows for multiple input functions and complex meshes.

4 Learning neural operators

In this section, we discuss various technical aspects involved in training neural operators, such as the data acquisition of forcing terms and solutions, the amount of training data required in practice, and the optimization algorithms and loss functions used to train neural operators.

4.1 Data acquisition

This section focuses on the data acquisition process for learning neural operators. In real-world applications, one may not have control over the distribution of source terms and solutions or locations of the sensors to measure the solutions at specific points in the domain. Therefore, we consider an idealized setting where one is interested in generating synthetic data using numerical PDE solvers to develop neural operator architectures. In this case, one has complete control over the distribution of source terms and solutions, as well as the locations of the sensors.

4.1.1 Distribution of source terms

The source terms $\{f_j\}_{j=1}^N$ used to generate pairs of training data to train neural operators are usually chosen to be random functions, sampled from a Gaussian random field (Lu et al., 2021a). Let $\Omega \subset \mathbb{R}^d$ be a domain, then a stochastic process $\{X_x, x \in \Omega\}$ indexed by Ω, is Gaussian if, for every finite set of indices $x_1, \ldots, x_n \in \Omega$, the vector of random variables $(X_{x_1}, \ldots, X_{x_n})$ follows a multivariate Gaussian distribution. The Gaussian process (GP) distribution is completely determined by the following mean and covariance functions (Adler, 2010, Sec. 1.6):

$$\mu(x) = \mathbb{E}\{X_x\}, \quad K(x, y) = \mathbb{E}\{[X_x - \mu(x)]^\top [X_y - \mu(y)]\}, \quad x \in \Omega.$$

In the rest of the paper, we will denote a Gaussian process with mean μ and covariance kernel K by $\mathcal{GP}(\mu, K)$. The mean function μ is usually chosen to be zero, while K is symmetric and positive-definite.

When K is continuous Mercer's theorem (Mercer, 1909) states that there exists an orthonormal basis of eigenfunctions $\{\psi_j\}_{j=1}^\infty$ of $L^2(\Omega)$, and nonnegative eigenvalues $\lambda_1 \geq \lambda_2 \geq \cdots > 0$ such that

$$K(x, y) = \sum_{j=1}^{\infty} \lambda_j \psi_j(x) \psi_j(y), \quad x, y \in \Omega,$$

where the sum is absolutely and uniformly convergent (Hsing and Eubank, 2015, Thm. 4.6.5). Here, the eigenvalues and eigenfunctions of the kernel are defined as solutions to the associated Fredholm integral equation:

$$\int_\Omega K(x, y) \psi_j(y) \, \mathrm{d}y = \lambda_j \psi_j(x), \quad x \in \Omega.$$

Then, the Karhunen–Loève theorem (Karhunen, 1946; Loève, 1946) ensures that a zero mean square-integrable Gaussian process X_x with continuous covariance function K admits the following representation:

$$X_x = \sum_{j=1}^{\infty} \sqrt{\lambda_j} c_j \psi_j(x), \quad c_j \sim \mathcal{N}(0, 1), \quad x \in \Omega, \tag{10}$$

where c_j are independent and identically distributed (i.i.d.) random variables, and the convergence is uniform in $x \in \Omega$. Suppose the eigenvalue decomposition of the covariance function is known. In that case, one can sample a random function from the associated GP, $\mathcal{GP}(0, K)$, by sampling the coefficients c_j in Eq. (10) from a standard Gaussian distribution and truncated the series up to the desired resolution. Under suitable conditions, one can relate the covariance function K's smoothness to the random functions sampled from the associated GP (Adler, 2010, Sec. 3). Moreover, the decay rate of the eigenvalues provides information about the smoothness of the underlying kernel (Ritter et al., 1995; Zhu et al., 1998). In practice, the number of eigenvalues greater than machine precision dictates the dimension of the finite-dimensional vector space spanned by the random functions sampled from $\mathcal{GP}(0, K)$.

One of the most common choices of covariance functions for neural operator learning include the squared-exponential kernel (Lu et al., 2021a; Boullé et al., 2022a), which is defined as

$$K(x, y) = \exp(-|x - y|^2/(2\ell^2)), \quad x, y \in \Omega,$$

where $\ell > 0$ is the length-scale parameter, which roughly characterizes the distance at which two point values of a sampled random function become uncorrelated (Rasmussen and Williams, 2006, Chapt. 5). Moreover, eigenvalues of the squared-exponential kernel decay exponentially fast at a rate that depends on the choice of ℓ (Zhu et al., 1998; Boullé and Townsend, 2022). After a random function f has been sampled from the GP, one typically discretizes it by performing a piecewise linear interpolation at sensor points $x_1, \ldots, x_m \in \Omega$ by evaluating f as these points. The interpolant can then solve the underlying PDE or train a neural operator. The number of sensors is chosen to resolve the underlying random functions and depends on their smoothness. Following the analysis by Lu et al. (2021a, Suppl. Inf. S4), in one dimension, the error between f and its piecewise linear interpolant is of order $\mathcal{O}(1/(m^2\ell^2))$, and one should choose $m \geq 1/\ell$. A typical value of ℓ lies in the range $\ell \in [0.01, 0.1]$ with $m = 100$ sensors (Lu et al., 2021a; Boullé et al., 2022a). We illustrate the eigenvalues of the squared-exponential kernel on $\Omega = [0, 1]$ with length-scale parameters $\ell \in \{0.1, 0.05, 0.01\}$, along with the corresponding random functions sampled from the associated GP in Fig. 9. As the length-scale parameter ℓ decreases, the eigenvalues decay faster, and the sampled random functions become smoother.

Another possible choice of covariance functions for neural operator learning (Benitez et al., 2023; Zhu et al., 2023) comes from the Matérn class of covariance functions (Rasmussen and Williams, 2006; Stein, 1999):

$$K(x, y) = \frac{2^{1-\nu}}{\Gamma(\nu)} \left(\frac{\sqrt{2\nu}|x - y|}{\ell} \right)^{\nu} K_\nu \left(\frac{\sqrt{2\nu}|x - y|}{\ell} \right), \quad x, y \in \Omega,$$

where Γ is the Gamma function, K_ν is a modified Bessel function, and ν, ℓ are positive parameters that enable the control of the smoothness of the sampled

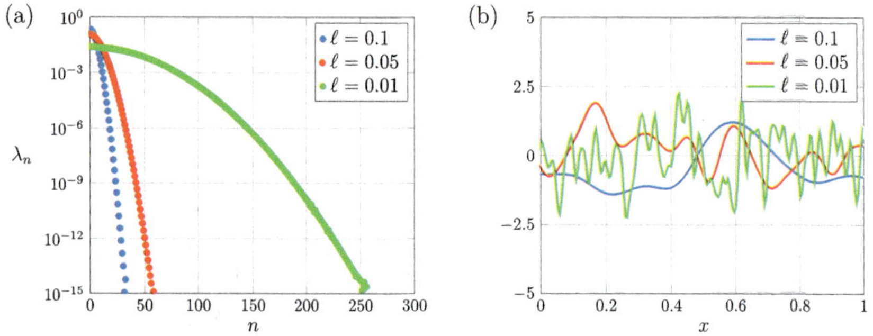

FIGURE 9 (a) Eigenvalues of the squared-exponential kernel on $\Omega = [0, 1]$ with length-scale parameters $\ell \in \{0.1, 0.05, 0.01\}$. (b) Random functions sampled from the associated Gaussian process $\mathcal{GP}(0, K)$, where K is the squared-exponential kernel with length-scale parameters $\ell \in \{0.1, 0.05, 0.01\}$.

random functions. Hence, the resulting Gaussian process is $\lceil \nu \rceil - 1$ times mean-squared differentiable (Rasmussen and Williams, 2006, Sec. 4.2). Moreover, the Matérn kernel converges to the squared-exponential covariance function as $\nu \to \infty$. We refer to the book by Rasmussen and Williams (2006) for a detailed analysis of other standard covariance functions in Gaussian processes.

Finally, Li et al. (2021a); Kovachki et al. (2023) propose to use a Green kernel associated with a differential operator, which is a power of the Helmholtz equation, as a covariance function:

$$K = A(-\nabla^2 + cI)^{-\nu}.$$

Here, A, $\nu > 0$, and $c \geq 0$ are parameters that respectively govern the scaling of the Gaussian process, the algebraic decay rate of the spectrum, and the frequency of the first eigenfunctions of the covariance function. One motivation for this choice of distribution is that it allows the enforcement of prior information about the underlying model, such as the order of the differential operator, directly into the source terms. A similar behavior has been observed in a randomized linear algebra context when selecting the distribution of random vectors for approximating matrices from matrix-vector products using the randomized SVD (Boullé and Townsend, 2022). For example, the eigenvalues associated with this covariance kernel decay at an algebraic rate, implying that random functions sampled from this GP would be more oscillatory. This could lead to higher performance of the neural operator on high-frequency source terms. However, one downside of this approach is that a poor choice of the parameters can affect the training and approximation error of the neural operator. Hence, the studies (Li et al., 2021a; Kovachki et al., 2023) employ different parameter choices in each of the reported numerical experiments, suggesting that the covariance hyperparameters have been heavily optimized.

4.1.2 Numerical PDE solvers

After choosing the covariance function and generating the random source terms from the Gaussian process, one must solve the underlying PDE to generate the corresponding solutions. In general, the PDE is unknown, and one only has access to an oracle (such as physical experiments or black-box numerical solver) that outputs solutions u to the PDE from input source terms f as $\mathcal{L}u = f$. However, generating synthetic data from known mathematical models to design or evaluate neural operator architectures is often convenient. One can use numerical PDE solvers to generate the corresponding solutions in this case. This section briefly describes the different numerical methods that can be used to solve the underlying PDEs, along with their key attributes summarized in Table 3. We want to emphasize that these methods have many variations, and we refer to the most standard ones.

TABLE 3 Summary of the different properties of standard finite difference, finite element, and spectral methods for solving PDEs.

Property	Finite differences	Finite elements	Spectral methods
Domain geometry	Simple	Complex	Simple
Approximation	Local	Local	Global
Linear system	Large sparse	Large sparse	Small dense
Convergence rate	Algebraic	Algebraic	Spectral

When the PDE does not depend on time, the most common techniques for discretizing and solving it are finite difference methods (FDM), finite element methods (FEM), and spectral methods. The finite difference method consists of discretizing the computational domain Ω into a grid and approximating spatial derivatives of the solution u from linear combinations of the values of u at the grid points using finite difference operators (Iserles, 2009, Chap. 8). This approach is based on a local Taylor expansion of the solution and is very easy to implement on rectangular geometries. However, it usually requires a uniform grid approximation of the domain, which might not be appropriate for complex geometries and boundary conditions. The finite element method (Süli and Mayers, 2003, Chap. 14) employs a different approach than FDM and considers the approximation of the solution u on a finite-dimensional vector space spanned by basis functions with local support on Ω. The spatial discretization of the domain Ω is performed via a mesh representation. The basis functions are often chosen as piecewise polynomials supported on a set of elements, which are adjacent cells in the mesh. This approach is more flexible than FDM and can be used to solve PDEs on complex geometries and boundary conditions. However, it is more challenging to implement and requires the construction of a mesh of the domain Ω, which can be computationally expensive. We highlight that the finite difference and finite element methods lead to large, sparse, and highly structured linear algebra systems, which can be solved efficiently using iterative methods.

Two commonly used finite element software for generating training data for neural operators are FEniCS (Alnæs et al., 2015) and Firedrake (Rathgeber et al., 2016; Ham et al., 2023). These open-source software exploit the Unified Form Language (UFL) developed by Alnæs et al. (2014) to define the weak form of the PDE in a similar manner as in mathematics and automatically generate the corresponding finite element assembly code before exploiting fast linear and nonlinear solvers through a Python interface with the high-performance PETSc library (Balay et al., 2023).

Finally, spectral methods (Iserles, 2009; Gottlieb and Orszag, 1977; Trefethen, 2000) are based on the approximation of the solution u on a finite-dimensional vector space spanned by basis functions with global support on Ω, usually Chebyshev polynomials or trigonometric functions. Spectral methods are motivated by the fact that the solution u of a PDE defined on a 1D interval is often smooth if the source term is smooth and so can be well-approximated by a Fourier series if it is periodic or a Chebyshev series if it is not periodic. Hence, spectral methods lead to exponential convergence, also called spectral accuracy, to analytic solutions with respect to the number of basis functions, unlike FDM and FEM, which only converge at an algebraic rate. The fast convergence rate of spectral methods implies that a small number of basis functions is usually required to achieve a given accuracy. Therefore, the matrices associated with the resulting linear algebra systems are much smaller than for FEM but are dense. In summary, spectral methods have a competitive advantage over FDM and FEM on simple geometries, such as tensor-product domains, and when the solution is smooth. At the same time, FEM might be difficult to implement but is more flexible. A convenient software for solving simple PDEs using spectral methods is the Chebfun software system (Driscoll et al., 2014), an open-source package written in MATLAB®.

For time-dependent PDEs, one typically starts by performing a time-discretization using a time-stepping scheme, such as backward differentiation schemes (e.g. backward Euler) and Runge–Kutta methods (Iserles, 2009; Süli and Mayers, 2003), and then employ a spatial discretization method, such as the techniques described before in this section, to solve the resulting stationary PDE at each time-step.

4.1.3 Amount of training data

Current neural operator approaches typically require a relatively small amount of training data, in the order of a thousand input-output pairs, to approximate solution operators associated with PDEs (Lu et al., 2021a; Goswami et al., 2023; Kovachki et al., 2023; Boullé et al., 2023). This contrasts with the vast amount of data used to train neural networks for standard supervised learning tasks, such as image classification, which could require hundreds of millions of labeled samples (LeCun et al., 2015). This difference can be explained by the fact that solution operators are often highly structured, which can be exploited to design data-efficient neural operator architectures (see Section 3).

Recent numerical experiments have shown that the rate of convergence of neural operators with respect to the number of training samples evolves in two regimes (Lu et al., 2021a, Fig. S10). In the first regime, we observe a fast decay of the testing error at an exponential rate (Boullé et al., 2023). Then, the testing error decays at a slower algebraic rate in the second regime for a larger amount of samples and saturates due to discretization error and optimization issues, such as convergence to a suboptimal local minimum.

On the theoretical side, several works derived sample complexity bounds that characterize the amount of training data required to learn solution operators associated with certain classes of linear PDEs to within a prescribed accuracy $0 < \epsilon < 1$. In particular, Boullé et al. (2023); Schäfer and Owhadi (2021) focus on approximating Green's functions associated with uniformly elliptic PDEs in divergence form:

$$-\operatorname{div}(A(x)\nabla u) = f, \quad x \in \Omega \subset \mathbb{R}^d, \tag{11}$$

where $A(x)$ is a symmetric bounded coefficient matrix (see Eq. (7)). These studies construct data-efficient algorithms that converge exponentially fast with respect to the number of training pairs. Hence, they can recover the Green's function associated with Eq. (11) to within ϵ using only $\mathcal{O}(\text{polylog}(1/\epsilon))$ sample pairs. The method employed by Boullé et al. (2023) consists of recovering the hierarchical low-rank structure satisfied by Green's function on well-separated subdomains (Bebendorf and Hackbusch, 2003; Lin et al., 2011; Levitt and Martinsson, 2022b) using a generalization of the rSVD to Hilbert–Schmidt operators (Boullé and Townsend, 2022, 2023; Halko et al., 2011; Martinsson and Tropp, 2020). Interestingly, the approach by Schäfer and Owhadi (2021) is not based on low-rank techniques but relies on the sparse Cholesky factorization of elliptic solution operators (Schäfer et al., 2021).

Some of the low-rank recovery techniques employed by Boullé et al. (2023) extend naturally to time-dependent parabolic PDEs (Boullé et al., 2022b) in spatial dimension $d \geq 1$ of the form:

$$\frac{\partial u}{\partial t} - \operatorname{div}(A(x,t)\nabla u) = f, \quad x \in \Omega \subset \mathbb{R}^d, \quad t \in (0, T], \tag{12}$$

where the coefficient matrix $A(x,t) \in \mathbb{R}^{d\times d}$ is symmetric positive definite with bounded coefficient functions in $L^\infty(\Omega \times [0, T])$, for some $0 < T < \infty$, and satisfies the uniform parabolicity condition (Evans, 2010, Sec. 7.1.1). Parabolic systems in the form of Eq. (12) model various time-dependent phenomena, including heat conduction and particle diffusion. Boullé et al. (2022b, Thm. 9) showed that the Green's function associated with Eq. (12) admits a hierarchical low-rank structure on well-separated subdomains, similarly to the elliptic case (Bebendorf and Hackbusch, 2003). Combining this with the pointwise bounds satisfied by the Green's function (Cho et al., 2012), one can construct an algorithm that recovers the Green's function to within ϵ using $\mathcal{O}(\text{poly}(1/\epsilon))$ sample pairs (Boullé et al., 2022b, Thm. 10).

Finally, other approaches (de Hoop et al., 2023; Jin et al., 2023; Stepaniants, 2023) derived convergence rates for a broader class of operators between infinite-dimensional Hilbert spaces, which are not necessarily associated with solution operators of PDEs. In particular, de Hoop et al. (2023) consider the problem of estimating the eigenvalues of an unknown, and possibly unbounded, self-adjoint operator assuming that the operator is diagonalizable in a known basis of eigenfunctions, and highlight the impact of varying the smoothness of training and test data on the convergence rates. Then, Stepaniants (2023); Jin et al. (2023) derive upper and lower bounds on the sample complexity of Hilbert–Schmidt operators between two reproducing kernel Hilbert spaces (RKHS) that depend on the smoothness of the input and output functions.

4.2 Optimization

Once the neural operator architecture has been selected and the training dataset is constituted, the next task is to train the neural operator by solving an optimization problem in the form of Eq. (1). The aim is to identify the best parameters of the underlying neural network so that the output $\hat{\mathcal{A}}(f;\theta)$ of the neural operator evaluated at a forcing term f in the training dataset fits the corresponding ground truth solution u. This section describes the most common choices of loss functions and optimization algorithms employed in current operator learning approaches. Later in Section 4.2.3, we briefly discuss how to measure the convergence and performance of a trained neural operator.

4.2.1 Loss functions

The choice of loss function in operator learning is a critical step, as it directs the optimization process and ultimately affects the model's performance. Different types of loss functions can be utilized depending on the task's nature, the operator's structure, and the function space's properties. A common choice of loss function in ML is the mean squared error (MSE), which is defined as

$$L_{\mathrm{MSE}} = \frac{1}{N}\sum_{i=1}^{N}\frac{1}{m}\sum_{j=1}^{m}|\hat{A}(f_i)(x_j) - u_i(x_j)|^2 \approx \frac{1}{N}\sum_{i=1}^{N}\|\hat{A}(f_i) - u_i\|^2_{L^2(\Omega)}, \tag{13}$$

and is employed in the original DeepONet study (Lu et al., 2021a). Here, N is the number of training samples, m is the number of sensors, f_i is the i-th forcing term, u_i is the corresponding ground truth solution, $\hat{A}(f_i)$ is the output of the neural operator evaluated at f_i, and x_j is the j-th sensor location. This loss function discretizes the squared L^2 error between the output of the neural operator and the ground truth solution at the sensor locations. When the sensor grid is regular, one can employ a higher-order quadrature rule to discretize the L^2 norm. Moreover, in most cases, it may be beneficial to use a relative error, especially when the magnitudes of the output function can vary widely. Then,

Boullé et al. (2022a) use the following relative squared L^2 loss function

$$L = \frac{1}{N}\sum_{i=1}^{N} \frac{\|\hat{A}(f_i) - u_i\|_{L^2(\Omega)}^2}{\|u_i\|_{L^2(\Omega)}^2}, \tag{14}$$

which is discretized using a trapezoidal rule. The most common loss function in operator learning is the relative L^2 error employed in Fourier neural operator techniques (Li et al., 2021a):

$$L_2 = \frac{1}{N}\sum_{i=1}^{N} \frac{\|\hat{A}(f_i) - u_i\|_{L^2(\Omega)}}{\|u_i\|_{L^2(\Omega)}}. \tag{15}$$

Kovachki et al. (2023) observed a better normalization of the model when using a relative loss function, and that the choice of Eq. (15) decreases the testing error by a factor of two compared to Eq. (14).

For tasks that require robustness to outliers or when it is important to measure the absolute deviation, the L^1 loss can be employed. It is defined as

$$L_1 = \frac{1}{N}\sum_{i=1}^{N} \frac{\|\hat{A}(f_i) - u_i\|_{L^1(\Omega)}}{\|u_i\|_{L^1(\Omega)}}. \tag{16}$$

This loss function tends to be less sensitive to large deviations than the L^2 loss (Alpak et al., 2023; Lyu et al., 2023; Zhao et al., 2024). Furthermore, Sobolev norms can also be used as a loss function when the unknown operator $\mathcal{A}$ involves functions in Sobolev spaces (Evans, 2010, Chapt. 5), particularly when the derivatives of the input and output functions play a role (Son et al., 2021; Yu et al., 2023; O'Leary-Roseberry et al., 2024). For example, one could perform training with a relative H^1 loss to enforce the smoothness of the neural operator output. Finally, when the underlying PDE is known, one can enforce it as a weak constraint when training the neural operator by adding a PDE residual term to the loss function (Li et al., 2021b; Wang et al., 2021b), similarly to physics-informed neural networks (Raissi et al., 2019).

4.2.2 Optimization algorithms and implementation

The training procedure of neural operators is typically performed using Adam optimization algorithm (Kingma and Ba, 2015; Kovachki et al., 2023; Lu et al., 2021a; Li et al., 2021a; Goswami et al., 2023) or one of its variants such as AdamW (Loshchilov and Hutter, 2019; Hao et al., 2023a). Hence, the work introducing DeepONets by Lu et al. (2021a) employed Adam algorithm to train the neural network architecture with a default learning rate of 0.001. In contrast, Kovachki et al. (2023) incorporate learning rate decay throughout the optimization of Fourier neural operators. One can also employ a two-step training approach by minimizing the loss function using Adam algorithm for a fixed number of iterations and then fine-tuning the neural operator using the L-BFGS algorithm (Byrd et al., 1995; Cuomo et al., 2022; He et al., 2020; Mao et al., 2020; Boullé et al., 2022a). This approach has been shown to improve the convergence rate of the optimization when little data is available in PINNs applications (He et al., 2020). Popular libraries for implementing and training neural operators include PyTorch (Paszke et al., 2019) and TensorFlow (Abadi et al., 2016).

Thus far, there has been limited focus on the theoretical understanding of convergence and optimization of neural operators. Since neural operators are a generalization of neural networks in infinite dimensions, existing convergence results of physics-informed neural networks (Wang et al., 2021a, 2022b) based on the neural tangent kernel (NTK) framework (Jacot et al., 2018; Du et al., 2019; Allen-Zhu et al., 2019) should naturally extend to neural operators. One notable exception is the study by Wang et al. (2022a), which analyzes the training of physics-informed DeepONets (Wang et al., 2021b) and derives a weighting scheme guided by NTK theory to balance the data and the PDE residual terms in the loss function.

4.2.3 Measuring convergence and superresolution

After training a neural operator, one typically measures its performance by evaluating the testing error, such as the relative L^2-error, on a set of unseen data generated using the procedure described in Section 4. In general, state-of-the-art neural operator architectures report a relative testing error of $1\% - 10\%$ depending on the problems considered (Kovachki et al., 2023; Lu et al., 2021a; Li et al., 2021a).

However, it is essential to note that the testing error may not be a good measure of the performance of a neural operator, as it does not provide any information about the generalization properties of the model. Hence, the testing forcing terms are usually sampled from the same distribution as the training forcing terms, so they lie on the same finite-dimensional function space, determined by the spectral decay of the GP covariance kernel eigenvalues (see Section 4.1). Moreover, in real applications, the testing source terms could have different distributions than the ones used for training ones, and extrapolation of the neural operator may be required. Zhu et al. (2023) investigate the extrapolation properties of DeepONet with respect to the length-scale parameter of the underlying source term GP. They observe that the testing error increases when the length-scale parameter corresponding to the test data decreases. At the same time, the neural operator can extrapolate to unseen data with a larger length scale than the training dataset, i.e., smoother functions.

One attractive property of neural operators is their resolution invariance to perform predictions at finer spatial resolutions than the training dataset on which they have been trained. This is usually called zero-shot superresolution (Kovachki et al., 2023, Sec. 7.2.3). To investigate this property, we reproduce the numerical examples of Kovachki et al. (2023, Sec. 7.2) and train a Fourier neu-

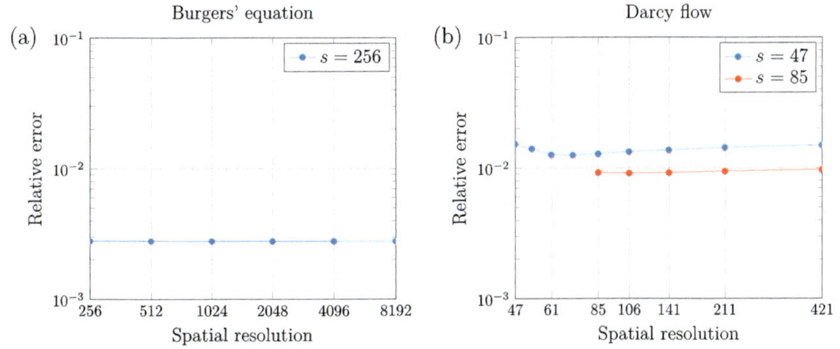

FIGURE 10 (a) Relative test error at different spatial resolutions of the Fourier neural operator trained to approximate the solution operator of the 1D Burgers' equation (17) with trained resolution of $s = 256$. (b) FNO trained on 2D Darcy flow a resolution of $s = 47$ (blue) and $s = 85$ (red) and evaluated at higher spatial resolutions.

ral operator to approximate the solution operator of Burgers' equation and Darcy flow at a low-resolution data and evaluate the operator at higher resolutions. We consider the one-dimensional Burgers' equation:

$$\frac{\partial}{\partial t}u(x,t) + \frac{1}{2}\frac{\partial}{\partial_x}(u(x,t)^2) = \nu\frac{\partial^2}{\partial_{x^2}}u(x,t), \quad x \in (0, 2\pi), \quad t \in [0, 1], \tag{17}$$

with periodic boundary conditions and viscosity $\nu = 0.1$. We are interested in learning the solution operator $\mathcal{A} : L^2_{\text{per}}((0, 2\pi)) \to H^1_{\text{per}}((0, 2\pi))$, which maps initial conditions $u_0 \in L^2_{\text{per}}((0, 2\pi))$ to corresponding solutions $u(\cdot, 1) \in H^1_{\text{per}}((0, 2\pi))$ to Eq. (17) at time $t = 1$. We then discretize the source and solution training data on a uniform grid with spatial resolution $s = 256$ and evaluate the trained neural operator at finer spatial resolutions $s \in \{512, 1024, 2048\}$. We observe in Fig. 10(a) that the relative testing error of the neural operator is independent of the spatial resolutions, as reported by Kovachki et al. (2023, Sec. 7.2).

Next, we consider the two-dimensional Darcy flow equation (18) with constant source term $f = 1$ and homogeneous Dirichlet boundary conditions on a unit square domain $\Omega = [0, 1]^2$:

$$-\operatorname{div}(a(x)\nabla u) = 1, \quad x \in [0, 1]^2. \tag{18}$$

We train a Fourier neural operator to approximate the solution operator, mapping the coefficient function a to the associated solution u to Eq. (18). We reproduce the numerical experiment in Kovachki et al. (2023, Sec. 6.2), where the random coefficient functions a are piecewise constant. The random functions a are generated as $a \sim T \circ f$, where $f \sim \mathcal{GP}(0, C)$, with $C = (-\Delta + 9I)^{-2}$ and

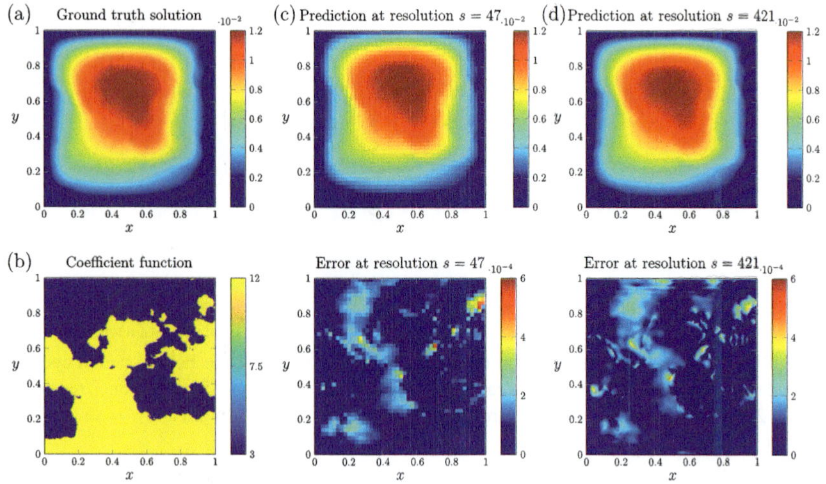

FIGURE 11 (a) Ground truth solution u to the 2D Darcy flow equation (18) corresponding to the coefficients function plotted in (b). (c)-(d) Predicted solution and approximation error at $s = 47$ and $s = 421$ by a Fourier neural operator trained on a Darcy flow dataset with spatial resolution of $s = 47$.

$T : \mathbb{R} \to \mathbb{R}^+$ is defined as

$$T(x) = \begin{cases} 12, & \text{if } x \geq 0, \\ 3, & \text{if } x < 0. \end{cases}$$

We discretize the coefficient and solution training data on a $s \times s$ uniform grid with spatial resolution $s = 47$ and evaluate the trained neural operator at higher spatial resolutions in Fig. 10. The relative testing error does not increase as the spatial resolution increases. Moreover, training the neural operator on a higher spatial resolution dataset can decrease the testing error. We also plot the ground truth solution u to Eq. (18) in Fig. 11(a) corresponding to the coefficient function plotted in panel (b), along with the predicted solutions and approximation errors at $s = 47$ and $s = 421$ by the Fourier neural operator in panels (c) and (d). We want to point the reader interested in the discretization properties of neural operators to the recent perspective on representation equivalent neural operators (ReNO) by Bartolucci et al. (2023).

5 Conclusions and future challenges

In this paper, we provided a comprehensive overview of the recent developments in neural operator learning, a new paradigm at the intersection of scientific computing and ML for learning solution operators of PDEs. Given the recent surge of interest in this field, a key question concerns the choice of neural architectures for different PDEs. Most theoretical studies in the field analyze and compare neural operators through the prism of approximation theory. We proposed a framework based on numerical linear algebra and matrix recovery problems for interpreting the type of neural operator architectures that can be used to learn solution operators of PDEs. Hence, solution operators associated with linear PDEs can often be written as integral operators with a Green's function and recover by a one-layer neural operator, which after discretization is equivalent to a matrix recovery problem.

Moreover, the choice of architectures, such as FNO or DeepONet, enforces or preserves different properties of the PDE solution operator, such as being translation invariant, low-rank, or off-diagonal low-rank (see Table 2). We then focused on the data acquisition process. We highlighted the importance of the distribution of source terms, usually sampled from a Gaussian process with a tailored covariance kernel, on the resulting performance of the neural operator. Following recent works on elliptic and parabolic PDEs and numerical experiments, we also discussed the relatively small amount of training data needed for operator learning. Finally, we studied the different choices of optimization algorithms and loss functions and highlighted the superresolution properties of neural operators, i.e., their ability to be evaluated at higher resolution than the training dataset with a minor impact on the performance. There are, however, several remaining challenges in the field.

Distribution of probes

Most applications of neural operators employ source terms that are globally supported on the domain, sampled from a Gaussian process, and whose distribution is fixed before training. However, this might now always apply to real-world engineering or biological systems, where source terms could be localized in space and time. A significant problem is to study the impact of the distribution of locally supported source terms on the performance of neural operators, both from a practical and theoretical viewpoint. Hence, recent sample complexity works on elliptic and parabolic PDEs exploit structured source terms (Boullé et al., 2023, 2022b; Schäfer and Owhadi, 2021). Another area of future research is to employ adaptive source terms to fine-tune neural operators for specific applications. This could lead to higher performance by selecting source terms that maximize the training error or allow efficient transfer learning between different applications without retraining a large neural operator.

Software and datasets

An essential step towards democratizing operator learning involves the development of open-source software and datasets for training and comparing neural operators, similar to the role played by the MNIST (LeCun et al., 1998) and ImageNet (Deng et al., 2009) databases in the improvement of computer vision techniques. However, due to the fast emerging methods in operator learning, there have been limited attempts beyond (Lu et al., 2022) to standardize the datasets and software used in the field. Establishing a list of standard PDE problems across different scientific fields, such as fluid dynamics, quantum mechanics, and epidemiology, with other properties (e.g. linear/nonlinear, steady/time-dependent, low/high dimensional, smooth/rough solutions, simple/complex geometry) would allow researchers to compare and identify the neural operator architectures that are the most appropriate for a particular task. A recent benchmark has been proposed to evaluate the performance of physics-informed neural networks for solving PDEs (Hao et al., 2023b).

Real-world applications

Neural operators have been successfully applied to perform weather forecasting and achieve spectacular performance in terms of accuracy and computational time to solutions compared to traditional numerical weather prediction techniques while being trained on historical weather data (Kurth et al., 2023; Lam et al., 2023). An exciting development in the field of operator learning would be to expand the scope of applications to other scientific fields and train the models on real datasets, where the underlying PDE governing the data is unknown to discover new physics.

Theoretical understanding

Following the recent works on the approximation theory of neural operators and sample complexity bounds for different classes of PDEs, there is a growing need for a theoretical understanding of convergence and optimization. In particular, an exciting area of research would be to extend the convergence results of physics-informed neural networks and the neural tangent kernel framework to neural operators. This would enable the derivation of rigorous convergence rates for different types of neural operator architectures and loss functions and new schemes for initializing the weight distributions in the underlying neural networks.

Physical properties

Most neural operator architectures are motivated by obtaining a good approximation of the solution operator of a PDE. However, the resulting neural operator is often highly nonlinear, difficult to interpret mathematically, and might not satisfy the physical properties of the underlying PDE, such as conservation laws or symmetries (Olver, 1993b). There are several promising research directions in operator learning related to symmetries and conservation laws (Otto et al., 2023). One approach would be to enforce known physical properties when training neural operators, either strongly through structure preserving architectures (Richter-Powell et al., 2022), or weakly by adding a residual term in the loss function (Li et al., 2021b; Wang et al., 2021b). Another direction is to discover new physical properties of the underlying PDEs from the trained neural operator. While (Boullé et al., 2022a) showed that symmetries of linear PDEs can be recovered from the learned Green's function, this approach has not been extended to nonlinear PDEs. Finally, one could also consider using reinforcement learning techniques for enforcing physical constraints after the optimization procedure, similar to recent applications in large language models (Ouyang et al., 2022).

Acknowledgments

The work of both authors was supported by the Office of Naval Research (ONR), under grant N00014-23-1-2729. N.B. was supported by an INI-Simons Postdoctoral Research Fellowship. A.T. was supported by National Science Foundation grants DMS-2045646 and a Weiss Junior Fellowship Award.

References

Abadi, M., Barham, P., Chen, J., Chen, Z., Davis, A., Dean, J., Devin, M., Ghemawat, S., Irving, G., Isard, M., et al., 2016. Tensorflow: a system for large-scale machine learning. In: 12th USENIX Symposium on Operating Systems Design and Implementation, pp. 265–283.

Adler, R.J., 2010. The Geometry of Random Fields. SIAM.

Allen-Zhu, Z., Li, Y., Song, Z., 2019. A convergence theory for deep learning via over-parameterization. In: International Conference on Machine Learning, pp. 242–252.

Alnæs, M., Blechta, J., Hake, J., Johansson, A., Kehlet, B., Logg, A., Richardson, C., Ring, J., Rognes, M.E., Wells, G.N., 2015. The FEniCS project version 1.5. Arch. Numer. Softw. 3 (100).

Alnæs, M.S., Logg, A., Ølgaard, K.B., Rognes, M.E., Wells, G.N., 2014. Unified form language: a domain-specific language for weak formulations of partial differential equations. ACM Trans. Math. Softw. 40 (2), 1–37.

Alpak, F.O., Vamaraju, J., Jennings, J.W., Pawar, S., Devarakota, P., Hohl, D., 2023. Augmenting deep residual surrogates with Fourier neural operators for rapid two-phase flow and transport simulations. SPE J., 1–22.

Arridge, S., Maass, P., Öktem, O., Schönlieb, C.-B., 2019. Solving inverse problems using data-driven models. Acta Numer. 28, 1–174.

Balay, S., Abhyankar, S., Adams, M.F., et al., 2023. PETSc Users Manual. Argonne National Laboratory.

Bartolucci, F., de Bézenac, E., Raonić, B., Molinaro, R., Mishra, S., Alaifari, R., 2023. Are neural operators really neural operators? Frame theory meets operator learning. arXiv preprint. arXiv: 2305.19913.

Bebendorf, M., Hackbusch, W., 2003. Existence of $\mathcal{H}$-matrix approximants to the inverse FE-matrix of elliptic operators with L^∞-coefficients. Numer. Math. 95 (1), 1–28.

Benitez, J.A.L., Furuya, T., Faucher, F., Kratsios, A., Tricoche, X., de Hoop, M.V., 2023. Out-of-distributional risk bounds for neural operators with applications to the Helmholtz equation. arXiv preprint. arXiv:2301.11509.

Bhattacharya, K., Hosseini, B., Kovachki, N.B., Stuart, A.M., 2021. Model reduction and neural networks for parametric PDEs. SMAI J. Comput. Math. 7, 121–157.

Boullé, N., Townsend, A., 2022. A generalization of the randomized singular value decomposition. In: International Conference on Learning Representations.

Boullé, N., Townsend, A., 2023. Learning elliptic partial differential equations with randomized linear algebra. Found. Comput. Math. 23 (2), 709–739.

Boullé, N., Nakatsukasa, Y., Townsend, A., 2020. Rational neural networks. In: Advances in Neural Information Processing Systems, vol. 33, pp. 14243–14253.

Boullé, N., Earls, C.J., Townsend, A., 2022a. Data-driven discovery of Green's functions with human-understandable deep learning. Sci. Rep. 12 (1), 4824.

Boullé, N., Kim, S., Shi, T., Townsend, A., 2022b. Learning Green's functions associated with time-dependent partial differential equations. J. Mach. Learn. Res. 23 (218), 1–34.

Boullé, N., Halikias, D., Townsend, A., 2023. Elliptic PDE learning is provably data-efficient. Proc. Natl. Acad. Sci. USA 120 (39), e2303904120.

Bronstein, M.M., Bruna, J., Cohen, T., Veličković, P., 2021. Geometric deep learning: grids, groups, graphs, geodesics, and gauges. arXiv preprint. arXiv:2104.13478.

Brown, T., Mann, B., Ryder, N., Subbiah, M., Kaplan, J.D., Dhariwal, P., Neelakantan, A., Shyam, P., Sastry, G., Askell, A., et al., 2020. Language models are few-shot learners. In: Advances in Neural Information Processing Systems, vol. 33, pp. 1877–1901.

Bruno, O.P., Han, Y., Pohlman, M.M., 2007. Accurate, high-order representation of complex three-dimensional surfaces via Fourier continuation analysis. J. Comput. Phys. 227 (2), 1094–1125.

Brunton, S.L., Proctor, J.L., Kutz, J.N., 2016. Discovering governing equations from data by sparse identification of nonlinear dynamical systems. Proc. Natl. Acad. Sci. USA 113 (15), 3932–3937.

Byrd, R.H., Lu, P., Nocedal, J., Zhu, C., 1995. A limited memory algorithm for bound constrained optimization. SIAM J. Sci. Comput. 16 (5), 1190–1208.

Canzani, Y., 2013. Analysis on Manifolds via the Laplacian. Harvard University.

Cao, S., 2021. Choose a transformer: Fourier or Galerkin. In: Advances in Neural Information Processing Systems, vol. 34, pp. 24924–24940.

Champion, K., Lusch, B., Kutz, J.N., Brunton, S.L., 2019. Data-driven discovery of coordinates and governing equations. Proc. Natl. Acad. Sci. USA 116 (45), 22445–22451.

Chen, T., Chen, H., 1995. Universal approximation to nonlinear operators by neural networks with arbitrary activation functions and its application to dynamical systems. IEEE Trans. Neural Netw. 6 (4), 911–917.

Cho, S., Dong, H., Kim, S., 2008. On the Green's matrices of strongly parabolic systems of second order. Indiana Univ. Math. J. 57 (4), 1633–1677.

Cho, S., Dong, H., Kim, S., 2012. Global estimates for Green's matrix of second order parabolic systems with application to elliptic systems in two dimensional domains. Potential Anal. 36 (2), 339–372.

Cooley, J.W., Tukey, J.W., 1965. An algorithm for the machine calculation of complex Fourier series. Math. Comput. 19 (90), 297–301.

Cuomo, S., Di Cola, V.S., Giampaolo, F., Rozza, G., Raissi, M., Piccialli, F., 2022. Scientific machine learning through physics–informed neural networks: where we are and what's next. J. Sci. Comput. 92 (3), 88.

Cybenko, G., 1989. Approximation by superpositions of a sigmoidal function. Math. Control Signals Syst. 2 (4), 303–314.

de Hoop, M.V., Huang, D.Z., Qian, E., Stuart, A.M., 2022. The cost-accuracy trade-off in operator learning with neural networks. arXiv preprint. arXiv:2203.13181.

de Hoop, M.V., Kovachki, N.B., Nelsen, N.H., Stuart, A.M., 2023. Convergence rates for learning linear operators from noisy data. SIAM/ASA J. Uncertain. Quantificat. 11 (2), 480–513.

De Ryck, T., Mishra, S., 2022. Generic bounds on the approximation error for physics-informed (and) operator learning. In: Advances in Neural Information Processing Systems, vol. 35, pp. 10945–10958.

Deng, B., Shin, Y., Lu, L., Zhang, Z., Karniadakis, G.E., 2022. Approximation rates of DeepONets for learning operators arising from advection–diffusion equations. Neural Netw. 153, 411–426.

Deng, J., Dong, W., Socher, R., Li, L.-J., Li, K., Fei-Fei, L., 2009. Imagenet: a large-scale hierarchical image database. In: Conference on Computer Vision and Pattern Recognition. IEEE, pp. 248–255.

DeVore, R.A., 1998. Nonlinear approximation. Acta Numer. 7, 51–150.

Di Leoni, P.C., Lu, L., Meneveau, C., Karniadakis, G.E., Zaki, T.A., 2023. Neural operator prediction of linear instability waves in high-speed boundary layers. J. Comput. Phys. 474, 111793.

Dong, H., Kim, S., 2009. Green's matrices of second order elliptic systems with measurable coefficients in two dimensional domains. Trans. Am. Math. Soc. 361 (6), 3303–3323.

Dosovitskiy, A., Beyer, L., Kolesnikov, A., Weissenborn, D., Zhai, X., Unterthiner, T., Dehghani, M., Minderer, M., Heigold, G., Gelly, S., et al., 2020. An image is worth 16x16 words: transformers for image recognition at scale. arXiv preprint. arXiv:2010.11929.

Driscoll, T.A., Hale, N., Trefethen, L.N., 2014. Chebfun Guide. Pafnuty Publications. http://www.chebfun.org/docs/guide/.
Du, S., Lee, J., Li, H., Wang, L., Zhai, X., 2019. Gradient descent finds global minima of deep neural networks. In: International Conference on Machine Learning, pp. 1675–1685.
E, W., Yu, B., 2018. The deep Ritz method: a deep learning-based numerical algorithm for solving variational problems. Commun. Math. Stat. 6 (1), 1–12.
Evans, L.C., 2010. Partial Differential Equations, 2nd edition. American Mathematical Society.
Fanaskov, V., Oseledets, I., 2022. Spectral neural operators. arXiv preprint. arXiv:2205.10573.
Feliu-Faba, J., Fan, Y., Ying, L., 2020. Meta-learning pseudo-differential operators with deep neural networks. J. Comput. Phys. 408, 109309.
Gilmer, J., Schoenholz, S.S., Riley, P.F., Vinyals, O., Dahl, G.E., 2017. Neural message passing for quantum chemistry. In: International Conference on Machine Learning, pp. 1263–1272.
Gin, C.R., Shea, D.E., Brunton, S.L., Kutz, J.N., 2021. DeepGreen: deep learning of Green's functions for nonlinear boundary value problems. Sci. Rep. 11 (1), 1–14.
Goswami, S., Yin, M., Yu, Y., Karniadakis, G.E., 2022. A physics-informed variational deeponet for predicting crack path in quasi-brittle materials. Comput. Methods Appl. Mech. Eng. 391, 114587.
Goswami, S., Bora, A., Yu, Y., Karniadakis, G.E., 2023. Physics-informed deep neural operator networks. In: Machine Learning in Modeling and Simulation: Methods and Applications. Springer, pp. 219–254.
Gottlieb, D., Orszag, S.A., 1977. Numerical Analysis of Spectral Methods: Theory and Applications. SIAM.
Greengard, L., Rokhlin, V., 1997. A new version of the fast multipole method for the Laplace equation in three dimensions. Acta Numer. 6, 229–269.
Grüter, M., Widman, K.-O., 1982. The Green function for uniformly elliptic equations. Manuscr. Math. 37 (3), 303–342.
Gupta, G., Xiao, X., Bogdan, P., 2021. Multiwavelet-based operator learning for differential equations. In: Advances in Neural Information Processing Systems, vol. 34, pp. 24048–24062.
Hackbusch, W., Khoromskij, B.N., Kriemann, R., 2004. Hierarchical matrices based on a weak admissibility criterion. Computing 73 (3), 207–243.
Halikias, D., Townsend, A., 2023. Structured matrix recovery from matrix-vector products. Numer. Linear Algebra Appl., e2531.
Halko, N., Martinsson, P.-G., Tropp, J.A., 2011. Finding structure with randomness: probabilistic algorithms for constructing approximate matrix decompositions. SIAM Rev. 53 (2), 217–288.
Ham, D.A., Kelly, P.H.J., Mitchell, L., Cotter, C.J., Kirby, R.C., Sagiyama, K., Bouziani, N., Vorderwuelbecke, S., Gregory, T.J., Betteridge, J., Shapero, D.R., Nixon-Hill, R.W., Ward, C.J., Farrell, P.E., Brubeck, P.D., Marsden, I., Gibson, T.H., Homolya, M., Sun, T., McRae, A.T.T., Luporini, F., Gregory, A., Lange, M., Funke, S.W., Rathgeber, F., Bercea, G.-T., Markall, G.R., 2023. Firedrake User Manual, 1st edition. Imperial College London and University of Oxford and Baylor University and University of Washington.
Hao, Z., Wang, Z., Su, H., Ying, C., Dong, Y., Liu, S., Cheng, Z., Song, J., Zhu, J., 2023a. GNOT: a general neural operator transformer for operator learning. In: International Conference on Machine Learning, pp. 12556–12569.
Hao, Z., Yao, J., Su, C., Su, H., Wang, Z., Lu, F., Xia, Z., Zhang, Y., Liu, S., Lu, L., et al., 2023b. PINNacle: a comprehensive benchmark of physics-informed neural networks for solving PDEs. arXiv preprint. arXiv:2306.08827.
He, Q., Barajas-Solano, D., Tartakovsky, G., Tartakovsky, A.M., 2020. Physics-informed neural networks for multiphysics data assimilation with application to subsurface transport. Adv. Water Resour. 141, 103610.
Ho, J., Jain, A., Abbeel, P., 2020. Denoising diffusion probabilistic models. In: Advances in Neural Information Processing Systems, vol. 33, pp. 6840–6851.
Hofmann, S., Kim, S., 2004. Gaussian estimates for fundamental solutions to certain parabolic systems. Publ. Mat., 481–496.

Hornik, K., 1991. Approximation capabilities of multilayer feedforward networks. Neural Netw. 4 (2), 251–257.
Hsing, T., Eubank, R., 2015. Theoretical Foundations of Functional Data Analysis, with an Introduction to Linear Operators. John Wiley & Sons.
Iserles, A., 2009. A First Course in the Numerical Analysis of Differential Equations. Cambridge University Press.
Jacot, A., Gabriel, F., Hongler, C., 2018. Neural tangent kernel: convergence and generalization in neural networks. In: Advances in Neural Information Processing Systems, vol. 31.
Jin, J., Lu, Y., Blanchet, J., Ying, L., 2023. Minimax optimal kernel operator learning via multilevel training. In: International Conference on Learning Representations.
Jumper, J., Evans, R., Pritzel, A., Green, T., Figurnov, M., Ronneberger, O., Tunyasuvunakool, K., Bates, R., Žídek, A., Potapenko, A., et al., 2021. Highly accurate protein structure prediction with AlphaFold. Nature 596 (7873), 583–589.
Karhunen, K., 1946. Über lineare methoden in der wahrscheinlichkeitsrechnung. Ann. Acad. Sci. Fenn., Ser. A I 37, 3–79.
Karniadakis, G.E., Kevrekidis, I.G., Lu, L., Perdikaris, P., Wang, S., Yang, L., 2021. Physics-informed machine learning. Nat. Rev. Phys. 3 (6), 422–440.
Kingma, D.P., Ba, J., 2015. Adam: a method for stochastic optimization. In: Proc. 3rd International Conference on Learning Representation.
Kissas, G., Seidman, J.H., Guilhoto, L.F., Preciado, V.M., Pappas, G.J., Perdikaris, P., 2022. Learning operators with coupled attention. J. Mach. Learn. Res. 23 (1), 9636–9698.
Kovachki, N., Lanthaler, S., Mishra, S., 2021. On universal approximation and error bounds for Fourier neural operators. J. Mach. Learn. Res. 22, 1–76.
Kovachki, N., Li, Z., Liu, B., Azizzadenesheli, K., Bhattacharya, K., Stuart, A., Anandkumar, A., 2023. Neural operator: learning maps between function spaces with applications to PDEs. J. Mach. Learn. Res. 24 (89), 1–97.
Krizhevsky, A., Sutskever, I., Hinton, G.E., 2012. Imagenet classification with deep convolutional neural networks. In: Advances in Neural Information Processing Systems, vol. 25.
Kurth, T., Subramanian, S., Harrington, P., Pathak, J., Mardani, M., Hall, D., Miele, A., Kashinath, K., Anandkumar, A., 2023. Fourcastnet: accelerating global high-resolution weather forecasting using adaptive Fourier neural operators. In: Proceedings of the Platform for Advanced Scientific Computing Conference, pp. 1–11.
Lam, R., Sanchez-Gonzalez, A., Willson, M., Wirnsberger, P., Fortunato, M., Alet, F., Ravuri, S., Ewalds, T., Eaton-Rosen, Z., Hu, W., et al., 2023. Learning skillful medium-range global weather forecasting. Science, eadi2336.
Lanthaler, S., Mishra, S., Karniadakis, G.E., 2022. Error estimates for DeepONets: a deep learning framework in infinite dimensions. Trans. Math. Appl. 6 (1).
LeCun, Y., Bottou, L., Bengio, Y., Haffner, P., 1998. Gradient-based learning applied to document recognition. Proc. IEEE 86 (11), 2278–2324.
LeCun, Y., Bengio, Y., Hinton, G., 2015. Deep learning. Nature 521 (7553), 436–444.
Levitt, J., Martinsson, P.-G., 2022a. Linear-complexity black-box randomized compression of hierarchically block separable matrices. arXiv preprint. arXiv:2205.02990.
Levitt, J., Martinsson, P.-G., 2022b. Randomized compression of rank-structured matrices accelerated with graph coloring. arXiv preprint. arXiv:2205.03406.
Li, Z., Kovachki, N., Azizzadenesheli, K., Liu, B., Bhattacharya, K., Stuart, A., Anandkumar, A., 2020a. Neural operator: graph kernel network for partial differential equations. arXiv preprint. arXiv:2003.03485.
Li, Z., Kovachki, N., Azizzadenesheli, K., Liu, B., Stuart, A., Bhattacharya, K., Anandkumar, A., 2020b. Multipole graph neural operator for parametric partial differential equations. In: Advances in Neural Information Processing Systems, vol. 33, pp. 6755–6766.
Li, Z., Kovachki, N., Azizzadenesheli, K., Liu, B., Bhattacharya, K., Stuart, A., Anandkumar, A., 2021a. Fourier neural operator for parametric partial differential equations. In: International Conference on Learning Representations.

Li, Z., Zheng, H., Kovachki, N., Jin, D., Chen, H., Liu, B., Azizzadenesheli, K., Anandkumar, A., 2021b. Physics-informed neural operator for learning partial differential equations. arXiv preprint. arXiv:2111.03794.

Li, Z., Huang, D.Z., Liu, B., Anandkumar, A., 2022. Fourier neural operator with learned deformations for PDEs on general geometries. arXiv preprint. arXiv:2207.05209.

Li, Z., Kovachki, N.B., Choy, C., Li, B., Kossaifi, J., Otta, S.P., Nabian, M.A., Stadler, M., Hundt, C., Azizzadenesheli, K., et al., 2023a. Geometry-informed neural operator for large-scale 3D PDEs. arXiv preprint. arXiv:2309.00583.

Li, Z., Peng, W., Yuan, Z., Wang, J., 2023b. Long-term predictions of turbulence by implicit U-Net enhanced Fourier neural operator. Phys. Fluids 35 (7).

Lin, G., Chen, F., Hu, P., Chen, X., Chen, J., Wang, J., Shi, Z., 2023. BI-GreenNet: learning Green's functions by boundary integral network. Commun. Math. Stat. 11 (1), 103–129.

Lin, L., Lu, J., Ying, L., 2011. Fast construction of hierarchical matrix representation from matrix–vector multiplication. J. Comput. Phys. 230 (10), 4071–4087.

Loève, M., 1946. Fonctions aleatoire de second ordre. Rev. Sci. 84, 195–206.

Loshchilov, I., Hutter, F., 2019. Decoupled weight decay regularization. In: International Conference on Learning Representations.

Lu, L., Jin, P., Pang, G., Zhang, Z., Karniadakis, G.E., 2021a. Learning nonlinear operators via DeepONet based on the universal approximation theorem of operators. Nat. Mach. Intell. 3 (3), 218–229.

Lu, L., Meng, X., Mao, Z., Karniadakis, G.E., 2021b. DeepXDE: a deep learning library for solving differential equations. SIAM Rev. 63 (1), 208–228.

Lu, L., Meng, X., Cai, S., Mao, Z., Goswami, S., Zhang, Z., Karniadakis, G.E., 2022. A comprehensive and fair comparison of two neural operators (with practical extensions) based on fair data. Comput. Methods Appl. Mech. Eng. 393, 114778.

Lyu, Y., Zhao, X., Gong, Z., Kang, X., Yao, W., 2023. Multi-fidelity prediction of fluid flow based on transfer learning using Fourier neural operator. Phys. Fluids 35 (7).

Mao, S., Dong, R., Lu, L., Yi, K.M., Wang, S., Perdikaris, P., 2023. PPDONet: deep operator networks for fast prediction of steady-state solutions in disk–planet systems. Astrophys. J. Lett. 950 (2), L12.

Mao, Z., Jagtap, A.D., Karniadakis, G.E., 2020. Physics-informed neural networks for high-speed flows. Comput. Methods Appl. Mech. Eng. 360, 112789.

Martinsson, P.-G., 2011. A fast randomized algorithm for computing a hierarchically semiseparable representation of a matrix. SIAM J. Matrix Anal. Appl. 32 (4), 1251–1274.

Martinsson, P.-G., Tropp, J.A., 2020. Randomized numerical linear algebra: foundations and algorithms. Acta Numer. 29, 403–572.

Mathieu, M., Henaff, M., LeCun, Y., 2014. Fast training of convolutional networks through FFTs. In: International Conference on Learning Representations.

Mercer, J., 1909. Functions of positive and negative type, and their connection with the theory of integral equations. Philos. Trans. R. Soc. A 209, 415–446.

Minakshisundaram, S., Pleijel, Å., 1949. Some properties of the eigenfunctions of the Laplace-operator on Riemannian manifolds. Can. J. Math. 1 (3), 242–256.

Moya, C., Zhang, S., Lin, G., Yue, M., 2023. Deeponet-grid-uq: a trustworthy deep operator framework for predicting the power grid's post-fault trajectories. Neurocomputing 535, 166–182.

O'Leary-Roseberry, T., Chen, P., Villa, U., Ghattas, O., 2024. Derivative-informed neural operator: an efficient framework for high-dimensional parametric derivative learning. J. Comput. Phys. 496, 112555.

Olver, P.J., 1993a. Applications of Lie Groups to Differential Equations. Springer Science & Business Media.

Olver, P.J., 1993b. Applications of Lie Groups to Differential Equations, 2nd edition. Springer-Verlag.

Otto, S.E., Zolman, N., Kutz, J.N., Brunton, S.L., 2023. A unified framework to enforce, discover, and promote symmetry in machine learning. arXiv preprint. arXiv:2311.00212.

Ouyang, L., Wu, J., Jiang, X., Almeida, D., Wainwright, C., Mishkin, P., Zhang, C., Agarwal, S., Slama, K., Ray, A., et al., 2022. Training language models to follow instructions with human feedback. In: Advances in Neural Information Processing Systems, vol. 35, pp. 27730–27744.
Paszke, A., Gross, S., Massa, F., Lerer, A., Bradbury, J., Chanan, G., Killeen, T., Lin, Z., Gimelshein, N., Antiga, L., et al., 2019. Pytorch: an imperative style, high-performance deep learning library. In: Advances in Neural Information Processing Systems, vol. 32.
Peng, R., Dong, J., Malof, J., Padilla, W.J., Tarokh, V., 2023. Deep generalized Green's functions. arXiv preprint. arXiv:2306.02925.
Peng, W., Yuan, Z., Wang, J., 2022. Attention-enhanced neural network models for turbulence simulation. Phys. Fluids 34 (2).
Raissi, M., Perdikaris, P., Karniadakis, G.E., 2019. Physics-informed neural networks: a deep learning framework for solving forward and inverse problems involving nonlinear partial differential equations. J. Comput. Phys. 378, 686–707.
Raonic, B., Molinaro, R., Rohner, T., Mishra, S., de Bezenac, E., 2023. Convolutional neural operators. In: ICLR 2023 Workshop on Physics for Machine Learning.
Rasmussen, C.E., Williams, C., 2006. Gaussian Processes for Machine Learning. MIT Press.
Rathgeber, F., Ham, D.A., Mitchell, L., Lange, M., Luporini, F., McRae, A.T., Bercea, G.-T., Markall, G.R., Kelly, P.H., 2016. Firedrake: automating the finite element method by composing abstractions. ACM Trans. Math. Softw. 43 (3), 1–27.
Richter-Powell, J., Lipman, Y., Chen, R.T., 2022. Neural conservation laws: a divergence-free perspective. In: Advances in Neural Information Processing Systems, vol. 35, pp. 38075–38088.
Ritter, K., Wasilkowski, G.W., Woźniakowski, H., 1995. Multivariate integration and approximation for random fields satisfying Sacks-Ylvisaker conditions. Ann. Appl. Probab., 518–540.
Scarselli, F., Gori, M., Tsoi, A.C., Hagenbuchner, M., Monfardini, G., 2008. The graph neural network model. IEEE Trans. Neural Netw. 20 (1), 61–80.
Schäfer, F., Owhadi, H., 2021. Sparse recovery of elliptic solvers from matrix-vector products. arXiv preprint. arXiv:2110.05351.
Schäfer, F., Sullivan, T.J., Owhadi, H., 2021. Compression, inversion, and approximate PCA of dense kernel matrices at near-linear computational complexity. Multiscale Model. Simul. 19 (2), 688–730.
Schmidt, M., Lipson, H., 2009. Distilling free-form natural laws from experimental data. Science 324 (5923), 81–85.
Searson, D.P., Leahy, D.E., Willis, M.J., 2010. GPTIPS: an open source genetic programming toolbox for multigene symbolic regression. In: Proceedings of the International Multiconference of Engineers and Computer Scientists, vol. 1. Citeseer, pp. 77–80.
Sirignano, J., Spiliopoulos, K., 2018. DGM: a deep learning algorithm for solving partial differential equations. J. Comput. Phys. 375, 1339–1364.
Sohl-Dickstein, J., Weiss, E., Maheswaranathan, N., Ganguli, S., 2015. Deep unsupervised learning using nonequilibrium thermodynamics. In: International Conference on Machine Learning, pp. 2256–2265.
Son, H., Jang, J.W., Han, W.J., Hwang, H.J., 2021. Sobolev training for physics informed neural networks. arXiv preprint. arXiv:2101.08932.
Song, Y., Sohl-Dickstein, J., Kingma, D.P., Kumar, A., Ermon, S., Poole, B., 2021. Score-based generative modeling through stochastic differential equations. In: International Conference on Learning Representations.
Stein, M.L., 1999. Interpolation of Spatial Data: Some Theory for Kriging. Springer Science & Business Media.
Stepaniants, G., 2023. Learning partial differential equations in reproducing kernel Hilbert spaces. J. Mach. Learn. Res. 24 (86), 1–72.
Stuart, A.M., 2010. Inverse problems: a Bayesian perspective. Acta Numer. 19, 451–559.
Süli, E., Mayers, D.F., 2003. An Introduction to Numerical Analysis. Cambridge University Press.
Sun, J., Liu, Y., Wang, Y., Yao, Z., Zheng, X., 2023. BINN: a deep learning approach for computational mechanics problems based on boundary integral equations. Comput. Methods Appl. Mech. Eng. 410, 116012.

Trefethen, L.N., 2000. Spectral Methods in MATLAB. SIAM.
Tripura, T., Chakraborty, S., 2022. Wavelet neural operator: a neural operator for parametric partial differential equations. arXiv preprint. arXiv:2205.02191.
Udrescu, S.-M., Tegmark, M., 2020. AI Feynman: a physics-inspired method for symbolic regression. Sci. Adv. 6 (16), eaay2631.
Udrescu, S.-M., Tan, A., Feng, J., Neto, O., Wu, T., Tegmark, M., 2020. AI Feynman 2.0: Pareto-optimal symbolic regression exploiting graph modularity. In: Advances in Neural Information Processing Systems, vol. 33, pp. 4860–4871.
Vaswani, A., Shazeer, N., Parmar, N., Uszkoreit, J., Jones, L., Gomez, A.N., Kaiser, Ł., Polosukhin, I., 2017. Attention is all you need. In: Advances in Neural Information Processing Systems, vol. 30.
Venturi, S., Casey, T., 2023. Svd perspectives for augmenting deeponet flexibility and interpretability. Comput. Methods Appl. Mech. Eng. 403, 115718.
Wang, S., Wang, H., Perdikaris, P., 2021a. On the eigenvector bias of Fourier feature networks: from regression to solving multi-scale PDEs with physics-informed neural networks. Comput. Methods Appl. Mech. Eng. 384, 113938.
Wang, S., Wang, H., Perdikaris, P., 2021b. Learning the solution operator of parametric partial differential equations with physics-informed DeepONets. Sci. Adv. 7 (40), eabi8605.
Wang, S., Wang, H., Perdikaris, P., 2022a. Improved architectures and training algorithms for deep operator networks. J. Sci. Comput. 92 (2), 35.
Wang, S., Yu, X., Perdikaris, P., 2022b. When and why PINNs fail to train: a neural tangent kernel perspective. J. Comput. Phys. 449, 110768.
Wang, S., Sankaran, S., Wang, H., Perdikaris, P., 2023. An expert's guide to training physics-informed neural networks. arXiv preprint. arXiv:2308.08468.
Weyl, H., 1911. Über die asymptotische verteilung der eigenwerte. Nachr. Ges. Wiss. Gött., Math.-Phys. Kl. 1911, 110–117.
Wu, Z., Pan, S., Chen, F., Long, G., Zhang, C., Philip, S.Y., 2020. A comprehensive survey on graph neural networks. IEEE Trans. Neural Netw. Learn. Syst. 32 (1), 4–24.
Yarotsky, D., 2017. Error bounds for approximations with deep ReLU networks. Neural Netw. 94, 103–114.
Ying, L., Biros, G., Zorin, D., 2004. A kernel-independent adaptive fast multipole algorithm in two and three dimensions. J. Comput. Phys. 196 (2), 591–626.
You, H., Zhang, Q., Ross, C.J., Lee, C.-H., Yu, Y., 2022. Learning deep implicit Fourier neural operators (IFNOs) with applications to heterogeneous material modeling. Comput. Methods Appl. Mech. Eng. 398, 115296.
Yu, A., Yang, Y., Townsend, A., 2023. Tuning frequency bias in neural network training with nonuniform data. In: International Conference on Learning Representations.
Zhao, X., Chen, X., Gong, Z., Zhou, W., Yao, W., Zhang, Y., 2024. RecFNO: a resolution-invariant flow and heat field reconstruction method from sparse observations via Fourier neural operator. Int. J. Therm. Sci. 195, 108619.
Zheng, H., Nie, W., Vahdat, A., Azizzadenesheli, K., Anandkumar, A., 2023. Fast sampling of diffusion models via operator learning. In: International Conference on Machine Learning, pp. 42390–42402.
Zhou, J., Cui, G., Hu, S., Zhang, Z., Yang, C., Liu, Z., Wang, L., Li, C., Sun, M., 2020. Graph neural networks: a review of methods and applications. AI Open 1, 57–81.
Zhu, H., Williams, C.K., Rohwer, R., Morciniec, M., 1998. Gaussian regression and optimal finite dimensional linear models. In: Neural Networks and Machine Learning. Springer-Verlag.
Zhu, M., Zhang, H., Jiao, A., Karniadakis, G.E., Lu, L., 2023. Reliable extrapolation of deep neural operators informed by physics or sparse observations. Comput. Methods Appl. Mech. Eng. 412, 116064.

Chapter 4

The multiverse of dynamic mode decomposition algorithms

Matthew J. Colbrook

Department of Applied Mathematics and Theoretical Physics, University of Cambridge, Cambridge, United Kingdom

e-mail address: m.colbrook@damtp.cam.ac.uk

Contents

Handbook of Numerical Analysis, Volume 25, ISSN 1570-8659, https://doi.org/10.1016/bs.hna.2024.05.004

Abstract

Dynamic Mode Decomposition (DMD) is a popular data-driven analysis technique used to decompose complex, nonlinear systems into a set of modes, revealing underlying patterns and dynamics through spectral analysis. This review presents a comprehensive and pedagogical examination of DMD, emphasizing the role of Koopman operators in transforming complex nonlinear dynamics into a linear framework. A distinctive feature of this review is its focus on the relationship between DMD and the spectral properties of Koopman operators, with particular emphasis on the theory and practice of DMD algorithms for spectral computations. We explore the diverse "multiverse" of DMD methods, categorized into three main areas: linear regression-based methods, Galerkin approximations, and structure-preserving techniques. Each category is studied for its unique contributions and challenges, providing a detailed overview of significant algorithms and their applications as outlined in Table 1. We include a MATLAB® package with examples and applications to enhance the practical understanding of these methods. This review serves as both a practical guide and a theoretical reference for various DMD methods, accessible to both experts and newcomers, and enabling readers to explore their areas of interest in the expansive field of DMD.

Keywords

Dynamical systems, Koopman operator, Data-driven discovery, Dynamic mode decomposition, Spectral theory, Spectral computations

MSC Codes
37A30, 37Mxx, 37Nxx, 47A10, 47B33, 65J10, 65P99

1 Introduction

Dynamical systems provide a powerful framework for modeling the evolution of various scientific and engineering systems over time. They are crucial for understanding complex phenomena ranging from weather patterns and population growth to stock market fluctuations. We consider discrete-time dynamical systems represented as:

$$\mathbf{x}_{n+1} = \mathbf{F}(\mathbf{x}_n), \qquad n = 0, 1, 2, \ldots, \tag{1.1}$$

where $\mathbf{x} \in \Omega$ denotes the state of the system, and $\Omega \subseteq \mathbb{R}^d$ is the state space. The function $\mathbf{F} : \Omega \to \Omega$ governs the system's evolution. The classical approach to analyzing such systems, tracing back over a century to the seminal work of Poincaré (1899), is geometric. It involves local analysis of fixed points, periodic orbits, and stable or unstable manifolds. While Poincaré's framework has significantly advanced our understanding of dynamical systems, it faces two main challenges in modern applications:

- **Global understanding of nonlinear dynamics:** Unlike linear systems, there is no comprehensive mathematical framework for nonlinear systems. The principle of linear superposition is not applicable in this context. Local models can predict long-term dynamics near fixed points and attracting manifolds but have limited predictive power for other initial conditions. Consequently, the global understanding of nonlinear dynamics in state space is predominantly qualitative.
- **Incomplete knowledge of evolution:** Many systems cannot be analytically described due to their complexity or our incomplete understanding. Typically, our knowledge is limited to discrete-time snapshots of the system, i.e., a finite dataset

$$\left\{\mathbf{x}^{(m)}, \mathbf{y}^{(m)}\right\}_{m=1}^{M} \quad \text{such that} \quad \mathbf{y}^{(m)} = \mathbf{F}(\mathbf{x}^{(m)}), \quad m = 1, \ldots, M.$$

  We concisely write this data in the form of snapshot matrices

$$\mathbf{X} = \begin{pmatrix}\mathbf{x}^{(1)} & \mathbf{x}^{(2)} & \cdots & \mathbf{x}^{(M)}\end{pmatrix} \in \mathbb{R}^{d\times M},$$
$$\mathbf{Y} = \begin{pmatrix}\mathbf{y}^{(1)} & \mathbf{y}^{(2)} & \cdots & \mathbf{y}^{(M)}\end{pmatrix} \in \mathbb{R}^{d\times M}. \tag{1.2}$$

  Advances in measurement technologies have significantly enhanced our ability to collect detailed multimodal and multi-fidelity snapshot data. Data could be collected from one long trajectory or multiple shorter trajectories. It can come from experimental observations or numerical simulations. The question becomes how to use this data to meaningfully study the dynamical system in (1.1).

The advent of big data (Hey et al., 2009), coupled with strides in modern statistical learning (Hastie et al., 2009) and machine learning (Mohri et al., 2018), has heralded a new era of data-driven algorithms to address these issues. This review will focus on one of the most prominent of these algorithms, *Dynamic Mode Decomposition* (DMD), closely connected with *Koopman operators*.

Koopman Operators — In 1931, Koopman introduced his operator-theoretic approach to dynamical systems, initially to describe Hamiltonian systems (Koopman, 1931). This theory was further expanded by Koopman and von Neumann (1932) to include systems with continuous spectra. Koopman operators offer a powerful alternative to the classical geometric view of dynamical systems by addressing the fundamental issue of *nonlinearity*. We lift a nonlinear system (1.1) into an infinite-dimensional space of observable functions $g : \Omega \to \mathbb{C}$ using a Koopman operator $\mathcal{K}$:

$$[\mathcal{K}g](\mathbf{x}) = g(\mathbf{F}(\mathbf{x})), \quad \text{so that} \quad [\mathcal{K}g](\mathbf{x}_n) = g(\mathbf{x}_{n+1}).$$

Through this approach, the evolution dynamics become linear, enabling the use of generic solution techniques based on spectral decompositions. Initially, the primary application of Koopman operators was in ergodic theory (Eisner et al., 2015), notably playing a pivotal role in proving the ergodic theorem by von Neumann (Neumann, 1932) and Birkhoff (Birkhoff, 1931; Birkhoff and Koopman, 1932). More recently, they have been extensively used in data-driven methods for studying dynamical systems.

Dynamic Mode Decomposition — A significant objective of modern Koopman operator theory is to identify a coordinate transformation under which even strongly nonlinear dynamics may be approximated by a linear system. This coordinate system is related to the spectrum of the Koopman operator. DMD was initially developed by Schmid (2009, 2010) (see also Schmid and Sesterhenn, 2008) in the context of fluid dynamics. Mezić (2005) introduced the Koopman mode decomposition, providing a theoretical basis for Rowley et al. (2009) to connect DMD with Koopman operators. This connection validated DMD's application in nonlinear systems and offered a powerful yet straightforward, data-driven approach for approximating Koopman operators. The fusion of contemporary Koopman theory with an efficient numerical algorithm has led to significant advancements and a surge in research. DMD is now the central algorithm for computational approximations of Koopman operators with applications in various fields beyond fluid mechanics, such as neuroscience, disease modeling, robotics, video processing, power grids, financial markets, and plasma physics. The simplicity and effectiveness of DMD have led to numerous innovations, giving rise to a diverse array of DMD methods, playfully described here as a "multiverse", aimed at addressing specific challenges.

This Review — We provide a comprehensive tour of this "multiverse" of DMD methods. Our primary focus is on the interplay between DMD, the spectral properties of Koopman operators, and their numerical computations. At the time of writing, these methods can be broadly categorized into three main areas:

- DMD methods based on linear regression;
- DMD methods utilizing Galerkin approximations;
- DMD methods aimed at preserving structures or symmetries of (1.1).

These distinctions are not rigid, and some methods encompass multiple flavors. This review navigates these key areas and variants, summarized in Table 1, where we also highlight the unique challenges each algorithm addresses (see also Section 2.4). We provide detailed summaries and examples of these algorithms in action. Accompanying this review is a MATLAB package:

https://github.com/MColbrook/DMD-Multiverse

featuring user-friendly implementations and examples from the paper, most of which are new. We aim for readers to utilize this paper as a practical manual for various DMD methods. Although extensive, the review is structured modularly, enabling readers to selectively engage with DMD versions and topics that interest them most.

Differentiating itself from prior reviews, this review specifically focuses on the theory and practice of DMD algorithms for computing spectral properties of Koopman operators, complementing other reviews on the subject. Mezić (2013) and the more recent review of Schmid (2022) (see also Taira et al., 2017, 2020) focus on developments associated with applications in fluid dynamics. While the initial applications of Koopman and DMD techniques were in fluid problems, their utility has been demonstrated in a broader range of fields. We also briefly explore the applications of DMD in control theory, and readers seeking further exploration in this area are encouraged to read (Otto and Rowley, 2021). An excellent early review of "Applied Koopmanism" is presented by Budišić et al. (2012). More recently, Brunton et al. (2022) have provided a broad overview of Koopman operators and their applications, with connections to other fields.

This review is organized into several sections. In Section 2, we introduce the basic DMD algorithm, offering a concise introduction to Koopman operators and their spectral properties, before presenting the fundamental DMD algorithm and its two key interpretations: regression and projection. We discuss three canonical examples, followed by an examination of the goals and challenges of DMD. Section 3 focuses on variants from the regression viewpoint, including noise reduction, compression, randomized linear algebra, multiscale dynamics, and control. The connection with Koopman operators is further explored in Section 4, where we discuss nonlinear observables, time-delay embedding methods, and methods for controlling the infinite-dimensional projection error of DMD (e.g., to ensure convergence). In Section 5, we review recent methods that preserve the structures of dynamical systems. These methods often lead to greater noise resistance, improved generalization, and reduced data demands for training. We conclude in Section 6 by discussing further connections and open

TABLE 1 Executive summary of the DMD methods discussed in detail in this review. Numerous others are also discussed. The bold horizontal lines separate the different flavors of regression (top), Galerkin (middle), and structure-preserving (bottom). The fundamental DMD algorithm, exact DMD, is given in Algorithm 1. **NB:** For measure-preserving systems, discretizations that preserve the measure are crucial for convergence, recovering the correct dynamical behavior, stability, robustness to noise, and improved qualitative and long-time behavior.

DMD Method	Challenges Overcome	Key Insight/Development	Key Reference(s)
Forward-Backward DMD	Sensor noise bias.	Take geometric mean of forward and backward propagators for the data.	Dawson et al. (2016)
Total Least-Squares DMD	Sensor noise bias.	Replace least-squares problem with total least-squares problem.	Hemati et al. (2017) Dawson et al. (2016)
Optimized DMD Bagging Optimized DMD	Sensor noise bias. Optimal collective processing of snapshots.	Exponential fitting problem, solve using variable projection method. Statistical bagging sampling strategy.	Chen et al. (2012) Askham and Kutz (2018) Sashidhar and Kutz (2022)
Compressed Sensing	Computational efficiency. Temporal or spatial undersampling.	Unitary invariance of DMD extended to settings of compressed sensing (e.g., RIP, sparsity-promoting regularizers).	Tu et al. (2014a) S.L. Brunton et al. (2016b) Erichson et al. (2019a)
Randomized DMD	Computational efficiency. Memory usage.	Sketch data matrix for computations in reduced-dimensional space.	Erichson et al. (2019b)
Multiresolution DMD	Multiscale dynamics.	Filtered decomposition across scales.	Kutz et al. (2016b)
DMD with Control	Separation of unforced dynamics and actuation.	Generalized regression for globally linear control framework.	Proctor et al. (2016)

continued on next page

TABLE 1 *(continued)*

DMD Method	Challenges Overcome	Key Insight/Development	Key Reference(s)
Extended DMD	Nonlinear observables.	Arbitrary (nonlinear) dictionaries, recasting of DMD as a Galerkin method.	Williams et al. (2015a)
Hankel DMD	Delay-embedding for ergodic systems. Convergence under invariant subspace assumption.	Connection with Krylov subspace methods and Birkhoff's ergodic theorem.	Arbabi and Mezić (2017a)
HAVOK	Lack of closed linear models for chaotic systems.	Delay-embedding with chaos as forcing.	Brunton et al. (2017)
Residual DMD	Infinite-dimensional projection errors, verification (general systems). Computation of Koopman spectra (general systems). Spectral measures (measure-preserving systems).	Append EDMD with additional matrix (available from the snapshot data) to compute infinite-dimensional residuals and overcome the nonconvergence of EDMD.	Colbrook and Townsend (2023) Colbrook et al. (2023a)
Physics-Informed DMD	Preserving structure of dynamical systems. Numerous instances given in general framework.	Restrict the least-squares optimization to lie on a matrix manifold.	Baddoo et al. (2023)
Measure-Preserving Extended DMD	Measure-preserving discretizations of system. Convergence to Koopman spectral properties (including continuous spectra/spectral measures).	Alter EDMD to be measure-preserving with respect to a learned inner product. (via a polar decomposition of EDMD)	Colbrook (2023)
Compactification Methods	Continuous-time generator of measure-preserving system. Convergence to Koopman spectral properties (including continuous spectra/spectral measures). Conditioning of dictionary.	Compactification of generator or its resolvent using kernel integral operators, dictionary of kernel eigenvectors.	Das et al. (2021) Valva and Giannakis (2023)

problems. I hope the reader enjoys this tour of the DMD multiverse as much as I have enjoyed writing it!

Due to the sheer breadth and thousands of papers written on DMD, it is impossible for this review to cover every version of DMD in great detail. Significant DMD algorithms that are not discussed in their own sections are still discussed in some detail. If the reader searches this paper, they will find dozens of DMD algorithms. I have included all significant references I am aware of, but many others may not have been included. I apologize for that in advance and encourage all readers to inform the author about results that deserve more discussion.

2 The basics of DMD

To understand the DMD "multiverse", we must first study the basic DMD algorithm. We begin with Koopman operator theory, the theoretical underpinning of DMD, before moving on to the fundamental DMD algorithm and two important viewpoints. We then provide three canonical examples and discuss the goals and challenges of DMD.

2.1 The underlying theory: Koopman operators and spectra

In this section, we recall the definition of Koopman operators and equip the reader with a crash course on their relevant spectral properties. At its core, DMD is an algorithm that uses the snapshot data in (1.2) to approximate the spectral properties of Koopman operators.

2.1.1 *What is a Koopman operator?*

To define a Koopman operator, we begin with a space $\mathcal{F}$ of functions $g : \Omega \to \mathbb{C}$, where Ω is the state space of our dynamical system. The functions g, referred to as *observables*, serve as tools for indirectly measuring the state of the system described in (1.1). Specifically, $g(\mathbf{x}_n)$ indirectly measures the state $\mathbf{x}_n$. Koopman operators enable us to capture the time evolution of these observables through a linear operator framework. For a suitable domain $\mathcal{D}(\mathcal{K}) \subset \mathcal{F}$, we define the Koopman operator via the composition formula:

$$[\mathcal{K}g](\mathbf{x}) = [g \circ \mathbf{F}](\mathbf{x}) = g(\mathbf{F}(\mathbf{x})), \qquad g \in \mathcal{D}(\mathcal{K}). \tag{2.1}$$

In this context, $[\mathcal{K}g](\mathbf{x}_n) = g(\mathbf{F}(\mathbf{x}_n)) = g(\mathbf{x}_{n+1})$ represents the measurement of the state one time step ahead of $g(\mathbf{x}_n)$. This process effectively captures the dynamic progression of the system. The overarching concept is summarized in Fig. 1.

The key property of the Koopman operator $\mathcal{K}$ is its *linearity*. This linearity holds irrespective of whether the system's dynamics, as represented in (1.1), are linear or nonlinear. Consequently, the spectral properties of $\mathcal{K}$ become a powerful tool in analyzing the dynamical system's behavior. To study spectra, we

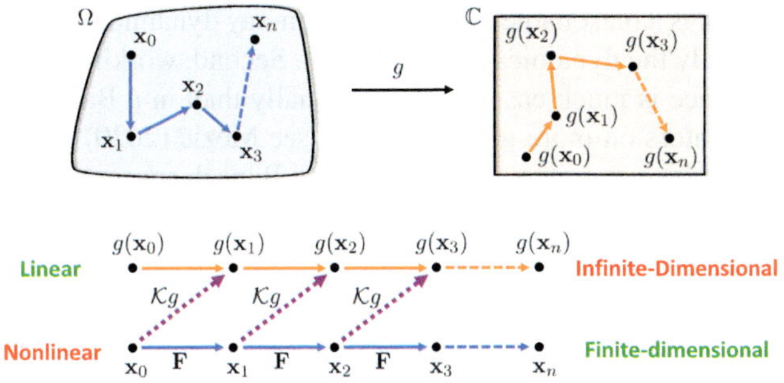

FIGURE 1 Summary of the idea of Koopman operators. By lifting to a space of observables, we trade a nonlinear finite-dimensional system for a linear infinite-dimensional system.

assume that $\mathcal{F}$ is a Banach space.[1] For the spectrum of $\mathcal{K}$ to be meaningful and nontrivial, we assume that its domain, $\mathcal{D}(\mathcal{K})$, is dense in $\mathcal{F}$ and that $\mathcal{K}$ itself is a closed operator.[2] If these conditions are not met, the spectrum would encompass the entirety of $\mathbb{C}$. It is crucial to recognize that the Koopman operator is not uniquely defined by the dynamical system in (1.1); rather, it is fundamentally dependent on the choice of the space of observables $\mathcal{F}$. In this review, we focus on cases where $\mathcal{F}$ is defined as the following Hilbert space:

$$\mathcal{F} = L^2(\Omega, \omega) \quad \text{with inner product} \quad \langle g_1, g_2 \rangle = \int_\Omega g_1(\mathbf{x})\overline{g_2(\mathbf{x})}\, \mathrm{d}\omega(\mathbf{x}) \quad \text{and}$$
$$\text{norm } \|g\| = \sqrt{\langle g, g \rangle},$$

for some positive measure ω.[3] In going from a pointwise definition in (2.1) to the space $L^2(\Omega, \omega)$, a little care is needed since $L^2(\Omega, \omega)$ consists of equivalence classes of functions. We assume that the map $\mathbf{F}$ is nonsingular with respect to ω, meaning that

$$\omega(E) = 0 \quad \text{implies that} \quad \omega(\{\mathbf{x} : \mathbf{F}(\mathbf{x}) \in E\}) = 0.$$

This ensures that the Koopman operator is well-defined since $g_1(\mathbf{x}) = g_2(\mathbf{x})$ for ω-almost every $\mathbf{x}$ implies that $g_1(\mathbf{F}(\mathbf{x})) = g_2(\mathbf{F}(\mathbf{x}))$ for ω-almost every $\mathbf{x}$. The above Hilbert space setting is standard in most of the Koopman literature for two

---

1 A Banach space is a normed vector space that is complete, i.e., every Cauchy sequence converges. Thus, a Banach space has no 'holes' in it. We have deliberately kept the background functional analysis to a minimum in this review.

2 An operator being 'closed' means that its graph $\{(g, \mathcal{K}g) : g \in \mathcal{D}(\mathcal{K})\}$ is a closed subset within the product space $\mathcal{F} \times \mathcal{F}$.

3 We do not assume that this measure is invariant. For Hamiltonian systems, a common choice of ω is the standard Lebesgue measure, for which the Koopman operator is unitary on $L^2(\Omega, \omega)$. For other systems, we can select ω according to the region where we wish to study the dynamics, such as a Gaussian measure. In many applications, ω corresponds to an unknown ergodic physical measure on an attractor.

reasons. First, it is a reasonable assumption for many dynamical systems, particularly if we study the dynamics on an attractor. Second, working with operators in a Hilbert space is much easier computationally than in a Banach space. For Koopman operators on more general spaces, see Mezić (2020). Practical algorithms for Koopman operators on more general Banach spaces remain a largely open problem (see Section 6.4).

Since $\mathcal{K}$ acts on an *infinite-dimensional* function space, we have exchanged the nonlinearity in (1.1) for an infinite-dimensional linear system. This means that the spectral properties of $\mathcal{K}$ can be significantly more complex than those of a finite matrix, making them more challenging to compute. While this might seem disheartening, as we will explore in Section 4, methodologies exist that enable the analysis of infinite-dimensional spectral properties through a series of finite-dimensional approximations.

2.1.2 Crash course on spectral properties of Koopman operators

We will now review the relevant spectral properties of $\mathcal{K}$. Readers primarily interested in applying DMD algorithms will still find the dynamical interpretations of these properties insightful. The sole assumption made throughout this paper is that $\mathcal{K}$ is a closed and densely defined operator. Specifically, unless stated otherwise, we do not presuppose that $\mathcal{K}$ possesses a nontrivial[4] finite-dimensional invariant subspace, nor do we assume it has an eigenvector basis. These two assumptions are often implicitly (and sometimes wrongly) assumed in DMD papers and can lead to confusion if care is not taken.

Koopman spectra

If $g \in L^2(\Omega, \omega)$ is an *eigenfunction* of $\mathcal{K}$ with *eigenvalue* λ, then g exhibits perfect coherence[5] with

$$g(\mathbf{x}_n) = [\mathcal{K}^n g](\mathbf{x}_0) = \lambda^n g(\mathbf{x}_0) \quad \forall n \in \mathbb{N}. \tag{2.2}$$

The oscillation and decay/growth of the observable g are dictated by the complex argument and absolute value of the eigenvalue λ, respectively. In infinite dimensions, the appropriate generalization of the set of eigenvalues of $\mathcal{K}$ is the *spectrum*, denoted by $\mathrm{Sp}(\mathcal{K})$, and defined as

$$\mathrm{Sp}(\mathcal{K}) = \left\{ z \in \mathbb{C} : (\mathcal{K} - zI)^{-1} \text{ does not exist as a bounded operator} \right\} \subset \mathbb{C}.$$

Here, I denotes the identity operator. The spectrum $\mathrm{Sp}(\mathcal{K})$ includes the set of eigenvalues of $\mathcal{K}$, but in general, $\mathrm{Sp}(\mathcal{K})$ contains points that are not eigenvalues.

---

[4] If the measure ω is finite, then the constant function $g(\mathbf{x}) = 1$ is a trivial eigenfunction with eigenvalue 1. It is deemed trivial because the dynamics of a constant observable lack useful information.

[5] In the setting of dynamical systems, coherent sets or structures are subsets of the phase space where elements (e.g., particles, agents, etc.) exhibit similar behavior over some time interval. This behavior remains relatively consistent despite potential perturbations or the chaotic nature of the system.

This is because there are more ways for $(\mathcal{K}-zI)^{-1}$ to not exist in infinite dimensions than in finite dimensions. For example, we may have continuous spectra. The standard Lorenz system on the Lorenz attractor gives rise to a Koopman operator that has no nontrivial eigenvalues, yet the spectrum is the whole unit circle!

In general, we cannot numerically approximate an eigenfunction perfectly. Moreover, the operator $\mathcal{K}$ may not have any nontrivial eigenfunctions, for instance, if the system is weakly mixing. Instead, the so-called *approximate point spectrum* is the following subset of $\mathrm{Sp}(\mathcal{K})$:

$$\mathrm{Sp}_{\mathrm{ap}}(\mathcal{K}) = \Big\{\lambda \in \mathbb{C} : \exists \{g_n\}_{n\in\mathbb{N}} \subset L^2(\Omega,\omega) \\ \text{such that } \|g_n\| = 1,\ \lim_{n\to\infty} \|(\mathcal{K}-\lambda I)g_n\| = 0\Big\} \subset \mathbb{C}.$$

An observable g with $\|g\| = 1$ and $\|(\mathcal{K}-\lambda I)g\| \leq \epsilon$ for $\lambda \in \mathbb{C}$ is known as ϵ-pseudoeigenfunction. Such observables are important for the dynamical system (1.1) since

$$\|\mathcal{K}^n g - \lambda^n g\| = \mathcal{O}(n\epsilon) \quad \forall n \in \mathbb{N}.$$

In other words, λ describes an approximate coherent oscillation and decay/growth of the observable g with time. The pseudoeigenfunctions and $\mathrm{Sp}_{\mathrm{ap}}(\mathcal{K})$ encode information about the underlying dynamical system (Mezić, 2021). For example, the level sets of certain eigenfunctions determine the invariant manifolds (Mezić, 2015) and isostables (Mauroy et al., 2013), and the global stability of equilibria (Mauroy and Mezić, 2016) and ergodic partitions (Budišić et al., 2012; Mezić and Wiggins, 1999) can be characterized by pseudoeigenfunctions and $\mathrm{Sp}_{\mathrm{ap}}(\mathcal{K})$.

Koopman pseudospectra

Approximate point spectra and pseudoeigenfunctions are related to the notion of *pseudospectra* (Trefethen and Embree, 2005). For a finite matrix $A \in \mathbb{C}^{n\times n}$ and $\epsilon > 0$, the ϵ-pseudospectrum of A is the set[6]

$$\mathrm{Sp}_\epsilon(A) = \Big\{\lambda \in \mathbb{C} : \|(A-\lambda I)^{-1}\| \geq 1/\epsilon\Big\} = \bigcup_{B\in\mathbb{C}^{n\times n},\, \|B\|\leq\epsilon} \mathrm{Sp}(A+B).$$

The ϵ-pseudospectra of A are regions in the complex plane enclosing the eigenvalues of A. These regions tell us how far an ϵ-sized perturbation can perturb an eigenvalue. Pseudospectra of Koopman operators must be defined with some care because $\mathcal{K}$ may be an unbounded operator and hence the resolvent norm $\|(\mathcal{K}-\lambda I)^{-1}\|$ can be constant on open subsets of $\mathbb{C}\backslash\mathrm{Sp}(\mathcal{K})$ (Shargorodsky,

6 Some authors use a strict inequality in the definition of ϵ-pseudospectra. We prefer the given definition since then the pseudospectrum is a *closed* subset of $\mathbb{C}$.

2008). We define the ϵ-pseudospectrum of $\mathcal{K}$ as (Roch and Silbermann, 1996, Prop. 4.15)[7]:

$$\mathrm{Sp}_\epsilon(\mathcal{K}) = \mathrm{Cl}\left(\{\lambda \in \mathbb{C} : \|(\mathcal{K} - \lambda I)^{-1}\| > 1/\epsilon\}\right) = \mathrm{Cl}\left(\bigcup_{\|\mathcal{B}\|<\epsilon} \mathrm{Sp}(\mathcal{K} + \mathcal{B})\right), \tag{2.3}$$

where Cl denotes the closure of a set. To see the connection with $\mathrm{Sp}_{\mathrm{ap}}(\mathcal{K})$, note that if $\|(\mathcal{K} - \lambda I)g\| \le \epsilon$ for an observable g with $\|g\| = 1$, then $\|(\mathcal{K} - \lambda I)^{-1}\| \ge 1/\epsilon$. We care about pseudospectra for several reasons, but two stand out as the most important:

- Pseudospectra allow us to determine which regions of computed spectra are accurate and trustworthy. This could be in terms of the numerical stability, but also pseudospectra aid in detecting so-called *spectral pollution* (see Figs. 7 and 9). These are spurious eigenvalues arising from discretization that are unrelated to the underlying Koopman operator. The term spectral pollution refers to the accumulation of these spurious eigenvalues at points outside the spectrum of $\mathcal{K}$ as the discretization size increases (Lewin and Séré, 2010). This occurs even when $\mathcal{K}$ is a normal operator (see Fig. 8). It is essential to realize that spectral pollution leads to spurious modes that are not linked to stability issues but are a consequence of discretizing the infinite-dimensional operator $\mathcal{K}$ to a finite matrix.
- If the Koopman operator is nonnormal, the system's transient behavior can differ significantly from the asymptotic behavior captured by $\mathrm{Sp}(\mathcal{K})$. Pseudospectra can be employed to detect and quantify transients not represented by the spectrum (Trefethen et al., 1993) (Trefethen and Embree, 2005, Section IV).

Pseudospectra also provide a means of computing spectra since $\lim_{\epsilon \downarrow 0} \mathrm{Sp}_\epsilon(\mathcal{K}) = \mathrm{Sp}(\mathcal{K})$. This convergence occurs in the so-called Attouch–Wets metric space (Beer, 1993), which roughly means that we obtain uniform convergence on any compact region of $\mathbb{C}$. This observation goes beyond Koopman operators and has been behind some recent breakthroughs in the computation of spectra in infinite dimensions (Ben-Artzi et al., 2020; Colbrook, 2020, 2022; Colbrook and Hansen, 2022; Colbrook et al., 2019).

Koopman mode decompositions and spectral theorems beyond eigenvalues

One of the most useful features of Koopman operators is the *Koopman Mode Decomposition* (KMD) (Mezić, 2005). The KMD expresses the state $\mathbf{x}$ or an observable $g(\mathbf{x})$ as a linear combination of dominant coherent structures. It can be considered a diagonalization of the Koopman operator. As a result, the KMD

[7] While Roch and Silbermann (1996, Prop. 4.15) consider bounded operators, it can be adjusted to cover unbounded operators (Trefethen and Embree, 2005, Thm. 4.3).

is invaluable for tasks such as dimensionality and model reduction. It generalizes the space-time separation of variables typically achieved through the Fourier transform or singular value decomposition (SVD). It is crucial to realize that an exact KMD is rigorously justified only if $\mathcal{K}$ possesses some form of spectral theorem, which extends the concept of diagonalization to infinite dimensions. Nevertheless, obtaining an *approximate* KMD is still possible even without a spectral theorem.

For example, suppose that the system (1.1) is *measure-preserving* with respect to the positive measure ω. This means that $\omega(E) = \omega(\{\mathbf{x} : \mathbf{F}(\mathbf{x}) \in E\})$ for any measurable set $E \subset \Omega$. In other words, the dynamical system preserves a volume. Measure-preserving systems encompass many systems of interest such as Hamiltonian flows (Arnold, 1989), geodesic flows (Dubrovin et al., 1984), Bernoulli schemes (Shields, 1973), physical systems in equilibrium (Hill, 1986), and ergodic systems (Walters, 2000). Furthermore, many dynamical systems either admit invariant measures (Kryloff and Bogoliouboff, 1937) or exhibit measure-preserving post-transient behavior (Mezić, 2005). In fact, if Ω is a compact metric space and $\mathbf{F}$ is continuous, then there is an invariant measure Mañé (1987, Prop. 8.1).[8] For a measure-preserving system, the Koopman operator $\mathcal{K}$ is an isometry, i.e., $\|\mathcal{K}g\| = \|g\|$ for all observables $g \in \mathcal{D}(\mathcal{K}) = L^2(\Omega, \omega)$. For simplicity, we further assume that $\mathcal{K}$ is unitary, implying that it is normal (it commutes with its adjoint).[9]

Under these conditions, the spectral theorem (Conway, 2007, Thm. X.4.11) allows us to diagonalize the Koopman operator $\mathcal{K}$. There is a *projection-valued measure* $\mathcal{E}$ supported on $\mathrm{Sp}(\mathcal{K})$. For readers unfamiliar with the spectral theorem, Halmos (1963) provides an excellent and readable introduction. In our example, $\mathcal{K}$ is unitary, which implies that $\mathrm{Sp}(\mathcal{K})$ lies within the unit circle $\mathbb{T}$. The measure $\mathcal{E}$ associates an orthogonal projector with each Borel measurable subset of $\mathbb{T}$. For such a subset $S \subset \mathbb{T}$, $\mathcal{E}(S)$ is a projection onto the spectral elements of $\mathcal{K}$ inside S. For any observable $g \in L^2(\Omega, \omega)$,

$$g = \left(\int_{\mathbb{T}} \mathrm{d}\mathcal{E}(\lambda)\right) g \qquad \text{and} \qquad \mathcal{K}g = \left(\int_{\mathbb{T}} \lambda \, \mathrm{d}\mathcal{E}(\lambda)\right) g.$$

The essence of this formula is the decomposition of g according to the spectral content of $\mathcal{K}$. The projection-valued measure $\mathcal{E}$ simultaneously decomposes the space $L^2(\Omega, \omega)$ and diagonalizes the Koopman operator. For example, we have

$$g(\mathbf{x}_n) = [\mathcal{K}^n g](\mathbf{x}_0) = \left[\left(\int_{\mathbb{T}} \lambda^n \, \mathrm{d}\mathcal{E}(\lambda)\right) g\right](\mathbf{x}_0). \tag{2.4}$$

---

[8] Of course, whether or not this is useful or whether our chosen ω is invariant is another matter.

[9] A Koopman operator that is an isometry need not be unitary, e.g., the Koopman operator associated with the tent map. However, an isometry can always be extended to a unitary operator, and the spectral measures associated with forward-time dynamics are independent of the chosen extension (Colbrook, 2023).

This directly extends (2.2). The spectral theorem can be perceived as offering a custom Fourier-type transform specifically for the operator $\mathcal{K}$ that extracts coherent features. Of particular interest are *scalar-valued* spectral measures. Given a normalized observable $g \in L^2(\Omega, \omega)$ with $\|g\| = 1$, the scalar-valued spectral measure of $\mathcal{K}$ with respect to g is a probability measure defined as

$$\mu_g(S) = \langle \mathcal{E}(S)g, g \rangle.$$

These measures can be further refined using Lebesgue's decomposition into a pure point part, supported on the eigenvalues of $\mathcal{K}$, and a continuous part. The continuous part can further be decomposed into an absolutely continuous part with a density function and a singular continuous part. The moments of the measure μ_g are the correlations

$$\langle \mathcal{K}^n g, g \rangle = \int_{\mathbb{T}} \lambda^n \, \mathrm{d}\mu_g(\lambda), \quad n \in \mathbb{Z}.$$

For example, if our system corresponds to the dynamics on an attractor, these statistical properties allow comparison of complex dynamics (Mezić and Banaszuk, 2004). More generally, the spectral measure of $\mathcal{K}$ with respect to $g \in L^2(\Omega, \omega)$ is a signature for the forward-time dynamics of (1.1).

Going one step further, $\mathcal{E}$ leads to a decomposition of $L^2(\Omega, \omega)$ into parts associated with quasiperiodic evolution and weak-mixing dynamics. Namely, we have the following orthogonal decomposition into two $\mathcal{K}$-invariant subspaces (Halmos, 2017)

$$L^2(\Omega, \omega) = \mathcal{H}_{\mathrm{pp}} \oplus \mathcal{H}_{\mathrm{c}}.$$

Here, the subspace $\mathcal{H}_{\mathrm{pp}}$ consists of the closure of the linear span of eigenvectors and admits an orthonormal basis of eigenvectors $\{\phi_j\}$ of $\mathcal{K}$ with eigenvalues $\{\lambda_j\}$. This means that we can write

$$\mathcal{K}^n g = \sum_j \lambda_j^n \langle g, \phi_j \rangle \phi_j \quad \forall g \in \mathcal{H}_{\mathrm{pp}}, n \in \mathbb{N}. \tag{2.5}$$

The spectrum of $\mathcal{K} \restriction_{\mathcal{H}_{\mathrm{pp}}}$ need not be a discrete subset of $\mathbb{T}$. For example, an ergodic rotation on the circle has eigenvalues that densely fill $\mathbb{T}$. In contrast to (2.5), observables in the continuous part $\mathcal{H}_{\mathrm{c}}$ exhibit a decay of correlations that is typical of chaotic systems. Namely, for any $\epsilon > 0$ (Katznelson, 2004, p.45),

$$\lim_{n\to\infty} \frac{1}{n} \sum_{j=1}^{n} \left| \langle \mathcal{K}^j g, f \rangle \right|^{\epsilon} = 0 \quad \forall g \in \mathcal{H}_{\mathrm{c}}, f \in L^2(\Omega, \omega).$$

This result says that $|\langle \mathcal{K}^j g, f \rangle|$ converges to zero in density, that is, for any $\delta > 0$, the proportion in all sufficiently large intervals of integers j such that $\left|\langle \mathcal{K}^j g, f \rangle\right| > \delta$ is arbitrarily small.

The above dichotomy is an example of how the decomposition of $\mathcal{E}$ into atomic and continuous parts often characterizes a dynamical system. For example, suppose that **F** is measure-preserving and bijective, and ω is a probability measure. Then, the dynamical system is (Halmos, 2017)

- **Ergodic** if and only if $\lambda = 1$ is a simple eigenvalue of $\mathcal{K}$,
- **Weakly mixing** if and only if $\lambda = 1$ is a simple eigenvalue of $\mathcal{K}$ and there are no other eigenvalues,
- **Mixing** if $\lambda = 1$ is a simple eigenvalue of $\mathcal{K}$, and $\mathcal{K}$ has absolutely continuous spectrum on $\text{span}\{1\}^{\perp}$.

Different spectral types find interpretations across various applications, including fluid mechanics (Mezić, 2013), anomalous transport (Zaslavsky, 2002), and the analysis of invariants/exponents related to trajectories (Kantz and Schreiber, 2006). The approximation of $\mathcal{E}$ is critical in many applications. For example, the approximate spectral projections provide reduced-order models (Mezić and Banaszuk, 2004; Mezić, 2005). The computation of spectral measures is discussed in Section 6.2.

2.2 The fundamental DMD algorithm

With the definition of a Koopman operator in hand, we can now present the fundamental DMD algorithm and two interpretations. The first interpretation of DMD is as a linear regression. The second is as a projection method. Both interpretations are instrumental, and understanding their interplay is often key to unlocking the power of DMD.

2.2.1 The linear regression interpretation

The simplest and historically first interpretation of DMD is as a linear regression. Given the snapshot matrices $\mathbf{X}, \mathbf{Y} \in \mathbb{C}^{d \times M}$ in (1.2), we seek a matrix $\mathbf{K}_{\text{DMD}}$ such that $\mathbf{Y} \approx \mathbf{K}_{\text{DMD}}\mathbf{X}$. We can think of this as constructing a *linear and approximate* dynamical system. To find a suitable matrix $\mathbf{K}_{\text{DMD}}$, we consider the minimization problem

$$\min_{\mathbf{K}_{\text{DMD}} \in \mathbb{C}^{d \times d}} \|\mathbf{Y} - \mathbf{K}_{\text{DMD}}\mathbf{X}\|_{\text{F}}, \tag{2.6}$$

where $\|\cdot\|_{\text{F}}$ denotes the Frobenius norm. Similar optimization problems will be at the heart of the various DMD-type algorithms we consider in this review. A solution to the problem in (2.6) is

$$\mathbf{K}_{\text{DMD}} = \mathbf{Y}\mathbf{X}^{\dagger} \in \mathbb{C}^{d \times d},$$

where $\dagger$ denotes the Moore–Penrose pseudoinverse. Often, the matrices **X** and **Y** are tall and skinny, meaning that $d \gg M$. In this scenario, we typically first project onto a low-dimensional subspace to reconstruct the leading nonzero eigenvalues and eigenvectors of the matrix $\mathbf{K}_{\text{DMD}}$ without explicitly computing it. The standard DMD algorithm does this using an SVD and is summarized

Algorithm 1 The exact DMD algorithm (Tu et al., 2014b), which has become the workhorse DMD algorithm.

Input: Snapshot data $\mathbf{X} \in \mathbb{C}^{d\times M}$ and $\mathbf{Y} \in \mathbb{C}^{d\times M}$, rank $r \in \mathbb{N}$.

1: Compute a truncated SVD of the data matrix $\mathbf{X} \approx \mathbf{U\Sigma V}^*$, $\mathbf{U} \in \mathbb{C}^{d\times r}$, $\mathbf{\Sigma} \in \mathbb{R}^{r\times r}$, $\mathbf{V} \in \mathbb{C}^{M\times r}$. The columns of $\mathbf{U}$ and $\mathbf{V}$ are orthonormal and $\mathbf{\Sigma}$ is diagonal.

2: Compute the compression $\tilde{\mathbf{K}}_{\mathrm{DMD}} = \mathbf{U}^*\mathbf{YV\Sigma}^{-1} \in \mathbb{C}^{r\times r}$.

3: Compute the eigendecomposition $\tilde{\mathbf{K}}_{\mathrm{DMD}}\mathbf{W} = \mathbf{W\Lambda}$.
The columns of $\mathbf{W}$ are eigenvectors and $\mathbf{\Lambda}$ is a diagonal matrix of eigenvalues.

4: Compute the modes $\mathbf{\Phi} = \mathbf{YV\Sigma}^{-1}\mathbf{W}$.

Output: The eigenvalues $\mathbf{\Lambda}$ and modes $\mathbf{\Phi} \in \mathbb{C}^{d\times r}$.

in Algorithm 1, where we have assumed that the projected DMD matrix is diagonalizable.[10] Algorithm 1 is known as *exact DMD* (Tu et al., 2014b) and often the modes are further scaled by Λ^{-1}. There are several remarks about this algorithm that are worth mentioning:

- The rank r is usually chosen based on the decay of singular values of $\mathbf{X}$. If low-dimensional structure is present in the data (Udell and Townsend, 2019), the singular values decrease rapidly, and small r captures the dominant modes. Moreover, the lowest energy modes may be corrupted by noise, and low-dimensional projection is a form of spectral filtering which has the positive effect of dampening the influence of noise (Hansen et al., 2006).[11] The question of how best to truncate is difficult to answer and is often performed heuristically. If the measurement error is additive white noise, there are algorithmic choices (Gavish and Donoho, 2014). In the context of Koopman operators, r is equivalent to the size of the space spanned by basis functions, and a good choice depends on the chosen observables. For example, we shall see below that Algorithm 1 corresponds to a linear set of basis functions, which may not capture the relevant nonlinear dynamics. Hence, a larger r may be suitable for other basis choices. Often, the choice of r is modest, meaning that randomized methods (Halko et al., 2011) for computing the SVD can significantly reduce the computational cost. We will explore this and other compression methods in Section 3.2.

[10] We make this assumption about various matrices throughout. Mathematically, a Jordan decomposition may be substituted for an eigendecomposition, and the modes corresponding to a single Jordan block can be considered as interacting modes. However, computing a Jordan block should be avoided. A stable alternative is a Schur decomposition that provides an orthogonal set of interacting modes (in sharp contrast with what is typically considered a DMD mode) or a block-diagonal Schur decomposition with nearly confluent eigenvalues grouped together.

[11] Low-energy modes can be important though, for example, in optimized control (Rowley, 2005; Rowley et al., 2006).

- We can interpret the algorithm as constructing a linear model of the dynamical system on projected coordinates $\tilde{\mathbf{x}} = \mathbf{U}^*\mathbf{x}$. Namely, $\tilde{\mathbf{x}}_{n+1} \approx \tilde{\mathbf{K}}_{\text{DMD}}\tilde{\mathbf{x}}_n$. The left singular vectors $\mathbf{U}$ are known as proper orthogonal decomposition (POD) modes (Berkooz et al., 1993).
- If the SVD is exact, so that $\mathbf{X} = \mathbf{U}\boldsymbol{\Sigma}\mathbf{V}^*$, then

$$\mathbf{K}_{\text{DMD}} = \mathbf{Y}\mathbf{V}\boldsymbol{\Sigma}^{-1}\mathbf{U}^*.$$

  Using this relation, we have

$$\mathbf{K}_{\text{DMD}}[\mathbf{Y}\mathbf{V}\boldsymbol{\Sigma}^{-1}\mathbf{W}] = \mathbf{Y}\mathbf{V}\boldsymbol{\Sigma}^{-1}\underbrace{\mathbf{U}^*\mathbf{Y}\mathbf{V}\boldsymbol{\Sigma}^{-1}}_{\tilde{\mathbf{K}}_{\text{DMD}}}\mathbf{W} = [\mathbf{Y}\mathbf{V}\boldsymbol{\Sigma}^{-1}\mathbf{W}]\boldsymbol{\Lambda},$$

  and hence Algorithm 1 computes exact eigenvalues and eigenvectors of $\mathbf{K}_{\text{DMD}}$. Moreover, one can show that this process identifies all of the nonzero eigenvalues of $\mathbf{K}_{\text{DMD}}$ (Tu et al., 2014b, Thm. 1). It is common to call $\mathbf{Y}\mathbf{V}\boldsymbol{\Sigma}^{-1}\mathbf{W}$ *exact modes* and $\mathbf{U}\mathbf{W}$ *projected modes*.
- Originally, DMD was developed in connection with Krylov subspaces and the Arnoldi algorithm. In this version, it is assumed that data is gathered along a single trajectory. The SVD version, on the other hand, is capable of handling more general trajectory data. Strategies for using this flexibility to reduce computational cost and average snapshot data noise are given in Tu et al. (2014b). The SVD version is also more numerically stable. Drmač has carefully analyzed the stability of DMD (Drmač et al., 2018, 2019; Drmač, 2020; Drmač et al., 2020).
- Centering the data before applying DMD can be helpful if the mean-subtracted data have linearly dependent columns, especially if the dynamics are perturbations about an equilibrium (Hirsh et al., 2020). This is equivalent to including an affine term in the linear regression. However, computing the DMD of centered data can have undesirable consequences (Chen et al., 2012).

The above interpretation of DMD is simple and intuitive. However, using DMD to analyze nonlinear dynamics globally seems dubious, as there is an underlying assumption of approximately linear dynamics in (2.6). Nevertheless, we shall now demonstrate that DMD can be interpreted as an approximation to Koopman spectral analysis. This provides a solid theoretical foundation for applying DMD in analyzing nonlinear dynamics, which will be further elaborated upon in Section 4.

2.2.2 The Galerkin interpretation: connecting to Koopman operators

The connection between Algorithm 1 and Koopman operators is revealed once we interpret DMD as a Galerkin method. Consider the two correlation matrices

$$\mathbf{G} = \frac{1}{M}\overline{\mathbf{U}^*\mathbf{X}}(\mathbf{U}^*\mathbf{X})^\top, \quad \mathbf{A} = \frac{1}{M}\overline{\mathbf{U}^*\mathbf{X}}(\mathbf{U}^*\mathbf{Y})^\top.$$

We can think of the jth row of the POD matrix $\mathbf{U}^*\mathbf{X}$ as an affine function u_j on the state space Ω evaluated at the snapshot data:

$$u_j(\mathbf{x}) = [\mathbf{U}_{:,j}]^*\mathbf{x}, \quad u_j(\mathbf{x}^{(m)}) = [\mathbf{U}^*\mathbf{X}]_{jm}.$$

It follows that $\mathbf{G}$ can be interpreted as a Gram matrix with respect to the positive semidefinite Hermitian form induced by the probability measure with equal point masses at the $\{\mathbf{x}^{(m)}\}$. Namely,

$$\mathbf{G}_{jk} = \frac{1}{M}\sum_{m=1}^{M} \overline{u_j(\mathbf{x}^{(m)})} u_k(\mathbf{x}^{(m)}) = \int_\Omega \overline{u_j(\mathbf{x})} u_k(\mathbf{x})\, \mathrm{d}\omega_M(\mathbf{x}),$$
$$\omega_M = \frac{1}{M}\sum_{m=1}^{M} \delta_{\mathbf{x}^{(m)}}.$$

Writing $\langle\cdot,\cdot\rangle_M$ for the form induced by ω_M, we can argue similarly for $\mathbf{A}$ and succinctly write

$$\mathbf{G}_{jk} = \langle u_k, u_j\rangle_M,$$
$$\mathbf{A}_{jk} = \frac{1}{M}\sum_{m=1}^{M} \overline{u_j(\mathbf{x}^{(m)})} u_k(\mathbf{y}^{(m)}) = \int_\Omega \overline{u_j(\mathbf{x})} u_k(\mathbf{F}(\mathbf{x}))\, \mathrm{d}\omega_M(\mathbf{x}) = \langle \mathcal{K}u_k, u_j\rangle_M.$$

Assuming that the matrix $\mathbf{U}^*\mathbf{X}$ is of rank r, and using $\mathbf{U}^*\mathbf{X} = \mathbf{\Sigma}\mathbf{V}^*$, we can write

$$\tilde{\mathbf{K}}_{\mathrm{DMD}}^{\top} = \mathbf{\Sigma}^{-1}\mathbf{V}^{\top}\mathbf{Y}^{\top}\overline{\mathbf{U}} = (\mathbf{X}^{\top}\overline{\mathbf{U}})^{\dagger}\mathbf{Y}^{\top}\overline{\mathbf{U}} = \mathbf{G}^{-1}\mathbf{A}.$$

The matrix $\mathbf{G}^{-1}\mathbf{A}$ is an approximation of the action of $\mathcal{K}$ on the subspace spanned by the functions $\{u_j\}_{j=1}^{r}$. Namely, if g is an observable that can be expressed as the linear combination

$$g(\mathbf{x}) = \sum_{j=1}^{r} u_j(\mathbf{x})\mathbf{g}_j, \quad \text{for some} \quad \mathbf{g} \in \mathbb{C}^r,$$

then

$$[\mathcal{K}g](\mathbf{x}) \approx \sum_{j=1}^{r} u_j(\mathbf{x})(\mathbf{G}^{-1}\mathbf{A}\mathbf{g})_j = \sum_{j=1}^{r} u_j(\mathbf{x})(\tilde{\mathbf{K}}_{\mathrm{DMD}}^{\top}\mathbf{g})_j.$$

In other words, $\tilde{\mathbf{K}}_{\mathrm{DMD}}^{\top}$ is a matrix that approximates the action of the Koopman operator on expansion coefficients. More precisely, it is a Galerkin method corresponding to $\mathcal{K}$ and the form $\langle\cdot,\cdot\rangle_M$. This connection is explored more deeply in Section 4.1.

If $\langle\cdot,\cdot\rangle_M$ converges to $\langle\cdot,\cdot\rangle$ in the large data limit $M \to \infty$, then DMD can be considered to be a numerical approximation to Koopman spectral analysis. The terms *DMD mode* and *Koopman mode* are often used interchangeably in the literature. It is important to note that the Koopman modes and eigenfunctions are distinct mathematical objects, requiring different approaches for approximation. The right eigenvectors of $\tilde{\mathbf{K}}_{\text{DMD}}$ give rise to time-invariant directions in the state space $\mathbf{x}$, whereas the left-eigenvectors give rise to Koopman eigenfunctions, which are similarly time-invariant directions in the space of observables.

2.2.3 The Koopman mode decomposition

We can now connect DMD with the spectral expansions discussed in Section 2.1.2. First, we approximate an initial condition $\mathbf{x}_0$ in the eigenvector coordinates via

$$\mathbf{x}_0 \approx \mathbf{\Phi}\mathbf{b}, \quad \mathbf{b} = \mathbf{\Phi}^{\dagger}\mathbf{x}_0.$$

This is not the only choice, but it is the simplest. The KMD then provides an approximation of the dynamics by

$$\mathbf{x}_n \approx \mathbf{K}_{\text{DMD}}^n \mathbf{x}_0 \approx \mathbf{K}_{\text{DMD}}^n \mathbf{\Phi}\mathbf{b} = \mathbf{\Phi}\mathbf{\Lambda}^n \mathbf{b}, \tag{2.7}$$

which echoes (2.4). Since zero eigenvalues do not contribute to the dynamics, this decomposition further justifies the compression in Algorithm 1. The KMD has also been related to other decompositions in various situations, particularly those that have arisen in the fluid dynamics community (Taira et al., 2017, 2020). These include POD (Towne et al., 2018), optimal mode decomposition (Wynn et al., 2013), and resolvent analysis (Sharma et al., 2016; Herrmann et al., 2021). Under suitable conditions, the KMD converges as we increase the dimension of the projected Koopman operator (see Section 4.1.3 and Section 5.2.2).

2.3 Three canonical examples

Having grasped the notion of Koopman operators and the basic DMD algorithm, it is time for some examples. As a warm-up for the reader, we consider three canonical well-studied examples of Algorithm 1, each with a unitary Koopman operator:

- The flow past a cylinder wake at $Re = 100$ with a state space dimension $d = 160{,}000$ that corresponds to the number of spatial measurement points in the flow. The associated Koopman operator has a pure point spectrum consisting of powers of a fundamental eigenvalue.
- The Lorenz system on the Lorenz attractor with a state space dimension $d = 3$. The associated Koopman operator possesses no eigenvalues, except for the simple eigenvalue $\lambda = 1$ whose eigenfunction is the constant function. The rest of the spectrum is continuous.
- The Duffing oscillator with a state space dimension $d = 2$. The associated Koopman operator possesses no eigenvalues, except for $\lambda = 1$, whose eigenspace now corresponds to the conserved Hamiltonian energy of the system and indicator functions associated with invariant sets of positive area.

Despite its large ambient space dimension, the first example is the easiest to address using Algorithm 1. This is because the cylinder wake exhibits an attracting limit cycle, and the Koopman operator has a basic spectrum. The other two examples demonstrate three difficulties of DMD: noise, projection error (which can lead to spurious modes and missing parts of the spectrum), and continuous spectra. These and further challenges are discussed in Section 2.4.

2.3.1 Flow past a cylinder wake

We first consider the classic DMD example of low Reynolds number flow past a circular cylinder. Due to its simplicity and relevance in engineering, this is one of the most studied examples in modal-analysis techniques (Rowley and Dawson, 2017, Table 3) (Chen et al., 2012; Taira et al., 2020). $Re = 100$ is chosen so that it is larger than the critical Reynolds number at which the flow undergoes a supercritical Hopf bifurcation, resulting in laminar vortex shedding (Jackson, 1987; Zebib, 1987). This limit cycle is stable and is representative of the three-dimensional flow (Noack and Eckelmann, 1994; Noack et al., 2003). The Koopman operator of the post-transient flow has a pure point spectrum with a lattice structure on the unit circle (Bagheri, 2013).

To collect snapshot data, we numerically compute the velocity field of a flow around a circular cylinder of diameter $D = 1$ using an incompressible, two-dimensional lattice-Boltzmann solver (Józsa et al., 2016; Szőke et al., 2017). The temporal resolution of the flow is chosen so that approximately 24 snapshots of the flow field correspond to the period of vortex shedding. The computational domain size is $18D$ in length and $5D$ in height, with a 800×200 grid resolution. The cylinder is positioned $2D$ downstream of the inlet at the mid-height of the domain. The cylinder side walls are defined as bounce-back and no-slip walls, and a parabolic velocity profile is given at the inlet of the domain. The outlet is defined as a nonreflecting outflow. After simulations converge to steady-state vortex shedding, we collect $M = 120$ snapshots for the DMD algorithm and a further 880 snapshots to test the prediction of the KMD. One should think of this as training data and test data, respectively. Letting $\mathbf{V}_x(t)$ denote the vectorized horizontal velocity field at time t, our snapshot matrices have the form

$$\mathbf{X} = \begin{pmatrix} \mathbf{V}_x(0) & \mathbf{V}_x(\Delta t) & \cdots & \mathbf{V}_x(119\Delta t) \end{pmatrix},$$
$$\mathbf{Y} = \begin{pmatrix} \mathbf{V}_x(\Delta t) & \mathbf{V}_x(2\Delta t) & \cdots & \mathbf{V}_x(120\Delta t) \end{pmatrix}.$$

We use a rank of $r = 47$ to recover the trivial mode corresponding to $\lambda = 1$ and 24 conjugate pairs of modes up to the timescale of vortex shedding. The eigenvalues come in conjugate pairs due to processing real-valued data $\mathbf{X}$ and $\mathbf{Y}$.

Fig. 2 shows the output of Algorithm 1. In the left panel, we see the lattice structure of the DMD modes correctly identified by Algorithm 1. In the middle plot, we show the predictive error of (2.7). The relative error is computed by taking the 2-norm of the error in the velocity field $\mathbf{V}_x$ and normalizing it by the 2-norm of the mean-subtracted flow at each time step. Due to the periodic

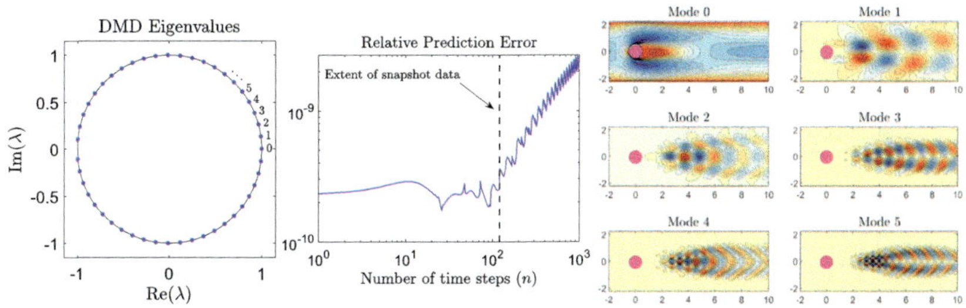

FIGURE 2 Output of Algorithm 1 for the flow past a cylinder wake. Left: The DMD eigenvalues. Middle: The total relative prediction error of (2.7). Right: Example modes of the horizontal component of the velocity field. The magenta disc corresponds to the cylinder. The zeroth mode corresponds to the time-averaged flow.

nature of the flow, there is excellent agreement between the KMD and flow, with slow algebraic growth of the error beyond the snapshot data time window. The right panel of Fig. 2 shows the real part of some of the Koopman modes for the horizontal velocity field.

2.3.2 *Lorenz system*

The Lorenz (63) system (Lorenz, 1963) is the following three coupled ordinary differential equations:

$$\dot{x} = 10\,(y - x)\,, \quad \dot{y} = x\,(28 - z) - y, \quad \dot{z} = xy - 8z/3.$$

We consider the dynamics of $\mathbf{x} = (x, y, z)$ on the Lorenz attractor. The system is chaotic and strongly mixing (Luzzatto et al., 2005). It follows that $\lambda = 1$ is the only eigenvalue of $\mathcal{K}$, corresponding to a constant eigenfunction, and that this eigenvalue is simple. We consider a discrete-time dynamical system by sampling with a time-step $\Delta t = 0.001$. We use time-delay embedding, which is a popular method for DMD-type algorithms (Susuki and Mezić, 2015; Arbabi and Mezić, 2017a; Brunton et al., 2017; Das and Giannakis, 2019; Kamb et al., 2020; Pan and Duraisamy, 2020a) and corresponds to building a Krylov subspace. This technique is justified through Takens' embedding theorem (Takens, 2006), which says that under certain technical conditions, delay embedding a signal coordinate of the system can reconstruct the attractor of the original system up to a diffeomorphism. In this example, we augment $\mathbf{x}$ by $N - 1$ further time-delays of length $\Delta t' = 0.2$ and consider $M = 5 \times 10^5$ snapshots along a single trajectory. Specifically, our snapshot matrices have the form

$$\mathbf{X} = \begin{pmatrix} \mathbf{x}(0) & \mathbf{x}(\Delta t) & \cdots & \mathbf{x}((M-1)\Delta t) \\ \mathbf{x}(\Delta t') & \mathbf{x}(\Delta t'+\Delta t) & \cdots & \mathbf{x}(\Delta t'+(M-1)\Delta t) \\ \vdots & \vdots & \vdots & \vdots \\ \mathbf{x}((N-1)\Delta t') & \mathbf{x}((N-1)\Delta t'+\Delta t) & \cdots & \mathbf{x}((N-1)\Delta t'+(M-1)\Delta t) \end{pmatrix}$$
$$\in \mathbb{R}^{3N \times M},$$

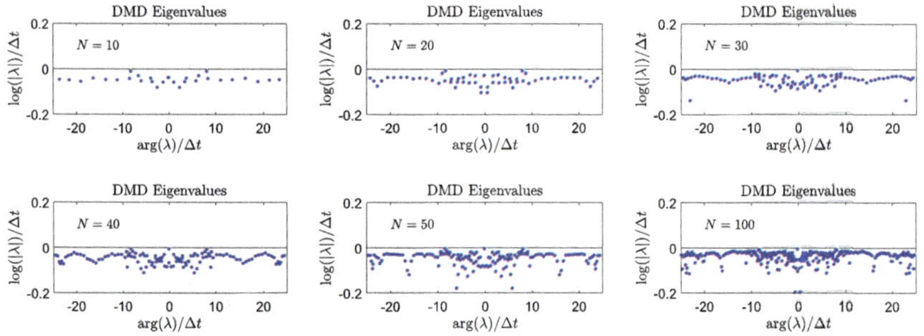

FIGURE 3 DMD eigenvalues for the Lorenz system and different choices of N (the number of eigenvalues is $3N$). The logarithm of the eigenvalues is plotted to align with the continuous-time system. As N increases, the eigenvalues cluster to approximate the continuous spectrum.

$$\mathbf{Y} = \begin{pmatrix} \mathbf{x}(\Delta t) & \mathbf{x}(2\Delta t) & \cdots & \mathbf{x}(M\Delta t) \\ \mathbf{x}(\Delta t' + \Delta t) & \mathbf{x}(\Delta t' + 2\Delta t) & \cdots & \mathbf{x}(\Delta t' + M\Delta t) \\ \vdots & \vdots & \vdots & \vdots \\ \mathbf{x}((N-1)\Delta t' + \Delta t) & \mathbf{x}((N-1)\Delta t' + 2\Delta t) & \cdots & \mathbf{x}((N-1)\Delta t' + M\Delta t) \end{pmatrix}$$
$$\in \mathbb{R}^{3N \times M}.$$

We use the `ode45` command in MATLAB to collect the data after an initial burn-in time to ensure that the initial point $\mathbf{x}(0)$ is (approximately) on the Lorenz attractor. The system is chaotic, so we cannot hope to integrate for long periods accurately numerically. However, convergence is still obtained in the large data limit $M \to \infty$ due to an effect known as shadowing.

In addition to a discrete time-step Δt, we can consider Koopman operators associated with continuous-time dynamical systems. The continuous-time infinitesimal generator is defined by

$$\mathcal{L}g = \lim_{\Delta t \downarrow 0} \frac{\mathcal{K}_{\Delta t} g - g}{\Delta t}, \tag{2.8}$$

where $\mathcal{K}_{\Delta t}$ is the Koopman operator corresponding to a time-step Δt. The generator satisfies

$$\mathcal{K}_{\Delta t} = \exp(\Delta t \mathcal{L}),$$

which can be made precise through the theory of semigroups (Pazy, 2010). Hence, in this example, we consider the following time-scaled logarithms of the eigenvalues:

$$\log(\lambda)/\Delta t = \log(|\lambda|)/\Delta t + i \arg(\lambda)/\Delta t.$$

Fig. 3 shows the DMD eigenvalues for various choices of N. The horizontal line corresponds to a portion of the spectrum of the Koopman operator. The DMD

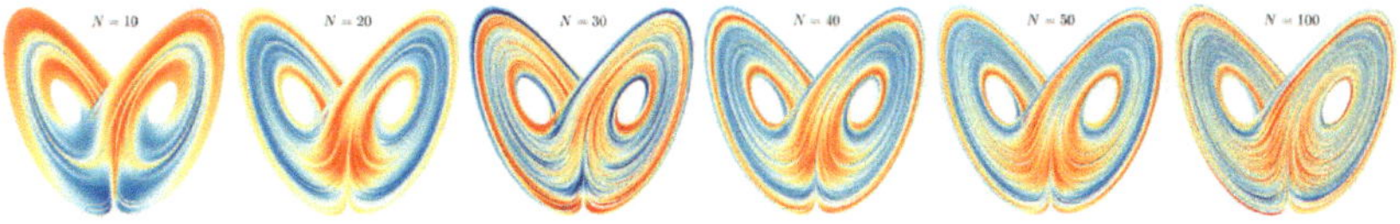

FIGURE 4 DMD eigenfunctions corresponding to $\log(\lambda)/(i\Delta t) \approx \pm 8.2$ rad/s. These are similar to the singularity in spectral measures detected in Korda et al. (2020).

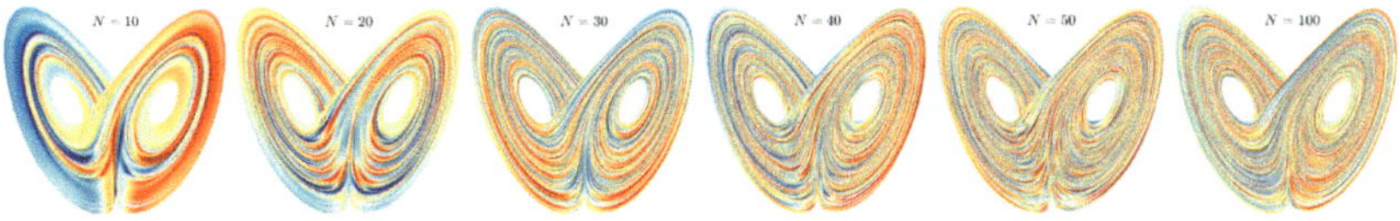

FIGURE 5 DMD eigenfunctions corresponding to $\log(\lambda)/(i\Delta t) \approx 6$ rad/s. We see increasing oscillations as N gets larger and the eigenfunctions resemble unstable periodic orbits, see also Colbrook and Townsend (2023).

eigenvalues fall below this line, corresponding to a dampening effect in the dynamics encapsulated by DMD. For these choices of parameters, this error is largely due to the finite amount of trajectory data and noise in the data matrices from the numerical solver. In Section 3.1.5, we shall see that this effect can be reduced using DMD methods designed to be robust to noise and by increasing M. Another reason for the dampening of the eigenvalues is the projection error (from projecting onto a finite matrix). Another interesting feature of Fig. 3 is the clustering of the DMD eigenvalues with increasing N as they attempt to approximate the continuous spectrum. Recall from above that the Koopman operator for this example has no eigenvalues except the trivial eigenvalue $\lambda = 1$. In Section 5.2, we shall see that *Measure-Preserving Extended DMD* (Colbrook, 2023) can deal with continuous spectra (see also the discussion in Section 6.2 for further methods).

We next show DMD eigenfunctions, but associated with the matrix $\mathbf{X}^{\dagger}\mathbf{Y}$. The discussion in Section 2.2.2 shows that these correspond to pseudoeigenfunctions of $\mathcal{K}$. Note that these pseudoeigenfunctions do not approximate true eigenfunctions - since eigenfunctions do not exist for this system! For visualization over the attractor, we plot function values along the trajectory of snapshot data. In Fig. 3 there are DMD eigenvalues close to the horizontal line with $\log(\lambda)/(i\Delta t) \approx \pm 8.2$ rad/s. These correspond to an apparent singularity in the spectral measure detected in Korda et al. (2020). Fig. 4 shows the corresponding pseudoeigenfunctions. These bear a striking resemblance to the local spectral projections in Korda et al. (2020, Figure 13), which the authors attributed to an almost-periodic motion of the z component during the time that the state resides in either of the two lobes of the Lorenz attractor. In Fig. 5, we plot the pseudoeigenfunctions corresponding to the DMD eigenvalue with $\log(\lambda)/(i\Delta t)$ closest to 6 rad/s. In this case, we see increasing oscillations as N gets larger, and the pseudoeigenfunctions resemble unstable periodic orbits, which in a sense, form a backbone of the attractor (Eckmann and Ruelle, 1985;

Tufillaro et al., 1993). For further examples of these kinds of pseudoeigenfunctions, see Colbrook and Townsend (2023).

2.3.3 Duffing oscillator

We now consider the Hamiltonian system

$$\dot{x} = y, \quad \dot{y} = x - x^3,$$

known as the (undamped nonlinear) Duffing oscillator with state $\mathbf{x} = (x, y) \in \Omega = \mathbb{R}^2$. This dynamical system has three fixed points at $\mathbf{x} = (0, 0)$ (a saddle), and $\mathbf{x} = (\pm 1, 0)$ (centers). The Hamiltonian for this system is $H = y^2 - x^2/2 + x^4/2$. We consider the corresponding discrete-time dynamical system by sampling with a time-step $\Delta t = 0.25$. Instead of using the state $\mathbf{x}$ or time-delay embedding to form our snapshot matrices, we consider an example of *Extended DMD* (Williams et al., 2015a), discussed in more detail in Section 4.1. Specifically, we consider 10^3 random points sampled uniformly in $[-2, 2]^2$, and then the trajectory of these points for 50 times steps. This leads to $M = 5 \times 10^4$ snapshots $\{\mathbf{x}^{(m)}, \mathbf{y}^{(m)}\}_{m=1}^{M}$. We then partition these into N clusters using k-means, and use these as centers $\mathbf{c}_j$ for N radial basis functions of the form

$$\psi_j(\mathbf{x}) = \exp(-\gamma \|\mathbf{x} - \mathbf{c}_j\|),$$

where γ is the squared reciprocal of the average ℓ^2-norm of the snapshot data after it is shifted to mean zero. Our snapshot matrices are then given by

$$\mathbf{X} = \begin{pmatrix} \psi_1(\mathbf{x}^{(1)}) & \psi_1(\mathbf{x}^{(2)}) & \cdots & \psi_1(\mathbf{x}^{(M)}) \\ \psi_2(\mathbf{x}^{(1)}) & \psi_2(\mathbf{x}^{(2)}) & \cdots & \psi_2(\mathbf{x}^{(M)}) \\ \vdots & \vdots & \vdots & \vdots \\ \psi_N(\mathbf{x}^{(1)}) & \psi_N(\mathbf{x}^{(2)}) & \cdots & \psi_N(\mathbf{x}^{(M)}) \end{pmatrix},$$

$$\mathbf{Y} = \begin{pmatrix} \psi_1(\mathbf{y}^{(1)}) & \psi_1(\mathbf{y}^{(2)}) & \cdots & \psi_1(\mathbf{y}^{(M)}) \\ \psi_2(\mathbf{y}^{(1)}) & \psi_2(\mathbf{y}^{(2)}) & \cdots & \psi_2(\mathbf{y}^{(M)}) \\ \vdots & \vdots & \vdots & \vdots \\ \psi_N(\mathbf{y}^{(1)}) & \psi_N(\mathbf{y}^{(2)}) & \cdots & \psi_N(\mathbf{y}^{(M)}) \end{pmatrix}.$$

Fig. 6 shows some of the Koopman eigenfunctions corresponding to $\lambda = 1$ and computed using $N = 1000$. For visualization, the values of the functions are plotted at the data points. We see that the level sets of the eigenfunctions correspond to trajectories, as expected. However, moving away from $\lambda = 1$ in the spectral plane becomes more challenging. Fig. 7 shows the DMD eigenvalues for various choices of N, along with the unit circle, which is the spectrum of the Koopman operator. Most of the DMD eigenvalues are spurious and correspond

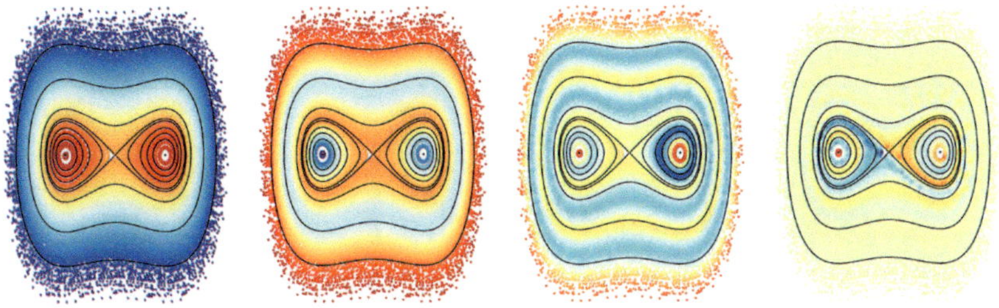

FIGURE 6 Some examples of computed eigenfunctions of the Duffing oscillator corresponding to $\lambda = 1$. The black lines show trajectory orbits and correspond to level sets of the eigenfunctions, which are invariant in time.

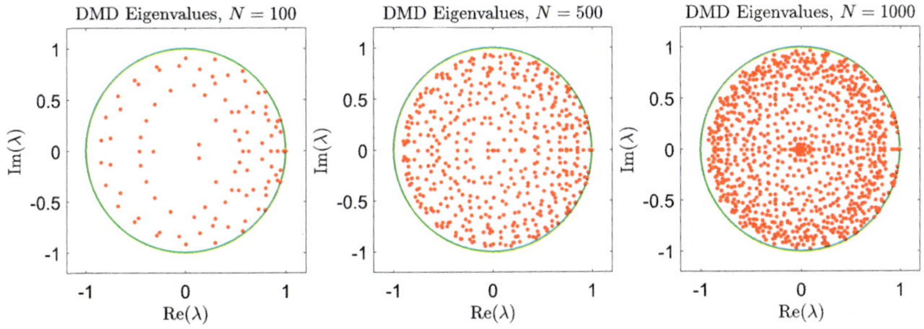

FIGURE 7 DMD eigenvalues (red dots) computed for various choices of N. The spectrum is the unit circle (green); hence, most eigenvalues are spurious. This occurs because a projection error occurs when the Koopman operator $\mathcal{K}$ is approximated by a finite DMD matrix.

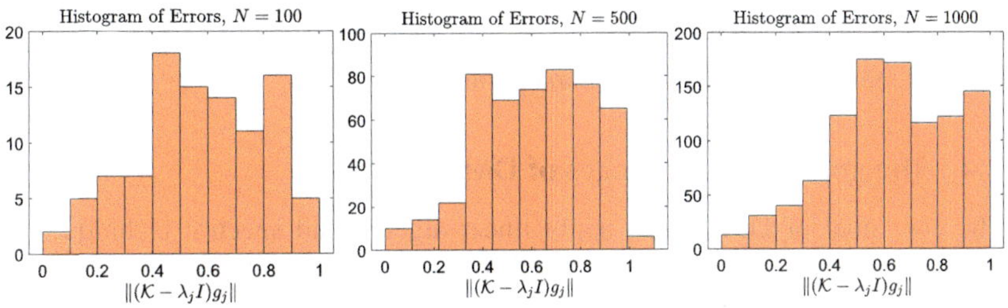

FIGURE 8 Histograms of the errors of the DMD eigenpairs. The errors are computed using ResDMD in Algorithm 11 and show the persistence of heavy spectral pollution of DMD as N increases.

to spectral pollution. This occurs because we have approximated the infinite-dimensional Koopman operator $\mathcal{K}$, by a finite matrix. These errors persist, even as we increase N.

To measure the errors, we can use *Residual DMD* (ResDMD) (Colbrook and Townsend, 2023; Colbrook et al., 2023a) to compute the error $\|(\mathcal{K} - \lambda_j I)g_j\|$ associated with a DMD eigenfunction g_j and eigenvalue λ_j. In other words, we can compute the projection error of DMD. This is detailed in Section 4.3. Fig. 8 shows the histograms of these projection errors. Only a small proportion of reliable DMD eigenvalues persist as N increases. Finally, we can also use ResDMD

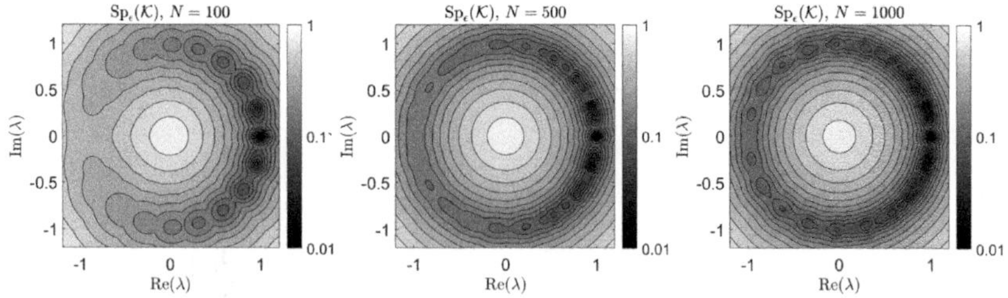

FIGURE 9 Pseudospectra (see (2.3)) computed using ResDMD (see Algorithm 12) and visualized by plotting several contour plots of ϵ on a logarithmic scale. The pseudospectra demonstrate the heavy spectral pollution present in Fig. 7. Note also that these pseudospectra are computed using the same snapshot data and dictionary used for Figs. 6 to 8. As $N \to \infty$, the algorithm converges to the pseudospectra.

to compute pseudospectra. Algorithm 12 converges to the pseudospectrum as $N \to \infty$. The output is shown in Fig. 9, where we visualize pseudospectra by plotting several contour plots of ϵ on a logarithmic scale. For this example, the pseudospectra are the annuli $\mathrm{Sp}_\epsilon(\mathcal{K}) = \{\lambda \in \mathbb{C} : ||\lambda| - 1| \leq \epsilon\}$. The projection errors are computed using the same snapshot data and dictionary used for Figs. 6 to 8. ResDMD allows us to compute and minimize projection errors directly in infinite dimensions to avoid spectral pollution and spurious modes.

In summary, DMD can suffer from closure issues (projection errors) associated with approximating the infinite-dimensional Koopman operator by a finite-dimensional matrix. This phenomenon is well-known (S.L. Brunton et al., 2016a; Kaiser et al., 2021). Nevertheless, by computing projection errors, we can avoid difficulties (such as spurious modes) associated with the infinite-dimensional nature of the Koopman operator.

2.4 The goals and challenges of DMD

The core goal of DMD is to apply linear algebra and spectral techniques to the analysis, prediction, and control of nonlinear dynamical systems. However, DMD often faces several challenges (Kutz et al., 2016a), many of which are discussed in Table 1. These challenges have been a driving force for the many versions of the DMD algorithm that have appeared.

For example, the KMD in (2.7) highlights the potential usefulness of DMD in forecasting. In instances where DMD is applied to noise-free data, such as in generating reduced-order models from high-fidelity numerical simulations (Kutz et al., 2016b; Alla and Kutz, 2017; Lu and Tartakovsky, 2020b), DMD proves effective for both reconstruction and accurate forecasting of the solutions. However, practitioners familiar with DMD's performance in noisy conditions recognize its shortcomings; the algorithm often fails to forecast and reconstruct even the time series it was trained on. In particular, the prediction error in Fig. 2 is somewhat misleading of what a user might expect in the general case. Even after over a decade, the application of DMD for forecasting or reconstructing time-series data remains limited, typically restricted to high-quality, low-noise scenarios. In Section 3.1, we will focus on methods that mitigate the effect of noise in snapshot data. Many of the structure-preserving methods we discuss in Section 5 have an in-built robustness to noise.

Generally speaking, the error of DMD and its approximate KMD can be split into three types:

- The *projection error* is due to projecting/truncating the Koopman operator onto a finite-dimensional space of observables. This is linked to the issue of closure and lack of (or lack of knowledge of) nontrivial finite-dimensional Koopman invariant subspaces.
- The *estimation error* is due to estimating the matrices that represent the projected Koopman operator from a finite set of potentially noisy trajectory data.
- *Numerical errors* (e.g., roundoff, stability, further compression, etc.) incurred when processing the finite DMD matrix.

In particular, Wu et al. (2021) highlight the issues of robustness to noise and closure/projection errors as the two fundamental challenges for DMD methods. In Section 4 we will consider methods that directly connect DMD with Koopman operators through the Galerkin perspective. Ways of controlling and measuring the projection error are discussed in Section 4.3.

DMD's primary value has been as a diagnostic tool, and the interpretability of DMD modes and frequencies is crucial to this role. Most DMD papers focus on analyzing DMD modes and eigenvalues. This emphasis shapes much of this review. The KMD approximated by DMD modes and eigenvalues facilitates dimensionality reduction and model simplification, analogous to classical methods like the Fourier transform or SVD (Brunton and Kutz, 2022). There are numerous software packages for DMD methods, including https://github.com/dynamicslab/pykoopman, https://github.com/mathLab/PyDMD, and https://github.com/decargroup/pykoop. Moreover, there are numerous repositories connected to the papers cited below. Drmač has implemented the DMD algorithm and extensions in LAPACK (Drmač, 2022a,b).

3 Variants from the regression perspective

This section gives the reader a flavor of DMD variants from the regression perspective.[12] We focus on four key aspects that have proved influential over the last decade or so:

- Noise reduction;
- Compression and randomized linear algebra;
- Multiscale dynamics; and
- Control.

[12] As we shall see, there is less convergence theory (e.g., in the large data limit $M \to \infty$ or as the number of observables increases) for DMD methods based on this viewpoint than for those based on the Galerkin viewpoint in Section 4. This is due to a looser connection with Koopman operators.

The methods we discuss are only some of the variants - it is impossible to do justice to the breadth of techniques! Notable omissions include the following. *Bayesian DMD* (Takeishi et al., 2017a) transfers the Bayesian formulation into DMD. *Higher order DMD* (Le Clainche and Vega, 2017) applies time-delay embedding to build a larger state space after projecting onto POD modes. *Parametric DMD* (Huhn et al., 2023) performs DMD independently per parameter realization and interpolates the resulting Koopman operators. See also Andreuzzi et al. (2023). *Refined Rayleigh–Ritz data driven modal decomposition* (Drmač et al., 2018) produces refined Ritz pairs of the finite DMD matrix $\mathbf{K}_{\text{DMD}}$. *Spatio-temporal Koopman decomposition* (Vega and Le Clainche, 2021) approximates spatio-temporal data as a linear combination of (possibly growing or decaying exponentially) standing or traveling waves. Klus et al. (2018b); Klus and Schütte (2016) develop tensor-based DMD methods for computing eigenfunctions of the Koopman operator. For example, *tensor-based DMD* exploits low-rank tensor decompositions of the data matrices to improve efficiency and memory use. There are extensions of this approach based on reproducing kernel Hilbert spaces (RKHSs) (Fujii and Kawahara, 2019) and EDMD (Nüske et al., 2021). Recent work has also explored connections between DMD and tensor factorizations (Redman, 2021).

3.1 Increasing robustness to noise

A challenge of DMD is that the computed eigenvalues are biased in the presence of sensor noise. Noise typically dampens the eigenvalues, meaning that for discrete-time systems, the absolute values of the eigenvalues are decreased, and the Koopman modes become distorted. For studies of this effect in various physical systems, see Duke et al. (2012); Bagheri (2014); Pan et al. (2015). Dawson et al. (2016) provide an exceptionally clear discussion of this topic. The bias occurs because standard algorithms treat the data "snapshot to snapshot" rather than as a whole and favor one direction (forward in time). Several variants of DMD aim to address this bias. In addition to the methods presented below, other techniques include utilizing Kalman filters (Nonomura et al., 2018, 2019; Jiang and Liu, 2022), adapting DMD to online data (Hemati et al., 2014, 2016), robust principal component analysis (Scherl et al., 2020), and using a second set of noisy observables that meet some independence requirements (Wanner and Mezić, 2022). Moreover, the structure-preserving methods we discuss in Section 5 often have an inbuilt robustness to noise. Finally, in Section 6.3, we discuss the stochastic Koopman operator, which can handle both system and sensor noise.

3.1.1 The problem of noise

We can understand the bias often encountered in DMD as follows. Assume that the snapshots come with additive sensor noise that affects only our measurements of a given system and does not interact with the true dynamics. This means that we have access to noisy data matrices

$$\mathbf{X}_s = \mathbf{X} + \mathbf{N}_X, \quad \mathbf{Y}_s = \mathbf{Y} + \mathbf{N}_Y,$$

where $\mathbf{N}_X$ and $\mathbf{N}_Y$ are random matrices representing sensor noise, and $\mathbf{X}$ and $\mathbf{Y}$ are the noise-free snapshots. We then represent the data in a truncated POD mode basis so that

$$\tilde{\mathbf{X}}_s = \tilde{\mathbf{X}} + \tilde{\mathbf{N}}_X, \quad \tilde{\mathbf{Y}}_s = \tilde{\mathbf{Y}} + \tilde{\mathbf{N}}_Y,$$

and assume that a subset of POD modes has been selected so that $\tilde{\mathbf{X}}_s\tilde{\mathbf{X}}_s^*$ is invertible. Assuming the noise is sufficiently small, the DMD matrix can be expanded as

$$\begin{aligned}
\tilde{\mathbf{K}}_{\mathrm{DMD}} &= \tilde{\mathbf{Y}}_s\tilde{\mathbf{X}}_s^\dagger = \tilde{\mathbf{Y}}_s\tilde{\mathbf{X}}_s^*(\tilde{\mathbf{X}}_s\tilde{\mathbf{X}}_s^*)^{-1} \\
&= (\tilde{\mathbf{Y}} + \tilde{\mathbf{N}}_Y)(\tilde{\mathbf{X}} + \tilde{\mathbf{N}}_X)^*\left[(\tilde{\mathbf{X}} + \tilde{\mathbf{N}}_X)(\tilde{\mathbf{X}} + \tilde{\mathbf{N}}_X)^*\right]^{-1} \\
&= (\tilde{\mathbf{Y}} + \tilde{\mathbf{N}}_Y)(\tilde{\mathbf{X}} + \tilde{\mathbf{N}}_X)^*(\tilde{\mathbf{X}}\tilde{\mathbf{X}}^*)^{-1} \\
&\quad \times \left[\mathbf{I} - (\tilde{\mathbf{N}}_X\tilde{\mathbf{X}}^* + \tilde{\mathbf{X}}\tilde{\mathbf{N}}_X^* + \tilde{\mathbf{N}}_X\tilde{\mathbf{N}}_X^*)(\tilde{\mathbf{X}}\tilde{\mathbf{X}}^*)^{-1} + \cdots\right].
\end{aligned}$$

Dawson et al. (2016) discard high-order terms in the expectation of this expansion to arrive at

$$\mathbb{E}(\tilde{\mathbf{K}}_{\mathrm{DMD}}) \approx \tilde{\mathbf{Y}}\tilde{\mathbf{X}}^{-1}(\mathbf{I} - \mathbb{E}(\tilde{\mathbf{N}}_X\tilde{\mathbf{N}}_X^*)(\tilde{\mathbf{X}}\tilde{\mathbf{X}}^*)^{-1}). \tag{3.1}$$

This indicates that DMD has an inherent bias due to sensor noise, causing a dampening effect. Interestingly, this bias depends only on $\tilde{\mathbf{N}}_X$ and not $\tilde{\mathbf{N}}_Y$. The reason is that the least squares problem in (2.6) is optimal only when assuming that all of the noise is in $\tilde{\mathbf{Y}}$, but not in $\tilde{\mathbf{X}}$. Another way of seeing this is that the expression $\tilde{\mathbf{Y}}\tilde{\mathbf{X}}^{-1}$ is linear in $\tilde{\mathbf{Y}}$, but not in $\tilde{\mathbf{X}}$, which is why perturbations to $\tilde{\mathbf{X}}$ do not have to propagate through the equation in an unbiased manner.

If the noise structure is known, DMD can be adjusted using a method called *noise-corrected DMD* (ncDMD) (Dawson et al., 2016). However, it is preferable to have methods that correct for noise without requiring explicit knowledge of its structure. We will now outline three popular DMD variants that address this bias without specific assumptions about the noise. The first two can be executed directly using SVDs. The final method requires an iterative method for solving an optimization problem and is more expensive yet more robust.

3.1.2 Forward-backward dynamic mode decomposition (fbDMD)

Forward-Backward DMD (fbDMD) can be considered a correction to the unidirectional bias of Algorithm 1 (Dawson et al., 2016). Let $\mathbf{X} = \mathbf{U}_X\mathbf{\Sigma}_X\mathbf{V}_X^*$ and $\mathbf{Y} = \mathbf{U}_Y\mathbf{\Sigma}_Y\mathbf{V}_Y^*$ be truncated SVDs of the matrices $\mathbf{X}$ and $\mathbf{Y}$, respectively. We define

$$\tilde{\mathbf{K}}_f = \mathbf{U}_X^*\mathbf{Y}\mathbf{V}_X\mathbf{\Sigma}_X^{-1}, \quad \tilde{\mathbf{K}}_b = \mathbf{U}_Y^*\mathbf{X}\mathbf{V}_Y\mathbf{\Sigma}_Y^{-1},$$

Algorithm 2 Forward-backward DMD (Dawson et al., 2016).

Input: Snapshot data $\mathbf{X} \in \mathbb{C}^{d\times M}$ and $\mathbf{Y} \in \mathbb{C}^{d\times M}$, rank $r \in \mathbb{N}$.

1: Compute a truncated SVD $\mathbf{X} \approx \mathbf{U}\boldsymbol{\Sigma}\mathbf{V}^*$, $\mathbf{U} \in \mathbb{C}^{d\times r}$, $\boldsymbol{\Sigma} \in \mathbb{R}^{r\times r}$, $\mathbf{V} \in \mathbb{C}^{M\times r}$.

2: Compute the projected data matrices $\tilde{\mathbf{X}} = \mathbf{U}^*\mathbf{X}$, $\tilde{\mathbf{Y}} = \mathbf{U}^*\mathbf{Y}$ and their economized SVDs $\tilde{\mathbf{X}} = \mathbf{U}_X\boldsymbol{\Sigma}_X\mathbf{V}_X^*$, $\tilde{\mathbf{Y}} = \mathbf{U}_Y\boldsymbol{\Sigma}_Y\mathbf{V}_Y^*$.

3: Compute the forward and backward matrices $\tilde{\mathbf{K}}_f = \mathbf{U}_X^*\tilde{\mathbf{Y}}\mathbf{V}_X\boldsymbol{\Sigma}_X^{-1}$, $\tilde{\mathbf{K}}_b = \mathbf{U}_Y^*\tilde{\mathbf{X}}\mathbf{V}_Y\boldsymbol{\Sigma}_Y^{-1}$.

4: Compute the matrices $\mathbf{S}_f = \tilde{\mathbf{Y}}\mathbf{V}_X\boldsymbol{\Sigma}_X^{-1}$, $\mathbf{S}_b = \tilde{\mathbf{X}}\mathbf{V}_Y\boldsymbol{\Sigma}_Y^{-1}$, and $\mathbf{K}_f = \mathbf{S}_f\tilde{\mathbf{K}}_f\mathbf{S}_f^\dagger$, $\mathbf{K}_b = \mathbf{S}_b\tilde{\mathbf{K}}_b\mathbf{S}_b^\dagger$.

5: Compute the DMD matrix $\tilde{\mathbf{K}} = \left(\mathbf{K}_f\mathbf{K}_b^{-1}\right)^{1/2}$ and its eigendecomposition $\tilde{\mathbf{K}}\mathbf{W} = \mathbf{W}\boldsymbol{\Lambda}$.

6: Compute the modes $\boldsymbol{\Phi} = \mathbf{Y}\mathbf{V}\boldsymbol{\Sigma}^{-1}\mathbf{W}$.

Output: The eigenvalues $\boldsymbol{\Lambda}$ and modes $\boldsymbol{\Phi} \in \mathbb{C}^{d\times r}$.

which represent forward and backward propagators for the data, analogous to Algorithm 1. Assuming the system's dynamics are invertible and $\tilde{\mathbf{K}}_b$ is also invertible, the matrix

$$\tilde{\mathbf{K}} = \left(\tilde{\mathbf{K}}_f\tilde{\mathbf{K}}_b^{-1}\right)^{1/2}$$

provides a debiased estimate of the forward propagator. The method is presented in Algorithm 2. Nonetheless, caution is required due to the nonuniqueness of the matrix square root (Higham, 2008). Dawson et al. (2016) suggest selecting the square root that is closest to $\tilde{\mathbf{K}}_f$ in norm, although this can be computationally costly. A more economical alternative involves measuring closeness in the computed eigencoordinates. Sometimes, the nonuniqueness can be avoided. For instance, if the samples are snapshots from a continuous system whose signal has a bandwidth of λ_B and the time-step satisfies $\Delta t < \pi/(2\lambda_B)$, then the discrete eigenvalues expected to be recovered will have a positive real part, which resolves the ambiguity mentioned previously. The square root issue is further analyzed in Drmač et al. (2018, Section 5.4). Finally, Askham and Kutz (2018) recommend first projecting onto r POD modes before applying fbDMD, an alteration that has demonstrated superior performance in practice. For a variational problem involving forward and backward dynamics, see *Consistent DMD* (Azencot et al., 2019).

3.1.3 Total least-squares dynamic mode decomposition (tlsDMD)

Total Least-Squares DMD (tlsDMD) addresses the asymmetric treatment of noise in $\mathbf{X}$ and $\mathbf{Y}$ by Algorithm 1. The least-squares problem in (2.6) can be formulated as

$$\min_{\mathbf{K}} \|\mathbf{E}_Y\|_{\mathrm{F}} \quad \text{such that} \quad \mathbf{Y} + \mathbf{E}_Y = \mathbf{K}\mathbf{X}.$$

Algorithm 3 Total least-squares DMD (Dawson et al., 2016; Hemati et al., 2017).

Input: Snapshot data $\mathbf{X} \in \mathbb{C}^{d \times M}$ and $\mathbf{Y} \in \mathbb{C}^{d \times M}$, rank $r \in \mathbb{N}$.

1: Compute a truncated SVD $\mathbf{X} \approx \mathbf{U}\mathbf{\Sigma}\mathbf{V}^*$, $\mathbf{U} \in \mathbb{C}^{d \times r}$, $\mathbf{\Sigma} \in \mathbb{R}^{r \times r}$, $\mathbf{V} \in \mathbb{C}^{M \times r}$.
2: Compute the projected data matrices $\tilde{\mathbf{X}} = \mathbf{U}^*\mathbf{X}$, $\tilde{\mathbf{Y}} = \mathbf{U}^*\mathbf{Y}$.
3: Form the matrix $\mathbf{Z} = \begin{pmatrix} \tilde{\mathbf{X}}^\top & \tilde{\mathbf{Y}}^\top \end{pmatrix}^\top$ and compute its reduced SVD $\mathbf{Z} = \mathbf{U}_Z\mathbf{\Sigma}_Z\mathbf{V}_Z^*$.
4: Set $\mathbf{U}_1 = \mathbf{U}_Z(1:r, 1:r)$, $\mathbf{U}_2 = \mathbf{U}_Z(r+1:2r, 1:r)$.
5: Compute the DMD matrix $\tilde{\mathbf{K}} = \mathbf{U}_2\mathbf{U}_1^{-1}$ and its eigendecomposition $\tilde{\mathbf{K}}\mathbf{W} = \mathbf{W}\mathbf{\Lambda}$.
6: Compute the modes $\mathbf{\Phi} = \mathbf{Y}\mathbf{V}\mathbf{\Sigma}^{-1}\mathbf{W}$.

Output: The eigenvalues $\mathbf{\Lambda}$ and modes $\mathbf{\Phi} \in \mathbb{C}^{d \times r}$.

Considering the reverse time direction, as in fbDMD, leads to the problem

$$\min_{\mathbf{K}} \|\mathbf{E}_X\|_\mathrm{F} \quad \text{such that} \quad \mathbf{Y} = \mathbf{K}(\mathbf{X} + \mathbf{E}_X).$$

While fbDMD accounts for both directions of error, a more direct approach utilizes the total least-squares problem (Van Huffel and Vandewalle, 1991):

$$\min_{\mathbf{K}} \left\| \begin{pmatrix} \mathbf{E}_X \\ \mathbf{E}_Y \end{pmatrix} \right\|_\mathrm{F} \quad \text{such that} \quad \mathbf{Y} + \mathbf{E}_Y = \mathbf{K}(\mathbf{X} + \mathbf{E}_X).$$

This problem can be solved via an SVD, and we follow the version presented by Dawson et al. (2016), which is similar in spirit to that of Hemati et al. (2017). First, we project $\mathbf{X}$ and $\mathbf{Y}$ onto $r < M/2$ POD modes to obtain $\tilde{\mathbf{X}}$ and $\tilde{\mathbf{Y}}$. We then define

$$\mathbf{Z} = \begin{pmatrix} \tilde{\mathbf{X}} \\ \tilde{\mathbf{Y}} \end{pmatrix}$$

and compute its reduced SVD $\mathbf{Z} = \mathbf{U}_Z\mathbf{\Sigma}_Z\mathbf{V}_Z^*$. The matrix $\tilde{\mathbf{K}} = \mathbf{U}_Z(r+1:2r, 1:r)\mathbf{U}_Z(1:r, 1:r)^{-1}$ then provides a debiased estimate of the forward propagator. The method is summarized in Algorithm 3.

3.1.4 Optimized dynamic mode decomposition (optDMD)

Optimized DMD (optDMD) is a variation of DMD that processes all data snapshots collectively (Chen et al., 2012). This approach reduces much of the bias associated with exact DMD. Nonetheless, it necessitates solving a nonlinear optimization problem, initially thought to hinder its practical application. However, Askham and Kutz (2018) demonstrated that an approximate solution to the optimization problem can be efficiently computed using the variable projection method (Golub and Pereyra, 1973). In this framework, DMD is reformulated as an exponential data fitting problem (Pereyra and Scherer, 2010), which brings an additional advantage: the data snapshots do not have to be equidistant in time. For further DMD methodologies tailored for data with irregular time intervals, see Tu et al. (2014a); Guéniat et al. (2015); Leroux and Cordier (2016).

Initially, we project onto r POD modes to construct the data matrix $\tilde{\mathbf{X}} = [\mathbf{z}_0\ \mathbf{z}_1 \cdots \mathbf{z}_M]$, which corresponds to the projected data at times $t_0, t_1, \ldots, t_M$. Depending on the data structure, the projected matrix $\mathbf{Y}$ may also be incorporated into this matrix. We posit that the data represents the solution to a linear system of differential equations, expressed as

$$\mathbf{z}(t) \approx \mathbf{S}e^{\mathbf{\Lambda}t}\mathbf{S}^{\dagger}\mathbf{z}_0\,,$$

where $\mathbf{S} \in \mathbb{C}^{r\times r}$ and $\mathbf{\Lambda} \in \mathbb{C}^{r\times r}$. This representation can be reformulated to

$$\tilde{\mathbf{X}}^{\top} \approx \mathbf{\Phi}(\boldsymbol{\alpha})\mathbf{B}, \quad \mathbf{B}_{i,j} = \mathbf{S}_{j,i}\left(\mathbf{S}^{\dagger}\mathbf{z}_0\right)_i,$$

where $\mathbf{\Phi}(\boldsymbol{\alpha}) \in \mathbb{C}^{(m+1)\times r}$, whose elements are $\Phi(\boldsymbol{\alpha})_{i,j} = \exp(\boldsymbol{\alpha}_j t_i)$. From this, we arrive at an exponential fitting problem:

$$\min_{\boldsymbol{\alpha}\in\mathbb{C}^r,\mathbf{B}\in\mathbb{C}^{r\times r}} \left\|\tilde{\mathbf{X}}^{\top} - \mathbf{\Phi}(\boldsymbol{\alpha})\mathbf{B}\right\|_F.$$

The optimized DMD eigenvalues are determined by $\lambda_j = \boldsymbol{\alpha}_j$. This optimization problem is solved using the variable projection method, which exploits the specific structure of the exponential data fitting problem to eliminate many of the variables from the optimization process. A summary is provided in Algorithm 4, with practical details given in Askham and Kutz (2018), including strategies for selecting the initial guess (e.g., employing an alternate DMD algorithm).

While this nonlinear, nonconvex optimization problem is not guaranteed to be solved globally, and the method may be computationally intensive due to its iterative nature, optDMD often yields significant enhancements over traditional DMD approaches. Moreover, optDMD's efficacy can be further heightened by employing Breiman's statistical bagging sampling strategy (Breiman et al., 2017), which assembles a collection of models to reduce model variance, mitigate overfitting, and facilitate uncertainty quantification. This augmented method is referred to as *bagging optimized DMD* (bopDMD) (Sashidhar and Kutz, 2022).

3.1.5 Examples

For simplicity, we focus on the error associated with the approximated eigenvalues. Other error metrics related to the modes or the accuracy of the decomposition in fitting the data or forecasting are also frequently considered in the literature, often yielding similar results.

Algorithm 4 Optimized DMD, algorithmic details are given in Askham and Kutz (2018).

Input: Snapshot data $\mathbf{X} \in \mathbb{C}^{d\times(M+1)}$, rank $r \in \mathbb{N}$, initial guess for $\boldsymbol{\alpha}$.

1: Compute a truncated SVD $\mathbf{X} \approx \mathbf{U\Sigma V}^*$, $\mathbf{U} \in \mathbb{C}^{d\times r}$, $\boldsymbol{\Sigma} \in \mathbb{R}^{r\times r}$, $\mathbf{V} \in \mathbb{C}^{(M+1)\times r}$.

2: Compute the projected data matrix $\tilde{\mathbf{X}} = \mathbf{U}^*\mathbf{X}$.

3: Solve the problem

$$\min_{\boldsymbol{\alpha}\in\mathbb{C}^r,\mathbf{B}\in\mathbb{C}^{r\times r}} \left\|\tilde{\mathbf{X}}^\top - \boldsymbol{\Phi}(\boldsymbol{\alpha})\mathbf{B}\right\|_{\mathrm{F}}$$

using a variable projection algorithm.

4: Set $\lambda_j = \boldsymbol{\alpha}_j$ and $\boldsymbol{\Phi}(:,i) = (\|\mathbf{UB}^\top(:,i)\|_{\ell^2})^{-1}\mathbf{UB}^\top(:,i)$.

Output: The eigenvalues $\boldsymbol{\Lambda}$ and modes $\boldsymbol{\Phi} \in \mathbb{C}^{d\times r}$.

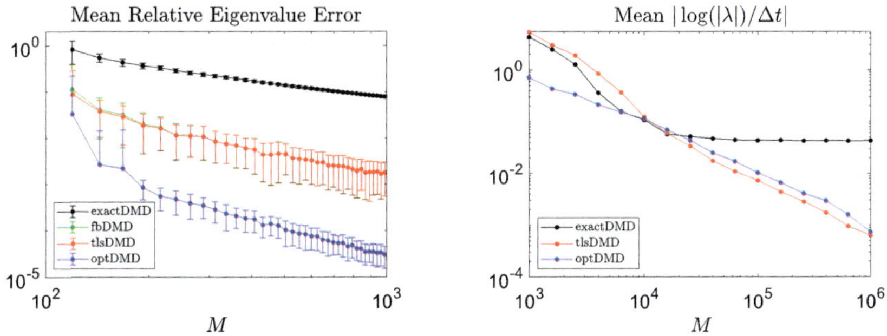

FIGURE 10 Left: Mean error (error bars correspond to the standard deviation across noise realizations) in the first 11 eigenvalues of the cylinder example. Right: Mean value of $|\log(|\lambda|)/\Delta t|$ for the DMD eigenvalues for the Lorenz system. We have not shown the results for fbDMD since they are almost identical to tlsDMD.

Noisy cylinder wake

We revisit the example of flow past a cylinder from Section 2.3.1. We center and normalize the data grid-wise before adding 40% Gaussian random noise to the measurements. Fig. 10 (left) shows the mean relative ℓ^2 error of the first 11 eigenvalues (see Fig. 2), averaged over 100 realizations of random noise. The errors are calculated by comparison with eigenvalues computed from noise-free snapshots that have converged in terms of both the size of the truncated SVD and the number of snapshots. The error bars represent one standard deviation from the mean. All methods exhibit a decreasing error as M increases, which is largely attributable to the truncation in the SVD used in DMD. As often noted in the literature, the fbDMD and tlsDMD methods perform comparably. However, optDMD demonstrates a significantly smaller error.

Lorenz system

We revisit the Lorenz system example from Section 2.3.2. Since the spectrum is continuous (apart from the trivial eigenvalue $\lambda = 1$), measuring the error of individual DMD eigenvalues is meaningless unless methods such as the residual in Section 4.3 are used. However, $|\log(|\lambda|)/\Delta t|$ vanishes on the spectrum of the Koopman operator. Therefore, we select $N = 10$ and compute the mean value of $|\log(|\lambda|)/\Delta t|$ over the DMD eigenvalues. Fig. 10 (right) shows the results, averaged over 50 randomly selected initial conditions on the attractor for the initial value $\mathbf{x}(0)$. For exact DMD, this error metric plateaus as M increases. Generally, the eigenvalues computed using DMD with delay embedding are damped and lie strictly within the unit disk (Korda et al., 2020, Corollary 2). Consequently, their logarithms are in the left-half plane, corresponding to positions below the horizontal line in Fig. 3. Conversely, the eigenvalues computed by tlsDMD and optDMD approach the unit disk with increasing M and exhibit greater robustness to noise in the measurements.

3.2 Compression and randomized linear algebra

With ever-increasing volumes of measurement data from simulations and experiments, modal extraction algorithms such as DMD can become prohibitively expensive, particularly for online or real-time analysis. Dynamics often evolve on low-dimensional attractors, indicating sparsity in a suitable coordinate system or an intrinsic low rankness. However, the SVD used in Algorithm 1 scales with the dimension of the measurements, not with the intrinsic dimension of the data. This section explores two principles aimed at mitigating this computational cost:

- **Compressed sensing** (Donoho, 2006; Candes et al., 2006) facilitates the reconstruction of sparse signals from a limited number of measurements, allowing for undersampling below traditional Shannon–Nyquist limits (Nyquist, 1928; Shannon, 1948). Applying compressed sensing to DMD can substantially improve computational efficiency, particularly during the SVD step of the algorithm. Acquiring high-resolution, time-resolved measurements can be challenging. Nevertheless, temporally and spatially sparse signals may be sampled less frequently than traditionally expected, which is crucial if data acquisition is costly.
- **Randomized numerical linear algebra** (Martinsson and Tropp, 2020) offers a way to solve certain linear algebra problems much faster than classical methods. The randomized SVD is a fast and straightforward technique for computing an approximate low-rank SVD (Halko et al., 2011). It is robust and amenable to parallelization and can benefit from GPU architectures. When coupled with randomized SVD, DMD scales with the intrinsic rank of the data matrices rather than the measurement dimension. The approximation error is manageable through oversampling and power iterations, providing a balance between computational speed and accuracy. Moreover, it can accom-

modate large datasets that exceed the capacity of fast memory by using a blocked matrix approach.

Beyond the methods detailed here, several DMD variants are based on related principles. For instance, due to the nonorthogonality of DMD modes, choosing an appropriate low-rank representation can be difficult (Kou and Zhang, 2017). *Sparsity-Promoting DMD* (Jovanović et al., 2014) aims to strike a balance between accuracy and the number of modes by identifying a sparse subset of modes. Other techniques for selecting dominant modes include ranking each DMD mode's importance by time integration (Kou and Zhang, 2017) or by assessing the time-averaged modal energy contribution (Tissot et al., 2014). Furthermore, one can apply DMD recursively to achieve orthogonality (Noack et al., 2016), a method termed *Recursive DMD*, which blends the principles of POD and DMD. Additionally, regularization terms can be imposed to encourage sparsity in the Koopman matrix (Sinha et al., 2019).

Finally, it is crucial to recognize that the usefulness of the methods in this section presumes the dynamics are evolving on a low-dimensional subspace characterized by a quickly decaying singular value spectrum. While not a fundamental limitation of DMD, this is a common underlying assumption which may not hold for all dynamical systems. Erichson et al. (2019b) provide a turbulent flow example that demonstrates the limits of the approaches in this section when this assumption does not hold.

3.2.1 Compressed sensing meets DMD (cDMD and csDMD)

A full description of the extensive field of compressed sensing is beyond the scope of this review. We outline the key points to understand its interplay with DMD. The reader is encouraged to consult the excellent textbooks (Foucart and Rauhut, 2013; Adcock and Hansen, 2021) for a comprehensive understanding or (Candes and Wakin, 2008) for a concise introductory tutorial. Compressed sensing is founded on two central principles: *sparsity*, which pertains to the signals of interest, and *incoherence*, which relates to the sensing methodology.

Consider a signal $\mathbf{x} \in \mathbb{C}^d$ that is approximately sparse in some basis $\mathbf{B} \in \mathbb{C}^{d\times d}$, meaning that $\mathbf{x} = \mathbf{B}\mathbf{z}$, where the vector $\mathbf{z}$ can be well approximated by a sparse vector. Many natural signals, such as images and audio, are approximately sparse in specific bases like the Fourier or wavelet bases. When we transform an image using Fourier or wavelet transformations, most coefficients are small and can be disregarded while still retaining the quality of the image. We assume that we have access to measurements:

$$\mathbf{x}_c = \mathbf{C}\mathbf{x} = \mathbf{C}\mathbf{B}\mathbf{z},$$

where $\mathbf{C} \in \mathbb{C}^{p\times d}$ is a measurement matrix with $p < d$. Compressed sensing theory implies that, under suitable conditions, we can recover an accurate approximation of $\mathbf{z}$ (and hence $\mathbf{x}$) from the subsampled measurements $\mathbf{x}_c$. For

Algorithm 5 Compressed DMD (S.L. Brunton et al., 2016b), suitable when given access to the full snapshots.

Input: Snapshot data $\mathbf{X} \in \mathbb{C}^{d \times M}$ and $\mathbf{Y} \in \mathbb{C}^{d \times M}$, rank $r \in \mathbb{N}$, and measurement matrix $\mathbf{C} \in \mathbb{C}^{p \times d}$.

1: Compress $\mathbf{X}$ and $\mathbf{Y}$ to $\mathbf{X}_c = \mathbf{CX}$ and $\mathbf{Y}_c = \mathbf{CY}$.
2: Apply Algorithm 1 with input $\mathbf{X}_c$ and $\mathbf{Y}_c$ and outputs $\mathbf{\Lambda}_c$, $\mathbf{W}_c$, $\mathbf{V}_c$ and $\mathbf{\Sigma}_c$.
3: Reconstruct full-state modes via $\mathbf{\Phi} = \mathbf{Y}\mathbf{V}_c\mathbf{\Sigma}_c^{-1}\mathbf{W}_c$.

Output: The eigenvalues $\mathbf{\Lambda}_c$ and DMD modes $\mathbf{\Phi} \in \mathbb{C}^{d \times r}$.

example, consider the ℓ^1-minimization problem

$$\min \|\mathbf{z}\|_{\ell^1} \quad \text{subject to} \quad \mathbf{x}_c = \mathbf{CBz}. \tag{3.2}$$

Specifically, the measurement matrix $\mathbf{C}$ must be incoherent with respect to the sparse basis $\mathbf{B}$, meaning that the rows of $\mathbf{C}$ are uncorrelated with the columns of $\mathbf{B}$. If the matrix $\mathbf{CB}$ satisfies a restricted isometry property (RIP)[13]:

$$(1-\delta_k)\|\mathbf{z}\|_{\ell^2}^2 \leq \|\mathbf{CBz}\|_{\ell^2}^2 \leq (1+\delta_k)\|\mathbf{z}\|_{\ell^2}^2 \quad \text{for } k\text{-sparse vectors } \mathbf{z},$$

then we can prove results about how close solutions of (3.2) are to the true $\mathbf{z}$, how issues such as only approximately numerically solving (3.2) affect the solution, robustness to noise, and so forth. Beyond the above ℓ^1-minimization problem, many successful optimization problems and algorithms approximate their solutions in compressed sensing.

Tu et al. (2014a) combine *temporal* compressed sensing with ideas from DMD to recover POD modes. For the remainder of this section, we focus instead on *spatial* compressed sensing, following the methods of S.L. Brunton et al. (2016b). In essence, S.L. Brunton et al. (2016b) demonstrated that the unitary invariance of the DMD algorithm can be extended to approximate invariance under transformations satisfying a RIP, provided that the data is sparse in a basis that is incoherent with respect to the measurements. For compressed data matrices

$$\mathbf{X}_c = \mathbf{CX}, \quad \mathbf{Y}_c = \mathbf{CY},$$

there are essentially two approaches, depending on whether one has access to the matrix $\mathbf{Y}$ or not. Algorithm 5 illustrates *compressed DMD* (cDMD) (see also Erichson et al., 2019a), where one performs the standard DMD algorithm on the compressed data matrices, and then reconstructs the full-state modes using $\mathbf{Y}$.

[13] There are no known large matrices with bounded restricted isometry constants since computing these constants is NP-hard and hard to approximate. Typically, one builds random matrices so that the RIP holds with overwhelming probability. For example, Bernoulli and Gaussian random measurement matrices satisfy the RIP for a generic basis $\mathbf{B}$ with high probability (Candes and Tao, 2006).

Algorithm 6 Compressed sensing DMD (S.L. Brunton et al., 2016b), suitable when given access to only compressed data. The ℓ^1-minimization can be replaced with a plethora of similar minimization problems from the compressed sensing literature.

Input: Compressed snapshot data $\mathbf{X}_c \in \mathbb{C}^{p\times M}$ and $\mathbf{Y}_c \in \mathbb{C}^{p\times M}$, measurement matrix $\mathbf{C} \in \mathbb{C}^{p\times d}$, and basis $\mathbf{B} \in \mathbb{C}^{d\times d}$.

1: Apply Algorithm 1 with input $\mathbf{X}_c$ and $\mathbf{Y}_c$ and outputs $\mathbf{\Lambda}_c$ and $\mathbf{\Phi}_c$.
2: Apply ℓ^1-minimization (3.2) columnwise to reconstruct modes $\mathbf{\Phi}_s \in \mathbb{C}^{d\times r}$.
3: Recover full-state modes via $\mathbf{\Phi} = \mathbf{B}\mathbf{\Phi}_s$.

Output: The eigenvalues $\mathbf{\Lambda}_c$ and DMD modes $\mathbf{\Phi} \in \mathbb{C}^{d\times r}$.

If access to $\mathbf{Y}$ is not available, we can use an optimization problem such as (3.2) to recover the modes in the sparse basis $\mathbf{\Phi}_s$, and then reconstruct the full-state modes. This approach, known as *compressed sensing DMD* (csDMD), is outlined in Algorithm 6.

3.2.2 Randomized dynamic mode decomposition (rDMD)

Early uses of DMD with randomized SVD include Erichson and Donovan (2016), who utilized it to expedite DMD applications in video background subtraction, and Bistrian and Navon (2017), who applied it as a component of a reduced-order model for two-dimensional fluid flows. Although this method is reliable and robust to noise, it only accelerates the computation of the SVD, with subsequent computational steps in the DMD algorithm remaining costly. Instead, Erichson et al. (2019b) developed a *randomized DMD* (rDMD) algorithm. This algorithm relies on sketching the range of $\mathbf{X}$ and executing the entire DMD process in a reduced-dimensional space, ultimately recovering the DMD of the original system at the end.

The idea is to use randomness as a computational strategy to find a smaller representation, known as a *sketch*. This smaller matrix sketch can be used to compute an approximate low-rank factorization for the high-dimensional data matrix. rDMD utilizes the off-the-shelf probabilistic framework proposed in the seminal work of Halko et al. (2011). Given a target rank r, the aim is to compute a near-optimal basis $\mathbf{Q} \in \mathbb{C}^{d\times r}$ for the input matrix $\mathbf{X}$ such that $\mathbf{X} \approx \mathbf{Q}\mathbf{Q}^*\mathbf{X}$. A test matrix $\mathbf{\Omega} \in \mathbb{R}^{M\times r}$ is drawn from a normal Gaussian distribution to sample the range of $\mathbf{X}$ via

$$\mathbf{Z} = \mathbf{X}\mathbf{\Omega}.$$

To mitigate the $\mathcal{O}(dMr)$ cost of dense matrix multiplication, more sophisticated random test matrices, such as the subsampled randomized Hadamard transform, can also be used, leading to a complexity of $\mathcal{O}(dM\log(r))$. The orthonormal basis $\mathbf{Q}$ is then obtained via QR decomposition of $\mathbf{Z}$. In practice, we slightly oversample the desired rank r by a constant factor (typically, 10 suffices). A

Algorithm 7 Randomized range finder. Other choices of random test matrices may be employed for computational efficiency. The QR algorithm is written using MATLAB notation.

Input: Snapshot data $\mathbf{X} \in \mathbb{C}^{d \times M}$, target rank $r \in \mathbb{N}$, oversampling factor $p \in \mathbb{N}$, and power iteration factor $q \in \mathbb{N} \cup \{0\}$.

1: Generate a random Gaussian matrix $\mathbf{\Omega} \in \mathbb{R}^{M \times (r+p)}$ and form the matrix $\mathbf{Z} = \mathbf{X\Omega}$.
2: **for** $j = 1, \ldots, q$ **do**
3: $[\mathbf{Q}, \sim] = \texttt{qr}(\mathbf{Z}, \text{'econ'})$
4: $[\mathbf{C}, \sim] = \texttt{qr}(\mathbf{X}^*\mathbf{Q}, \text{'econ'})$
5: $\mathbf{Z} = \mathbf{XC}$
6: **end for**
7: $[\mathbf{Q}, \sim] = \texttt{qr}(\mathbf{Z}, \text{'econ'})$.

Output: Range matrix $\mathbf{Q} \in \mathbb{C}^{d \times (r+p)}$.

Algorithm 8 Randomized DMD (Erichson et al., 2019b).

Input: Snapshot data $\mathbf{X} \in \mathbb{C}^{d \times M}$ and $\mathbf{Y} \in \mathbb{C}^{d \times M}$, target rank $r \in \mathbb{N}$, oversampling factor $p \in \mathbb{N}$, and power iteration factor $q \in \mathbb{N} \cup \{0\}$.

1: Run Algorithm 7 to generate the matrix $\mathbf{Q}$.
2: Compress $\mathbf{X}$ and $\mathbf{Y}$ to $\mathbf{X}_c = \mathbf{Q}^*\mathbf{X}$ and $\mathbf{Y}_c = \mathbf{Q}^*\mathbf{Y}$.
3: Apply Algorithm 1 with input $\mathbf{X}_c$ and $\mathbf{Y}_c$ and outputs $\mathbf{\Lambda}_c$ and $\mathbf{\Phi}_c$.
4: Reconstruct full-state modes via $\mathbf{\Phi} = \mathbf{Q\Phi}_c$.

Output: The eigenvalues $\mathbf{\Lambda}_c$ and DMD modes $\mathbf{\Phi} \in \mathbb{C}^{d \times r}$.

second strategy to improve performance involves power iterations (Rokhlin et al., 2010; Gu, 2015). Particularly, a slowly decaying singular value spectrum of the input matrix can significantly affect the quality of the approximated basis matrix $\mathbf{Q}$. Power iterations are employed to preprocess the input matrix to promote a more rapidly decaying spectrum. The sampling matrix obtained is

$$\mathbf{Z} = (\mathbf{XX}^*)^q \mathbf{X\Omega},$$

and as few as $q = 2$ power iterations can considerably improve the approximation quality, even when the singular values of the input matrix decay slowly. This procedure is outlined in Algorithm 7, and we direct the reader to Martinsson and Tropp (2020, Section 11) for probabilistic error bounds. With the matrix $\mathbf{Q}$ in hand, we can perform DMD on the lower-dimensional space, as summarized in Algorithm 8. One can also simultaneously sketch the range and corange of $\mathbf{X}$. This method, known as *sketchy DMD*, was proposed by Ahmed et al. (2022).

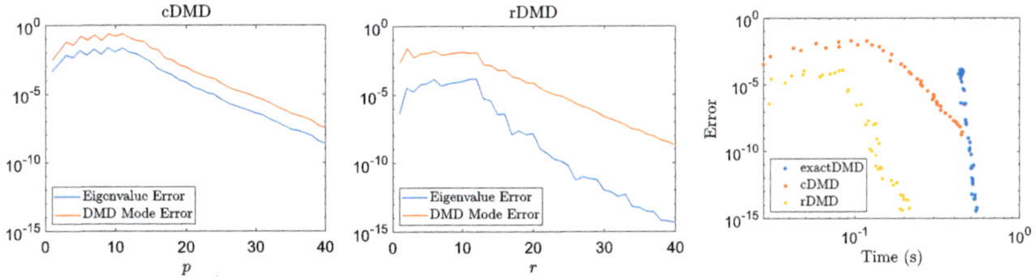

FIGURE 11 Left: Errors of cDMD (Algorithm 5) for the first 6 eigenvalues and DMD modes. Middle: Errors of rDMD (Algorithm 8) for the first six eigenvalues and DMD modes. Right: Computational times vs eigenvalue error.

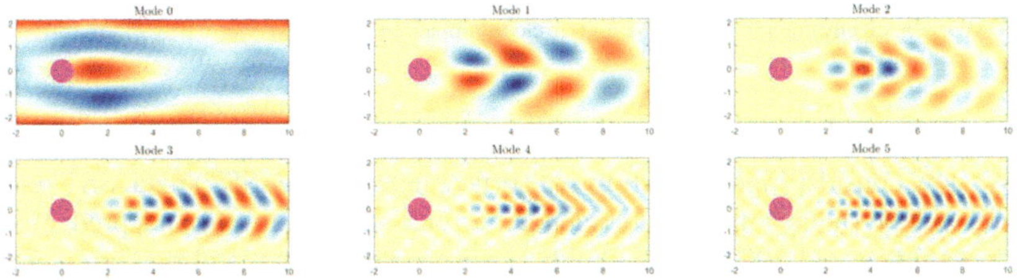

FIGURE 12 DMD modes constructed using csDMD (Algorithm 6) with a 99.5% compression in the dimension of the measurements.

3.2.3 Examples

Cylinder wake

We return to the cylinder wake discussed in Section 2.3.1 as an illustrative example. For cDMD and csDMD, we use the (inverse) two-dimensional discrete Fourier transform as our basis $\mathbf{B}$ and Gaussian random measurements $\mathbf{C}$. Fig. 11 shows the results for cDMD and csDMD when recovering the first six modes plotted in Fig. 2. In the left and middle panels, we have shown the mean eigenvalue error and the DMD mode error (computed as a subspace angle) averaged over 20 random realizations. The errors quickly become negligible for $p, r = \mathcal{O}(10)$. In the right panel, we have displayed the mean execution times on a standard laptop (without GPU) against the eigenvalue error. Even for this simple example with rapidly decreasing singular values, cDMD performs better than exact DMD, while rDMD is the clear winner.

Fig. 12 shows the modes computed by csDMD with $p = 800$, which corresponds to a 99.5% compression in the dimension of the snapshot matrices. We employ the CoSaMP algorithm (Needell and Tropp, 2009) to perform the ℓ^1-minimization step. When only compressed measurements are available, it is still possible to reconstruct full-state modes using compressed sensing. However, this typically requires more measurements and computational resources than cDMD or rDMD.

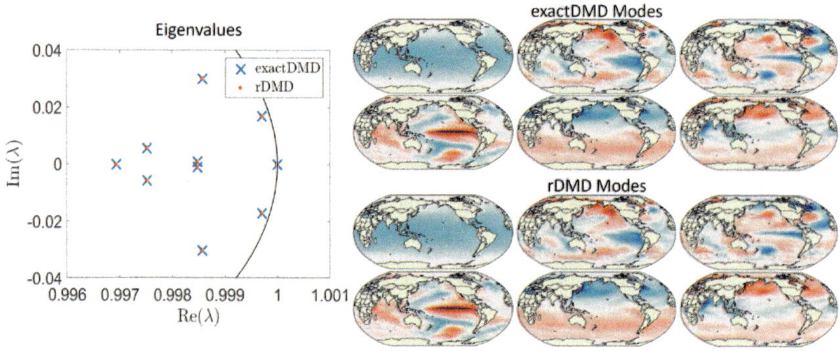

FIGURE 13 Eigenvalues and dynamic modes of the SST data set computed using exact DMD and rDMD. The eigenvalue plot shows the unit circle as a black line.

Sea surface data

We now consider high-resolution sea surface temperature (SST) data. SST data are widely studied in climate science for climate monitoring and prediction (Reynolds et al., 2007, 2002; Smith and Reynolds, 2005), and measurements are constructed by combining infrared satellite data with observations provided by ships and buoys. The data are available from the National Oceanic and Atmospheric Administration at https://www.esrl.noaa.gov/psd/ for the years 1981 to 2023, with a grid resolution of 0.25°. Omitting data over land results in $d = 691,150$ spatial grid points. The following experiments, similar in spirit to Erichson et al. (2019b), were performed using a system with Intel(R) Xeon(R) Gold 6126 CPU at 2.60GHz (48 cores) and 767GiB system memory.

We first consider a temporal resolution of one day and a data matrix $\mathbf{X} \in \mathbb{R}^{691,150 \times 15,097}$. Fig. 13 shows the eigenvalues and dynamic modes computed using exact DMD and rDMD (with $r = 10$), demonstrating the accuracy of rDMD. The bottom left mode is reminiscent of an El Niño mode generated from the El Niño-Southern Oscillation (ENSO). El Niño is the warm phase of the ENSO cycle. It is associated with a band of warm ocean water that develops in the central and east-central equatorial Pacific, including off the Pacific coast of South America (see also Fig. 15).

Next, we compare the accuracy and computational times for a temporal resolution of one week and a data matrix $\mathbf{X} \in \mathbb{R}^{691,150 \times 2156}$ of weekly averages. Fig. 14 shows the relative error in the Frobenius norm of the reconstructed data matrix and time taken for exact DMD, rDMD, and blocked rDMD (using four blocks). We observe substantial gains in computational time when using rDMD while maintaining an accuracy similar to the full deterministic exact DMD.

3.3 Multiresolution dynamic mode decomposition (mrDMD)

Multiscale systems are widespread across various scientific disciplines. Modeling the interactions between microscale and macroscale phenomena, which may

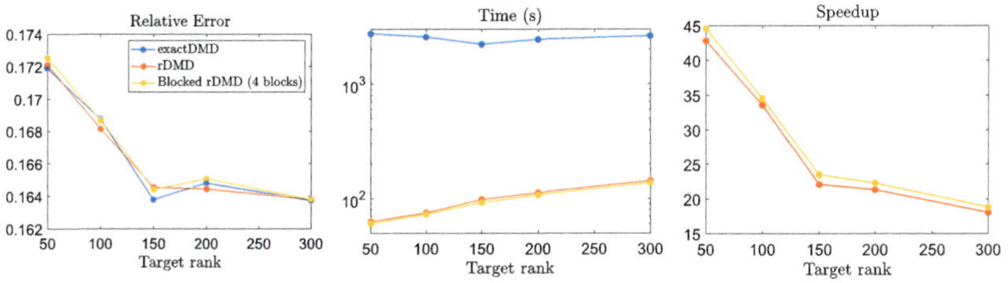

FIGURE 14 Left: Accuracy of exact DMD, rDMD, and batched rDMD for the SST data set. Middle: The computational times of each method. Right: The speedup of rDMD and batched rDMD compared to exact DMD. All plots show the mean of 10 independent runs.

differ by orders of magnitude either spatially or temporally, poses a considerable challenge. Wavelet-based methods and windowed Fourier transforms are well-suited for multiresolution analysis (MRA), as they systematically isolate temporal or spatial features through recursive refinement when sampling from the targeted data (Daubechies, 1992). Typically, MRA is employed separately in either space or time, but it is seldom applied to both simultaneously.

Multiresolution DMD (mrDMD) (Kutz et al., 2016b) integrates DMD with core principles from wavelet theory and MRA. It adjusts the sampling window of the data collection process in line with wavelet theory, filtering information across various scales. The process is iteratively refined through progressively shorter snapshot sampling windows, leading to the recursive extraction of DMD modes from slow to rapidly changing timescales. The benefits of this approach include enhanced prediction of the near-future state of the system, which is vital for control; effective management of transient phenomena; and improved handling of moving (translating/rotating) structures within the data. The latter two points underscore significant challenges inherent in standard DMD methods. The mrDMD algorithm has led to practical applications such as determining optimal sensor placement (Manohar et al., 2019).

3.3.1 The algorithm

When using mrDMD, it is typical to work with an M such that a full-rank approximation with $r = M$ in Algorithm 1 is feasible and such that high- and low-frequency content is present. We assume that data is collected along a single time trajectory with time-step Δt and express the eigenvalues in terms of their time-scaled logarithms $\eta = \log(\lambda)/\Delta t$. In the first pass, mrDMD separates the approximation in (2.7) into slow modes and fast modes:

$$\mathbf{x}(t) \approx \underbrace{\sum_{|\eta_k| \leq \tau} \boldsymbol{\phi}_k^{(1)} \exp(\eta_k t)\mathbf{b}(k)}_{\text{slow modes}} + \underbrace{\sum_{|\eta_k| > \tau} \boldsymbol{\phi}_k^{(1)} \exp(\eta_k t)\mathbf{b}(k)}_{\text{fast modes}}, \tag{3.3}$$

where $\boldsymbol{\phi}_k^{(1)} = \boldsymbol{\Phi}(:, k)$, and the superscript (1) indicates the level. The first sum in the expression (3.3) represents the slow-mode dynamics, whereas the second sum is everything else. How to choose the slow modes is important in the practical implementation. In the original mrDMD paper, it is suggested to set the threshold τ to select eigenvalues whose temporal behavior allows for at most one wavelength to fit within the sampling window.

The fast modes in (3.3) can be collected into a data matrix $\mathbf{X}_{M/2}$, where we let m_1 denote the number of slow modes in (3.3). The matrix $\mathbf{X}_{M/2}$ is now split into two matrices, where the first matrix contains the first $M/2$ snapshots, and the second matrix contains the remaining $M/2$ snapshots. The process is now repeated, where m_2 slow modes are collected at the second level and computed separately in the first and second intervals of snapshots. This process is repeated to obtain the decomposition

$$\mathbf{x}(t) \approx \sum_{k=1}^{m_1} b_k^{(1)} \boldsymbol{\phi}_k^{(1)} \exp(\eta_k^{(1)} t) + \sum_{k=1}^{m_2} b_k^{(2)} \boldsymbol{\phi}_k^{(2)} \exp(\eta_k^{(2)} t)$$
$$+ \sum_{k=1}^{m_3} b_k^{(3)} \boldsymbol{\phi}_k^{(3)} \exp(\eta_k^{(3)} t) + \cdots,$$

where the $\boldsymbol{\phi}_k^{(\ell)}$ and $\eta_k^{(\ell)}$ are the DMD modes and eigenvalues at the ℓth level of the decomposition, the $b_k^{(\ell)}$ are the initial projections of the data onto the time interval of interest, and the m_ℓ are the number of slow modes retained at each level. The idea is that different spatiotemporal DMD modes are used to represent key multiresolution features. Thus, no single set of modes dominates the SVD and potentially marginalizes features at other time scales.

We can make the mrDMD more precise, letting L denote the number of levels of the decomposition. The solution is a sum with ℓ indexing the level, $j = 1, \ldots, 2^{(\ell-1)}$ indexing the time bins $[t_j^{(\ell)}, t_{j+1}^{(\ell)}]$ in each level and $k = 1, \ldots, m_\ell$ indexing the modes extracted at each level. To simplify the sum, define the following indicator function

$$f_{\ell,j}(t) = \begin{cases} 1, & \text{if } t \in [t_j^{(\ell)}, t_{j+1}^{(\ell)}] \\ 0, & \text{otherwise} \end{cases},$$

$$\text{so that} \quad \mathbf{x}_{\text{mrDMD}}(t) = \sum_{\ell=1}^{L} \sum_{j=1}^{2^{\ell-1}} \sum_{k=1}^{m_\ell} f_{\ell,j}(t) b_k^{(\ell,j)} \boldsymbol{\phi}_k^{(\ell,j)} \exp(\eta_k^{(\ell,j)} t).$$

In particular, each mode is represented in its respective time bin and level. Alternatively, this solution can be interpreted as yielding the least-squares fit to the dynamics within a given time bin at each level of the decomposition.

Numerous innovations enhance the practical implementation of mrDMD. Since only slow modes matter within a window, we can limit sampling to a fixed number of points per window, reducing the data matrix size for more manageable SVD computations. The sampling window locations are flexible, and

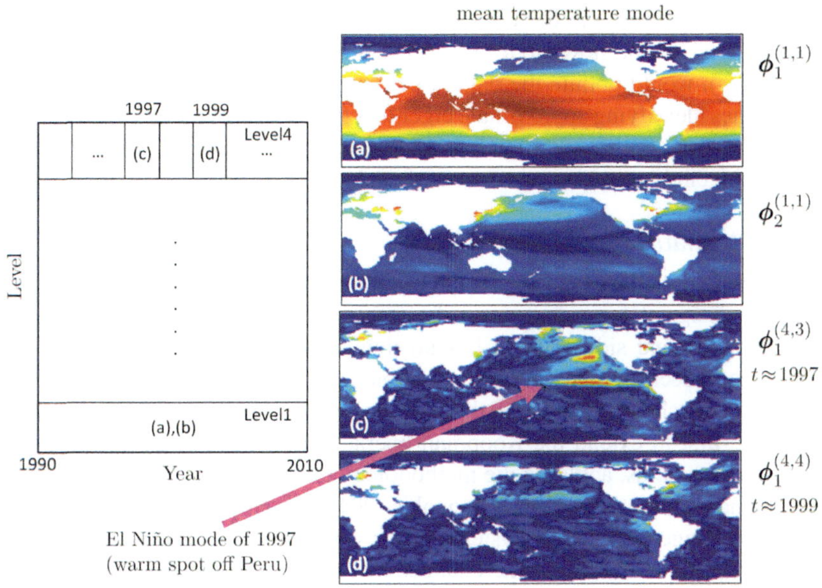

FIGURE 15 Application of mrDMD on sea surface temperature data from 1990 to 2010. The left panel illustrates the process for a 4-level decomposition. At each level, the slowest modes are extracted. Mode (c) clearly shows the El Niño mode of interest that develops in the central and east-central equatorial Pacific. The El Niño mode was absent in 1999, as is clear from mode (d). Reproduced with permission from Kutz et al. (2016b), copyright © 2016 Society for Industrial and Applied Mathematics, all rights reserved.

smoothing their edges can prevent the Gibbs phenomenon due to abrupt data cutoffs. One can also employ wavelet functions, like Haar, Daubechies, or Mexican Hat, for the sifting function $f_{\ell,j}(t)$. Finally, overlapping windows prevent data loss during sampling. A sliding window approach can robustly track data features, enabling pattern correlation across windows akin to a Gabor transform-generated spectrogram.

3.3.2 *Example*

The following example is from Kutz et al. (2016b). We consider the global sea surface temperature (see Section 3.2.3) with data that spans 20 years from 1990 to 2010. Fig. 15 illustrates the outcomes of employing a 4-level mrDMD decomposition. At the first level, mrDMD identifies two modes: the mean ocean temperature, denoted as $\boldsymbol{\phi}_1^{(1,1)}$, and an annual cycle, represented by $\boldsymbol{\phi}_2^{(1,1)}$. Intriguingly, at the fourth level, the approximate zero mode of the sampling window uncovers noteworthy phenomena; specifically, it isolates the 1997 El Niño event. In contrast, when the same sampling window is applied to the year 1999, the El Niño mode is absent, aligning with the recognized oceanic patterns of that year. These insights would not have been obtainable using traditional DMD without preselecting the correct sampling windows. Moreover, even if such a step were taken, the slow modes identified at the first level, as shown in Fig. 15 (a) and (b), would pollute the data at the level of investigation. For additional techniques identifying approximate eigenfunctions that provide a rectified representation of the ENSO and function as (approximate) semiconjugacies or factor maps with circle rotations, see Froyland et al. (2021).

3.3.3 Nonautonomous systems

The use of mrDMD to detect transient behavior is tantalizing! Most DMD methods are designed for autonomous dynamical systems, where the function **F** on the right-hand side of (1.1) has no time dependence. However, there has been some recent initial work on nonautonomous systems, and we expect this area to grow significantly over the next few years. Mezić and Surana (2016) were the first to extend the Koopman operator framework to nonautonomous dynamical systems, applying the methodology to linear-periodic and quasiperiodic nonautonomous systems. Giannakis (2019) developed a strategy inspired by time-changed dynamical systems that involves rescaling the generator; this can be applied to a class of time-changed mixing systems. Further development of this approach, using delay-coordinate maps for recovering the dynamical system on tori with multiple time scales, is presented in Das and Giannakis (2019). The extraction of spatiotemporal patterns using an extension of approximation techniques developed in Giannakis (2019) on the space-time manifold defined as a skew-product structure is considered in Giannakis and Das (2020); Giannakis et al. (2019). For the online computation of windowed DMD using rank-one updates, see *Online DMD* (Zhang et al., 2019). Maćešić et al. (2018) provides an error analysis for DMD with moving stencils. For extensions to actuated systems, see Williams et al. (2016); Bai et al. (2020). Redman et al. (2023) develop an episodic memory approach that saves spectral objects associated with temporally local approximations of the Koopman operator, and utilizes this information to make new predictions. Nonautonomous systems have also been studied using transfer operators, which are the dual of Koopman operators (see the discussion in Section 6.1), and space-time manifolds (Froyland and Koltai, 2023).

3.4 Control

One of the most successful applications of the Koopman operator framework lies in control (Mauroy et al., 2020; Otto and Rowley, 2021), with demonstrated successes in various challenging applications. These include fluid dynamics (Arbabi et al., 2018; Peitz and Klus, 2020), robotics (Abraham et al., 2017; Bruder et al., 2019; Mamakoukas et al., 2019; Haggerty et al., 2023), power grids (Korda et al., 2018; Netto and Mili, 2018), biology (Hasnain et al., 2020), and chemical processes (Narasingam and Kwon, 2019). The key point is that Koopman operators represent nonlinear dynamics within a globally linear framework. This approach leads to tractable convex optimization problems and circumvents theoretical and computational limitations associated with nonlinearity. Moreover, it is amenable to data-driven, model-free approaches (Proctor et al., 2016; Williams et al., 2016; Korda and Mezić, 2018a; Proctor et al., 2018; Surana, 2016; Kaiser et al., 2021, 2018a; Peitz and Klus, 2019; Abraham and Murphey, 2019). The resulting models reveal insights into global stability properties (Sootla and Mauroy, 2016; Mauroy and Mezić, 2016), observability/controllability (Vaidya, 2007; Goswami and Paley, 2017; Yeung et al., 2018), and sensor/actuator placement (Sinha et al., 2016; Sharma et al., 2019) for the underlying nonlinear systems.

Koopman operator theory was first extended to actuated systems by Mezić and Banaszuk (2004), with stochastic forcing interpreted as actuation. Proctor et al. (2016) developed the first control schemes based on DMD. A significant strength of DMD is the ability to describe complex and high-dimensional dynamical systems with a few dominant modes. Reducing the system's dimensionality enables faster and lower-latency prediction and estimation, leading to high-performance, robust controllers.

3.4.1 Dynamic mode decomposition with control (DMDc)

We will focus on the *DMD with control* (DMDc) algorithm (Proctor et al., 2016). DMDc extends DMD to disambiguate between unforced dynamics and the effect of actuation. The DMD regression of Section 2.2.1 is generalized to

$$\mathbf{x}_{n+1} = \mathbf{F}(\mathbf{x}_n, \mathbf{u}_n) \approx \mathbf{A}\mathbf{x}_n + \mathbf{B}\mathbf{u}_n,$$

where $\mathbf{u}_n \in \mathbb{C}^q$ is a vector of control inputs for each time-step. Here $\mathbf{A} \in \mathbb{C}^{d\times d}$ and $\mathbf{B} \in \mathbb{C}^{d\times q}$ are unknown matrices. Snapshot triplets of the form $\{\mathbf{x}^{(m)}, \mathbf{y}^{(m)}, \mathbf{u}^{(m)}\}_{m=1}^{M}$ are collected, where we assume that

$$\mathbf{y}^{(m)} \approx \mathbf{F}(\mathbf{x}^{(m)}, \mathbf{u}^{(m)}), \quad m = 1, \ldots, M.$$

The control portion of the snapshots is arranged into the matrix $\boldsymbol{\Upsilon} = \begin{pmatrix}\mathbf{u}^{(1)} \, \mathbf{u}^{(2)} \, \cdots \, \mathbf{u}^{(M)}\end{pmatrix} \in \mathbb{C}^{q\times M}$. The optimization problem in (2.6) is replaced by

$$\min_{(\mathbf{A}\ \mathbf{B})} \|\mathbf{Y} - (\mathbf{A}\ \mathbf{B})\boldsymbol{\Omega}\|_{\mathrm{F}}^2, \quad \text{where} \quad \boldsymbol{\Omega} = \begin{pmatrix}\mathbf{X} \\ \boldsymbol{\Upsilon}\end{pmatrix} \in \mathbb{C}^{(d+q)\times M}.$$

A solution is given as $(\mathbf{A}\ \mathbf{B}) = \mathbf{Y}\boldsymbol{\Omega}^\dagger$. In practice, we seek a reduced-order model by performing a truncated SVD on both the input and output space. The full algorithm is summarized in Algorithm 9 and is an extension of Algorithm 1. DMDc has been used with Model-Predictive Control (MPC) for enhanced control of nonlinear systems in Korda and Mezić (2018a); Kaiser et al. (2018b), with the DMDc method performing surprisingly well, even for strongly nonlinear systems. Extensions are discussed in Section 3.4.3.

Algorithm 9 DMD with control (Proctor et al., 2016).

Input: Snapshot data $\mathbf{X} \in \mathbb{C}^{d\times M}$, $\mathbf{Y} \in \mathbb{C}^{d\times M}$ and $\boldsymbol{\Upsilon} \in \mathbb{C}^{q\times M}$, ranks $r, p \in \mathbb{N}$.

1: Compute a truncated SVD of the input matrix $\begin{pmatrix}\mathbf{X}\\ \boldsymbol{\Upsilon}\end{pmatrix} \approx \tilde{\mathbf{U}}\tilde{\boldsymbol{\Sigma}}\tilde{\mathbf{V}}^*$, $\tilde{\mathbf{U}} \in \mathbb{C}^{(d+q)\times p}$, $\tilde{\boldsymbol{\Sigma}} \in \mathbb{R}^{p\times p}$, $\tilde{\mathbf{V}} \in \mathbb{C}^{M\times p}$. Break up the matrix $\tilde{\mathbf{U}}$ into $\tilde{\mathbf{U}}^* = [\tilde{\mathbf{U}}_1^*\ \tilde{\mathbf{U}}_2^*]$ where $\tilde{\mathbf{U}}_1 \in \mathbb{C}^{d\times p}$ and $\tilde{\mathbf{U}}_2 \in \mathbb{C}^{q\times p}$.

2: Compute a truncated SVD of $\mathbf{Y} \approx \widehat{\mathbf{U}}\widehat{\boldsymbol{\Sigma}}\widehat{\mathbf{V}}^*$, $\widehat{\mathbf{U}} \in \mathbb{C}^{d\times r}$, $\widehat{\boldsymbol{\Sigma}} \in \mathbb{R}^{r\times r}$, $\widehat{\mathbf{V}} \in \mathbb{C}^{M\times r}$.

3: Compute the compressions $\tilde{\mathbf{A}} = \widehat{\mathbf{U}}^*\mathbf{Y}\tilde{\mathbf{V}}\tilde{\boldsymbol{\Sigma}}^{-1}\tilde{\mathbf{U}}_1^*\widehat{\mathbf{U}} \in \mathbb{C}^{r\times r}$ and $\tilde{\mathbf{B}} = \widehat{\mathbf{U}}^*\mathbf{Y}\tilde{\mathbf{V}}\tilde{\boldsymbol{\Sigma}}^{-1}\tilde{\mathbf{U}}_2^* \in \mathbb{C}^{r\times q}$.

4: Compute the eigendecomposition $\tilde{\mathbf{A}}\mathbf{W} = \mathbf{W}\boldsymbol{\Lambda}$. The columns of $\mathbf{W}$ are eigenvectors and $\boldsymbol{\Lambda}$ is a diagonal matrix of eigenvalues.

5: Compute the modes $\boldsymbol{\Phi} = \mathbf{Y}\tilde{\mathbf{V}}\tilde{\boldsymbol{\Sigma}}^{-1}\tilde{\mathbf{U}}_1^*\widehat{\mathbf{U}}\mathbf{W}$.

Output: The eigenvalues $\boldsymbol{\Lambda}$ and modes $\boldsymbol{\Phi} \in \mathbb{C}^{d\times r}$.

3.4.2 *Example*

We illustrate DMDc for system identification on a high-dimensional, linear system with spectral sparsity following Proctor et al. (2016, Section 4.3). We consider a two-dimensional torus discretized by a 128×128 equispaced grid such that $\mathbf{x} \in \mathbb{R}^{128\times 128} \cong \mathbb{R}^{16,384}$. Taking the two-dimensional discrete Fourier transform of $\mathbf{x}$, we obtain $\hat{\mathbf{x}}$. The system evolves according to

$$\hat{\mathbf{x}}_{n+1} = \hat{\mathbf{A}}\hat{\mathbf{x}}_n + \hat{\mathbf{B}}\hat{\mathbf{u}}_n.$$

Here, $\hat{\mathbf{A}}$ is a diagonal matrix with five nonzero entries representing the modes, each with a randomly chosen frequency and small damping. The random input signal, $\hat{\mathbf{u}}$, is one-dimensional and directly influences the sparse modes, resulting in a localized negative control input when transformed back to the spatial domain. This back-transformation yields our dynamical system in physical space. The system is constructed by sampling a continuous-time system at time steps of $\Delta t = 0.01$. We collect $M = 400$ snapshots of the data for our analysis. Further details of this system can be found in S.L. Brunton et al. (2016b).

Fig. 16 displays the true eigenvalues alongside those computed by DMDc and exact DMD. Ten eigenvalues are present in conjugate pairs due to processing real-valued data. DMDc demonstrates greater accuracy than exact DMD, which inaccurately estimates some eigenvalues and generates unstable modes. The true DMD modes for this system appear at the top Fig. 17. The DMDc modes in the middle row align almost perfectly with the true modes. The subspace angle between the true modes and the DMDc-computed modes is on the order of machine precision. In contrast, the modes produced by exact DMD show significant distortion.

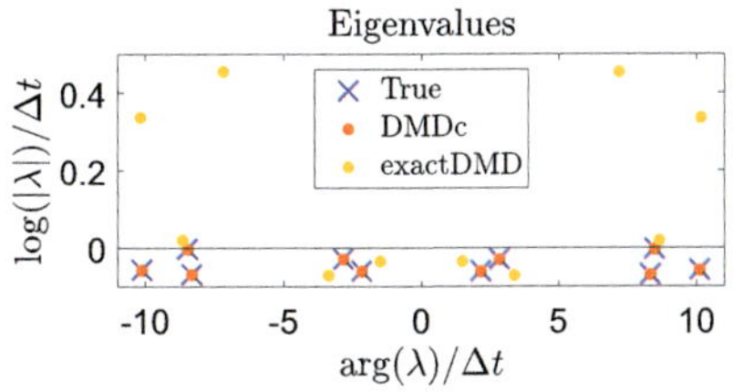

FIGURE 16 True eigenvalues of the torus example and those computed by DMDc and exact DMD. The logarithm of the eigenvalues are plotted to align with the continuous-time system.

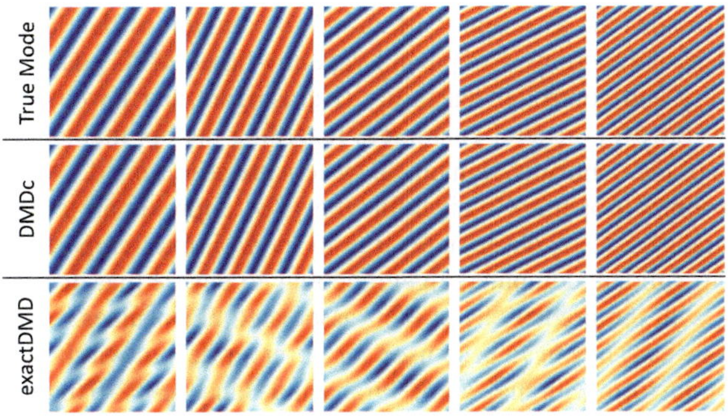

FIGURE 17 The true DMD modes for the torus example, alongside those computed by DMDc and exact DMD. The modes obtained from DMDc are accurate to machine precision, whereas those computed using exact DMD are significantly distorted.

3.4.3 Extensions and connection with Koopman operators

Koopman theory has been used in combination with the Linear Quadratic Regulator (LQR) (S.L. Brunton et al., 2016a; Mamakoukas et al., 2019, 2021), state-dependent LQR (Kaiser et al., 2021), and MPC (Korda and Mezić, 2018a; Kaiser et al., 2018b). Other noteworthy directions include optimal control for switching control problems (Peitz and Klus, 2019, 2020), Lyapunov-based stabilization (Huang et al., 2018, 2020), eigenstructure assignment (Hemati and Yao, 2017), and active learning (Abraham and Murphey, 2019). Additionally, deep learning architectures have been employed to represent the nonlinear observables in combination with MPC (Li et al., 2019), see also (Liu et al., 2018; Han et al., 2020), and (Peitz and Klus, 2019; Peitz et al., 2020; Klus et al., 2020b) for parametrized models.

Koopman theory is closely related to Carleman linearization (Carleman, 1932), which embeds finite-dimensional dynamics into infinite-dimensional linear systems using a polynomial basis. Carleman linearization has been used for decades to obtain truncated linear (and bilinear) state estimators (Krener, 1974; Brockett, 1976) and to examine stability, observability, and controllability of the underlying nonlinear system (Loparo and Blankenship, 1978).

The DMDc framework may be extended to nonlinear observables using EDMD (see Section 4), an approach called eDMDc (Williams et al., 2016). Korda and Mezić (2018a) integrated eDMDc into MPC. Here, the Koopman operator is characterized as an autonomous operator on the extended state vector $(\mathbf{x}^\top, \mathbf{u}^\top)^\top$, with observables that may be nonlinear functions of both the state and the input. In practical applications, simplifications are employed to ensure the control problem remains convex (Korda and Mezić, 2018a; Proctor et al., 2018). This method has been applied for control in the coordinates of Koopman eigenfunctions (Kaiser et al., 2021, 2018a; Folkestad et al., 2020) and in interpolated Koopman models (Peitz, 2018; Peitz et al., 2020). Convergence can be established under the assumption of an infinite amount of data and an infinite number of basis functions. Koopman Lyapunov-based MPC guarantees closed-loop stability and controller feasibility (Narasingam and Kwon, 2019; Son et al., 2020). However, general guarantees regarding the optimality, stability, and robustness of the controlled dynamical system are still limited.

The Koopman operator's eigenfunctions (or approximate eigenfunctions) are a natural choice of observables due to their simple temporal behavior. It is crucial to validate computed eigenfunctions to ensure that their evolution is consistent with the predictions of their associated eigenvalues, particularly for prediction tasks. They have been used for observer design within the Koopman canonical transform (Surana, 2016; Surana and Banaszuk, 2016) and within the Koopman reduced-order nonlinear identification and control framework (Kaiser et al., 2021), which both typically yield a global bilinear representation of the underlying system. Subsequent research has focused on directly identifying Koopman eigenfunctions (Korda and Mezić, 2020; Pan et al., 2021) and approximate invariant subspaces (Haseli and Cortés, 2023).

The efficacy of Koopman-based MPC is currently at odds with the difficulties of approximating the Koopman operator and its spectra. Only a limited number of systems with a known Koopman-invariant subspace and verifiable eigenfunctions exist for model analysis and evaluation. Furthermore, the linearity of Koopman eigenfunctions is seldom validated. Nevertheless, Koopman-based MPC demonstrates remarkable resilience with models of marginal predictive ability. Despite notable successes, understanding how well the Koopman operator is actually approximated and producing error bounds remains largely incomplete.

4 Variants from the Galerkin perspective

We now explore variants of DMD from the Galerkin (or projection) perspective, building on the connection established in Section 2.2.2. This approach particularly focuses on addressing the infinite-dimensional nature of Koopman operators. Given that a Koopman operator transforms a finite-dimensional nonlinear system into an infinite-dimensional linear one, a significant part of this section will address nonlinear observables. We will concentrate on three methods designed to tackle these challenges:

- **Extended DMD:** This represents a fundamental extension of DMD that treats it as a Galerkin method.[14] In particular, it introduces nonlinear observables to form a dictionary, which generates a subspace within $L^2(\Omega, \omega)$. Adopting the Galerkin perspective enables the application of numerical tools for addressing infinite-dimensional spectral problems. However, the well-studied challenges of infinite-dimensional spectral computations are significant. Generally, EDMD will not converge to the spectral properties of the Koopman operator, either theoretically or practically (see Section 4.1.3 and common pitfalls in Section 4.3).
- **Time-delay Embedding:** This technique is commonly used to construct a dictionary of observables for EDMD and generates a Krylov subspace. Our focus will be on two methods: Hankel-DMD, which is a widely used technique suitable for ergodic systems that have a low-dimensional attractor, and HAVOK (Hankel Alternative View Of Koopman) analysis, which produces a linear model using the leading delay coordinates and includes forcing terms represented by low-energy delay coordinates.
- **Residual DMD:** This algorithm computes verified spectral properties of Koopman operators via an infinite-dimensional residual corresponding to the projection error of (E)DMD. This residual is computed from the snapshot data by augmenting EDMD with an additional matrix. This leads to the computation of spectra and pseudospectra without spectral pollution (general systems) and can be used to compute spectral measures (measure-preserving systems). Since the algorithms have error control, ResDMD allows a posteriori verification of spectral quantities, Koopman mode decompositions, and learned dictionaries.[15]

4.1 Nonlinear observables: extended dynamic mode decomposition (EDMD)

The standard DMD algorithm can accurately characterize periodic and quasi-periodic behaviors in nonlinear systems. However, DMD models based on linear observables generally fail to capture truly nonlinear phenomena. To address this limitation, Williams et al. (2015a) introduced *Extended DMD* (EDMD), which also elucidated the interpretation of DMD as a Galerkin method. Specifically, they demonstrated that EDMD converges to the numerical approximation obtained by a Galerkin method in the limit of large data sets. Prior research in a similar vein includes (Tu et al., 2014b). Moreover, the connection between EDMD and the earlier variational approach of conformation dynamics (Noé and Nüske, 2013; Nüske et al., 2014) from molecular dynamics is explored in Wu et al. (2017); Klus et al. (2018c).

---

[14] Though once nonlinear observables have been chosen, one can also apply the regression interpretation of Section 3.

[15] One can often show that a priori error control is impossible (Colbrook et al., 2019).

4.1.1 The algorithm

Following the discussion of Koopman operators in Section 2.1, the objective of EDMD is to approximate the Koopman operator with a matrix. For the sake of simplicity, the initial formulation of EDMD assumes that the columns of the snapshot matrix $\mathbf{X}$ are independently sampled from the distribution ω. In our discussion, we extend EDMD to accommodate any given snapshot matrices

$$\mathbf{X} = \begin{pmatrix} \mathbf{x}^{(1)} & \mathbf{x}^{(2)} & \dots & \mathbf{x}^{(M)} \end{pmatrix} \quad \text{and} \quad \mathbf{Y} = \begin{pmatrix} \mathbf{y}^{(1)} & \mathbf{y}^{(2)} & \dots & \mathbf{y}^{(M)} \end{pmatrix},$$

and consider the $\mathbf{x}^{(m)}$ as *quadrature nodes* used for integration with respect to ω. This adaptability permits the application of various quadrature weights tailored to the specific scenario. It will be shown that EDMD generalizes the setup of Section 2.2.2.

One first chooses a dictionary $\{\psi_1, \dots, \psi_N\}$, i.e., a list of observables, in the space $L^2(\Omega, \omega)$. These observables form a finite-dimensional subspace $V_N = \operatorname{span}\{\psi_1, \dots, \psi_N\}$. EDMD selects a matrix $\mathbf{K} \in \mathbb{C}^{N \times N}$ that approximates the action of $\mathcal{K}$ confined to this subspace. We desire that

$$[\mathcal{K}\psi_j](\mathbf{x}) = \psi_j(\mathbf{F}(\mathbf{x})) \approx \sum_{i=1}^{N} \mathbf{K}_{ij} \psi_i(\mathbf{x}), \quad 1 \le j \le N.$$

Define the vector-valued feature map

$$\Omega \ni \mathbf{x} \mapsto \mathbf{\Psi}(\mathbf{x}) = \begin{bmatrix} \psi_1(\mathbf{x}) & \cdots & \psi_N(\mathbf{x}) \end{bmatrix} \in \mathbb{C}^{1 \times N}.$$

Any $g \in V_N$ can be written as $g(\mathbf{x}) = \sum_{j=1}^{N} \psi_j(\mathbf{x})\mathbf{g}_j = \mathbf{\Psi}(\mathbf{x})\,\mathbf{g}$ for some vector $\mathbf{g} \in \mathbb{C}^N$. Hence

$$[\mathcal{K}g](\mathbf{x}) = \mathbf{\Psi}(\mathbf{F}(\mathbf{x}))\,\mathbf{g} = \mathbf{\Psi}(\mathbf{x})(\mathbf{K}\mathbf{g}) + \underbrace{\left(\sum_{j=1}^{N} \psi_j(\mathbf{F}(\mathbf{x}))\mathbf{g}_j - \mathbf{\Psi}(\mathbf{x})(\mathbf{K}\mathbf{g}) \right)}_{=:R(\mathbf{g},\mathbf{x})}.$$

Typically, V_N is not an invariant subspace of $\mathcal{K}$. Hence, there is no choice of $\mathbf{K}$ that makes $R(\mathbf{g}, \mathbf{x})$ zero for all $g \in V_N$ and ω-almost every $\mathbf{x} \in \Omega$. Instead, it is natural to select $\mathbf{K}$ as a solution of

$$\min_{\mathbf{K} \in \mathbb{C}^{N \times N}} \left\{ \int_{\Omega} \max_{\mathbf{g} \in \mathbb{C}^N, \|\mathbf{C}\mathbf{g}\|_{\ell^2} = 1} |R(\mathbf{g}, \mathbf{x})|^2 \, \mathrm{d}\omega(\mathbf{x}) \right.$$
$$\left. = \int_{\Omega} \left\| \mathbf{\Psi}(\mathbf{F}(\mathbf{x}))\mathbf{C}^{-1} - \mathbf{\Psi}(\mathbf{x})\mathbf{K}\mathbf{C}^{-1} \right\|_{\ell^2}^2 \, \mathrm{d}\omega(\mathbf{x}) \right\}. \qquad (4.1)$$

Here, $\|\cdot\|_{\ell^2}$ denotes the standard Euclidean norm of a vector, and $\mathbf{C}$ is a positive self-adjoint matrix that controls the size of $g = \mathbf{\Psi}\mathbf{g}$. One should think of this $\mathbf{C}$ as choosing an appropriate norm. This is important since not all norms on an infinite-dimensional vector space are equivalent (for an example in DMD analysis of fluid flow, see Colbrook, 2023, Figure 7).

In practical, data-driven contexts, it is not possible to directly evaluate the integral in (4.1). Instead, we approximate it via a quadrature rule with nodes $\{\mathbf{x}^{(m)}\}_{m=1}^{M}$ and weights $\{w_m\}_{m=1}^{M}$. For notational convenience, let $\mathbf{D} = \mathrm{diag}(w_1, \ldots, w_M)$ and

$$\mathbf{\Psi}_X = \begin{pmatrix} \mathbf{\Psi}(\mathbf{x}^{(1)}) \\ \vdots \\ \mathbf{\Psi}(\mathbf{x}^{(M)}) \end{pmatrix} \in \mathbb{C}^{M\times N}, \quad \mathbf{\Psi}_Y = \begin{pmatrix} \mathbf{\Psi}(\mathbf{y}^{(1)}) \\ \vdots \\ \mathbf{\Psi}(\mathbf{y}^{(M)}) \end{pmatrix} \in \mathbb{C}^{M\times N}. \tag{4.2}$$

The discretized version of (4.1) is the following weighted least-squares problem:

$$\begin{aligned} \min_{\mathbf{K}\in\mathbb{C}^{N\times N}} \Bigg\{ & \sum_{m=1}^{M} w_m \left\| \mathbf{\Psi}(\mathbf{y}^{(m)})\mathbf{C}^{-1} - \mathbf{\Psi}(\mathbf{x}^{(m)})\mathbf{K}\mathbf{C}^{-1} \right\|_{\ell^2}^2 \\ & = \left\| \mathbf{D}^{1/2}\mathbf{\Psi}_Y\mathbf{C}^{-1} - \mathbf{D}^{1/2}\mathbf{\Psi}_X\mathbf{K}\mathbf{C}^{-1} \right\|_{\mathrm{F}}^2 \Bigg\}, \end{aligned} \tag{4.3}$$

where we remind the reader that $\|\cdot\|_{\mathrm{F}}$ denotes the Frobenius norm. By reducing the size of the dictionary if necessary, we may assume without loss of generality that $\mathbf{D}^{1/2}\mathbf{\Psi}_X$ has rank N. For example, we can do this in DMD by projecting onto POD modes. Regularization through a truncated singular value decomposition may also be considered. A solution to (4.3) is

$$\mathbf{K} = (\mathbf{D}^{1/2}\mathbf{\Psi}_X)^{\dagger}\mathbf{D}^{1/2}\mathbf{\Psi}_Y = (\mathbf{\Psi}_X^*\mathbf{D}\mathbf{\Psi}_X)^{\dagger}\mathbf{\Psi}_X^*\mathbf{D}\mathbf{\Psi}_Y,$$

where '$\dagger$' denotes the pseudoinverse. Note that this solution is independent of the matrix $\mathbf{C}$. However, a suitable choice of $\mathbf{C}$ is vital once we add constraints to the optimization problem in (4.1), see Section 5.2.1. As observed in Section 2.2.2, if the quadrature weights are equal and $\mathbf{\Psi} = \begin{bmatrix} u_1 & \cdots & u_r \end{bmatrix}$ constitutes an appropriate linear dictionary, then $\mathbf{K}$ is the transpose of the DMD matrix. Conceptually, DMD can be regarded as a particular instance of EDMD employing a set of linear basis functions.

We now generalize Section 2.2.2 by defining the two correlation matrices

$$\mathbf{G} = \mathbf{\Psi}_X^*\mathbf{D}\mathbf{\Psi}_X = \sum_{m=1}^{M} w_m \mathbf{\Psi}(\mathbf{x}^{(m)})^*\mathbf{\Psi}(\mathbf{x}^{(m)}),$$

Algorithm 10 The EDMD algorithm (Williams et al., 2015a).

Input: Snapshot data $\mathbf{X} \in \mathbb{C}^{d\times M}$ and $\mathbf{Y} \in \mathbb{C}^{d\times M}$, quadrature weights $\{w_m\}_{m=1}^{M}$, and a dictionary of functions $\{\psi_j\}_{j=1}^{N}$.

1: Compute the matrices $\mathbf{\Psi}_X$ and $\mathbf{\Psi}_Y$ defined in (4.2) and $\mathbf{D} = \operatorname{diag}(w_1, \ldots, w_M)$.
2: Compute the EDMD matrix $\mathbf{K} = (\mathbf{D}^{1/2}\mathbf{\Psi}_X)^{\dagger}\mathbf{D}^{1/2}\mathbf{\Psi}_Y \in \mathbb{C}^{N\times N}$.
3: Compute the eigendecomposition $\mathbf{KV} = \mathbf{V\Lambda}$.
The columns of $\mathbf{V}$ are eigenvector coefficients and $\mathbf{\Lambda}$ is a diagonal matrix of eigenvalues.

Output: The eigenvalues $\mathbf{\Lambda}$ and eigenvector coefficients $\mathbf{V} \in \mathbb{C}^{N\times N}$.

$$\mathbf{A} = \mathbf{\Psi}_X^*\mathbf{D}\mathbf{\Psi}_Y = \sum_{m=1}^{M} w_m \mathbf{\Psi}(\mathbf{x}^{(m)})^* \mathbf{\Psi}(\mathbf{y}^{(m)}). \tag{4.4}$$

If we consider the discrete measure $\omega_M = \sum_{m=1}^{M} w_m \delta_{\mathbf{x}^{(m)}}$, then

$$\mathbf{G}_{jk} = \int_\Omega \overline{\psi_j(\mathbf{x})}\psi_k(\mathbf{x})\, d\omega_M(\mathbf{x}), \quad \mathbf{A}_{jk} = \int_\Omega \overline{\psi_j(\mathbf{x})}\psi_k(\mathbf{F}(\mathbf{x}))\, d\omega_M(\mathbf{x}).$$

If the quadrature converges, then

$$\lim_{M\to\infty} \mathbf{G}_{jk} = \langle \psi_k, \psi_j\rangle \quad \text{and} \quad \lim_{M\to\infty} \mathbf{A}_{jk} = \langle \mathcal{K}\psi_k, \psi_j\rangle, \tag{4.5}$$

where $\langle\cdot,\cdot\rangle$ is the inner product associated with $L^2(\Omega, \omega)$. Hence, in the large data limit, $\mathbf{K} = \mathbf{G}^{\dagger}\mathbf{A}$ approaches a matrix representation of $\mathcal{P}_{V_N}\mathcal{K}\mathcal{P}_{V_N}^*$, where $\mathcal{P}_{V_N}$ denotes the orthogonal projection onto V_N. In essence, EDMD is a Galerkin method. The EDMD eigenvalues thus approach the spectrum of $\mathcal{P}_{V_N}\mathcal{K}\mathcal{P}_{V_N}^*$, and EDMD is an example of the so-called *finite section method* (Böttcher and Silbermann, 1983) (Mezić, 2022, Section 4). Since the finite section method can suffer from spectral pollution (spurious modes), spectral pollution is a concern for EDMD (Williams et al., 2015a). We saw an explicit example in Section 2.3.3. See also Mezić (2022, Example 2) for the worked example $\mathbf{F}(\mathbf{x}) = \mathbf{x}^2$ on the unit circle.

Algorithm 10 summarizes the procedure for computing eigenvalues and eigenvectors. We can also use EDMD to compute Koopman modes. Given an observable $g = \mathbf{\Psi}\mathbf{g} \in V_N$, we may expand g in terms of the eigenvectors of $\mathbf{K}$ as

$$g = \mathbf{\Psi}\mathbf{g} = \mathbf{\Psi}\mathbf{V}\left[\mathbf{V}^{-1}\mathbf{g}\right], \tag{4.6}$$

where $\mathbf{V}$ is the matrix of eigenvectors of $\mathbf{K}$ with eigenvalues $\{\lambda_j\}_{j=1}^N$. Similarly, for general $g \in L^2(\Omega, \omega) \backslash V_N$, we obtain an approximate expansion

$$g \approx \mathbf{\Psi V} \left[\mathbf{V}^{-1} (\mathbf{D}^{1/2} \mathbf{\Psi}_X)^\dagger \mathbf{D}^{1/2} \left(g(\mathbf{x}^{(1)}), \ldots, g(\mathbf{x}^{(M)}) \right)^\top \right]. \tag{4.7}$$

This expansion is called the KMD of g.[16] With an abuse of notation, if $g \notin V_N$, we set

$$\mathbf{g} = (\mathbf{D}^{1/2} \mathbf{\Psi}_X)^\dagger \mathbf{D}^{1/2} \left(g(\mathbf{x}^{(1)}), \ldots, g(\mathbf{x}^{(M)}) \right)^\top.$$

As $M \to \infty$, assuming that the quadrature rule underlying EDMD converges, the approximation $g \approx \mathbf{\Psi g}$ converges to the projected observable $\mathcal{P}_{V_N} g$. As a particular case, we can vectorize and obtain

$$\mathbf{x} \approx \mathbf{\Psi}(\mathbf{x}) \mathbf{V} \left[\mathbf{V}^{-1} (\mathbf{D}^{1/2} \mathbf{\Psi}_X)^\dagger \mathbf{D}^{1/2} \left(\mathbf{x}^{(1)}, \ldots, \mathbf{x}^{(M)} \right)^\top \right].$$

The jth row of the matrix in square brackets is known as the jth Koopman mode, which we denote as $\boldsymbol{\xi}_j \in \mathbb{C}^{1 \times d}$. Generalizing (2.7), the KMD provides an approximation of the dynamics by

$$\mathbf{x}_n \approx \mathbf{\Psi}(\mathbf{x}_0) \mathbf{K}^n \mathbf{V} \begin{pmatrix} \boldsymbol{\xi}_1 \\ \vdots \\ \boldsymbol{\xi}_N \end{pmatrix} = \mathbf{\Psi}(\mathbf{x}_0) \mathbf{V} \mathbf{\Lambda}^n \begin{pmatrix} \boldsymbol{\xi}_1 \\ \vdots \\ \boldsymbol{\xi}_N \end{pmatrix} = \sum_{j=1}^N \mathbf{\Psi}(\mathbf{x}_0) \mathbf{V}(:, j) \lambda_j^n \boldsymbol{\xi}_j.$$

Similarly, for general g, we obtain

$$g(\mathbf{x}_n) \approx \mathbf{\Psi}(\mathbf{x}_0) \mathbf{K}^n \mathbf{V} \mathbf{V}^{-1} \mathbf{g} = \mathbf{\Psi}(\mathbf{x}_0) \mathbf{V} \lambda^n \mathbf{V}^{-1} \mathbf{g} = \sum_{j=1}^N \mathbf{\Psi}(\mathbf{x}_0) \mathbf{V}(:, j) \lambda_j^n [\mathbf{V}^{-1} \mathbf{g}]_j,$$

which includes the triple of Koopman eigenvectors, eigenvalues, and modes.

4.1.2 Choices of dictionary

We have already met two examples of EDMD in this review: the Lorenz system discussed in Section 2.3.2, where a dictionary was constructed from delay embedding, and the Duffing oscillator discussed in Section 2.3.3, where we utilized a dictionary of radial basis functions. The selection of the dictionary significantly affects the efficacy of EDMD. In their original formulation, Williams et al. (2015a) suggest various dictionary choices, such as polynomials, Fourier

16 Unfortunately, there are numerous meanings of the term KMD in the literature. There is the KMD of Mezić (2005), which we discussed in Section 2.1.2 and is based on the spectral theorem for unitary Koopman operators. There is also the (typically approximate) KMD produced by DMD and EDMD.

modes, spectral elements, and radial basis functions. Subsequent extensions have primarily focused on addressing the challenges of large state-space dimensions and mitigating the curse of dimensionality.

Kernelized EDMD (Williams et al., 2015b) (developed in parallel in Kawahara, 2016) uses the kernel trick (Scholkopf, 2001) to perform EDMD with a choice of dictionary determined implicitly by a choice of a kernel function. This approach can help circumvent the curse of dimensionality and can be very effective when the state-space dimension d is large. Numerous papers have been written on the approximation of Koopman operators in a RKHS (Klus et al., 2018a; Fujii and Kawahara, 2019; DeGennaro and Urban, 2019; Alexander and Giannakis, 2020; Das and Giannakis, 2020; Klus et al., 2020c,a; Mezić, 2020; Burov et al., 2021; Baddoo et al., 2022; Kostic et al., 2022; Khosravi, 2023; Philipp et al., 2023). This also includes methods for continuous-time dynamical systems (Das et al., 2021; Rosenfeld et al., 2022). A challenge associated with RKHS techniques is that a general RKHS does not exhibit invariance under the action of the Koopman operator. This situation renders the selection of a reproducing kernel a delicate task. Ideally, one should choose the kernel so that the Koopman operator on the RKHS is not only densely defined but also closable. Finding such a kernel is generally nontrivial, as indicated in Ikeda et al. (2022).

Kernel analog forecasting (KAF) (Zhao and Giannakis, 2016) is a kernel method used for nonparametric statistical forecasting of dynamically generated time series data. Under measure-preserving and ergodic dynamics, KAF consistently approximates the conditional expectation of observables that are acted upon by the Koopman operator of the dynamical system and are conditioned on the observed data at forecast initialization (Alexander and Giannakis, 2020). KAF yields optimal predictions in the sense of minimal root mean square error with respect to the invariant measure in the asymptotic limit of large data. This connection facilitates the analysis of generalization error and uncertainty quantification. KAF has been used with streaming kernel regression (Giannakis et al., 2023) and for multiscale systems (Burov et al., 2021).

Diffusion forecasting (Berry et al., 2015) uses the diffusion maps algorithm (Coifman and Lafon, 2006) to construct a data-driven basis. Leveraging spectral convergence results for kernel integral operators (García Trillos et al., 2020; von Luxburg et al., 2008), this approach produces a well-conditioned and consistent approximation as both the amount of training data and the number of basis functions increase. Giannakis et al. (2015); Giannakis (2019) use the diffusion forecasting technique in a framework that approximates the generator $\mathcal{L}$ of measure-preserving ergodic flows on manifolds by an advection-diffusion operator $\mathcal{L}_\tau = \mathcal{L} - \tau\Delta$, where τ is a regularization parameter, and Δ is a Laplace-type diffusion operator. A Galerkin method was developed for the eigenvalue problem of $\mathcal{L}_\tau$, which was observed to perform efficiently for systems with a pure point spectrum, such as ergodic rotations on tori. The most straightforward case for analyzing the spectral properties of diffusion-regularized generators arises when the regularizing operator Δ commutes with $\mathcal{L}$. Das and Giannakis

(2019); Giannakis (2019) demonstrated that a commuting operator Δ can be derived from the infinite-delay limit of a family of kernel integral operators constructed using time-delay embedding.

Another prevalent method involves training *neural networks* as a suitable dictionary to construct Koopman forecasts, as demonstrated in several studies (Li et al., 2017; Takeishi et al., 2017b; Wehmeyer and Noé, 2018; Yeung et al., 2019; Azencot et al., 2020; Eivazi et al., 2021; Li and Jiang, 2021; Alford-Lago et al., 2022). This approach is typically implemented in two ways: by identifying a few key latent variables or by lifting to a higher-dimensional input space. Variational autoencoders (VAMPnets) have been employed for stochastic dynamical systems such as in molecular dynamics (Mardt et al., 2018; Wehmeyer and Noé, 2018), wherein the mapping back to the physical configuration space from the latent variables is probabilistic. The integration of Koopman analysis with graph convolutional neural networks has been explored to learn the dynamics of atoms within materials (Xie et al., 2019). Lusch et al. (2018) employ an auxiliary network to parameterize the continuously varying spectral parameter, enabling a network structure that offers both parsimony and interpretability. A notable challenge when incorporating EDMD with neural networks is the trade-off between representing data accurately and the potential for overfitting, particularly with limited data. To address this issue, Otto and Rowley (2019) proposed an architecture that combines an autoencoder with linear recurrent dynamics in the encoded space. Beyond employing neural networks for learning Koopman embeddings, Koopman theory has also been applied to understand the behavior of neural networks themselves (Manojlović et al., 2020; Dogra and Redman, 2020) and algorithms more broadly (Dietrich et al., 2020; Redman et al., 2022).

4.1.3 Convergence theory

We now outline the convergence theory for EDMD. Unfortunately, the type of convergence (in the strong operator topology) is too weak to ensure the convergence of spectral properties. To take into account the snapshot data and dictionary, we let $\mathbf{K}_{N,M}$ denote the EDMD matrix. When considering the convergence of EDMD and related methods, there are two limits of interest:

- The large-data limit which corresponds to $M \to \infty$;
- The large-subspace limit which corresponds to $N \to \infty$.

To compute the spectral properties of $\mathcal{K}$, a double limit

$$\lim_{N\to\infty} \lim_{M\to\infty} \mathbf{K}_{N,M}$$

must be considered. Generally, these limits do not commute. More broadly (and beyond the above specific interpretations of the two limits), the use of successive limits is a common occurrence in spectral problems and other areas of scientific computation and cannot be overcome regardless of the choice of algorithm (Colbrook, 2020, 2022; Colbrook and Hansen, 2022; Ben-Artzi et al., 2020).

We saw above that if the quadrature rule converges, i.e., (4.5) holds, then $\lim_{M\to\infty} \mathbf{K}_{N,M} = \mathbf{K}_N$ is a Galerkin matrix. There are essentially three options for the quadrature rule:

- **Random sampling:** We may draw $\mathbf{x}^{(m)}$ at random according to a probability measure that is absolutely continuous with respect to ω, and select the quadrature weights according to the corresponding Radon–Nikodym derivative. This was essentially observed in (Williams et al., 2015a). Convergence holds with probability one (Klus et al., 2016, Section 3.4) provided that ω is not supported on a zero level set that is a linear combination of the dictionary (Korda and Mezić, 2018b, Section 4). The convergence rate is typically $\mathcal{O}(M^{-1/2})$ (Caflisch, 1998), but is a practical approach if the state-space dimension is large. One could also consider quasi-Monte Carlo integration, which can achieve a faster rate of $\mathcal{O}(M^{-1})$ (up to logarithmic factors) under suitable conditions (Caflisch, 1998).
- **Ergodic sampling:** If the system is ergodic, then we can replace the strong law of large numbers with Birkhoff's Ergodic theorem (Birkhoff, 1931):

$$\lim_{n\to\infty} \frac{1}{n}\sum_{j=0}^{n-1}[\mathcal{K}^j g](\mathbf{x}_0) = \lim_{n\to\infty} \frac{1}{n}\sum_{j=0}^{n-1} g(\mathbf{x}_j)$$
$$= \int_\Omega g(\mathbf{x})\,\mathrm{d}\omega(\mathbf{x}) \quad \forall g \in L^1(\Omega,\omega). \tag{4.8}$$

We may select $\mathbf{x}^{(m)} = \mathbf{x}_{m-1}$ from a single trajectory starting at ω-almost any initial condition $\mathbf{x}_0$ and $w_m = 1/M$. Often, the measures are 'physical,' meaning that the set of initial points with convergence has a positive Lebesgue measure.[17] For example, taking $g = [\mathcal{K}\psi_k]\cdot\overline{\psi_j}$ in (4.8), we obtain

$$\underbrace{\lim_{M\to\infty} \frac{1}{M}\sum_{n=0}^{M-1} \psi_k(\mathbf{x}_{n+1})\overline{\psi_j(\mathbf{x}_n)}}_{=\mathbf{A}_{jk}} = \langle \mathcal{K}\psi_k, \psi_j\rangle.$$

Convergence in this scenario is analyzed in Arbabi and Mezić (2017a); Korda and Mezić (2018b). The convergence rate in M is problem dependent (Kachurovskii, 1996; Mezić and Sotiropoulos, 2002). For periodic and quasiperiodic attractors, the error of approximating the inner products is generally $\mathcal{O}(M^{-1})$. For strongly mixing systems, the rate of convergence slows down to $\mathcal{O}(M^{-1/2})$. However, convergence rates cannot be established for the general class of ergodic systems. For convergence rates of von Neumann's ergodic theorem in the context of Koopman operators, see Aloisio et al. (2022).

[17] There is also the notion of SRB measure, which often coincides. For a survey of these measures and their definitions, see Young (2002).

- **High-order quadrature:** If the state space dimension d is not too large and Ω is sufficiently simple, it can be effective to choose $\{(\mathbf{x}^{(m)}, w_m)\}$ according to a high-order quadrature rule. Even changing the weights $\{w_m\}$ for a fixed set of sample points $\{\mathbf{x}^{(m)}\}$ can lead to a considerable acceleration of the convergence (Colbrook and Townsend, 2023).

Colbrook et al. (2023b) provide concentration bounds on the error of the finite M EDMD matrix. Mollenhauer et al. (2022) provide a rigorous analysis of kernel autocovariance operators, including nonasymptotic error bounds under classical ergodic and mixing assumptions. Nüske et al. (2023) presented the first rigorously derived probabilistic bounds on the finite-data approximation error for the truncated Koopman generator of stochastic differential equations (SDEs) and nonlinear control systems. Two settings were analyzed: independent and identically distributed sampling and ergodic sampling, where it was assumed that the Koopman semigroup is exponentially stable for the latter. Lu and Tartakovsky (2020b) provide bounds for parabolic PDEs.

Suppose the quadrature rule converges and we have passed to the limit $M \to \infty$. Korda and Mezić (2018b) show that under a natural density assumption of V_N as $N \to \infty$, $\mathbf{K}_N$ converges strongly to $\mathcal{K}$ for bounded Koopman operators. This means that

$$\lim_{N\to\infty} \|\mathcal{K}g - \mathbf{\Psi}\mathbf{K}_N\mathcal{P}_{V_N}g\|_{L^2(\Omega,\omega)} = 0 \quad \forall g \in L^2(\Omega,\omega), \tag{4.9}$$

where, with an abuse of notation, $\mathcal{P}_{V_N}g$ denotes the vector of coefficients of $\mathcal{P}_{V_N}g$. It is straightforward to drop the assumption that $\mathcal{K}$ is bounded by making natural assumptions on the dictionary and considering g in the domain of $\mathcal{K}$ (Colbrook and Townsend, 2023). Unfortunately, strong convergence is insufficient to ensure that the spectral properties of $\mathbf{K}_N$ converge to that of $\mathcal{K}$ - Mezić (2022) provides an explicit example. We also saw an example of this effect in Section 2.3.3. In Section 4.3, we will show how Residual DMD provides convergence and error control in the final limit $N \to \infty$.

4.1.4 Infinitesimal generators

Several methods have also been proposed for continuous-time systems and approximating the Koopman infinitesimal generator defined in (2.8). For example, *generator EDMD* (gEDMD) (Klus et al., 2020b) uses time derivatives of the dictionary to extend EDMD to compute the generator, see also Klus et al. (2020a); Rosenfeld et al. (2022). Other methods include computing the matrix logarithm of the Koopman operator (Mauroy and Goncalves, 2020; Drmač et al., 2021), approximating the Koopman operator family, and using finite-differences to compute the Lie derivative of the Koopman operator (Giannakis, 2021; Sechi et al., 2021). Finally, Das and Giannakis (2019); Giannakis (2019); Giannakis and Das (2020) approach the problem of approximating both the Koopman and its generator as a manifold-learning problem on a space-time manifold. This

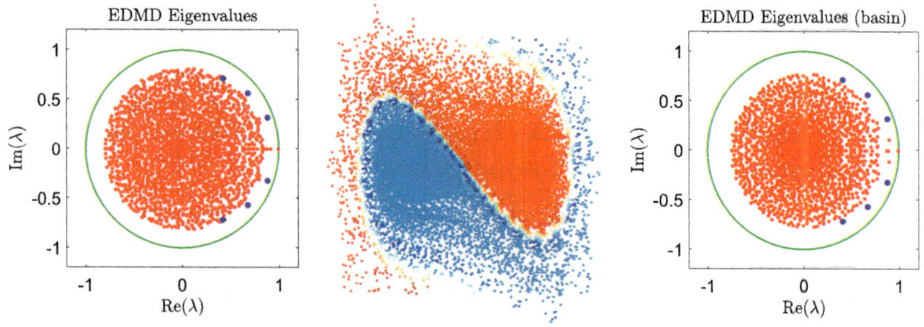

FIGURE 18 Left: EDMD eigenvalues computed over the entire state space. Middle: Eigenfunction that parametrizes the basins of attraction. Right: EDMD eigenvalues obtained after restricting the process to one basin of attraction. The powers of the dominant damped mode are shown in blue.

challenge was successfully addressed for ergodic dynamical systems, such as those evolving on a chaotic attractor.

4.1.5 Example

As an example of EDMD, we revisit the Duffing oscillator from Section 2.3.3 but follow the experiment of Williams et al. (2015a) closely. Namely, we consider the damped system:

$$\dot{x} = y, \quad \dot{y} = -0.5y + x - x^3.$$

In this regime, there are two stable spirals at $(\pm 1, 0)$ and a saddle at the origin. Almost every initial condition, except those on the stable manifold of the saddle, is drawn to one of the spirals. We collect trajectory data and form the dictionary of observables $\{\psi_j\}_{j=1}^{N}$ in the same manner as before. Fig. 18 (left) shows the eigenvalues computed using EDMD with $N = 2000$. The system is now damped, and there is a lattice structure of dominant but damped modes, $\{\lambda_1^n, \overline{\lambda_1}^n : n \in \mathbb{N}, \lambda_1 \approx 0.8831 + 0.3203i\}$ shown in blue. The lattice structure can be understood as follows: if g and f are eigenfunctions of $\mathcal{K}$ corresponding to eigenvalues λ and μ, respectively, and if the product fg is within the function space that forms the domain of $\mathcal{K}$, then

$$[\mathcal{K}(fg)](\mathbf{x}) = f(\mathbf{F}(\mathbf{x}))g(\mathbf{F}(\mathbf{x})) = [\mathcal{K}f](\mathbf{x})[\mathcal{K}g](\mathbf{x}) = \lambda\mu f(\mathbf{x})g(\mathbf{x}).$$

Namely, further eigenvalues and eigenfunctions can be constructed by taking products.

For this system, the eigenspace corresponding to $\lambda = 1$ is spanned by the constant function and the indicator function of the invariant set corresponding to the two basins of attraction. This is illustrated in the middle of Fig. 18. Utilizing the level sets of this eigenfunction, we limit the data to the basin of $(-1, 0)$ and rerun the process to compute a new dictionary. The resulting EDMD eigenvalues are displayed on the right side of Fig. 18, where the eigenvalue λ_1^3 is

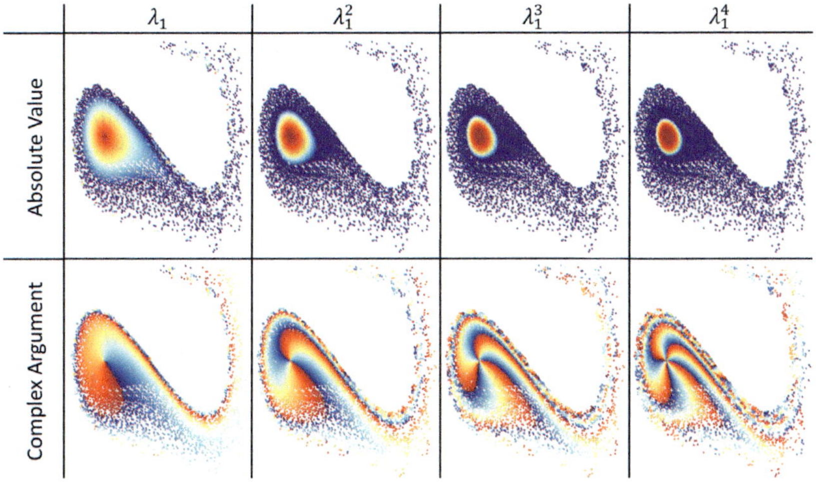

FIGURE 19 Top row: Absolute value of eigenfunction. Bottom row: Complex argument of eigenfunction. Left to right: Eigenfunctions corresponding to powers λ_1, λ_1^2, λ_1^3 and λ_1^4 of the fundamental eigenvalue λ_1.

now more distinctly observable. In Fig. 19, we plot the eigenfunctions corresponding to the powers λ_1, λ_1^2, λ_1^3, and λ_1^4 of the fundamental eigenvalue. Note that the eigenfunctions are successive powers of one another. Furthermore, the amplitude and phase of a Koopman eigenfunction are analogous to an 'action–angle' parametrization of the basin of attraction. The level sets of the absolute values of the eigenfunctions are the so-called isostables, while the level sets of the arguments of the eigenfunctions are termed isochrons (Mauroy et al., 2013).

4.2 Time-delay embedding

In many applications, only partial observations of the system are available, leading to hidden or latent variables. Additionally, the explicit construction of a robust nonlinear dictionary can be challenging, particularly when the system evolves on a low-dimensional attractor that may be unknown or fractal. Nevertheless, it is often feasible to utilize time-delayed measurements of the system to construct an augmented state vector. This approach yields an intrinsic coordinate system that is hoped to form an approximate invariant subspace. This technique was discussed in Section 2.3.2, where it was justified by Takens' embedding theorem (Takens, 2006). Mezić and Banaszuk (2004) established the connection between delay embeddings and the Koopman operator via a statistical Takens' embedding theorem.

Employing the same time step for both the delay interval and the frequency of measurements results in a data matrix with a Hankel structure. Hankel matrices have been used in system identification for decades, as seen in the eigensystem realization algorithm (Juang and Pappa, 1985) and singular spectrum analysis (Broomhead and Jones, 1989). Although these early algorithms were initially developed for linear systems, they have frequently been applied to weakly nonlinear systems as well. The practice of computing DMD on a Hankel matrix was introduced by Tu et al. (2014b) and subsequently utilized in the field of neuroscience (B.W. Brunton et al., 2016). In this section, we focus on two prevalent methods: Hankel-DMD, which is essentially EDMD applied to a dictionary created from time-delay embedding, and the Hankel Alternative View of Koopman (HAVOK) framework, which enhances the DMD model by incorporating a forcing term.

4.2.1 Hankel dynamic mode decomposition (Hankel-DMD)

Hankel-DMD, introduced by Arbabi and Mezić (2017a) and closely related to the Prony approximation of the KMD (Susuki and Mezić, 2015), represents a specialized instance of EDMD where the dictionary is constructed through time-delay embedding. This approach is particularly effective for ergodic systems that exhibit low-dimensional attractors. We saw a slightly generalized variant of this algorithm in Section 2.3.2, where we employed distinct time steps for sampling trajectories and the lengths of the time delays. Typically, Hankel-DMD utilizes the same time steps for delay embedding and trajectory data collection.

Suppose the map $\mathbf{F}$ in (1.1) is ergodic. We can construct a dictionary by starting with an observable g and forming the *Krylov subspace*

$$V_N = \operatorname{span}\{g, \mathcal{K}g, \mathcal{K}^2 g, \ldots, \mathcal{K}^{N-1} g\}.$$

Given a single trajectory of the observable, $\{g(\mathbf{x}_0), g(\mathbf{x}_1), \ldots, g(\mathbf{x}_{M+N-1})\}$, the matrices $\boldsymbol{\Psi}_X$ and $\boldsymbol{\Psi}_Y$ in (4.2) are given explicitly by the Hankel matrices

$$\boldsymbol{\Psi}_X = \begin{pmatrix} g(\mathbf{x}_0) & g(\mathbf{x}_1) & \cdots & g(\mathbf{x}_{N-1}) \\ g(\mathbf{x}_1) & g(\mathbf{x}_2) & \cdots & g(\mathbf{x}_N) \\ \vdots & \vdots & \vdots & \vdots \\ g(\mathbf{x}_{M-1}) & g(\mathbf{x}_M) & \cdots & g(\mathbf{x}_{M+N-2}) \end{pmatrix},$$

$$\boldsymbol{\Psi}_Y = \begin{pmatrix} g(\mathbf{x}_1) & g(\mathbf{x}_2) & \cdots & g(\mathbf{x}_N) \\ g(\mathbf{x}_2) & g(\mathbf{x}_3) & \cdots & g(\mathbf{x}_{N+1}) \\ \vdots & \vdots & \vdots & \vdots \\ g(\mathbf{x}_M) & g(\mathbf{x}_{M+1}) & \cdots & g(\mathbf{x}_{M+N-1}) \end{pmatrix}. \tag{4.10}$$

Applying Birkhoff's ergodic theorem (4.8), we obtain the convergence specified in (4.5). A common simplifying assumption in Hankel-DMD is the existence of a finite-dimensional $\mathcal{K}$-invariant subspace V of $L^2(\Omega, \omega)$ generated by g. $\mathcal{K}$-invariance means that $\mathcal{K}V \subset V$ and allows us to study a portion of the spectral properties of $\mathcal{K}$ by restricting to the finite-dimensional subspace V. Suppose such a subspace exists and has dimension k, then $V_k = V$. We can identify this

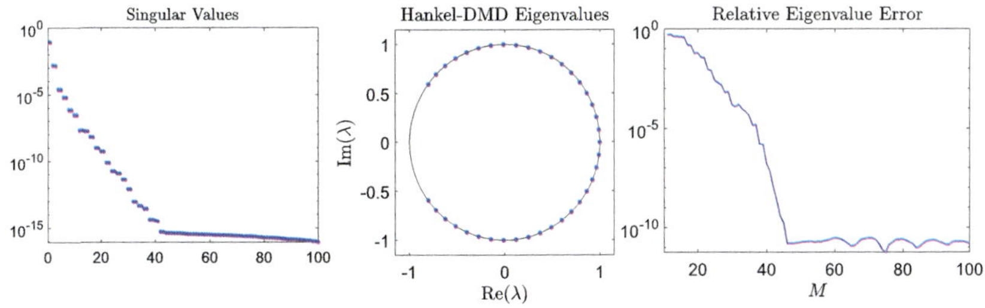

FIGURE 20 Left: Singular values of Ψ_X for the Hankel-DMD algorithm applied to the cylinder wake with $N = 100$ and $M = 120$. In essence, we are building a Krylov subspace by measuring the horizontal component of the velocity field at a single point. Middle: Eigenvalues computed using Hankel-DMD. Right: Convergence to the first 11 eigenvalues with increasing amount of data M.

invariant subspace as $M \to \infty$ by selecting $N = k$ and employing the aforementioned dictionary. This is proven in Arbabi and Mezić (2017a) and is derived from the ergodic theorem in conjunction with the quadrature interpretation of EDMD. These findings also apply when constructing a Krylov subspace from multiple initial observables $g_1, \ldots, g_p$. Nonetheless, the existence of such a subspace is not guaranteed, and even if it is, the dimension k is typically unknown. Practically, one postulates an *approximate invariant subspace* and truncates to $r \leq N$ modes for the basis by executing an SVD.

As an example, we revisit the cylinder wake discussed in Section 2.3.1. For the observable g, we choose the horizontal velocity at a *single point* in the middle of the channel, situated $4D$ downstream from the center of the cylinder. Initially setting $N = 100$ and $M = 120$, we plot the singular values of the data matrix Ψ_X on the left side of Fig. 20. It is crucial to recognize that although the spectrum is pure point in this example, g does not generate a finite-dimensional invariant subspace since g projects nontrivially onto each eigenspace. Guided by these singular values, we apply Algorithm 1 with $r = 39$, using the transposes of Ψ_X and Ψ_Y as the snapshot matrices. The eigenvalues are illustrated in the middle of Fig. 20 and should be compared with a subset of the eigenvalues from Fig. 2. On the right side of Fig. 20, we present the relative ℓ^2 error for the first 11 eigenvalues as a function of M. The convergence is remarkable. Nonetheless, we must stress that this example is rather straightforward. Systems such as the Lorenz system, as discussed in Section 2.3.2 and tackled in Arbabi and Mezić (2017a, Section 4.1), pose a substantially more significant challenge. This is further exemplified by a slow decay of singular values in the data matrices.

4.2.2 Hankel alternative view of Koopman (HAVOK)

Consider the (truncated) SVD of the transpose of the matrix Ψ_X in (4.10),

$$\Psi_X^\top \approx \mathbf{U\Sigma V}^*, \quad \mathbf{U} \in \mathbb{C}^{N\times r}, \quad \mathbf{\Sigma} \in \mathbb{R}^{r\times r}, \quad \mathbf{V} \in \mathbb{C}^{M\times r}.$$

We can view the columns of the matrix $\mathbf{V}$ as coordinates for a state $\mathbf{v} = [v_1 \ \ v_2 \ \ \cdots \ \ v_r]$. If our discrete-time dynamical system corresponds to sampling

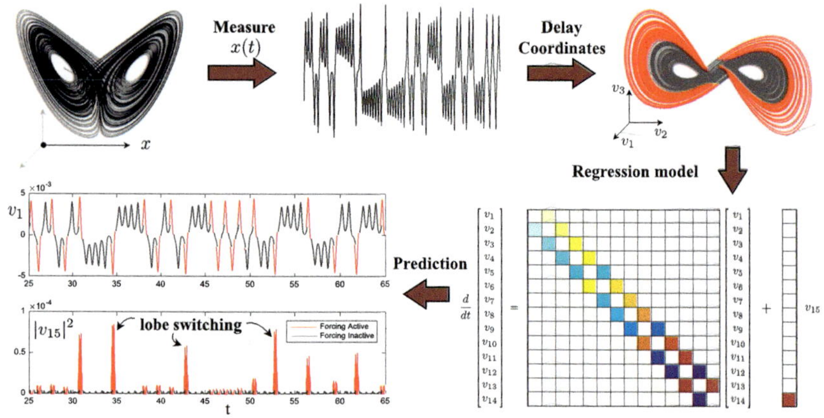

FIGURE 21 Decomposition of chaos into a linear system with forcing. A time series $x(t)$ is stacked into a Hankel matrix, the SVD of which yields a hierarchy of *eigen* time series that produce a delay-embedded attractor. A best-fit linear regression model is obtained on the delay coordinates; the linear fit for the first $r-1$ variables is excellent, but the last coordinate v_r is not well-modeled as linear. Instead, v_r is an input that forces the first $r-1$ variables. Rare forcing events correspond to lobe switching in the chaotic dynamics. Reproduced with permission from Brunton et al. (2022), copyright © 2022 Society for Industrial and Applied Mathematics, all rights reserved. This in turn was adapted from Brunton et al. (2017).

a continuous-time dynamical system, DMD/EDMD results in a linear regression model

$$\frac{d\mathbf{v}}{dt} = \hat{\mathbf{K}}\mathbf{v}, \quad \text{for some matrix} \quad \hat{\mathbf{K}} \in \mathbb{C}^{r\times r}.$$

This can be very effective for weakly nonlinear systems (Champion et al., 2019) and if r is sufficiently large to capture an almost invariant subspace (Arbabi and Mezić, 2017a). However, it can be challenging to identify a small (approximately) closed linear model for chaotic systems.

An alternative, known as the *Hankel Alternative View of Koopman* (HAVOK) framework, proposed by Brunton et al. (2017), is to build a linear model on the first $r-1$ variables $\tilde{\mathbf{v}} = [v_1 \;\; v_2 \;\; \cdots \;\; v_{r-1}]$ and impose the last variable, v_r, as a forcing term:

$$\frac{d\tilde{\mathbf{v}}}{dt} = \tilde{\mathbf{K}}\tilde{\mathbf{v}} + \mathbf{B}v_r, \quad \text{for some matrices} \quad \tilde{\mathbf{K}} \in \mathbb{C}^{r-1\times r-1}, \mathbf{B} \in \mathbb{C}^{r-1\times 1}.$$

Here, v_r acts as an input forcing to the linear dynamics of the model, which approximates the nonlinear dynamics of the original system. Typically, the statistics of v_r are non-Gaussian. For instance, in Fig. 21, we summarize the results for the Lorenz system. The long tails in the statistics of v_r correspond to rare-event forcing that drives lobe switching. For strategies on using HAVOK in systems with multiple time scales, see Champion et al. (2019). Hirsh et al. (2021) established connections between HAVOK and the Frenet–Serret frame from differential geometry, motivating a more accurate computational modeling approach.

4.3 Controlling projection errors: residual dynamic mode decomposition (ResDMD)

We saw in Section 4.1 that EDMD builds a finite matrix approximation of the Koopman operator. In particular, for a dictionary $\{\psi_1, \ldots, \psi_N\}$ forming a finite-dimensional subspace $V_N = \text{span}\{\psi_1, \ldots, \psi_N\}$, the EDMD matrix corresponds to the projected operator $\mathcal{P}_{V_N}\mathcal{K}\mathcal{P}_{V_N}^*$. Care must be taken when discretizing or truncating an infinite-dimensional operator to a finite matrix to compute spectral properties. In general, several well-studied pitfalls include:

- **Spectral Pollution:** This term describes false eigenvalues that accumulate at points not in the spectrum as the discretization size increases.
- **Spectral Invisibility:** Discretizing an operator can cause us to miss parts of its spectrum, even as the size of the discretization increases.
- **Lack of Verification:** Even if a method converges as the discretization parameter grows, how much of the output can we trust for a finite discretization size?
- **Continuous Spectra:** Discretizing to a finite matrix results in a discrete set of eigenvalues. How can we recover continuous spectra?

We have already encountered these effects in this review (e.g., spectral pollution as discussed in Fig. 7), and they are well-known throughout the Koopman literature. In the following section, we will consider strategies to mitigate these issues, focusing on controlling projection errors when transitioning from $\mathcal{K}$ to $\mathcal{P}_{V_N}\mathcal{K}\mathcal{P}_{V_N}^*$. The algorithm that does this is *Residual DMD* (ResDMD), introduced by Colbrook and Townsend (2023).

4.3.1 The algorithm

The main idea behind ResDMD is to compute an infinite-dimensional residual. We follow the notation of Section 4.1 that described EDMD. Consider an observable $g = \mathbf{\Psi}\mathbf{g} \in V_N$, which we aim to be an approximate eigenfunction of $\mathcal{K}$ with an approximate eigenvalue λ. For now, the method of determining the pair (λ, g) is left unspecified. In connection with pseudospectra and the approximate point spectrum discussed in Section 2.1.2, a way to measure the suitability of the candidate pair (λ, g) is through the relative residual

$$\begin{aligned}\frac{\|(\mathcal{K}-\lambda I)g\|}{\|g\|} &= \sqrt{\frac{\int_\Omega |[\mathcal{K}g](\mathbf{x}) - \lambda g(\mathbf{x})|^2 \,\mathrm{d}\omega(\mathbf{x})}{\int_\Omega |g(\mathbf{x})|^2 \,\mathrm{d}\omega(\mathbf{x})}} \\ &= \sqrt{\frac{\langle \mathcal{K}g, \mathcal{K}g\rangle - \lambda\langle g, \mathcal{K}g\rangle - \overline{\lambda}\langle \mathcal{K}g, g\rangle + |\lambda|^2\langle g, g\rangle}{\langle g, g\rangle}}. \end{aligned} \tag{4.11}$$

For instance, if $\mathcal{K}$ is a normal operator (one that commutes with its adjoint), then

$$\operatorname{dist}(\lambda, \operatorname{Sp}(\mathcal{K})) = \inf_f \frac{\|(\mathcal{K}-\lambda I)f\|}{\|f\|} \leq \frac{\|(\mathcal{K}-\lambda I)g\|}{\|g\|}.$$

In the case of a nonnormal $\mathcal{K}$, the residual in (4.11) is closely related to the concept of pseudospectra. Adopting the quadrature interpretation of EDMD, we can define a finite data approximation of the relative residual as:

$$\operatorname{res}(\lambda, g) = \|(\mathbf{D}^{1/2}\mathbf{\Psi}_Y - \lambda \mathbf{D}^{1/2}\mathbf{\Psi}_X)\mathbf{g}\|_{\ell^2} / \|\mathbf{D}^{1/2}\mathbf{\Psi}_X\mathbf{g}\|_{\ell^2}.$$

We then have

$$\begin{aligned}
[\operatorname{res}(\lambda, g)]^2 &= \frac{\mathbf{g}^*\left[\mathbf{\Psi}_Y^*\mathbf{D}\mathbf{\Psi}_Y - \lambda\mathbf{\Psi}_Y^*\mathbf{D}\mathbf{\Psi}_X - \overline{\lambda}\mathbf{\Psi}_X^*\mathbf{D}\mathbf{\Psi}_Y + |\lambda|^2\mathbf{\Psi}_X^*\mathbf{D}\mathbf{\Psi}_X\right]\mathbf{g}}{\mathbf{g}^*\mathbf{\Psi}_X^*\mathbf{D}\mathbf{\Psi}_X\mathbf{g}} \\
&= \frac{\mathbf{g}\left[\mathbf{\Psi}_Y^*\mathbf{D}\mathbf{\Psi}_Y - \lambda\mathbf{A}^* - \overline{\lambda}\mathbf{A} + |\lambda|^2\mathbf{G}\right]\mathbf{g}}{\mathbf{g}^*\mathbf{G}\mathbf{g}},
\end{aligned} \tag{4.12}$$

where $\mathbf{G}$ and $\mathbf{A}$ are the same matrices from (4.4). The right-hand side of (4.12) has an additional matrix $\mathbf{L} := \mathbf{\Psi}_Y^*\mathbf{D}\mathbf{\Psi}_Y$. Under the assumption that the quadrature rule converges, this matrix approximates $\mathcal{K}^*\mathcal{K}$:

$$\lim_{M\to\infty} \mathbf{L}_{jk} = \langle \mathcal{K}\psi_k, \mathcal{K}\psi_j \rangle. \tag{4.13}$$

Comparing (4.11) and the square-root of (4.12), we observe that

$$\lim_{M\to\infty} \operatorname{res}(\lambda, g) = \|(\mathcal{K}-\lambda I)g\| / \|g\|.$$

Note that there is no approximation or projection on the right-hand side of this equation. Consequently, we can compute an *infinite-dimensional residual* directly using finite matrices, achieving exactness in the limit of large data sets. ResDMD leverages this residual in a suite of algorithms to compute various spectral properties of $\mathcal{K}$, two of which are the focus of the subsequent discussion.

As a first approach, we can implement the EDMD algorithm (Algorithm 10) to generate candidate eigenpairs (λ, g), followed by the computation of residuals. This process is outlined in Algorithm 11. Importantly, this approach is not more computationally demanding than EDMD itself. Additionally, it is worth noting that one is not restricted to using EDMD exclusively for selecting candidate eigenpairs; any suitable method can be employed. We can avoid spectral pollution by setting a threshold to discard residuals that exceed a certain tolerance. This also serves as a validation mechanism for the computations. However, it is crucial to recognize that Algorithm 11, when relying on EDMD for computing candidate eigenpairs, does not inherently circumvent the issue of spectral

Algorithm 11 ResDMD for computing residuals (Colbrook and Townsend, 2023).

Input: Snapshot data $\mathbf{X} \in \mathbb{C}^{d\times M}$ and $\mathbf{Y} \in \mathbb{C}^{d\times M}$, quadrature weights $\{w_m\}_{m=1}^{M}$, and a dictionary of functions $\{\psi_j\}_{j=1}^{N}$.

1: Compute the matrices $\mathbf{\Psi}_X$ and $\mathbf{\Psi}_Y$ defined in (4.2) and $\mathbf{D} = \mathrm{diag}(w_1, \ldots, w_M)$.
2: Compute the EDMD matrix $\mathbf{K} = (\mathbf{D}^{1/2}\mathbf{\Psi}_X)^{\dagger}\mathbf{D}^{1/2}\mathbf{\Psi}_Y \in \mathbb{C}^{N\times N}$.
3: Compute the eigendecomposition $\mathbf{KV} = \mathbf{V\Lambda}$. The columns of $\mathbf{V} = [\mathbf{v}_1 \cdots \mathbf{v}_n]$ are eigenvector coefficients and $\mathbf{\Lambda}$ is a diagonal matrix of eigenvalues $\lambda_1, \ldots, \lambda_n$.
4: For each eigenpair $(\lambda_j, \mathbf{v}_j)$ compute $\mathrm{res}(\lambda_j, \mathbf{\Psi}\mathbf{v}_j) = \|(\mathbf{D}^{1/2}\mathbf{\Psi}_Y - \lambda_j \mathbf{D}^{1/2}\mathbf{\Psi}_X)\mathbf{v}_j\|_{\ell^2} / \|\mathbf{D}^{1/2}\mathbf{\Psi}_X\mathbf{v}_j\|_{\ell^2}$.

Output: The eigenvalues $\mathbf{\Lambda}$, eigenvector coefficients $\mathbf{V} \in \mathbb{C}^{N\times N}$ and residuals $\{\mathrm{res}(\lambda_j, \mathbf{\Psi}\mathbf{v}_j)\}$.

invisibility. To address this, we need to consider approaches that approximate the pseudospectrum.

For computing pseudospectra, working in the standard ℓ^2 norm is beneficial instead of the norm induced by the matrix $\mathbf{G}$. We compute an economy QR decomposition of the data matrix

$$\mathbf{D}^{1/2}\mathbf{\Psi}_X = \mathbf{QR}, \quad \mathbf{Q} \in \mathbb{C}^{M\times N}, \mathbf{R} \in \mathbb{C}^{N\times N},$$

where $\mathbf{Q}$ has orthonormal columns and $\mathbf{R}$ is upper triangular with positive diagonals. Letting $\mathbf{w} = \mathbf{Rg}$, we have

$$\|\mathbf{D}^{1/2}\mathbf{\Psi}_X\mathbf{g}\|_{\ell^2}^2 = \mathbf{g}^*\mathbf{R}^*\mathbf{Q}^*\mathbf{QRg} = \mathbf{g}^*\mathbf{R}^*\mathbf{Rg} = \mathbf{w}^*\mathbf{w} = \|\mathbf{w}\|_{\ell^2}^2.$$

Consequently, the residual can be expressed as:

$$\mathrm{res}(z, g) = \|(\mathbf{D}^{1/2}\mathbf{\Psi}_Y\mathbf{R}^{-1} - z\mathbf{Q})\mathbf{w}\|_{\ell^2} / \|\mathbf{w}\|_{\ell^2}. \tag{4.14}$$

For a given $z \in \mathbb{C}$, our objective is to minimize this residual, which corresponds to finding the smallest singular value of the matrix $(\mathbf{D}^{1/2}\mathbf{\Psi}_Y\mathbf{R}^{-1} - z\mathbf{Q}) \in \mathbb{C}^{M\times N}$. Denoting the smallest singular value by $\sigma_{\inf}$, we must do this for various values of z. Given that $M > N$, a computational advantage is gained by considering the $N \times N$ matrix $(\mathbf{D}^{1/2}\mathbf{\Psi}_Y\mathbf{R}^{-1} - z\mathbf{Q})^*(\mathbf{D}^{1/2}\mathbf{\Psi}_Y\mathbf{R}^{-1} - z\mathbf{Q})$ and computing

$$\sqrt{\sigma_{\inf}((\mathbf{D}^{1/2}\mathbf{\Psi}_Y\mathbf{R}^{-1} - z\mathbf{Q})^*(\mathbf{D}^{1/2}\mathbf{\Psi}_Y\mathbf{R}^{-1} - z\mathbf{Q}))} = \sigma_{\inf}(\mathbf{D}^{1/2}\mathbf{\Psi}_Y\mathbf{R}^{-1} - z\mathbf{Q}).$$

Typically, computing singular values in this manner is not recommended due to the potential loss of precision owing to the square root. However, in most applications, the resulting error is significantly smaller than the errors inherent in

Algorithm 12 ResDMD for computing pseudospectra (Colbrook and Townsend, 2023). One can also compute the singular values directly (of a $\mathbb{C}^{M\times N}$ matrix) without the square root.

Input: Snapshot data $\mathbf{X}\in\mathbb{C}^{d\times M}$ and $\mathbf{Y}\in\mathbb{C}^{d\times M}$, quadrature weights $\{w_m\}_{m=1}^{M}$, dictionary of functions $\{\psi_j\}_{j=1}^{N}$, accuracy goal $\epsilon>0$, and grid of points $\{z_\ell\}_{\ell=1}^{k}\subset\mathbb{C}$.

1: Compute the matrices $\mathbf{\Psi}_X$ and $\mathbf{\Psi}_Y$ defined in (4.2) and $\mathbf{D}=\mathrm{diag}(w_1,\ldots,w_M)$.
2: Compute an economy QR decomposition $\mathbf{D}^{1/2}\mathbf{\Psi}_X=\mathbf{QR}$, where $\mathbf{Q}\in\mathbb{C}^{M\times N}$, $\mathbf{R}\in\mathbb{C}^{N\times N}$.
3: Compute $\mathbf{C}_2=(\mathbf{R}^*)^{-1}\mathbf{\Psi}_Y^*\mathbf{D}\mathbf{\Psi}_Y\mathbf{R}^{-1}$ and $\mathbf{C}_1=\mathbf{Q}^*\mathbf{D}^{1/2}\mathbf{\Psi}_Y\mathbf{R}^{-1}$.
4: Compute $\tau_\ell=\sigma_{\inf}(\mathbf{C}_2-z_\ell\mathbf{C}_1^*-\overline{z_\ell}\mathbf{C}_1+|z_\ell|^2\mathbf{I})$ for $\ell=1,\ldots,k$ ($\sigma_{\inf}$ is smallest singular value).
(If wanted, compute the corresponding right-singular vectors $\mathbf{w}_\ell$ and set $\mathbf{v}_j=\mathbf{R}^{-1}\mathbf{w}_j$.)

Output: Estimate of the pseudospectrum $\{z_\ell:\tau_\ell<\epsilon\}$ (if wanted, corresponding pseudoeigenfunctions $\{\mathbf{\Psi}\mathbf{v}_\ell:\tau_\ell<\epsilon\}$).

the data matrices or the quadrature approximation of inner products. If precision becomes a concern, $\sigma_{\inf}(\mathbf{D}^{1/2}\mathbf{\Psi}_Y\mathbf{R}^{-1}-\lambda\mathbf{Q})$ can be computed directly in subsequent algorithms. Since $\mathbf{Q}^*\mathbf{Q}=\mathbf{I}$, we have

$$
\begin{aligned}
&(\mathbf{D}^{1/2}\mathbf{\Psi}_Y\mathbf{R}^{-1}-z\mathbf{Q})^*(\mathbf{D}^{1/2}\mathbf{\Psi}_Y\mathbf{R}^{-1}-z\mathbf{Q})\\
&\quad=(\mathbf{R}^*)^{-1}\mathbf{\Psi}_Y^*\mathbf{D}\mathbf{\Psi}_Y\mathbf{R}^{-1}-z(\mathbf{R}^*)^{-1}\mathbf{\Psi}_Y^*\mathbf{D}^{1/2}\mathbf{Q}-\overline{z}\mathbf{Q}^*\mathbf{D}^{1/2}\mathbf{\Psi}_Y\mathbf{R}^{-1}+|z|^2\mathbf{I}.
\end{aligned}
$$

The minimum singular values of this matrix are then computed across a grid of z values. This procedure is detailed in Algorithm 12, where the approximation of the ϵ-pseudospectrum is defined as the set of grid points where the minimized residual falls below ϵ. If required, the algorithm can also be extended to compute ϵ-pseudoeigenfunctions (discussed in Section 2.1.2).

When $M<N$

In the above discussion, we assumed that $M\geq N$. If $M<N$, then there are two options. One can consider two subsets of snapshot data (training and test) (Colbrook and Townsend, 2023). Or, as developed in Colbrook (2024), in particular for exact DMD and kernelized EDMD, one can compute dual residuals.

4.3.2 Convergence theory

Colbrook and Townsend (2023) present several convergence results concerning ResDMD. We have already discussed that if the quadrature rule underlying

EDMD converges,

$$\lim_{M\to\infty} \mathrm{res}(\lambda, g) = \|(\mathcal{K} - \lambda I)g\|/\|g\|.$$

Therefore, we can avoid spectral pollution in the large data limit by selecting eigenpairs with small residuals (as computed in Algorithm 11). Let $\Gamma_{N,M}^{\epsilon}$ be the output $\{z_\ell : \tau_\ell < \epsilon\}$ of Algorithm 12. With a minor modification for the boundary case where $\tau = \epsilon$,

$$\lim_{M\to\infty} \Gamma_{N,M}^{\epsilon} =: \Gamma_N^{\epsilon} \subset \mathrm{Sp}_\epsilon(\mathcal{K}).$$

In other words, ResDMD provides verified approximations of pseudospectra. Moreover, under mild conditions on the dictionary and an N-dependent grid $\{z_\ell\}_{\ell=1}^{k}$,

$$\lim_{N\to\infty} \Gamma_N^{\epsilon} = \mathrm{Cl}\left(\left\{\lambda \in \mathbb{C} : \exists g \in L^2(\Omega, \omega) \text{ such that } \|g\| = 1, \|(\mathcal{K} - \lambda I)g\| < \epsilon\right\}\right).$$

As $\epsilon \downarrow 0$, the set on the right-hand side converges to the approximate point spectrum $\mathrm{Sp}_{\mathrm{ap}}(\mathcal{K})$. Thus, ResDMD allows us to compute $\mathrm{Sp}_{\mathrm{ap}}(\mathcal{K})$ via a convergent algorithm. Colbrook and Townsend (2023) further discuss alterations that allow the computation of the full pseudospectrum $\mathrm{Sp}_\epsilon(\mathcal{K})$, and consequently the complete spectrum $\mathrm{Sp}(\mathcal{K})$. In summary, ResDMD addresses the challenges of spectral pollution and spectral invisibility, providing a method for verified spectral computations of general Koopman operators.

A careful reader will note that a few of these algorithms require us to take several parameters successively to infinity. This was also the case for EDMD, as discussed in Section 4.1.3. These limits do not generally commute, and it may be impossible to rewrite them with fewer limits or develop a different algorithm that uses fewer limits. This is a generic feature of infinite-dimensional spectral problems (Colbrook, 2020) and has given rise to the *Solvability Complexity Index* (Hansen, 2011; Ben-Artzi et al., 2020; Colbrook, 2022; Colbrook and Hansen, 2022). We do not go into the details, but there are many open questions on the foundations of computing spectral properties of Koopman operators. In particular, lower bounds on the number of successive limits needed to compute spectra of Koopman operators are an ongoing research problem (see Section 6.4).

We have yet to discuss continuous spectra, the final pitfall mentioned in the bullet point list at the start of this section. Using ResDMD, Colbrook and Townsend (2023) also provide an algorithm that computes spectral measures of Koopman operators associated with generic measure-preserving systems. This approach and others for spectral measures are discussed in Section 6.2.

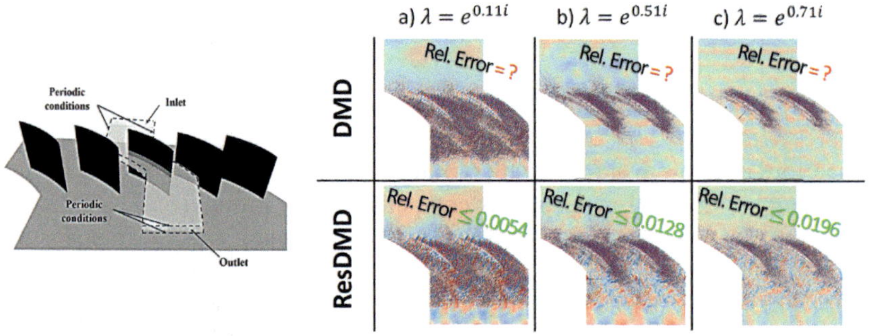

FIGURE 22 Left: Large-scale wall-resolved turbulent flow past a periodic cascade of aerofoils. Right: Comparison of computed Koopman modes using ResDMD and DMD across various frequencies. ResDMD highlights stronger acoustic waves between cascades and larger-scale turbulent fluctuations past the trailing edge, which is crucial for understanding acoustic interactions with engine turbines and nearby structures. The residuals in ResDMD underscore its capability to capture nonlinear dynamics accurately and verifiably. Reproduced with permission from Colbrook and Townsend (2023).

4.3.3 Examples

We have already seen an example of ResDMD in action in Section 2.3.3. Here, we present some examples from Colbrook and Townsend (2023); Colbrook et al. (2023a).

The first example we consider is a large-scale wall-resolved turbulent flow past a periodic cascade of aerofoils depicted on the left in Fig. 22. This setup is motivated by ongoing efforts to mitigate noise sources from aerial vehicles. The data is collected from a high-fidelity simulation solving the fully nonlinear Navier–Stokes equations (Koch et al., 2021), with a Reynolds number of 3.88×10^5 and a Mach number of 0.07. A two-dimensional slice of the pressure field is recorded at 295,122 points across trajectories of length 798 and sampled every 2×10^{-5} seconds. ResDMD can be used with kernelized EDMD, and we use $N = 250$ functions in our dictionary. Fig. 22 (right) shows the computed Koopman modes for a range of representative frequencies. We also show the corresponding Koopman modes computed using DMD. For the first column, ResDMD shows stronger acoustic waves between the cascades. Detecting these vibrations is essential as they can damage engine turbines (Parker, 1984). ResDMD shows larger-scale turbulent fluctuations past the trailing edge for the second and third columns. This can be crucial for understanding acoustic interactions with nearby structures such as subsequent blade rows (Woodley and Peake, 1999). The residuals for ResDMD are small, particularly given the enormous state-space dimension. This example demonstrates two benefits of ResDMD compared with DMD: (1) ResDMD can capture the nonlinear dynamics (just like EDMD), and (2) it computes residuals, thus providing an accuracy certificate.

With its capability to verifiably compute spectra, ResDMD can be employed for validating dictionaries in methods like EDMD and verifying the efficacy

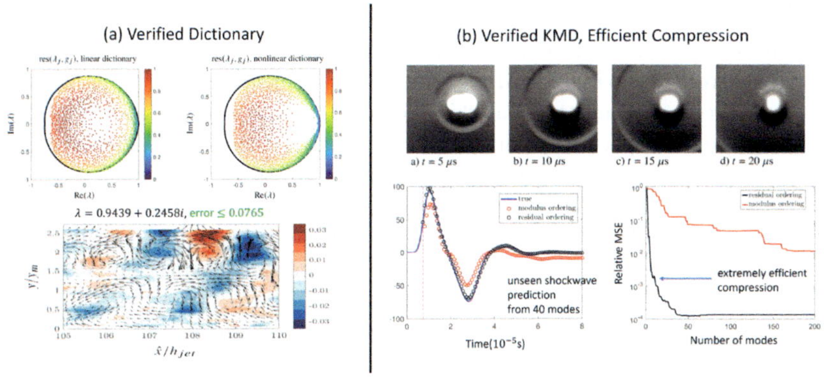

FIGURE 23 ResDMD applications in validating EDMD and KMD. Left: Comparison of two dictionaries (linear: truncated SVD, nonlinear: kernel method) for turbulent boundary layer flow analysis, with the nonlinear dictionary showing smaller residuals and revealing verified transient modes (bottom of the figure). Right: Demonstration of KMD's ability to capture a highly nonlinear shockwave, where ordering modes by residual values enables efficient data compression and precise shockwave prediction. Reproduced with permission from Colbrook et al. (2023a).

of KMD itself. We present two illustrative examples in Fig. 23, adapted from Colbrook et al. (2023a). The comparison of two dictionaries used for analyzing turbulent boundary layer flow is shown on the left. Here, the nonlinear dictionary demonstrates smaller residuals, leading to the identification of verified transient modes, as depicted at the bottom of the figure. On the right, the figure illustrates the proficiency of KMD in capturing a highly nonlinear shockwave. By ordering modes based on their residual values, we achieve efficient data compression and accurate prediction of the shockwave dynamics.

5 Variants that preserve structure

One of the most exciting recent developments in DMD is the introduction of methods that preserve structures of the underlying dynamical system in (1.1). When studying a system from a data-driven perspective, it is often the case that one possesses partial knowledge of the system's underlying physics. Methods that leverage this structure typically exhibit a greater resistance to noise, better generalization, and demand less data for training. Structure-preserving algorithms have a deep-rooted history in geometric integration (Hairer et al., 2010) and have recently gained traction in data-driven methods (Celledoni et al., 2021; Greydanus et al., 2019; Hernández et al., 2021; Hesthaven et al., 2022; Karniadakis et al., 2021; Loiseau and Brunton, 2018; Otto et al., 2023b). In the context of DMD, this area is burgeoning. We will concentrate on three methods:

- **Physics-Informed DMD:** This provides a framework for incorporating symmetries into DMD through additional constraints in the least-squares problem (2.6). The original paper focused on five fundamental physical principles:

conservation, self-adjointness, localization, causality, and shift-equivariance. The idea is far more general and has ushered in a new wave of DMD methods.

- **Measure-Preserving EDMD:** This enforces measure-preserving EDMD truncations, leading to a Galerkin method whose eigendecomposition converges to the spectral quantities of Koopman operators (including spectral measures and continuous spectra) for general measure-preserving dynamical systems. Like EDMD, it can be used with any dictionary of observables and with different data types. Preserving the measure is crucial for convergence, recovering the correct dynamical behavior, stability, robustness to noise, and improved qualitative and long-time behavior.
- **Compactification:** These methods for continuous-time measure-preserving systems are based on the compactification of the Koopman generator or its resolvent. They automatically lead to skew-adjoint approximations whose spectral properties converge to that of the Koopman generator. Additionally, approximations are expressed in a well-conditioned basis of kernel eigenvectors computed from trajectory data.

Subsequently, we will discuss additional DMD methods based on preserving structure. The methods we discuss open the door to future extensions to more general structure-preserving methods for Koopman operators and data-driven dynamical systems.

5.1 Physics-informed dynamic mode decomposition (piDMD)

5.1.1 The framework

Physics-Informed DMD (piDMD), introduced by Baddoo et al. (2023), provides an overarching framework for integrating physical principles – such as symmetries, invariances and conservation laws – into DMD. The idea is to replace the optimization problem in (2.6) by a *constrained* optimization problem

$$\min_{\mathbf{K}_{\text{piDMD}} \in \mathcal{M}} \left\| \mathbf{Y} - \mathbf{K}_{\text{piDMD}} \mathbf{X} \right\|_{\mathrm{F}} . \tag{5.1}$$

The matrix manifold $\mathcal{M}$ is dictated by the known physics of the system in (1.1). One selects $\mathcal{M}$ so that its members satisfy certain symmetries of the system. The optimization problem in (5.1) is known as a Procrustes problem,[18] which comprises of finding the optimal transformation between two matrices subject to certain constraints. Numerous exact solutions exist for Procrustes problems, including the notable cases of orthogonal matrices (Schönemann, 1966), and symmetric matrices (Higham, 1988). When exact solutions are not possible, algorithmic solutions can be effective (Boumal et al., 2014). Procrustes analysis finds relevance in many fields, as detailed in the monograph of Gower and Dijksterhuis (2004).

[18] In Greek mythology, Procrustes was a bandit who would stretch or amputate the limbs of his victims to force them to fit onto his bed. Herein, $\mathbf{X}$ plays the role of Procrustes' victim, $\mathbf{Y}$ is the bed, and $\mathbf{K}_{\text{piDMD}}$ is the 'treatment' (stretching or amputation).

To apply piDMD, we first identify the system's known or suspected physical properties. Once the physical principles we wish to enforce are determined, these laws must be translated into the matrix manifold where the linear model will be constrained. With a defined target matrix manifold, we numerically solve the relevant Procrustes problem in (5.1). The concluding step encompasses extracting physical information from the refined model. For instance, one might analyze the spectrum, DMD modes, and the related KMD.

Baddoo et al. (2023) focus on five fundamental physical principles: conservation, self-adjointness, localization, causality, and shift-equivariance. Several closed-form solutions and efficient algorithms for the corresponding piDMD optimizations are derived. With fewer degrees of freedom, piDMD models are typically less prone to overfitting, require less training data, and are often less computationally expensive to build than standard DMD models. This reduction in the size of required training data is connected with the problem of matrix recovery from matrix-vector products, whereby enforcing structures reduces the number of queries needed (Halikias and Townsend, 2023). A fundamental issue related to the DMD algorithm is the fact that low-rank matrices are not provably recoverable from snapshot pairs (without access to adjoints) until there are at least as many pairs as state dimensions (Otto et al., 2023a, Thm. 2.5).

5.1.2 Examples

To showcase the breadth of piDMD, Fig. 24 shows six physical examples. Baddoo et al. (2023) provide full experimental details for each example. Each row corresponds to a different system, and the corresponding constraint is listed in the second column. Exact DMD (Algorithm 1) is compared to piDMD in terms of the computed matrix **K** and the eigenvalues. In general, constraining the matrix **K** to lie on the appropriate manifold $\mathcal{M}$ leads to more accurate approximations of the eigenvalues. The advantage of preserving structure is striking!

5.1.3 Future work

We have only started to tap the potential of adding constraints in the optimization problem in (2.6). This idea will likely be an active research area over the next few years. With that in mind, it is worth mentioning several challenges and directions of future work pointed out by Baddoo et al. (2023):

- **Knowing the physics:** In some scenarios where the physics is poorly understood, determining suitable physical laws to impose on the model can be challenging. Is it possible to learn symmetries and then incorporate them as constraints?
- **Complicated manifolds:** For problems with intricate geometries and multiple dimensions, interpreting the physical principle as a matrix manifold can be a roadblock, as the manifold can become exceedingly complicated.
- **Regularizers:** In many applications, such as when the data are very noisy or the physical laws and constraints are only approximately understood, it

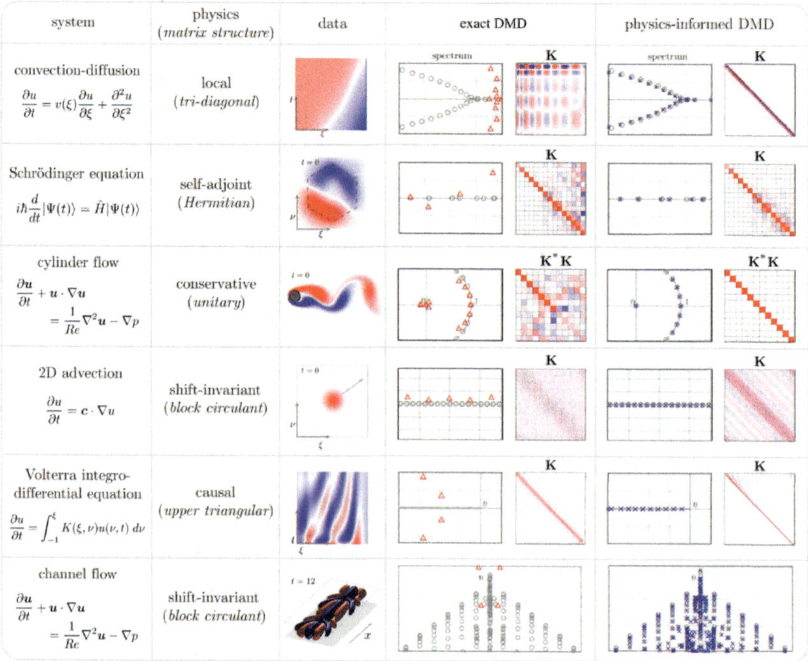

FIGURE 24 A comparison of the models learned by exact DMD (Algorithm 1) and piDMD for a range of applications. The structure of the model matrices is also illustrated. In the spectrum subplots, the true eigenvalues are shown as ○, the DMD eigenvalues as △, and the piDMD eigenvalues as ×. In each case, the eigenvalues of piDMD are more accurate than exact DMD. Reproduced with permission from Baddoo et al. (2023).

may be more appropriate to merely encourage **K** towards $\mathcal{M}$, e.g., through a regularizer.

- **Nonlinear observables:** It is not always clear how to extend the approach of piDMD to nonlinear observables and EDMD. Such an approach is crucial for strongly nonlinear systems to maintain the connection with Koopman operators. For example, if the dictionary consists of the state vector **x**, an upper triangular matrix **K** can have a clear meaning in terms of causality. But how should one incorporate causality into other choices of dictionaries? The manifold can depend on the chosen dictionary in a highly complex manner.
- **Convergence:** In connection with the previous point in this list, proving the convergence of piDMD in the large data limit or large dictionary limit typically requires a Galerkin interpretation as in Section 4. This connection is not always immediate.

We expect these last two points, in particular, to lead to many exciting future works.

5.2 Measure-preserving extended dynamic mode decomposition (mpEDMD)

Measure-Preserving EDMD (mpEDMD), introduced by Colbrook (2023), enforces that the EDMD approximation is measure-preserving. The system being measure-preserving is equivalent to the Koopman operator $\mathcal{K}$ being an isometry. Namely, $\|\mathcal{K}g\| = \|g\|$ for any observable $g \in L^2(\Omega, \omega)$. The mpEDMD algorithm is simple and robust, with no tuning parameters. We outline the method, discuss its convergence properties, and end with two examples. Note that we do not need to assume the system is ergodic or invertible.

5.2.1 The algorithm

We follow the notation of Section 4.1 that described EDMD. Recall that we have a dictionary $\{\psi_1, \ldots, \psi_N\}$, i.e., a list of observables, in the space $L^2(\Omega, \omega)$. These observables form a finite-dimensional subspace $V_N = \operatorname{span}\{\psi_1, \ldots, \psi_N\}$. Our starting point is the observation that the Gram matrix $\mathbf{G} = \mathbf{\Psi}_X^* \mathbf{D} \mathbf{\Psi}_X$ in (4.4) provides an approximation of the inner product $\langle \cdot, \cdot \rangle$ on $L^2(\Omega, \omega)$. Namely, we have the following inner product induced by $\mathbf{G}$:

$$\mathbf{h}^* \mathbf{G} \mathbf{g} = \sum_{j,k=1}^{N} \overline{\mathbf{h}_j} \mathbf{g}_k \mathbf{G}_{j,k} \approx \sum_{j,k=1}^{N} \overline{\mathbf{h}_j} \mathbf{g}_k \langle \psi_k, \psi_j \rangle = \langle \mathbf{\Psi} \mathbf{g}, \mathbf{\Psi} \mathbf{h} \rangle. \tag{5.2}$$

If the convergence in (4.5) holds, then the left-hand side of (5.2) converges to the right-hand side as $M \to \infty$. Hence, if $g = \mathbf{\Psi}\mathbf{g} \in V_N$ and we approximate the action of $\mathcal{K}$ on V_N by a matrix $\mathbf{K}$,

$$\|g\|^2 \approx \mathbf{g}^* \mathbf{G} \mathbf{g}, \quad \|\mathcal{K}g\|^2 \approx \|\mathbf{\Psi}\mathbf{K}\mathbf{g}\|^2 \approx \mathbf{g}^* \mathbf{K}^* \mathbf{G} \mathbf{K} \mathbf{g}.$$

Since $\mathcal{K}$ is an isometry, $\|g\|^2 = \|\mathcal{K}g\|^2$. Therefore, it is natural to enforce

$$\mathbf{g}^* \mathbf{G} \mathbf{g} = \mathbf{g}^* \mathbf{K}^* \mathbf{G} \mathbf{K} \mathbf{g} \quad \forall \mathbf{g} \in \mathbb{C}^N.$$

This condition holds if and only if $\mathbf{K}^* \mathbf{G} \mathbf{K} = \mathbf{G}$. Returning to the optimization problem in (4.1), we now make two changes. First, we set $\mathbf{C} = \mathbf{G}^{1/2}$ so that $\|\mathbf{C}\mathbf{g}\|_{\ell^2} = \sqrt{\mathbf{g}^* \mathbf{G} \mathbf{g}} \approx \|g\|$. Second, we enforce the additional constraint $\mathbf{K}^* \mathbf{G} \mathbf{K} = \mathbf{G}$. This leads us to the optimization problem

$$\min_{\substack{\mathbf{K} \in \mathbb{C}^{N \times N} \\ \mathbf{K}^* \mathbf{G} \mathbf{K} = \mathbf{G}}} \int_{\Omega} \left\| \mathbf{\Psi}(\mathbf{F}(\mathbf{x})) \mathbf{G}^{-1/2} - \mathbf{\Psi}(\mathbf{x}) \mathbf{K} \mathbf{G}^{-1/2} \right\|_{\ell^2}^2 \, d\omega(\mathbf{x}). \tag{5.3}$$

In a nutshell, we enforce that our Galerkin approximation is an isometry with respect to the learned, data-driven inner product induced by $\mathbf{G}$. After applying the quadrature rule we used for EDMD, the discretized version of (5.3) is

$$\min_{\substack{\mathbf{K}\in\mathbb{C}^{N\times N}\\ \mathbf{K}^*\mathbf{G}\mathbf{K}=\mathbf{G}}} \sum_{m=1}^{M} w_m \left\| \boldsymbol{\Psi}(\mathbf{y}^{(m)})\mathbf{G}^{-1/2} - \boldsymbol{\Psi}(\mathbf{x}^{(m)})\mathbf{K}\mathbf{G}^{-1/2} \right\|_{\ell^2}^2. \tag{5.4}$$

Letting $\mathbf{K}=\mathbf{G}^{-1/2}\mathbf{B}\mathbf{G}^{1/2}$ for some matrix $\mathbf{B}$, the problem in (5.4) is equivalent to

$$\min_{\substack{\mathbf{B}\in\mathbb{C}^{N\times N}\\ \mathbf{B}^*\mathbf{B}=\mathbf{I}}} \left\| \mathbf{D}^{1/2}\boldsymbol{\Psi}_X\mathbf{G}^{-1/2}\mathbf{B} - \mathbf{D}^{1/2}\boldsymbol{\Psi}_Y\mathbf{G}^{-1/2} \right\|_{\mathrm{F}}^2, \tag{5.5}$$

where $\mathbf{I}$ denotes the identity matrix. The optimization problem in (5.5) is known as the *orthogonal Procrustes problem* (Schönemann, 1966; Arun, 1992). The predominant method for computing a solution is via an SVD. First, we compute an SVD of

$$\mathbf{G}^{-1/2}\boldsymbol{\Psi}_Y^*\mathbf{D}\boldsymbol{\Psi}_X\mathbf{G}^{-1/2} = \mathbf{G}^{-1/2}\mathbf{A}^*\mathbf{G}^{-1/2} = \mathbf{U}_1\boldsymbol{\Sigma}\mathbf{U}_2^*,$$

where $\mathbf{A}=\boldsymbol{\Psi}_X^*\mathbf{D}\boldsymbol{\Psi}_Y$ is the matrix from (4.4). A solution of (5.5) is then $\mathbf{B}=\mathbf{U}_2\mathbf{U}_1^*$ and we take $\mathbf{K}=\mathbf{G}^{-1/2}\mathbf{U}_2\mathbf{U}_1^*\mathbf{G}^{1/2}$.

Since $\mathbf{K}$ is similar to a unitary matrix, its eigenvalues lie along the unit circle. For stability purposes, the best way to compute the eigendecomposition of $\mathbf{K}$ is to do so for the unitary matrix $\mathbf{U}_2\mathbf{U}_1^*$. To numerically ensure an orthonormal basis of eigenvectors, we use MATLAB's `schur` command in the examples of this paper. It is also beneficial to replace the square root $\mathbf{G}^{1/2}$ with a suitable upper triangular matrix $\mathbf{R}$ such that $\mathbf{G}=\mathbf{R}^*\mathbf{R}$. Such an upper triangular matrix can be computed using an economy QR decomposition of the data matrix as

$$\mathbf{D}^{1/2}\boldsymbol{\Psi}_X = \mathbf{Q}\mathbf{R}, \quad \mathbf{Q}\in\mathbb{C}^{M\times N}, \mathbf{R}\in\mathbb{C}^{N\times N},$$

where $\mathbf{Q}$ has orthonormal columns and $\mathbf{R}$ is upper triangular with positive diagonals. This leads to a mathematically equivalent algorithm but is faster and more numerically robust in practice.[19] The computation of $\mathbf{K}$ and its eigendecomposition is summarized in Algorithm 13. Arguing as we did for EDMD, we obtain a KMD via

$$g \approx \boldsymbol{\Psi}\mathbf{V}\left[\mathbf{V}^{-1}(\mathbf{D}^{1/2}\boldsymbol{\Psi}_X)^{\dagger}\mathbf{D}^{1/2}\left(g(\mathbf{x}^{(1)}),\ldots,g(\mathbf{x}^{(M)})\right)^{\top}\right], \quad g\in L^2(\Omega,\omega).$$

Explicitly applied to the state vector $\mathbf{x}$, we have (transposed) Koopman modes

$$\boldsymbol{\Phi}^{\top} = \mathbf{V}^{-1}(\mathbf{D}^{1/2}\boldsymbol{\Psi}_X)^{\dagger}\mathbf{D}^{1/2}\left(\mathbf{x}^{(1)},\ldots,\mathbf{x}^{(M)}\right)^{\top} \in \mathbb{C}^{N\times d}.$$

Note that mpEDMD can be used with generic choices of dictionary that generate $\mathbf{G}$ and $\mathbf{A}$.

19 I am indebted to Zlatko Drmač for pointing this out.

Algorithm 13 The mpEDMD algorithm (Colbrook, 2023).

Input: Snapshot data $\mathbf{X} \in \mathbb{C}^{d\times M}$ and $\mathbf{Y} \in \mathbb{C}^{d\times M}$, quadrature weights $\{w_m\}_{m=1}^{M}$, and a dictionary of functions $\{\psi_j\}_{j=1}^{N}$.

1: Compute the matrices $\boldsymbol{\Psi}_X$ and $\boldsymbol{\Psi}_Y$ defined in (4.2) and $\mathbf{D} = \mathrm{diag}(w_1, \ldots, w_M)$.

2: Compute an economy QR decomposition $\mathbf{D}^{1/2}\boldsymbol{\Psi}_X = \mathbf{QR}$, where $\mathbf{Q} \in \mathbb{C}^{M\times N}$, $\mathbf{R} \in \mathbb{C}^{N\times N}$.

3: Compute an SVD of $(\mathbf{R}^{-1})^*\boldsymbol{\Psi}_Y^*\mathbf{D}^{1/2}\mathbf{Q} = \mathbf{U}_1\boldsymbol{\Sigma}\mathbf{U}_2^*$.

4: Compute the eigendecomposition $\mathbf{U}_2\mathbf{U}_1^* = \hat{\mathbf{V}}\boldsymbol{\Lambda}\hat{\mathbf{V}}^*$ (via a Schur decomposition).

5: Compute $\mathbf{K} = \mathbf{R}^{-1}\mathbf{U}_2\mathbf{U}_1^*\mathbf{R}$ and $\mathbf{V} = \mathbf{R}^{-1}\hat{\mathbf{V}}$.

Output: Koopman matrix $\mathbf{K}$, with eigenvectors $\mathbf{V}$ and eigenvalues $\boldsymbol{\Lambda}$.

Finally, the relationship between mpEDMD and piDMD is worth commenting on. For conservative systems, piDMD enforces the DMD matrix in (2.6) to be orthogonal and uses linear observables. This implicitly assumes that these linear observables (and the coordinates used) are orthonormal in $L^2(\Omega, \omega)$, an assumption that typically does not hold. In contrast, mpEDMD works in a data-driven inner product space induced by $\mathbf{G}$. The resulting Gram matrix of the observables must be included in a measure-preserving discretization; otherwise, the wrong measure may be preserved (see the example of turbulent flow in Colbrook, 2023 where mpEDMD and piDMD are contrasted). Thus, we can think of the relationship between mpEDMD and piDMD as akin to the relationship between EDMD and DMD (see the discussion in Sections 2.2.2 and 4.1.1), with an additional difference arising from the use of the inner product arising from the Gram matrix $\mathbf{G}$.

5.2.2 *Convergence theory*

Several convergence results for mpEDMD are proven in Colbrook (2023). First, echoing Section 4.1.3, we can consider the two limits $M \to \infty$ and $N \to \infty$. Assuming that the quadrature rule underlying EDMD converges, i.e., (4.5) holds, the EDMD matrix corresponds to $\mathcal{P}_{V_N}\mathcal{K}\mathcal{P}_{V_N}^*$. In contrast, the mpEDMD matrix corresponds to the unitary part of a *polar decomposition* of $\mathcal{P}_{V_N}\mathcal{K}\mathcal{P}_{V_N}^*$. Call this matrix $\mathbf{K}_N$. Under a natural density assumption of V_N as $N \to \infty$, $\mathbf{K}_N$ converges strongly to $\mathcal{K}$, meaning that (4.9) holds.

We can consider the spectral measures from Section 2.1.2 for a measure-preserving system. These spectral measures provide a diagonalization of the Koopman operator $\mathcal{K}$ and form the foundation of the KMD. As the dictionary $\{\psi_1, \ldots, \psi_N\}$ used in EDMD becomes richer, the spectral measures computed by EDMD do not typically converge in any sense to that of $\mathcal{K}$.[20] This contrasts

[20] Even more fundamentally, the eigenvalues of EDMD typically lie within and accumulate within the unit disk. So, the measures are not even on the same space.

the convergence of the spectral measures of mpEDMD (Colbrook, 2023). The critical step in making this convergence work is that mpEDMD provides a unitary Galerkin approximation of $\mathcal{K}$.

The mpEDMD algorithm leads to the following approximations of spectral measures, where $\mathbf{K}$, $\mathbf{V} = [\mathbf{v}_1 \cdots \mathbf{v}_N]$ and $\mathbf{\Lambda} = \text{diag}(\lambda_1, \ldots, \lambda_N)$ are the outputs of Algorithm 13. To approximate the spectral measure $\mathcal{E}$, we consider the spectral measure, $\mathcal{E}_{N,M}$, of the matrix $\mathbf{K}$ on the Hilbert space $\mathbb{C}^N$ with the inner product in (5.2) induced by G:

$$d\mathcal{E}_{N,M}(\lambda) = \sum_{j=1}^{N} \mathbf{v}_j \mathbf{v}_j^* \mathbf{G} \delta(\lambda - \lambda_j)\, d\lambda.$$

Let $g \in L^2(\Omega, \omega)$ with $\|g\| = 1$. We approximate μ_g by $\mu_{\mathbf{g}}^{(N,M)}$, where

$$d\mu_{\mathbf{g}}^{(N,M)}(\lambda) = \sum_{j=1}^{N} \delta(\lambda - \lambda_j)|\mathbf{v}_j^* \mathbf{G}\mathbf{g}|^2 d\lambda$$

and $\mathbf{g}$ is normalized so that $\mathbf{g}^*\mathbf{G}\mathbf{g} = 1$. Since $\{\mathbf{G}^{1/2}\mathbf{v}_j\}_{j=1}^N$ is an orthonormal basis for $\mathbb{C}^N$, $\mu_{\mathbf{g}}^{(N,M)}$ is a probability measure on the unit circle $\mathbb{T}$.

The most natural way for measures to converge is in a *weak sense*. We say that a sequence of measures μ_n converges weakly to a measure μ on $\mathbb{T}$ if for any continuous function $\phi : \mathbb{T} \to \mathbb{C}$,

$$\lim_{n\to\infty} \int_{\mathbb{T}} \phi(\lambda)\, d\mu_n(\lambda) = \int_{\mathbb{T}} \phi(\lambda)\, d\mu(\lambda).$$

This convergence is captured by the so-called Wasserstein 1 metric between probability measures:

$$W_1(\mu, \nu) := \sup\left\{\int_{\mathbb{T}} \phi(\lambda)\, d(\mu - \nu)(\lambda) : \phi : \mathbb{T} \to \mathbb{R} \text{ Lip. cts., Lip. constant} \leq 1\right\}.$$

Under mild conditions on the dictionary as $N \to \infty$, mpEDMD has the following convergence properties. If $\lim_{N\to\infty} \mathbf{\Psi}\mathbf{g}_N = g$ and $\phi : \mathbb{T} \to \mathbb{C}$ is continuous, then

$$\lim_{N\to\infty} \limsup_{M\to\infty} \left\| \int_{\mathbb{T}} \phi(\lambda)\, d\mathcal{E}(\lambda) g - \mathbf{\Psi} \int_{\mathbb{T}} \phi(\lambda)\, d\mathcal{E}_{N,M}(\lambda)\mathbf{g}_N \right\| = 0.$$

Moreover, for the scalar-valued spectral measures,

$$\lim_{N\to\infty} \limsup_{M\to\infty} W_1\left(\mu_g, \mu_{\mathbf{g}}^{(N,M)}\right) = 0. \tag{5.6}$$

Furthermore, if $\{g, \mathcal{K}g, \ldots, \mathcal{K}^{L_N-1}g\} \subset V_N$ (time-delay embedding), then

$$\limsup_{M\to\infty} W_1\left(\mu_g, \mu_{\mathbf{g}}^{(N,M)}\right) \lesssim \log(L_N)/L_N. \tag{5.7}$$

The bound in (5.7) provides an explicit convergence rate in the number of delays used.

Further properties of mpEDMD proven in Colbrook (2023) include respecting invariance subspace properties of $\mathcal{K}$, well-conditioning of the matrix $\mathbf{K}$ and its eigendecomposition (these do not hold for EDMD in general), convergence of KMDs, and

$$\lim_{N\to\infty} \limsup_{M\to\infty} \sup_{\lambda\in\mathrm{Sp}_{\mathrm{ap}}(\mathcal{K})} \mathrm{dist}(\lambda, \{\lambda_1, \ldots, \lambda_N\}) = 0.$$

In other words, we avoid spectral invisibility and do not miss parts of the spectrum. We can also combine with the techniques of Section 4.3 to avoid spectral pollution. Finally, in connection with Section 3.1, the solution to the orthogonal Procrustes problem (5.5) is also the solution to the corresponding constrained total least squares problem (Arun, 1992). Hence, in a similar vein to tlsDMD in Section 3.1.3, mpEDMD is optimally robust when noise is present in both data matrices in (5.5) (Van Huffel and Vandewalle, 1991).

5.2.3 Examples

We consider two examples of mpEDMD. The first shows the convergence to spectral measures for a system with continuous spectra. The second shows the conservation of energy and statistics for a turbulent boundary layer flow, where the snapshots are collected experimentally.

Convergence to continuous spectra

We first revisit the Lorenz system from Section 2.3.2, but now with a discrete time step of $\Delta t = 0.1$. An arbitrary observable is chosen as

$$g(\mathbf{x}) = g(x, y, z) = c\tanh((xy - 3z)/5),$$

where c is a normalization constant ensuring $\|g\| = 1$. We employ delay-embedding to construct a Krylov subspace $V_N = \{g, \mathcal{K}g, \ldots, \mathcal{K}^{N-1}g\}$. The matrices $\mathbf{\Psi}_X$ and $\mathbf{\Psi}_Y$ are computed by evaluating g pointwise at the snapshot matrices of $\mathbf{x}$. A set of $M = 10^4$ snapshots is collected along a single trajectory following an initial burn-in period. It is important to recall that the spectrum of the Koopman operator is continuous, featuring an embedded trivial eigenvalue at $\lambda = 1$. Therefore, we demonstrate the convergence of the mpEDMD approximation of μ_g. For visualization purposes, we transition from variables in $\mathbb{T}$ to complex-arguments in the interval $[-\pi, \pi)$. For a probability measure μ on $\mathbb{T}$, its cumulative distribution function (cdf) on $[-\pi, \pi)$ is defined as

$$F_\mu(\theta) = \mu\left(\{\exp(it) : -\pi \le t \le \theta\}\right).$$

One can express the metric W_1 in terms of these cdfs (Hundrieser et al., 2022).

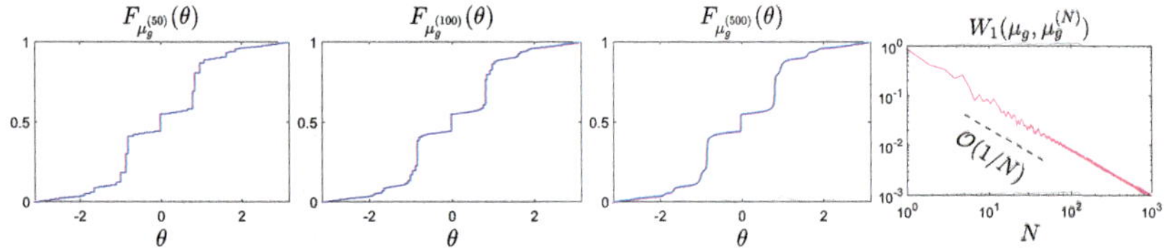

FIGURE 25 First three subplots: The cdfs computed by mpEDMD for various values of N. Far-right: The W_1 metric between the spectral measure computed by mpEDMD, $\mu_g^{(N)}$, and the spectral measure of the Koopman operator, μ_g. The W_1 distance to μ_g is computed by comparing to an approximation with larger N selected large enough to have a negligible effect on the shown errors.

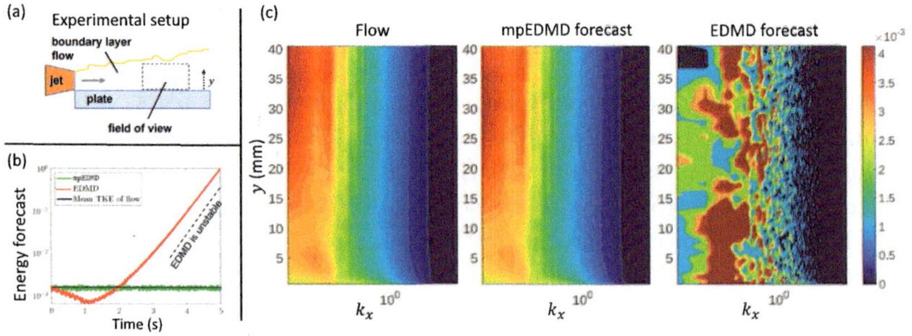

FIGURE 26 (a) Experimental wall-jet boundary layer flow setup with Reynolds number 6.4×10^4. (b) Horizontal averages of the forecasts for turbulent kinetic energy, which show the stability of mpEDMD. (c) Wavenumber spectra measure the energy content of various turbulent structures as a function of their size, thus providing an efficient measure of a flow reconstruction method's performance over various spatial scales. This demonstrates the importance of structure-preserving discretizations (mpEDMD). Reproduced with permission from Colbrook (2023),

Fig. 25 displays the cdfs of $\mu_g^{(N)} = \mu_{\mathbf{g}}^{(N,10^4)}$ for various choices of N, illustrating the convergence of spectral measures. Notably, there is a discontinuity in the cdfs at $\theta = 0$, corresponding to the eigenvalue at $\lambda = 1$. Away from this value, the cdfs show pointwise convergence. The error measured in the Wasserstein 1 metric is depicted on the far right of the plot. Consistent with (5.7), this error decreases as $\mathcal{O}(1/N)$. For similar analyses and examples regarding the projection-valued spectral measures, see Colbrook (2023).

Conservation of energy and statistics for turbulent boundary layer flow

We now examine the boundary layer formed by a thin jet injecting air onto a smooth, flat wall, as depicted in panel (a) of Fig. 26. The experiments are conducted in the wind tunnel at Virginia Tech (Szőke et al., 2021). We utilize a two-component, time-resolved particle image velocimetry system to capture 10^3 snapshots of the two-dimensional velocity field of the wall-jet flow. These snapshots are taken over a spatial grid and a period of 1 second. The jet velocity is set at $U_{\text{jet}} = 50$ m/s, corresponding to a jet Reynolds number of 6.4×10^4.

The field-of-view spans approximately 75 mm by 40 mm, with a spatial resolution of approximately $\Delta x = \Delta y \approx 0.24$ mm. This setup leads to a dimension $d = 102{,}300$ in (1.1). An SVD of the data matrix is employed to create a dictionary with $N = 1000$. The flow exhibits zero pressure gradient turbulent boundary layer characteristics within the region between the wall and the velocity profile peak at approximately $y = 15.5$ mm. Above this region, the flow is dominated by a two-dimensional shear layer characterized by large, energetic flow structures. This scenario presents a significant challenge for conventional DMD approaches due to the multiple turbulent scales within the boundary layer.

We investigate the conservation of energy and flow statistics using the KMD for future state predictions. We consider the velocity profiles predicted by mpEDMD and EDMD over 5 seconds, five times the observation window, starting from an initial state $\mathbf{x}_0$ randomly selected from the trajectory data. The results are averaged over 100 such random initializations. Panel (b) of Fig. 26 shows the turbulent kinetic energy (TKE) of the predictions, averaged in the homogeneous horizontal direction and normalized by U_{jet}^2. The instability of EDMD is evident. In contrast, mpEDMD preserves the inner product associated with the TKE.

To examine the statistics of the predictions, panel (c) of Fig. 26 presents the wavenumber spectrum. This spectrum is computed by applying the Fourier transform to the spatial autocorrelations of the predictions in the horizontal direction, as detailed in Glegg and Devenport (2017, Chapter 8). The wavenumber spectrum provides insights into the energy content of various turbulent structures based on their size. It also serves as an efficient measure of a flow-reconstruction method's performance across different spatial scales. The wavenumber spectrum derived from mpEDMD shows remarkable alignment with the actual flow, demonstrating its efficacy. In contrast, EDMD completely fails to capture the correct turbulent statistics.

5.3 Compactification methods for continuous-time systems

For continuous-time invertible measure-preserving systems, the Koopman generator $\mathcal{L}$, as defined in (2.8), is skew-adjoint. A sophisticated suite of methods exists aimed at approximating such generators through compactification. Working in continuous time presents at least two advantages. First, the generator $\mathcal{L}$ is skew-adjoint, while the Koopman operators $\mathcal{K}_{\Delta t}$ are unitary. Developing projection methods that preserve skew-adjointness is generally much more straightforward than preserving unitarity (although we have seen that mpEDMD leads to an appropriate unitary discretization). Second, by computing the spectral properties of the generator $\mathcal{L}$, we are no longer constrained by the need to select a specific discrete time step. We gain comprehensive spectral information for the entire family of Koopman operators $\{\mathcal{K}_{\Delta t} : \Delta t > 0\}$.

Das et al. (2021) developed an approach based on a one-parameter family of reproducing kernels, $\{p_\tau : \tau > 0\}$, satisfying mild regularity assumptions.

This method utilizes corresponding integral operators to perturb the Koopman generator $\mathcal{L}$ to a compact operator on the corresponding RKHS, $\mathcal{H}_\tau$. Assuming ergodic flow, Das et al. (2021) constructed a one-parameter family of skew-adjoint compact operators, $W_\tau : \mathcal{H}_\tau \to \mathcal{H}_\tau$, where $W_\tau = P_\tau \mathcal{L} P_\tau^*$ and $P_\tau : L^2(\Omega, \omega) \to \mathcal{H}_\tau$ is the integral operator defined by

$$[P_\tau g](\mathbf{x}') = \int_\Omega p_\tau(\mathbf{x}', \mathbf{x}) g(\mathbf{x}) \, \mathrm{d}\omega(\mathbf{x}).$$

The operators W_τ are unitarily equivalent to $\mathcal{L}_\tau = G_\tau^{1/2} \mathcal{L} G_\tau^{1/2}$ acting on $L^2(\Omega, \omega)$, with $G_\tau = P_\tau^* P_\tau$. The operators $\mathcal{L}_\tau$ are compact, skew-adjoint, and converge in the strong resolvent sense to the generator $\mathcal{L}$ as $\tau \to 0$. Since each $\mathcal{L}_\tau$ is compact, its spectrum can be computed by projection onto finite-dimensional subspaces without spectral pollution and without missing parts of the spectrum in the limit of infinite discretization size (Ben-Artzi et al., 2020). This procedure yields approximate Koopman eigenvalues and eigenfunctions, which have been demonstrated to lie within the ϵ-pseudospectrum of the Koopman operator, with the value of ϵ dependent on an RKHS-induced Dirichlet energy functional. In particular, approximate eigenfunctions with small Dirichlet energy as $\tau \to 0$ are approximately cyclical, slowly decorrelating observables under potentially mixing dynamics. It is important to note that two limits are implicitly involved here: the first concerns the parameter that controls the projection size used to approximate spectra of $\mathcal{L}_\tau$, and the second is as τ approaches zero. Another potential limitation of this method is its use of finite-difference schemes on time-ordered data. Although the error from these approximations can be controlled in the limit of a vanishing sampling interval via RKHS regularity, finite differencing generally reduces noise robustness.

Another approach involves the *resolvent* of the generator, $(\mathcal{L} - zI)^{-1}$, where $z \in \mathbb{C} \backslash i\mathbb{R}$. By taking the Laplace transform of the Koopman semigroup, we can observe that (Susuki et al., 2021)

$$(\mathcal{L} - zI)^{-1} = -\int_0^\infty e^{-zt} \mathcal{K}_{\Delta t} \, \mathrm{d}t, \quad \mathrm{Re}(z) > 0. \tag{5.8}$$

Valva and Giannakis (2023) combine the compactification approach from Das et al. (2021) and the integral representation of the resolvent used in Susuki et al. (2021) to construct a compact operator that acts as the resolvent of a skew-adjoint operator. The result is a family of skew-adjoint unbounded operators with compact resolvents, whose spectral measures converge weakly to those of $\mathcal{L}$. This method not only preserves skew-adjointness but also eliminates the need for finite-difference approximations of the generator by using a quadrature approximation for the integral in (5.8). It offers a flexible framework that allows for the control of approximation accuracy by varying z in relation to the sampling interval and the timespan of the training data. Additionally, the finite-rank operators are expressed in a well-conditioned basis of kernel eigenvectors, computed from trajectory data with convergence assurances in the large-data limit (Das et al., 2021). These basis vectors are particularly well-suited to invariant measures supported on sets with complex geometries (e.g., fractal attractors) that are embedded in high-dimensional ambient spaces.

5.3.1 Example

The Rössler system (Rössler, 1976) consists of the following three coupled ordinary differential equations:

$$\dot{x} = -y - z, \quad \dot{y} = x + 0.1y, \quad \dot{z} = 0.1 + z(x - 14).$$

We consider the dynamics of $\mathbf{x} = (x, y, z)$ on the Rössler attractor. The Rössler system is often viewed as a simplified analog of the Lorenz (63) system. However, despite the simplicity of its governing equations, it exhibits complex dynamical characteristics. Theorems on the existence and measure-theoretic mixing properties of the Rössler system analogous to those for the Lorenz system have not been established. Nevertheless, the system has been studied extensively through analytical and numerical techniques, supporting the hypothesis that the Rössler system is mixing (Peifer et al., 2005). Assuming this, it follows that 0 is the only eigenvalue of $\mathcal{L}$, corresponding to a constant eigenfunction and is simple. The integral in (5.8) is approximated by truncating the domain of integration (taking advantage of the exponential decay in the integrand) and Simpson's quadrature rule. Full algorithmic details of the method are given in Valva and Giannakis (2023, Algorithm 1). Data is collected along a single trajectory of length 64,000 with time-step $\Delta t = 0.04$. We use MATLAB's `ode45` command to collect the data after an initial burn-in time to ensure that the initial point is (approximately) on the attractor. The dictionary consists of 2001 data-driven kernel eigenfunctions, and the smoothing parameter is set as $\tau = 2 \times 10^{-6}$.

Fig. 27 shows three approximate eigenfunctions along with the corresponding values of σ, so that $i\sigma$ lies in the spectrum of $\mathcal{L}$. To the figure's right, we illustrate the trajectory of these approximate eigenfunctions. The chaotic behavior of the Rössler system predominantly occurs in the $(r = \sqrt{x^2 + y^2}, z)$ coordinates, while the evolution of the azimuthal angle θ in the $z = 0$ plane proceeds at a near-constant angular frequency, approximately equal to one in natural time units. We suspect this distinction contributes to the challenge of capturing the approximate eigenfunctions in Fig. 27. The first two approximate eigenfunctions are highly coherent, predominantly functions of the azimuthal phase angle, and evolve near-periodically over several Lyapunov times. The third approximate eigenfunction (bottom row) exhibits manifest radial variability in state space and amplitude-modulated time series. Additionally, resolving the radial direction may be more challenging in a data-driven basis, as most variability in the input data occurs in the azimuthal or vertical directions.

We have also presented the Dirichlet energies, indicative of the function's variability or roughness. The approximate eigenfunction corresponding to a

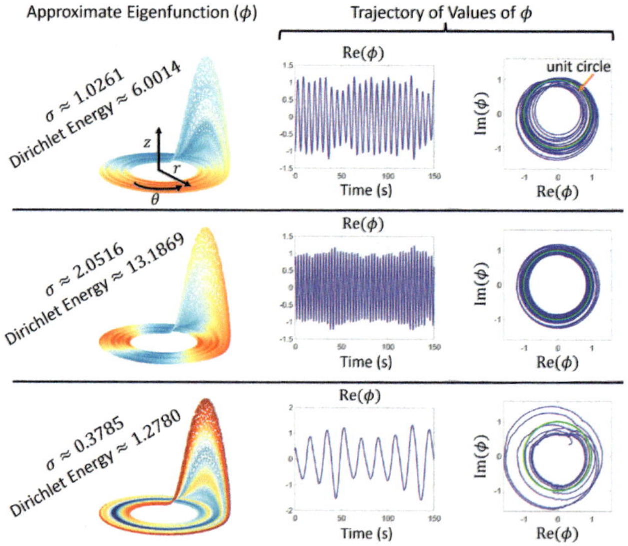

FIGURE 27 Approximate eigenfunctions of the Rössler system computed using the method of Valva and Giannakis (2023). On the right, we show the trajectories of these functions. A trajectory on the unit circle corresponds to coherent periodic behavior. The bottom row displays an approximate eigenfunction with radial variability and larger deviations from the unit circle.

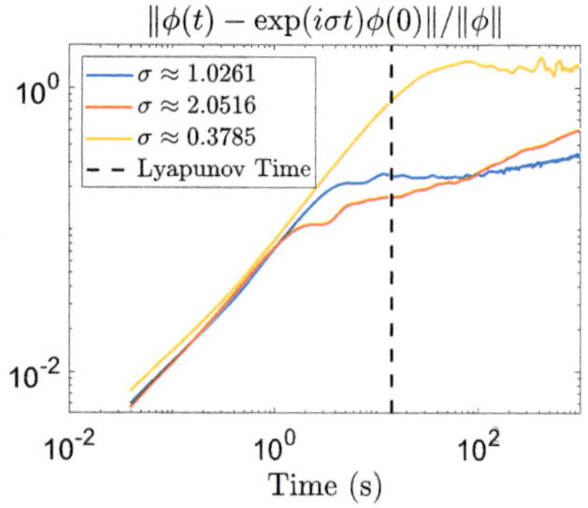

FIGURE 28 Relative residuals of the approximate eigenfunctions in Fig. 27 plotted as a function of time. The residuals increase steadily up to the characteristic Lyapunov timescale.

frequency of $\sigma = 0.3785$ demonstrates relatively low variability compared to the others. Furthermore, the increase in energy from the eigenfunction with frequency 1.0261 to its harmonic with frequency 2.0516 also mirrors this variability increase. Denoting each approximate eigenfunction's trajectory by $\phi(t)$, Fig. 28 displays the relative residual $\|\phi(t) - \exp(i\sigma t)\phi(0)\|/\|\phi\|$. These residuals steadily increase up to the characteristic Lyapunov time of the system and exhibit a larger residual for the third approximate eigenfunction. To summarize, as an a posteriori metric, Dirichlet energy is generally independent from pseudospectral residuals. The information it provides can be useful in supervised learning tasks, e.g., when performing out-of-sample evaluation of the eigenfunctions in prediction problems. In practice, eigenfunctions with small Dirichlet energy also tend to have small pseudospectral residuals, though the precise ordering obtained from the two approaches may differ.

5.4 Further methods

Huang and Vaidya (2018) were among the first to enforce structure in DMD by introducing *Naturally Structured DMD*. This variant ensures positivity and offers the added option of the Markov property. In another early paper, Salova et al. (2019) investigated dynamical systems with symmetries characterized by a finite group. Utilizing representation theory, the authors demonstrated that the Koopman operator and its EDMD approximations can be block diagonalized using a symmetry basis. This basis respects the isotypic component structure related to the underlying symmetry group and the actions of its elements, providing insights into suitable dictionaries. However, the data must align with the system's symmetries to achieve an exact block-diagonal approximation matrix. In an earlier work, Sharma et al. (2016) connected the spatiotemporal symmetries of the Navier–Stokes equation with its spatial and temporal Koopman operators. Kaiser et al. (2018a) presented a method to detect conservation laws using Koopman operator approximations, which can subsequently be employed to control Hamiltonian systems.

We saw that mpEDMD and compactification methods are well-suited to measure-preserving systems. Govindarajan et al. (2019, 2021) provide another approach that is similar to the Ulam approximation of the Perron–Frobenius operator (Ulam, 1960; Li, 1976). They proposed *periodic approximations* for Koopman operators under conditions where Ω is compact, ω is absolutely continuous with respect to the Lebesgue measure, and the system is both measure-preserving and invertible (Govindarajan et al., 2019). This framework was developed into an algorithm for systems on tori and extended to continuous-time systems in Govindarajan et al. (2021). The technique hinges on constructing a periodic approximation of the dynamics via a state-space partition, thus enabling the approximation of the Koopman operator's action through a permutation. The concept of periodic approximations has roots in the works of Halmos (1944) and Lax (1971). This method yields measures that converge weakly to the spectral measures of the Koopman operator. Furthermore, periodic approximations are positive operators and uphold the multiplicative structure of the Koopman operator, i.e., $\mathcal{K}(fg) = (\mathcal{K}f)(\mathcal{K}g)$. A significant unresolved question is how these results can be generalized to handle systems that are not necessarily invariant with respect to a Lebesgue absolutely continuous measure, such as those defined on intricate domains like chaotic attractors, and how to develop efficient schemes in high dimension. Boullé and Colbrook (2024a) introduced *Multiplicative DMD* which enforces the multiplicative structure of Koopman operators, improving the selection of observables and enabling efficient matrix approximation to better reflect spectral properties, with demonstrated robustness to noise across several examples.

DMD fails in translational problems, such as wave-like phenomena, moving interfaces, and moving shocks (Kutz et al., 2016a). This limitation can be attributed to the dominant advection behavior propagating through the entire high-dimensional domain. This propagation makes establishing a global spatiotemporal basis challenging within a low-dimensional subspace. Drawing inspiration from Lagrangian POD (Mojgani and Balajewicz, 2017), Lu and Tartakovsky (2020a) introduced *Lagrangian DMD* that constructs a reduced-order model within the Lagrangian framework. Temporally evolving characteristic lines are selected as a central observable, and a low-dimensional structure in the Lagrangian framework is identified. *Port-Hamiltonian DMD* (Morandin et al., 2023) adapts the DMD within the port-Hamiltonian systems framework to ensure the system satisfies a dissipation inequality. *Symmetric DMD* (Cohen et al., 2020) mandates the dynamics matrix to be symmetric. *Constrained DMD* (Krake et al., 2022) ensures the presence of specific frequencies by incorporating constraints into DMD. *Hermitian DMD* (Baddoo et al., 2023; Drmač, 2022b) provides a self-adjoint approximation that converges for self-adjoint Koopman operators (Boullé and Colbrook, 2024b).

On the transfer operator side, Mehta et al. (2006) address symmetries of the Perron–Frobenius operator in relation to the admissible symmetry properties of attractors. *Constrained Ulam DMD* (Goswami et al., 2018) uses a minimization problem with constraints that guarantee a positive operator with a row sum equal to one. Beyond DMD, Mardt et al. (2020) developed deep learning Markov and Koopman models with physical constraints. Pan and Duraisamy (2020b) learn continuous-time Koopman operators with deep neural networks and enforce stability by ensuring that eigenvalues have nonpositive real parts. Hirsh et al. (2021) presented a theoretical connection between time-delay embedding models and the Frenet–Serret frame (intrinsic coordinates formed by applying the Gram–Schmidt procedure to the derivatives of the trajectory) from differential geometry. This was used to develop structured HAVOK models.

6 Further topics and open problems

We conclude this review of DMD with further topics that are connected with DMD and Koopman operators, followed by outlining some future challenges in the field.

6.1 Transfer operators

Perron–Frobenius operators, also known as Ruelle (Ruelle, 1968) or transfer operators, act on measures through pullbacks. When considering appropriately chosen spaces of observables and measures, the Koopman and Perron–

Frobenius operators emerge as dual pairs, thereby offering equivalent information. In the context of ergodic dynamical systems, natural spaces of observables are typically L^2 spaces of complex-valued scalar functions associated with invariant probability measures, while natural spaces of measures involve complex measures with L^2 densities. The operational distinction between Koopman and Perron–Frobenius operators has led to the development of two distinct families of approximation techniques. However, recent works, such as Klus et al. (2016), have started to bridge this gap.

For data-driven techniques employing Perron–Frobenius operators, see the seminal work of Dellnitz and Junge (1999). A prevalent approach in these methods involves approximating the Perron–Frobenius operator's spectrum using Ulam's method (Ulam, 1960; Li, 1976). This method involves partitioning the state space into a finite set of disjoint subsets. The transition probabilities between these subsets are then estimated by counting transitions observed in extensive simulations or experimental data. The derived transition probability matrix essentially serves as a Galerkin projection of a smoothed compact transfer operator, slightly perturbed by noise. The matrix's eigenvectors, corresponding to eigenvalues at or near the unit circle, are instrumental in identifying coherent sets. Subsequent research by Dellnitz, Froyland, and colleagues (Dellnitz et al., 2000; Froyland and Dellnitz, 2003; Froyland, 2007, 2008; Froyland et al., 2014a) focused on specific system classes with quasicompact Perron–Frobenius operators. Their work rigorously demonstrated Ulam's method's efficacy in accurately approximating isolated Perron–Frobenius eigenvalues and their associated eigenfunctions.

The Perron–Frobenius operator has been used to analyze the global behavior of dynamical systems across various fields. Its applications span molecular dynamics (Schütte and Sarich, 2013; Schütte et al., 2023), fluid dynamics (Froyland et al., 2014b, 2016), meteorology and atmospheric sciences (Tantet et al., 2015, 2018; Froyland et al., 2021), as well as engineering (Vaidya et al., 2010; Ober-Blöbaum and Padberg-Gehle, 2015). Various toolboxes, such as GAIO (Dellnitz et al., 2001), can compute almost invariant sets or metastable states. These toolboxes utilize adaptive box discretizations of the state space to approximate the system's behavior efficiently. However, it is important to note that this approach is generally more suited to low-dimensional problems.

6.2 Continuous spectra and spectral measures

In Section 2.1.2, we saw how Koopman operators associated with measure-preserving systems have spectral measures that provide a KMD. In the course of this review, we have met several methods that converge to spectral measures: mpEDMD in Section 5.2 (see Fig. 25 for an example), methods based on compactification in Section 5.3, and partitioning of the state space to obtain periodic approximations in Section 5.4. There are other methods that are not based on the eigenvalues of a finite matrix. Korda et al. (2020) approximate the moments of spectral measures of ergodic systems using (4.8), and then use the Christoffel–Darboux kernel to analyze the atomic and absolutely continuous parts of the spectrum. They also compute the spectral projection on a given segment of the unit circle. See also Arbabi and Mezić (2017b), who use harmonic averaging and Welch's method (Welch, 1967) to compute the discrete and continuous spectrum of the Koopman operator for post-transient flows. Using the resolvent operator and ResDMD, Colbrook and Townsend (2023) compute smoothed approximations of spectral measures associated with general measure-preserving dynamical systems. They prove explicit high-order convergence theorems for the computation of spectral measures in various senses, including computing the density of the continuous spectrum, spectral projections of subsets of the unit circle, and the discrete spectrum. These smoothing techniques can also be used for self-adjoint operators (Colbrook et al., 2021). *Rigged DMD* (Colbrook et al., 2024) takes this even further and computes generalized eigenfunctions of Koopman operators.

However, we end this discussion with the following warning to the reader about recovering atomic parts of spectral measures that should be kept in mind for all of the above methods. As soon as the spectral measure μ_g has atoms (i.e., $\mathcal{K}$ has eigenvalues, and g is not orthogonal to all the eigenspaces), the map $\lambda \mapsto \mu_g(\{\lambda\})$ is discontinuous. One can prove that, in general, separating the point spectrum from the rest of the spectrum, either in terms of spectral measures or spectral sets, is impossible for *any* algorithm. This holds even for simple classes of operators (Colbrook, 2021), unless we know a priori that the spectrum is discrete in a region of interest (Colbrook et al., 2021, Section 7.3). An excellent example and discussion of this point is provided on the second page of Govindarajan et al. (2019). This is one reason the above methods can only compute spectral measures in a weak or setwise sense. It also helps explain why many of these techniques involve some form of smoothing.

6.3 Stochastic dynamical systems

Stochastic dynamical systems are widely used to model and study systems that evolve under the influence of both deterministic and random effects. It is common to replace (1.1) with a discrete-time Markov process

$$\mathbf{x}_{n+1} = \mathbf{F}(\mathbf{x}_n, \tau_n), \qquad n = 0, 1, 2, \ldots, \tag{6.1}$$

where $\{\tau_n\} \in \Omega_s$ are independent and identically distributed random variables with distribution ρ supported on Ω_s, and $\mathbf{F} : \Omega \times \Omega_s \to \Omega$ is a function. The stochastic Koopman operator (also called the Kolmogorov operator) is the expectation:

$$[\mathcal{K}g](\mathbf{x}) = \int_{\Omega_s} g(\mathbf{F}(\mathbf{x}, \tau))\, d\rho(\tau). \tag{6.2}$$

In contrast to the deterministic case, stochastic Koopman operators typically have discrete spectra due to diffusion. A primary focus has been the challenge of noisy observables in EDMD-type methods (Takeishi et al., 2017c; Wanner and Mezić, 2022), and debiasing DMD (Hemati et al., 2017; Dawson et al., 2016; Takeishi et al., 2017c). Črnjarić-Žic et al. (2020) developed a stochastic Hankel-DMD algorithm for numerical approximations of the stochastic Koopman operator. Klus et al. (2020b) used gEDMD to derive models for SDEs with applications in control. Sinha et al. (2020) provided an explicit optimization-based approximation of stochastic Koopman operators. Wu and Noé (2020) developed a variational approach for Markov processes that finds optimal feature mappings and optimal Markovian models of the dynamics from the top singular components of the Koopman operator.

The definition in (6.2) involves an expectation. Colbrook et al. (2023b) demonstrated the benefits and necessity of going beyond expectations of trajectories. They incorporated the concept of variance into the Koopman framework, establishing its relationship with batched Koopman operators. This led to an extension of ResDMD, resulting in convergence to the spectral properties of stochastic Koopman operators, a Koopman analog of a variance-bias decomposition, and the concept of variance-pseudospectra as a measure of statistical coherency.

6.4 Some open problems

There has been substantial interest in Koopmanism over the last decade, and we expect this interest only to grow. This is an exciting time to be working in this field, which is at the crossroads of dynamical systems theory, data analysis, spectral theory, and computational analysis. We end with some open problems that the author believes will lead to important breakthroughs in the coming years:

- **Banach (and other function) spaces:** We have focused on Koopman operators defined on the Hilbert space $L^2(\Omega, \omega)$. In some cases, it is more appropriate to consider function spaces that are Banach spaces (Mohr and Mezić, 2014; Mezić, 2020). Computational tools for infinite-dimensional spectral problems on Banach spaces are less developed than those for Hilbert spaces. An exception is the transfer operator community, which has developed methods for quasicompact Perron–Frobenius operators (see Section 6.1). Another challenge is the development of the theory of the KMD in the absence of spectral theorems. These issues are expected to be particularly significant in applying the Koopman framework and DMD to transient and off-attractor dynamics.

  For some chaotic systems under appropriate conditions, the eigenvalues in the large subspace limit of EDMD correspond to the eigenvalues of transfer operators in suitable function spaces, as discussed in Slipantschuk et al. (2020); Wormell (2023); Bandtlow et al. (2023). Reconciling these generalized eigenfunctions, the L^2 flavor of DMD methods, and appropriate Banach spaces is a key open problem. These questions could be crucial for understanding and improving the effectiveness of EDMD, including guidance on the choice of dictionary.

- **Choice of dictionary:** One of the most significant open problems in EDMD is the selection of observables or dictionaries. At present, this process can be considered more of an art than a science. While well-conditioned bases can be constructed for some systems, as outlined in Section 5.3, this task often presents a substantial challenge. This is also true for methods based on delay-embedding, where the choice of delay itself is a classical problem with many available heuristics.
- **Foundations:** All of the convergence results for DMD and Koopman operators rely on algorithms that depend on several parameters, with successive limits of these parameters taken to achieve convergence. This is not accidental and occurs generically in infinite-dimensional spectral computations (Ben-Artzi et al., 2020). It is possible to classify the difficulty of computational problems, including data-driven ones. To date, the Koopman community has only provided upper bounds, i.e., algorithms that converge for specific classes of problems. A significant open problem is the development of lower bounds, i.e., universal impossibility results that indicate an intrinsic difficulty in a problem that cannot be overcome by any algorithm. Such results are beginning to emerge in the world of deep learning, particularly regarding the existence vs. trainability of neural networks (Colbrook et al., 2022). We expect them to be equally fruitful in the Koopman context.
  Lower bounds are essential for several reasons. First, they prevent futile searches for nonexistent algorithms. Second, they often elucidate why certain algorithms cannot exist. When combined with upper bounds, this knowledge can lead to natural assumptions about the dynamical systems or the data required to achieve our computational goals.
- **Further structure-preserving methods:** We have already outlined several open problems stimulated by piDMD in Section 5.1.3. Understanding the relationships between structures or symmetries in dynamical systems and their manifestation in the Koopman spectrum lies at the forefront of current knowledge. A crucial challenge is extending constraints applicable to DMD with linear observables to EDMD with nonlinear observables, which will be instrumental in applying structure-preserving methods to nonlinear systems effectively. Another related open problem is establishing the convergence of constrained DMD methods to the spectral properties of Koopman operators and the convergence of KMDs.
- **Verified control:** An exciting development area in modern Koopman theory is its use for controlling nonlinear systems. In Section 3.4.3, we discussed how the practice of Koopman control currently surpasses the theoretical understanding. Only a limited number of systems with a known Koopman-invariant subspace and verifiable eigenfunctions exist. Thus, devel-

oping methods to validate Koopman models for control purposes is a crucial problem. Successfully addressing this issue will likely lead to further insights and enhancements in the practice of Koopman control.

Acknowledgments

I am grateful to numerous friends for their support and stimulating discussions during the completion of this work. The following is in no particular order. Thank you Steven Brunton and Nathan Kutz, for your thoughts and perspectives on an early version of this manuscript, in particular, Table 1. Thank you Igor Mezić, for our discussions regarding spectral theory and computation. Thank you Benjamin Erichson, for running the numerical experiments behind Figs. 13 and 14 and discussing rDMD with me. Thank you Claire Valva and Dimitrios Giannakis, for discussing compactification methods with me, and thank you Claire for running the numerical experiments behind Fig. 27. Thank you Zlatko Drmač, for our discussions regarding the numerical linear algebra aspects of DMD. Thank you Julia Slipantschuk and Caroline Wormell for our discussions regarding numerical methods for transfer operators. Thank you Zlatko Drmač, Dimitrios Giannakis, Igor Mezić, and Samuel Otto, for your insightful feedbacks on an initial draft.

References

Abraham, I., de la Torre, G., Murphey, T., 2017. Model-based control using Koopman operators. In: Robotics: Science and Systems XIII. RSS2017, Robotics: Science and Systems Foundation. MIT Press Journals.

Abraham, I., Murphey, T.D., 2019. Active learning of dynamics for data-driven control using Koopman operators. IEEE Transactions on Robotics 35 (5), 1071–1083.

Adcock, B., Hansen, A.C., 2021. Compressive Imaging: Structure, Sampling, Learning. Cambridge University Press.

Ahmed, S.E., Dabaghian, P.H., San, O., Bistrian, D.A., Navon, I.M., 2022. Dynamic mode decomposition with core sketch. Physics of Fluids 34 (6).

Alexander, R., Giannakis, D., 2020. Operator-theoretic framework for forecasting nonlinear time series with kernel analog techniques. Physica D. Nonlinear Phenomena 409, 132520.

Alford-Lago, D.J., Curtis, C.W., Ihler, A.T., Issan, O., 2022. Deep learning enhanced dynamic mode decomposition. Chaos 32 (3).

Alla, A., Kutz, J.N., 2017. Nonlinear model order reduction via dynamic mode decomposition. SIAM Journal on Scientific Computing 39 (5), B778–B796.

Aloisio, M., Carvalho, S.L., de Oliveira, C.R., Souza, E., 2022. On spectral measures and convergence rates in von Neumann's ergodic theorem. arXiv preprint. arXiv:2209.05290.

Andreuzzi, F., Demo, N., Rozza, G., 2023. A dynamic mode decomposition extension for the forecasting of parametric dynamical systems. SIAM Journal on Applied Dynamical Systems 22 (3), 2432–2458.

Arbabi, H., Korda, M., Mezić, I., 2018. A data-driven Koopman model predictive control framework for nonlinear partial differential equations. In: 2018 IEEE Conference on Decision and Control (CDC). IEEE, pp. 6409–6414.

Arbabi, H., Mezić, I., 2017a. Ergodic theory, dynamic mode decomposition, and computation of spectral properties of the Koopman operator. SIAM Journal on Applied Dynamical Systems 16 (4), 2096–2126.

Arbabi, H., Mezić, I., 2017b. Study of dynamics in post-transient flows using Koopman mode decomposition. Physical Review Fluids 2 (12), 124402.

Arnold, V.I., 1989. Mathematical Methods of Classical Mechanics. Springer New York.

Arun, K.S., 1992. A unitarily constrained total least squares problem in signal processing. SIAM Journal on Matrix Analysis and Applications 13 (3), 729–745.

Askham, T., Kutz, J.N., 2018. Variable projection methods for an optimized dynamic mode decomposition. SIAM Journal on Applied Dynamical Systems 17 (1), 380–416.

Azencot, O., Erichson, N.B., Lin, V., Mahoney, M., 2020. Forecasting sequential data using consistent Koopman autoencoders. In: International Conference on Machine Learning. PMLR, pp. 475–485.

Azencot, O., Yin, W., Bertozzi, A., 2019. Consistent dynamic mode decomposition. SIAM Journal on Applied Dynamical Systems 18 (3), 1565–1585.

Baddoo, P.J., Herrmann, B., McKeon, B.J., Brunton, S.L., 2022. Kernel learning for robust dynamic mode decomposition: linear and nonlinear disambiguation optimization. Proceedings of the Royal Society A. Mathematical, Physical and Engineering Sciences 478 (2260), 20210830.

Baddoo, P.J., Herrmann, B., McKeon, B.J., Nathan Kutz, J., Brunton, S.L., 2023. Physics-informed dynamic mode decomposition. Proceedings of the Royal Society A. Mathematical, Physical and Engineering Sciences 479 (2271), 20220576.

Bagheri, S., 2013. Koopman-mode decomposition of the cylinder wake. Journal of Fluid Mechanics 726, 596–623.

Bagheri, S., 2014. Effects of weak noise on oscillating flows: linking quality factor, Floquet modes, and Koopman spectrum. Physics of Fluids 26 (9).

Bai, Z., Kaiser, E., Proctor, J.L., Kutz, J.N., Brunton, S.L., 2020. Dynamic mode decomposition for compressive system identification. AIAA Journal 58 (2), 561–574.

Bandtlow, O.F., Just, W., Slipantschuk, J., 2023. EDMD for expanding circle maps and their complex perturbations. arXiv preprint. arXiv:2308.01467.

Beer, G., 1993. Topologies on Closed and Closed Convex Sets, vol. 268. Springer Netherlands.

Ben-Artzi, J., Colbrook, M.J., Hansen, A.C., Nevanlinna, O., Seidel, M., 2020. Computing spectra - on the solvability complexity index hierarchy and towers of algorithms. arXiv preprint. arXiv: 1508.03280.

Berkooz, G., Holmes, P., Lumley, J.L., 1993. The proper orthogonal decomposition in the analysis of turbulent flows. Annual Review of Fluid Mechanics 25 (1), 539–575.

Berry, T., Giannakis, D., Harlim, J., 2015. Nonparametric forecasting of low-dimensional dynamical systems. Physical Review E 91 (3), 032915.

Birkhoff, G.D., 1931. Proof of the ergodic theorem. Proceedings of the National Academy of Sciences 17 (12), 656–660.

Birkhoff, G.D., Koopman, B.O., 1932. Recent contributions to the ergodic theory. Proceedings of the National Academy of Sciences 18 (3), 279–282.

Bistrian, D.A., Navon, I.M., 2017. Randomized dynamic mode decomposition for nonintrusive reduced order modelling. International Journal for Numerical Methods in Engineering 112 (1), 3–25.

Böttcher, A., Silbermann, B., 1983. The finite section method for Toeplitz operators on the quarter-plane with piecewise continuous symbols. Mathematische Nachrichten 110 (1), 279–291.

Boullé, N., Colbrook, M.J., 2024a. Multiplicative dynamic mode decomposition. arXiv preprint. arXiv:2405.05334.

Boullé, N., Colbrook, M.J., 2024b. On the convergence of hermitian dynamic mode decomposition. arXiv preprint. arXiv:2401.03192.

Boumal, N., Mishra, B., Absil, P.-A., Sepulchre, R., 2014. Manopt, a MATLAB toolbox for optimization on manifolds. Journal of Machine Learning Research 15 (1), 1455–1459.

Breiman, L., Friedman, J.H., Olshen, R.A., Stone, C.J., 2017. Classification and Regression Trees. Routledge.

Brockett, R.W., 1976. Volterra series and geometric control theory. Automatica 12 (2), 167–176.

Broomhead, D.S., Jones, R., 1989. Time-series analysis. Proceedings of the Royal Society of London. Series A, Mathematical and Physical Sciences 423 (1864), 103–121.

Bruder, D., Gillespie, B., David Remy, C., Vasudevan, R., 2019. Modeling and control of soft robots using the Koopman operator and model predictive control. In: Robotics: Science and Systems XV. RSS2019, Robotics: Science and Systems Foundation.

Brunton, B.W., Johnson, L.A., Ojemann, J.G., Kutz, J.N., 2016. Extracting spatial-temporal coherent patterns in large-scale neural recordings using dynamic mode decomposition. Journal of Neuroscience Methods 258, 1–15.

Brunton, S.L., Brunton, B.W., Proctor, J.L., Kaiser, E., Kutz, J.N., 2017. Chaos as an intermittently forced linear system. Nature Communications 8 (1), 1–9.

Brunton, S.L., Brunton, B.W., Proctor, J.L., Kutz, J.N., 2016a. Koopman invariant subspaces and finite linear representations of nonlinear dynamical systems for control. PLoS ONE 11 (2), e0150171.

Brunton, S.L., Budišić, M., Kaiser, E., Kutz, J.N., 2022. Modern Koopman theory for dynamical systems. SIAM Review 64 (2), 229–340.

Brunton, S.L., Kutz, J.N., 2022. Data-Driven Science and Engineering: Machine Learning, Dynamical Systems, and Control. Cambridge University Press.

Brunton, S.L., Proctor, J.L., Tu, J.H., Kutz, J.N., 2016b. Compressed sensing and dynamic mode decomposition. Journal of Computational Dynamics 2 (2), 165–191.

Budišić, M., Mohr, R., Mezić, I., 2012. Applied Koopmanism. Chaos 22 (4), 047510.

Burov, D., Giannakis, D., Manohar, K., Stuart, A., 2021. Kernel analog forecasting: multiscale test problems. Multiscale Modeling & Simulation 19 (2), 1011–1040.

Caflisch, R.E., 1998. Monte Carlo and quasi-Monte Carlo methods. Acta Numerica 7, 1–49.

Candes, E.J., Romberg, J., Tao, T., 2006. Robust uncertainty principles: exact signal reconstruction from highly incomplete frequency information. IEEE Transactions on Information Theory 52 (2), 489–509.

Candes, E.J., Tao, T., 2006. Near-optimal signal recovery from random projections: universal encoding strategies? IEEE Transactions on Information Theory 52 (12), 5406–5425.

Candes, E.J., Wakin, M.B., 2008. An introduction to compressive sampling. IEEE Signal Processing Magazine 25 (2), 21–30.

Carleman, T., 1932. Application de la théorie des équations intégrales linéaires aux systèmes d'équations différentielles non linéaires. Acta Mathematica 59, 63–87.

Celledoni, E., Ehrhardt, M.J., Etmann, C., McLachlan, R.I., Owren, B., Schönlieb, C.-B., Sherry, F., 2021. Structure-preserving deep learning. European Journal of Applied Mathematics 32 (5), 888–936.

Champion, K.P., Brunton, S.L., Kutz, J.N., 2019. Discovery of nonlinear multiscale systems: sampling strategies and embeddings. SIAM Journal on Applied Dynamical Systems 18 (1), 312–333.

Chen, K.K., Tu, J.H., Rowley, C.W., 2012. Variants of dynamic mode decomposition: boundary condition, Koopman, and Fourier analyses. Journal of Nonlinear Science 22 (6), 887–915.

Cohen, I., Azencot, O., Lifshits, P., Gilboa, G., 2020. Mode decomposition for homogeneous symmetric operators. Preprint arXiv.

Coifman, R.R., Lafon, S., 2006. Diffusion maps. Applied and Computational Harmonic Analysis 21 (1), 5–30.

Colbrook, M., Horning, A., Townsend, A., 2021. Computing spectral measures of self-adjoint operators. SIAM Review 63 (3), 489–524.

Colbrook, M.J., 2020. The Foundations of Infinite-Dimensional Spectral Computations. PhD thesis. University of Cambridge.

Colbrook, M.J., 2021. Computing spectral measures and spectral types. Communications in Mathematical Physics 384 (1), 433–501.

Colbrook, M.J., 2022. On the computation of geometric features of spectra of linear operators on Hilbert spaces. Foundations of Computational Mathematics, 1–82.

Colbrook, M.J., 2023. The mpEDMD algorithm for data-driven computations of measure-preserving dynamical systems. SIAM Journal on Numerical Analysis 61 (3), 1585–1608.

Colbrook, M.J., Antun, V., Hansen, A.C., 2022. The difficulty of computing stable and accurate neural networks: on the barriers of deep learning and Smale's 18th problem. Proceedings of the National Academy of Sciences 119 (12), e2107151119.

Colbrook, M.J., Ayton, L.J., Szőke, M., 2023a. Residual dynamic mode decomposition: robust and verified Koopmanism. Journal of Fluid Mechanics 955, A21.
Colbrook, M.J., Hansen, A.C., 2022. The foundations of spectral computations via the solvability complexity index hierarchy. Journal of the European Mathematical Society 25 (12), 4639–4728.
Colbrook, M.J., Li, Q., Raut, R.V., Townsend, A., 2023b. Beyond expectations: residual dynamic mode decomposition and variance for stochastic dynamical systems. arXiv preprint. arXiv:2308.10697.
Colbrook, M.J., Roman, B., Hansen, A.C., 2019. How to compute spectra with error control. Physical Review Letters 122 (25), 250201.
Colbrook, M.J., Townsend, A., 2023. Rigorous data-driven computation of spectral properties of Koopman operators for dynamical systems. Communications on Pure and Applied Mathematics 77 (1), 221–283.
Colbrook, M.J., 2024. Another look at residual dynamic mode decomposition in the regime of fewer snapshots than dictionary size. arXiv preprint. arXiv:2403.05891.
Colbrook, M.J., Drysdale, C., Horning, A., 2024. Rigged dynamic mode decomposition: data-driven generalized eigenfunction decompositions for Koopman operators. arXiv preprint. arXiv:2405.00782.
Conway, J.B., 2007. A Course in Functional Analysis, vol. 96, 2 edn. Springer New York.
Črnjarić-Žic, N., Maćešić, S., Mezić, I., 2020. Koopman operator spectrum for random dynamical systems. Journal of Nonlinear Science 30 (5), 2007–2056.
Das, S., Giannakis, D., 2019. Delay-coordinate maps and the spectra of Koopman operators. Journal of Statistical Physics 175 (6), 1107–1145.
Das, S., Giannakis, D., 2020. Koopman spectra in reproducing kernel Hilbert spaces. Applied and Computational Harmonic Analysis 49 (2), 573–607.
Das, S., Giannakis, D., Slawinska, J., 2021. Reproducing kernel Hilbert space compactification of unitary evolution groups. Applied and Computational Harmonic Analysis 54, 75–136.
Daubechies, I., 1992. Ten Lectures on Wavelets. Society for Industrial and Applied Mathematics.
Dawson, S.T.M., Hemati, M.S., Williams, M.O., Rowley, C.W., 2016. Characterizing and correcting for the effect of sensor noise in the dynamic mode decomposition. Experiments in Fluids 57 (3), 1–19.
DeGennaro, A.M., Urban, N.M., 2019. Scalable extended dynamic mode decomposition using random kernel approximation. SIAM Journal on Scientific Computing 41 (3), A1482–A1499.
Dellnitz, M., Froyland, G., Junge, O., 2001. The Algorithms Behind GAIO - Set Oriented Numerical Methods for Dynamical Systems. Springer Berlin Heidelberg, pp. 145–174.
Dellnitz, M., Froyland, G., Sertl, S., 2000. On the isolated spectrum of the Perron–Frobenius operator. Nonlinearity 13 (4), 1171–1188.
Dellnitz, M., Junge, O., 1999. On the approximation of complicated dynamical behavior. SIAM Journal on Numerical Analysis 36 (2), 491–515.
Dietrich, F., Thiem, T.N., Kevrekidis, I.G., 2020. On the Koopman operator of algorithms. SIAM Journal on Applied Dynamical Systems 19 (2), 860–885.
Dogra, A.S., Redman, W., 2020. Optimizing neural networks via Koopman operator theory. Advances in Neural Information Processing Systems 33, 2087–2097.
Donoho, D.L., 2006. Compressed sensing. IEEE Transactions on Information Theory 52 (4), 1289–1306.
Drmač, Z., 2020. Dynamic Mode Decomposition - a Numerical Linear Algebra Perspective. Springer International Publishing, pp. 161–194.
Drmač, Z., 2022a. A LAPACK implementation of the dynamic mode decomposition I.
Drmač, Z., 2022b. A LAPACK implementation of the dynamic mode decomposition II. The symmetric/Hermitian DMD (xSYDMD/xHEDMD).
Drmač, Z., Mezić, I., Mohr, R., 2018. Data driven modal decompositions: analysis and enhancements. SIAM Journal on Scientific Computing 40 (4), A2253–A2285.
Drmač, Z., Mezić, I., Mohr, R., 2019. Data driven Koopman spectral analysis in Vandermonde–Cauchy form via the DFT: numerical method and theoretical insights. SIAM Journal on Scientific Computing 41 (5), A3118–A3151.

Drmač, Z., Mezić, I., Mohr, R., 2020. On least squares problems with certain Vandermonde–Khatri–Rao structure with applications to DMD. SIAM Journal on Scientific Computing 42 (5), A3250–A3284.

Drmač, Z., Mezić, I., Mohr, R., 2021. Identification of nonlinear systems using the infinitesimal generator of the Koopman semigroup - a numerical implementation of the Mauroy–Goncalves method. Mathematics 9 (17), 2075.

Dubrovin, B.A., Fomenko, A.T., Novikov, S.P., 1984. Modern Geometry - Methods and Applications Part I. The Geometry of Surfaces, Transformation Groups, and Fields, vol. 104. Springer New York.

Duke, D., Soria, J., Honnery, D., 2012. An error analysis of the dynamic mode decomposition. Experiments in Fluids 52 (2), 529–542.

Eckmann, J.-P., Ruelle, D., 1985. Ergodic theory of chaos and strange attractors. Reviews of Modern Physics 57 (3), 617–656.

Eisner, T., Farkas, B., Haase, M., Nagel, R., 2015. Operator Theoretic Aspects of Ergodic Theory, vol. 272. Springer International Publishing.

Eivazi, H., Guastoni, L., Schlatter, P., Azizpour, H., Vinuesa, R., 2021. Recurrent neural networks and Koopman-based frameworks for temporal predictions in a low-order model of turbulence. International Journal of Heat and Fluid Flow 90, 108816.

Erichson, N.B., Brunton, S.L., Kutz, J.N., 2019a. Compressed dynamic mode decomposition for background modeling. Journal of Real-Time Image Processing 16 (5), 1479–1492.

Erichson, N.B., Donovan, C., 2016. Randomized low-rank dynamic mode decomposition for motion detection. Computer Vision and Image Understanding 146, 40–50.

Erichson, N.B., Mathelin, L., Kutz, J.N., Brunton, S.L., 2019b. Randomized dynamic mode decomposition. SIAM Journal on Applied Dynamical Systems 18 (4), 1867–1891.

Folkestad, C., Pastor, D., Mezić, I., Mohr, R., Fonoberova, M., Burdick, J., 2020. Extended dynamic mode decomposition with learned Koopman eigenfunctions for prediction and control. In: 2020 American Control Conference (ACC). IEEE, pp. 3906–3913.

Foucart, S., Rauhut, H., 2013. A Mathematical Introduction to Compressive Sensing. Springer New York.

Froyland, G., 2007. On Ulam approximation of the isolated spectrum and eigenfunctions of hyperbolic maps. Discrete and Continuous Dynamical Systems 17 (3), 671–689.

Froyland, G., 2008. Unwrapping eigenfunctions to discover the geometry of almost-invariant sets in hyperbolic maps. Physica D. Nonlinear Phenomena 237 (6), 840–853.

Froyland, G., Dellnitz, M., 2003. Detecting and locating near-optimal almost-invariant sets and cycles. SIAM Journal on Scientific Computing 24 (6), 1839–1863.

Froyland, G., Giannakis, D., Lintner, B.R., Pike, M., Slawinska, J., 2021. Spectral analysis of climate dynamics with operator-theoretic approaches. Nature Communications 12 (1).

Froyland, G., González-Tokman, C., Quas, A., 2014a. Detecting isolated spectrum of transfer and Koopman operators with Fourier analytic tools. Journal of Computational Dynamics 1 (2), 249–278.

Froyland, G., González-Tokman, C., Watson, T.M., 2016. Optimal mixing enhancement by local perturbation. SIAM Review 58 (3), 494–513.

Froyland, G., Koltai, P., 2023. Detecting the birth and death of finite-time coherent sets. Communications on Pure and Applied Mathematics 76 (12), 3642–3684.

Froyland, G., Stuart, R.M., van Sebille, E., 2014b. How well-connected is the surface of the global ocean? Chaos 24 (3), 033126.

Fujii, K., Kawahara, Y., 2019. Dynamic mode decomposition in vector-valued reproducing kernel Hilbert spaces for extracting dynamical structure among observables. Neural Networks 117, 94–103.

García Trillos, N., Gerlach, M., Hein, M., Slepčev, D., 2020. Error estimates for spectral convergence of the graph Laplacian on random geometric graphs toward the Laplace–Beltrami operator. Foundations of Computational Mathematics 20 (4), 827–887.

Gavish, M., Donoho, D.L., 2014. The optimal hard threshold for singular values is $4/\sqrt{3}$. IEEE Transactions on Information Theory 60 (8), 5040–5053.

Giannakis, D., 2019. Data-driven spectral decomposition and forecasting of ergodic dynamical systems. Applied and Computational Harmonic Analysis 47 (2), 338–396.

Giannakis, D., 2021. Delay-coordinate maps, coherence, and approximate spectra of evolution operators. Research in the Mathematical Sciences 8 (1), 1–33.

Giannakis, D., Das, S., 2020. Extraction and prediction of coherent patterns in incompressible flows through space-time Koopman analysis. Physica D. Nonlinear Phenomena 402, 132211.

Giannakis, D., Henriksen, A., Tropp, J.A., Ward, R., 2023. Learning to forecast dynamical systems from streaming data. SIAM Journal on Applied Dynamical Systems 22 (2), 527–558.

Giannakis, D., Ourmazd, A., Slawinska, J., Zhao, Z., 2019. Spatiotemporal pattern extraction by spectral analysis of vector-valued observables. Journal of Nonlinear Science 29 (5), 2385–2445.

Giannakis, D., Slawinska, J., Zhao, Z., 2015. Spatiotemporal feature extraction with data-driven Koopman operators. In: Feature Extraction: Modern Questions and Challenges. PMLR, pp. 103–115.

Glegg, S., Devenport, W., 2017. Aeroacoustics of Low Mach Number Flows: Fundamentals, Analysis, and Measurement. Academic Press, London, England.

Golub, G.H., Pereyra, V., 1973. The differentiation of pseudo-inverses and nonlinear least squares problems whose variables separate. SIAM Journal on Numerical Analysis 10 (2), 413–432.

Goswami, D., Paley, D.A., 2017. Global bilinearization and controllability of control-affine nonlinear systems: a Koopman spectral approach. In: 2017 IEEE 56th Annual Conference on Decision and Control (CDC). IEEE, pp. 6107–6112.

Goswami, D., Thackray, E., Paley, D.A., 2018. Constrained Ulam dynamic mode decomposition: approximation of the Perron–Frobenius operator for deterministic and stochastic systems. IEEE Control Systems Letters 2 (4), 809–814.

Govindarajan, N., Mohr, R., Chandrasekaran, S., Mezić, I., 2019. On the approximation of Koopman spectra for measure preserving transformations. SIAM Journal on Applied Dynamical Systems 18 (3), 1454–1497.

Govindarajan, N., Mohr, R., Chandrasekaran, S., Mezic, I., 2021. On the approximation of Koopman spectra of measure-preserving flows. SIAM Journal on Applied Dynamical Systems 20 (1), 232–261.

Gower, J.C., Dijksterhuis, G.B., 2004. Procrustes Problems, vol. 30. Oxford University Press.

Greydanus, S., Dzamba, M., Yosinski, J., 2019. Hamiltonian neural networks. Advances in Neural Information Processing Systems 32.

Gu, M., 2015. Subspace iteration randomization and singular value problems. SIAM Journal on Scientific Computing 37 (3), A1139–A1173.

Guéniat, F., Mathelin, L., Pastur, L.R., 2015. A dynamic mode decomposition approach for large and arbitrarily sampled systems. Physics of Fluids 27 (2).

Haggerty, D.A., Banks, M.J., Kamenar, E., Cao, A.B., Curtis, P.C., Mezić, I., Hawkes, E.W., 2023. Control of soft robots with inertial dynamics. Science Robotics 8 (81), eadd6864.

Hairer, E., Lubich, C., Wanner, G., 2010. Geometric Numerical Integration, second edition, first softcover print edn. Springer Series in Computational Mathematics, vol. 31. Springer, Berlin.

Halikias, D., Townsend, A., 2023. Structured matrix recovery from matrix-vector products. Numerical Linear Algebra with Applications 31 (1).

Halko, N., Martinsson, P.G., Tropp, J.A., 2011. Finding structure with randomness: probabilistic algorithms for constructing approximate matrix decompositions. SIAM Review 53 (2), 217–288.

Halmos, P.R., 1944. Approximation theories for measure preserving transformations. Transactions of the American Mathematical Society 55 (1), 1–18.

Halmos, P.R., 1963. What does the spectral theorem say? The American Mathematical Monthly 70 (3), 241–247.

Halmos, P.R., 2017. Lectures on Ergodic Theory. Dover Books on Mathematics. Unabridged republication of the work originally published in 1956 edn, Dover Publications, Mineola, New York.

Han, Y., Hao, W., Vaidya, U., 2020. Deep learning of Koopman representation for control. In: 2020 59th IEEE Conference on Decision and Control (CDC). IEEE, pp. 1890–1895.

Hansen, A., 2011. On the solvability complexity index, the n-pseudospectrum and approximations of spectra of operators. Journal of the American Mathematical Society 24 (1), 81–124.

Hansen, P.C., Nagy, J.G., O'Leary, D.P., 2006. Deblurring Images: Matrices, Spectra, and Filtering. Society for Industrial and Applied Mathematics.

Haseli, M., Cortés, J., 2023. Modeling nonlinear control systems via Koopman control family: universal forms and subspace invariance proximity. arXiv preprint. arXiv:2307.15368.

Hasnain, A., Boddupalli, N., Balakrishnan, S., Yeung, E., 2020. Steady state programming of controlled nonlinear systems via deep dynamic mode decomposition. In: 2020 American Control Conference (ACC). IEEE, pp. 4245–4251.

Hastie, T., Tibshirani, R., Friedman, J., 2009. The Elements of Statistical Learning. Springer New York.

Hemati, M., Deem, E., Williams, M., Rowley, C.W., Cattafesta, L.N., 2016. Improving separation control with noise-robust variants of dynamic mode decomposition. In: 54th AIAA Aerospace Sciences Meeting. American Institute of Aeronautics and Astronautics, p. 1103.

Hemati, M., Yao, H., 2017. Dynamic mode shaping for fluid flow control: new strategies for transient growth suppression. In: 8th AIAA Theoretical Fluid Mechanics Conference. American Institute of Aeronautics and Astronautics, p. 3160.

Hemati, M.S., Rowley, C.W., Deem, E.A., Cattafesta, L.N., 2017. De-biasing the dynamic mode decomposition for applied Koopman spectral analysis of noisy datasets. Theoretical and Computational Fluid Dynamics 31 (4), 349–368.

Hemati, M.S., Williams, M.O., Rowley, C.W., 2014. Dynamic mode decomposition for large and streaming datasets. Physics of Fluids 26 (11).

Hernández, Q., Badías, A., González, D., Chinesta, F., Cueto, E., 2021. Structure-preserving neural networks. Journal of Computational Physics 426, 109950.

Herrmann, B., Baddoo, P.J., Semaan, R., Brunton, S.L., McKeon, B.J., 2021. Data-driven resolvent analysis. Journal of Fluid Mechanics 918, A10.

Hesthaven, J.S., Pagliantini, C., Rozza, G., 2022. Reduced basis methods for time-dependent problems. Acta Numerica 31, 265–345.

Hey, A.J.G., Tansley, S., Tolle, K.M. (Eds.), 2009. The Fourth Paradigm, vol. 1. Microsoft Research, Redmond, Wash.

Higham, N.J., 1988. The symmetric procrustes problem. BIT 28 (1), 133–143.

Higham, N.J., 2008. Functions of Matrices: Theory and Computation. Society for Industrial and Applied Mathematics.

Hill, T.L., 1986. An Introduction to Statistical Thermodynamics. Courier Corporation.

Hirsh, S.M., Harris, K.D., Kutz, J.N., Brunton, B.W., 2020. Centering data improves the dynamic mode decomposition. SIAM Journal on Applied Dynamical Systems 19 (3), 1920–1955.

Hirsh, S.M., Ichinaga, S.M., Brunton, S.L., Nathan Kutz, J., Brunton, B.W., 2021. Structured time-delay models for dynamical systems with connections to Frenet–Serret frame. Proceedings of the Royal Society A. Mathematical, Physical and Engineering Sciences 477 (2254), 20210097.

Huang, B., Ma, X., Vaidya, U., 2018. Feedback stabilization using Koopman operator. In: 2018 IEEE Conference on Decision and Control (CDC). IEEE, pp. 6434–6439.

Huang, B., Ma, X., Vaidya, U., 2020. Data-Driven Nonlinear Stabilization Using Koopman Operator. Springer International Publishing, pp. 313–334.

Huang, B., Vaidya, U., 2018. Data-driven approximation of transfer operators: naturally structured dynamic mode decomposition. In: 2018 Annual American Control Conference (ACC). IEEE, pp. 5659–5664.

Huhn, Q.A., Tano, M.E., Ragusa, J.C., Choi, Y., 2023. Parametric dynamic mode decomposition for reduced order modeling. Journal of Computational Physics 475, 111852.

Hundrieser, S., Klatt, M., Munk, A., 2022. The Statistics of Circular Optimal Transport. Springer Nature Singapore, pp. 57–82.

Ikeda, M., Ishikawa, I., Schlosser, C., 2022. Koopman and Perron–Frobenius operators on reproducing kernel Banach spaces. Chaos 32 (12).

Jackson, C.P., 1987. A finite-element study of the onset of vortex shedding in flow past variously shaped bodies. Journal of Fluid Mechanics 182 (1), 23–45.

Jiang, L., Liu, N., 2022. Correcting noisy dynamic mode decomposition with Kalman filters. Journal of Computational Physics 461, 111175.

Jovanović, M.R., Schmid, P.J., Nichols, J.W., 2014. Sparsity-promoting dynamic mode decomposition. Physics of Fluids 26 (2).

Józsa, T., Szőke, M., Teschner, T.-R., Könözsy, L.Z., Moulitsas, I., 2016. Validation and verification of a 2D lattice Boltzmann solver for incompressible fluid flow. In: Proceedings of the VII European Congress on Computational Methods in Applied Sciences and Engineering (ECCOMAS Congress 2016). ECCOMAS Congress 2016, Institute of Structural Analysis and Antiseismic Research School of Civil Engineering National Technical University of Athens (NTUA) Greece.

Juang, J.-N., Pappa, R.S., 1985. An eigensystem realization algorithm for modal parameter identification and model reduction. Journal of Guidance, Control, and Dynamics 8 (5), 620–627.

Kachurovskii, A.G., 1996. The rate of convergence in ergodic theorems. Russian Mathematical Surveys 51 (4), 653–703.

Kaiser, E., Kutz, J.N., Brunton, S.L., 2018a. Discovering conservation laws from data for control. In: 2018 IEEE Conference on Decision and Control (CDC). IEEE, pp. 6415–6421.

Kaiser, E., Kutz, J.N., Brunton, S.L., 2018b. Sparse identification of nonlinear dynamics for model predictive control in the low-data limit. Proceedings of the Royal Society A. Mathematical, Physical and Engineering Sciences 474 (2219), 20180335.

Kaiser, E., Kutz, J.N., Brunton, S.L., 2021. Data-driven discovery of Koopman eigenfunctions for control. Machine Learning: Science and Technology 2 (3), 035023.

Kamb, M., Kaiser, E., Brunton, S.L., Kutz, J.N., 2020. Time-delay observables for Koopman: theory and applications. SIAM Journal on Applied Dynamical Systems 19 (2), 886–917.

Kantz, H., Schreiber, T., 2006. Nonlinear Time Series Analysis, second edition edn. Cambridge Nonlinear Science Series. Cambridge Univ. Press, Cambridge.

Karniadakis, G.E., Kevrekidis, I.G., Lu, L., Perdikaris, P., Wang, S., Yang, L., 2021. Physics-informed machine learning. Nature Reviews Physics 3 (6), 422–440.

Katznelson, Y., 2004. An Introduction to Harmonic Analysis. Cambridge University Press.

Kawahara, Y., 2016. Dynamic mode decomposition with reproducing kernels for Koopman spectral analysis. Advances in Neural Information Processing Systems 29.

Khosravi, M., 2023. Representer theorem for learning Koopman operators. IEEE Transactions on Automatic Control 68 (5), 2995–3010.

Klus, S., Bittracher, A., Schuster, I., Schütte, C., 2018a. A kernel-based approach to molecular conformation analysis. Journal of Chemical Physics 149 (24), 244109.

Klus, S., Gelß, P., Peitz, S., Schütte, C., 2018b. Tensor-based dynamic mode decomposition. Nonlinearity 31 (7), 3359–3380.

Klus, S., Koltai, P., Schütte, C., 2016. On the numerical approximation of the Perron-Frobenius and Koopman operator. Journal of Computational Dynamics 3 (1), 1–12.

Klus, S., Nüske, F., Hamzi, B., 2020a. Kernel-based approximation of the Koopman generator and Schrödinger operator. Entropy 22 (7), 722.

Klus, S., Nüske, F., Koltai, P., Wu, H., Kevrekidis, I., Schütte, C., Noé, F., 2018c. Data-driven model reduction and transfer operator approximation. Journal of Nonlinear Science 28 (3), 985–1010.

Klus, S., Nüske, F., Peitz, S., Niemann, J.-H., Clementi, C., Schütte, C., 2020b. Data-driven approximation of the Koopman generator: model reduction, system identification, and control. Physica D. Nonlinear Phenomena 406, 132416.

Klus, S., Schuster, I., Muandet, K., 2020c. Eigendecompositions of transfer operators in reproducing kernel Hilbert spaces. Journal of Nonlinear Science 30 (1), 283–315.

Klus, S., Schütte, C., 2016. Towards tensor-based methods for the numerical approximation of the Perron–Frobenius and Koopman operator. Journal of Computational Dynamics 3 (2), 139–161.

Koch, R., Sanjosé, M., Moreau, S., 2021. Large-eddy simulation of a linear compressor cascade with tip gap: aerodynamic and acoustic analysis. In: AIAA Aviation 2021 Forum. American Institute of Aeronautics and Astronautics, p. 2312.

Koopman, B.O., 1931. Hamiltonian systems and transformation in Hilbert space. Proceedings of the National Academy of Sciences 17 (5), 315–318.

Koopman, B.O., von Neumann, J., 1932. Dynamical systems of continuous spectra. Proceedings of the National Academy of Sciences 18 (3), 255–263.

Korda, M., Mezić, I., 2018a. Linear predictors for nonlinear dynamical systems: Koopman operator meets model predictive control. Automatica 93, 149–160.

Korda, M., Mezić, I., 2018b. On convergence of extended dynamic mode decomposition to the Koopman operator. Journal of Nonlinear Science 28 (2), 687–710.

Korda, M., Mezić, I., 2020. Optimal construction of Koopman eigenfunctions for prediction and control. IEEE Transactions on Automatic Control 65 (12), 5114–5129.

Korda, M., Putinar, M., Mezić, I., 2020. Data-driven spectral analysis of the Koopman operator. Applied and Computational Harmonic Analysis 48 (2), 599–629.

Korda, M., Susuki, Y., Mezić, I., 2018. Power grid transient stabilization using Koopman model predictive control. IFAC-PapersOnLine 51 (28), 297–302.

Kostic, V., Novelli, P., Maurer, A., Ciliberto, C., Rosasco, L., Pontil, M., 2022. Learning dynamical systems via Koopman operator regression in reproducing kernel Hilbert spaces. Advances in Neural Information Processing Systems 35, 4017–4031.

Kou, J., Zhang, W., 2017. An improved criterion to select dominant modes from dynamic mode decomposition. European Journal of Mechanics. B, Fluids 62, 109–129.

Krake, T., Klötzl, D., Eberhardt, B., Weiskopf, D., 2022. Constrained dynamic mode decomposition. IEEE Transactions on Visualization and Computer Graphics 29 (1), 1–11.

Krener, A.J., 1974. Linearization and bilinearization of control systems. In: Proceedings of the 1974 Allerton Conference on Circuit and Systems Theory, Urbana III.

Kryloff, N., Bogoliouboff, N., 1937. La théorie générale de la mesure dans son application à l'étude des systèmes dynamiques de la mécanique non linéaire. Annals of Mathematics 38 (1), 65–113.

Kutz, J.N., Brunton, S.L., Brunton, B.W., Proctor, J.L., 2016a. Dynamic Mode Decomposition: Data-Driven Modeling of Complex Systems. Society for Industrial and Applied Mathematics.

Kutz, J.N., Fu, X., Brunton, S.L., 2016b. Multiresolution dynamic mode decomposition. SIAM Journal on Applied Dynamical Systems 15 (2), 713–735.

Lax, P.D., 1971. Approximation of measure preserving transformations. Communications on Pure and Applied Mathematics 24 (2), 133–135.

Le Clainche, S., Vega, J.M., 2017. Higher order dynamic mode decomposition. SIAM Journal on Applied Dynamical Systems 16 (2), 882–925.

Leroux, R., Cordier, L., 2016. Dynamic mode decomposition for non-uniformly sampled data. Experiments in Fluids 57 (5), 94.

Lewin, M., Séré, É., 2010. Spectral pollution and how to avoid it. Proceedings of the London Mathematical Society 100 (3), 864–900.

Li, M., Jiang, L., 2021. Deep learning nonlinear multiscale dynamic problems using Koopman operator. Journal of Computational Physics 446, 110660.

Li, Q., Dietrich, F., Bollt, E.M., Kevrekidis, I.G., 2017. Extended dynamic mode decomposition with dictionary learning: a data-driven adaptive spectral decomposition of the Koopman operator. Chaos 27 (10), 103111.

Li, T.-Y., 1976. Finite approximation for the Frobenius-Perron operator. A solution to Ulam's conjecture. Journal of Approximation Theory 17 (2), 177–186.

Li, Y., He, H., Wu, J., Katabi, D., Torralba, A., 2019. Learning compositional Koopman operators for model-based control. arXiv preprint. arXiv:1910.08264.

Liu, Z., Kundu, S., Chen, L., Yeung, E., 2018. Decomposition of nonlinear dynamical systems using Koopman Gramians. In: 2018 Annual American Control Conference (ACC). IEEE, pp. 4811–4818.

Loiseau, J.-C., Brunton, S.L., 2018. Constrained sparse Galerkin regression. Journal of Fluid Mechanics 838, 42–67.

Loparo, K., Blankenship, G., 1978. Estimating the domain of attraction of nonlinear feedback systems. IEEE Transactions on Automatic Control 23 (4), 602–608.

Lorenz, E.N., 1963. Deterministic nonperiodic flow. Journal of the Atmospheric Sciences 20 (2), 130–141.

Lu, H., Tartakovsky, D.M., 2020a. Lagrangian dynamic mode decomposition for construction of reduced-order models of advection-dominated phenomena. Journal of Computational Physics 407, 109229.

Lu, H., Tartakovsky, D.M., 2020b. Prediction accuracy of dynamic mode decomposition. SIAM Journal on Scientific Computing 42 (3), A1639–A1662.

Lusch, B., Kutz, J.N., Brunton, S.L., 2018. Deep learning for universal linear embeddings of nonlinear dynamics. Nature Communications 9 (1), 1–10.

Luzzatto, S., Melbourne, I., Paccaut, F., 2005. The Lorenz attractor is mixing. Communications in Mathematical Physics 260 (2), 393–401.

Maćešić, S., Črnjarić-Žic, N., Mezić, I., 2018. Koopman operator family spectrum for nonautonomous systems. SIAM Journal on Applied Dynamical Systems 17 (4), 2478–2515.

Mamakoukas, G., Castano, M., Tan, X., Murphey, T., 2019. Local Koopman operators for data-driven control of robotic systems. In: Robotics: Science and Systems XV. RSS2019, Robotics: Science and Systems Foundation.

Mamakoukas, G., Castano, M.L., Tan, X., Murphey, T.D., 2021. Derivative-based Koopman operators for real-time control of robotic systems. IEEE Transactions on Robotics 37 (6), 2173–2192.

Mañé, R., 1987. Ergodic Theory and Differentiable Dynamics. Springer Berlin Heidelberg.

Manohar, K., Kaiser, E., Brunton, S.L., Kutz, J.N., 2019. Optimized sampling for multiscale dynamics. Multiscale Modeling & Simulation 17 (1), 117–136.

Manojlović, I., Fonoberova, M., Mohr, R., Andrejčuk, A., Drmač, Z., Kevrekidis, Y., Mezić, I., 2020. Applications of Koopman mode analysis to neural networks. arXiv preprint. arXiv:2006.11765.

Mardt, A., Pasquali, L., Noé, F., Wu, H., 2020. Deep learning Markov and Koopman models with physical constraints. In: Mathematical and Scientific Machine Learning. PMLR, pp. 451–475.

Mardt, A., Pasquali, L., Wu, H., Noé, F., 2018. VAMPnets for deep learning of molecular kinetics. Nature Communications 9 (1), 1–11.

Martinsson, P.-G., Tropp, J.A., 2020. Randomized numerical linear algebra: foundations and algorithms. Acta Numerica 29, 403–572.

Mauroy, A., Goncalves, J., 2020. Koopman-based lifting techniques for nonlinear systems identification. IEEE Transactions on Automatic Control 65 (6), 2550–2565.

Mauroy, A., Mezić, I., 2016. Global stability analysis using the eigenfunctions of the Koopman operator. IEEE Transactions on Automatic Control 61 (11), 3356–3369.

Mauroy, A., Mezić, I., Moehlis, J., 2013. Isostables, isochrons, and Koopman spectrum for the action–angle representation of stable fixed point dynamics. Physica D. Nonlinear Phenomena 261, 19–30.

Mauroy, A., Susuki, Y., Mezić, I., 2020. Koopman Operator in Systems and Control. Lecture Notes in Control and Information Sciences Ser., vol. 484. Springer International Publishing AG, Cham.

Mehta, P.G., Hessel-von Molo, M., Dellnitz, M., 2006. Symmetry of attractors and the Perron–Frobenius operator. Journal of Difference Equations and Applications 12 (11), 1147–1178.

Mezić, I., 2005. Spectral properties of dynamical systems, model reduction and decompositions. Nonlinear Dynamics 41 (1), 309–325.

Mezić, I., 2013. Analysis of fluid flows via spectral properties of the Koopman operator. Annual Review of Fluid Mechanics 45 (1), 357–378.

Mezić, I., 2015. On applications of the spectral theory of the Koopman operator in dynamical systems and control theory. In: 2015 54th IEEE Conference on Decision and Control (CDC). IEEE, pp. 7034–7041.

Mezić, I., 2020. Spectrum of the Koopman operator, spectral expansions in functional spaces, and state-space geometry. Journal of Nonlinear Science 30 (5), 2091–2145.

Mezić, I., 2021. Koopman operator, geometry, and learning of dynamical systems. Notices of the American Mathematical Society 68 (07), 1.

Mezić, I., 2022. On numerical approximations of the Koopman operator. Mathematics 10 (7), 1180.

Mezić, I., Banaszuk, A., 2004. Comparison of systems with complex behavior. Physica D. Nonlinear Phenomena 197 (1–2), 101–133.

Mezić, I., Sotiropoulos, F., 2002. Ergodic theory and experimental visualization of invariant sets in chaotically advected flows. Physics of Fluids 14 (7), 2235–2243.

Mezić, I., Surana, A., 2016. Koopman mode decomposition for periodic/quasi-periodic time dependence. IFAC-PapersOnLine 49 (18), 690–697.

Mezić, I., Wiggins, S., 1999. A method for visualization of invariant sets of dynamical systems based on the ergodic partition. Chaos 9 (1), 213–218.

Mohr, R., Mezić, I., 2014. Construction of eigenfunctions for scalar-type operators via Laplace averages with connections to the Koopman operator. arXiv preprint. arXiv:1403.6559.

Mohri, M., Rostamizadeh, A., Talwalkar, A., 2018. Foundations of Machine Learning, Adaptive Computation and Machine Learning, second edition edn. The MIT Press, Cambridge, Massachusetts.

Mojgani, R., Balajewicz, M., 2017. Lagrangian basis method for dimensionality reduction of convection dominated nonlinear flows. arXiv preprint. arXiv:1701.04343.

Mollenhauer, M., Klus, S., Schütte, C., Koltai, P., 2022. Kernel autocovariance operators of stationary processes: estimation and convergence. Journal of Machine Learning Research 23 (327), 1–34.

Morandin, R., Nicodemus, J., Unger, B., 2023. Port-Hamiltonian dynamic mode decomposition. SIAM Journal on Scientific Computing 45 (4), A1690–A1710.

Narasingam, A., Kwon, J.S.-I., 2019. Koopman Lyapunov-based model predictive control of nonlinear chemical process systems. AIChE Journal 65 (11), e16743.

Needell, D., Tropp, J.A., 2009. CoSaMP: iterative signal recovery from incomplete and inaccurate samples. Applied and Computational Harmonic Analysis 26 (3), 301–321.

Netto, M., Mili, L., 2018. A robust data-driven Koopman Kalman filter for power systems dynamic state estimation. IEEE Transactions on Power Systems 33 (6), 7228–7237.

Neumann, J.v., 1932. Proof of the quasi-ergodic hypothesis. Proceedings of the National Academy of Sciences 18 (1), 70–82.

Noack, B.R., Afanasiev, K., Morzynski, M., Tadmor, G., Thiele, F., 2003. A hierarchy of low-dimensional models for the transient and post-transient cylinder wake. Journal of Fluid Mechanics 497, 335–363.

Noack, B.R., Eckelmann, H., 1994. A global stability analysis of the steady and periodic cylinder wake. Journal of Fluid Mechanics 270, 297–330.

Noack, B.R., Stankiewicz, W., Morzyński, M., Schmid, P.J., 2016. Recursive dynamic mode decomposition of transient and post-transient wake flows. Journal of Fluid Mechanics 809, 843–872.

Noé, F., Nüske, F., 2013. A variational approach to modeling slow processes in stochastic dynamical systems. Multiscale Modeling & Simulation 11 (2), 635–655.

Nonomura, T., Shibata, H., Takaki, R., 2018. Dynamic mode decomposition using a Kalman filter for parameter estimation. AIP Advances 8 (10).

Nonomura, T., Shibata, H., Takaki, R., 2019. Extended-Kalman-filter-based dynamic mode decomposition for simultaneous system identification and denoising. PLoS ONE 14 (2), e0209836.

Nüske, F., Gelß, P., Klus, S., Clementi, C., 2021. Tensor-based computation of metastable and coherent sets. Physica D. Nonlinear Phenomena 427, 133018.

Nüske, F., Keller, B.G., Pérez-Hernández, G., Mey, A.S.J.S., Noé, F., 2014. Variational approach to molecular kinetics. Journal of Chemical Theory and Computation 10 (4), 1739–1752.

Nüske, F., Peitz, S., Philipp, F., Schaller, M., Worthmann, K., 2023. Finite-data error bounds for Koopman-based prediction and control. Journal of Nonlinear Science 33 (1), 14.

Nyquist, H., 1928. Certain topics in telegraph transmission theory. Transactions of the American Institute of Electrical Engineers 47 (2), 617–644.

Ober-Blöbaum, S., Padberg-Gehle, K., 2015. Multiobjective optimal control of fluid mixing. PAMM 15 (1), 639–640.

Otto, S.E., Padovan, A., Rowley, C.W., 2023a. Model reduction for nonlinear systems by balanced truncation of state and gradient covariance. SIAM Journal on Scientific Computing 45 (5), A2325–A2355.

Otto, S.E., Rowley, C.W., 2019. Linearly recurrent autoencoder networks for learning dynamics. SIAM Journal on Applied Dynamical Systems 18 (1), 558–593.

Otto, S.E., Rowley, C.W., 2021. Koopman operators for estimation and control of dynamical systems. Annual Review of Control, Robotics, and Autonomous Systems 4 (1), 59–87.

Otto, S.E., Zolman, N., Kutz, J.N., Brunton, S.L., 2023b. A unified framework to enforce, discover, and promote symmetry in machine learning. arXiv preprint. arXiv:2311.00212.

Pan, C., Xue, D., Wang, J., 2015. On the accuracy of dynamic mode decomposition in estimating instability of wave packet. Experiments in Fluids 56 (8), 1–15.

Pan, S., Arnold-Medabalimi, N., Duraisamy, K., 2021. Sparsity-promoting algorithms for the discovery of informative Koopman-invariant subspaces. Journal of Fluid Mechanics 917, A18.

Pan, S., Duraisamy, K., 2020a. On the structure of time-delay embedding in linear models of nonlinear dynamical systems. Chaos 30 (7), 073135.

Pan, S., Duraisamy, K., 2020b. Physics-informed probabilistic learning of linear embeddings of nonlinear dynamics with guaranteed stability. SIAM Journal on Applied Dynamical Systems 19 (1), 480–509.

Parker, R., 1984. Acoustic resonances and blade vibration in axial flow compressors. Journal of Sound and Vibration 92 (4), 529–539.

Pazy, A., 2010. Semigroups of Linear Operators and Applications to Partial Differential Equations, 3 edn. Applied Mathematical Sciences, vol. 44. Springer, New York.

Peifer, M., Schelter, B., Winterhalder, M., Timmer, J., 2005. Mixing properties of the Rössler system and consequences for coherence and synchronization analysis. Physical Review E 72 (2), 026213.

Peitz, S., 2018. Controlling nonlinear PDEs using low-dimensional bilinear approximations obtained from data. arXiv preprint. arXiv:1801.06419.

Peitz, S., Klus, S., 2019. Koopman operator-based model reduction for switched-system control of PDEs. Automatica 106, 184–191.

Peitz, S., Klus, S., 2020. Feedback Control of Nonlinear PDEs Using Data-Efficient Reduced Order Models Based on the Koopman Operator. Springer, pp. 257–282.

Peitz, S., Otto, S.E., Rowley, C.W., 2020. Data-driven model predictive control using interpolated Koopman generators. SIAM Journal on Applied Dynamical Systems 19 (3), 2162–2193.

Pereyra, V., Scherer, G., 2010. Exponential data fitting and its applications. Bentham eBooks, Sharjah, UAE.

Philipp, F., Schaller, M., Worthmann, K., Peitz, S., Nüske, F., 2023. Error bounds for kernel-based approximations of the Koopman operator. arXiv preprint. arXiv:2301.08637.

Poincaré, H., 1899. Les méthodes nouvelles de la mécanique céleste. Il Nuovo Cimento 10 (1), 128–130.

Proctor, J.L., Brunton, S.L., Kutz, J.N., 2016. Dynamic mode decomposition with control. SIAM Journal on Applied Dynamical Systems 15 (1), 142–161.

Proctor, J.L., Brunton, S.L., Kutz, J.N., 2018. Generalizing Koopman theory to allow for inputs and control. SIAM Journal on Applied Dynamical Systems 17 (1), 909–930.

Redman, W.T., 2021. On Koopman mode decomposition and tensor component analysis. Chaos 31 (5).

Redman, W.T., Fonoberova, M., Mohr, R., Kevrekidis, I.G., Mezić, I., 2022. Algorithmic (semi-) conjugacy via Koopman operator theory. In: 2022 IEEE 61st Conference on Decision and Control (CDC). IEEE.

Redman, W.T., Huang, D., Fonoberova, M., Mezić, I., 2023. Koopman learning with episodic memory. arXiv preprint. arXiv:2311.12615.
Reynolds, R.W., Rayner, N.A., Smith, T.M., Stokes, D.C., Wang, W., 2002. An improved in situ and satellite SST analysis for climate. Journal of Climate 15 (13), 1609–1625.
Reynolds, R.W., Smith, T.M., Liu, C., Chelton, D.B., Casey, K.S., Schlax, M.G., 2007. Daily high-resolution-blended analyses for sea surface temperature. Journal of Climate 20 (22), 5473–5496.
Roch, S., Silbermann, B., 1996. C^*-algebra techniques in numerical analysis. Journal of Operator Theory, 241–280.
Rokhlin, V., Szlam, A., Tygert, M., 2010. A randomized algorithm for principal component analysis. SIAM Journal on Matrix Analysis and Applications 31 (3), 1100–1124.
Rosenfeld, J.A., Kamalapurkar, R., Gruss, L.F., Johnson, T.T., 2022. Dynamic mode decomposition for continuous time systems with the Liouville operator. Journal of Nonlinear Science 32 (1), 1–30.
Rössler, O.E., 1976. An equation for continuous chaos. Physics Letters A 57 (5), 397–398.
Rowley, C.W., 2005. Model reduction for fluids, using balanced proper orthogonal decomposition. International Journal of Bifurcation and Chaos 15 (03), 997–1013.
Rowley, C.W., Dawson, S.T.M., 2017. Model reduction for flow analysis and control. Annual Review of Fluid Mechanics 49 (1), 387–417.
Rowley, C.W., Mezić, I., Bagheri, S., Schlatter, P., Henningson, D.S., 2009. Spectral analysis of nonlinear flows. Journal of Fluid Mechanics 641, 115–127.
Rowley, C.W., Williams, D.R., Colonius, T., Murray, R.M., Macmynowski, D.G., 2006. Linear models for control of cavity flow oscillations. Journal of Fluid Mechanics 547 (1), 317–330.
Ruelle, D., 1968. Statistical mechanics of a one-dimensional lattice gas. Communications in Mathematical Physics 9 (4), 267–278.
Salova, A., Emenheiser, J., Rupe, A., Crutchfield, J.P., D'Souza, R.M., 2019. Koopman operator and its approximations for systems with symmetries. Chaos 29 (9).
Sashidhar, D., Kutz, J.N., 2022. Bagging, optimized dynamic mode decomposition for robust, stable forecasting with spatial and temporal uncertainty quantification. Philosophical Transactions of the Royal Society A: Mathematical, Physical and Engineering Sciences 380 (2229), 20210199.
Scherl, I., Strom, B., Shang, J.K., Williams, O., Polagye, B.L., Brunton, S.L., 2020. Robust principal component analysis for modal decomposition of corrupt fluid flows. Physical Review Fluids 5 (5), 054401.
Schmid, P., Sesterhenn, J., 2008. Dynamic mode decomposition of numerical and experimental data. Bulletin of the American Physical Society 53.
Schmid, P.J., 2009. Dynamic mode decomposition of experimental data. In: 8th International Symposium on Particle Image Velocimetry (PIV09).
Schmid, P.J., 2010. Dynamic mode decomposition of numerical and experimental data. Journal of Fluid Mechanics 656, 5–28.
Schmid, P.J., 2022. Dynamic mode decomposition and its variants. Annual Review of Fluid Mechanics 54 (1), 225–254.
Scholkopf, B., 2001. The kernel trick for distances. Advances in Neural Information Processing Systems, 301–307.
Schönemann, P.H., 1966. A generalized solution of the orthogonal procrustes problem. Psychometrika 31 (1), 1–10.
Schütte, C., Klus, S., Hartmann, C., 2023. Overcoming the timescale barrier in molecular dynamics: transfer operators, variational principles and machine learning. Acta Numerica 32, 517–673.
Schütte, C., Sarich, M., 2013. Metastability and Markov State Models in Molecular Dynamics, vol. 24. American Mathematical Society.
Sechi, R., Sikorski, A., Weber, M., 2021. Estimation of the Koopman generator by Newton's extrapolation. Multiscale Modeling & Simulation 19 (2), 758–774.
Shannon, C.E., 1948. A mathematical theory of communication. The Bell System Technical Journal 27 (3), 379–423.

Shargorodsky, E., 2008. On the level sets of the resolvent norm of a linear operator. Bulletin of the London Mathematical Society 40 (3), 493–504.

Sharma, A.S., Mezić, I., McKeon, B.J., 2016. Correspondence between Koopman mode decomposition, resolvent mode decomposition, and invariant solutions of the Navier-Stokes equations. Physical Review Fluids 1 (3), 032402.

Sharma, H., Vaidya, U., Ganapathysubramanian, B., 2019. A transfer operator methodology for optimal sensor placement accounting for uncertainty. Building and Environment 155, 334–349.

Shields, P.C., 1973. The Theory of Bernoulli Shifts. University of Chicago Press, Chicago.

Sinha, S., Huang, B., Vaidya, U., 2020. On robust computation of Koopman operator and prediction in random dynamical systems. Journal of Nonlinear Science 30 (5), 2057–2090.

Sinha, S., Vaidya, U., Rajaram, R., 2016. Operator theoretic framework for optimal placement of sensors and actuators for control of nonequilibrium dynamics. Journal of Mathematical Analysis and Applications 440 (2), 750–772.

Sinha, S., Vaidya, U., Yeung, E., 2019. On computation of Koopman operator from sparse data. In: 2019 American Control Conference (ACC). IEEE, pp. 5519–5524.

Slipantschuk, J., Bandtlow, O.F., Just, W., 2020. Dynamic mode decomposition for analytic maps. Communications in Nonlinear Science and Numerical Simulation 84, 105179.

Smith, T.M., Reynolds, R.W., 2005. A global merged land–air–sea surface temperature reconstruction based on historical observations (1880–1997). Journal of Climate 18 (12), 2021–2036.

Son, S.H., Narasingam, A., Kwon, J.S.-I., 2020. Handling plant-model mismatch in Koopman Lyapunov-based model predictive control via offset-free control framework. arXiv preprint. arXiv:2010.07239.

Sootla, A., Mauroy, A., 2016. Properties of isostables and basins of attraction of monotone systems. In: 2016 American Control Conference (ACC). IEEE, pp. 7365–7370.

Surana, A., 2016. Koopman operator based observer synthesis for control-affine nonlinear systems. In: 2016 IEEE 55th Conference on Decision and Control (CDC). IEEE, pp. 6492–6499.

Surana, A., Banaszuk, A., 2016. Linear observer synthesis for nonlinear systems using Koopman operator framework. IFAC-PapersOnLine 49 (18), 716–723.

Susuki, Y., Mauroy, A., Mezić, I., 2021. Koopman resolvent: a Laplace-domain analysis of nonlinear autonomous dynamical systems. SIAM Journal on Applied Dynamical Systems 20 (4), 2013–2036.

Susuki, Y., Mezić, I., 2015. A Prony approximation of Koopman mode decomposition. In: 2015 54th IEEE Conference on Decision and Control (CDC). IEEE.

Szőke, M., Jozsa, T.I., Koleszár, Á., Moulitsas, I., Könözsy, L., 2017. Performance evaluation of a two-dimensional lattice Boltzmann solver using CUDA and PGAS UPC based parallelisation. ACM Transactions on Mathematical Software 44 (1), 1–22.

Szőke, M., Nurani Hari, N., Devenport, W.J., Glegg, S.A., Teschner, T.-R., 2021. Flow field analysis around pressure shielding structures. In: AIAA Aviation 2021 Forum. American Institute of Aeronautics and Astronautics, p. 2293.

Taira, K., Brunton, S.L., Dawson, S.T.M., Rowley, C.W., Colonius, T., McKeon, B.J., Schmidt, O.T., Gordeyev, S., Theofilis, V., Ukeiley, L.S., 2017. Modal analysis of fluid flows: an overview. AIAA Journal 55 (12), 4013–4041.

Taira, K., Hemati, M.S., Brunton, S.L., Sun, Y., Duraisamy, K., Bagheri, S., Dawson, S.T.M., Yeh, C.-A., 2020. Modal analysis of fluid flows: applications and outlook. AIAA Journal 58 (3), 998–1022.

Takeishi, N., Kawahara, Y., Tabei, Y., Yairi, T., 2017a. Bayesian dynamic mode decomposition. In: Proceedings of the Twenty-Sixth International Joint Conference on Artificial Intelligence. IJCAI-2017, International Joint Conferences on Artificial Intelligence Organization, pp. 2814–2821.

Takeishi, N., Kawahara, Y., Yairi, T., 2017b. Learning Koopman invariant subspaces for dynamic mode decomposition. Advances in Neural Information Processing Systems 30.

Takeishi, N., Kawahara, Y., Yairi, T., 2017c. Subspace dynamic mode decomposition for stochastic Koopman analysis. Physical Review E 96 (3), 033310.

Takens, F., 2006. Detecting strange attractors in turbulence. In: Dynamical Systems and Turbulence, Warwick 1980: Proceedings of a Symposium Held at the University of Warwick 1979/80. Springer, pp. 366–381.

Tantet, A., Lucarini, V., Lunkeit, F., Dijkstra, H.A., 2018. Crisis of the chaotic attractor of a climate model: a transfer operator approach. Nonlinearity 31 (5), 2221–2251.

Tantet, A., van der Burgt, F.R., Dijkstra, H.A., 2015. An early warning indicator for atmospheric blocking events using transfer operators. Chaos 25 (3).

Tissot, G., Cordier, L., Benard, N., Noack, B.R., 2014. Model reduction using dynamic mode decomposition. Comptes Rendus. Mécanique 342 (6–7), 410–416.

Towne, A., Schmidt, O.T., Colonius, T., 2018. Spectral proper orthogonal decomposition and its relationship to dynamic mode decomposition and resolvent analysis. Journal of Fluid Mechanics 847, 821–867.

Trefethen, L.N., Embree, M., 2005. Spectra and Pseudospectra: The Behavior of Nonnormal Matrices and Operators. Princeton University Press.

Trefethen, L.N., Trefethen, A.E., Reddy, S.C., Driscoll, T.A., 1993. Hydrodynamic stability without eigenvalues. Science 261 (5121), 578–584.

Tu, J.H., Rowley, C.W., Kutz, J.N., Shang, J.K., 2014a. Spectral analysis of fluid flows using sub-Nyquist-rate PIV data. Experiments in Fluids 55 (9), 1–13.

Tu, J.H., Rowley, C.W., Luchtenburg, D.M., Brunton, S.L., Nathan Kutz, J., 2014b. On dynamic mode decomposition: theory and applications. Journal of Computational Dynamics 1 (2), 391–421.

Tufillaro, N.B., Abbot, T., Reilly, J., Hickey, F.R., 1993. An experimental approach to nonlinear dynamics and chaos. American Journal of Physics 61 (10), 958–959.

Udell, M., Townsend, A., 2019. Why are big data matrices approximately low rank? SIAM Journal on Mathematics of Data Science 1 (1), 144–160.

Ulam, S.M., 1960. Problems in Modern Mathematics. Science Editions.

Vaidya, U., 2007. Observability Gramian for nonlinear systems. In: 2007 46th IEEE Conference on Decision and Control. IEEE, pp. 3357–3362.

Vaidya, U., Mehta, P.G., Shanbhag, U.V., 2010. Nonlinear stabilization via control Lyapunov measure. IEEE Transactions on Automatic Control 55 (6), 1314–1328.

Valva, C., Giannakis, D., 2023. Consistent spectral approximation of Koopman operators using resolvent compactification. arXiv preprint. arXiv:2309.00732.

Van Huffel, S., Vandewalle, J., 1991. The Total Least Squares Problem: Computational Aspects and Analysis. Society for Industrial and Applied Mathematics.

Vega, J.M., Le Clainche, S., 2021. Spatio-Temporal Koopman Decomposition, vol. 28. Elsevier, pp. 121–157.

von Luxburg, U., Belkin, M., Bousquet, O., 2008. Consistency of spectral clustering. The Annals of Statistics 36 (2), 555–586.

Walters, P., 2000. An Introduction to Ergodic Theory, 1 softcover printing edn. Graduate Texts in Mathematics, vol. 79. Springer, New York.

Wanner, M., Mezić, I., 2022. Robust approximation of the stochastic Koopman operator. SIAM Journal on Applied Dynamical Systems 21 (3), 1930–1951.

Wehmeyer, C., Noé, F., 2018. Time-lagged autoencoders: deep learning of slow collective variables for molecular kinetics. Journal of Chemical Physics 148 (24).

Welch, P., 1967. The use of fast Fourier transform for the estimation of power spectra: a method based on time averaging over short, modified periodograms. IEEE Transactions on Audio and Electroacoustics 15 (2), 70–73.

Williams, M.O., Hemati, M.S., Dawson, S.T.M., Kevrekidis, I.G., Rowley, C.W., 2016. Extending data-driven Koopman analysis to actuated systems. IFAC-PapersOnLine 49 (18), 704–709.

Williams, M.O., Kevrekidis, I.G., Rowley, C.W., 2015a. A data–driven approximation of the Koopman operator: extending dynamic mode decomposition. Journal of Nonlinear Science 25 (6), 1307–1346.

Williams, M.O., Rowley, C.W., Kevrekidis, I.G., 2015b. A kernel-based method for data-driven Koopman spectral analysis. Journal of Computational Dynamics 2 (2), 247–265.

Woodley, B.M., Peake, N., 1999. Resonant acoustic frequencies of a tandem cascade. Part 2. Rotating blade rows. Journal of Fluid Mechanics 393, 241–256.

Wormell, C.L., 2023. Orthogonal polynomial approximation and extended dynamic mode decomposition in chaos. arXiv preprint. arXiv:2305.08074.

Wu, H., Noé, F., 2020. Variational approach for learning Markov processes from time series data. Journal of Nonlinear Science 30 (1), 23–66.

Wu, H., Nüske, F., Paul, F., Klus, S., Koltai, P., Noé, F., 2017. Variational Koopman models: slow collective variables and molecular kinetics from short off-equilibrium simulations. Journal of Chemical Physics 146 (15).

Wu, Z., Brunton, S.L., Revzen, S., 2021. Challenges in dynamic mode decomposition. Journal of the Royal Society Interface 18 (185), 20210686.

Wynn, A., Pearson, D.S., Ganapathisubramani, B., Goulart, P.J., 2013. Optimal mode decomposition for unsteady flows. Journal of Fluid Mechanics 733, 473–503.

Xie, T., France-Lanord, A., Wang, Y., Shao-Horn, Y., Grossman, J.C., 2019. Graph dynamical networks for unsupervised learning of atomic scale dynamics in materials. Nature Communications 10 (1), 2667.

Yeung, E., Kundu, S., Hodas, N., 2019. Learning deep neural network representations for Koopman operators of nonlinear dynamical systems. In: 2019 American Control Conference (ACC). IEEE, pp. 4832–4839.

Yeung, E., Liu, Z., Hodas, N.O., 2018. A Koopman operator approach for computing and balancing Gramians for discrete time nonlinear systems. In: 2018 Annual American Control Conference (ACC). IEEE, pp. 337–344.

Young, L.-S., 2002. What are SRB measures, and which dynamical systems have them? Journal of Statistical Physics 108 (5/6), 733–754.

Zaslavsky, G.M., 2002. Chaos, fractional kinetics, and anomalous transport. Physics Reports 371 (6), 461–580.

Zebib, A., 1987. Stability of viscous flow past a circular cylinder. Journal of Engineering Mathematics 21 (2), 155–165.

Zhang, H., Rowley, C.W., Deem, E.A., Cattafesta, L.N., 2019. Online dynamic mode decomposition for time-varying systems. SIAM Journal on Applied Dynamical Systems 18 (3), 1586–1609.

Zhao, Z., Giannakis, D., 2016. Analog forecasting with dynamics-adapted kernels. Nonlinearity 29 (9), 2888–2939.

Chapter 5

Deep learning variational Monte Carlo for solving the electronic Schrödinger equation

Leon Gerard[a], Philipp Grohs[a,b,*], and Michael Scherbela[a]

[a]*Faculty of Mathematics, University of Vienna, Vienna, Austria,* [b]*Johann Radon Institute for Computational and Applied Mathematics, Austrian Academy of Sciences, Linz, Austria*

**Corresponding author: e-mail address: philipp.grohs@univie.ac.at*

Contents

Abstract

The electronic Schrödinger equation is one of the most fundamental models in physics, due to its capability of accurately predicting all properties of molecules. Having efficient numerical methods for its solution would revolutionize drug- or material discovery – among many other fields. However, while the equation itself is easily stated, its efficient and accurate numerical solution poses formidable challenges. This has sparked decades-long research on the development of a plethora of different highly specialized numerical schemes. In recent years, methods based on Deep Learning have been introduced and shown to outperform the previous state-of-the-art in terms of accuracy, allowing for increasingly large systems to be computable to within chemical accuracy. In this paper we survey these exciting developments from the perspective of a (numerical) mathematician. To this end, we first provide an introduction into the mathematical theory of the electronic Schrödinger equation and then outline numerical methods to partially overcome some of its considerable challenges. Finally we survey recent work on Deep Learning-based Variational Monte Carlo methods and showcase some numerical results.

Keywords
Electronic Schrödinger equation, Deep learning, Variational Monte Carlo, Fermionic neural networks

MSC Codes
68T07, 81-08, 81Q10, 65-02, 46N50, 35-02, 35Q40, 65C05

1 Introduction

> *The underlying physical laws necessary for the mathematical theory of a large part of physics and the whole of chemistry are thus completely known, and the difficulty is only that the exact application of these laws leads to equations much too complicated to be soluble. It therefore becomes desirable that approximate practical methods of applying quantum mechanics should be developed, which can lead to an explanation of the main features of complex atomic systems without too much computation.*
>
> P. Dirac (1929)

This famous quote taken from Dirac (1929) refers to the Schrödinger equation of electronic systems, which is also the subject of the present survey article. The Schrödinger equation represents a partial differential equation capable of accurately describing all nonrelativistic properties of atoms and molecules. This means that one can – theoretically – simulate all properties of molecules from first principles without having to resort to expensive and time-consuming experiments. This has in principle the potential to lower cost and enable the search for new materials on a much greater scale than ever before. In view of this, the importance of developing efficient numerical methods for the electronic Schrödinger equation is hard to overstate and solving it can be considered the "holy grail" of computational chemistry.

Solving the equation – i.e., finding a ground-state wavefunction for a given molecule – is computationally challenging. Analytical solutions are only known for atoms with a single electron (i.e., a single Hydrogen atom) and thus any system of practical interest must be solved numerically. It has furthermore been shown that for model-Hamiltonians such as the Hubbard model, finding the ground-state wavefunction is a QMA-hard problem (Troyer and Wiese, 2005), making it at least as hard as any NP-complete problem.

This has motivated a diverse range of computational methods, including renowned techniques like Density Functional Theory (DFT), which was awarded the Nobel Prize in 1998. On the one hand, there exist methods like DFT or Hartree Fock that can handle systems comprising hundreds of atoms with relatively crude but efficient approximations. On the other hand, methods such as CCSD(T), often named the gold standard when it comes to precise approximations, tend to exhibit scaling behavior typically in the order of $O(n_{el}^7)$, where n_{el} represents the number of electrons considered. In the case of the configuration interaction singles, doubles, triples, quadruples (CISDTQ) method the scaling can even go up to $O(n_{el}^{10})$ (Scherbela et al., 2022), limiting the method to rather small system sizes. For an introduction to computational methods see Szabo and Ostlund (1996).

In more recent years, methods from machine learning have made a significant impact in improving the tradeoff between accuracy and computational cost. These approaches can be roughly categorized into supervised and unsupervised. In the supervised regime one typically starts from a data set of high accuracy calculations for several different compounds. This data set is then used as training data for a machine learning regressor. Once trained, such a regressor is capable of speeding up calculations by many orders of magnitude while retaining the accuracy of the training data set. We refer to Schütt et al. (2020) for an overview focused on material discovery.

The supervised learning approach leaves open the key problem of generating highly accurate solutions to the Schrödinger equation that can be subsequently used as training data. To address this issue, and due to their excellent approximation properties (Elbrächter et al., 2021), deep neural networks have been explored as ansatz functions for electronic wave functions. These neural networks are then trained in an unsupervised fashion to approximate wave functions of electronic ground states. This approach – pioneered in Carleo and Troyer (2017); Pfau et al. (2020); Hermann et al. (2020) and coined *Deep Learning Variational Monte Carlo (DL-VMC)* – has been shown to significantly outperform classical methods in terms of accuracy. We refer to Hermann et al. (2022) for a summary of recent results in this direction. With its apparent ability to better describe wavefunctions of molecules when compared to classical representations and the great progress that could already be achieved within the span of only a few years, DL-VMC holds the potential to revolutionize the field of computational chemistry.

Broadly speaking, DL-VMC can be regarded as a specific instance of so-called PINNs (physics-informed-neural-networks (De Ryck and Mishra, 2024)) applied to the Schrödinger equation. While the analysis and development of PINN methods for solving PDEs constitutes a highly active field of research, these methods often lag behind more standard algorithms (such as finite elements or low rank methods) in terms of accuracy (Chuang and Barba, 2022) or sample complexity (Bayer et al., 2023; Berner et al., 2023). DL-VMC is a rare instance of a PINN-like method that actually surpasses conventional algorithms in terms of accuracy, making it all the more interesting.

The present paper aims to provide a survey of these exciting developments with a specific focus on a readership coming from the field of mathematical/numerical analysis. While empirical results of DL-VMC are convincing, the mathematical theory behind these successes has yet to be developed. We therefore hope that our work can serve as an invitation to members of the mathematical community to contribute to this vibrant field.

1.1 The molecular and electronic Schrödinger equations

We start with an informal description of the molecular and electronic Schrödinger equations. A molecule can be described by n_{nuc} nuclei with $\mathbf{R} = \left(R_1, \ldots, R_{n_{\text{nuc}}}\right) \in \mathbb{R}^{n_{\text{nuc}} \times 3}$ denoting the nuclear coordinates and $\mathbf{Z} = \left(Z_1, \ldots, Z_{n_{\text{nuc}}}\right) \in \mathbb{N}^{n_{\text{nuc}}}$ denoting their respective charges. These nuclei are surrounded by a cloud of n_{el} electrons with $\mathbf{r} = \left(r_1, \ldots, r_{n_{\text{el}}}\right) \in \mathbb{R}^{n_{\text{el}} \times 3}$ denoting the electron coordinates. In atomic units (meaning that electron mass, elementary charge, Planck's constant $\hbar$ and permittivity are equal to 1) we assume that the mass of the I-th nucleus is equal to M_I. The electrons and nuclei interact with each other through Coulomb attraction and repulsion forces, which leads to the molecular Hamiltonian

$$\begin{aligned}\mathcal{H}^{\text{mol}} = &-\frac{1}{2}\sum_{I=1}^{n_{\text{nuc}}} \frac{1}{M_I}\nabla_{R_I}^2 - \frac{1}{2}\sum_{i=1}^{n_{\text{el}}} \nabla_{r_i}^2 + \sum_{i=1}^{n_{\text{el}}-1}\sum_{j=i+1}^{n_{\text{el}}} \frac{1}{|r_i - r_j|} \\ &+ \sum_{I=1}^{n_{\text{nuc}}-1}\sum_{J=I}^{n_{\text{nuc}}} \frac{Z_I Z_J}{|R_I - R_J|} - \sum_{i=1}^{n_{\text{el}}}\sum_{I=1}^{n_{\text{nuc}}} \frac{Z_I}{|r_i - R_I|}. \end{aligned} \tag{1}$$

Here the term

$$-\frac{1}{2}\sum_{I=1}^{n_{\text{nuc}}} \frac{1}{M_I}\nabla_{R_I}^2 - \frac{1}{2}\sum_{i=1}^{n_{\text{el}}} \nabla_{r_i}^2$$

represents the kinetic energy and the term

$$\sum_{i=1}^{n_{\text{el}}-1}\sum_{j=i+1}^{n_{\text{el}}} \frac{1}{|r_i - r_j|} + \sum_{I=1}^{n_{\text{nuc}}-1}\sum_{J=I}^{n_{\text{nuc}}} \frac{Z_I Z_J}{|R_I - R_J|} - \sum_{i=1}^{n_{\text{el}}}\sum_{I=1}^{n_{\text{nuc}}} \frac{Z_I}{|r_i - R_I|},$$

which should be understood as a multiplication operator, represents the potential energy induced by Coulomb forces. The state of a molecule can be quantum mechanically described by its molecular wave function $\Psi = \Psi(\mathbf{R}, \mathbf{r})$. The time evolution of such a state is governed by the *time-dependent Schrödinger Equation*

$$i\frac{\partial}{\partial t}\Psi(t) = \mathcal{H}^{\text{mol}}\Psi(t). \tag{2}$$

While (2) in principle allows for a full description of a given molecule, it is often too complicated to be solved, even by numerical methods.

A common simplification is given by the *Born-Oppenheimer approximation* which essentially assumes that the positions of the nuclei stay fixed so that the term corresponding to the kinetic energy of the nuclei can be omitted in the Hamiltonian. Heuristically, this approximation is justified since nuclei are much heavier than electrons and therefore the motion of the nuclei $\mathbf{R}$ can be assumed to occur on a much slower timescale than the motion of the electrons. Hence (to within a certain accuracy) the motion of the nuclei can be neglected. For a fixed geometrical conformation described by nuclear coordinates and charges $(\mathbf{R}, \mathbf{Z})$ this leads to the new Hamiltonian

$$\mathcal{H}^{\mathrm{BO}}_{(\mathbf{R},\mathbf{Z})} = -\frac{1}{2}\sum_{i=1}^{n_{\mathrm{el}}} \nabla^2_{r_i} + \sum_{i=1}^{n_{\mathrm{el}}-1}\sum_{j=i+1}^{n_{\mathrm{el}}} \frac{1}{|r_i - r_j|} + \sum_{I=1}^{n_{\mathrm{nuc}}-1}\sum_{J=I}^{n_{\mathrm{nuc}}} \frac{Z_I Z_J}{|R_I - R_J|} - \sum_{i=1}^{n_{\mathrm{el}}}\sum_{I=1}^{n_{\mathrm{nuc}}} \frac{Z_I}{|r_i - R_I|}, \tag{3}$$

which is now applied to *electronic wavefunctions* $\Psi^{\mathrm{e}} : \mathbb{R}^{n_{\mathrm{el}} \times 3} \ni (r_1, \ldots, r_{n_{\mathrm{el}}}) \to \mathbb{C}$, depending on the electron coordinates only, whereas the geometric conformation $(\mathbf{R}, \mathbf{Z})$ is treated as a parameter. We have therefore reduced the dimensionality by $3 \cdot n_{\mathrm{nuc}}$. Studying the simplified quantum system described by the Hamiltonian (3) is then referred to as the Born-Oppenheimer approximation.

Of specific interest is the *time-independent electronic Schrödinger Equation*, which amounts to solving the eigenvalue problem

$$\mathcal{H}^{\mathrm{BO}}_{(\mathbf{R},\mathbf{Z})} \Psi^{\mathrm{e}} = \lambda_{(\mathbf{R},\mathbf{Z})} \Psi^{\mathrm{e}}, \quad \lambda_{(\mathbf{R},\mathbf{Z})} \in \mathbb{R}. \tag{4}$$

The Eigenvalues $\lambda_{(\mathbf{R},\mathbf{Z})}$ correspond to the energies that the electronic system can assume without breaking apart (Hunziker and Sigal, 2000). The corresponding Eigenvectors describe precisely those electronic states Ψ^{e} whose energy $E = \lambda_{(\mathbf{R},\mathbf{Z})} = \langle \mathcal{H}^{\mathrm{BO}} \Psi^{\mathrm{e}}, \Psi^{\mathrm{e}} \rangle$ can be measured without uncertainty. Their time evolution under the time-dependent electronic Schrödinger equation

$$i \frac{\partial}{\partial t} \Psi^{\mathrm{e}}(t) = \mathcal{H}^{\mathrm{BO}} \Psi^{\mathrm{e}}(t)$$

can be easily calculated to yield $\Psi^{\mathrm{e}}(t) = e^{-iEt} \Psi^{\mathrm{e}}(0)$. Therefore, knowledge of the eigenstates allows for a simple solution of the time-dependent electronic Schrödinger equation whenever the initial state is given as a superposition of eigenstates.

The eigenvectors $\Psi^{\mathrm{e},0}_{(\mathbf{R},\mathbf{Z})}$ corresponding to the smallest possible eigenvalue $\lambda^0_{(\mathbf{R},\mathbf{Z})}$ in (4) are called *electronic ground states* and generally represent the most stable and likely electronic states. The eigenvectors corresponding to higher eigenvalues are called *excited states*. For a fixed number of nuclei n_{nuc} and fixed nuclear charges $\mathbf{Z}$, the mapping $\mathcal{E}_{\mathbf{Z}} : \mathbb{R}^{n_{\mathrm{nuc}} \times 3} \ni \mathbf{R} \mapsto \lambda^0_{(\mathbf{R},\mathbf{Z})}$ is called the *potential energy surface (PES)*.

The PES contains a wealth of information about the chemical properties of a given molecule. For example, it can be used as an approximation of the potential energy of the nuclear coordinates $\mathbf{R}$ leading to the force field $F_{\mathbf{Z}}(\mathbf{R}) := -\nabla_{\mathbf{R}}\mathcal{E}_{\mathbf{Z}}(\mathbf{R})$ acting on the nuclei. Fig. 1 shows a calculation of a PES of hydrogen chains using a deep learning ansatz (Scherbela et al., 2022). Using these forces, the dynamics of the nuclei can be approximately simulated classically by solving Newton's equations

$$M\ddot{\mathbf{R}}(t) = F_{\mathbf{Z}}(\mathbf{R}(t)), \quad M = \operatorname{diag}(M_1, \ldots, M_{n_{\mathrm{nuc}}}) \in \mathbb{R}^{n_{\mathrm{nuc}} \times n_{\mathrm{nuc}}}.$$

This in turn allows for the determination of the structure of molecules or the ab initio simulation of chemical reactions.

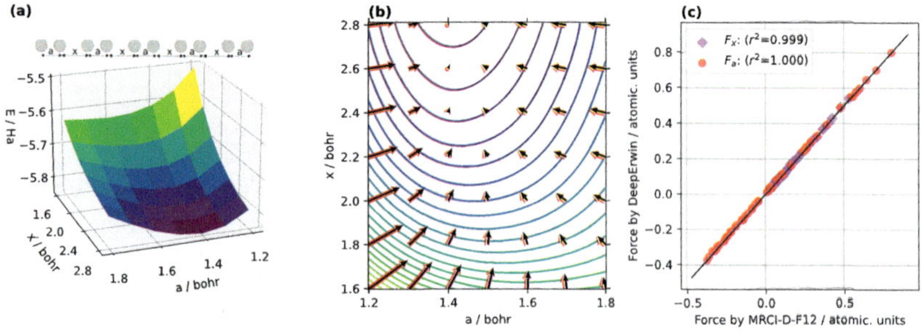

FIGURE 1 (a) Potential energy surface (PES) for the H_{10} chain. The variable a is representing the distance between two H-atoms and x is the distance between two H_2 molecules. The lowest ground state of the PES describes the dimerization. (b) Cubic interpolation of the PES with force vectors for the H_{10} chain. Red arrows depict force vectors computed by DeepErwin via the Hellmann-Feynman theorem, whereas black arrows represent numerical gradients that are based on finite differences of MRCI-F12 reference calculations. (c) Forces computed by DeepErwin plotted against the respective forces obtained from finite differences of the MRCI-F12 reference calculation. Figure is taken from Scherbela et al. (2022).

Furthermore, one can approximate the molecular ground state $\Psi^0(\mathbf{R}, \mathbf{r})$ by the separated product $\Psi^0(\mathbf{R}, \mathbf{r}) \sim \Psi^{\mathrm{n},0}_{\mathbf{Z}}(\mathbf{R}) \cdot \Psi^{\mathrm{e},0}_{(\mathbf{R},\mathbf{Z})}(\mathbf{r})$, where $\Psi^{\mathrm{n},0}_{\mathbf{Z}}(\mathbf{R})$ solves the time-independent nuclear Schrödinger Equation

$$\Psi^{\mathrm{n}} = \lambda \left(-\sum_{I}^{n_{\mathrm{nuc}}} \frac{1}{2M_I} \nabla^2_{\mathbf{R}_I} + \mathcal{E}_{\mathbf{Z}} \right) \Psi^{\mathrm{n}}, \quad \lambda \in \mathbb{R}$$

with minimal eigenvalue. The degree to which these approximations are valid depends on the structure of the eigenvalues of the electronic Hamiltonian, see for example Jecko (2014) and the references therein. At any rate, it should be clear that the PES is an extremely important quantity since it allows for the computation of properties of molecules without having to conduct actual experiments.

A grand goal of computational chemistry is thus as follows.

Find efficient and accurate algorithms for evaluating the PES $(\mathbf{R}, \mathbf{Z}) \mapsto \mathcal{E}_{\mathbf{Z}}(\mathbf{R})$

Efficient evaluation of the PES allows, for example, to perform a computational structure search or to compute chemical reaction rates. However, for these types of tasks, highly accurate estimates of the ground-state energy become essential. For example, a study by Barone et al. (2013) demonstrated that the geometrical conformations of the smallest amino acid, Glycine, are partitioned by transition barriers of only ~20 millihartree (mHa). To put the scale of the transition barrier into perspective, the ground-state energy is typically several Hartree (see Fig. 2 for a visual representation).

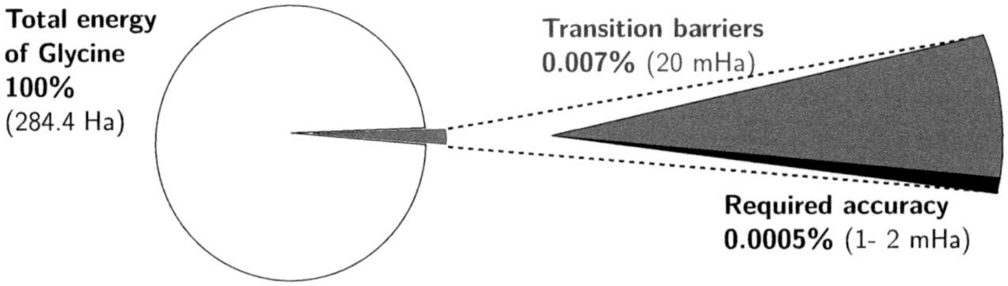

FIGURE 2 Conceptual visualization of the relevant energy-scales on the example of Glycine. Total energies are often hundreds of Hartrees, whereas required accuracy to predict experiments is often only milli-Hartrees. Figure adapted from Gerard et al. (2022).

The requirement of having to achieve very high accuracy in combination with the high dimensionality of the problem (4) (the dimension being $3\times$ the number of electrons) renders the efficient evaluation of the PES a formidable challenge. An additional complication is given by the *Pauli Exclusion Principle*, which posits that the electronic wave function must satisfy certain antisymmetry conditions, see Section 2.3 below.

As already mentioned, a plethora of methods have been developed over the last decades to solve the Schrödinger Equation approximately. Some methods such as the Hartree-Fock (HF) method are computationally cheap and scale well with system size, but are only accurate for a limited class of systems. Other methods such as Configuration Interaction (CI) and Coupled Cluster (CC) often yield highly-accurate results that are in good agreement with experiments, but scale poorly with system size and are thus limited to small systems (cf. Fig. 3). Density Functional Theory (DFT) has emerged as a breakthrough, and has been awarded the Nobel Prize in chemistry in 1998, since it yields surprisingly high accuracy while scaling well with system size. However, DFT requires the use of an essentially uncontrolled approximation, which can fail for many systems of interest and is thus not universally applicable.

Broadly speaking, the aim of DL-VMC is to represent the wave function as a neural network Ψ_θ and aim to approximate the electronic ground state by minimizing the Raleigh-Ritz quotient corresponding to the Eigenvalue problem (4), which amounts to minimizing the loss

$$\mathcal{L}(\theta) := \frac{\langle \mathcal{H}^{\mathrm{BO}} \Psi_\theta, \Psi_\theta \rangle}{\langle \Psi_\theta, \Psi_\theta \rangle}.$$

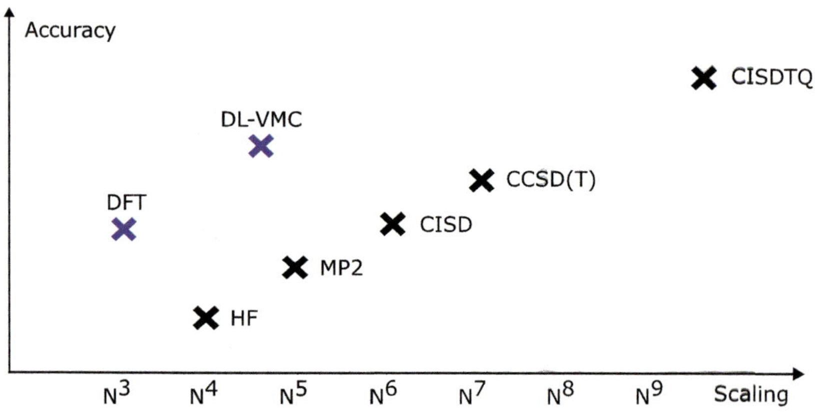

FIGURE 3 A conceptual visualization of the computational scaling with respect to the estimated accuracy for different Quantum Chemistry methods. The abbreviations stand for DFT: Density Functional Theory; DL-VMC: Deep-Learning-based Variational Monte Carl; HF: Hartree-Fock; MP2: Møller-Plesset 2nd order; CISD: Configuration Interaction (doubles); CCSD(T): Coupled-Cluster; CISDTQ: Configuration Interaction (quadruples). The figure is adapted from Hermann et al. (2022).

In recent years it has been demonstrated empirically (see for example Fig. 8) that a judiciously chosen DL-VMC ansatz is capable of achieving considerably more accurate ground state energies than previous methods, bringing us closer to the aforementioned "grand goal".

1.2 Outline

In Section 2 we provide a mathematical introduction into the spectral theory of the electronic Schrödinger equation. In particular, we will see that the electronic Hamiltonian is self-adjoint and induces a bounded and (up to a translation) coercive bilinear form on H^1, characterizes its spectrum, and elaborates on the Pauli exclusion principle.

Section 3 provides a concise introduction into variational Monte Carlo (VMC) and introduces ansatz spaces of functions that satisfy the Pauli exclusion principle. We furthermore present results and techniques related to efficient Markov chain Monte Carlo (MCMC) quadrature methods and preconditioning methods for optimization.

In Section 4 we describe various neural network architectures used in DL-VMC. This includes constructions that are transferable among different geometric conformations.

Finally, in Section 5 we review numerical results.

1.3 Notation

We denote with $L^2(\mathbb{R}^d)$, $H^k(\mathbb{R}^d)$ the usual Lebesgue space, resp. Sobolev spaces, see Evans (2022). The symbols $\|\cdot\|$ and $\langle\cdot,\cdot\rangle$ typically denote the norm and inner product in L^2, unless stated otherwise. The inner product of two vectors $v, w \in \mathbb{R}^d$ will be denoted by $v \cdot w$ and the norm of $v \in \mathbb{R}^d$ by $|v|$.

2 Mathematical preliminaries

So far the main problem (4) does not stand on a rigorous mathematical footing. Specifically, it is a priori not clear in which sense the eigenvalue equation (4) is well-posed, what would be the correct domain of definition of $\mathcal{H}^{\mathrm{BO}}$, whether this operator is self adjoint and whether an isolated minimal eigenvalue exists. Addressing these issues requires studying the spectral theory of the electronic Hamiltonian $\mathcal{H}^{\mathrm{BO}}_{(\mathbf{R},\mathbf{Z})}$. This will be done in Section 2.2 after a brief introduction into some mathematical foundations of quantum mechanics in Section 2.1.

Furthermore, electrons carry an additional variable, namely the *spin*, which can assume the values $\pm\frac{1}{2}$. Therefore, the electronic wave function actually depends on *spin coordinates*

$$\left((\mathbf{r}_1, \sigma_1), \ldots, (\mathbf{r}_{n_{\mathrm{el}}}, \sigma_{n_{\mathrm{el}}})\right), \quad \sigma = (\sigma_1, \ldots, \sigma_{n_{\mathrm{el}}}) \in \left\{-\frac{1}{2}, \frac{1}{2}\right\}^{n_{\mathrm{el}}}.$$

The *Pauli Exclusion Principle* states that every admissible wave function must be antisymmetric with respect to permutations of different spin coordinates. We elaborate on the consequences in Section 2.3.

For notational convenience we will often drop sub- and superscripts such as in $\mathcal{H}^{\mathrm{BO}}$. However, it is important to note that all objects that follow will depend on the geometric conformation $(\mathbf{R}, \mathbf{Z})$.

2.1 Basic mathematical setting

In classical mechanics, the state of a single particle in $\mathbb{R}$ is completely determined as a point (x, p) in phase-space $\mathbb{R}^2$. Its position is uniquely determined as x, its momentum as p and, if it has mass m and is subjected to a force field $F(x) = -\nabla V(x)$, its energy as

$$E = \frac{p^2}{2m} + V(x). \tag{5}$$

Its motion is governed by Newton's law $m\ddot{x}(t) = F(x(t))$, which conserves its energy.

Quantum states are modeled as elements of a complex Hilbert space H. Two elements $\Psi_1, \Psi_2 \in H$ represent the same state if there is $c \in \mathbb{C} \setminus \{0\}$ with $\Psi_1 = c\Psi_2$ and we can therefore assume that states are represented by unit vectors in H.

The position of a single particle

As an example we may consider $H = L^2(\mathbb{R})$, where each $\Psi \in L^2(\mathbb{R})$ with $\|\Psi\| = 1$ describes the wave function of a single particle. Similar to classical mechanics, we would like to infer properties of quantum states. For example, for a wave function $\Psi \in L^2(\mathbb{R})$ of a single particle we may be interested in determining its position, its momentum or its energy, etc. However, as it is well-known, properties of quantum states are typically not fully determined. What quantum mechanics provides instead is a probability distribution that models the likelihood of a particular observable assuming its value in a specific set. Returning to our example of $\Psi \in L^2(\mathbb{R})$ modeling a single particle in $\mathbb{R}$ we may define the probability that the position of Ψ lies in a measurable set $E \subset \mathbb{R}$ as

$$\mathbb{P}\left[\text{The position of } \Psi \text{ is observed in } E\right] = \int_E |\Psi(x)|^2 dx,$$

implying that the expected position is given as

$$\mathbb{E}\left[\text{Position of } \Psi\right] = \int_{\mathbb{R}} x|\Psi(x)|^2 dx = \langle \mathcal{X}\Psi, \Psi \rangle$$

with $\mathcal{X}\Psi := (x \mapsto x \cdot \Psi(x))$ denoting the *position operator*, which is defined on the dense subset $\mathcal{D}(\mathcal{X}) := \left\{\Psi \in L^2(\mathbb{R}) : \ x \cdot \Psi(x) \in L^2(\mathbb{R})\right\} \subset L^2(\mathbb{R})$. For $E \subset \mathbb{R}$ measurable define the projection operator $\mu^{\mathcal{X}}(E) : \Psi(x) \mapsto \chi_E(x)\Psi(x)$ with χ_E denoting the indicator function of E. The mapping $E \mapsto \mu^{\mathcal{X}}(E)$ is called *projection-valued measure*. This projection-valued measure allows to formally decompose $\mathcal{X}$ as

$$\mathcal{X} = \int_{\mathbb{R}} \lambda d\mu^{\mathcal{X}}(\lambda) := \lim_{h \to 0} \sum_{i \in \mathbb{Z}} hi\mu^{\mathcal{X}}([ih, (i+1)h]). \tag{6}$$

It is not hard to verify that this Riemann sum converges in the sense that

$$\lim_{h \to 0} \left\langle \sum_{i \in \mathbb{Z}} hi\mu^{\mathcal{X}}([ih, (i+1)h])\Psi, \Psi \right\rangle = \langle \mathcal{X}\Psi, \Psi \rangle \quad \forall \Psi \in \mathcal{D}(\mathcal{X}).$$

Given our projection-valued measure we also have

$$\mathbb{P}[\mathcal{X} \in E] = \langle \mu^{\mathcal{X}}(E)\Psi, \Psi \rangle. \tag{7}$$

The decomposition (6) also defines a functional calculus that allows for the study of functions of $\mathcal{X}$: indeed for a measurable function $f : \mathbb{R} \to \mathbb{R}$ we may define the operator

$$f(\mathcal{X}) := \int_{\mathbb{R}} f(\lambda) d\mu^{\mathcal{X}}(\lambda). \tag{8}$$

In our example it can be seen that $f(\mathcal{X})\Psi(x) = f(x) \cdot \Psi(x)$, as expected. Finally we have that

$$\mathbb{E}[f\,(\text{Position of } \Psi)] = \langle f(\mathcal{X})\Psi, \Psi \rangle =: \langle f(\mathcal{X}) \rangle_{\Psi} \tag{9}$$

which provides a simple expression for moments of the distribution of the position of Ψ. For instance, the variance of the position of Ψ is given by $\mathbb{V}\,[\text{Position of }\Psi] = \left\langle (\mathcal{X} - \langle \mathcal{X} \rangle_\Psi)^2 \right\rangle_\Psi$.

Self-adjoint operators

The properties of having a decomposition (6) which defines a probability distribution (7) and a functional calculus (8) are enjoyed by every self-adjoint operator.

Definition 2.1. Let $\mathcal{A}$ be a linear operator on a Hilbert space H, defined on a dense subset $\mathcal{D}(\mathcal{A}) \subset H$. Let

$$\mathcal{D}(\mathcal{A}^*) := \{\Phi \in H : \exists C \in [0, \infty) : \forall \Psi \in \mathcal{D}(\mathcal{A}) : |\langle \Phi, \mathcal{A}\Psi \rangle| \leqslant C \cdot \|\Psi\|\}. \tag{10}$$

In other words, $\mathcal{D}(\mathcal{A}^*)$ consists of those $\Phi \in H$ such that the linear functional $L_\Phi : \Psi \mapsto \langle \Phi, \mathcal{A}\Psi \rangle$ is bounded on the dense set $\mathcal{D}$. The boundedness implies that L_Φ can be extended to a bounded linear functional on all of H and by the Riesz representation theorem there exists a unique $\chi \in H$ with $L_\Phi(\cdot) = \langle \chi, \cdot \rangle$. We can then define the *adjoint operator* by $\mathcal{A}^*\Psi = \chi$. $\mathcal{A}$ is called *symmetric* if $\mathcal{A} = \mathcal{A}^*$ on $\mathcal{D}(\mathcal{A})$ and *self-adjoint* if additionally $\mathcal{D}(\mathcal{A}) = \mathcal{D}(\mathcal{A}^*)$. Any self-adjoint operator is called a *quantum observable*.

Remark 2.2. Note that for every symmetric operator it holds that $\mathcal{D}(\mathcal{A}) \subset \mathcal{D}(\mathcal{A}^*)$. This is because for $\Phi, \Psi \in \mathcal{D}(\mathcal{A})$ we have that

$$\langle \Phi, \mathcal{A}\Psi \rangle = \langle \mathcal{A}\Phi, \Psi \rangle \leqslant \|\mathcal{A}\Phi\| \cdot \|\Psi\|.$$

Self-adjointness is a somewhat subtle definition since it involves a correct specification of the domain of definition of $\mathcal{A}$. It is however the right concept in the context of quantum mechanics, as evidenced by the spectral theorem that we only state in an informal way. For more details please consult the excellent monograph by Hall (2013).

Definition 2.3. The *resolvent set* $\rho(\mathcal{A})$ of an operator $\mathcal{A} : \mathcal{D}(\mathcal{A}) \to \mathcal{H}$ consists of all $\lambda \in \mathbb{C}$ such that $\mathcal{A} - \lambda : \mathcal{D}(\mathcal{A}) \to H$ is boundedly invertible. The spectrum $\sigma(\mathcal{A})$ of $\mathcal{A}$ is defined as

$$\sigma(\mathcal{A}) = \mathbb{C} \setminus \rho(\mathcal{A}).$$

$\lambda \in \mathbb{C}$ is called an Eigenvalue of $\mathcal{A}$ if there is $\Psi \in \mathcal{D}(\mathcal{A}) \setminus \{0\}$ with

$$\mathcal{A}\Psi = \lambda\Psi. \tag{11}$$

The subspace $E_\lambda \subset H$ containing all $\Psi \in H$ such that (11) holds is called the Eigenspace of $\mathcal{A}$ corresponding to the Eigenvalue λ. All $\Psi \in E_\lambda \setminus \{0\}$ are called Eigenvector of $\mathcal{A}$ corresponding to the Eigenvalue λ. The dimension $m_\lambda := \dim(E_\lambda)$ of E_λ is called the multiplicity of the Eigenvalue λ and λ is of finite multiplicity if $m_\lambda \in \mathbb{N}$. An Eigenvalue λ is called isolated if there is an open neighborhood B of λ in $\mathbb{C}$ with $\sigma(\mathcal{A}) \cap B = \{\lambda\}$. If $\mathcal{A}$ is self-adjoint, the discrete spectrum $\sigma_{\text{disc}}(\mathcal{A})$ consists of all isolated Eigenvalues with finite multiplicity and the essential spectrum equals $\sigma_{\text{ess}}(\mathcal{A}) := \sigma(\mathcal{A}) \setminus \sigma_{\text{disc}}(\mathcal{A})$.

If $\mathcal{A}$ is self-adjoint, then the spectrum has the following intuitive characterization of consisting of those values λ for which there are states that are "almost" Eigenvectors.

Lemma 2.4. *For a self-adjoint operator $\mathcal{A}$ it holds that*

$$\lambda \in \sigma(\mathcal{A}) \Leftrightarrow \forall \varepsilon \in (0, \infty)\ \exists \Psi \in \mathcal{D}(\mathcal{A}):\ \|(\mathcal{A} - \lambda)\, \Psi\| \leqslant \varepsilon \|\Psi\|. \tag{12}$$

Proof. If the right hand side of (12) holds true, then $\mathcal{A} - \lambda$ cannot be boundedly invertible.

On the other hand, suppose that λ does not satisfy the right hand side of (12). Then there exists $\delta \in (0, \infty)$ with

$$\|(\mathcal{A} - \lambda)\Psi\| \geqslant \delta \|\Psi\| \quad \forall \Psi \in \mathcal{D}(\mathcal{A}). \tag{13}$$

This clearly implies that $\mathcal{A} - \lambda$ is injective. It also has closed range because if we have a Cauchy sequence $\Phi_n = (\mathcal{A} - \lambda)\Psi_n$ in the range of $\mathcal{A} - \lambda$ with $\Phi = \lim_{n\to\infty} \Phi_n$, then by (13) the limit $\Psi = \lim_{n\to\infty} \Psi_n$ exists.

We now show that $\Psi \in \mathcal{D}(\mathcal{A} - \lambda)$ and $(\mathcal{A} - \lambda)\Psi = \Phi$. Let $\mathcal{B} := \mathcal{A} - \lambda$. Then $\mathcal{B}$ is self-adjoint. Let $\Xi \in \mathcal{D}(\mathcal{A})$ be arbitrary. Then

$$\langle \Psi, \mathcal{B}\Xi \rangle = \lim_{n\to\infty} \langle \Psi_n, \mathcal{B}\Xi \rangle = \lim_{n\to\infty} \langle \mathcal{B}\Psi_n, \Xi \rangle = \langle \Phi, \Xi \rangle$$

Therefore it holds that $\Psi \in \mathcal{D}(\mathcal{B}^*) = \mathcal{D}(\mathcal{B})$ and $\mathcal{B}\Psi = \mathcal{B}^*\Psi = \Phi$.

We still need to show that $\mathcal{B}$ is surjective. To this end suppose that $\Xi \in \text{range}(\mathcal{B})^\perp$. Then

$$\|\mathcal{B}\Xi\|^2 = \langle \mathcal{B}\Xi, \mathcal{B}\Xi \rangle = \langle \Xi, \mathcal{B}^*\mathcal{B}\Xi \rangle = 0.$$

By (13) this implies that $\Xi = 0$. Therefore, the range of $\mathcal{B}$ is dense and closed and hence $\mathcal{B}$ is surjective. By (13) we conclude by the Closed Graph Theorem that $\mathcal{B} = \mathcal{A} - \lambda$ is boundedly invertible. □

Remark 2.5. An examination of the proof of Lemma 2.4 shows the useful property that every self-adjoint operator is *closed*, that is, the set $\{(\Psi, \mathcal{A}\Psi) : \Psi \in \mathcal{D}(\mathcal{A})\}$ is closed in $H \times H$.

Furthermore, one can easily deduce from Lemma 2.4 that the spectrum is always closed.

By working a bit harder one can obtain a more detailed characterization of the essential spectrum of self-adjoint operators.

Theorem 2.6. *For a self-adjoint operator $\mathcal{A}$ it holds that*

$$\lambda \in \sigma_{ess}(\mathcal{A}) \Leftrightarrow \forall n \in \mathbb{N}\ \exists \Psi^n \in \mathcal{D}(\mathcal{A}): \ \left\|(\mathcal{A}-\lambda)\,\Psi^n\right\| \leqslant \frac{1}{n}\cdot \|\Psi^n\| \wedge$$
$$\forall \Xi \in H: \ \lim_{n\to\infty} \langle \Psi^n, \Xi\rangle = 0. \tag{14}$$

Proof. $\Rightarrow$: Suppose that $\lambda \in \sigma_{ess}(\mathcal{A})$.

We distinguish three cases.

Case 1: λ is an Eigenvalue with infinite multiplicity. Then there exist countably many orthonormal corresponding Eigenvectors $(\Psi^n)_{n\in\mathbb{N}}$. Let $\Xi \in H$. Then $\sum_{n=1}^{\infty}\langle \Psi^n, \Xi\rangle^2 \leqslant \|\Xi\|^2$ which implies that $\lim_{n\to\infty}\langle \Psi^n, \Xi\rangle = 0$.

Case 2: λ is not an Eigenvalue. Since $\lambda \in \sigma(\mathcal{A})$ there exists by Lemma 2.4 a sequence $\Psi^n \in H$ with $\|\Psi^n\| = 1$ and $\|(\mathcal{A}-\lambda)\Psi^n\| \leqslant \frac{1}{n}$. Since the Ψ^n's are bounded, by the Banach-Alaoglu Theorem there exists a weak limit (possibly after passing to a subsequence), e.g. some $\Omega \in H$ with $\lim_{n\to\infty}\langle \Psi^n, \Xi\rangle = \langle \Omega, \Xi\rangle$ for all $\Xi \in H$. It holds for all $\Xi \in \mathcal{D}(\mathcal{A}-\lambda)$ that

$$\langle \Omega, (\mathcal{A}-\lambda)\Xi\rangle = \lim_{n\to\infty}\langle \Psi^n, (\mathcal{A}-\lambda)\Xi\rangle = \lim_{n\to\infty}\langle (\mathcal{A}-\lambda)\Psi^n, \Xi\rangle = 0.$$

Therefore it holds that $\Omega \in \mathcal{D}((\mathcal{A}-\lambda)^*) = \mathcal{D}(\mathcal{A}-\lambda)$ and $(\mathcal{A}-\lambda)^*\Omega = (\mathcal{A}-\lambda)\Omega = 0$. This implies that either Ω is an Eigenvalue or $\Omega = 0$. Since we already excluded the former, it must hold that $\Omega = 0$ which is what we wanted to show.

Case 3: λ is an Eigenvalue with finite multiplicity that is not isolated. W.l.o.g. assume $\lambda \neq 0$. Let $\mu^{\mathcal{A}}(\{\lambda\})$ be the orthogonal projection on the corresponding Eigenspace. Consider $\tilde{\mathcal{A}} := \mathcal{A} - \lambda\mu^{\mathcal{A}}(\{\lambda\})$. Then λ cannot be an Eigenvalue of $\tilde{\mathcal{A}}$. Since it holds that $\sigma(\mathcal{A})\setminus\{\lambda, 0\} = \sigma(\tilde{\mathcal{A}})\setminus\{\lambda, 0\}$ and due to the assumption that λ is an accumulation point of $\sigma(\mathcal{A})$, λ must be an accumulation point of $\sigma(\tilde{\mathcal{A}})$. Therefore (since the spectrum is always closed) it must hold that $\lambda \in \sigma(\tilde{\mathcal{A}})$. By Case 2 there must thus exist Ψ^n with $\|\Psi^n\| = 1$, $\|(\tilde{\mathcal{A}}-\lambda)\Psi^n\| \leqslant \frac{1}{n}$ and $\forall \Xi \in H: \ \lim_{n\to\infty}\langle \Psi^n, \Xi\rangle = 0$. The last condition implies that for all $\Xi \in H: \ \lim_{n\to\infty}\langle \mu^{\mathcal{A}}(\{\lambda\})\Psi^n, \Xi\rangle = \lim_{n\to\infty}\langle \Psi^n, \mu^{\mathcal{A}}(\{\lambda\})\Xi\rangle = 0$ and, since the range of $\mu^{\mathcal{A}}$ is finite-dimensional, together with the fact that on finite dimensional spaces strong and weak convergence coincide, it follows that $\lim_{n\to\infty}\|\mu^{\mathcal{A}}(\{\lambda\})\Psi^n\| = 0$. This implies that $\lim_{n\to\infty}\|(\mathcal{A}-\lambda)\Psi^n\| = 0$ which proves that the right hand side of (14) holds true.

$\Leftarrow$: For the other direction we assume that $\lambda \in \sigma_{\text{disc}}(\mathcal{A})$ and suppose w.l.o.g that $\lambda \neq 0$. Define again $\tilde{\mathcal{A}} := \mathcal{A} - \lambda\mu^{\mathcal{A}}(\{\lambda\})$. Since λ is an isolated Eigenvalue of $\mathcal{A}$, it holds that $\lambda \notin \sigma(\tilde{\mathcal{A}})$. This can be argued as follows. If λ was in the spectrum of $\tilde{\mathcal{A}}$, then λ would necessarily have to be isolated. This implies that λ would be an Eigenvalue of $\tilde{\mathcal{A}}$ (the reader can verify this as an exercise. It will probably be convenient to use the Spectral Theorem 2.11). But this would imply that there exists an Eigenvector of $\mathcal{A}$ that lies in the orthogonal complement of the Eigenspace range$(\mu^{\mathcal{A}}(\{\lambda\}))$ of $\mathcal{A}$ w.r.t. λ, which is a contradiction. Therefore, the right hand side of (14) cannot hold for the operator $\tilde{\mathcal{A}}$ and λ.

Now assume that the right hand side of (14) holds true for the operator $\mathcal{A}$ and λ. Then, using the fact that the range of $\mu^{\mathcal{A}}(\{\lambda\})$ is finite dimensional, we can argue exactly as in Case 3 above to deduce that the right hand side of (14) must also hold for $\tilde{\mathcal{A}}$ and λ, which would give a contradiction. □

Corollary 2.7. *For a self-adjoint operator $\mathcal{A}$ it holds that*

$$\sigma(\mathcal{A}) \subset \mathbb{R}. \tag{15}$$

Proof. Let $\lambda = a + ib$ with $b \neq 0$ and $\Psi \in \mathcal{D}(\mathcal{A})$. Then it holds that

$$\begin{aligned}\langle(\mathcal{A}-\lambda)\,\Psi, (\mathcal{A}-\lambda)\,\Psi\rangle &= \langle(\mathcal{A}-a)\,\Psi, (\mathcal{A}-a)\,\Psi\rangle + ib\,\langle\Psi, (\mathcal{A}-\lambda)\,\Psi\rangle \\ &\quad + ib\,\langle\Psi, (\mathcal{A}-\lambda)\,\Psi\rangle + b^2\|\Psi\|^2 \\ &= \langle(\mathcal{A}-a)\,\Psi, (\mathcal{A}-a)\,\Psi\rangle + b^2\|\Psi\|^2 \geqslant b^2\|\Psi\|^2,\end{aligned}$$

where the last equality follows from the self-adjointness of $\mathcal{A}$. Due to Lemma 2.4 it follows that λ cannot be in the spectrum. □

Remark 2.8. It might be instructive to go back to the position operator $\mathcal{X}$ of a single particle. It is not hard to check that $\mathcal{X}$ is self-adjoint. Its spectrum is all of $\mathbb{R}$ but there are no eigenvalues. Indeed, for $\lambda \in \mathbb{R}$, any Ψ with $(\mathcal{X}-\lambda)\Psi = 0$ would have to be vanished on all of $\mathbb{R}\setminus\{\lambda\}$ which is impossible. On the other hand, it is easy to construct function Ψ with $\|\Psi\| = 1$ which are arbitrarily well localized near λ in the sense of Ψ being supported on $[\lambda - \varepsilon/2, \lambda + \varepsilon/2]$. These functions then satisfy the right hand side of (12) and therefore λ must be in the spectrum of $\mathcal{X}$. Furthermore it is easy to see that these functions converge weakly to 0 as $\varepsilon \to 0$. Therefore, $\sigma(\mathcal{X}) = \sigma_{\text{ess}}(\mathcal{X}) = \mathbb{R}$.

The spectral theorem

We will now state the spectral theorem for unbounded self-adjoint operators. Recall the definition of a projection-valued measure.

Definition 2.9. A mapping μ from the Borel sigma algebra on a set $A \subset \mathbb{R}$ to the set of orthogonal projections on H is called a *projection-valued measure* supported on A if

1. $\mu(\emptyset) = 0$ and $\mu(A) = I$,
2. For all disjoint and measurable subsets $(E_i)_{i\in\mathbb{N}}$ of A it holds that $\mu(\bigcup_{i\in\mathbb{N}} E_i) = \sum_{i\in\mathbb{N}} \mu(E_i)$, and
3. For E_1, E_2 measurable we have $\mu(E_1 \cap E_2) = \mu(E_1)\mu(E_2)$.

For $\Psi \in H$ we denote $\mu_\Psi : E \mapsto \langle\mu(E)\Psi, \Psi\rangle$ and note that μ_Ψ is a positive Borel measure on A.

We have the following important result

Theorem 2.10. *For μ a projection-valued measure supported on A and $f : A \to \mathbb{C}$ measurable, let*

$$\mathcal{D}_f := \left\{ \Psi \in H : \int_A |f(\lambda)|^2 d\mu_\Psi(\lambda) < \infty \right\}.$$

Then $\mathcal{D}_f$ is dense in H and there exists a unique (unbounded) operator, denoted $\int_A f(\lambda) d\mu(\lambda)$ with $\mathcal{D}\left(\int_A f(\lambda) d\mu(\lambda)\right) = \mathcal{D}_f$ and

$$\left\langle \Psi, \int_A f(\lambda) d\mu(\lambda) \Psi \right\rangle = \int_A f(\lambda) d\mu_\Psi(\lambda). \tag{16}$$

Furthermore it holds that

$$\int_A f(\lambda) \cdot g(\lambda) d\mu(\lambda) = \int_A f(\lambda) d\mu(\lambda) \int_A g(\lambda) d\mu(\lambda) \tag{17}$$

and

$$\int_A \overline{f(\lambda)} d\mu(\lambda) = \left(\int_A f(\lambda) d\mu(\lambda) \right)^*. \tag{18}$$

Proof. See Hall (2013, Proposition 10.1 and Proposition 10.2). □

We can now state the spectral theorem.

Theorem 2.11 (Spectral Theorem for unbounded self-adjoint Operators). *For every self-adjoint operator $\mathcal{A}$ the spectrum $\sigma(\mathcal{A}) \subset \mathbb{R}$ there exists a unique spectral measure $\mu^{\mathcal{A}}$ supported on $\sigma(\mathcal{A})$ such that*

$$\mathcal{A} = \int_{\sigma(\mathcal{A})} \lambda d\mu^{\mathcal{A}}(\lambda). \tag{19}$$

Proof. The proof is quite long and involved. See Hall (2013, Section 10) for an excellent exposition. □

The spectral theorem readily allows for the definition of a functional calculus on self-adjoint operators.

Definition 2.12 (Functional calculus). Let $\mathcal{A}$ be self-adjoint with spectral measure $\mu^{\mathcal{A}}$. For $f : \sigma(\mathcal{A}) \to \mathbb{C}$ measurable we define

$$f(\mathcal{A}) := \int_{\sigma(\mathcal{A})} f(\lambda) d\mu^{\mathcal{A}}(\lambda).$$

We have now shown that all previous desirable properties of the position operator $\mathcal{X}$ can be established for every self-adjoint operator. This motivates the following definition.

Definition 2.13. A self-adjoint operator $\mathcal{A}$ on a Hilbert space H is called a *quantum observable* on H.

A quantum observable thus observes (or measures) a certain property of a state Ψ. The outcome of this measurement is random and the spectral measure represents the probability measure on the measurement outcome lying in a set E if the system is in the state Ψ with $\|\Psi\| = 1$:

$$\mathbb{P}[\text{The measurement } \mathcal{A} \text{ of a state } \Psi \text{ lies in } E] = \mu_{\Psi}^{\mathcal{A}}(E).$$

Remark 2.14. Note that a measurement can only be realized deterministically if the measure $\mu_{\Psi}^{\mathcal{A}}$ is concentrated in a single point λ. This can only occur if λ is an Eigenvalue of $\mathcal{A}$ and Ψ a corresponding Eigenvector.

We close this paragraph with the following result that will be used later to turn the Eigenvalue problem into an optimization problem.

Lemma 2.15. *For a self-adjoint operator $\mathcal{A}$ let $\mathcal{W}(\mathcal{A}) := \{\Psi \in H : \langle \mathcal{A}\Psi, \Psi\rangle < \infty\}$. Then it holds that*

$$\inf \sigma(\mathcal{A}) = \inf\{\langle \mathcal{A}\Psi, \Psi\rangle : \Psi \in \mathcal{W}(\mathcal{A}), \|\Psi\| = 1\}. \tag{20}$$

Proof. Let $\lambda \in \sigma(\mathcal{A})$. Then, by Lemma 2.4 for every $\varepsilon > 0$ there is $\Psi \in \mathcal{D}((A))$ with $\|\Psi\| = 1$ and $\|(\mathcal{A} - \lambda)\Psi\| \leqslant \varepsilon$. Therefore, by Cauchy-Schwarz

$$|\langle \mathcal{A}\Psi, \Psi\rangle - \lambda| = |\langle (\mathcal{A} - \lambda)\Psi, \Psi\rangle| \leqslant \varepsilon$$

This implies that

$$\begin{aligned} &\inf\{\langle \mathcal{A}\Psi, \Psi\rangle : \Psi \in \mathcal{W}(\mathcal{A}), \|\Psi\| = 1\} \\ &\qquad \leqslant \inf\{\langle \mathcal{A}\Psi, \Psi\rangle : \Psi \in \mathcal{D}(\mathcal{A}), \|\Psi\| = 1\} \leqslant \inf \sigma(\mathcal{A}). \end{aligned}$$

For the other direction we note that by the spectral theorem we have for $\Psi \in H$ with $\|\Psi\| = 1$ that

$$\langle \mathcal{A}\Psi, \Psi\rangle = \int_{\sigma(\mathcal{A})} \lambda d\mu_{\Psi}^{\mathcal{A}}(\lambda) \geqslant \inf \sigma(\mathcal{A}) \cdot \int_{\sigma(\mathcal{A})} 1 d\mu_{\Psi}^{\mathcal{A}}(\lambda) = \inf \sigma(\mathcal{A})$$

which implies that

$$\inf\{\langle \mathcal{A}\Psi, \Psi\rangle : \Psi \in \mathcal{W}(\mathcal{A}), \|\Psi\| = 1\} \geqslant \inf \sigma(\mathcal{A}).$$ □

Other observables and the Schrödinger equation

Returning to our example of a particle moving in $\mathbb{R}$ we would now like to find a quantum mechanical analog of other observables, such as momentum or energy. The momentum operator is defined as $\mathcal{P} := -i\frac{d}{dx}$. Intuitively this makes sense because $\mathcal{P}$ is simply the position operator in the Fourier domain. This also shows that $\mathcal{P}$ is self-adjoint. Having position and momentum operators we can now define the energy operator, or the Hamiltonian in analogy with (5) via

$$\mathcal{H} = \frac{\mathcal{P}^2}{2m} + V(\mathcal{X}) = -\frac{1}{2m}\frac{d^2}{dx^2} + V(\mathcal{X}).$$

Remark 2.16. The process of constructing self-adjoint quantum observables from classical observables $f(x, p)$ is called *quantization*. Due to the fact that the position and momentum operators do not commute, this is a highly nontrivial problem, see Hall (2013).

One can show that the Hamiltonian is self-adjoint (this is not completely trivial due to the conditions on the domains of self-adjoint operators!) and therefore, the energy of a state $\Phi \in \mathcal{D}(\mathcal{H})$ with $\|\Psi\| = 1$ is distributed according to $\mu_\Psi^{\mathcal{H}}$.

The evolution of quantum systems is governed by the Schrödinger equation

$$\dot{\Psi}(t) = \frac{1}{i}\mathcal{H}\Psi(t). \tag{21}$$

Using our functional calculus one can show that $\Phi(t) = e^{-it\mathcal{H}}\Psi(0)$ and that $U(t) = e^{-it\mathcal{H}}$ is a unitary one-parameter group which is defined on all of $L^2(\mathbb{R})$ (this is one part of the famous Stone-von Neumann Theorem (Stone, 1932; von Neumann, 1932)). Using (21) we can now model a quantum particle moving in $\mathbb{R}$.

Molecules consist of several particles, each with coordinates in $\mathbb{R}^3$. It is straightforward how the position and momentum operators can be extended to particles in $\mathbb{R}^3$. This naturally leads to the definition of the molecular Hamiltonian $\mathcal{H}^{\text{mol}}$ and $\mathcal{H}^{\text{BO}}$ from the introduction which is our main interest of study.

2.2 The Hamiltonian of the electronic Schrödinger equation

At this point it is not clear that the electronic Hamiltonian $\mathcal{H}^{\text{BO}}_{(\mathbf{R},\mathbf{Z})}$ is self-adjoint. Furthermore, we would like to know more about the precise nature of the spectrum of $\mathcal{H}^{\text{BO}}_{(\mathbf{R},\mathbf{Z})}$. Addressing these issues are the subject of this section. For notational convenience we will omit the subscript describing the geometric conformation and simply write $\mathcal{H} = \mathcal{H}^{\text{BO}}_{(\mathbf{R},\mathbf{Z})}$.

Our main tool in asserting the self-adjointness of $\mathcal{H}$ will be the Kato-Rellich theorem.

Theorem 2.17 (Kato-Rellich Theorem). *Let $\mathcal{A}$, $\mathcal{B}$ be self-adjoint with $\mathcal{D}(\mathcal{A}) \subset \mathcal{D}(\mathcal{B})$. Suppose that there are $a \in (0,1)$ and $b \in (0,\infty)$ with*

$$\|\mathcal{B}\Psi\| \leqslant a\|\mathcal{A}\Psi\| + b\|\Psi\| \quad \forall \Psi \in \mathcal{D}(\mathcal{A}).$$

Then $\mathcal{A} + \mathcal{B}$ is self-adjoint on $\mathcal{D}(\mathcal{A})$.

Proof. See Kato (2013, Section V, Theorem 4.13) or Hall (2013, Theorem 9.37). □

We rewrite the operator $\mathcal{H}$ of (3) as

$$\mathcal{H} = -\frac{1}{2}\Delta + V(\mathbf{r})$$

and aim to apply the Kato-Rellich Theorem with $\mathcal{A} = -\frac{1}{2}\Delta$ and $\mathcal{B} = V(\mathbf{r})$. To this end the following result will prove useful.

Proposition 2.18 (Hardy Inequality). *For any smooth and compactly supported function u on $\mathbb{R}^3$ it holds that*

$$\int_{\mathbb{R}^3} \frac{1}{|r|^2} u(r)^2 dr \leqslant 4 \cdot \int_{\mathbb{R}^3} |\nabla u(r)|^2 dr.$$

Proof. See Yserentant (2010, Lemma 4.1). □

Theorem 2.19. *The operator $\mathcal{H}$ is self-adjoint on $L^2(\mathbb{R}^{n_{el}\times 3})$ with $\mathcal{D}(\mathcal{H}) = H^2(\mathbb{R}^{n_{el}\times 3})$.*

Proof. We will show that the operators $\mathcal{A} = -\frac{1}{2}\Delta$ and $\mathcal{B} = V(r)$ satisfy the conditions of the Kato-Rellich Theorem 2.17.

We will use the Fourier transform $\mathcal{F}$ on $\mathbb{R}^{n_{\text{el}}\times 3}$ and assume that the reader is familiar with its basic properties. First note that the Fourier transform of the Laplace operator is simply given by multiplication with $|\omega|^2$. Using this fact one can easily show that Δ is self-adjoint with domain given by H^2.

Hardy's inequality (Proposition 2.18) implies that there exists a constant C (depending on $(\mathbf{R}, \mathbf{Z})$) with

$$\|\mathcal{B}\Psi\| \leqslant C \cdot \|\nabla\Psi\|. \tag{22}$$

Therefore, by Plancherel's theorem and the fact that the Fourier transform of the gradient is a multiplication operator we have that

$$\|\mathcal{B}\Psi\|^2 \leqslant C \cdot \int_{\mathbb{R}^{n_{\text{el}}\times 3}} |\omega|^2 |\mathcal{F}\Psi(\omega)|^2 d\omega.$$

Let $\varepsilon > 0$ arbitrary and $D_\varepsilon > 0$ so that

$$|\omega|^2 \leqslant \varepsilon|\omega|^4 + D_\varepsilon \quad \forall \omega \in \mathbb{R}^{n_{\text{el}}\times 3}.$$

Then, we have that

$$\|\mathcal{B}\Psi\|^2 \leqslant C \cdot \varepsilon \int_{\mathbb{R}^{n_{\text{el}}\times 3}} |\omega|^4 |\mathcal{F}\Psi(\omega)|^2 d\omega + C \cdot D_\varepsilon \int_{\mathbb{R}^{n_{\text{el}}\times 3}} |\mathcal{F}\Psi(\omega)|^2 d\omega.$$

Using Plancherel's theorem again with the fact that the Laplace operator turns into multiplication with $|\omega|^2$ in Fourier space yields that

$$\|\mathcal{B}\Psi\|^2 \leqslant C \cdot \varepsilon \|\Delta\Psi\|^2 + C \cdot D_\varepsilon \|\Psi\|^2.$$

By choosing ε sufficiently small we can now satisfy the conditions of Theorem 2.17, which proves the desired statement. □

Having now established the self-adjointness of $\mathcal{H}$ we can now speak of the associated Schrödinger equation and its spectrum in a rigorous way.

First we establish that the spectrum of $\mathcal{H}$ is bounded from below.

Theorem 2.20. *It holds that*

$$\inf \sigma(\mathcal{H}) > -\infty.$$

Proof. The proof goes by establishing a Gårding inequality for $\mathcal{H}$. First note that by Lemma 2.15 it suffices to show that

$$\inf\{\langle \mathcal{H}\Psi, \Psi\rangle : \|\Psi\| = 1\} > -\infty.$$

Next we observe that the Hardy inequality Proposition 2.18 and the Cauchy-Schwartz Inequality imply that there is a constant $C \in (0, \infty)$ with

$$|\langle V\Psi, \Psi\rangle| \leqslant C\|\nabla\Psi\|\|\Psi\|. \tag{23}$$

This, together with the fact that $\langle \Delta\Psi, \Psi\rangle = \|\nabla\Psi\|^2$ implies that

$$\begin{aligned}
&2\langle(\mathcal{H}\Psi, \Psi) + \left(2C^2 + \frac{1}{2}\right)\langle\Psi, \Psi\rangle \\
&\geqslant \|\nabla\Psi\|^2 - 2C\|\Psi\|\|\nabla\Psi\| + \left(2C^2 + \frac{1}{2}\right)\|\Psi\|^2 \\
&= \frac{1}{2}\left(\|\nabla\Psi\|^2 + \|\Psi\|^2\right) + \frac{1}{2}\left(\|\nabla\Psi\|^2 - 4C\|\Psi\|\|\nabla\Psi\| + 4C^2\|\Psi\|^2\right) \\
&= \frac{1}{2}\left(\|\nabla\Psi\|^2 + \|\Psi\|^2\right) + \frac{1}{2}\left(\|\nabla\Psi\| - 2C\|\Psi\|\right)^2 \\
&\geqslant \frac{1}{2}(\|\Psi\| + \|\nabla\Psi\|^2) \geqslant 0.
\end{aligned}$$

In particular, it follows that

$$\langle \mathcal{H}\Psi, \Psi\rangle \geqslant -C^2 - \frac{1}{4} > -\infty$$

for every Ψ with $\|\Psi\| = 1$, which is what we wanted to show □

Remark 2.21. A careful examination of our proof of Theorem 2.20 yields a stronger statement. There exist constants $c_1, c_2 \in (0, \infty)$ such that

$$\langle \mathcal{H}\Psi, \Psi\rangle \geqslant c_1\|\Psi\|_{H^1}^2 - c_2\|\Psi\|^2 \quad \forall \Psi \in H^1. \tag{24}$$

Such an inequality is commonly referred to as *Gårding inequality* and shows that a shifted version of $\mathcal{H}$ is coercive on H^1. Furthermore, using (23), it is easy to see that there is another constant $c_3 \in (0, \infty)$ with

$$|\langle \mathcal{H}\Psi, \Phi\rangle| \leqslant c_3\|\Psi\|_{H^1}\|\Phi\|_{H^1} \quad \forall \Psi, \Phi \in H^1. \tag{25}$$

This shows that the Schrödinger equation can be formulated in a weak form on H^1 and studied using familiar tools from the theory of elliptic operators. This route is taken in Yserentant (2010).

Since $\mathcal{H}$ is an unbounded operator, its spectrum must be unbounded. In order to find out more about its precise structure, we follow Yserentant (2010, Section 5) and define for $T \in (0, \infty)$ the quantity

$$\Sigma(T) := \tag{26}$$
$$\inf\left\{\langle \mathcal{H}\Psi, \Psi\rangle : \ \|\Psi\| = 1 \ \wedge \ \forall \mathbf{r} \in \mathbb{R}^{n_{\mathrm{el}}\times 3} \setminus \{\mathbf{r} \in \mathbb{R}^{n_{\mathrm{el}}\times 3} : \ |\mathbf{r}| \leqslant T\} : \ \Psi(\mathbf{r}) = 0\right\}$$

Lemma 2.22. *For all $T > 0$ it holds that $\Sigma(T) \leqslant 0$.*

Proof. Observe that the Hardy inequality (Proposition 2.18), the Cauchy-Schwartz Inequality, and the fact that $\langle \Delta\Psi, \Psi\rangle = \|\nabla\Psi\|^2$ imply that there is a constant $C \in (0, \infty)$ with

$$\langle \mathcal{H}\Psi, \Psi\rangle \leqslant C \cdot \left(\|\nabla\Psi\|^2 + \|\nabla\Phi\|\|\Phi\|\right). \tag{27}$$

Let $\Phi \in \mathcal{S}$ be a function vanishing on the unit ball of $\mathbb{R}^{n_{\mathrm{el}}\times 3}$ with $\|\Phi\| = 1$. Then for $t \in (0, \infty)$ the function $\Psi_t := \frac{1}{t^{\frac{n_{\mathrm{el}}\times 3}{2}}}\Psi\left(\frac{\cdot}{t}\right)$ has unit norm and vanishes on a ball of radius t. It is not hard to see (for example by a scaling argument) that $\lim_{t\to\infty}\|\nabla\Phi_t\| = 0$ and therefore, since for $t > T$ the function Ψ_t can be considered in the infimum of (26), the estimate (27) implies that $\Sigma(T) \leqslant 0$. □

Definition 2.23. Define the *ionization threshold*

$$\Sigma := \lim_{T\to\infty} \Sigma(T) \leqslant 0. \tag{28}$$

Remark 2.24. By Lemma 2.22, the limit in (28) exists and is bounded above by 0.

We now come to the main result about the structure of $\sigma(\mathcal{H})$.

Theorem 2.25. *Assume that*

$$\inf \sigma(\mathcal{H}) < \Sigma. \tag{29}$$

Then it holds that

$$\inf \sigma_{ess}(\mathcal{H}) = \Sigma.$$

In particular, the essential spectrum is nonempty and it holds that

$$\sigma(\mathcal{H}) \cap [\inf \sigma(\mathcal{H}), \Sigma) \subset \sigma_{disc}(\mathcal{H}).$$

Proof. We follow the arguments in Yserentant (2010, Section 5, Theorem 5.6).

1. First we show that for all $T > 0$ and all $\lambda \in \sigma_{\text{ess}}(\mathcal{H})$ it holds that

$$\lambda \geqslant \Sigma(R). \tag{30}$$

To this end, for any $T > 0$ let Ω_1, Ω_2 be functions with $\Omega_1^2 + \Omega_2^2 = 1$ and $\text{supp}(\Omega_1) \subset \{\mathbf{r} : |\mathbf{r}| \geqslant T\}$. Since this implies that $|\nabla\Psi|^2 = |\nabla(\Omega_1\Psi)|^2 + |\nabla(\Omega_2\Psi)|^2 - \left(|\nabla\Omega_1|^2 + |\nabla\Omega_2|^2\right)\Psi^2$, we get that

$$\langle \mathcal{H}\Psi, \Psi\rangle = \langle \mathcal{H}\Omega_1\Psi, \Omega_1\Psi\rangle + \langle \mathcal{H}\Omega_2\Psi, \Omega_2\Psi\rangle - \int \left(|\nabla\Omega_1|^2 + |\nabla\Omega_2|^2\right)\Psi^2.$$

Due to (24) there is $\mu > 0$ with $\langle \mathcal{H}\Omega_2\Psi, \Omega_2\Psi\rangle \geqslant -\mu\|\Omega_2\Psi\|^2$ and hence we get that

$$\langle \mathcal{H}\Psi, \Psi\rangle \geqslant \langle \mathcal{H}\Omega_1\Psi, \Omega_1\Psi\rangle - \int \left(\mu\Omega_2^2 + |\nabla\Omega_1|^2 + |\nabla\Omega_2|^2\right)\Psi^2.$$

Since $\Omega_1\Psi$ is supported in a ball of radius T we furthermore get that $\langle \mathcal{H}\Omega_1\Psi, \Omega_1\Psi\rangle \geqslant \Sigma(T)\|\Omega_1\Psi\|^2$ which implies that

$$\langle \mathcal{H}\Psi, \Psi\rangle \geqslant \Sigma(T)\|\Omega_1\Psi\|^2 - \int \left(\mu\Omega_2^2 + |\nabla\Omega_1|^2 + |\nabla\Omega_2|^2\right)\Psi^2.$$

Since $\|\Omega_1\Psi\|^2 = \|\Psi\|^2 - \|\Omega_2\Psi\|^2$ it follows that

$$\langle \mathcal{H}\Psi, \Psi\rangle \geqslant \Sigma(T)\|\Psi\|^2 - \int \left((\Sigma(T) + \mu)\Omega_2^2 + |\nabla\Omega_1|^2 + |\nabla\Omega_2|^2\right)\Psi^2.$$

Define $\Xi := (\Sigma(T) + \mu)\Omega_2^2 + |\nabla\Omega_1|^2 + |\nabla\Omega_2|^2$. Then Ξ is compactly supported and for all $\Psi \in \mathcal{D}(\mathcal{H})$ it holds that

$$\langle \mathcal{H}\Psi, \Psi\rangle + \langle \Xi\Psi, \Psi\rangle \geqslant \Sigma(T)\|\Psi\|^2. \tag{31}$$

By the assumption that $\lambda \in \sigma_{\text{ess}}(\mathcal{H})$ and Theorem 2.6 there is a sequence Ψ^n with $\|\Psi^n\| = 1$, $\|(\mathcal{H} - \lambda)\Psi^n\| \leqslant \frac{1}{n}$ and Ψ^n converges weakly to 0.

First of all this implies that there is a constant C with $\|\Psi\|_{H^1} \leqslant C$ for all n. To see this we observe that the condition $\|(\mathcal{H} - \lambda)\Psi^n\| \leqslant \frac{1}{n}$ implies that there is a constant $C > 0$ with $\langle \mathcal{H}\Psi^n, \Psi^n \rangle \leqslant C$ for all n. By (24) this implies that the norms $\|\Psi^n\|_{H^1}$ must be uniformly bounded. Take the function Ξ from (31) and consider $D := \text{supp}(\Xi)$, which is a bounded subset of $\mathbb{R}^{n_{\text{el}} \times 3}$. Since $H^1(D)$ is compactly embedded into $L^2(D)$ by the Rellich-Kondrachov Theorem, the uniform H^1 boundedness of the Ψ^n's implies that there is a subsequence of the Ψ^n's that converges *strongly* in $L^2(D)$. Since we also know that the Ψ^n's converge weakly to 0 this implies (possibly after passing to a subsequence) that $\lim_{n\to\infty} \|\Psi^n\|_{L^2(D)} = 0$. This implies that $\lim_{n\to\infty}\langle \Xi\Psi^n, \Psi^n \rangle = 0$. This, together with the fact that $\lim_{n\to\infty}\langle \mathcal{H}\Psi^n, \Psi^n \rangle = \lambda$, the fact that $\|\Psi^n\| = 1$, as well as (31) implies that $\lambda \geqslant \Sigma(T)$. This proves (30).

2. In the other direction we show that any S with $(-\infty, S] \cap \sigma_{\text{ess}}(\mathcal{H}) = \varnothing$ satisfies that for every $\varepsilon > 0$ there exists T_ε with

$$S - \varepsilon \leqslant \Sigma(T_\varepsilon). \tag{32}$$

In particular, this implies that $\inf \sigma_{\text{ess}}(\mathcal{H}) \leqslant \Sigma$.

To prove (32) we assume that S satisfies $(-\infty, S] \cap \sigma_{\text{ess}}(\mathcal{H}) = \varnothing$. This means that the set $(-\infty, S] \cap \sigma(\mathcal{H})$ only contains isolated eigenvalues of finite multiplicity. By Theorem 2.20 there can only be finitely many such Eigenvalues. Let $\Psi^1, \ldots, \Psi^N$ be an orthonormal basis of the union of the finitely many corresponding Eigenspaces of finite dimension and let $\lambda_1, \ldots, \lambda_N$ denote the corresponding Eigenvalues. If $N = 0$, Lemma 2.15 implies that $S \leqslant \inf\{\langle \mathcal{H}\Psi, \Psi \rangle : \Psi \in \mathcal{D}(\mathcal{H})\} \leqslant \Sigma$ which implies (32). If $N > 0$ we define by $\mathcal{P}$ the orthogonal projection onto the span of the Ψ^n's, i.e. $\mathcal{P}\Psi = \sum_{i=1}^N \langle \Psi, \Psi^i \rangle \Psi^i$. The spectral Theorem 2.11 together with the fact that for every Ψ the spectral measure $\mu^{\mathcal{H}}_{\Psi - \mathcal{P}\Psi}$ is supported on $[S, \infty)$ implies that

$$\langle \mathcal{H}(\Psi - \mathcal{P}\Psi), \Psi - \mathcal{P}\Psi \rangle \geqslant S\|\Psi - \mathcal{P}\Psi\|^2$$

for all $\Psi \in \mathcal{D}(\mathcal{H})$. Furthermore, the fact that the span of the Ψ^n's is $\mathcal{H}$-invariant yields that

$$\langle \mathcal{H}\Psi, \Psi \rangle = \langle \mathcal{H}(\Psi - \mathcal{P}\Psi), \Psi - \mathcal{P}\Psi \rangle + \sum_{i=1}^N \lambda_i \langle \Psi, \Psi^i \rangle^2$$

Taken together the previous two inequalities yield that

$$\langle \mathcal{H}\Psi, \Psi \rangle \geqslant S\|\Psi - \mathcal{P}\Psi\|^2 + \sum_{i=1}^N \lambda_i \langle \Psi, \Psi^i \rangle^2 = S\|\Psi\|^2 - \sum_{i=1}^N (S - \lambda_i) \langle \Psi, \Psi^i \rangle^2$$

Now let Ω_T be the characteristic function of the set $\{\mathbf{r} \in \mathbb{R}^{n_{\text{el}} \times 3} : |\mathbf{r}| \geqslant T\}$ and $\Psi \in \mathcal{D}(\mathcal{H})$ with $\|\Psi\| = 1$ and supported on $\{\mathbf{r} \in \mathbb{R}^{n_{\text{el}} \times 3} : |\mathbf{r}| \geqslant T\}$. Then

$\Psi = \Omega_T \Psi$ and taking the infimum of the previous inequality yields

$$\Sigma(T) \geqslant S - \sum_{i=1}^{N} (S - \lambda_i) \langle \Psi, \Omega_T \Psi^i \rangle^2$$

Since $\lim_{T \to \infty} \|\Omega_T \Psi^i\| = 0$, $(S - \lambda_i) \geqslant 0$, and $|\langle \Psi, \Omega_T \Psi^i \rangle| \leqslant \|\Omega_T \Psi^i\|$ for all $i = 1, \ldots, N$ it follows that for every $\varepsilon > 0$ there is T_ε with $\sum_{i=1}^{N} (S - \lambda_i) \langle \Psi, \Omega_T \Psi^i \rangle^2 \leqslant \varepsilon$. For this T_ε it then holds that

$$\Sigma(T_\varepsilon) \geqslant S - \varepsilon,$$

as claimed. □

Remark 2.26. We comment on the assumption (29). Looking at the definition of the ionization threshold we notice that Σ represents precisely the infimum of energies that can be assumed by states whose electron positions are arbitrarily far removed from the nuclei. This is called ionization. Assumption (29) posits that it is energetically advantageous for electrons to stay bounded to the nuclei, which is certainly a natural thing to assume.

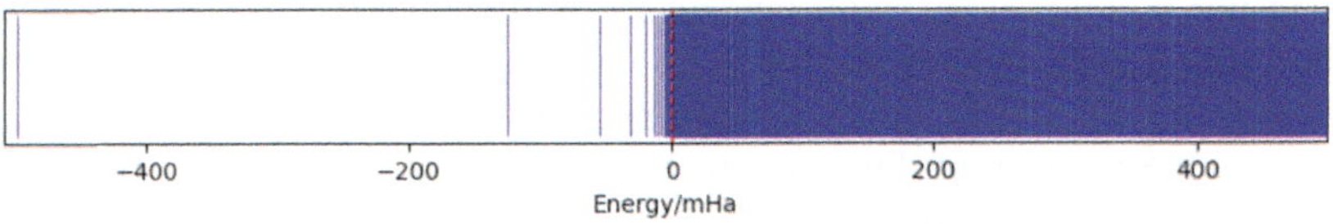

FIGURE 4 Spectrum of a Hydrogen Atom. Negative energies correspond to isolated Eigenvalues which cluster around the ionization threshold 0. The essential spectrum is made up by the positive reals.

Fig. 4 plots the spectrum of the hydrogen atom. Theorem 2.25 shows that for every molecule (that is capable of binding its electrons to the nuclei) the spectrum qualitatively looks the same.

Remark 2.27. There is much more to say about the spectral properties of $\mathcal{H}$. For example, for $\lambda \in \sigma_{\text{disc}}(\mathcal{H})$ it holds that $\|e^{\sqrt{2(\Sigma - \lambda)}|\cdot|}\Psi(\cdot)|\| < \infty$ for any corresponding Eigenfunction. This means that any Eigenfunction corresponding to an Eigenvalue below the ionization threshold decays exponentially. A proof can be found in Yserentant (2010, Theorem 5.17). These estimates are however not sharp. Since the potential does not decay equally fast in every direction, one can get improved anisotropic decay estimates that precisely characterize the decay of Eigenfunctions (Agmon, 2014).

Furthermore, one can show that the discrete spectrum is infinite (Hunziker and Sigal, 2000) and clusters at the ionization threshold.

Corollary 2.28. *Under the assumptions of Theorem 2.25 the quantity* $\mathcal{E}_{\mathbf{Z}}(\mathbf{R}) := \inf \sigma(\mathcal{H})$ *is called the* ground state energy. *It is an isolated Eigenvalue of* $\mathcal{H}$ *of finite multiplicity and corresponding Eigenvectors are called* ground states.

Furthermore it holds that

$$\mathcal{E}_{\mathbf{Z}}(\mathbf{R}) = \min_{\Psi \in H^1} \frac{\langle \mathcal{H}\Psi, \Psi \rangle}{\langle \Psi, \Psi \rangle} \tag{33}$$

and the minimizers are precisely the ground states.

Proof. This follows from Lemma 2.15 and by noting that $\mathcal{W}(\mathcal{H}) = H^1(\mathbb{R}^{n_{\mathrm{el}} \times 3})$. The fact that the infimum is a minimum that is attained by the ground states follows from the fact that the infimum of $\sigma(\mathcal{H})$ is an isolated Eigenvalue with finite multiplicity by Theorem 2.25. □

Our main interest will thus be to efficiently minimize the Raleigh quotient (33).

2.3 Spin and the Pauli exclusion principle

We will again write $\mathcal{H} = \mathcal{H}^{\mathrm{BO}}_{(\mathbf{R},\mathbf{Z})}$ for notational convenience.

There is another serious complication that we have not addressed thus far. The electrons $r_1, \ldots, r_{n_{\mathrm{el}}}$ are *indistinguishable*. This means that for every permutation π of the set $\{1, \ldots, n_{\mathrm{el}}\}$ the state $\Psi(\pi \circ \mathbf{r}) := \Psi(r_{\pi(1)}, \ldots, r_{\pi(n_{\mathrm{el}})})$ must be identical to the state Ψ. In other words, there must be $\tau(\pi) \in \{z \in \mathbb{C} : |z| = 1\}$ with $\Psi(\pi \circ \mathbf{r}) = \tau(\pi) \cdot \Psi(\mathbf{r})$. For two permutations π, η it clearly holds that $\tau(\pi \circ \eta) = \tau(\pi) \cdot \tau(\eta)$. Furthermore, if π is a transposition (e.g., a permutation that exchanges two elements and leaves all other elements fixed) it holds that $\pi \circ \pi = \mathrm{id}$, and therefore it holds that $\tau(\pi)^2 = \tau(\mathrm{id}) = 1$ which implies that $\tau(\pi) \in \{\pm 1\}$ if π is a transposition. Since for any two transpositions π, π' there exists a permutation η with $\pi = \eta^{-1} \circ \pi' \circ \eta$ it follows that the phase factor must be equal for all transpositions. Since all permutations arise as products of transpositions, this implies that either $\tau(\pi) = 1$ for all permutations or $\tau(\pi) = \mathrm{sign}(\pi)$ for all permutations. In the first case the particles are called *bosonic*, in the latter case *fermionic*. Electrons are fermionic which thus implies that the electronic wave function Ψ must be antisymmetric.

Furthermore, each electron has an additional property called *spin* which can assume values in $\left\{-\frac{1}{2}, \frac{1}{2}\right\}$. Therefore the electronic wave function really depends on *spin coordinates*

$$\left((r_1, s_1), \ldots, (r_{n_{\mathrm{el}}}, s_{n_{\mathrm{el}}})\right) \in \left(\mathbb{R}^3 \times \left\{-\frac{1}{2}, \frac{1}{2}\right\}\right)^{n_{\mathrm{el}}}.$$

Definition 2.29. The *Pauli exclusion principle* states that any admissible electronic wave function $\Psi : \left((r_1, s_1), \ldots, (r_{n_{\mathrm{el}}}, s_{n_{\mathrm{el}}})\right) \in \left(\mathbb{R}^3 \times \left\{-\frac{1}{2}, \frac{1}{2}\right\}\right)^{n_{\mathrm{el}}} \to \mathbb{C}$ must be antisymmetric with respect to permutations of the spin coordinates. In other words, for every permutation π and every admissible wavefunction Ψ it must hold that

$$\Psi\left((r_{\pi(1)}, s_{\pi(1)}), \ldots, (r_{\pi(n_{\mathrm{el}})}, s_{\pi(n_{\mathrm{el}})})\right) = \mathrm{sign}(\pi) \cdot \Psi\left((r_1, s_1), \ldots, (r_{n_{\mathrm{el}}}, s_{n_{\mathrm{el}}})\right). \tag{34}$$

Writing $\mathbf{s} = (s_1, \ldots, s_{n_{\mathrm{el}}})$ we succinctly write

$$\Psi(\pi \circ (\mathbf{r}, \mathbf{s})) = \mathrm{sign}(\pi) \cdot \Psi(\mathbf{r}, \mathbf{s}).$$

For each $\mathbf{s}$ and each $\Psi : \left(\mathbb{R}^3 \times \left\{-\frac{1}{2}, \frac{1}{2}\right\}\right)^{n_{\mathrm{el}}} \to \mathbb{C}$ we denote by $\Psi_{\mathbf{s}} : \mathbb{R}^{n_{\mathrm{el}} \times 3} \to \mathbb{C}$ the function defined by $\Psi_{\mathbf{s}}(\mathbf{r}) = \Psi(\mathbf{r}, \mathbf{s})$. The correct quantum Hilbert space for electronic systems is thus given by

$$H^{\mathrm{fermi}} := \left\{\Psi : \left(\mathbb{R}^3 \times \left\{-\frac{1}{2}, \frac{1}{2}\right\}\right)^{n_{\mathrm{el}}} \to \mathbb{C} : \ \Psi \text{ satisfies (34) and } \forall \mathbf{s} \in \left\{-\frac{1}{2}, \frac{1}{2}\right\}^{n_{\mathrm{el}}} : \ \Psi_{\mathbf{s}} \in L^2(\mathbb{R}^{n_{\mathrm{el}} \times 3})\right\}$$

with inner product

$$\langle \Psi, \Phi \rangle_{H^{\mathrm{fermi}}} := \sum_{\mathbf{s} \in \left\{-\frac{1}{2}, \frac{1}{2}\right\}^{n_{\mathrm{el}}}} \langle \Psi_s, \Phi_s \rangle_{L^2(\mathbb{R}^{n_{\mathrm{el}} \times 3})}. \tag{35}$$

The electronic Hamiltonian on H^{fermi} is defined componentwise as

$$(\mathcal{H}^{\mathrm{fermi}} \Psi)_s = \mathcal{H} \Psi_s, \tag{36}$$

for $\Psi \in H^{\mathrm{fermi}}$.

We still need to establish self-adjointness. This is done in the next result.

Theorem 2.30. *The operator $\mathcal{H}^{fermi}$ is self-adjoint on H^{fermi} with $\mathcal{D}(\mathcal{H}^{\mathrm{fermi}}) = \left[H^2(\mathbb{R}^{n_{el} \times 3})\right]^{2^{n_{el}}} \cap H^{\mathrm{fermi}}$.*

Proof. Let

$$H^{\mathrm{full}} := \left\{\Psi : \left(\mathbb{R}^3 \times \left\{-\frac{1}{2}, \frac{1}{2}\right\}\right)^{n_{\mathrm{el}}} \to \mathbb{C} : \forall \mathbf{s} \in \left\{-\frac{1}{2}, \frac{1}{2}\right\}^{n_{\mathrm{el}}} : \ \Psi_{\mathbf{s}} \in L^2(\mathbb{R}^{n_{\mathrm{el}} \times 3})\right\}.$$

We first show that $\mathcal{H}^{\mathrm{full}}$, which is simply the extension of $\mathcal{H}^{\mathrm{fermi}}$ to H^{full} is self-adjoint on H^{full}. Not that the operators in this notation do not change, but only their domain of definition.

Self-adjointness of $\mathcal{H}^{\mathrm{full}}$ follows from the fact that H^{full} is a direct sum of $2^{n_{\mathrm{el}}}$ copies of $L^2(\mathbb{R}^{n_{\mathrm{el}}})$ and $\mathcal{H}^{\mathrm{full}}$ is the direct sum of $2^{n_{\mathrm{el}}}$ copies of a self-adjoint operator (namely the Hamiltonian $\mathcal{H}$ on $\mathbb{R}^{n_{\mathrm{el}}\times 3}$). By Reed and Simon (1978, Theorem XIII.85) a direct sum of self-adjoint operators is self-adjoint and therefore $\mathcal{H}^{\mathrm{full}}$ is self-adjoint on H^{full}.

We want to show that the restriction of $\mathcal{H}^{\mathrm{full}}$ to the antisymmetric subspace H^{fermi} is still self-adjoint. We will make use of the following fact that is proven in Hall (2013, Proposition 9.23): For any Hilbert space J and any symmetric operator $\mathcal{B}$ on J it holds that

$$\mathcal{B} \text{ self-adjoint } \Leftrightarrow \operatorname{range}(\mathcal{B} \pm i) = J. \tag{37}$$

Consider the antisymmetrization operator

$$\mathcal{A}\Psi(\mathbf{r},\mathbf{s}) := \frac{1}{n_{\mathrm{el}}!} \sum_{\pi \in \mathfrak{S}_{n_{\mathrm{el}}}} \operatorname{sign}(\pi) \cdot \Psi(\pi \circ (\mathbf{r},\mathbf{s})), \tag{38}$$

where $\mathfrak{S}_{n_{\mathrm{el}}}$ denotes the group of permutations of n_{el} elements. Since $\mathcal{A}$ is the identity on H^{fermi}, $\mathcal{A}^2 = \mathcal{A}$, and $\mathcal{A}$ is symmetric (and therefore self-adjoint since it is bounded) it follows that $\mathcal{A}$ is the orthogonal projection from H^{full} onto H^{fermi}.

Observe that

$$\mathcal{A}\left(\mathcal{D}(\mathcal{H}^{\mathrm{full}})\right) \subset \mathcal{D}(\mathcal{H}^{\mathrm{full}}) \tag{39}$$

and

$$\forall \Phi \in \mathcal{D}(\mathcal{H}^{\mathrm{full}}):\ \mathcal{A}\mathcal{H}^{\mathrm{full}}\Phi = \mathcal{H}^{\mathrm{full}}\mathcal{A}\Phi. \tag{40}$$

In view of (37) we need to prove that $\operatorname{range}(\mathcal{H}^{\mathrm{fermi}} \pm i) = H^{\mathrm{fermi}}$. To this end let $\Psi \in H^{\mathrm{fermi}}$ be arbitrary. Since $\mathcal{H}^{\mathrm{full}}$ is self-adjoint, Eq. (37) implies that there exists $\Phi \in \mathcal{D}(\mathcal{H}^{\mathrm{full}})$ with $(\mathcal{H}^{\mathrm{full}} \pm i)\Phi = \Psi$. By (39) it holds that $\mathcal{A}\Phi \in \mathcal{D}(\mathcal{H}^{\mathrm{full}})$ and that $(\mathcal{H}^{\mathrm{full}} \pm i)\mathcal{A}\Phi = \mathcal{A}(\mathcal{H}^{\mathrm{full}} \pm i)\Phi = \mathcal{A}\Psi = \Psi$. Thus, $\mathcal{H}^{\mathrm{fermi}} \pm i$ are surjective on H^{fermi} which proves our desired claim. □

Remark 2.31. While the original spectral problem for the operator $\mathcal{H}$ is generally already very high-dimensional and extremely challenging, the Pauli exclusion principle adds another layer of intractability. Even evaluating the inner product according to the formula (35) requires the evaluation of $2^{n_{\mathrm{el}}}$ terms, each of which is given as an integral of a $3 \cdot n_{\mathrm{el}}$ dimensional function! This sounds daunting. Fortunately, we will see that the spectral problem of $\mathcal{H}^{\mathrm{fermi}}$ decouples into only n_{el} subproblems analogous to the spectral problem of $\mathcal{H}$ – one for each number of electrons with spin equal to $\frac{1}{2}$ – but with additional antisymmetry constraints.

In order to clarify the connection between functions $\Psi = (\Psi_s)_{s\in\{\pm 1/2\}^{n_{\mathrm{el}}}} \in H^{\mathrm{fermi}}$ and their components Ψ_s we need the following definition.

Definition 2.32. For $\mathbf{s} \in \left\{\pm\frac{1}{2}\right\}^{n_{\mathrm{el}}}$ define $L^2(\mathbb{R}^{n_{\mathrm{el}}\times 3})_{\mathbf{s}}$ as the subspace of $L^2(\mathbb{R}^{n_{\mathrm{el}}\times 3})$ consisting of all functions Ψ which are antisymmetric with respect to all permutations that leave $\mathbf{s}$ invariant, i.e.,

$$\Psi(\pi \circ \mathbf{r}) = \mathrm{sign}(\pi) \cdot \Psi(\mathbf{r}) \quad \forall \pi \in \mathfrak{S}_{n_{\mathrm{el}}} : \pi \circ \mathbf{s} = \mathbf{s}.$$

Analogously we define the spaces $H^k(\mathbb{R}^{n_{\mathrm{el}}\times 3})_{\mathbf{s}}$. Finally we denote by $\mathcal{H}_{\mathbf{s}}$ the restriction of $\mathcal{H}$ to $L^2(\mathbb{R}^{n_{\mathrm{el}}\times 3})_{\mathbf{s}}$ and its ionization threshold $\Sigma_{\mathbf{s}}$ in the same way as (28) but with the infimum (26) taken only over $L^2(\mathbb{R}^{n_{\mathrm{el}}\times 3})_{\mathbf{s}}$.

Remark 2.33. It is important to keep in mind that the operators $\mathcal{H}_{\mathbf{s}}$ are really independent of $\mathbf{s}$ and all equal to $\mathcal{H}$. The only thing that distinguishes them is their domain, which does depend on $\mathbf{s}$ and is given by $H^2(\mathbb{R}^{n_{\mathrm{el}}\times 3})_{\mathbf{s}}$ (see Theorem 2.34 below).

The operators $\mathcal{H}_{\mathbf{s}}$ behave similarly to the operators $\mathcal{H}$ in terms of their spectral properties.

Theorem 2.34. **1.** *For each* $\mathbf{s} \in \left\{-\frac{1}{2}, \frac{1}{2}\right\}^{n_{el}}$ *the operator* $\mathcal{H}_{\mathbf{s}}$ *is self adjoint with* $\mathcal{D}(\mathcal{H}_{\mathbf{s}}) = H^2(\mathbb{R}^{n_{el}\times 3})_{\mathbf{s}}$.

2. *For every permutation* $\pi \in \mathfrak{S}_{n_{el}}$ *and every* $\mathbf{s} \in \left\{-\frac{1}{2}, \frac{1}{2}\right\}$, *the operators* $\mathcal{H}_{\mathbf{s}}$ *and* $\mathcal{H}_{\pi\circ\mathbf{s}}$ *are unitarily equivalent.*

3. *Assume that*

$$\inf \sigma(\mathcal{H}_{\mathbf{s}}) < \Sigma_{\mathbf{s}}.$$

Then it holds that

$$\inf \sigma_{ess}(\mathcal{H}_{\mathbf{s}}) = \Sigma_{\mathbf{s}}.$$

In particular, the essential spectrum is nonempty and it holds that

$$\sigma(\mathcal{H}_{\mathbf{s}}) \cap [\inf \sigma(\mathcal{H}_{\mathbf{s}}), \Sigma_{\mathbf{s}}) \subset \sigma_{disc}(\mathcal{H}_{\mathbf{s}}).$$

Proof. The fact that the restriction $\mathcal{H}_{\mathbf{s}}$ of the self-adjoint operator $\mathcal{H}$ to $L^2_{\mathbf{s}}$ is self-adjoint can be argued in exactly the same way as the proof of Theorem 2.30. The second point follows from the fact that $\mathcal{H}_{\mathbf{s}}$ and $\mathcal{H}_{\pi\circ\mathbf{s}}$ are unitarily equivalent via the operator that permutes the electronic coordinates $\mathbf{r}$ via π. The proof of the third point is identical to the proof of Theorem 2.25. □

We now want to relate the spectral properties of the operators $\mathcal{H}_{\mathbf{s}}$ to properties of the full Hamiltonian $\mathcal{H}^{\mathrm{fermi}}$.

Lemma 2.35. *A function* $\Phi \in L^2(\mathbb{R}^{n_{el}\times 3})$ *is a component* $\Psi_{\mathbf{s}}$ *of a function* $\Psi = (\Psi_{\mathbf{s}})_{\mathbf{s}\in\{\pm 1/2\}^{n_{el}}} \in H^{\mathrm{fermi}}$ *if and only if* $\Phi \in L^2(\mathbb{R}^{n_{el}\times 3})_{\mathbf{s}}$.

Proof. First consider $\Psi = (\Psi_{\mathbf{s}})_{\mathbf{s}\in\{\pm 1/2\}^{n_{\mathrm{el}}}} \in H^{\mathrm{fermi}}$ and fix $\mathbf{s}$. Let π be a permutation with $\pi \circ \mathbf{s} = \mathbf{s}$.

Then

$$\Psi_{\mathbf{s}}(\pi \circ \mathbf{r}) = \mathrm{sign}(\pi) \cdot \Psi_{\pi \circ \mathbf{s}}(\pi \circ \mathbf{s}) = \mathrm{sign}(\pi) \cdot \Psi_{\mathbf{s}}(\pi \circ \mathbf{r}).$$

Therefore, $\Psi_{\mathbf{s}} \in L^2(\mathbb{R}^{n_{\mathrm{el}}\times 3})_{\mathbf{s}}$ which proves the "only if" part.

For the other direction we assume that for a fixed $\mathbf{s}$ it holds that $\Phi \in L^2(\mathbb{R}^{n_{\mathrm{el}}\times 3})_{\mathbf{s}}$. We need to find $\Psi = (\Psi_{\mathbf{t}})_{\mathbf{t}\in\{\pm 1/2\}^{n_{\mathrm{el}}}} \in H^{\mathrm{fermi}}$ with $\Psi_{\mathbf{s}} = \Phi$. Such a Ψ can be explicitly constructed as

$$\Psi_{\mathbf{t}}(\mathbf{r}) := \frac{\sum_{\pi\in\mathfrak{S}_{n_{\mathrm{el}}}} \mathrm{sign}(\pi)\Phi(\pi \circ \mathbf{r}) \cdot \delta_{\pi\circ\mathbf{t},s}}{\sum_{\pi\in\mathfrak{S}_{n_{\mathrm{el}}}} \delta_{\pi\circ\mathbf{t},\mathbf{t}}}. \tag{41}$$

□

Theorem 2.36. *It holds that*

$$\sigma(\mathcal{H}^{\mathrm{fermi}}) = \bigcup_{\mathbf{s}\in\left\{-\frac{1}{2},\frac{1}{2}\right\}^{n_{el}}} \sigma(\mathcal{H}_{\mathbf{s}}). \tag{42}$$

Proof. Let $\lambda \in \sigma(\mathcal{H}^{\mathrm{fermi}})$, then by Lemma 2.4 for every $n \in \mathbb{N}$ there is $\Psi^n \in H^{\mathrm{fermi}}$ with $\|(\mathcal{H}^{\mathrm{fermi}} - \lambda)\Psi^n\| \leqslant \frac{1}{n}\|\Psi^n\|$.

By definition this implies that

$$\sum_{\mathbf{s}\in\{\pm 1/2\}^{n_{\mathrm{el}}}} \|(\mathcal{H}_{\mathbf{s}} - \lambda)\Psi^n_{\mathbf{s}}\|^2 \leqslant \frac{1}{n^2} \sum_{\mathbf{s}\in\{\pm 1/2\}^{n_{\mathrm{el}}}} \|\Psi_{\mathbf{s}}\|^2$$

and therefore, for each $n \in \mathbb{N}$ there must exist $\mathbf{s}_n$ such that

$$\|(\mathcal{H}_{\mathbf{s}_n} - \lambda)\Psi^n_{\mathbf{s}_n}\| \leqslant \frac{1}{n}\|\Psi^n_{\mathbf{s}_n}\|.$$

Since there are only finitely many values that $\mathbf{s}_n$ can assume, we conclude that there exists $\mathbf{s}$ such that for infinitely many $n \in \mathbb{N}$ there is $\Psi^n_{\mathbf{s}} \in L^2(\mathbb{R}^{n_{\mathrm{el}}\times 3})_{\mathbf{s}}$ with

$$\|(\mathcal{H}_{\mathbf{s}} - \lambda)\Psi^n_{\mathbf{s}}\| \leqslant \frac{1}{n}\|\Psi^n_{\mathbf{s}}\|.$$

By Lemma 2.4 this implies that $\lambda \in \sigma(\mathcal{H}_{\mathbf{s}})$.

For the other direction suppose that $\lambda \in \sigma(\mathcal{H}_{\mathbf{s}})$ and for each $n \in \mathbb{N}$ let $\Phi^n \in L^2(\mathbb{R}^{n_{\mathrm{el}}\times 3})_{\mathbf{s}}$ satisfy that

$$\|(\mathcal{H}_s - \lambda)\Phi^n\| \leqslant \frac{1}{n}\|\Phi^n\|.$$

The existence of Φ^n follows from Lemma 2.4. Now construct $\Psi^n \in H^{\mathrm{fermi}}$ according to (41), which only contains components for spin vectors $\mathbf{t}$ that arise from permuting the fixed spin vector $\mathbf{s}$ in the sense that $s = \pi \circ \mathbf{t}$. By Item 2 of Theorem 2.34 the operators $\mathcal{H}_\mathbf{t}$ for such spin vectors are unitarily equivalent to $\mathcal{H}_\mathbf{s}$ and therefore it holds that

$$\|(\mathcal{H}_\mathbf{t} - \lambda)\Phi^n(\pi \circ \mathbf{r})\| = \|(\mathcal{H}_\mathbf{s} - \lambda)\Phi^n(\mathbf{r})\| \leqslant \frac{1}{n}\|\Psi_\mathbf{s}^n\| = \frac{1}{n}\|\Psi_\mathbf{t}^n\|$$

This implies that

$$\|(\mathcal{H}^{\text{fermi}} - \lambda)\Psi^n\| \leqslant \frac{1}{n}\|\Psi^n\|,$$

which by Lemma 2.4 implies that $\lambda \in \sigma(\mathcal{H}^{\text{fermi}})$ □

By Theorem 2.36 the spectrum of the full fermionic system $\mathcal{H}$ can be determined from solving the spectral problems of all operators $\mathcal{H}_\mathbf{s}$ for $\mathbf{s} \in \{-\frac{1}{2}, \frac{1}{2}\}^{n_{\text{el}}}$. Since operators $\mathcal{H}_\mathbf{s}$ and $\mathcal{H}_\mathbf{t}$ are unitarily equivalent if $\mathbf{s}$ is a permutation of $\mathbf{t}$, it suffices to only consider spin assignments of the form

$$\mathbf{s}_{n_\uparrow} := \Bigg(\underbrace{\frac{1}{2}, \ldots, \frac{1}{2}}_{n_\uparrow \text{ times}}, \underbrace{-\frac{1}{2}, \ldots, -\frac{1}{2}}_{n_{\text{el}} - n_\uparrow \text{ times}}\Bigg), \quad n_\uparrow \in \{0, \ldots, n_{\text{el}}\}, \tag{43}$$

where $n_\uparrow$ corresponds to the number of electrons with positive spin. Since the spectral problem does not change if positive spin electrons are changed into negative spin electrons and vice versa, it suffices to solve the spectral problems of the operators $\mathcal{H}_{\mathbf{s}_{n_\uparrow}}$ for $n_\uparrow \in \{0, \ldots, \lfloor \frac{n_{\text{el}}}{2} \rfloor\}$ denoting the number of spin-up electrons.

By Theorem 2.34 the lowest possible energy is given as an isolated eigenvalue (the ground state energy) and corresponding eigenvectors are the ground states. By Lemma 2.15, the ground states and the ground state energies (and thus the PES) can be determined as solutions of the minimization problems

$$\mathcal{E}_\mathbf{Z}^{n_\uparrow}(\mathbf{R}) = \min_{\Psi \in H^1_{\mathbf{s}_{n_\uparrow}}} \frac{\langle \mathcal{H}\Psi, \Psi \rangle}{\langle \Psi, \Psi \rangle}, \quad n_\uparrow \in \left\{0, \ldots, \left\lfloor \frac{n_{\text{el}}}{2} \right\rfloor\right\}. \tag{44}$$

Finally we note that, although the space $H^1_{\mathbf{s}_{n_\uparrow}}$ is a complex Hilbert space, we can without loss of generality minimize (44) over the space of real-valued antisymmetric H^1 functions. This is because the Hamiltonian acts separately on the real and imaginary part which implies that the real (or imaginary) part of an Eigenvector is still an Eigenvector.

Our goal is thus to

Find efficient and accurate algorithms for solving the minimization problems (44).

3 Introduction to variational Monte Carlo (VMC)

Given the problems (44) we would like to devise numerical algorithms that provide an accurate solution. The most straightforward idea to achieve this is to start with a parametrized class of functions

$$\mathcal{F} := \left\{ \Psi_\theta : \theta \in \mathbb{R}^{N_{\text{param}}} \right\} \subset H^1(\mathbb{R}^{n_{\text{el}} \times 3})_{\mathfrak{s}_{n_\uparrow}} \tag{45}$$

and (try to) solve the restricted minimization problem

$$\min_{\theta \in \mathbb{R}^{N_{\text{param}}}} \frac{\langle \mathcal{H}\Psi_\theta, \Psi_\theta \rangle}{\langle \Psi_\theta, \Psi_\theta \rangle}, \quad n_\uparrow \in \left\{ 0, \ldots, \left\lfloor \frac{n_{\text{el}}}{2} \right\rfloor \right\} \tag{46}$$

in the parameter θ.

This approach has several obvious benefits:

1. It is conceptually simple
2. It is *variational* in the sense that for any approximate solution Ψ_{θ^*} of (46) (or rather any $\Psi_{\theta^*} \in H^1(\mathbb{R}^{n_{\text{el}} \times 3})_{\mathfrak{s}_{n_\uparrow}}$) we always have the upper bound

$$\mathcal{E}_{\mathbf{Z}}^{n_\uparrow}(\mathbf{R}) \leqslant \frac{\langle \mathcal{H}\Psi_{\theta^*}, \Psi_{\theta^*} \rangle}{\langle \Psi_{\theta^*}, \Psi_{\theta^*} \rangle}, \quad n_\uparrow \in \left\{ 0, \ldots, \left\lfloor \frac{n_{\text{el}}}{2} \right\rfloor \right\}. \tag{47}$$

   This means that we can compare the quality of different algorithms a posteriori – the smaller the computed energy, the better.
3. Since by (24) and (25) the Hamiltonian is bounded and coercive on H^1, one can in principle treat the corresponding Eigenvalue problem within the well-developed mathematical framework of elliptic eigenvalue problems, see Yserentant (2010) for results in this direction.

In reality, it is however extremely challenging to come up with efficient and accurate algorithms. This is at least due to the following issues.

1. The problem is high-dimensional and potentially carries the curse of dimension (it may well be NP-hard (Troyer and Wiese, 2005)).
2. Enforcing the antisymmetry condition of Definition 2.32 is nonstandard and challenging.
3. The minimization problem (47) is nonconvex and therefore algorithms can get stuck on local minima.
4. The electronic ground state is not globally smooth, has cusps near the nuclei and for electron coordinates approaching each other (Kato, 1957) and complicated long-range interactions. It is not at all clear how to choose $\mathcal{F}$ so that such functions can be well approximated with a reasonable number of parameters.

3.1 Slater determinants

We address the question of how to enforce the antisymmetry condition of Definition 2.32.

Slater determinants

Suppose that there are $n_\uparrow$ electrons with spin $\frac{1}{2}$ and $n_{\mathrm{el}} - n_\uparrow$ electrons with spin $-\frac{1}{2}$. Then for $\mathbf{s}_{n_\uparrow}$ as in (43) a permutation π leaves $\mathbf{s}_{n_\uparrow}$ invariant if and only if π can be decomposed into $\pi = \pi_\uparrow \circ \pi_\downarrow$ where $\pi_\uparrow$ is any permutation of the first $n_\uparrow$ electron coordinates and $\pi_\downarrow$ is any permutation of the last $n - n_\uparrow$ electron coordinates.

Now take any set $\phi_\uparrow^i : \mathbb{R}^3 \to \mathbb{R}$ for $i = 1, \ldots, n_\uparrow$ and $\phi_\downarrow^i : \mathbb{R}^3 \to \mathbb{R}$ for $i = 1, \ldots, n_{\mathrm{el}} - n_\uparrow$ of $H^1(\mathbb{R}^3)$ functions. Then it is easy to see that the function

$$\mathfrak{D}\left(\phi_\uparrow^1, \ldots, \phi_\uparrow^{n_\uparrow}; \phi_\downarrow^1, \ldots, \phi_\downarrow^{n_{\mathrm{el}}-n_\uparrow}\right) := \frac{1}{\sqrt{n_\uparrow!}} \begin{vmatrix} \phi_\uparrow^1(r_1) & \ldots & \phi_\uparrow^{n_\uparrow}(r_1) \\ \vdots & \ddots & \vdots \\ \phi_\uparrow^1(r_{n_\uparrow}) & \ldots & \phi_\uparrow^{n_\uparrow}(r_{n_\uparrow}) \end{vmatrix} \cdot \frac{1}{\sqrt{(n_{\mathrm{el}} - n_\uparrow)!}} \begin{vmatrix} \phi_\downarrow^1(r_{n_\uparrow+1}) & \ldots & \phi_\downarrow^{n-n_\uparrow}(r_{n_\uparrow+1}) \\ \vdots & \ddots & \vdots \\ \phi_\downarrow^1(r_{n_{\mathrm{el}}}) & \ldots & \phi_\downarrow^{n-n_{\mathrm{el}}}(r_{n_{\mathrm{el}}}) \end{vmatrix} \tag{48}$$

is in $H^1_{\mathbf{s}_{n_\uparrow}}$. Here we denote by $\begin{vmatrix} * & \ldots & * \\ \vdots & \ddots & \vdots \\ * & \ldots & * \end{vmatrix}$ denotes the determinant of a matrix. We also note that the determinant can be evaluated in cubic complexity using Gaussian elimination.

Functions of the Form (48) are called *Slater determinants*. A suitable approximation set $\mathcal{F}$ can now be constructed by choosing a *basis set* $\mathcal{B} = \chi_1, \ldots, \chi_{n_{\mathrm{basis}}} \in H^1(\mathbb{R}^3)$, writing

$$\phi(\mathcal{B}, \mathbf{b}) := \sum_{j=1}^{n_{\mathrm{basis}}} (\mathbf{b}^i)_j \chi_j.$$

Then one can define

$$\mathcal{F}_{\mathcal{B}} := \left\{ \mathfrak{D}\left(\phi(\mathcal{B}, \mathbf{b}_\uparrow^1), \ldots, \phi(\mathcal{B}, \mathbf{b}_\uparrow^{n_\uparrow}); \phi(\mathcal{B}, \mathbf{b}_\downarrow^1), \ldots, \phi(\mathcal{B}, \mathbf{b}_\downarrow^{n_{\mathrm{el}}-n_\uparrow})\right) : \mathbf{b}_\uparrow^i,\ \mathbf{b}_\downarrow^i \in \mathbb{R}^{n_{\mathrm{basis}}} \right\}$$

and solve the minimization problem (46). The resulting method is called the *Hartree Fock method* and it corresponds to an antisymmetrized rank-1 approximation. The Hartree Fock method is computationally relatively cheap. However, it disregards electron correlations (beyond the correlations caused by the antisymmetry constraint) and is therefore of very limited accuracy.

To improve accuracy one can try to use linear combinations of several different slater determinants which corresponds to an antisymmetrized low-rank approximation. The corresponding approximation set is then given as

$$\mathcal{F}_{n_{\text{det}},\mathcal{B}} := \left\{ \sum_{l=1}^{n_{\text{det}}} \mathfrak{D}\left(\phi(\mathcal{B}, \mathbf{b}_{\uparrow}^{1,l}), \ldots, \phi(\mathcal{B}, \mathbf{b}_{\uparrow}^{n_{\uparrow},l}); \phi(\mathcal{B}, \mathbf{b}_{\downarrow}^{1,l}), \ldots, \phi(\mathcal{B}, \mathbf{b}_{\downarrow}^{n_{\text{el}}-n_{\uparrow},l})\right) : \mathbf{b}_{\uparrow}^{i,l},\ \mathbf{b}_{\downarrow}^{i,l} \in \mathbb{R}^{n_{\text{basis}}} \right\}.$$

Remark 3.1. These constructions can be augmented by multiplying each Slater determinant with any function that leaves the antisymmetry property intact. This holds for instance for symmetric functions. Such multiplicative corrections are called *Jastrow factor*. Such Jastrow factors are important to capture the correct cusp behavior of the ground state.

Approximation results for sets of the form $\mathcal{F}_{n_{\text{det}},\mathcal{B}}$ are derived in Yserentant (2010) where it is shown that the H^1 approximation error for the electronic ground state decays as $C_{N_{\text{det}}} \cdot n_{\text{det}}^{-1}$. Unfortunately, a close inspection of the proofs in Yserentant (2010) reveals that the constant $C_{N_{\text{det}}}$ grows exponentially in n_{det} which makes approximation by $\mathcal{F}_{n_{\text{det}},\mathcal{B}}$ intractable. Overall it seems that accurate and tractable approximations of electronic ground states using Slater determinants are limited to quite small systems of up to around 15 electrons.

Generalized slater determinants

A key drawback of Slater determinants is their inability to efficiently model electron correlations. To a certain extent this can be remedied by considering *generalized Slater determinants* of the following form.

For $n \in \mathbb{N}, l \in \{1, \ldots, n\}$ denote the classes

$$\mathcal{O}_{\uparrow}^{l,n} := \left\{\phi \in H^1(\mathbb{R}^{n\times 3}) : \phi(r_1, \ldots, r_n) \text{ is symmetric in the variables } (r_2, \ldots, r_l) \text{ and } (r_{l+1}, \ldots, r_n)\right\} \tag{49}$$

and

$$\mathcal{O}_{\downarrow}^{l,n} := \left\{\phi \in H^1(\mathbb{R}^{n\times 3}) : \phi(r_1, \ldots, r_n) \text{ is symmetric in the variables } (r_{l+2}, \ldots, r_n) \text{ and } (r_2, \ldots, r_l)\right\}. \tag{50}$$

We also introduce the following notation

$$\{\mathbf{r}^{\uparrow}\} := \{r_1, \ldots, r_{n_\uparrow}\}, \quad \{\mathbf{r}^{\downarrow}\} := \{r_{n_\uparrow+1}, \ldots, r_{n_{\mathrm{el}}}\}, \quad \{\mathbf{r}^{\uparrow\downarrow}_{\setminus j}\} := \{\mathbf{r}^{\uparrow\downarrow}\} \setminus \{r_j\}. \tag{51}$$

Here, the symbol $\{\cdot\}$ should be understood as describing a *multiset*, meaning that it can contain equal elements several times. For example, if $r_1 = r_2 = 1$, $r_3 = 2$ and $n_\uparrow = 3$ it holds that

$$\{\mathbf{r}^{\uparrow}\} = \{1, 1, 2\} = \{1, 2, 1\} = \{2, 1, 1\} \neq \{1, 2\} = \{\mathbf{r}^{\uparrow}_{\setminus 1}\} = \{\mathbf{r}^{\uparrow}_{\setminus 2}\}.$$

Suppose that $\phi^1_\uparrow, \ldots, \phi^{n_\uparrow}_\uparrow \in \mathcal{O}^{n_\uparrow, n_{\mathrm{el}}}_\uparrow$ and $\phi^{n_\uparrow+1}_\downarrow, \ldots, \phi^{n_{\mathrm{el}}}_\downarrow \in \mathcal{O}^{n_\uparrow, n_{\mathrm{el}}}_\downarrow$.

Then, we can define

$$\Phi^{i,j}_\uparrow(\mathbf{r}) := \phi^i_\uparrow(r_j; \{\mathbf{r}^{\uparrow}_{\setminus j}\}; \{\mathbf{r}^{\downarrow}\}), \quad i, j = 1, \ldots, n_\uparrow, \tag{52}$$

and

$$\Phi^{i,j}_\downarrow(\mathbf{r}) := \phi^i_\downarrow(r_j; \{\mathbf{r}^{\uparrow}\}; \{\mathbf{r}^{\downarrow}_{\setminus j}\}), \quad i, j = n_\uparrow + 1, \ldots, n_{\mathrm{el}}. \tag{53}$$

Observe that due to the symmetry requirements of the functions $\phi^i_{\uparrow\downarrow}$ the set notation (51) in (52) and (53) is justified since the respective coordinates are independent of their order.

It is again easy to see that the function

$$\mathfrak{D}^{\mathrm{fermi}}\left(\varphi^1_\uparrow, \ldots, \varphi^{n_\uparrow}_\uparrow; \varphi^1_\downarrow, \ldots, \varphi^{n_{\mathrm{el}}-n_\uparrow}_\downarrow\right) := \\ \frac{1}{\sqrt{n_\uparrow!}} \begin{vmatrix} \Phi^{1,1}_\uparrow(\mathbf{r}) & \ldots & \Phi^{n_\uparrow,1}_\uparrow(\mathbf{r}) \\ \vdots & \ddots & \vdots \\ \Phi^{1,n_\uparrow}_\uparrow(\mathbf{r}) & \ldots & \Phi^{n_\uparrow,n_\uparrow}_\uparrow(\mathbf{r}) \end{vmatrix} \\ \cdot \frac{1}{\sqrt{(n_{\mathrm{el}} - n_\uparrow)!}} \begin{vmatrix} \Phi^{n_\uparrow+1,n_\uparrow+1}_\downarrow(\mathbf{r}) & \ldots & \Phi^{n_{\mathrm{el}},n_\uparrow+1}_\downarrow(\mathbf{r}) \\ \vdots & \ddots & \vdots \\ \Phi^{n_\uparrow+1,n_{\mathrm{el}}}_\downarrow(\mathbf{r}) & \ldots & \Phi^{n_{\mathrm{el}},n_{\mathrm{el}}}_\downarrow(\mathbf{r}) \end{vmatrix} \tag{54}$$

is in $H^1_{\mathfrak{S}_{n_\uparrow}}$. In Pfau et al. (2020) such determinant functions are called *generalized Slater determinants*.

Remark 3.2. The paper by Pfau et al. (2020, Appendix B) proves a universal approximation result that shows that one can approximate any function $L^\infty(\mathbb{R}^{n_{\mathrm{el}}\times 3})_{\mathfrak{S}_{n_\uparrow}}$ arbitrarily well in the L^∞ norm if one allows orbitals Φ^i that are discontinuous (and not in H^1). While this result is reassuring, an application to solving the Schrödinger equation would require such results in the H^1 norm.

It is an open question if such an approximation result from a single generalized Slater determinant holds true, see however Ye et al. (2024) for recent progress in this direction.

Given a pair of mappings

$$\phi_\uparrow : \mathbb{R}^{n_{\text{oparams}}} \to \mathcal{O}_\uparrow^{n_\uparrow, n_{\text{el}}} \quad \text{and} \quad \phi_\downarrow : \mathbb{R}^{n_{\text{oparams}}} \to \mathcal{O}_\downarrow^{n_\uparrow, n_{\text{el}}} \tag{55}$$

we can now again construct an approximation set by using parametrized orbitals $\phi_\uparrow^i, \phi_\downarrow^i$, e.g.,

$$\phi_\uparrow^i = \phi_\uparrow(\theta_\uparrow^i) \quad \text{for some } \theta_\uparrow^i \in \mathbb{R}^{n_{\text{oparams}}} \tag{56}$$

and

$$\phi_\downarrow^i = \phi_\downarrow(\theta_\downarrow^i) \quad \text{for some } \theta_\downarrow^i \in \mathbb{R}^{n_{\text{oparams}}}. \tag{57}$$

The corresponding approximation set using N_{det} generalized Slater determinants is then given as

$$\begin{aligned} \mathcal{F}^{\text{fermi}}_{n_{\text{det}}, \Phi_\uparrow, \Phi_\downarrow} :=& \\ &\left\{ \sum_{l=1}^{n_{\text{det}}} \mathfrak{D}^{\text{fermi}} \left(\phi_\uparrow(\theta_\uparrow^{1,l}), \ldots, \phi_\uparrow(\theta_\uparrow^{n_\uparrow,l}); \phi_\downarrow(\theta_\downarrow^{1,l}), \ldots, \phi_\downarrow(\theta_\downarrow^{n_{\text{el}}-n_\uparrow,l}) \right) : \right. \\ &\left. \theta_\uparrow^{i,l}, \ \theta_\downarrow^{i,l} \in \mathbb{R}^{n_{\text{oparams}}} \right\}. \end{aligned} \tag{58}$$

Remark 3.3. One can further generalize (54) by considering n_{el} spin-up orbitals of the form (56) and n_{el} spin-down orbitals of the form (57) and assembling the $n_{\text{el}} \times n_{\text{el}}$ matrix $(\Xi_{i,j})_{i,j=1}^{n_{\text{el}}}$ with

$$\Xi_{i,j} = \begin{cases} \Phi_\uparrow^{i,j} & j \in \{1, \ldots, n_\uparrow\} \\ \Phi_\uparrow^{i,j} & j \in \{n_\uparrow + 1, \ldots, n_{\text{el}}\}, \end{cases}$$

and taking its determinant. It is easy to see that this approach yields functions with the correct antisymmetry properties. Furthermore, the construction (54) arises as a special case by letting $\Phi_\uparrow^i = 0$ for all $i = n_\uparrow + 1, \ldots, n_{\text{el}}$ and $\Phi_\downarrow^i = 0$ for all $i = 1, \ldots, n_\uparrow$. This construction is sometimes referred to as *full Determinant* (Lin et al., 2023).

Furthermore, the construction (58) is typically extended by including a Jastrow factor, i.e., a symmetric function as multiplicative correction term, see Remark 3.1.

Having mappings (55) at hand we can thus construct corresponding parametrized classes of functions given by (58) and try to solve the minimization problem (44) in terms of the parameter $\theta = (\theta_\uparrow^{i,l})_{l=1,\ldots,n_{\text{det}},\ i=1,\ldots,n_\uparrow} \times$

$(\theta_{\downarrow}^{i,l})_{l=1,\ldots,N_{\text{det}},\ i=1,\ldots,n_{\text{el}}-n_{\uparrow}} \cong \mathbb{R}^{n_{\text{det}}\times n_{\text{el}}\times n_{\text{oparams}}}$ over this set. In this case we have that $n_{\text{params}} = n_{\text{det}} \times n_{\text{el}} \times n_{\text{oparams}}$ and for $\theta = (\theta_{\uparrow\downarrow}^{i,l})_{i,l,\uparrow\downarrow} \in \mathbb{R}^{n_{\text{params}}}$ it then holds that

$$\Psi_\theta := \sum_{l=1}^{n_{\text{det}}} \mathfrak{D}^{\text{fermi}}\left(\phi_\uparrow(\theta_\uparrow^{1,l}), \ldots, \phi_\uparrow(\theta_\uparrow^{n_\uparrow,l}); \phi_\downarrow(\theta_\downarrow^{1,l}), \ldots, \phi_\downarrow(\theta_\downarrow^{n_{\text{el}}-n_\uparrow,l})\right)$$
$$\in H^1(\mathbb{R}^{n_{\text{el}}\times 3})_{\mathfrak{S}_{n_\uparrow}}. \tag{59}$$

The key challenge is then to

Find parametrized orbitals (55) such that the sets (58) are maximally expressive.

3.2 Sampling using the Metropolis-Hastings algorithm

For solving the optimization problem (44) we need to be able to efficiently evaluate the $3n_{\text{el}}$-dimensional integrals $\frac{\langle \mathcal{H}\Psi_\theta, \Psi_\theta\rangle}{\langle \Psi_\theta, \Psi_\theta}$. Therefore, during optimization of the parameters θ and to get a final prediction of the ground-state energy, it is essential to have an efficient method for calculating the high-dimensional integrals. In Variational Monte Carlo, one commonly employs *Markov chain Monte Carlo* (MCMC), utilizing algorithms like *Metropolis-Hastings*. To perform Monte Carlo integration, one needs to rewrite the Rayleigh-Ritz Quotient

$$\frac{\langle \mathcal{H}\Psi_\theta, \Psi_\theta\rangle}{\langle \Psi_\theta, \Psi_\theta\rangle} = \int \frac{|\Psi_\theta(\mathbf{r})|^2}{\langle \Psi_\theta, \Psi_\theta\rangle} \frac{\mathcal{H}\Psi_\theta(\mathbf{r})}{\Psi_\theta(\mathbf{r})} d\mathbf{r} \approx \frac{1}{n_{\text{s}}} \sum_{a=1}^{n_{\text{s}}} \frac{\mathcal{H}\Psi_\theta(\mathbf{r}_a)}{\Psi_\theta(\mathbf{r}_a)} \tag{60}$$

with n_{s} electron samples $\mathbf{r}_a \sim p_\theta(\mathbf{r}) := \frac{|\Psi_\theta(\mathbf{r})|^2}{\langle \Psi_\theta, \Psi_\theta\rangle}$.

The density p_θ we sample from with MCMC gets updated during optimization due to the dependence on θ (cf. Sec. 3.3). To ensure the sampled electron positions follow the correct distribution, we have to perform multiple consecutive steps of MCMC, whereas a single step with the Metropolis-Hastings algorithm can be broken down into the following stages, as outlined in Algorithm 1. Firstly, starting with an initial set of electron positions, either from a previous step or randomly initialized, represented as $\mathbf{r}_n \in \mathbb{R}^{n_{\text{el}}\times 3}$ a proposed state $\mathbf{r}_{\text{p}} \in \mathbb{R}^{n_{\text{el}}\times 3}$ is generated by sampling from a proposal function q. Subsequently, the acceptance probability is computed, and the proposed state is accepted if this probability exceeds a uniformly distributed random value. Thus, given a suitable proposal function q, the algorithm produces a sample $\mathbf{r}_n$ that is distributed according to p_θ in the limit of $n \to \infty$. Intuitively, over multiple steps n the sample tends to move towards high-probability regions because proposals towards higher probability are always accepted, while proposals towards lower-probability regions are often rejected. Because there is some chance to accept proposals towards low-probability regions, the algorithm does not only return the value with highest probability, but a distribution of samples. Fig. 5a depicts the convergence of the distribution of samples towards the target distribution p_θ for a simple 1D example.

Algorithm 1 Metropolis-Hastings sampling

Require: Probability density $p_\theta(\mathbf{r})$, proposal distribution $q(\mathbf{r}_p|\mathbf{r}_n)$, initial configurations $\mathbf{r}_0$, number of steps N

for $n = 0$ to $N-1$ **do**

  $\mathbf{r}_p \sim q(\mathbf{r}_p|\mathbf{r}_n)$ ▷ Propose new configuration $\mathbf{r}_p$

  $a = \min\left(1, \frac{p_\theta(\mathbf{r}_p)q(\mathbf{r}_n|\mathbf{r}_p)}{p_\theta(\mathbf{r}_n)q(\mathbf{r}_p|\mathbf{r}_n)}\right)$ ▷ Compute acceptance probability a

  if $a \geqslant$ RandomUniform(0, 1) **then**

    $\mathbf{r}_{n+1} \leftarrow \mathbf{r}_p$ ▷ Accept the proposal with probability a

  else

    $\mathbf{r}_{n+1} \leftarrow \mathbf{r}_n$ ▷ Reject the proposal with probability $1-a$

  end if

end for

return $\mathbf{r}_N$ ▷ In the limit of $N \to \infty$, $\mathbf{r}_N$ is distributed according to $p_\theta(\mathbf{r})$

A common choice for the proposal function $q(\mathbf{r}_p|\mathbf{r}_n)$ is a multivariate Gaussian distribution centered around $\mathbf{r}_n$ and variance s^2

$$q(\mathbf{r}_p|\mathbf{r}_n) \propto \exp\left(-\frac{1}{2s^2}|\mathbf{r}_p - \mathbf{r}_n|^2\right) \tag{61}$$

where s is a tuneable parameter known as the stepsize. It is a valid choice, because in principle any configuration can be reached from any other configuration in a single step (since the Gaussian distribution has support on the whole domain) and thus the proposal satisfies ergodicity. Furthermore the fact that the Gaussian is symmetric in $\mathbf{r}_p$ and $\mathbf{r}_n$ allows to omit the q-ratio on the calculation of the acceptance rate since it is always 1. A slight modification to Algorithm 1 is to divide the number of electrons into multiple blocks of electrons and separately accepting / rejecting a subset of electron positions. This potentially reduces the number of steps one needs to perform to reach more decorrelated electron positions (von Glehn et al., 2023).

One important aspect of the Metropolis-Hastings algorithm is that its acceptance criterion only depends on the ratio of probability densities but not on p_θ directly. Therefore, one can consider instead the unnormalized density $|\Psi_\theta|^2$.

An alternative to a Gaussian Proposal distribution is to bias proposals towards increasing probability density to increase the probability of acceptance. This is known as Metropolis Adjusted Langevin Algorithm (MALA) and increases sampling efficiency (Schätzle et al., 2023) at the expensive of higher computational cost to evaluate ∇p_θ.

$$\mathbf{r}_p = \mathbf{r}_n + \tau \nabla_{\mathbf{r}} \log p_\theta(\mathbf{r}_n) + s\mathcal{N}(\mathbf{0}, 1) \tag{62}$$

The choice of the stepsize s is important to achieve fast convergence and mixing of the Markov Chain: Choosing a very small stepsize only allows very small changes in $\mathbf{r}$, leading to slow convergence. Choosing a very large stepsize leads to proposed configurations $\mathbf{r}_p$ that are far from the original configuration $\mathbf{r}_n$ and are very often rejected, thus not moving at all. To address this issue, one can set a target acceptance rate of around 50% and automatically adjust s to approximately reach this acceptance rate.

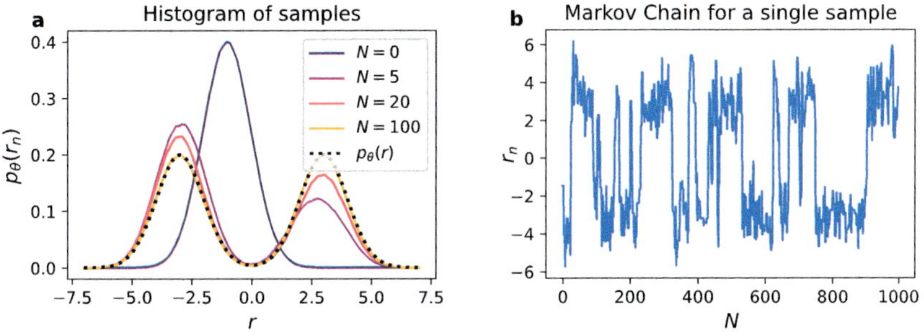

FIGURE 5 1D example of MCMC on a 1D density $p_\theta(\mathbf{r})$ consisting of two Gaussians. The initial configurations $\mathbf{r}_0$ are drawn from a single Gaussian distribution. **a**: Histogram of samples after different number of MCMC steps N. After $N \approx 100$ steps the distribution of $\mathbf{r}_N$ aligns with the target distribution p_θ. **b**: Path of a single sample. Subsequent samples are strongly correlated, depicting two distinct time-scales: A short time-scale corresponding to moves within a density peak and a long time scale corresponding to moves between the two peaks.

In principle any initial distribution can be used for the samples $\mathbf{r}_0$, but choosing an initial distribution that resembles the target distribution is obviously advantageous. Therefore the typical approach is to first take a large number of steps $N_{\text{burn-in}}$ for the samples to converge to the target distribution. Then, to obtain more samples one does not start again from the initial distribution, but uses the latest sample as starting point for the next N_{intermed} steps to obtain a new sample. In practice $N_{\text{burn-in}} \gg N_{\text{intermed}}$, for sampling from a wavefunction of a small molecule $N_{\text{burn-in}} \approx 10^3$, while $N_{\text{intermed}} \approx 10^1$. A disadvantage of this approach is that subsequent samples are not fully independent of each other, but can still be correlated if N_{intermed} is too small. Fig. 5b shows the trajectory of a single sample as a function of Metropolis-Hastings steps, clearly showing correlations between subsequent samples. This issue is particularly pronounced when p_θ has multiple maxima that are separated by regions of low probability, because it takes many steps to transition between these maxima.

Besides using Metropolis-Hastings to sample from $|\Psi_\theta|^2$, one can in principle also design models which allow direct sampling from the probability distribution.

One option are Normalizing Flows, a type of model that maps an easy to sample probability distribution (e.g. a Gaussian) to the target probability distribution p_θ. It has been applied to the Schrödinger Equation, but only in the substantially simplified 1D case (Thiede et al., 2023).

Another option are autoregressive models, which generate a full configuration of electrons one electron at a time, by conditioning the probability distribution on all previously added electrons:

$$p_\theta(\mathbf{r}_1, \ldots \mathbf{r}_{n_{\text{el}}}) = p_\theta(\mathbf{r}_1)\, p_\theta(\mathbf{r}_2|\mathbf{r}_1)\, p_\theta(\mathbf{r}_3|\mathbf{r}_1, \mathbf{r}_2) \ldots p_\theta(\mathbf{r}_{n_{\text{el}}}|\mathbf{r}_1, \ldots \mathbf{r}_{n_{\text{el}}-1}) \tag{63}$$

Instead of sampling from the probability distribution of the left hand side of Eq. (63), which is $3 \times n_{\text{el}}$-dimensional, all at once, one samples n_{el} times from a 3-dimensional probability distribution (each term on the right hand side of Eq. (63)). This structure is the currently dominant paradigm in large language models, which autoregressively sample one token / word at a time, from a probability distribution which is conditioned on the previously generated tokens (Radford et al., 2021). This approach has also been applied to wavefunctions, but so far only for model Hamiltonians (Hibat-Allah et al., 2020) and molecules in second quantization (Barrett et al., 2022). In both cases the state space is discretized, simplifying the sampling from the low-dimensional conditional probability distributions.

3.3 Optimization

After discussing the variational principle and the sampling technique for the high-dimensional integral, we can now delve into the problem of optimizing for the parameters θ. Applying the variational principle, we can utilize the formulation provided in Eq. (44) as our loss function:

$$L(\theta) = \langle E \rangle_{\mathbf{r}\sim\Psi_\theta^2} := \mathbb{E}_{\mathbf{r}\sim\Psi_\theta^2}\left[\frac{\mathcal{H}\Psi_\theta(\mathbf{r})}{\Psi_\theta(\mathbf{r})}\right]. \tag{64}$$

To perform gradient descent, the computation of the gradient with respect to the parameters would in general require third derivatives: second derivatives with respect to $\mathbf{r}$ for the kinetic energy and first derivatives with respect to θ. Furthermore, stochastic gradient descent potentially leads to a biased estimator because the sampling process depends on the parameterized wave function Ψ_θ. Fortunately, one can exploit the hermiticity of the Hamiltonian to rewrite the gradient with respect to the parameters of the loss function as:

$$\nabla_\theta L(\theta) = 2\left\langle\left(\frac{\mathcal{H}\Psi_\theta(\mathbf{r})}{\Psi_\theta(\mathbf{r})} - L(\theta)\right)\nabla_\theta \log|\Psi_\theta|\right\rangle_{\mathbf{r}\sim p_\theta(\mathbf{r})}. \tag{65}$$

Eq. (65) allows the computation of an unbiased estimator for stochastic gradient descent with, at most, second derivatives. Additionally, in the case that Ψ_θ represents the true wave function, the Monte Carlo estimator has zero variance due to the local energy $\frac{H\Psi_\theta(\mathbf{r})}{\Psi_\theta(\mathbf{r})}$ being spatially constant and equal to $L(\theta)$. A full derivation of the gradient can be found for example in Inui et al. (2021). The energy can be minimized using gradient based optimizers, such as Stochastic

Gradient Descent (SGD) or the Adam optimizer (Kingma and Ba, 2017), used in many deep-learning applications. The update rule for SGD with learning rate λ is given by

$$\theta_{t+1} = \theta_t - \lambda \nabla_\theta L. \tag{66}$$

Convergence of optimization can be substantially accelerated by not using the energy gradient directly as in (66), but rather preconditioning it with the following matrix $\mathbf{S} \in \mathbb{R}^{n_{\text{param}} \times n_{\text{param}}}$:

$$S_{\mu\nu} := \left\langle \frac{\partial \log|\Psi|}{\partial \theta_\mu} \frac{\partial \log|\Psi|}{\partial \theta_\nu} \right\rangle - \left\langle \frac{\partial \log|\Psi|}{\partial \theta_\mu} \right\rangle \left\langle \frac{\partial \log|\Psi|}{\partial \theta_\nu} \right\rangle \tag{67}$$

and then using this preconditioned gradient for stochastic gradient descent

$$\theta_{t+1} = \theta_t - \lambda \mathbf{S}^{-1} \nabla_\theta L. \tag{68}$$

Here, and in what follows, we use the notation

$$\langle \Phi \rangle := \mathbb{E}_{\mathbf{r} \sim \Psi_\theta^2}[\Phi]$$

for any $\Phi : \mathbb{R}^{n_{\text{el}} \times 3} \to \mathbb{R}$.

The update rule (68) is known as *Stochastic Reconfiguration* in the physics community (Becca and Sorella, 2017) (where $\mathbf{S}$ is then referred to as the *Quantum Geometric Tensor*) and is very closely related to *Natural Gradient Descent* in the machine learning community (Martens and Grosse, 2015) (where an object closely related to $\mathbf{S}$ is referred to as the Fisher information matrix). The following argument, adapted from Becca and Sorella (2017), should give some perspective on why using $\mathbf{S}$ as a preconditioner is a sensible choice.

Stochastic reconfiguration as a local metric

When performing SGD, a crucial choice is the stepsize λ. One way of formulating this is to consider at every step as loss $\mathcal{L}$ the original loss $L(\theta)$ plus an additional regularization term, which penalizes large changes δ in parameter space.

$$\mathcal{L}^{\text{SGD}}(\delta) = L(\theta + \delta) + \frac{\lambda}{2} \delta^T \delta \tag{69}$$

$$\delta := \theta_{t+1} - \theta_t \tag{70}$$

When minimizing Eq. (69) with respect to all parameter updates δ_μ the classical SGD update rule is recovered:

$$\frac{\partial \mathcal{L}^{\text{SGD}}}{\partial \delta_\mu} = \frac{\partial L}{\partial \delta_\mu} + \lambda \delta_\mu \stackrel{!}{=} 0 \tag{71}$$

$$\delta = -\lambda \nabla_\theta L. \tag{72}$$

SGD with a given learning-rate λ therefore minimizes the energy, while at the same time minimizing the Euclidean norm of the parameter update. While this is not an unreasonable choice per se, it would be better to minimize the energy, while making *minimal changes to the wavefunction*. After all, the wavefunction might be very sensitive to some parameters and insensitive to others. We would therefore like to make small steps for sensitive parameters and larger steps for insensitive parameters.

The following metric can be used to assess the distance between two unnormalized real-valued wavefunctions Ψ and Φ:

$$s(\Phi, \Psi)^2 = 1 - \frac{\langle \Psi, \Phi \rangle^2}{\langle \Psi, \Psi \rangle \langle \Phi, \Phi \rangle} \tag{73}$$

Eq. (73) corresponds to 1 minus the squared overlap of the normalized wavefunctions and is thus 0 for $\Phi \equiv \Psi$ and 1 for $\Phi \perp \Psi$. Using Eq. (73) as a metric to regularize the loss yields

$$\mathcal{L}^{\text{SR}}(\delta) = L(\theta + \delta) + \lambda \left(1 - \frac{\langle \Psi_\theta, \Psi_{\theta+\delta} \rangle^2}{\langle \Psi_\theta, \Psi_\theta \rangle \langle \Psi_{\theta+\delta}, \Psi_{\theta+\delta} \rangle} \right). \tag{74}$$

The updated wavefunction $\Psi_{\theta+\delta}$ can be expressed as Taylor expansion up to first order,

$$\Psi_{\theta+\delta} \approx \Psi_\theta + \delta^T \nabla_\theta \Psi_\theta, \tag{75}$$

yielding (subscripts θ omitted for clarity):

$$\mathcal{L}^{\text{SR}}(\delta) = \langle E \rangle + \lambda \left(1 - \frac{\langle \Psi, \Psi + \delta^T \nabla_\theta \Psi \rangle^2}{\langle \Psi, \Psi \rangle \langle \Psi + \delta^T \nabla_\theta \Psi, \Psi + \delta^T \nabla_\theta \Psi \rangle} \right). \tag{76}$$

Expanding the regularization term, dividing the denominator and enumerator by $\langle \Psi, \Psi \rangle^2$, and introducing O yields

$$O := \frac{\nabla_\theta \Psi}{\Psi} = \nabla_\theta \log |\Psi| \tag{77}$$

$$\mathcal{L} = \langle E \rangle + \lambda \left(1 - \frac{\left(1 + \delta^T \frac{\langle \Psi, \nabla_\theta \Psi \rangle}{\langle \Psi, \Psi \rangle} \right)^2}{1 + 2\delta^T \frac{\langle \Psi, \nabla_\theta \Psi \rangle}{\langle \Psi, \Psi \rangle} + \delta^T \frac{\langle \nabla_\theta \Psi, \nabla_\theta \Psi \rangle}{\langle \Psi, \Psi \rangle} \delta} \right) \tag{78}$$

$$= \langle E \rangle + \lambda \left(1 - \frac{\left(1 + \langle \delta^T O \rangle \right)^2}{1 + 2\delta^T \langle O \rangle + \delta^T \langle O O^T \rangle \delta} \right) + \mathcal{O}(|\delta|^3). \tag{79}$$

Expanding the denominator up to second order in δ (using $(1+x)^{-1} \approx 1 - x + x^2$) and multiplying all terms finally yields

$$\mathcal{L}^{\text{SR}}(\delta) = \langle E \rangle + \lambda \delta^T \mathbf{S} \delta + \mathcal{O}(|\delta|^3) \tag{80}$$

$$\mathbf{S} = \left\langle O\,O^T \right\rangle - \langle O \rangle \langle O \rangle^T \tag{81}$$

The regularized loss in Eq. (80) has the same structure as (69): The original loss + a quadratic regularization term – the only difference being that this time the metric is given by **S** instead of the Euclidean-norm. When minimizing this regularized loss one obtains the stochastic reconfiguration update rule (up to a factor of 2 in the learning rate)

$$\delta = -2\lambda \mathbf{S}^{-1} \nabla_\theta L. \tag{82}$$

Note that **S** has been motivated here using (73) as distance metric. The same result (up to a constant factor) can be obtained by expanding the Kullback-Leibler divergence – a well known divergence to measure the distance between probability distribution – between the probability distributions $|\Psi_\theta^2|$ and $|\Psi_{\theta+\delta}|^2$. If the probability distributions are normalized then $\langle O \rangle \equiv 0$, simplifying (81) to the first term and the corresponding update rule of *natural gradient descent.*

Toy example: SR for 2-parameter system

Fig. 6 demonstrates the effect of this preconditioning on a 1D-example (a single particle in a parabolical potential) with a wavefunction that has only two parameters:

$$\Psi(x) = e^{-\frac{1}{2}\left(\frac{x-\theta_2}{2\sigma(\theta_1)}\right)^2}, \tag{83}$$

with the sigmoid function $\sigma(\theta) = \frac{1}{1+e^{-\theta}}$. This system has its ground-state at $\theta_1 = \theta_2 = 0$, depicted as Ψ_{GS} in Fig. 6a and the corresponding point in parameter-space θ_{GS} in Fig. 6b.

Starting from an arbitrary initial wavefunction Ψ_0 (and corresponding parameters θ_0), two distinct new wavefunctions (and corresponding parameters) are depicted. The parameters θ_{SGD} (and its wavefunction Ψ_{SGD}) are obtained from the gradient descent update rule $\theta_{\mathrm{SGD}} = \theta_0 - \lambda_{\mathrm{SGD}} \nabla_\theta E$. The parameters θ_{SR} (and its wavefunction Ψ_{SR}) are obtained from the stochastic-reconfiguration update rule $\theta_{\mathrm{SR}} = \theta_0 - \lambda_{\mathrm{SR}} \mathbf{S}^{-1} \nabla_\theta E$. The learning rates λ_{SGD} and λ_{SR} are chosen such that the Euclidean distance in parameter space is identical in both case (as depicted in Fig. 6c). However, the change of the wavefunction is markedly different: The SR-update rule leads to a much smaller change in the wavefunction (compared to the SGD update rule). This can be seen from the smaller S-distance in Fig. 6d, as well as visually when comparing Ψ_{SR} and Ψ_{SGD} in Fig. 6a. Overall this leads to a lower energy after the update step (as can be seen in Fig. 6b) and will lead to overall faster convergence towards the ground-state when iterating. The effect of the preconditioning is that the SR-update skews the update step towards larger updates along the θ_1 parameter, which is less sensitive in this point of the parameter space.

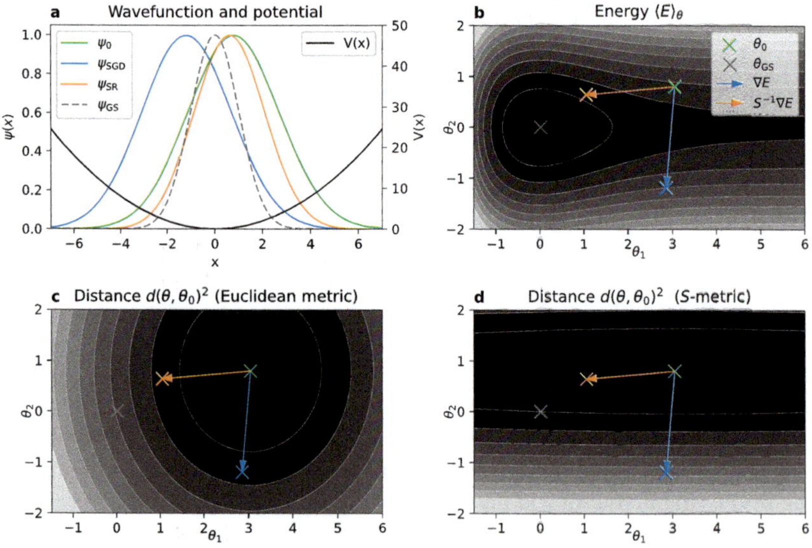

FIGURE 6 1D toy example for 2-parameter wavefunction: a) Plot of ground-state wavefunction Ψ_{GS}, initial wavefunction Ψ_0, and the resulting wavefunctions after update steps according to stochastic gradient descent (SGD) and stochastic reconfiguration (SR). b-d) Contour-plots as a function of wavefunction parameters θ_1 and θ_2 (darker colors correspond to lower values). b) Energy expectation value of corresponding wavefunction. c,d) Distance from initial parameter vector θ_0 measured in Euclidean metric and the metric induced by the preconditioner S.

Practical considerations for stochastic reconfiguration

Computing S^{-1} can be practically challenging. The first complication is that when $\mathbf{S}$ is being estimated from N_s samples it is at most of rank N_s:

$$S_{\mu\nu} = \langle O_\mu O_\nu \rangle - \langle O_\mu \rangle \langle O_\nu \rangle \tag{84}$$

$$= \Big\langle \Big(O_\mu - \langle O_\mu \rangle\Big)\Big(O_\nu - \langle O_\nu \rangle\Big)\Big\rangle \tag{85}$$

$$\approx \sum_{n=1}^{N_s} \Big(O_\mu(\mathbf{r}_n) - \bar{O}_\mu\Big)\Big(O_\nu(\mathbf{r}_n) - \bar{O}_\nu\Big) \tag{86}$$

Therefore for typical values of $N_s \approx 10^3$ and $n_{\text{param}} \approx 10^6$, $\mathbf{S}$ is rank deficient and cannot be inverted. The latter problem is typically addressed via Tikhonov regularization with a small damping constant ϵ:

$$\mathbf{S}_{\text{reg}} = \mathbf{S} + \epsilon\, \mathbb{1}_{N_s} \tag{87}$$

Another approach is to estimate $\mathbf{S}$ not only from the current batch, but as a moving average of the estimates from past batches, thus increasing the rank of the estimator. This helps to reduce Monte Carlo noise in the estimation but typically still requires regularization to avoid a singular matrix **S**. The second complication arises due to the size of **S**, which is of dimension $n_{\text{param}} \times n_{\text{param}}$. Therefore, for a neural network wavefunction with $\approx 10^6$ parameters, even storing this matrix with $\approx 10^{12}$ elements becomes impossible. Even worse, this large matrix must be inverted, an operation that has computational cost $\mathcal{O}({n_{\text{param}}}^3)$ using Gaussian elimination. There are two viable routes in practice: Find a (sparse) approximation of **S** and invert it exactly, or find a way to approximately invert **S** without fully materializing **S**.

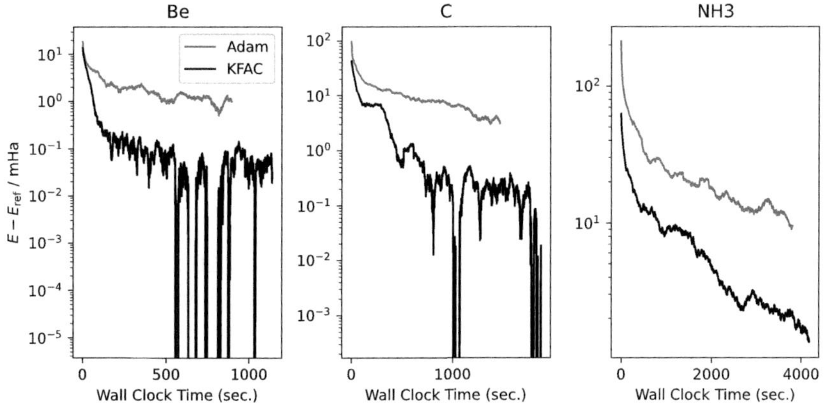

FIGURE 7 Comparing the optimizers Adam (grey) and KFAC (black) for three molecules (Beryllium, Carbon, and Ammonia). The x-axis represents the wall clock time in seconds, and the y-axis (log-scale) the energy error with respect to a highly accurate reference calculation in millihartree (mHa).

KFAC (Kronecker-Factored Approximation of Curvature) (Martens and Grosse, 2015) is of the first type, making two approximations to **S**. First it assumes that there are no dependencies between parameters belonging to different layers of the neural network, effectively assuming **S** to be block-diagonal. Second it assumes that each remaining block can be expressed as an eponymous Kronecker product of two smaller matrices. This allows inversion of the approximated **S** via inversion of many small matrices, which is computationally feasible even for networks involving millions of parameters. KFAC has first been applied to Neural Network wavefunctions by Pfau et al. (2020) and since been used widely throughout the neural wavefunction community. Compared to nonpreconditioned methods it yields substantially faster rates of convergence as depicted in Fig. 7. While it is computationally efficient and can yield good results, it has two downsides in practice. First, it involves approximations that cannot be systematically improved upon. Second, the optimizer does not only require access to the wavefunction, energies and their gradients, but also requires access to intermediate activations and gradients of the model. This can introduce substantial complexity for practical implementation and leads to some unwanted coupling between the wavefunction model and the optimizer.

An alternative approach is to make no approximations to $\mathbf{S}$, but to only invert it approximately. One common approach is to use the Conjugate Gradient (CG) method to compute $\mathbf{S}^{-1}\nabla E$ without materializing $\mathbf{S}$. CG only requires the repeated evaluation of the matrix-vector product $\mathbf{Sx}$ for arbitrary vectors $\mathbf{x}$. This can be obtained by a vector-jacobian-product (VJP) followed by a jacobian-vector-product (VJP), which are implemented using back-propagation and forward-mode differentiation respectively and don't require materializing the full jacobian.

Another approach is to use the fact that the regularized $\mathbf{S}_{\text{reg}}$ is a sum of a full-rank, but easy to invert diagonal matrix ($\epsilon\,\mathbb{1}$), and a rank-N_s matrix $\mathbf{S}$. Inversion of $\mathbf{S}_{\text{reg}}$ can therefore be done using the Sherman-Morrison-Woodbury formula (Woodbury, 1950), which only requires the inversion of a $N_s \times N_s$ matrix. This forms the basis for the MinSR (Rende et al., 2023) and SPRING optimizer (Goldshlager et al., 2024).

Supervised pretraining

In addition to the variational optimization, Pfau et al. (2020) proposed to perform a supervised optimization phase with respect to a reference method. This step, referred to as *supervised pretraining* in the following, minimizes the difference between the neural network orbitals and reference orbitals (we omit the spin dependence in the notation)

$$L^{\text{pre}}(\theta) = \mathbb{E}_{\mathbf{r}\sim p_\theta(\mathbf{r})}\left[\sum_{k=1}^{n_{\text{el}}}\sum_{i=1}^{n_{\text{el}}}\left(\phi_k^{\text{ref}}(r_i) - \phi(\theta_i)(r_i)\right)^2\right], \tag{88}$$

whereas the neural network-based orbitals are calculated as described in (119) and (120). As reference calculation, a mean field solution such as Hartree Fock, is usually used, where each orbital depends on a single electron. Pfau et al. (2020) argued that supervised pretraining can improve convergence and numerical stability of the subsequent variational optimization. Compared to variational optimization, supervised pretraining is computationally cheaper because it does not require evaluation of the Hamiltonian; in particular it avoids the derivatives associated with the kinetic energy. The downside of supervised pretraining is that it unlike variational optimization it requires a reference method. Furthermore excessive pretraining can bias the network initialization, leading to less accurate results after subsequent variational optimization (Gerard et al., 2022).

4 Deep learning VMC

In this section we introduce the DL-VMC method. In particular, we review how neural networks can be used to construct suitable parametrized orbitals (55) that can serve as a numerical ansatz in the VMC problem (44).

4.1 Multilayer perceptrons

The multi layer perceptron (MLP) forms the basic building block of all deep-learning based architectures. It consists of L layers, alternating between affine transformations and elementwise nonlinear functions. Given an input $x^0 \in \mathbb{R}^{d_0}$, the output of each subsequent layer $l = 1 \ldots L$ is computed as

$$y_n^l = \sum_m w_{nm}^l x_m^{l-1} + b_m^l \qquad l = 1 \ldots L \tag{89}$$

$$x_n^l = \sigma(y_n^l) \qquad l = 1 \ldots L-1 \tag{90}$$

$$\text{MLP}\left(x^0\right) := y^L, \tag{91}$$

with trainable weights $w^l \in \mathbb{R}^{d_l \times d_{l-1}}$, $b^l \in d_l$ for every layer l and a nonlinear function σ, referred to as *activation function*. Common choices for this activation function include

$$\sigma(x) = \tanh(x) \tag{92}$$

$$\sigma(x) = \text{SiLU}(x) = \frac{x}{1+e^{-x}} \tag{93}$$

Note that $\text{ReLU}(x) := \max(x, 0)$, which is used extensively in many deep-learning applications, is not used for neural wavefunctions, because its first derivative is discontinuous and its second derivative (required for the kinetic energy of $\mathcal{H}^{\text{BO}}$) is zero everywhere.

4.2 Overall structure of neural network wavefunctions

To model the functions $\Phi_{\uparrow}^{k,i}(\mathbf{r})$ (52) and $\Phi_{\downarrow}^{k,i}(\mathbf{r})$ (53), several different architectures have been proposed, which all follow the following structure. All functions in this section can depend on parameters θ, but we suppress this index for clarity.

1. **Input features**: Compute features $x_i \in \mathbb{R}^{n_{\text{feat}}^{\text{el}}}$ for single-electrons and $p_{ij} \in \mathbb{R}^{n_{\text{feat}}^{\text{el-el}}}$ for pairs of electrons i, j from the coordinates $\mathbf{r}$ and spins σ.

$$x_i = f^{\text{el}}(r_i, \sigma_i, (\mathbf{R}, \mathbf{Z})) \qquad i = 1 \ldots n_{\text{el}} \tag{94}$$

$$p_{ij} = f^{\text{el-el}}(r_i, \sigma_i, r_j, \sigma_j) \qquad i, j = 1 \ldots n_{\text{el}} \tag{95}$$

Analogous to the multisets $\{\mathbf{r}^{\uparrow}\}$, $\{\mathbf{r}^{\downarrow}\}$ defined in (51) we define the following multisets of features:

$$\{p_i^{\uparrow}\} := \{(x_1, p_{i,1}), \ldots, (x_{n_\uparrow}, p_{i,n_\uparrow})\},$$
$$\{p_i^{\downarrow}\} := \{(x_{n_\uparrow+1}, p_{i,n_\uparrow+1}), \ldots, (x_{n_{\text{el}}}, p_{i,n_{\text{el}}})\} \tag{96}$$

$$\{p_{i,\setminus i}^{\uparrow}\} := \{p_i^{\uparrow}\} \setminus \{(x_i, p_{i,i})\},$$
$$\{p_{i,\setminus i}^{\downarrow}\} := \{p_i^{\downarrow}\} \setminus \{(x_i, p_{i,i})\} \tag{97}$$

2. **Embedding**: Compute a high-dimensional embedding $h_i \in \mathbb{R}^{d_{\text{emb}}}$ for each electron i. These embeddings not only depend on the input features of electron i but also on the multisets of input features of all other electrons.

$$h_i = h\left(x_i, \{p_{i,\setminus i}^{\uparrow}\}, \{p_{i,\setminus i}^{\downarrow}\}\right), \qquad i = 1 \dots n_{\text{el}} \tag{98}$$

3. **Orbitals**: Compute entries $\phi_{\uparrow\downarrow}^{d,k,i}$ from the embeddings h_i for each orbital k and determinant d.

$$\phi_{\uparrow}^{d,k,i} = \phi_{dk}^{\uparrow}(h_i) \qquad i = 1 \dots n_{\uparrow}, \quad k = 1 \dots n_{\text{el}}, \quad d = 1 \dots n_{\text{det}} \tag{99}$$

$$\phi_{\downarrow}^{d,k,i} = \phi_{dk}^{\downarrow}(h_i) \quad i = n_{\uparrow} + 1 \dots n_{\text{el}}, \quad k = 1 \dots n_{\text{el}}, \quad d = 1 \dots n_{\text{det}} \tag{100}$$

4. **Slater determinant**: Compute Ψ as a sum of Slater determinants of these orbitals $\phi_{\uparrow}^{d,k,i}$ (cf. (54)) and optionally multiply it with a Jastrow factor J that is invariant under permutation of electrons with the same spin.

$$\begin{aligned} \Psi = {} & J\left(\{h_i\}_{i=1\dots n_{\uparrow}}, \{h_i\}_{i=n_{\uparrow}+1\dots n_{\text{el}}}\right) \\ & \times \sum_d \det\left(\phi_{\uparrow}^{d,1}, \dots, \phi_{\uparrow}^{d,n_{\uparrow}}, \phi_{\downarrow}^{d,n_{\uparrow}+1}, \dots, \phi_{\downarrow}^{d,n_{\text{el}}}\right) \end{aligned} \tag{101}$$

Note that by construction the embeddings h_i are equivariant under permutation of electrons of the same spin and therefore the whole ansatz satisfies antisymmetry. In the following we discuss a few common choices for each of these four steps.

4.3 Input features

Pairwise features

For the pairwise features p_{ij}, typically 3D difference vectors $r_i - r_j$ and the pairwise distances $|r_i - r_j|$ are being used. Some ansätze (Pfau et al., 2020; Gerard et al., 2022; Gao and Günnemann, 2022) use a simple concatenation (denoted by $[\cdot, \cdot]$) of these features

$$p_{ij}^{\text{concat}} = \left[r_i - r_j, |r_i - r_j|\right], \qquad p_{ij}^{\text{concat}} \in \mathbb{R}^4. \tag{102}$$

Others additionally include trainable functions of these distances and difference vectors. For example, Gao and Günnemann (2023a) proposed

$$\begin{aligned} p_{ij,\nu}^{\text{MOON}} &= \text{MLP}_{\nu}(r_i - r_j) \sum_{\mu} W_{\nu\mu} \exp\left(-\frac{|r_i - r_j|^2}{\zeta_{\mu}}\right) \\ \nu &= 1 \dots n_{\text{feat}}^{\text{el-el}}, \ \mu = 1 \dots n_{\text{filters}} \end{aligned} \tag{103}$$

It may seem redundant to include both $r_i - r_j$ and $|r_i - r_j|$ as an input feature, because the norm is just a function of $r_i - r_j$. But including $|r_i - r_j|$ which has discontinuous derivatives at $r_i = r_j$ allows the network to model functions that are not smooth at $r_i = r_j$, which is required to satisfy the Kato cusp conditions (Kato, 1957). Early approaches (Hermann et al., 2020) used only the distances $|r_i - r_j|$ as input features, but this has been recognized to be insufficiently expressive (Gerard et al., 2022).

For large molecules the interparticle distances $|r_i - r_j|$ can become very large, making training of neural networks using them numerically challenging. von Glehn et al. (2023) proposed to logarithmically scale the interparticle distances and differences to alleviate this problem:

$$\widetilde{r}_{ij} = \log\left(1 + |r_i - r_j|\right) \tag{104}$$

$$p_{ij}^{\log} = \left[\frac{\widetilde{r}_{ij}}{|r_i - r_j|}\left(r_i - r_j\right), \widetilde{r}_{ij}\right], \qquad p_{ij}^{\log} \in \mathbb{R}^4. \tag{105}$$

Single electron features

The single electron features x_i are typically computed as functions of the electron-nuclei distances and differences. FermiNet (Pfau et al., 2020) proposed a simple concatenation of all electron-nucleus pairs

$$x_i^{\text{concat}} = \left[r_i - R_1, |r_i - R_1|, \ldots, r_i - R_{n_{\text{nuc}}}, |r_i - R_{n_{\text{nuc}}}|\right], \qquad p_i^{\text{concat}} \in \mathbb{R}^{4n_{\text{nuc}}}, \tag{106}$$

whereas Gao and Günnemann (2022); Gerard et al. (2022) use sums of trainable functions of the differences and distances.

$$x_i^{\text{MLP}} = \sum_{J=1}^{n_{\text{nuc}}} \text{MLP}\Big([r_i - R_J, |r_i - R_J|]\Big) \tag{107}$$

While some architectures (von Glehn et al., 2023) encode spin explicitly as a feature $x_i = \sigma_i$, many others do not encode spin explicitly, but instead opt for different subsequent embedding functions depending on spin.

4.4 Embedding

The role of the embedding network is to take simple input features x_i and p_{ij} and compute embeddings h_i that form expressive basis functions for the subsequent many-body orbitals. To do this, the embedding network must on the one hand be able to incorporate information from all other electrons $i \neq j$, and on the other hand be able to represent arbitrary functions of a single electron i. These two requirements are typically addressed by interleaving two kinds of computation over multiple rounds l: A function f that gathers information from other electrons and function that acts only on a single electron (typically implemented as MLP or a single affine transformation). Most embedding networks thus follow the following structure:

$$h_i^0 = x_i \qquad \text{Initialization} \tag{108}$$

$$m_i^{\uparrow,l} = \sum_{j\in\{\uparrow\backslash i\}} f^{\uparrow}(h_i^{l-1}, h_j^{l-1}, p_{ij}) \qquad \text{Gather information across electrons}$$
$$m_i^{\downarrow,l} = \sum_{j\in\{\downarrow\backslash i\}} f^{\downarrow}(h_i^{l-1}, h_j^{l-1}, p_{ij}) \tag{109}$$

$$h_i^l = \text{MLP}\left(\left[h_i^{l-1},\ m_i^{\uparrow,l},\ m_i^{\downarrow,l}\right]\right) \qquad \text{Single-electron computation} \tag{110}$$

The sets $\{\uparrow \backslash i\}$ and $\{\downarrow \backslash i\}$ correspond to the electron-indices of all electrons of a given spin excluding i:

$$\{\uparrow \backslash i\} := \{1 \dots n_\uparrow\} \setminus \{i\}, \qquad \{\downarrow \backslash i\} := \{n_\uparrow + 1 \dots n_{\text{el}}\} \setminus \{i\} \tag{111}$$

The initial embeddings are $h^0 \in \mathbb{R}^{n_{\text{feat}}^{\text{el}}}$, and $h^l \in \mathbb{R}^{d_{\text{emb}}}$ for $l > 0$.

After iterating Eq. (109) and (110) for $l = 1 \dots L$, the final embeddings are given by the output of the last layer, i.e. $h_i = h_i^L$. A few design considerations are worth discussing:

- The *messages* $m_i^{\uparrow\downarrow,l}$ in Eq. (109) are a sum over all particles of a given spin. Since the sum is a permutation invariant operation, the resulting message is invariant under permutation of two electrons of the same spin. This ensures that h_i has the structure defined in (98), ultimately ensuring wavefunction antisymmetry.
- Not all parts of the network have the same impact on computational cost. While the functions $f(h_i, h_j, p_{ij})$ in Eq. (109) are evaluated for every pair of electrons – and thus have computational cost scaling as $\mathcal{O}({n_{\text{el}}}^2)$ – the MLP in Eq. (110) is only evaluated for every electron, thus scaling as $\mathcal{O}(n_{\text{el}})$. This difference in scaling is typically reflected by the fact that most architectures use wide (and thus costly) MLPs for the one-electron computations (Eq. (110)), and computationally cheaper functions for $f^\uparrow$ and $f^\downarrow$ (Eq. (109)).
- Some architectures differentiate between messages from up- and down-electrons (as denoted in Eq. (109)), while others differentiate between messages from spin-parallel or spin-antiparallel electrons. The latter choice enforces invariance w.r.t. exchanging all spin-up particles with spin-down particles and has been shown to be advantageous for closed-shell systems (Gao and Günnemann, 2023b).

Given this very general framework, the key difference between the various embedding architectures lies therefore in the message functions $f^\uparrow$, $f^\downarrow$, with a few common choices outlined below.

Hartree-Fock

If no information from other electrons is gathered (e.g. $f^{\uparrow} \equiv f^{\downarrow} \equiv 0$), the final embedding h_i only depends on the input features of that electron x_i. The network cannot capture any correlation effects and thus the best possible accuracy is Hartree-Fock.

FermiNet

In FermiNet (Pfau et al., 2020; Spencer et al., 2020) the message-function f simply concatenates the feature vectors of embedding h_j and an MLP of p_{ij}.

$$f^{\uparrow}(h_j, p_{ij}) = f^{\downarrow}(h_j, p_{ij}) = \left[h_j,\ \mathrm{MLP}\left(p_{ij}\right)\right] \tag{112}$$

FermiNet clearly improves upon a simple noninteracting embedding, by including in every layer information about all other electron embeddings as well as their relative positions. Note however that in FermiNet all electron embeddings h_j contribute equally to the message m_i, irrespective of the distance between electron i and j. This runs against physical intuition, which suggests that electrons at large separations would have a smaller influence.

Graph convolutional neural networks

In a Graph Convolutional Neural Network (GCN) (Zhou et al., 2020), the contributions of each electron j to the message m_i are weighted by their relative geometric positions encoded in p_{ij}. This weighting is commonly achieved using an elementwise product along the feature dimension, denoted by $\odot$:

$$f^{\uparrow}(h_j, p_{ij}) = \mathrm{MLP_h}\left(h_j\right) \odot \mathrm{MLP_p^{\uparrow}}\left(p_{ij}\right), \tag{113}$$

$$f^{\downarrow}(h_j, p_{ij}) = \mathrm{MLP_h}\left(h_j\right) \odot \mathrm{MLP_p^{\downarrow}}\left(p_{ij}\right). \tag{114}$$

Empirically, including graph convolutions can improve the accuracy and convergence of the ansatz compared to the purely MLP-based FermiNet (Gerard et al., 2022). One potential reason for the improved performance is that it enables the network to put larger weight on closer neighboring electrons than electrons which are far apart. Some approaches (Gao and Günnemann, 2023a) enforce this prior knowledge explicitly, by multiplying the $\mathrm{MLP}\left(p_{ij}\right)$ with functions that explicitly decay as a function of the distance between electrons i and j.

Self-attention based

Neither in FermiNet nor the GCN embedding does the message m_i explicitly depend on the message receiver h_i, but instead only depends on the message sender h_j and the pairwise features p_{ij}. Self-attention is an approach where the weighting of each message j is computed as an inner product between a query vectors q_i (derived from the receiving embedding h_i) and a key vector k_j (derived from the sending embedding h_j). For embeddings $h \in \mathbb{R}^{n_{\mathrm{el}} \times d_{\mathrm{emb}}}$ and trainable matrices $W^{\mathrm{q}}, W^{\mathrm{k}}, W^{\mathrm{v}} \in \mathbb{R}^{d_{\mathrm{emb}} \times d_{\mathrm{attn}}}$ the weights $w \in \mathbb{R}^{n_{\mathrm{el}} \times n_{\mathrm{el}}}$ and the corresponding message function are given as

$$q = hW^{\mathrm{q}}, \qquad k_j = hW^{\mathrm{k}}, \qquad v_j = hW^{\mathrm{q}}, \qquad q, k, v \in \mathbb{R}^{n_{\mathrm{el}} \times d_{\mathrm{attn}}} \tag{115}$$

$$\hat{w} = \exp\left(\frac{qk^T}{\sqrt{d_{\mathrm{attn}}}}\right), \qquad \hat{w} \in \mathbb{R}^{n_{\mathrm{el}} \times n_{\mathrm{el}}} \tag{116}$$

$$w_{ij} = \frac{\hat{w}_{ij}}{\sum_j' \hat{w}_{ij'}} \tag{117}$$

$$f^{\uparrow}(h_i, h_j) = f^{\downarrow}(h_i, h_j) = w_{ij} v_j. \tag{118}$$

The message m_i explicitly depends on the embedding for electron i and j, but no longer explicitly depends on their pairwise features p_{ij}. This geometric information must be inferred from the inner product of q_i and k_j, and thus requires that the single-electron input features contain information about their absolute positions. This architecture was implemented by von Glehn et al. (2023) and has been empirically shown to be among the most expressive ansätze.

4.5 Orbitals

Given permutation equivariant embeddings h_i, the elements $\phi_{\uparrow\downarrow}^{d,k,i}$ of the slater matrix are typically computed as

$$\phi_{\uparrow}^{d,k,i} = \left(W_{dk}^{\uparrow} \cdot h_i\right) \widetilde{\varphi}_{dk}^{\uparrow}(r_i), \qquad i = 1 \dots n_{\uparrow} \tag{119}$$

$$\phi_{\downarrow}^{d,k,i} = \left(W_{dk}^{\downarrow} \cdot h_i\right) \widetilde{\varphi}_{dk}^{\downarrow}(r_i), \qquad i = n_{\uparrow} + 1 \dots n_{\mathrm{el}} \tag{120}$$

where $W^{\uparrow\downarrow}$ are trainable matrices $W^{\uparrow}, W^{\downarrow} \in \mathbb{R}^{n_{\mathrm{det}} \times n_{\mathrm{el}} \times d_{\mathrm{emb}}}$, and $\widetilde{\varphi}$ is an envelope function enforcing that $\phi \to 0$ as $r_i \to \infty$.

For molecules the envelope function is typically expressed as a sum over nuclei, leading to

$$\widetilde{\varphi}_{dk}(r_i) = \sum_{J=1}^{n_{\mathrm{nuc}}} \varphi_{dkJ}(r_i). \tag{121}$$

The most common choice for the envelope function, originally proposed by Pfau et al. (2020) and simplified by Spencer et al. (2020) is *exponential envelopes*

$$\varphi_{dkJ}(r_i) = \pi_{dkJ}\, e^{-\alpha_{dkJ}|r_i - R_J|}, \tag{122}$$

with trainable parameters $\pi, \alpha \in \mathbb{R}^{n_{\mathrm{el}} \times n_{\mathrm{det}} \times n_{\mathrm{nuc}}}$.

An alternative proposed by Hermann et al. (2020) is using the single-particle orbitals from a Hartree-Fock calculation

$$\varphi_{dkJ}(r_i) = \varphi_{kJ}^{\mathrm{HF}}(r_i) = \sum_{\mu=1}^{n_{\mathrm{basis}}} c_{kJ\mu}\, b_\mu(r_i - R_J), \tag{123}$$

where $b_\mu : \mathbb{R}^3 \to \mathbb{R}$ are atom-centered basis functions and $c_{kJ\mu} \in \mathbb{R}$ are the expansion coefficients of orbital φ_k^{HF} in this basis.

Given that Hartree-Fock yields a good approximation of the groundstate wavefunction, one might think that using Hartree-Fock orbitals as envelopes provides a useful prior and good starting point for optimization. However, in practice the exponential envelopes are not only simpler to implement but also lead to substantially more accurate results (Gerard et al., 2022), potentially due to a bias introduced by the Hartree-Fock orbitals.

Even though using the HF-envelopes directly can decrease accuracy – and several groups that originally used them (Hermann et al., 2020; Scherbela et al., 2022), replaced them in later work with exponential envelopes (Gerard et al., 2022; Schätzle et al., 2023) – there is still information in the HF-orbitals which can be used: First, different HF-orbitals typically have different length-scales: Some orbitals (known by chemists as core orbitals) are tightly localized on an atom, whereas other orbitals (known by chemists as valence orbitals) are somewhat delocalized. This can be quantified and used to initialize the exponents α of the exponential envelopes, using large values for α to initialize strongly localized core orbitals and small values to initialize delocalized valence electrons. Numerical experiments show that this initialization accelerates wavefunction optimization (Gerard et al., 2022), in particular for heavy atoms where the length-scale of core orbitals differs by an order of magnitude from the length-scale of the valence electrons. Second, one can use the expansion coefficients of an HF-orbital as a descriptor of that orbital. This can be useful when designing a transferable wavefunction (cf. Sec. 4.7).

4.6 Jastrow factor

The wavefunction can be multiplied with a function $J(\mathbf{r}) : \mathbb{R}^{n_{\mathrm{el}} \times 3} \to \mathbb{R}$, which is invariant under permutations of electrons with the same spin, without affecting the total antisymmetry of the wavefunction. It is common to use a Jastrow-factor that does not alter the sign of Ψ, by enforcing $J > 0$ via $J = \exp(\hat{J})$ with an arbitrary permutation invariant function $\hat{J}$.

The Jastrow factor generally serves two purposes: Increasing expressivity of the wavefunction ansatz and enforcing the Kato cusp conditions (Kato, 1957). The first can be achieved by a permutation invariant pooling of the embeddings h_i, e.g. as

$$J = \exp\left(\sum_{i=1}^{n_\uparrow} \mathrm{MLP}^{\uparrow}(h_i) + \sum_{i=n_\uparrow+1}^{n_{\mathrm{el}}} \mathrm{MLP}^{\downarrow}(h_i) \right). \tag{124}$$

The latter refers to cusps that are present in the groundstate wavefunction whenever the positions of two particles coincide. The local energy $\frac{H\Psi}{\Psi}$ of the groundstate wavefunction is constant, but the individual terms in the Hamiltonian are not. In particular the potential energy terms in Eq. (3) diverge whenever the distance between two particles approaches zero. To obtain a constant local energy, the kinetic energy – given by the curvature of the wavefunction – must diverge with opposite sign, leading to discontinuous first derivatives of the wavefunction Ψ whenever two particles coincide. These cusps of high electron density (when an electron approaches a nucleus) or low electron density (when an electron approaches another electron) can be represented by an ansatz that has discontinuous derivatives at distance $|r_i - r_j| = 0$. A choice used by Hermann et al. (2020); von Glehn et al. (2023) is:

$$J^{\text{cusp}} = \exp\left(\sum_{i=1}^{n_{\text{el}}} \sum_{j=i+1}^{n_{\text{el}}} \frac{a}{b + |r_i - r_j|}\right) \tag{125}$$

with parameters a, b.

The full wavefunction Ψ is then given as

$$\Psi = J^{\text{cusp}} J \sum_{d=1}^{n_{\text{det}}} \det[\Phi_d] \tag{126}$$

4.7 Architectures for transferable wavefunctions

In many instances it is advantageous to have an ansatz which not only accurately represents $\Psi_{(\mathbf{R},\mathbf{Z})}(\mathbf{r})$ for a fixed geometry $(\mathbf{R}, \mathbf{Z})$, but which explicitly depends on the molecule and yields accurate groundstate wavefunctions across molecules. For example, when computing a Potential Energy Surface $\mathcal{E}_{\mathbf{Z}}(\mathbf{R})$, which requires finding the minimum eigenvalue for many instances of $\mathcal{H}^{\text{BO}}$, it can be more efficient to train a single transferable ansatz for all geometries, rather than training a separate ansatz for every geometry.

The architectures described in this section so far parameterize a wavefunction $\Psi_{(\mathbf{R},\mathbf{Z})}(\mathbf{r})$ which explicitly depends on the electron coordinates, but only implicitly depends on the molecular geometry contained in $(\mathbf{R}, \mathbf{Z})$. For example the nuclear coordinates $\mathbf{R}$ are used explicitly to compute input features in most architectures, but in many architectures the wavefunction Ψ does not explicitly depend on the nuclear charges $\mathbf{Z}$. Any realization of a wavefunction with parameters θ obtained through variational optimization will still implicitly depend on $\mathbf{Z}$ – because $\mathcal{H}^{\text{BO}}$ used throughout optimization depends on $\mathbf{Z}$ – but evaluating it for a molecule with different $\mathbf{Z}$, will not yield the correct groundstate wavefunction for this new molecule.

Several approaches for such transferable wavefunctions have been proposed by Gao and Günnemann (2022, 2023b,a); Scherbela et al. (2024, 2023), which make the following changes to the architecture outline in Sec. 4.2:

Input features and embedding

The input features explicitly depend on $(\mathbf{R}, \mathbf{Z})$. To encode the nuclear charges $\mathbf{Z}$, one-hot-encodings are typically used. The embeddings h_i no longer depend only on x_i, $\{p^{\uparrow}_{i,\setminus i}\}$, $\{p^{\downarrow}_{i,\setminus i}\}$, but also explicitly depend on the multiset of all nuclear positions and their nuclear charges:

$$h_i = h\left(x_i,\ \{p^{\uparrow}_{i,\setminus i}\},\ \{p^{\downarrow}_{i,\setminus i}\},\ \{(\mathbf{R}, \mathbf{Z})\}\right) \tag{127}$$

$$\{(\mathbf{R}, \mathbf{Z})\} := \{(R_1, Z_1),\ \ldots,\ (R_{n_{\text{nuc}}}, Z_{n_{\text{nuc}}})\} \tag{128}$$

This extra dependence is typically implemented in a similar fashion as the dependence on the set of other electrons, for example using graph convolutional networks or self-attention.

Orbitals

The widely used computation of orbitals in (119), (120) poses a challenge in designing transferable architectures, which generalize not only across different geometries $\mathbf{R}$, but also across molecules with different number of electrons. Because the dimensions of the trainable matrices $W^{\uparrow}, W^{\downarrow} \in \mathbb{R}^{n_{\text{el}} \times n_{\text{det}} \times d_{\text{emb}}}$ explicitly depend on n_{el} (corresponding to the number of orbitals in the Slater determinant), they cannot be transferred across different molecules with varying n_{el}. This is in contrast to other parts of the architecture (e.g. the GCN embedding) where the number of parameters does not depend on n_{el}, because all parameterized functions are only applied to objects corresponding to single electrons or pairs of electrons. The computational cost grows with system size – because the functions must be evaluated for more electrons / pairs of electrons – but the number of parameters is independent of n_{el}.

A solution proposed by Gao and Günnemann (2023a); Scherbela et al. (2024) is to compute W_k as a function of some features $c_k \in \mathbb{R}^{n^{\text{orb}}_{\text{feat}}}$ of each orbital k. The features c_k in turn are evaluated using a conventional method that yields orbitals for a given molecule $(\mathbf{R}, \mathbf{Z})$. Gao and Günnemann (2023a) propose a heuristic based on chemical bonds to obtain orbital positions and ultimately features c_k. Scherbela et al. (2024) propose to use the Hartree-Fock orbitals – which are typically already computed for the purpose of supervised pretraining (cf. Eq. (88)) – to provide atom-wise orbital features c_{kJ}. Given Hartree-Fock orbitals $\varphi^{\text{HF}}_k(r_i)$, expanded in atom-centered basis functions $b_\mu : \mathbb{R}^3 \to \mathbb{R}$

$$\varphi^{\text{HF}}_k(r_i) = \sum_{J=1}^{n_{\text{nuc}}} \sum_{\mu=1}^{n_{\text{basis}}} c_{kJ\mu}\, b_\mu(r_i - R_J), \tag{129}$$

with corresponding expansion coefficients $c \in \mathbb{R}^{n_{\text{el}} \times n_{\text{nuc}} \times n_{\text{basis}}}$, these expansion coefficients can be used as orbital features with $n^{\text{orb}}_{\text{feat}} = n_{\text{basis}}$. To this end the expansion coefficients are first mapped to $\hat{W}_{dkJ} \in \mathbb{R}^{d_{\text{emb}}}$ and $\hat{a}_{dkJ} \in \mathbb{R}$ using MLPs and then used to evaluate the orbitals analogous to (119) and (122). Omitting spins for clarity, this leads to

$$\hat{W}_{dkJ} = \mathrm{MLP_W}\left(c_{kJ}\right) \qquad \mathrm{MLP_W}: \mathbb{R}^{n_\mathrm{basis}} \to \mathbb{R}^{d_\mathrm{emb}} \tag{130}$$

$$\hat{a}_{dkJ} = \mathrm{MLP_a}\left(c_{kJ}\right) \qquad \mathrm{MLP_a}: \mathbb{R}^{n_\mathrm{basis}} \to \mathbb{R} \tag{131}$$

$$\phi^{d,k,i} = \sum_{J=1}^{n_\mathrm{nuc}} \left(\hat{W}_{dkJ} \cdot h_i\right) e^{-\hat{a}_{dkJ}|r_i - R_j|}, \tag{132}$$

which no longer requires a number of trainable parameters dependent on n_el.

5 Results

In this section we survey numerical results that have been achieved using DL-VMC.

5.1 Highly accurate variational energies

In general architectures like FermiNet introduced in Sec. 4.2 are capable of accurately representing the ground-state wavefunction and allow for highly accurate ground-state energy predictions. For example, Fig. 8 compares a neural network architecture proposed in 2022 against other computational approaches for a set of molecules up to 42 electrons (Gerard et al., 2022). The work finds that across a range of molecules, Deep Learning-based Variational Monte Carlo is able to reach lower energy predictions than conventional approaches and,

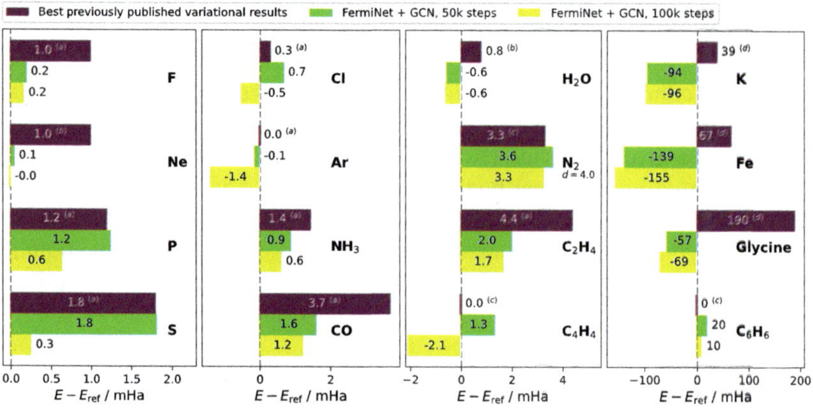

FIGURE 8 Energies relative to the previously known best estimate, (lower is better). Blue bars depict best published variational energies, footnotes mark the method: a: FermiNet VMC (Pfau et al., 2020; Spencer et al., 2020), b: Conventional DMC (Seth et al., 2011; Nemec et al., 2010; Clark et al., 2011), c: FermiNet DMC (Ren et al., 2023), d: MRCI-F12. Note that E_ref is not necessary variational and thus may underestimate the true energy. Figure and caption is taken from Gerard et al. (2022).

therefore, better estimates for the ground state, due to the variational principle. The figure distinguishes between the best estimate and other variational energies, whereas the best estimate can include methods such as CCSD(T), which is widely considered the gold standard for highly accurate ground-state energy predictions. However, a notable drawback of CCSD(T) lies in its nonvariational nature, potentially leading to an underestimation of the ground-state energy and offering no uncertainty guarantees. In contrast, DL-VMC exceeds all conventional variational approaches, including Diffusion Monte Carlo, except for the case of Benzene, the largest system tested, where another deep learning-based approach outperforms the proposed architecture. Since the architecture's initial publication, further improvements have been made in incorporating interparticle correlation, enhancing the presented results even further (Li et al., 2023; von Glehn et al., 2023; Gao and Günnemann, 2023a). These advancements underscore the potential of Variational Monte Carlo and its capabilities in the field.

5.2 Transfer learning for ground-state energy predictions

A potential direction to improve the method's efficiency and reduce training cost, is to use techniques such as deep transfer learning (Devlin et al., 2018; Alayrac et al., 2022). The idea is to pretrain a neural network ansatz on a specific set of molecules to predict the ground-state energy and subsequently transfer this pretrained model to novel, previously unencountered molecules. As discussed in the preceding Sec. 4.7, a common challenge with the proposed architectures lies in the inherent dependence of the ansatz's parameter count on the system size due to the unique construction of the orbital matrix. To address this issue, Scherbela et al. (2023) proposed an approach to map computationally cheap orbital descriptors from methods such as Hartree Fock to highly accurate deep-learning-based orbitals.

To assess the transfer capabilities of a pretrained model, Scherbela et al. pretrained a single neural network ansatz on a diverse set of molecules, comprising approximately 100 molecules, each containing up to four heavy atoms (counted as the number of nonhydrogen atoms). In Fig. 9, the model was evaluated on test sets, each containing four randomly perturbed molecules, grouped by the number of nonhydrogen atoms with up to 7 heavy atoms. To prevent any potential train/test leakage, none of the molecules in the test set were included in the training set. Utilizing a pretrained model, the authors achieved CCSD(T) accuracy with a 2Z basis set without the need for additional variational optimization steps (zero-shot) for molecules containing up to 6 heavy atoms. Additional optimization steps further improved accuracy (refer to Fig. 9 b). To benchmark the performance against other state-of-the-art Deep Learning-based Variational Monte Carlo methods, the energy error as a function of optimization steps, was compared for molecules with three heavy atoms. Specifically, an attention-based approach (von Glehn et al., 2023) without pretraining was used as a baseline. In summary, a pretrained model can generally achieve certain levels of accuracy orders of magnitude faster but may be outperformed after more prolonged

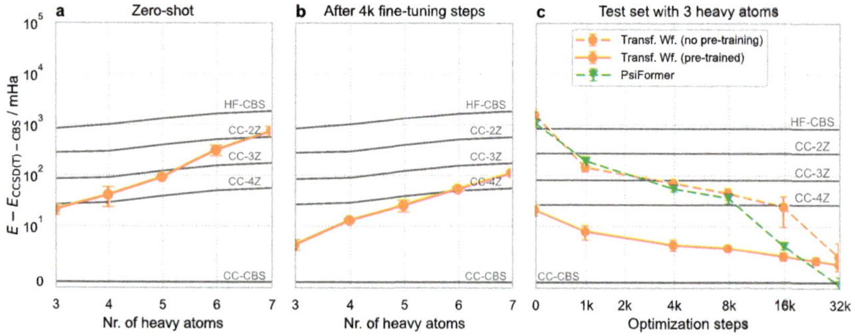

FIGURE 9 **Absolute energies**: Energies relative to CCSD(T)-CBS (complete basis set limit) when re-using the pretrained model on molecules of varying size without optimization (a) and after fine-tuning (b). (c) depicts energy for the test set containing 3 heavy atoms as a function of optimization steps and compares against SOTA method. Solid lines are with pretraining, dashed lines without. Gray lines correspond to conventional methods: Hartree-Fock in the complete basis set limit (HF-CBS), and CCSD(T) with correlation consistent basis sets of double to quadruple valence (CC-nZ). Figure and caption is taken from Scherbela et al. (2023).

optimization. This phenomenon could be related to the orbital construction, potentially suggesting that the proposed transferable ansatz might be inherently less expressive (Scherbela et al., 2023).

The more challenging task of relative energies was evaluated in a second experiment. By relative energies we denote the energy difference between different geometrical conformations of the same molecule. The authors show that by using only a few additional variational optimization steps qualitatively and quantitatively correct results can be achieved.

For instance, in Fig. 10a, the relative energy of five conformers to the equilibrium-state geometry of Bicyclobutane is illustrated. This system is of particular interest because CCSD(T) tends to inaccurately predict the relative energy for the "dis-TS" conformer, significantly underestimating the energy by approximately 60 mHa. While the pretrained deep learning model correctly predicts the sign of the relative energies without the need for additional optimization steps, it does yield quantitatively different relative energies. Therefore, additional optimization steps are necessary to bring the model into closer alignment with the Diffusion Monte Carlo reference method and FermiNet. With just 700 optimization steps per geometry, the pretrained model achieved a maximum deviation of 2.1 millihartree (mHa) to the reference method compared to FermiNet, which required 10,000 steps for a maximum deviation of 7.1 mHa.

Another frequently encountered test case involves the dissociation curve of the Nitrogen dimer. The dimer serves as another example wherein methods like CCSD(T) commonly struggle by notably overestimating the energy for conformations in the bond-breaking regime (i.e. a distance of 3-4 Bohr between the Nitrogen atoms). In this scenario, the pretrained model also tended to overesti-

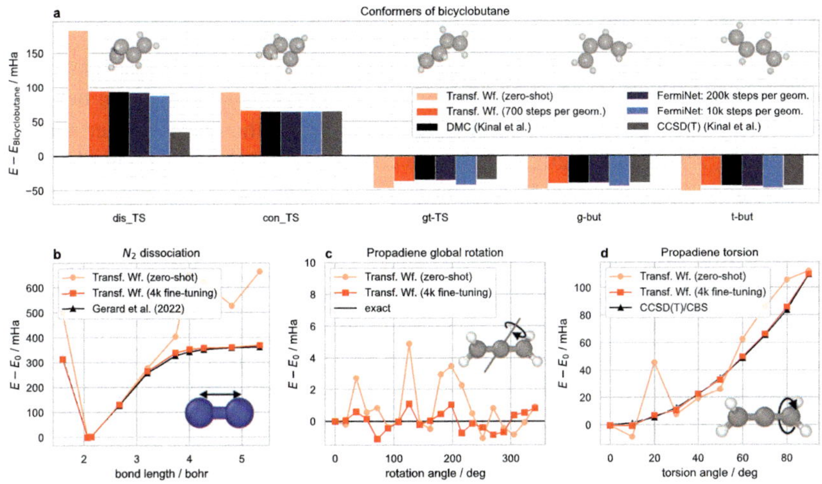

FIGURE 10 **Challenging relative energies**: Relative energies obtained with and without fine-tuning on 4 distinct, challenging systems, compared against high-accuracy reference methods. a) Relative energy of bicyclobutane conformers vs. the energy of bicyclobutane; b) Potential energy surface (PES) of N_2; c) global rotation of propadiene; d) relative energy of twisted vs. untwisted propadiene. Figure and caption is taken from Scherbela et al. (2023).

mate the energy significantly. However, it only required a few optimization steps to better align with the reference method (cf. Fig. 9 c).

The ground-state energy in general is invariant under global rotation of the molecule. Although the energy is invariant, this is in general not the case for the wavefunction. Enforcing complete invariance of the wavefunction under global rotation would be overly restrictive, as discussed in the work by Gao and Günnemann (2022). To address this, Scherbela et al. incorporated data augmentation during pretraining, involving random rotations of the entire molecules. While this proved to be a suitable proxy, with energy variations of approximately 1 mHa for propadiene, it was not entirely sufficient, leading to outliers with variations reaching up to 5 mHa. A short variational optimization phase once again helped to mitigate larger energy errors, achieving chemical accuracy as depicted in Fig. 10c.

For the final evaluation, the transition barrier of propadiene twisted around the C=C bond was examined. The equilibrium and transition states are differentiated by an energy difference of approximately 110 mHa. The model was intentionally pretrained on twisted molecules, encompassing equilibrium conformations, transition conformations, and one intermediate twist. However, as depicted in Fig. 10, this proves insufficient to accurately predict the complete transition path without the incorporation of additional optimization steps. However, again only a short amount of fine-tuning allowed for an accurate prediction of the whole path.

5.3 Literature overview

Research into Deep Learning-based Variational Monte Carlo has expanded rapidly over the last years, rendering a comprehensive overview of the field impossible. For further reviews on this subject we refer the reader to Hermann et al. (2022); Zhang et al. (2023); Schätzle et al. (2023); Medvidović and Moreno (2024) and highlight several advances in key areas below.

Embedding architectures

FermiNet (Pfau et al., 2020; Spencer et al., 2020) and PauliNet (Hermann et al., 2020) have been the first two neural network architectures to successfully demonstrate the potential of neural network-based wavefunctions for molecules in first quantization. As discussed in Sec. 4.2, a significant portion of research has been invested in improving the originally proposed embedding architectures by incorporating attention-based techniques (von Glehn et al., 2023; Pescia et al., 2023; Li et al., 2023) and graph-based approaches (Gerard et al., 2022; Gao and Günnemann, 2023a). Currently, the two state-of-the-art architectures, LapNet (Li et al., 2023) and PsiFormer (von Glehn et al., 2023), in terms of accuracy are based on attention mechanisms.

Antisymmetrization

In terms of antisymmetrization, as previously discussed, a common technique is to employ Slater determinants. From a scaling perspective, the determinant is the leading factor with cubical scaling. Therefore, efforts have been made to reduce the theoretical scaling using sorting algorithms (Richter-Powell et al., 2023) or products of two-particle functions (Han et al., 2019). Although the results are partially promising, they are still at a proof-of-concept stage, and the Slater determinant remains the most commonly used antisymmetrization method. However, for certain systems, such as for the case of superfluids, it was shown that approaches like antisymmetric geminal powered wavefunctions (Lou et al., 2024) or Pfaffian wavefunctions (Kim et al., 2023) can be beneficial. A recent preprint (Ye et al., 2024) proposed antisymmetrization using Vandermonde-like determinants, and showed that any continuous antisymmetric function can be represented by $\mathcal{O}(n_{\mathrm{el}})$ of these objects, potentially yielding another approach to antisymmetrization with favorable scaling of computational cost.

Generalization across molecules

By taking advantage of regularities within the geometrical conformation space of molecules, several approaches have been proposed to learn neural network-based wavefunctions simultaneously across a range of geometrical conformations. Either by only sharing parts of the neural network architecture (Scherbela et al., 2022) or by using a meta neural network to predict the linear mappings within the orbital construction (Gao and Günnemann, 2022, 2023b). Although they achieve faster evaluation of the potential energy surface of a molecule by an order of magnitude, they don't allow the efficient transfer to new molecules. A key reason for this is the explicit dependence of the parameter count of the architecture on the system size due to the unique construction of the orbital matrix. Therefore, Gao and Günnemann (2023a) and Scherbela et al. (2024, 2023) generalized the existing approach to allow for efficient optimization of a single neural network across a diverse range of molecules. This again allowed for a significant reduction in optimization steps, as discussed in Sec. 5.2.

Optimization

Another active area of research involves enhancing neural network optimization techniques. In Deep Learning-based Variational Monte Carlo, it is common to employ second-order methods like Natural Gradient Descent. Consequently, there arises the need to invert a preconditioner matrix with dimensions $n_{\text{params}} \times n_{\text{params}}$. FermiNet has proposed the use of KFAC as an approximation technique, which relies on the assumption that the matrix is block-diagonal. Another very recent line of work showed that it is possible to convert the problem of inverting the matrix of shape $n_{\text{params}} \times n_{\text{params}}$ to a problem of inverting a matrix of shape $n_{\text{samples}} \times n_{\text{samples}}$, whereas $n_{\text{samples}} \ll n_{\text{params}}$ represent the number of Monte Carlo samples (Goldshlager et al., 2024; Rende et al., 2023). On the other hand, Neklyudov et al. (2023) interpreted the optimization in Variational Monte Carlo as a gradient flow and by improving the underlying metric of the distribution space the author reached empirically faster convergence.

Observables and applications

Besides the improvements to the neural network-based architecture and the computation of the ground-state energy for molecules in first quantization, the method was also applied to a plethora of other observables and systems. For example the method has been applied to the computation of forces (Qian et al., 2022; Scherbela et al., 2022) and excited states (Entwistle et al., 2023; Pfau et al., 2023), as well as to other systems such as solids including real-solids and the homogeneous electron gas (Pescia et al., 2023; Cassella et al., 2023; Li et al., 2022a), superfluids (Lou et al., 2024; Kim et al., 2023), positron-molecule complexes (Cassella et al., 2024) and discrete systems (Carleo and Troyer, 2017). Also, techniques like effective core potentials (Li et al., 2022b) or Diffusion Monte Carlo (Ren et al., 2023) have been explored in the context of Deep Learning-based methods for the electronic Schrödinger equation to improve the scaling and accuracy further.

References

Agmon, Shmuel, 2014. Lectures on Exponential Decay of Solutions of Second-Order Elliptic Equations: Bounds on Eigenfunctions of N-Body Schrodinger Operations. (MN-29), vol. 29. Princeton University Press.

Alayrac, Jean-Baptiste, et al., 2022. Flamingo: a visual language model for few-shot learning. In: Advances in Neural Information Processing Systems, vol. 35. Curran Associates, Inc., pp. 23716–23736.

Barone, Vincenzo, et al., 2013. Accurate structure, thermodynamic and spectroscopic parameters from CC and CC/DFT schemes: the challenge of the conformational equilibrium in glycine. Physical Chemistry Chemical Physics 15 (25), 10094–10111. https://doi.org/10.1039/C3CP50439E.

Barrett, Thomas D., Malyshev, Aleksei, Lvovsky, A.I., 2022. Autoregressive neural-network wavefunctions for ab initio quantum chemistry. Nature Machine Intelligence 4 (4), 351–358. https://doi.org/10.1038/s42256-022-00461-z.

Bayer, Christian, et al., 2023. Pricing high-dimensional Bermudan options with hierarchical tensor formats. SIAM Journal on Financial Mathematics 14 (2), 383–406.

Becca, Federico, Sorella, Sandro, 2017. Quantum Monte Carlo Approaches for Correlated Systems. Cambridge University Press.

Berner, Julius, Grohs, Philipp, Voigtlaender, Felix, 2023. Training ReLU networks to high uniform accuracy is intractable. In: Proceedings ICLR 2023.

Carleo, Giuseppe, Troyer, Matthias, 2017. Solving the quantum many-body problem with artificial neural networks. Science 355 (6325), 602–606. https://doi.org/10.1126/science.aag2302.

Cassella, G., et al., 2024. Neural network variational Monte Carlo for positronic chemistry. arXiv preprint. arXiv:2310.05607.

Cassella, Gino, et al., 2023. Discovering quantum phase transitions with fermionic neural networks. Physical Review Letters 130 (3), 036401. https://doi.org/10.1103/PhysRevLett.130.036401.

Chuang, Pi-Yueh, Barba, Lorena A., 2022. Experience report of physics-informed neural networks in fluid simulations: pitfalls and frustration. arXiv preprint. arXiv:2205.14249.

Clark, Bryan K., et al., 2011. Computing the energy of a water molecule using multideterminants: a simple, efficient algorithm. Journal of Chemical Physics 135 (24), 244105. https://doi.org/10.1063/1.3665391.

De Ryck, Tim, Mishra, Siddhartha, 2024. Numerical analysis of physics-informed neural networks and related models in physics-informed machine learning. arXiv preprint. arXiv:2402.10926.

Devlin, Jacob, et al., 2018. Bert: pre-training of deep bidirectional transformers for language understanding. arXiv:1810.04805.

Dirac, Paul Adrien Maurice, 1929. Quantum mechanics of many-electron systems. Proceedings of the Royal Society of London. Series A, Containing Papers of a Mathematical and Physical Character 123 (792), 714–733.

Elbrächter, Dennis, et al., 2021. Deep neural network approximation theory. IEEE Transactions on Information Theory 67 (5), 2581–2623.

Entwistle, Mike.T., et al., 2023. Electronic excited states in deep variational Monte Carlo. Nature Communications 14 (1), 274. https://doi.org/10.1038/s41467-022-35534-5.

Evans, Lawrence C., 2022. Partial Differential Equations, vol. 19. American Mathematical Society.

Gao, Nicholas, Günnemann, Stephan, 2022. Ab-initio potential energy surfaces by pairing GNNs with neural wave functions. In: International Conference on Learning Representations.

Gao, Nicholas, Günnemann, Stephan, 2023a. Generalizing neural wave functions. In: Proceedings of the 40th International Conference on Machine Learning. In: Proceedings of Machine Learning Research, vol. 202. PMLR, pp. 10708–10726.

Gao, Nicholas, Günnemann, Stephan, 2023b. Sampling-free inference for ab-initio potential energy surface networks. In: The Eleventh International Conference on Learning Representations.

Gerard, Leon, et al., 2022. Gold-standard solutions to the Schrödinger equation using deep learning: how much physics do we need? In: Advances in Neural Information Processing Systems, vol. 35. Curran Associates, Inc., pp. 10282–10294.

Goldshlager, Gil, Abrahamsen, Nilin, Lin, Lin, 2024. A Kaczmarz-inspired approach to accelerate the optimization of neural network wavefunctions. arXiv preprint. arXiv:2401.10190.

Hall, Brian C., 2013. Quantum Theory for Mathematicians. Graduate Texts in Mathematics, vol. 136.

Han, Jiequn, Zhang, Linfeng, E, Weinan, 2019. Solving many-electron Schrödinger equation using deep neural networks. Journal of Computational Physics 399, 108929. https://doi.org/10.1016/j.jcp.2019.108929.

Hermann, Jan, et al., 2022. Ab-initio quantum chemistry with neural-network wavefunctions. arXiv preprint. arXiv:2208.12590.

Hermann, Jan, Schätzle, Zeno, Noé, Frank, 2020. Deep-neural-network solution of the electronic Schrödinger equation. Nature Chemistry 12 (10), 891–897. https://doi.org/10.1038/s41557-020-0544-y.

Hibat-Allah, Mohamed, et al., 2020. Recurrent neural network wave functions. Physical Review Research 2 (2), 023358. https://doi.org/10.1103/PhysRevResearch.2.023358.

Hunziker, Walter, Sigal, Israel Michael, 2000. The quantum N-body problem. Journal of Mathematical Physics 41 (6), 3448–3510.

Inui, Koji, Kato, Yasuyuki, Motome, Yukitoshi, 2021. Determinant-free fermionic wave function using feed-forward neural networks. Physical Review Research 3 (4), 043126. https://doi.org/10.1103/PhysRevResearch.3.043126.

Jecko, Thierry, 2014. On the mathematical treatment of the Born-Oppenheimer approximation. Journal of Mathematical Physics 55, 5.

Kato, Tosio, 1957. On the eigenfunctions of many-particle systems in quantum mechanics. Communications on Pure and Applied Mathematics 10 (2), 151–177.

Kato, Tosio, 2013. Perturbation Theory for Linear Operators, vol. 132. Springer Science & Business Media.

Kim, Jane, et al., 2023. Neural-network quantum states for ultra-cold Fermi gases. arXiv preprint. arXiv:2305.08831.

Kingma, Diederik P., Ba, Jimmy, 2017. Adam: a method for stochastic optimization. arXiv:1412.6980.

Li, Ruichen, et al., 2023. Forward Laplacian: a new computational framework for neural network-based variational Monte Carlo. arXiv:2307.08214.

Li, Xiang, et al., 2022b. Fermionic neural network with effective core potential. Physical Review Research 4, 013021. https://doi.org/10.1103/PhysRevResearch.4.013021.

Li, Xiang, Li, Zhe, Chen, Ji, 2022a. Ab initio calculation of real solids via neural network ansatz. Nature Communications 13 (1), 7895. https://doi.org/10.1038/s41467-022-35627-1.

Lin, Jeffmin, Goldshlager, Gil, Lin, Lin, 2023. Explicitly antisymmetrized neural network layers for variational Monte Carlo simulation. Journal of Computational Physics 474, 111765.

Lou, Wan Tong, et al., 2024. Neural wave functions for superfluids. arXiv preprint. arXiv:2305.06989.

Martens, James, Grosse, Roger, 2015. Optimizing neural networks with Kronecker-factored approximate curvature. arXiv:1503.05671.

Medvidović, Matija, Moreno, Javier Robledo, 2024. Neural-network quantum states for many-body physics. arXiv preprint. arXiv:2402.11014.

Neklyudov, Kirill, et al., 2023. Wasserstein quantum Monte Carlo: a novel approach for solving the quantum many-body Schrödinger equation. arXiv preprint. arXiv:2307.07050.

Nemec, Norbert, Towler, Michael D., Needs, R.J., 2010. Benchmark all-electron ab initio quantum Monte Carlo calculations for small molecules. Journal of Chemical Physics 132 (3), 034111. https://doi.org/10.1063/1.3288054.

Pescia, Gabriel, et al., 2023. Message-passing neural quantum states for the homogeneous electron gas. arXiv:2305.07240.

Pfau, David, et al., 2020. Ab initio solution of the many-electron Schrödinger equation with deep neural networks. Physical Review Research 2 (3), 033429. https://doi.org/10.1103/PhysRevResearch.2.033429.

Pfau, David, et al., 2023. Natural quantum Monte Carlo computation of excited states. arXiv:2308.16848.

Qian, Yubing, et al., 2022. Interatomic force from neural network based variational quantum Monte Carlo. Journal of Chemical Physics 157 (16), 164104. https://doi.org/10.1063/5.0112344.

Radford, Alec, et al., 2021. Improving Language Understanding by Generative Pre-Training.
Reed, M., Simon, B., 1978. Analysis of Operators. Methods of Modern Mathematical Physics, vol. 4. Academic, New York.
Ren, Weiluo, et al., 2023. Towards the ground state of molecules via diffusion Monte Carlo on neural networks. Nature Communications 14 (1), 1860. https://doi.org/10.1038/s41467-023-37609-3.
Rende, Riccardo, et al., 2023. A simple linear algebra identity to optimize large-scale neural network quantum states. arXiv preprint. arXiv:2310.05715.
Richter-Powell, Jack, et al., 2023. Sorting out quantum Monte Carlo. arXiv preprint. arXiv:2311.05598.
Schätzle, Zeno, et al., 2023. DeepQMC: an open-source software suite for variational optimization of deep-learning molecular wave functions. Journal of Chemical Physics 159 (9), 094108. https://doi.org/10.1063/5.0157512.
Scherbela, Michael, et al., 2022. Solving the electronic Schrödinger equation for multiple nuclear geometries with weight-sharing deep neural networks. Nature Computational Science 2 (5), 331–341. https://doi.org/10.1038/s43588-022-00228-x.
Scherbela, Michael, Gerard, Leon, Grohs, Philipp, 2023. Variational Monte Carlo on a budget - fine-tuning pre-trained neural wavefunctions. In: Thirty-Seventh Conference on Neural Information Processing Systems.
Scherbela, Michael, Gerard, Leon, Grohs, Philipp, 2024. Towards a transferable fermionic neural wavefunction for molecules. Nature Communications 15 (1), 120. https://doi.org/10.1038/s41467-023-44216-9.
Schütt, Kristof T., et al., 2020. Machine Learning Meets Quantum Physics. Lecture Notes in Physics.
Seth, P., López Ríos, P., Needs, R.J., 2011. Quantum Monte Carlo study of the first-row atoms and ions. Journal of Chemical Physics 134 (8), 084105. https://doi.org/10.1063/1.3554625.
Spencer, James S., et al., 2020. Better, faster fermionic neural networks. arXiv:2011.07125.
Stone, Marshall H., 1932. On one-parameter unitary groups in Hilbert space. Annals of Mathematics, 643–648.
Szabo, Attila, Ostlund, Neil S., 1996. Modern Quantum Chemistry: Introduction to Advanced Electronic Structure Theory. Dover Publications. ISBN 9780486691862.
Thiede, Luca, Sun, Chong, Aspuru-Guzik, Alán, 2023. Waveflow: enforcing boundary conditions in smooth normalizing flows with application to fermionic wave functions. arXiv preprint. arXiv:2211.14839.
Troyer, Matthias, Wiese, Uwe-Jens, 2005. Computational complexity and fundamental limitations to fermionic quantum Monte Carlo simulations. Physical Review Letters 94 (17), 170201.
von Glehn, Ingrid, Spencer, James S., Pfau, David, 2023. A self-attention ansatz for ab-initio quantum chemistry. In: The Eleventh International Conference on Learning Representations.
von Neumann, John, 1932. Über einen Satz von Herrn MH Stone. Annals of Mathematics 33 (3), 567–573.
Woodbury, Max A., 1950. Inverting Modified Matrices. Department of Statistics, Princeton University.
Ye, Haotian, Li, Ruichen, Gu, Yuntian, Lu, Yiping, He, Di, Wang, Liwei, 2024. $\widetilde{O}(N^2)$ representation of general continuous anti-symmetric function. https://doi.org/10.48550/arXiv.2402.15167. arXiv:2402.15167 [quant-ph]. http://arxiv.org/abs/2402.15167 (visited on 03/01/2024).
Yserentant, Harry, 2010. Regularity and Approximability of Electronic Wave Functions. Springer.
Zhang, Xuan, et al., 2023. Artificial intelligence for science in quantum, atomistic, and continuum systems. arXiv:2307.08423.
Zhou, Jie, et al., 2020. Graph neural networks: a review of methods and applications. AI Open (ISSN 2666-6510) 1, 57–81. https://doi.org/10.1016/j.aiopen.2021.01.001. https://www.sciencedirect.com/science/article/pii/S2666651021000012 (visited on 02/13/2024).

Chapter 6

Theoretical foundations of physics-informed neural networks and deep neural operators

A brief review

Yeonjong Shin[a,*], **Zhongqiang Zhang**[b], **and George Em Karniadakis**[c]
[a]*Department of Mathematics, North Carolina State University, Raleigh, NC, United States,* [b]*Department of Mathematical Sciences, Worcester Polytechnic Institute, Worcester, MA, United States,* [c]*Division of Applied Mathematics, Brown University, Providence, RI, United States*
**Corresponding author: e-mail address: yeonjong_shin@ncsu.edu*

Contents

Handbook of Numerical Analysis, Volume 25, ISSN 1570-8659, https://doi.org/10.1016/bs.hna.2024.05.008

Abstract

This chapter presents a brief review of the theoretical foundations of physics-informed neural networks (PINNs) and deep neural operators. PINN is one of the most popular deep learning approaches for solving both forward and inverse problems of partial differential equations (PDEs). It provides seamless ways of embedding laws of physics into deep neural networks (DNNs) by leveraging auto-differentiation. At the same time, operator learning emerged as a new learning paradigm for learning nonlinear operators, particularly ones relevant to PDEs. Deep Operator Network (DeepONet) is one of the first pioneering models whose architecture is inspired by the universal approximation theorem. DeepONets can generate reliable real-time responses when they are pretrained with a large amount of data pairs of inputs and outputs. Topics to be covered include mathematical formulations, approximation error estimates, approximation theory of DNNs, and training/optimization methods.

Keywords

Physics-informed neural networks, Deep operator networks, Deep learning, Optimization, Approximation theory, Partial differential equations

MSC Codes

65-00, 90-00, 41-00

1 Introduction

In this chapter, we present a review of the theoretical development of physics-informed neural networks (PINNs, e.g. in Lagaris et al., 1998, and Raissi et al., 2019) and deep operator networks (DeepONets, e.g., in Lu et al., 2021a) for partial differential equations.

In physics-informed neural networks, partial differential equations including forward and inverse problems are reformulated as minimization of least-squares of the residuals of the equations and their constraints at random points (often uniformly distributed). Since the neural networks are nonconvex and nonlinear in their trainable parameters, we often use stochastic gradient descent methods to find approximate solutions to the minimization problems. Advances in PINNs include choices of loss functions, distribution of sampling points, architectures of neural networks, efficient training methods, and many techniques addressing issues in various applications.

Compared to least-squares finite element methods (Bochev and Gunzburger, 1998; Bramble and Schatz, 1970) and least-squares collocations, the approximators are networks and thus are highly nonlinear and nonconvex. Correspondingly, different formulations and solvers for the least-squares problems have been applied and some error estimates of PINNs have been established.

While PINNs may need no data or little data, DeepONets require a large amount of data but DeepONets learn maps/operators from a function input to a function output. In the vanilla DeepONets, a regression problem is formulated as the empirical L^2-norm of the differences among the output of neural networks at input data and corresponding output data. Developments of DeepONets include but are not limited to those in architectures of neural networks, efficient training methods and error estimates.

In Section 2, we briefly introduce a special class of neural networks-feedforward neural networks, although many other networks work as well in most of the settings in this work. In Section 3, we introduce an abstract problem and present some classical forward and inverse problems to solve with PINNs. We also introduce the strong formulations and weak formulations used in PINNs. Various aspects of PINNs are addressed with various techniques to solve many types of problems. In Section 4, we discuss the error estimates of PINNs for conditionally stable problems. The estimates are based on the conditional stability of problems. We present some developments of training methods in Section 5 for the loss functions, which may not be limited to loss functions in PINNs. In Section 6, we present a special neural network where the weights and biases are small. This network is motivated by expressing functions with discontinuity or large derivatives, such as in hyperbolic problems. In Section 7, we briefly discuss PINNs when observational data are available, e.g. in inverse problems. In Section 8, we present error estimates for DeepONets for continuous operators and three examples of solution operators arising from linear and nonlinear advection-diffusion-reaction equations.

2 Neural networks

Many neural network architectures can be used for scientific machine learning, such as radial basis function networks, convolution neural networks, residual neural networks, recurrent Neural networks, U-nets, etc. Let's focus on feed-forward neural networks. A L-layer feed-forward neural network (NN) $u_{\rm NN}$ is a function $\mathbb{R}^{d_{\rm in}} \mapsto \mathbb{R}^{d_{\rm out}}$ defined by

$$u_{\rm NN}(x;\theta) = W^L u^{L-1}(x) + b^L, \tag{2.1}$$

where u^{L-1} is recursively computed by

$$u^\ell(x) = \phi(W^\ell u^{\ell-1}(x) + b^\ell), \quad 1 \le \ell < L, \qquad u^0(x) = x.$$

Here ϕ is a nonlinear activation function that applies element-wise. Some popular activation functions include the hyperbolic tangent (tanh) and the rectified linear unit (ReLU). L is a positive integer greater than or equal to 2, referred to as the depth. For notational convenience, the input and output dimensions are often denoted by $n_0 = d_{\rm in}$ and $n_L = d_{\rm out}$, respectively and these are assumed to be given for every learning task. The weight matrix and bias vector of the ℓ-th hidden layer are $W^\ell \in \mathbb{R}^{n_\ell \times n_{\ell-1}}$ and $b^\ell \in \mathbb{R}^{n_\ell}$, respectively. The collection θ of all the weight matrices and bias vectors determines $u_{\rm NN}$ and is called the network parameter. The vector $\vec{n}_L = (n_1, \ldots, n_{L-1}) \in \mathbb{N}^{L-1}$ is referred to as the architecture, which determines the size of θ. To explicitly express the dependency of the architecture, the network parameter is often denoted by $\theta(\vec{n}_L)$, whose size and magnitude are given by

$$|\vec{n}_L| = \sum_{\ell=1}^{L} n_\ell(n_{\ell-1}+1), \quad |\theta|_\infty = \max_{1\le\ell\le L} \max\{\|W^\ell\|_{\max}, \|b^\ell\|_{\max}\}.$$

A class of L-layer NNs is then given by

$$V_L(M) = \left\{u_{\mathrm{NN}}(\cdot;\theta(\vec{n}_L)) : |\theta(\vec{n}_L)|_\infty \le M, \vec{n}_L \in \mathbb{N}^{L-1}\right\}. \tag{2.2}$$

The class V_L does not form a linear space as it is not closed under addition, i.e., for any $u, v \in V_L$, $u+v \notin V_L$. However, it has been well known that V_L is universal. For example, for any continuous function f in a compact domain $\Omega \subset \mathbb{R}^{d_{\mathrm{in}}}$ and any $\epsilon > 0$, there exist M and $\vec{n}_L$ such that $\|f(x) - u_{\mathrm{NN}}(x, \theta(\vec{n}_L))\| < \epsilon$ for any $x \in \Omega$.

To provide a mathematical explanation of the empirical success of NNs, many works have focused on quantifying the approximation ability of NNs using the number of nonzero network parameters, the magnitudes, the depth, and the width, to name a few. The works of Petersen and Voigtlaender (2018) proved universal approximation theorems of rectified linear units (ReLU) networks, showing the advantage of the depth. On the other hand, the class of two-layer neural networks has been shown to effectively approximate certain functions, and such a class is now known as the Barron class.

3 Mathematical formulations

Consider the following problem

$$\begin{aligned} \mathcal{D}[u](x) &= f(x) \quad x \in \Omega \\ \mathcal{B}[u](x) &= g(x) \quad x \in \Gamma, \end{aligned} \tag{3.1}$$

where $\mathcal{D}: X \to Y$ $\mathcal{B}: X \to Z$ are appropriate differential operators. Here X, Y and Z are Banach spaces. Here $\Omega \subset \mathbb{R}^d$ is a computational (compact) domain and Γ is a subset of $\mathbb{R}^d$. This abstract problem can include many interesting problems. Below, we list some typical benchmark problems when developing deep learning methods for partial differential equations (PDEs).

Example 3.1 (Poisson equation with Dirichlet boundary condition, forward problem). Let $\mathcal{D} = -\sum_{j=1}^{d} \partial_{x_i}^2 := \Delta$ and $\mathcal{B} = \mathrm{Id}$ (identity operator). Let $\Omega = (0,1)^2$, $\Gamma = \partial\Omega = \{x = (x_1, x_2) \in \Omega | x{=}(0, x_2) \text{ or } (1, x_2), \text{ or } (x_1, 0), \text{ or } (x_1, 1)\}$. The problem is usually written as

$$-\Delta u(x) = f(x), \quad x \in \Omega; \qquad u(x) = g(x), \quad x \in \Gamma = \partial\Omega. \tag{3.2}$$

Here f and g are given, we seek the solution u on the Ω.

Example 3.2 (Data assimilation of Poisson's equation, inverse problem). Let $\Omega \subset \mathbb{R}^d$ be a Lipschitz domain. Let $\Omega_0 \subsetneq \Omega$ be open. Consider the following problem

$$-\Delta u = f, \quad x \in \Omega; \qquad u = g, \quad x \in \Omega_0.$$

Here f and g are given, we seek the solution u on Ω and $\partial\Omega$. This problem is often called a *continuation* problem.

Example 3.3 (Cauchy problem of Poisson's equation, inverse problem). Let $\Omega \subset \mathbb{R}^d$ be a Lipschitz domain. Let $\Gamma_D \subsetneq \partial\Omega$ be an open subset with a positive area. Consider the following Cauchy problem

$$-\Delta u = f, \quad x \in \Omega; \qquad u = g, \quad x \in \Gamma_D; \qquad \nabla u \cdot \mathbf{n} = \varphi, \quad x \in \Gamma_D.$$

In this problem, we only know partial data on the boundary and thus we seek the solution as well as the data on the whole boundary. This problem is often called a Cauchy problem for the Poisson equation.

Example 3.4 (Coefficient inverse problem). Let $D \subset \mathbb{R}^n$ be a smooth bounded domain. Consider the initial boundary value problem on $\Omega = D \times (0, T)$ (T>0),

$$\begin{aligned}&\partial_t^2 u = \nabla \cdot (p(x)\nabla u), \quad (x,t) \in D \times (0,T);\\ &u(x,0) = g_0(x),\ \partial_t u(x,0) = 0,\ x \in D;\ u = g(x,t),\ (x,t) \in S_T = \partial D \times (0,T).\end{aligned}$$

Here the coefficient $p(x) \in C^1(\bar{D})$ and $p(x) \geq p_0 > 0$. Here the functions g_0 and g are known, while The goal is to find the coefficient $p(x)$ in D. We also know the normal derivative of u $\nabla u \cdot \mathbf{n} = h(x,t)$ on S_T, where $\mathbf{n}$ is the outward normal vector at the cylindrical surface S_T.

3.1 Stability

We assume that the problem (3.1) is conditionally stable. We use the following definition of conditional stability, see e.g. in Burman and Oksanen (2018) and Dahmen et al. (2023).

Definition 3.5 (Conditional stability). Let X and Y be Banach spaces. Let $\mathcal{D} : X \mapsto Y$ and $\mathcal{B} : X \mapsto Z$ be linear maps. Assume that there exists a linear operator $L : X \to H$ where H is a Banach space such that there is a positive function $\mathsf{N} : \hat{X} \to (0, \infty)$[1] and a nondecreasing function $\rho_C : [0, \infty] \to [0, \infty]$ with $\lim_{t \searrow 0} \rho_C(t) = 0$ for any $C > 0$, such that for $v = v_1 - v_2 \in X$ $(v_1, v_2 \in X)$ with $\|Lv\|_H \leq C$, it holds that

$$\mathsf{N}(v_1 - v_2) \leq \rho_C(\|\mathcal{D}v_1 - \mathcal{D}v_2\|_Y + \|\mathcal{B}v_1 - \mathcal{B}v_2\|_Z). \tag{3.3}$$

Here N is usually a norm or a seminorm on a Banach space V which $\widehat{X}$ can be embedded into. We then call the problem (3.1) conditionally stable.

---

1 Here $\widehat{X}$ is a Banach space, which often coincides with $\widehat{X}$.

Here L is often called regularization. When N is a norm, the problem is well-posed in the Hadamard sense: the solution exists and is unique and stable with respect to inputs. In this case, the problem is called unconditionally stable.

A special case of the conditional stability is linear stability. In *linear stability*, N is a norm and ρ_C is a linear function. For example, in Example 3.1, we can derive the following stability[2]

$$\|u\|_{L^2(\Omega)} \leq C(\|\Delta u\|_{H^{-3/2}(\Omega)} + \|u\|_{L^2(\partial\Omega)}) \leq C(\|\Delta u\|_{L^2(\Omega)} + \|u\|_{L^2(\partial\Omega)}). \tag{3.4}$$

Here H^r is the Sobolev-Hilbert space with elements and their r-th weak derivatives being square integrable.

The conditional stability for the Problem 3.2 is stated as follows.

Theorem 3.6 (Conditional stability for the Poisson continuation problem, Burman and Oksanen, 2018). *Let $f \in L^2(\Omega)$ and let $u \in H^1(\Omega)$ such that the equation in the Problem 3.2 holds. Let $g \in H^1(\Omega_0)$. Then for every open simply connected set $E \subset \Omega$ such that* $\text{dist}(E, \partial D) > 0$, *there holds*

$$\|u\|_{H^1(E)} \leq C\left(\|u\|_{H^1(\Omega)}\right)^{1-r} \omega\left(\|f\|_{H^{-1}(\Omega)} + \|g\|_{L^2(\Omega_0)}\right)^r.$$

Here $r \in (0, 1)$, depending on the set E. On the domain Ω, we have

$$\|u\|_{H^1(\Omega)} \leq C\left(\|u\|_{H^1(\Omega)}\right)^r \left[\log((\|f\|_{H^{-1}(\Omega)} + \|g\|_{L^2(\Omega_0)})\right]^{-r}.$$

The conditional stability of the Problem 3.3 can be found in Burman and Oksanen (2018). The inverse problem in Example 3.4 is Lipschitz stable, i.e., ρ_C is stable, see e.g., Klibanov and Yamamoto (2006).

Many problems have proven to be conditionally stable in literature, such as classical partial differential equations, integral equations and a large class of inverse problems.

Remark 3.7. Here the conditional stability is described for linear problems. From the definition, it can be further extended to nonlinear problems, where ρ_C may depend on v_1 and v_2.

Solving the problem (3.1) can be formulated as an optimization problem using the residual minimization

$$\inf_{v \in \widehat{X}} \mathcal{J}(v), \tag{3.5}$$

where the loss functional $\mathcal{J}_\tau(v)$ can be established in many ways. Depending on how the optimization problem is reformulated, different methods are obtained, e.g. deep Ritz methods (E and Yu, 2018) using energy minimization and physics-informed neural networks (PINNs) in Lagaris et al. (1998) and Raissi et al. (2019), which is based on residual minimization while approximate solutions being neural networks.

2 The L^2 norm on the left can be replaced by $\|u\|_{H^{1/2}(\Omega)}$.

3.2 Strong formulation

The most straightforward method is based on the strong form of the governing equation.

$$\mathcal{J}(v) = \|f - \mathcal{D}v\|_Y^2 + \|g - \mathcal{B}v\|_Z^2 + \epsilon^2 \|Lv\|_H^2 . \tag{3.6}$$

Here ϵ is usually a small parameter. This is one of the most popular approach used in practice. To implement this formulation, we enforce the approximation to satisfy the governing equation over a finite set of points. More precisely, let $\Omega_M \subset \Omega$ and $\Gamma_{M'} \subset \partial\Omega$ be such sets with $|\Omega_M| = M$ and $|\partial\Omega_{M'}| = M'$. Let $\omega_M(\cdot) = \frac{1}{M}\sum_{x\in\Omega_M} \delta_x(\cdot)$ and $\mu_{M'}(\cdot) = \frac{1}{M'}\sum_{x\in\Gamma_{M'}} \delta_x(\cdot)$ be empirical measures for Ω_M and $\Gamma_{M'}$, respectively, where δ_x is the Dirac measure. The naive PINN method (Lagaris et al., 1998; Raissi et al., 2019) defines an objective (also known as a physics-informed loss) functional as $\mathcal{L}(v) = \mathcal{L}_{\text{re}}(v) + \mathcal{L}_{\text{bc}}(v)$ where

$$\mathcal{L}_{\text{re}}(v) = \int_\Omega \mathrm{L}_{\text{re}}[v](x) d\omega_M(x), \quad \mathcal{L}_{\text{bc}}(v) = \int_\Gamma \mathrm{L}_{\text{bc}}[v](x) d\mu_{M'}(x), \tag{3.7}$$
$$\mathrm{L}_{\text{re}}[v](x) := [\mathcal{D}[v](x) - f(x)]^2, \mathrm{L}_{\text{bc}}[v](x) := [\mathcal{B}[v](x) - g(x)]^2 .$$

and seeks an approximation in V_n that minimizes the loss functional $\mathcal{L}$, i.e., $u_n^* = \text{argmin}_{v\in V_n} \mathcal{L}(v)$. Ideally, one wants to minimize the continuous loss functional defined by

$$\mathcal{L}_\infty(v) = \int_\Omega \mathrm{L}_{\text{re}}[v](x) d\omega(x) + \int_\Gamma \mathrm{L}_{\text{bc}}[v](x) d\mu(x), \tag{3.8}$$

where ω and μ are probability measures typically determined by the underlying PDEs and the corresponding solution space. However, since the exact calculations of the integrals are not available, one has to rely on reasonable discretization or approximation. The work of Sirignano and Spiliopoulos (2018) proposed an NN approach, namely, the deep Galerkin method, which aims at minimizing the ideal loss (3.8) by using the stochastic mini-batch gradient descent (see also Section 5.2).

The loss functional can be further generalized by introducing nonhomogeneous weights for the summands:

$$\mathcal{L}(v; \boldsymbol{\lambda}) = \sum_{x\in\Omega_M} \lambda_x \Big(\mathcal{D}[v](x) - f(x)\Big)^2 + \sum_{x\in\Gamma_{M'}} \lambda_x \Big(\mathcal{B}[v](x) - g(x)\Big)^2, \tag{3.9}$$

where $\boldsymbol{\lambda}$ is the collection of all the λ_x, which includes the application of Monte Carlo integration when $\lambda_x = \frac{1}{M}$ if $x \in \Omega_m$ and $\lambda_x = \frac{1}{M'}$ if $x \in \partial\Omega_{M'}$.

It is worth mentioning that a minimizer u_n^* depends on Ω_M, $\partial\Omega_{M'}$, $\boldsymbol{\lambda}$ and V_n.

3.3 Weak/variational formulations

A weak formulation of the general equation may be written as follows. Let V be a Hilbert space satisfying $\mathcal{B}[u] = g$ for all u. For general discussion, let us define

$$a(u, v) := \langle \mathcal{D}[u], v \rangle, \quad L(v) := \langle f, v \rangle,$$
$$\mathrm{L}_{\mathrm{re}}[v](x) := [\mathcal{D}[v](x) - f(x)]^2, \quad \mathrm{L}_{\mathrm{bc}}[v](x) := [\mathcal{B}[v](x) - g(x)]^2,$$

where $\langle \cdot, \cdot \rangle$ is an appropriate inner product. The variational formulation then seeks to find $u \in V$ such that

$$a(u, v) = L(v) \quad \forall v \in V.$$

The NN approaches seek to solve the variational problem with a subset of V, denoted by V_n of NNs, which is the set of approximate solutions.

Let us consider the Poisson equation to illustrate the variational approach. Let $\mathcal{D} = -\Delta$, $g = 0$ and $\Omega = (0, 1)$. Let $V = H_0^1(\Omega) = \{v \in H^1(\Omega) : v(0) = v(1) = 0\}$, $\langle u, v \rangle = \int_0^1 u(x)v(x)dx$ for all $u, v \in V$ and $f \in L^2(\Omega)$. It then follows from the integration by parts that

$$a(u, v) = \langle u', v' \rangle, \quad L(v) = \langle f, v \rangle.$$

If V_n is a linear space spanned by $\{\phi_j \in V : j = 1, .., n\}$, the above is equivalent to solving a linear system of equations of the form

$$Ac = b, \quad \text{where} \quad A_{ij} = a(\phi_i, \phi_j), \quad b_i = L(\phi_i). \tag{3.10}$$

The solution is then given by $u_n^* = \sum_{j=1}^n \hat{c}_j \phi_j$ where $\hat{c}$ is the solution to $Ac = b$, which basically describes the finite element method (FEM). If V_n is a class of neural networks, since V_n is not a linear space, the FEM-like approach cannot apply. Therefore, one needs new approaches to solve the variational problem using NNs.

Ritz-Galerkin approach. The work of Ainsworth and Dong (2021) proposed an NN approach based on a standard Galerkin approximation of the variation equation, namely Galerkin Neural Networks (GNNs). Let $a : V \times V \mapsto \mathbb{R}$ be a bounded symmetric, bilinear form satisfying $a(v, v) \geq 0$ and $a(v, v) = 0$ if and only if $v = 0$, which defines an associated norm given by $|||v||| := \sqrt{a(v, v)}$.

For a given basis $\{\phi_i\}_{i=1}^n$, let us define $u_n^* = \mathrm{GALERKIN}(a, L, \{\phi_i\}_{i=1}^n)$ where u_n^* is the solution obtained from (3.10). The GNN method seeks basis functions from a class of NNs and provides the corresponding Galerkin NN approximation. The GNN constructs one basis at a time and gradually augments the basis as the algorithm proceeds. The GNN mechanism is built based on the following result.

Proposition 3.8 (Proposition 2.3 of Ainsworth and Dong, 2021). *Let u be the solution to the variational formulation and $u_0 \in V$ be a given approximation. Let $\varphi_1 = \frac{u-u_0}{|||u-u_0|||}$ and let $r(u_0) : V \to \mathbb{R}$ be a functional given by the rule $r(u_0)[v] := L(v) - a(u_0, v)$. Then, $r(u_0)[\varphi_1] = \max_{v \in B} r(u_0)[v]$ where B is the closed unit ball in V.*

The proposition established the relationship between the residual $r(u_0)$ and the approximation error $u - u_0$. Based on this error indicator, the GNN commences with an initial guess u_0 and recursively augments the basis according to the following. For $k = 1, \ldots, n$,

$$v_k := \underset{v \in V_n, |||v||| \neq 0}{\operatorname{argmax}} \; r(u_{k-1})[\bar{v}] = \underset{v \in V_n, |||v||| \neq 0}{\operatorname{argmax}} \; \frac{L(v) - a(u_{k-1}, v)}{|||v|||},$$

and $u_k = \text{GALERKIN}(a, L, \{\varphi_i^{\text{NN}}\}_{i=0}^k)$ where $\varphi_i^{\text{NN}} = \frac{v_i}{|||v_i|||}$ with $v_0 = u_0$. The GNN approximation to the variational formulation is then given by u_n where n is a user-defined input representing the number of NN basis functions. The original work allows one to use different V_n for each v_k to further facilitate the training and improve the performance. The convergence of the GNN with respect to the number of basis functions is proved in Ainsworth and Dong (2021).

Minmax formulation. The work of Zang et al. (2020) followed the min-max formulation

$$\min_{u \in V_n} \left(\max_{v \in V_n} \frac{|a(u, v) - L(v)|^2}{\|v\|^2} \right),$$

and the approach was termed the weak adversarial network (WAN). In practice, the integration is approximated by a quadrature rule. Also, similar to the deep Ritz method, the boundary loss term is added. Let $\{x_k, w_k\}_{k=1}^M$ be a quadrature rule over $[0, 1]$. Then the loss functional is given by

$$\mathcal{L}(u, v) = \frac{\left| \sum_{k=1}^M w_k \cdot \left(u'(x_k) v'(x_k) - f(x_k) v(x_k) \right) \right|^2}{\left| \sum_{k=1}^M w_k \cdot v^2(x_k) \right|^2}.$$

To further improve the training, the WAN approach takes the log of the loss and aims to solve the minmax problem

$$\min_{u \in V_n} \left(\max_{v \in V_n} \log \mathcal{L}(u, v) + \mathcal{L}_{\text{bc}}(u) \right).$$

Petrov-Galerkin approach. Several works investigated NN approaches that use the Petrov-Galerkin (PG) formulation. Let W be a Hilbert space that differs from V. The PG seeks to find $u \in V$ such that $a(u, v) = L(v)$ for all $v \in W$. The corresponding NN approach uses a fixed finite-dimensional linear space of

W_n and seeks a solution in a class V_n of NNs. The works of Khodayi-Mehr and Zavlanos (2020); Shang et al. (2022) used linear finite element spaces and the work of Kharazmi et al. (2019) used sine functions, Legendre polynomials as test spaces. In Kharazmi et al. (2021), the *hp*-variational physics-informed neural networks were proposed as a general framework that combines the PG approach and *hp*-refinement via domain decomposition, where piecewise polynomials were used for test functions.

3.4 Extended PINN: domain decomposition

Extended PINNs (XPINNs) (Jagtap and Karniadakis, 2020) combine a domain decomposition method with PINNs. Let $\{\Omega_i\}_{i=1}^N$ be a partition of the domain Ω and consider a set of subproblems:

$$\begin{aligned} \mathcal{D}[u_i](x) &= f(x) \quad x \in \Omega_i \\ \mathcal{B}[u_i](x) &= g(x) \quad x \in \partial\Omega_i \cap \Gamma \\ u_i(x) &= u_j(x) \quad x \in \partial\Omega_i \cap \partial\Omega_j. \end{aligned} \tag{3.11}$$

Then, the solution to the original problem (3.1) is written in terms of u_j's as $u(x) = \sum_{j=1}^N u_j|_{\Omega_j}(x)$, where $u|_{\Omega_j}(x) = u(x)$ if $x \in \Omega_j$ and 0 if $x \notin \Omega_j$. The XPINN aims at approximating each u_j using a NN, say $v_j \in V_n$. While XPINNs can be formulated in many different ways (e.g. variational formulation), for ease of discussion, we focus on the original formulation based on the strong form. Let us define the interface term by

$$\begin{aligned} \mathrm{L}_{\mathrm{if}}[v, u](x) = \beta_0(v(x) - u(x))^2 + \beta_1 \|\nabla v(x) - \nabla u(x)\|_2^2 \\ + (\mathcal{D}[v](x) - \mathcal{D}[u](x))^2, \end{aligned}$$

for some $\beta_0, \beta_1 \geq 0$, where "if" stands for interface.

Let $\Omega_{M_j} \subset \Omega_j$, $\partial\Omega_{M_j'} \subset \partial\Omega_j$, $\Gamma_j = \Gamma \cap \partial\Omega_{M_j'}$, and $\partial\Omega_{M_{ij}'} \subset \partial\Omega_{M_i'} \cap \partial\Omega_{M_j'}$ be finite sets of interest and consider the corresponding empirical probability distributions: $\omega_j(\cdot) = \frac{1}{M_j} \sum_{x \in \Omega_{M_j}} \delta_x(\cdot)$, $\mu_j(\cdot) = \frac{1}{|\Gamma_j|} \sum_{x \in \Gamma_j} \delta_x(\cdot)$, and $\bar{\mu}_{ij} = \frac{1}{M_{ij}'} \sum_{x \in \partial\Omega_{M_{ij}'}} \delta_x(\cdot)$. Then, the loss functional for XPINNs is given by

$$\begin{aligned} \mathcal{L}[v_1, \ldots, v_N] = \sum_{j=1}^N \Bigg[\int_{\Omega_j} \mathrm{L}_{\mathrm{re}}[v_j](x) d\omega_j(x) + \int_{\Gamma_j} \mathrm{L}_{\mathrm{bc}}[v_j](x) d\mu_j(x) \\ + \sum_{i \neq j} \int_{\partial\Omega_i \cap \partial\Omega_j} \mathrm{L}_{\mathrm{if}}[v_i, v_j](x) d\bar{\mu}_{ij}(x) \Bigg], \end{aligned}$$

where v_j's are all in V_n (e.g. a class of neural networks). Let $\{u_j^*\}_{j=1}^N$ be a minimizer of the loss functional. Then, the NN approximation by the XPINN formulation is given by

$$u_{\mathrm{XPINN}}(x)=\sum_{j=1}^{N} u_j^*(x)\cdot \mathbb{I}_j(x), \qquad \mathbb{I}_j(x)=\begin{cases} 0 & \text{if } x\notin \Omega_j \\ \frac{1}{S_j(x)} & \text{if } x\in \partial\Omega_j \\ 1 & \text{otherwise,} \end{cases}$$

where $S_j(x)$ represents the number of subdomains whose boundary contains x. Note that the XPINN does not necessarily preserve the continuity at interfaces.

The performance of XPINN can be enhanced by adding the jump condition at the interface, e.g., in Jagtap et al. (2020) and De Ryck et al. (2023) in the loss function. In Hu et al. (2022), it is shown that XPINN is not necessarily performing better than PINN. The XPINN can have less complex parts and simple locally but less data in each part can lead to overfitting and thus affect the accuracy.

More aspects of domain decompositions techniques have been explored, such as in Sheng and Yang (2022) (variational formulation) and Shukla et al. (2021) addressing parallel computation, Meng et al. (2020) and Penwarden et al. (2023) for time domains, Kim and Yang (2023) using overlapping subdomains, and Kopaničáková et al. (2023) using preconditioning, and Sun et al. (2023) assigning different boundary conditions on different subdomains.

3.5 Useful techniques

We discuss some techniques and methods that may improve the performance of the aforementioned formulations.

Enforcement of boundary conditions. The aforementioned NN approaches do not satisfy the boundary conditions from the beginning and require an additional loss term which penalizes the discrepancy between the NN prediction on boundary and the given boundary conditions. This could be cumbersome when it comes to the PINN method as well as many variational approaches. In general, the loss functionals comprise multiple terms and each term may have significantly different scales. If this is the case, training algorithms would focus on minimizing the largest term while the other terms marginally change.

In principle, one can use the distance function and modify the NN structure. The distance function d to the boundary $\partial\Omega$ is defined to be the function that gives the shortest distance between any point x to $\partial\Omega$. Thus, for example,

$$\overline{u}_{\mathrm{NN}}(x;\theta)=\overline{g}(x)+\mathrm{d}(x)u_{\mathrm{NN}}(x;\theta),$$

where $\overline{g}$ is a smooth extension of g, will exactly satisfy the Dirichlet boundary condition as $\mathrm{d}(x)=0$ on $\partial\Omega$. However, the numerical solution here is only Lipschitz continuous as $\mathrm{d}(x)$ is. Several works followed this principle with some variations (see Lagaris et al., 1998; Berg and Nyström, 2018). This idea can be further generalized. In particular, the work of Sukumar and Srivastava (2022) used the theory of R-functions to construct an approximate distance function ξ that allows one to satisfy exactly inhomogeneous Dirichlet, Neumann, and Robin boundary conditions on complex geometries.

Constraints. Instead of modifying NN structures, one can formulate the problem as a constrained optimization problem via the augmented Lagrangian method. For example, the PINN loss may be written as

$$\min_{\theta} \mathcal{L}_{\text{re}}(\theta) \quad \text{subject to} \quad \mathcal{B}[u_{\text{NN}}(\cdot;\theta)](x) = g(x) \quad \forall x \in \Gamma_{M'},$$

which takes the boundary conditions as constraints. The method of Lagrange multipliers will then construct the Lagrangian defined by $\mathcal{L}(\theta,\lambda) = \mathcal{L}_{\text{re}}(\theta) + \langle \lambda, h(\theta)\rangle$ where $h(\theta) := (\mathcal{B}[u_{\text{NN}}(\cdot;\theta)](x) - g(x))_{x\in\Gamma_{M'}}$ and seek to solve the minmax problem of $\max_\lambda \min_\theta \mathcal{L}(\theta,\lambda)$. The augmented Lagrangian method can also apply, which seeks to solve the minmax problem of

$$\max_{\lambda} \min_{\theta} \mathcal{L}(\theta,\lambda) + \beta \|h(\theta)\|^2,$$

for some appropriately chosen $\beta > 0$. Since $\|h(\theta)\|^2 = M'\mathcal{L}_{\text{bc}}(\theta)$, this may be viewed as a combination of the original PINN loss and the Lagrangian. The work of Son et al. (2023) proposed this formulation, namely, the augmented Lagrangian relaxation method for PINNs. See also Lu et al. (2021b) where the augmented Lagrangian is used to enforce constraints.

Continuity equations. One can impose a certain structure into the NN. For example, we can define a divergence-free NN by using a curl field, e.g., in Richter-Powell et al. (2022).

Sobolev formulation. In the design of loss functions, we can take Sobolev spaces for Y and Z in (3.6) instead of L^2 or L^p spaces. For example, we can take $Y = H^1(\Omega)$, where the norm of $v \in Y$ is defined by $\sqrt{\|v\|_{L^2}^2 + \sum_{i=1}^d \|\partial_{x_i} v\|_{L^2}^2}$. Similarly, we can consider $Z = H^1(\Gamma)$ instead of $L^2(\Gamma)$. Then we can apply proper discretizations on the integrals and obtain a loss functional for training, see e.g., in Yu et al. (2022) and Son et al. (2021). Even higher derivatives of the residuals are considered in Yu et al. (2022) and Son et al. (2021).

Adaptive activation function. The motivation of adaptive activation functions (Jagtap et al., 2022b) is that different activation functions can significantly affect the spectral gaps of neural networks and thus the training results. Two extreme cases are sine or hyperbolic tangent (smooth) and the ReLu (nonsmooth). To appreciate different activation functions, one can set the form of the activation functions as $\sigma(\cdot) = \sum_{i=1}^r a_i \sigma_i(\cdot)$, where σ_i's are the commonly used activation functions which users prefer and a_i's are trainable logical values.

Separable PINNs. In separable PINNs (Cho et al., 2022), the following network architecture $\sum_{j=1}^r \prod_{i=1}^d N_{i,j}(x_i)$ is used, where $N_{i,j}(x_i)$ is a feedforward neural network in x_i only. Compared to the feedforward neural networks as a function of $(x_1, \ldots, x_d)$ in (2.1), PINNs can be trained faster by 10-100 times. The acceleration is due to the tensor-product nature of the network architecture, which facilitates auto-differentiation. Similar network architectures of low rank are considered in Wang et al. (2022a,b); Zhang et al. (2023a).

PINNs for high dimensions. A random batch method in dimensionality has been developed in Hu et al. (2023), where the computational cost of derivatives of feed-forward neural networks consists of computing the derivative in a few randomly picked dimensions in batches and thus the computational time becomes feasible even when the dimension is very high.

4 Approximation error for PINN in strong formulations

Motivated by the stability, we can directly derive an (posterior) error estimate. Let $u* \in X$ be a solution to (3.1).

Assume that f and g are continuous (otherwise we use their approximations which can be evaluated at the sampling points). Assume the neural network u_{NN} is smooth enough such that $\mathcal{D}u_{NN}$ and $\mathcal{B}u_{NN}$ is well-defined at sampling points. Suppose that $\epsilon = 0$, i.e., we have no regularization.

4.1 A posteriori estimate

Let $X = L^2(\Omega)$ and $Y = L^2(\Gamma)$. By the conditional stability (3.3), we obtain

$$\begin{aligned} \mathcal{N}(u_{NN} - u^*) &\leq \rho_C(\|\mathcal{D}u_{NN} - f\|_Y + \|\mathcal{B}u_{NN} - g\|_Z) \\ &\leq \rho_C\sqrt{2\,\|\mathcal{D}u_{NN} - f\|_Y^2 + 2\,\|\mathcal{B}u_{NN} - g\|_Z^2)} \\ &= \rho_C(\sqrt{2\mathcal{L}(v,\lambda)} + \sqrt{2\epsilon_{\text{appr, f}} + 2\epsilon_{\text{appr, g}}}) \end{aligned}$$

where we have applied the inequality $\sqrt{a+b} \leq \sqrt{a} + \sqrt{b}$ for $a, b \geq 0$ and denote

$$\epsilon_{\text{appr,f}} = \|\mathcal{D}u_{NN} - f\|_Y^2 - \sum_{x \in \Omega_M} \frac{1}{\#\Omega_M}(\mathcal{D}u_{NN} - f)^2(x),$$

$$\epsilon_{\text{appr,g}} = \|\mathcal{B}u_{NN} - g\|_Y^2 - \sum_{x \in \Gamma_M} \frac{1}{\#\Gamma_M}(\mathcal{B}u_{NN} - g)^2(x)$$

Both quantities can be estimated by the upper bounds of their corresponding Radmacher complexity. If the sampling points are quadrature points and weights are points-dependent, we may improve the bounds of these quantities by errors of quadrature points, if all functions are smooth. We refer to Shin et al. (2023) (applying Radmacher complexity) and Mishra and Molinaro (2023, 2022) (using quadrature bounds) for more detailed analysis.

4.2 A priori estimate

Let $\mathcal{S}u^*$ be a good mollifier of u^* so that one can make of the derivatives of u^* at the sampling points. By the conditional stability (3.3), we obtain

$$\begin{aligned}\mathcal{N}(u_{NN} - u^*) &\le \rho_C(\|\mathcal{D}u_{NN} - f\|_Y + \|\mathcal{B}u_{NN} - g\|_Z) \\ &\le \rho_C(\underbrace{\left\|\mathcal{D}(u_{NN} - \mathcal{S}u^*)\right\|_Y + \left\|\mathcal{B}(u_{NN} - \mathcal{S}u^*)\right\|_Z}_{\text{approximation}} \\ &\quad + \underbrace{\left\|f - \mathcal{D}\mathcal{S}u^*\right\|_Y + \left\|(g - \mathcal{B}\mathcal{S}u^*)\right\|_Z}_{\text{mollifying}}).\end{aligned}$$

Observe that $\|\mathcal{D}(u_{NN} - \mathcal{S}u^*)\|_Y + \|\mathcal{B}(u_{NN} - \mathcal{S}u^*)\|_Z$ can be bounded by certain norms of $u_{NN} - \mathcal{S}u^*$ and their derivatives. Thus this error can be small by the universal approximation theorem of the underlying neural networks and convergence rate can be quantified. The mollifying error can be estimated by the classical mollifier $\int_{\mathbb{R}^d} \epsilon^{-d}\varphi(\frac{x-y}{\epsilon})u(y)\,dy$, where $\epsilon^{-d}\varphi(\frac{x-y}{\epsilon})$ is a compactly supported function approximating the Dirac delta function in the distribution sense. With some mild moment conditions on φ, the convergence order of mollifier depends on the smoothness of the solution u^*. Also, $\mathcal{S}u^*$ can be some approximate solution by certain numerical methods. A similar idea is presented in Shin et al. (2023) in more details.

The strong formulation may require high smoothness which the solution may not admit. While the estimate here is insightful, it may not be useful in practice, especially when solutions are not smooth. This is a strong motivation of using weak formulations.

5 Training/optimization methods

The PINN methodology requires to solve an optimization problem. In practice, one needs to properly choose a numerical optimization algorithm to solve it. First-order gradient-based optimization algorithms have been popularly employed in this regard mainly due to their simple implementations and reasonably good empirical performance for diverse problems. We briefly discus several numerical optimization algorithms for the PINN methodology.

Let V_n be a class of neural networks, e.g., V_L defined in (2.2). Optimization problems defined on V_n are equivalently written as optimization problems on a finite-dimensional space, e.g., $\mathbb{R}^{|\vec{n}_L|}$. That is, by letting $\mathcal{L}(\theta) := \mathcal{L}(u_{\text{NN}}(\cdot;\theta))$ for any $u_{\text{NN}} \in V_L$, we are concerned with solving the following minimization problem

$$\min_{\theta \in \mathbb{R}^{|\vec{n}_L|}} \mathcal{L}(\theta). \tag{5.1}$$

In what follows, we briefly discuss popular optimization methods and strategies for the PINN methodology.

5.1 Initialization schemes

How to initialize hyperparameters of neural networks plays a pivotal role when it comes to iterative algorithms. As illustrated by the convergence analysis of Newton's method, a well-chosen initialization can dramatically improve the performance in practice. Several works have studied random initialization schemes for deep NNs. Popular initialization schemes use either Gaussian or uniform distributions for weight matrices and set bias vectors being zeros. See the table below. The underlying principle of many initialization schemes is to make the unit variance to avoid the so-called exploding/vanishing gradient, which refers to the phenomenon in which deep NNs yield either too big or too small gradients. Very large gradients will make training algorithms diverge and very small gradients will take forever for algorithms to converge. This requires one to compute the variance with respect to all the weights.

The Glorot initialization (Glorot and Bengio, 2010) was proposed based on the linearity assumption. Suppose $\phi(x) = x$, also known as the linear activation function). Since all the bias vectors are initialized to zeros, assuming W^{ℓ}'s are independently initialized whose element has mean zero and variance σ_{ℓ}^2, it can be checked that the variance of the i-th element of u^{ℓ} is

$$\mathrm{Var}[u_i^{\ell}(x)] = \prod_{j=1}^{\ell} (n_{j-1}\sigma_j^2) \cdot \|x\|_2^2 \quad \forall 1 \le i \le n_{\ell}.$$

Let $s^{\ell}(x) = W^{\ell} u^{\ell-1}(x) + b^{\ell}$. It follows from the chain rule that

$$\frac{\partial u_{\mathrm{NN}}}{\partial s_k^{\ell}} = \left(\frac{\partial s^{\ell+1}}{\partial s_k^{\ell}}\right)^{\top} \frac{\partial u_{\mathrm{NN}}}{\partial s^{\ell+1}} = \phi'(s^{\ell})(W_{:,k}^{\ell+1})^{\top} \frac{\partial u_{\mathrm{NN}}}{\partial s^{\ell+1}},$$

where $W_{:,k}^{\ell+1}$ is the k-th column of $W^{\ell+1}$. Since $\phi'(x) = 1$ and $\frac{\partial u_{\mathrm{NN}}}{\partial s^{\ell+1}} = (W^L \cdots W^{\ell+2})^{\top}$, we have

$$\mathrm{Var}\left[\frac{\partial \mathcal{L}}{\partial s_k^{\ell}}\right] = \prod_{j=\ell+1}^{L} (n_j \sigma_j^2) \cdot \|x\|_2^2 \quad \forall 1 \le k \le n_{\ell}.$$

The Glorot initialization aims at satisfying

$$\mathrm{Var}[u_k^{\ell}(x)] = \mathrm{Var}[u_1^1(x)], \quad \mathrm{Var}\left[\frac{\partial \mathcal{L}}{\partial s_k^{\ell}}\right] = \mathrm{Var}\left[\frac{\partial \mathcal{L}}{\partial s_1^{1}}\right], \quad \forall \ell, k,$$

which are equivalent to $n_{\ell}\sigma_{\ell}^2 = 1$ and $n_{\ell-1}\sigma_{\ell}^2 = 1$ for all ℓ. Since the two are met only if $n_{\ell} = n_{\ell-1}$, as a compromise, the Glorot initialization (Glorot and Bengio, 2010) sets

$$(\text{Glorot initialization}): \quad \sigma_{\ell}^2 = \frac{2}{n_{\ell} + n_{\ell-1}}.$$

The Kaiming initialization scheme (He et al., 2015) was developed based on the rectified linear unit (ReLU) activation function. Let $\phi(x) = \max\{x, 0\}$ and suppose that the bias vectors are initialized to zeros and the weight matrices are independently initialized from symmetric distributions around zero. Let σ_ℓ^2 be the variance of each component of W^ℓ. It follows from the symmetric distribution assumption that $\mathbb{E}[(u_k^{\ell-1}(x))^2] = \frac{1}{2}\text{Var}[s_k^{\ell-1}(x)]$ and thus

$$\begin{aligned}\text{Var}[s_i^\ell(x)] = \sigma_\ell^2\mathbb{E}[\|u^{\ell-1}(x)\|_2^2] &= \frac{1}{2}\sigma_\ell^2\sum_{k=1}^{n_{\ell-1}}\text{Var}[s_k^{\ell-1}(x)] \\ &= \frac{1}{2}\sigma_\ell^2 n_{\ell-1}\text{Var}[s_i^{\ell-1}(x)].\end{aligned}$$

The Kaiming initialization enforces $\text{Var}[s_i^\ell(x)] = \text{Var}[s_1^1(x)]$ for all ℓ and i, which results in the condition of

$$(\text{Kaiming initialization}): \quad \sigma_\ell^2 = \frac{2}{n_{\ell-1}}.$$

	weight matrix	bias vector	hyperparameters	note
Glorot Normal (Glorot and Bengio, 2010)	$\mathcal{N}(0, \sigma^2)$	0	$\sigma^2 = \frac{2}{n_\ell + n_{\ell-1}}$	Linear, Tanh
Glorot Uniform (Glorot and Bengio, 2010)	$\mathcal{U}(-a, a)$	0	$a^2 = \frac{6}{n_\ell + n_{\ell-1}}$	
Kaiming Normal (He et al., 2015)	$\mathcal{N}(0, \sigma^2)$	0	$\sigma^2 = \frac{2}{n_\ell}$	ReLU and variants
Kaiming Uniform (He et al., 2015)	$\mathcal{U}(-a, a)$	0	$a^2 = \frac{6}{n_\ell}$	
RAI (Lu et al., 2020)	$\mathcal{N}(0, \sigma^2)$ + Beta(2, 1)		$\sigma^2 = \frac{0.6007^2}{n_{\ell-1}}$	Preventing Dying ReLU

Dying ReLU. Dying ReLU refers to the phenomenon in which ReLU neurons are inactivated and return zeros for all inputs. It has been empirically observed that this could happen not only at the initialization but also during training. If a deep NN was initialized as a constant function, i.e., all the neurons were dead, the network cannot be trained by gradient-based algorithms as the gradient is zero. The work of Lu et al. (2020) estimated the probability of such an event (namely "born dead probability" in Lu et al. (2020)) with standard initialization schemes including Glorot (Glorot and Bengio, 2010) and He (He et al., 2015).

Theorem 5.1 (Theorem 3.2 of Lu et al., 2020). *Let $\vec{n} = (n_1, \ldots, n_{L-1})$ be an architecture. Suppose that all weights are independently initialized from symmetric continuous probability distributions around zero, and all biases are either drawn from a symmetric distribution or set to zero. Then, the probability that the initialized deep ReLU NN is born dead is upper bounded by $1 - \prod_{\ell=1}^{L-1}(1 - 2^{-n_\ell})$.*

The randomized asymmetric initialization (RAI) was proposed in Lu et al. (2020) to alleviate the dying ReLU by breaking the symmetry in distributions. Let $V^\ell = [W^\ell, b^\ell] \in \mathbb{R}^{n_\ell \times (n_{\ell-1}+1)}$. The RAI scheme works as follows: $W^1_{ij} \sim N(0, \frac{2}{d_{\text{in}}}$ and $b^1 = 0$ as is in the Kaiming normal initialization. For every $\ell \geq 2$ and $j \in \{1, \ldots, n_\ell\}$, (a) randomly select an index k^ℓ_j from $\{1, \ldots, n_{\ell-1}, n_{\ell-1}+1\}$, and (b) initialize $(V^\ell_j)_{-k^\ell_j} \sim N(0, \sigma^2_\ell I)$ and $(v^\ell_j)_{k^\ell_j} \sim \text{Beta}(2, 1)$ where $\sigma^2_\ell = \frac{0.6007^2}{n_{\ell-1}}$, V^ℓ_j is the j-th row of V^ℓ, $(V^\ell_j)_k$ is the k-th component of V^ℓ_j, $(V^\ell_j)_{-k}$ is a vector constructed from V^ℓ_j by omitting the k-th component. Here 0.6007 is an approximation of $-\frac{2\sqrt{2}}{3\sqrt{\pi}} + \sqrt{1 + \frac{8}{9\pi}}$, which is derived from the second moment analysis (Lu et al., 2020).

Data-dependent bias initialization. For two-layer NNs, the aforementioned schemes often do not perform well as these schemes were meant for deep NNs. In Shin and Karniadakis (2020), the data-dependent bias initialization scheme for two-layer ReLU NNs was proposed which located neurons at data points. Given W^1, the scheme initializes the bias vector b^1 according to $(b^1)_k = -w_k^\top x_{i_k}$ where x_{i_k} is a randomly selected input data point. By construction, the two-layer ReLU NN at initialization is written as

$$\text{(Data-dependent Bias Initialization)}: \quad u_{\text{NN}}(x) = \sum_{k=1}^{n_1} W^2_{:,k} \phi(W^1_k (x - x_{i_k})),$$

$$\text{(Zero Bias Initialization)}: \quad u_{\text{NN}}(x) = \sum_{k=1}^{n_1} W^2_{:,k} \phi(W^1_k x),$$

where $W^2_{:,k}$ is the k-th column of W^2. When the biases are initialized as zeros, all initialized neurons are clustered at the origin. Consequently, it would take a long time for training algorithms to distribute neurons over the training domain to achieve a small training loss. In the worst case, along the way of distributing neurons, it will get stuck at the local minimum. The data-dependent bias initialization ensures that all neurons are randomly distributed over the training data points.

5.2 Generic methods: stochastic gradient descent

The first-order gradient-based optimization algorithms have been popularly employed when it comes to (5.1) due to their simple implementation and reasonably good empirical performance.

The vanilla gradient descent (GD) algorithm commences with a random guess $\theta^{(0)}$ (following one of the initialization schemes) and updates the parameter according to the rule: for $k = 1, 2, \ldots,$

$$\text{(Full-batch GD)}: \quad \theta^{(k)} = \theta^{(k-1)} - \eta_k \cdot \nabla \mathcal{L}(\theta^{(k-1)}), \tag{5.2}$$

where η_k is the learning rate or the stepsize for the k-th iteration. This is also known as the full-batch training.

If the underlying computational domain is in either 3D or 4D (including time), the discrete loss evaluation may require a large number of points in computational domains. Consequently, the computation of the gradient becomes very expensive. To alleviate such difficulty, (stochastic) mini-batch GD approaches are popularly employed. There are two main approaches. The first one is the subsampling from a fixed set containing a large number of points. Let $\Omega_M \subset \Omega$ be a fixed set with $M \gg 1$. Let m be a reasonably small number satisfying $1 \le m \ll M$, which is referred to as the batch size. In the PINN formulation, the full gradient of $\mathcal{L}$ is written as

$$\nabla \mathcal{L}(\theta) = \frac{1}{M} \sum_{x \in \Omega_M} \nabla \mathrm{L}_{\mathrm{re}}[\theta](x) + \lambda \nabla \mathcal{L}_{\mathrm{bc}}(\theta),$$

where $\mathrm{L}_{\mathrm{re}}[\theta](x) := \mathrm{L}_{\mathrm{re}}[u_{\mathrm{NN}}(\cdot;\theta)](x)$. It can be seen that the full gradient requires M evaluations of $\nabla \mathrm{L}_{\mathrm{re}}[\theta](\cdot)$ which can be computationally demanding as $M \gg 1$. The computation cost grows linearly with respect to the number M of sample points. Thus, the mini-batch training is popularly used in practice. For the k-th iteration, let $\Omega_m^{(k)} \subset \Omega_M$ and define

$$\nabla \mathcal{L}(\theta^{(k-1)}; \Omega_m^{(k)}) := \frac{1}{|\Omega_m^{(k)}|} \sum_{x \in \Omega_m^{(k)}} \nabla \mathrm{L}_{\mathrm{re}}[\theta^{(k-1)}](x) + \lambda \nabla \mathcal{L}_{\mathrm{bc}}(\theta), \tag{5.3}$$

whose computation requires only m evaluations of $\nabla \mathrm{L}_{\mathrm{re}}[\theta](\cdot)$. The mini-batch gradient descent algorithm then updates the parameter according to the rule:

$$\text{(Mini-batch GD)}: \quad \theta^{(k)} = \theta^{(k-1)} - \eta_k \cdot \nabla \mathcal{L}(\theta^{(k-1)}; \Omega_m^{(k)}), \tag{5.4}$$

where the sampling set $\Omega_m^{(k)}$ varies over iterations. Depending on how these sampling sets are chosen, different mini-batch schemes are obtained. Let $\Omega_M = \{x_j\}_{j=1}^M \subset \Omega$ and let $M = qm + r$ where q is the quotient and $r \in \{1, \ldots, m\}$ is the remainder. Define

$$I_k = \begin{cases} \{1 + (k-1)m, \ldots, m + (k-1)m\} & \text{if } k \not\equiv 0 \pmod{q+1}, \\ \{1 + qm, \ldots, M\} & \text{if } k \equiv 0 \pmod{q+1}. \end{cases}$$

Let S_M be the permutation group of the set $\{1, \ldots, M\}$. Whenever $(k-1) = s(q+1)$ for some nonnegative integer s, let $\boldsymbol{\sigma}_s$ be a randomly uniformly chosen element from S_M and define

$$\sigma^{(k)} := \begin{cases} \boldsymbol{\sigma}_s & \text{if } k \equiv 1 \pmod{q+1}, \\ \sigma^{(k-1)} & \text{if } k \not\equiv 1 \pmod{q+1}. \end{cases}$$

We are now in a position to describe some popular mini-batch schemes:

$$\begin{aligned}
\text{(Mini-batch without shuffling)}: &\quad \Omega_m^{(k)} = \{x_j\}_{j\in I_k},\\
\text{(Mini-batch with shuffling)}: &\quad \Omega_m^{(k)} = \{x_{\sigma^{(k)}(j)}\}_{j\in I_k},\\
\text{(Mini-batch with replacement)}: &\quad \Omega_m^{(k)} \sim \mathrm{P},
\end{aligned}$$

where P is the uniform probability distribution over the set of all subsets containing m points from Ω_M. Among the above three schemes, the mini-batch without shuffling is considered as deterministic while the other two are stochastic. An epoch is a unit that refers to the number of iterations taken to sweep the entire M points ($q+1$ iterations). In the full batch training, an epoch is equivalent to an iteration.

The other is based on random sampling from the underlying probability measure ω defined on Ω. The distinct feature of this approach is that the reference set Ω_M needs not to be formed. Rather, every iteration requires newly sampled fresh m points from the underlying probability distribution ω. Specifically,

$$\begin{aligned}
\text{(Stochastic Mini-batch)}: \quad \Omega_m^{(k)} = \{x_j^{(k)}\}_{j=1}^m&\\
\text{where} \quad \{x_j^{(k)}\} \subset \Omega \text{ are i.i.d. samples from } \omega.&
\end{aligned}$$

At every iteration, one has the option to design an appropriate learning rate or stepsize η_k. It has been widely known that the learning rates should be chosen sufficiently small for convergence and, at the same time, sufficiently large for making meaningful progress within practically reasonable maximum epochs/iterations (e.g. 100K or 1M). From the optimization perspective, the optimal learning rate is possibly found by following the minimization principle. Given the current update direction $d^{(k-1)}$ (e.g. $d^{(k-1)} = -\nabla\mathcal{L}(\theta^{(k-1)}; \Omega_m^{(k)})$), the principle seeks the best learning rate that solves

$$\eta_k^* = \underset{\eta\in\mathbb{R}}{\operatorname{argmin}}\, \mathcal{L}\left(\theta^{(k-1)} + \eta d^{(k-1)}\right).$$

However, the optimization problem is difficult to solve in general and the optimum solution is only available for a small class of problems (e.g. convex, linear). In addition, the mini-batch version of the minimization rule remains elusive which further prevents its use in practice. Hence, practically popular designs heavily rely on users' trial-and-errors, which require problem-dependent hyperparameters to be tuned appropriately.

Let T be a positive integer representing the number of iterations on which the learning rate remains unchanged. While the initial learning rate η_0 is chosen in a problem-dependent manner, some popularly used values lie in the range of $[10^{-5}, 10^{-3}]$. The decay factor is denoted by γ and the maximum and minimum learning rates are denoted by $\eta_{\max}$ and $\eta_{\min}$, respectively. Some popular schedulers include

$$\begin{aligned}
\text{[Constant learning rate]} \quad & \eta_k = \eta_0, \\
\text{[Step Scheduler]} \quad & \eta_k = \eta_0 \gamma^{\lfloor \frac{k}{T} \rfloor}, \\
\text{[Exponential Scheduler]} \quad & \eta_k = \eta_0 \gamma^k, \\
\text{[Cosine Scheduler]} \quad & \eta_k = \eta_{\min} + \frac{\eta_{\max} - \eta_{\min}}{2}(1 + \cos(\pi k/T)).
\end{aligned}$$

In practice, especially when it comes to the PINN methodology, the vanilla (stochastic) GD is rarely used, rather, their variants are typically employed. One of the most popular variants is the adaptive moment estimation algorithm, widely known as `Adam`. The update rule of `Adam` is given as follows. For $\beta_1, \beta_2 \in [0, 1)$ whose default values are 0.9 and 0.999, respectively, let $m^{(0)} = 0$, $v^{(0)} = 0$ and $\epsilon = 10^{-8}$. Given an initialization $\theta^{(0)}$, the k-th iterated parameter is given by

$$(\texttt{Adam}): \quad \theta^{(k)} = \theta^{(k-1)} - \eta_k \cdot \hat{m}^{(k)} / (\sqrt{\hat{v}^{(k)}} + \epsilon),$$

where $\hat{m}^{(k)} = \frac{1}{1-\beta_1^k} m^{(k)}$, $\hat{v}^{(k)} = \frac{1}{1-\beta_2^k} v^{(k)}$, $m^{(k)} = \beta_1 m^{(k-1)} + (1 - \beta_1) g^{(k-1)}$, $v^{(k)} = \beta_2 v^{(k-1)} + (1 - \beta_2) g^{(k-1)} \otimes g^{(k-1)}$, and $g^{(k-1)} = \nabla \mathcal{L}(\theta^{(k-1)}; \Omega_m^{(k)})$. All the operations are understood element-wise.

5.3 Quasi-Newton methods of 1.5-order

While first-order gradient-based optimization algorithms are popularly used, they often suffer from either being stuck at local minima or slow convergence. Perhaps, a natural extension of the first-order methods is the optimization methods that (approximately) use the second-order information, i.e., Hessian. Newton's method is the representative one. While it could achieve the quadratic rate of convergence, the computation of Hessian is computationally very expensive if optimization problems are defined in a high dimension, which is the case for NNs as the number of NN parameters is large. Therefore, the Gauss-Newton type algorithms are popularly used which aim at approximating Hessian using only gradient information. These algorithms are called 1.5-order methods.

The underlying idea of higher-order algorithms is to appropriately design a local proxy function on which the minimum can be found easily. More precisely, let U be a neighborhood of the current parameter $\theta^{(k-1)}$ and one seeks to design an approximation $\tilde{\mathcal{L}}$ on U. The minimizer of the proxy function is then set to the update rule. A natural proxy function comes from the Taylor expansion:

$$\begin{aligned}
\mathcal{L}(\theta) \approx\ & \mathcal{L}(\theta^{(k-1)}) + \langle \nabla \mathcal{L}(\theta^{(k-1)}), \theta - \theta^{(k-1)} \rangle \\
& + \frac{1}{2}(\theta - \theta^{(k-1)})^\top H_{k-1}(\theta - \theta^{(k-1)}) + \dots,
\end{aligned}$$

where H_{k-1} is the Hessian of $\mathcal{L}$ at $\theta^{(k-1)}$. The Taylor approximation degree up to one yields

$$\tilde{\mathcal{L}}_1(\theta) = \mathcal{L}(\theta^{(k-1)}) + \langle \nabla\mathcal{L}(\theta^{(k-1)}), \theta - \theta^{(k-1)} \rangle.$$

Seeking the minimizer of $\tilde{\mathcal{L}}_1$ then gives the gradient descent (also known as the steepest decent) method, i.e., $\theta^* = \theta^{(k-1)} - \eta_k \cdot \nabla\mathcal{L}(\theta^{(k-1)})$. The Taylor approximation degree up to two provides the multivariate polynomial of degree up to two involving the Hessian

$$\begin{aligned}\tilde{\mathcal{L}}_2(\theta) = {} & \mathcal{L}(\theta^{(k-1)}) + \langle \nabla\mathcal{L}(\theta^{(k-1)}), \theta - \theta^{(k-1)} \rangle \\ & + \frac{1}{2\eta}(\theta - \theta^{(k-1)})^\top H_{k-1}(\theta - \theta^{(k-1)}).\end{aligned}$$

Assuming the Hessian is invertible, the minimizer of $\tilde{\mathcal{L}}_2$ is explicitly written as

$$\text{(Newton's method)} \quad \theta^{(k)} := \theta^* = \theta^{(k-1)} - \eta_k \cdot H_{k-1}^{-1} \nabla\mathcal{L}(\theta^{(k-1)}),$$

which defines Newton's update rule. Since the Hessian is expensive to compute, the 1.5-order optimization algorithms seek to find alternative proxy functions that do not require the Hessian. A generic proxy function may be written as

$$\begin{aligned}\tilde{\mathcal{L}}_{1.5}(\theta; B) = {} & \mathcal{L}(\theta^{(k-1)}) + \langle \nabla\mathcal{L}(\theta^{(k-1)}), \theta - \theta^{(k-1)} \rangle \\ & + \frac{1}{2\eta_k}(\theta - \theta^{(k-1)})^\top B(\theta - \theta^{(k-1)}),\end{aligned}$$

where B is an appropriately chosen matrix that does not require the second order partial derivatives. The 1.5-order method is then determined by the minimizer of $\tilde{\mathcal{L}}_{1.5}(\theta; B)$, i.e.,

$$\text{(1.5-order method)} \quad \theta^{(k)} := \theta^* = \theta^{(k-1)} - \eta_k \cdot B^{-1} \nabla\mathcal{L}(\theta^{(k-1)}).$$

The quasi-Newton type algorithms are based on the principle of seeking the Hessian approximation with the exact gradient. That is, it aims at finding B such that $\nabla\mathcal{L}(\theta^{(k-1)} + \mu) = \nabla\tilde{\mathcal{L}}_{1.5}(\theta^{(k-1)} + \mu; B)$ where the gradient is taken with respect to μ. It then can be checked that μ and B should satisfy the so-called secant equation

$$\text{(quasi-Newton)} \quad \nabla\mathcal{L}(\theta^{(k-1)} + \mu) = \nabla\mathcal{L}(\theta^{(k-1)}) + \eta_k^{-1} B\mu.$$

The quasi-Newton type algorithms iteratively find the solution μ_k, B_k to the secant equation. The update rule for μ_k is determined by the first-order optimality condition, i.e., $\nabla\mathcal{L}(\theta^{(k-1)}) + \eta_k^{-1} B_k \mu_k = 0 \iff \mu_k = -\eta_k \cdot B_k^{-1} \nabla\mathcal{L}(\theta^{(k-1)})$ where the learning rate η_k is typically chosen to satisfy the Wolfe conditions. Hence, the quasi-Newton algorithm reads $\theta^{(k)} = \theta^{(k-1)} + \mu_k$. The core part of the quasi-Newton algorithms is on the update formula for B_k. While several formulas exist, in the context of the PINN methodology, the BFGS (Broyden–Fletcher–Goldfarb–Shanno) algorithm is one of the most popular ones. The BFGS updates B_k according to $B_{k+1} = B_k + \frac{y_k y_k^\top}{y_k^\top \mu_k} - \frac{B_k \mu_k (B_k \mu_k)^\top}{\mu_k^\top B_k \mu_k}$, where $y_k = \nabla \mathcal{L}(\theta^{(k-1)} + \mu_k) - \nabla \mathcal{L}(\theta^{(k-1)})$.

For nonlinear least squares problems, which are equivalent to minimizing a sum of squared function values, the Gauss-Newton type algorithms are popularly used. These algorithms also follow the above principle and are obtained by appropriate designs of B. The Levenberg-Marquart algorithm is one of the most well-known algorithms of this type, whose design is given by

$$\text{(Levenberg-Marquart)} \quad B = \lambda I + \mathrm{J}_r^\top \mathrm{J}_r \quad \text{or} \quad \lambda \mathrm{diag}(\mathrm{J}_r^\top \mathrm{J}_r) + \mathrm{J}_r^\top \mathrm{J}_r.$$

6 Approximation theory with small weights

Both shallow and deep neural networks are universal approximators, however, their training behaviors may vary significantly. Since the NN parameters are initialized as small numbers, deep NNs may be preferred if they remain universal under the constraints on the magnitude of weights and biases. Fig. 1 shows the histogram of the magnitude of weights and biases of a 5-layer NN of width 100 at the initialization. We employ the Kaiming initialization (He et al., 2015), one of the most popular initialization schemes for ReLU NNs. This demonstrates that weights and biases are initialized to small values.

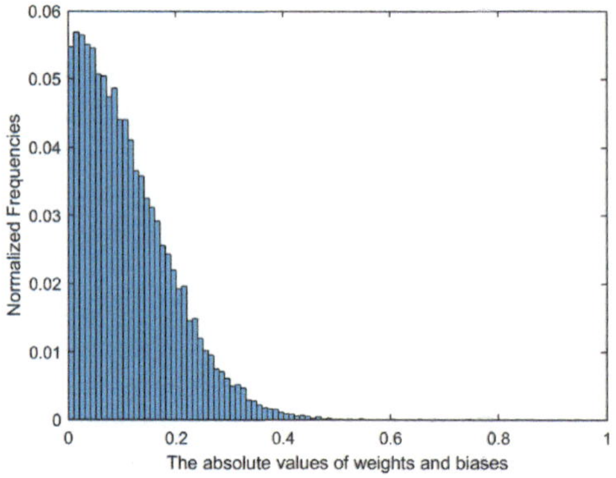

FIGURE 1 The histogram of the magnitude of randomly generated weights and biases for a 5-layer ReLU network of width 100. The Kaiming initialization (He et al., 2015) is employed.

This subsection establishes a universal approximation theorem of ReLU NNs with constraints on the magnitude of weights and biases. The primary focus will lie on the ReLU approximation to functions with (countably many) discontinuities. As a pedagogical example, let us consider a simple learning task of approximating a step function $f^*(x) = \mathbb{I}_{[-1,1]}(x)$ where $\mathbb{I}_A(x)$ is the indicator function where $\mathbb{I}_A(x) = 1$ if $x \in A$ and $\mathbb{I}_A(x) = 0$ if $x \notin A$. It is readily checked that for any $\epsilon > 0$, there exists a two-layer ReLU NN $u_{\text{NN}}(x)$ of width 4 that approximates f^* within the ℓ_1-error of ϵ, i.e., $\int_{\mathbb{R}} |f^*(x) - u_{\text{NN}}(x)| dx = \epsilon$. One such NN can explicitly be written as

$$u_{\text{NN}}(x) = \frac{1}{2\epsilon} \left[\phi(x+1+\epsilon) - \phi(x+1-\epsilon) - \phi(x-1+\epsilon) + \phi(x-1-\epsilon) \right],$$

where $\phi(x) = \max\{x, 0\}$. This indicates that if the desired accuracy is $\epsilon = 10^{-3}$ and a two-layer ReLU network, $\sum_{i=1}^{4} c_i \phi(w_i x + b_i)$, is used, the magnitudes of $c_i w_i$ and $c_i b_i$ should be at around 500. It may suggest that many gradient descent iterations are required to make all the $c_i w_i$, $b_i w_i$'s have the magnitude of 500. In Fig. 2, the histograms of NN parameters are plotted that illustrate the changes in magnitudes before and after the training. The top row shows the results with the two-layer (shallow) NNs at varying widths – 4, 16, 64. As expected, the magnitudes of parameters remain small as the width increases. Similar results are observed with the three-layer NNs whose results are shown in the bottom row.

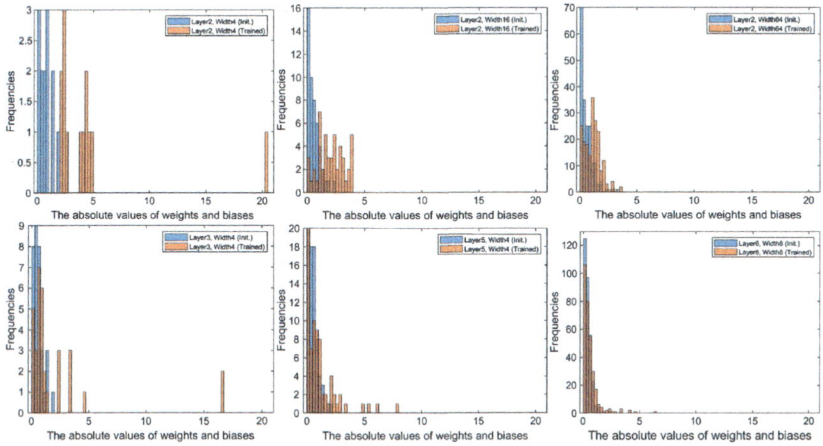

FIGURE 2 The histograms of weights and biases of networks before and after training for the task of approximating $f^*(x) = \mathbb{I}_{[-1,1]}(x)$. (Top row) Two-layer ReLU networks of width 4, 16, 64 (left to right) are employed. (Bottom row) L-layer ReLU networks of width W are employed where $(L, W) = (3, 4), (5, 4), (6, 8)$. Training is done by Adam over 300 equidistant data points from $[-3, 3]$. The maximum number of epochs is either 50,000 or 100,000.

Fig. 3 reports the training loss versus the number of iterations at varying NN architectures. It is seen that there is a tendency that the loss decays faster as the width gets larger and the depth gets larger.

In what follows, we present a universal approximation theorem of deep ReLU NNs with constraints on the magnitudes of NN parameters. We followed the proof structure of Petersen and Voigtlaender (2018), while we made modifications in each step to ensure that the magnitudes of the weights and biases

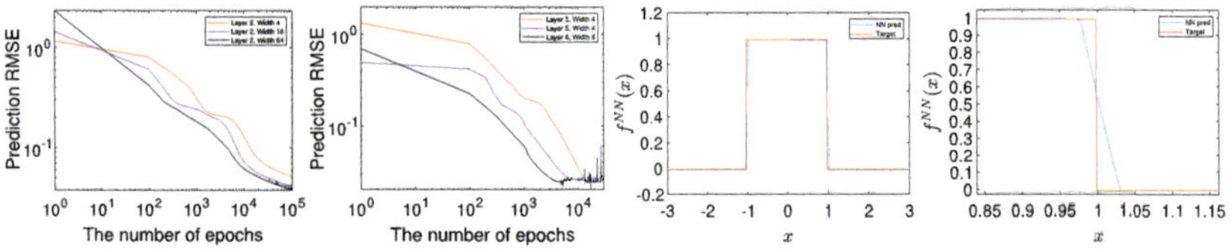

FIGURE 3 The prediction errors with respect to the number of epochs for the task of approximating $f^*(x) = \mathbb{I}_{[-1,1]}(x)$. (a) Two-layer ReLU networks of width 4, 16, 64 (left to right) are employed. (b) L-layer ReLU networks of width W are employed where $(L, W) = (3, 4), (5, 4), (6, 8)$. Training is done by Adam over 300 equidistant data points from $[-3, 3]$. The maximum number of epochs is 100,000.

were well controlled. Consequently, our new results will show the dependency of the magnitudes of NN parameters in the NN architecture.

The class of functions to be approximated contains functions defined on $\Omega = [-1/2, 1/2]^d$ that have the form of $\mathbb{I}_K(x)g(x)$ where g is a smooth function and K is an appropriate compact set in Ω on which $\mathbb{I}_K$ is locally a horizon function. To be more precise, let $r \in \mathbb{N}$, $\beta \in (0, \infty)$ such that $\beta = n + \sigma$ where $n \in \mathbb{N}_0$ and $\sigma \in (0, 1]$, $d \in \mathbb{N}_{\geq 2}$, and $B, p > 0$. Then, $\mathcal{E}^p_{r,\beta,d,B}$ is the target function class that is defined by

$$\mathcal{E}^p_{r,\beta,d,B} := \left\{\mathbb{I}_K(x)g(x) : K \in \mathcal{K}_{r,\beta,d,B},\ g \in \mathcal{F}_{\beta',d,B}\right\}, \tag{6.1}$$

where $\beta' = \frac{d\beta}{p(d-1+\beta)}$, $\mathcal{F}_{\beta',d,B}$ is a class of smooth functions and $\mathcal{K}_{r,\beta,d,B}$ is a class of tailored domains with smooth boundaries. Both classes will be defined shortly.

The smooth function class is defined as follows.

$$\mathcal{F}_{\beta,d,B} = \left\{ f \in C^n([-1/2, 1/2]^d) : \|f\|_{C^{0,\beta}} \leq B \right\},$$

where $\|f\|_{C^{0,\beta}} = \max\left\{\max_{|\mathbf{k}|\leq n} \|\partial^{\mathbf{k}} f\|_{C^0}, \max_{|\mathbf{k}|=n} \mathrm{Lip}_\sigma(\partial f)\right\}$ and $\mathrm{Lip}_\sigma(\partial f) := \sup_{x,y\in\Omega,x\neq y} \frac{|f(x)-f(y)|}{\|x-y\|^\sigma}$.

$$\|f\|_{C^{0,\beta}} = \max\left\{ \max_{|\mathbf{k}|\leq n} \|\partial^{\mathbf{k}} f\|_{C^0}, \max_{|\mathbf{k}|=n} \mathrm{Lip}_\sigma(\partial f) \right\},$$

$$\mathrm{Lip}_\sigma(\partial f) := \sup_{x,y\in\Omega,x\neq y} \frac{|f(x) - f(y)|}{\|x - y\|^\sigma}.$$

Next, let us introduce the class of horizon functions. Let $d \in \mathbb{N}_{\geq 2}$, $\mathbf{x}_{-1} = (x_2, \cdots, x_d) \in \mathbb{R}^{d-1}$ and $H := \mathbb{I}_{[0,\infty)\times\mathbb{R}^{d-1}}$ be the Heaviside function. We define

$$\mathcal{HF}_{\beta,d,B} = \{f \circ T \in L^\infty(\Omega) : f(x) = H(x_1 + \gamma(\mathbf{x}_{-1}), \mathbf{x}_{-1}),$$
$$\gamma \in \mathcal{F}_{\beta,d-1,B}, T \in \Pi(d, \mathbb{R})\},$$

where $\Pi(d,\mathbb{R})$ is the group of permutation matrices. We are now in a position to define a class of piecewise smooth functions of interest, which was originally introduced in Petersen and Voigtlaender (2018). Let us consider a set of domains with smooth boundaries. For $r \in \mathbb{N}$, let

$$\mathcal{K}_{r,\beta,d,B} := \big\{K \subset \Omega : \forall y \in \Omega, \exists f_y \in \mathcal{HF}_{\beta,d,B} \\ \text{such that } \mathbb{I}_K(x) = f_y(x) \text{ on } x \in \Omega \cap \overline{B}_{2^{-r}}(y)\big\},$$

where $\overline{B}_{2^{-r}}(y)$ is the closed ball centered at y with the radius of 2^{-r} in $\|\cdot\|_\infty$-norm. Note that every closed set K' in Ω whose boundary is locally the graph of a C^β function of all but one coordinate belongs to $\mathcal{K}_{r,\beta,d,B}$ for sufficiently large r and B.

To give a concrete idea of what functions belong to the class $\mathcal{E}^p_{r,\beta,d,B}$, here we present an example. Let $K = \{(x_1,x_2) \in \Omega : x_1 + 0.5\sin(2\pi x_2) \geq 0\}$ which corresponds to the yellow region shown on the left of Fig. 4. Since $\gamma = 0.5\sin(2\pi x) \in \mathcal{F}_{2,1,2\pi}$ as $\|\gamma\|_{C^{0,2}} = \pi$, K belongs to $\mathcal{K}_{r,\beta,d,B}$ with $r = 1$, $\beta = 2$, $d = 2$, $B = 100$. In particular, for all $y \in \Omega$ and for all $r \in \mathbb{N}$, we have $f_y(x_1,x_2) = H(x_1 + 0.5\sin(2\pi x_2), x_2) \in \mathcal{HF}_{\beta,d,B}$ which is equal to $\mathbb{I}_K$ and shows, in fact, r can be any integer. Lastly, let g be the Rosenbrock function defined by $g(x_1,x_2) = (1-0.5X_1)^2 + 100*(0.5x_2 - (0.5x_1).^2)^2$, which belongs to $\mathcal{E}^p_{r,\beta,d,B}$ where $\beta' = 2$. Then, the function $\mathbb{I}_K g$ belongs to the function class $\mathcal{E}^2_{1,2,2,100}$ whose graph is shown on the right of Fig. 4.

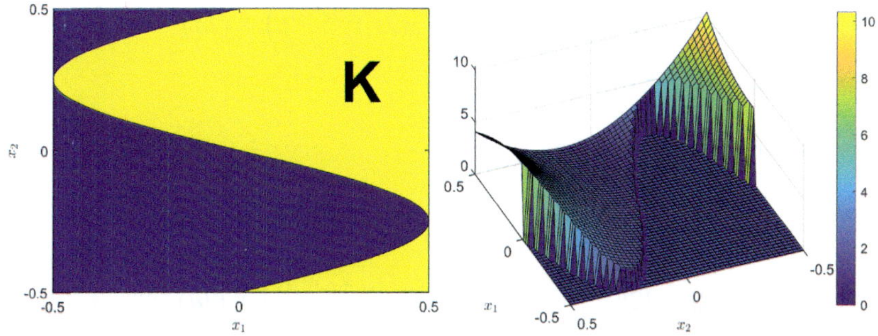

FIGURE 4 Left: The yellow region corresponds to $K = \{(x_1,x_2) \in \Omega : x_1 + 0.5\sin(2\pi x_2) \geq 0\}$ that belongs to the domain class $\mathcal{K}_{r,\beta,d,B}$ with $r = 1$, $\beta = 2$, $d = 2$, $B = 100$. Right: A piecewise smooth function of the form $\mathbb{I}_K(x)g(x) \in \mathcal{E}^p_{r,\beta,d,B}$ where $g(x)$ is the Rosenbrock function.

We now state the main theorem. Let r, n, σ, d, B, p be given with $\beta = n + \sigma$ where $r \in \mathbb{N}$, $n \in \mathbb{N}_0$, $\sigma \in (0,1]$, $d \in \mathbb{N}_{\geq 2}$, and $B, p > 0$, which define the function class $\mathcal{E}^p_{r,\beta,d,B}$ of interest.

Let $\mathrm{M} \geq 1$ be given and let $\mathfrak{n} = \lceil \log_2 n \rceil$. Let $m \in \mathbb{N}$ be sufficiently large to satisfy $(2m+1) \geq (r+1)\frac{p(d-1+\beta)}{\beta}$.

- Let $(k, L) \in \mathbb{N}^2$ such that $m = k(L-1) \geq \frac{1}{2}\log_2\left(\frac{\log_2 n}{2}\right)$.

- Let $(s', r') \in \mathbb{N}^2$ such that $\tilde{\gamma}(d-1, n, B) \le (s'\mathrm{M})^{r'-1}$ where $\tilde{\gamma}$ is defined in (6.B.1).
- Let $(s'', r'', r''') \in \mathbb{N}^3$ such that $2^{2(m+1)}\mathrm{M}^{-1} \le 0.5(s''\mathrm{M})^{r''-1}$ and $(1 + \frac{(d-1)^{n+\beta}}{n!})B \le (((d-1)s'+1)\mathrm{M})^{r'''-1}$.
- Let $(s_h, r_h) \in \mathbb{N}^2$ such that $2^{\frac{\beta}{d-1+\beta}(2m+1)}\mathrm{M}^{-1} \le 0.5(s_h\mathrm{M})^{r_h}$.
- Let $(s_I, r_I) \in \mathbb{N}^2$ such that $2^{\frac{\beta}{p(d-1+\beta)}(2m+1)}\mathrm{M}^{-1} \le 0.5(s_I\mathrm{M})^{r_I-1}$.
- Let $(s'_g, r'_g) \in \mathbb{N}^2$ such that $\tilde{\gamma}(d, n, B) \le (s'_g\mathrm{M})^{r'_g-1}$ where $\tilde{\gamma}$ is defined in (6.B.1).
- Let $(s''_g, r''_g, r'''_g) \in \mathbb{N}^3$ such that $2^{2(m+1)}\mathrm{M}^{-1} \le 0.5(s''\mathrm{M})^{r''-1}$ and $(1 + \frac{d^{n+\beta}}{n!})B \le ((ds'+1)\mathrm{M})^{r'''_g-1}$.

Let us define an architecture of θ_K by

$$\vec{n}_{\theta_K} = (2^{rd} \cdot (\vec{n}_{\theta_0} + 2^{\oplus L_{\theta_0}}, (2s_h)^{\oplus r_h}), (2^{rd+1}(ds_I+1))^{\oplus r_I}), \tag{6.2}$$

where L_{θ_0} is the length of $\vec{n}_{\theta_0}$, and

$$\vec{n}_{\theta_0} = (\vec{n}_{\theta_1}, 2(N^{d-1}+d-1), (2N^{d-1}((d-1)s''+1))^{\oplus(r''+r''')}),$$
$$N^{d-1} = 2^{\frac{d-1}{d-1+\beta}(2m+1)},$$

where $P_{d,n} = \binom{d+n}{n}$ and

$$\vec{n}_{\theta_1} = \left(P_{d-1,n} \cdot [2^{\mathrm{n}-1}, \ldots, 2^0] \otimes \vec{n}_{\tilde{\times}}, (2s')^{\oplus(r'-1)}\right) + (2d)^{\oplus(\mathrm{n}L+r'-1)},$$
$$\text{where} \quad \vec{n}_{\tilde{\times}} = (3(2^k+2))^{\oplus L}.$$

Also, define an architecture of θ_g by

$$\vec{n}_{\theta_g} = (\vec{n}_{\theta_2}, 2(N_g^d+d), (2N_g^d(ds''_g+1))^{\oplus(r''_g+r'''_g)}), \qquad N_g^d = 2^{\frac{d}{d+\beta}(2m+1)}, \tag{6.3}$$

where $\vec{n}_{\theta_2} = \left(P_{d,n} \cdot [2^{\mathrm{n}-1}, \ldots, 2^0] \otimes \vec{n}_{\tilde{\times}}, (2s'_g)^{\oplus(r'_g-1)}\right) + (2d)^{\oplus(\mathrm{n}L+r'_g-1)}$.

Theorem 6.1. *For any $\mathbb{I}_K(\cdot)g(\cdot) \in \mathcal{E}^p_{r,\beta,d,B}$, there exists a ReLU NN, Φ such that*

$$\|\mathcal{R}[\Phi](x) - \mathbb{I}_K(x)g(x)\|_{L_p} \le \tilde{C} \cdot 2^{-\frac{\beta}{p(d-1+\beta)+\beta}(2m+1)},$$

with $\vec{n}_\Phi = (\vec{n}_{\theta_K} + \vec{n}_{\theta_g}, \vec{n}_{\tilde{\times}})$ and $|\Phi|_\infty \le \mathrm{M}$. Here $\tilde{C}$ is a constant that depends only on r, β, d, B, p. Also, θ_K and θ_g are ReLU NNs whose architectures are given by (6.2) *and* (6.3)*, respectively.*

Proof. The proof can be found in Appendix 6.D. □

Remark 6.2. In the Appendix, we provided relevant approximation theorems with the controlled magnitude. The results indicate that one can approximate smooth functions with small weights, while discontinuous functions need delicate approaches to control the magnitude.

7 PINN with observational data

Assume we have a forward problem to solve, and have some observational data of the solution u, we can use the loss function by adding the data loss in the PINN loss (3.9).

$$\mathcal{L}(v;\boldsymbol{\lambda}) = \sum_{x\in\Omega_M} \lambda_x \Big(\mathcal{D}[v](x) - f(x)\Big)^2 + \sum_{x'\in\Gamma_{M'}} \lambda_{x'} \Big(\mathcal{B}[v](x) - g(x)\Big)^2 + \sum_{x_{\text{obs}}} \lambda_{x_{\text{obs}}} |v - u(x_{\text{obs}})|^2 \tag{7.1}$$

Then we solve the minimization problem $\min_\theta \mathcal{L}(u_{NN};\lambda)$.

If we have an inverse problem to solve, we may use a similar loss functional. For example, when we solve the data assimilation Problem 3.2 or the Cauchy problem of Poisson's equation 3.3, we can still use the above formulation. For the coefficient inverse Problem 3.4, we need to have two neural networks, one for the solution u, and the other for p. In this case, we denote the loss function at the continuous level by

$$\mathcal{L}(u_N, p_N;\lambda) = \left\|\partial_t^2 u_N - \nabla\cdot(p_N \nabla u_N)\right\|^2_{L^2(\Omega)} + \|u(x,0) - g(x)\|^2_{L^2} + \|\partial_t u(x,0)\|^2_{L^2} + \|u - g\|^2_{L^2(S_T)} + \underbrace{\|\nabla u\cdot\mathbf{n} - h(x,t)\|^2_W}_{\text{data}}.$$

Here the space W and the norm on it can be selected according to the Lipschitz stability in Klibanov and Yamamoto (2006). Suppose that Θ are the parameters for the network p_N, we then solve the following problem $\inf_{\theta,\Theta} \mathcal{L}(u_N, p_N;\lambda)$. While the observation data $u(x_{\text{obs}})$ are available, the choices of Ω_M, $\Gamma_{M'}$ may be adjusted from the formulation for forward problems.

In the PINNs framework, it is essentially the same to solve both forward and inverse problems. In both cases, we need stability of the problem to facilitate the setup of loss functions. The difference may lie in the complexity– we need two networks to solve inverse problems. Also, we may need some regularization to stabilize the formulation, as the stability may not be available theoretically.

PINNs have been applied to solve various inverse problems, such as elliptic and wave equations in Mishra and Molinaro (2022), supersonic problems in Jagtap et al. (2022a) and many others.

8 Deep operator networks

The PINN for PDEs usually solve one single problem at a time and can be time consuming. In particular training neural networks are performed for differential equations with fixed inputs, such as initial conditions, boundary conditions, forcing, and coefficients. If one input is changed, the training process has to be repeated. It is difficult to obtain outputs in real-time for systems that require various sets such in multiphysics.

Operator learning via neural networks has been developed to accommodate varying inputs and predict in real-time. In operator learning, a fixed-weighted/pretrained network approximates a continuous operator from the input(s) to the output(s), e.g. in the DeepONets (8.4). Once such networks are trained, we can predict/ evaluate at the desired input(s) in real-time.

To learn such networks, it often requires a huge amount of data to perform a least-squares regression for accurate approximations. For the learning of solution operators from PDEs, the data are usually generated by accurately solving the underlying PDEs with various inputs, which are usually modelled by stochastic processes with mild smoothness, such as in truncated Karhunen-Loeve expansions or a truncated Fourier expansion of stochastic processes.

8.1 Introduction

Let X and Y be Banach spaces. Consider a continuous operator $\mathcal{G} : X \to Y$ as follows. Let $\mathcal{G}(\cdot)$ be represented using a Schauder basis $\{e_k\}$ in the Banach space Y by $\mathcal{G}(u)(y) = \sum_{k=1}^{\infty} c_k(\mathcal{G}(u))e_k(y)$, where $c_k(\cdot)$ is a linear functional. We denote by $T_m u$ an approximation (encoder) or a parameterization of u in X and a truncation of the series

$$\mathcal{G}_{m,p}(T_m u)(y) = \sum_{k=1}^{p} c_k(\mathcal{G}(T_m u))e_k(y). \tag{8.1}$$

Assumption 8.1 (Modulus of continuity). Let $S \subset X$ and $w : [0,\infty) \to [0,\infty)$ be continuous at 0. There exists a constant $C > 0$ independent of $u, v \in S$.

$$\|\mathcal{G}(u) - \mathcal{G}(v)\|_Y \le C w(\|u - v\|_X), \quad \forall u, v \in S, \quad 0 < \alpha \le 1. \tag{8.2}$$

A particular useful choice of the function is $w(\cdot) = |\cdot|^{\alpha}$ for some $0 < \alpha \le 1$.

The error estimate of a network approximation can be derived as follows. We first split into three parts: approximation error from input, approximation error from output and approximation error from neural networks and then we can apply the modulus of continuity, the Schauder expansion and

$$\begin{aligned}\|\mathcal{G}(u) - \mathcal{G}_{\mathbb{N}}(T_m u)\|_Y \le \|\mathcal{G}(u) - \mathcal{G}(T_m u)\|_Y + \left\|\mathcal{G}(T_m u) - \mathcal{G}_{m,p}(T_m u)\right\| \\ + \left\|\mathcal{G}_{mp}(T_m u) - \mathcal{G}_{\mathbb{N}}(T_m u)\right\|_Y\end{aligned}$$

$$
\begin{aligned}
&\le Cw(\|u - T_m u\|_X) + \sum_{k=p+1}^{\infty} |c_k(\mathcal{G}(T_m u))| \, \|e_k\|_Y \\
&\quad + \left\|\mathcal{G}_{m,p}(T_m u) - \mathcal{G}_{\mathbb{N}}(T_m u)\right\|_Y \qquad (8.3) \\
&\le Cw(\|u - T_m u\|_X) \\
&\quad + \sum_{k=p+1}^{\infty} (|c_k(\mathcal{G}(T_m u) - \mathcal{G}(u))| + |c_k(\mathcal{G}(u))|) \, \|e_k\|_Y \\
&\quad + \left\|\mathcal{G}_{m,p}(T_m u) - \mathcal{G}_{\mathbb{N}}(T_m u)\right\|_Y .
\end{aligned}
$$

From this decomposition, we observe the factors of affecting the approximation error of an neural operator:

- How the input u is encoded/approximated. At least two typical approximations are used: interpolation, Deng et al. (2022) and Li et al. (2020); and truncated spectral expansion, Lanthaler et al. (2021) (from known basis) and Zhang et al. (2023b) (from Mercer's theorem).
- The choices of basis, which may significantly affect the efficiency of the approximation as in the spectral methods, empirical spectral methods (e.g., proper orthogonal decomposition (Lu et al., 2022)).
- The architecture of neural networks, e.g. DeepONets (Lu et al., 2019, 2021a) and extensions in Lu et al. (2022) and Zhang et al. (2023b), Fourier Neural operators (Li et al., 2020).

In practice, we also have errors from training/optimization and from *the mismatch of the inputs for training and evaluating*. Here we focus on the error from the mismatch. DeepONet takes function values at a set of fixed m points for the branch while the new input to evaluate at has function values at a different set of m' points. One way to accommodate this situation is to use a proper interpolation (or approximation) to derive a function (say $S_{m'}u$) so that one can evaluate the function at the set of m points that are required in the trained DeepONets. Note that this reformulation of the inputs does not require retraining DeepONets. The error for this accommodation can be bounded as follows.

$$
\begin{aligned}
\|\mathcal{G}_{\mathbb{N}}(T_m S_{m'}u) - \mathcal{G}(u)\|_Y &\le \|\mathcal{G}_{\mathbb{N}}(T_m S_{m'}u) - \mathcal{G}(S_{m'}u)\|_Y + \|\mathcal{G}(S_{m'}u) - \mathcal{G}(u)\|_Y \\
&\le \|\mathcal{G}_{\mathbb{N}}(T_m S_{m'}u) - \mathcal{G}(S_{m'}u)\|_Y + Cw(\|S_{m'}u - u\|_X),
\end{aligned}
$$

where we have applied the triangle inequality and the modulus continuity (8.2). The estimate of $w(\|S_{m'}u - u\|_X)$ is well studied in approximation theory and thus it is always assumed that the same set of m points is used for training and evaluating.

8.2 Vanilla DeepONets

In Chen and Chen (1995b); Lu et al. (2021a), the architecture of neural networks for operator learning is called Deep Operator Networks (DeepONets) in the following form

$$\mathcal{G}_{\mathbb{N}}(u) = \sum_{k=1}^{p} \underbrace{b^{\mathcal{N}}(T_m u; \Theta^{(k)})}_{\text{branch}} \underbrace{t^{\mathcal{N}}(y; \theta^{(k)})}_{\text{trunk}}, \tag{8.4}$$

where $T_m u = (u(x_1), u(x_2), \cdots, u(x_m))^\top$. Here the neural networks $f^{\mathcal{N}}$ and $g^{\mathcal{N}}$ can be any class of functions that satisfy the classical universal approximation theorem of continuous functions on compact sets.

Theorem 8.2 (Universal approximation theorem, Chen and Chen, 1995b; Lu et al., 2021a). *Let $K_1 \subset \mathbb{R}^{d_1}$ be a compact set. Let V be a compact set in $C(K_1)$, $K_2 \subset \mathbb{R}^d$ be a compact set and $Y = C(K_2)$. Assume that $\mathcal{G} : V \to Y$ is continuous. Then for any $\epsilon > 0$, there are positive integers p and m, neural networks $t^{\mathcal{N}}(\cdot; \theta^{(k)})$ and $b^{\mathcal{N}}(\cdot; \Theta^{(k)})$, $k = 1, \ldots, p$, $x_j \in K_1$, $j = 1, \ldots, m$, such that*

$$\sup_{u \in V} \sup_{y \in K_2} |\mathcal{G}(u)(y) - \mathcal{G}_{\mathbb{N}}(T_m u)(y)| < \epsilon, \tag{8.5}$$

where $T_m u$ is an interpolation of u at $x_1, \ldots, x_m$ using $(u(x_1), u(x_2), \cdots, u(x_m))^\top$.

The universal approximation theorem can be proved by emulating a truncated Fourier expansion as in Chen and Chen (1995b). However, the convergence order cannot be established therein as the branch networks is high-dimensional as a function of $(u(x_1), u(x_2), \cdots, u(x_m))$. Also, according to the proofs in Chen and Chen (1993, 1995a,b); Chen (1998), $C(K_1)$ can be replaced by $L^p(K_1)$, $p \geq 1$ and $C(K_2)$ can be replaced by $L^q(K_2)$, $q \geq 1$. But $T_m u$ should be replaced by averaged values $T_m u = (T_m u_h(x_1), \ldots, T_m u_h(x_m))^\top$, where $T_m u_h(x_i) = \int_{B(x_i,h)\cap K} u(t)\, dt / \mu(B(x_i, h) \cap K)$ while μ is the Lebesgue measure and $B(x_i, h)$ is the ball centered at x_i with radius h. More generally, these spaces may be replaced by Banach spaces with a Schauder basis.

8.3 Approximation rates for general Hölder operators

As indicated in (8.3), for a given operator, the approximation error of DeepONets depends on encoder/approximation of the input $T_m u$ and the choices of basis and architecture of branch and trunk networks. To establish approximation rates, we may use the current state-of-the-art where we emulate DeepONets by certain numerical methods.[3] Thus it is important to identify numerical methods which the architecture is analogy to and how the numerical methods behave.

Here we use some conclusions from Deng et al. (2022) on the approximation error of DeepONets using feedforward neural networks for Hölder continuous operators. The error is obtained by emulating truncated Fourier expansions in

[3] The emulation provides error estimates and thus insights on how one can obtain and improve architectures of networks for operator learning.

the L^q space. Let $\mathcal{N}(\cdot;\theta)$ be an feedforward neural network with parameters θ. We use $|\theta|$ to denote the size of is the total number of nonzero parameters; $N_{\mathcal{N}}$ for the width of $\mathcal{N}$ is the number of neurons in each layer; $L_{\mathcal{N}}$ for the depth or the number of layers.

Assume that

$$\|T_m u - u\|_X \leq C m^{-r} \|u\|_V, \forall u \in V. \quad r > 0. \tag{8.6}$$

Theorem 8.3 (Error estimates of DeepONets for Hölder continuous operators, Deng et al., 2022). *Assume that the conditions in Theorem 8.2 and* (8.6) *hold. Assume that T_m is Lipschitz continuous and $\mathcal{G}$ is Hölder continuous with exponent α. Let $Y = L^q(K_2)$, where $1 \leq q \leq \infty$, $K_2 \subset \mathbb{R}^d$ is compact. Then we have*

$$\|\mathcal{G}(u) - \mathcal{G}_{\mathbb{N}}(T_m u)\|_Y \leq \left\|\mathcal{G}(u) - \mathcal{G}_{m,p}(T_m u)\right\|_Y + \left\|\mathcal{G}_{m,p}(T_m u) - \mathcal{G}_{\mathbb{N}}(T_m u)\right\|_Y, \tag{8.7}$$

where $\mathcal{G}_{m,p}(T_m u)$ is defined in (8.4) *and*

$$\left\|\mathcal{G}(u) - \mathcal{G}_{m,p}(T_m u)\right\|_Y \leq C m^{-r\alpha} + C\omega_2(\mathcal{G}(T_m u), p^{-1/d})_q,$$
$$\left\|\mathcal{G}_{m,p}(T_m u) - \mathcal{G}_{\mathbb{N}}(T_m u)\right\|_Y \leq C\Big(p\sqrt{m} N_{b\mathcal{N}}^{-2\alpha/m} L_{b\mathcal{N}}^{-2\alpha/m} + p\exp(-|\theta|^{\frac{1}{1+d}})\Big),$$

where r is from (8.6). *The positive constant C is independent of m, p and $N_{b\mathcal{N}}$ is the number of neurons in each layer of the branch network, $|\theta|$ is the size (number of nonzero parameters) of each trunk network $t^{\mathcal{N}}$, and $L_{b\mathcal{N}}$ is the numbers of layers of the branch network.*

Let us denote the order/magnitude of '$\cdot$' by '$\sim \cdot$'. According to Theorem 8.3, in order to make the total error $\|\mathcal{G}(u) - \mathcal{G}_{\mathbb{N}}(T_m u)\|_Y < \varepsilon$, we need to set $m \sim \varepsilon^{-\frac{d}{\alpha}}$, $p \sim \varepsilon^{-\frac{d}{2}}$, $N_{g\mathcal{N}} L_{g\mathcal{N}} \sim \varepsilon^{-\frac{d}{\varepsilon}}$ and $|\theta| \sim \left(\frac{d+2}{2}\ln(\frac{1}{\varepsilon})\right)^{d+1}$. The low approximation rate and the high complexity is caused by the high dimensionality of the function (of $T_m u$) and Hölder continuous operators. If an operator is smoother, faster convergence can be obtained, e.g. exponential convergence in Lanthaler et al. (2021) and Marcati and Schwab (2023). In addition to the smoothness of the operator $\mathcal{G}$, the structure of the operators or approximate operator $\mathcal{G}_{m,p}$ is also important for the convergence rates of DeepONet. For example, for solution operators from partial differential equations we may have better convergence as we can obtain $\mathcal{G}_{m.p}$ using well-studied and efficient numerical methods and then neural networks can emulate these methods. In Deng et al. (2022), it is shown that there is no curse of dimensionality (exponential complexity in m) for several solution operators from PDEs. We will present such examples from Deng et al. (2022) in Section 8.4.

Below, we discuss some theoretical considerations when obtaining error estimates. *I. Physical domains of inputs and outputs K_1 and K_2*. We usually assume that the domains of K_1 and K_2 are cubes or balls, where approximations with convergence rates are well studied. If they are not cubes or balls, we use the Tietze-Urysohn-Brouwer extension theorem to extend the operator such that domains of input and output are cubes or balls (denoted by D). We denote the resulting operator (with also extension of V and image) by $\mathcal{G}(u)$ if no confusion arises. *II. Choices of function spaces for* $\mathcal{G}_{\mathbb{N}}$. Let's suppose that $\mathcal{G}_{\mathbb{N}}$ maps functions in V_m to functions in Y_p. Once we identify choices of the spaces V, V_m, Y, and Y_p, we are ready to calculate the approximation rate. In addition to the truncated Fourier expansion for both input and output, we list in Table 1 some typical choices of these spaces. Other choices are possible, e.g., V_m can be chosen as the set of networks with m parameters, which can well approximate functions in V while V_m is not a linear subspace of V. It is important to choose the best possible choices to parameterize the input and output as they determine the dimensionality of the input and output of the operator $\mathcal{G}_{m,p}(T_m u)$. The choices are highly problem dependent. In this work, we only consider two choices of piecewise polynomials and truncated Fourier space. We refer the interested readers for more discussion in Kovachki et al. (2021b), Deng et al. (2022), Lanthaler et al. (2021), and Marcati and Schwab (2023).

TABLE 1 List of some parameterizations of the input and output. Here RKHS refers to reproducing kernel Hilbert space.

$X(V)$ or Y	V_m or Y_p	$T_m u$	Ref.
L^p	piecewise polynomials	function values	Deng et al. (2022)
L^q	truncated Fourier space	Fourier coefficients	Deng et al. (2022) Lanthaler et al. (2021)
RKHS	truncated RKHS (first N term of the basis)	expansion coefficients	Lanthaler et al. (2021) Zhang et al. (2023b)
analytic functions	truncated orthogonal polynomials	expansion coefficients	Marcati and Schwab (2023)

8.4 Error estimates for solution operators from PDEs

We now consider error estimates of DeepONets for solution operators from linear and nonlinear advection-diffusion-reaction equations.

Example 8.4. Consider the following 1D advection-diffusion equation with Dirichlet boundary condition:

$$-u_{xx} + a(x)u_x = f(x), \quad x \in (0, L),\ 0 < L < \infty, \qquad u(0) = u(L) = 0, \quad (8.8)$$

where $a(x), f(x) \in L^\infty(0, L)$. The solution operator is defined from the coefficient a to the solution u by

$$u = \mathcal{G}^f(a). \quad (8.9)$$

Let $\{x_i\}_{i=1}^m$ be a partition of $(0, L)$. Define $\mathbf{a}_m = (a_1, \cdots, a_m)^\top$, where $a_i = a(x_i)$, $x_i \in (0, L)$, $i = 1, 2, \ldots, m$. From the solution (8.10), we can define a solution operator by

Theorem 8.5 (Error estimates of DeepONet for the solution operator (8.9), Deng et al., 2022). *For any given $f \in L^\infty$, let $\mathcal{G}^f(a)$ be the solution operator (8.9). Let $S = \{a(x) : a \in L^\infty(0, L),\ \partial_x a \in L^\infty(0, L)\}$. Then, there exist ReLU branch networks $b^N(\mathbf{a}_m; \Theta^{(i)})$ of size $|\Theta^{(i)}| = m^4 \ln(m)$ for $i = 1, \cdots, p$, and ReLU trunk networks $t^N(x; \theta^{(k)})$ of width $N_{t_N} = 3$ and depth $L_{t_N} = 1$, $k = 1, \cdots, p$, such that*

$$\sup_{a \in S} \left\| \mathcal{G}^f(a) - \mathcal{G}^f_{\mathbb{N}}(\mathbf{a}_m) \right\|_{L^\infty} \leq C\left(p^{-1} + m^{-1} + \left|\Theta^{(i)}\right|^{-\frac{1}{4}+\epsilon}\right),$$

where $\mathcal{G}^f_{\mathbb{N}}(\mathbf{a}_m)$ is of the form in (8.4), $\epsilon > 0$ is arbitrarily small and $C > 0$ is independent of m, p, $|\Theta^{(i)}|$ and $a(x)$.

The proof can be done by emulating the analytic solution as follows. The solution can be written as

$$u(x) = -\left(\mathcal{A}_- \circ \mathcal{A}_+\right)(f)(x) + \frac{\mathcal{A}_-(\mathbb{1})(x)}{\mathcal{A}_-(\mathbb{1})(L)}\left(\mathcal{A}_- \circ \mathcal{A}_+\right)(f)(L), \tag{8.10}$$

where $\mathbb{1}(x) = 1$ for $x \in [0, L]$ and $\mathcal{A}_+(g)(x) := \int_0^x A(y)g(y)dy$, $\mathcal{A}_-(g)(x) := \int_0^x A^{-1}(y)g(y)dy$ and $A(x) = \exp(-\int_0^x a(y)\,dy)$. We can utilize the analytical formulation (8.10) to show that $\left\|\mathcal{G}^f(a) - \mathcal{G}^f(\mathcal{I}_m^0 a)\right\|_{L^\infty} \leq C\left\|a - \mathcal{I}_m^0 a\right\|_{L^\infty}$, where $\mathcal{I}_m^0 v(x) = v(x_{i-1})$ (piecewise constant interpolation) on $[x_{i-1}, x_i)$. Define an approximation of $\mathcal{G}^f(a)$ as

$$\mathcal{G}^f_m(\mathbf{a}_m)(x) := -\left(\mathcal{A}^N_- \circ \mathcal{A}^N_+\right)(f)(x) + \frac{\mathcal{A}^N_-(\mathbb{1})(x)}{\mathcal{A}^N_-(\mathbb{1})(L)}\left(\mathcal{A}^N_- \circ \mathcal{A}^N_+\right)(f)(L), \tag{8.11}$$

where $\mathcal{A}^N_+(g)(x) := \int_0^x \mathcal{I}_m^0(Ag)(y)dy$, $\mathcal{A}^N_-(g)(x) := \int_0^x \mathcal{I}_m^0(A^{-1}g)(y)dy$. The approximation error introduced in this step can be calculated and the order is m^{-1}, which is the convergence order of piecewise linear interpolation. Then $\mathcal{G}(a)$ can be approximated by $\sum_{i=1}^p \mathcal{G}^f_m(\mathbf{a_m})(y_i)L_j^p(y)$ where $L_j^p(y)$ is the piecewise interpolation polynomial on a partition of $(0, L)$ and the resulting error is of order $p^{-1} + m^{-1}$. Approximating $\mathcal{G}^f_m(\mathbf{a_m})(y_i)$ by ReLU feedforward neural network $b^N(\mathbf{a_m}; \Theta^{(i)})$ and $L_j^p(y)$ by $t^N(y; \theta^{(k)})$ we then obtain a DeepONet in the form (8.4). The approximation error by neural networks can be bounded by observing that $\mathcal{G}^f_m(\mathbf{a_m})(y_i)$ is a rational polynomial in $\mathbf{V}_m = (v_1, v_2, \ldots, v_m)$ where $v_i = \exp(a_i(x_i - x_{i-1}))$ and $L_j^p(y)$ can be rewritten by a linear combination of two-layer ReLU neural networks and approximation error estimates in Telgarsky (2017).

Example 8.6. Consider the Burgers' equation (8.18). Define the solution operator by $u = \mathcal{G}(u_0)$ from the initial condition to the solution.

For $l \in \mathbb{Z}$, $x_j^l = x_j + 2\pi l$, $j = 0, 1, \cdots, m$, where $\Pi_m = \{-\pi = x_0 < x_1 < \cdots < x_m = \pi\}$ is a partition of $[-\pi, \pi]$. Let $M_0,\ M_1 > 0$. Define

$$\mathcal{S} = \mathcal{S}(M_0, M_1) := \{v|_{[-\pi,\pi)} : \left\|v|_{[-\pi,\pi)}\right\|_{L^\infty} \le M_0,\ \|\partial_x v|_{[-\pi,\pi)}\|_{L^\infty} \le M_1,$$
$$\bar{v} := \int_{-\pi}^{\pi} v(s)\,ds = 0\}. \tag{8.12}$$

Let $L_j^m(x)$ be the piecewise linear nodal basis over the partition Π_m, $h_j = x_j - x_{j-1}$, and $h = \max_{1\le j\le m} h_j$.

Theorem 8.7 (Error estimate of DeepONet for 1D Burgers equation with periodic boundary). *Let $\mathcal{G}(u_0)$ be the solution operator of the Burgers equation* (8.18). *Then, there exist ReLU branch networks $g^N(T_m u_{0,m}; \Theta^{(i)})$ of size $|\Theta^{(i)}| = O(m^2 \ln(m))$ for $i = 1, \cdots, p$, and ReLU trunk networks $f^N(x; \theta^{(k)})$ of size $O(1)$, $k = 1, \cdots, p$, such that for any $t > 0$,*

$$\sup_{u_0 \in \mathcal{S}} \|\mathcal{G}(u_0)(\cdot, t) - \mathcal{G}_{\mathbb{N}}(T_m u_0)(\cdot, t)\|_{L^\infty([-\pi,\pi))}$$
$$\le C\left(p^{-1} + m^{-1} + \left|\Theta^{(i)}\right|^{-\frac{1}{2}+\epsilon}\right),$$

where $\mathcal{G}_{\mathbb{N}}(T_m u_0)$ is of the form (8.4), *$\epsilon > 0$ is arbitrarily small and $C > 0$ is independent of m, p, $|\Theta^{(i)}|$ and the initial condition u_0.*

The proof of the theorem can be established by three steps. *First,* it can be shown that by (8.19), the solution operator $\mathcal{G} : X \to Y$ is Lipschitz continuous, where $X = Y = L^\infty([-\pi, \pi))$. *Second,* let

$$\mathcal{G}_{m,p}(T_m u_0)(x,t) = \sum_{k=1}^{p} \mathcal{G}_m(T_m u_0)(x_k, t) L_k^p(x), \quad x_k \in \Pi_p,$$

where $\mathcal{G}_m(T_m u_0)$ is defined by applying the piecewise constant and linear interpolation in (8.19)

$$\begin{aligned}\mathcal{G}_m(T_m u_0)(\mathbf{x}) &= \frac{-2\kappa \int_{\mathbb{R}} \partial_x \mathcal{K}(x,y,t)(\mathcal{I}_m^1 v_0)(y)dy}{\int_{\mathbb{R}} \mathcal{K}(x,y,t)(\mathcal{I}_m^0 v_0)(y)dy} \\ &= \frac{v_0^0 c_0^1(\mathbf{x}) + v_1^0 c_1^1(\mathbf{x}) + \cdots + v_{m-1}^0 c_{m-1}^1(\mathbf{x})}{v_0^0 c_0^2(\mathbf{x}) + v_1^0 c_1^2(\mathbf{x}) + \cdots + v_{m-1}^0 c_{m-1}^2(\mathbf{x})}, \quad \mathbf{x} = (x,t), \end{aligned} \tag{8.13}$$

where $\mathcal{I}_m^0 f$ and $\mathcal{I}_m^1 f$ be the piecewise constant interpolation and piecewise linear interpolation of f over Π_m, respectively, and $v_j^0 = v_0(x_j) = \prod_{i=0}^{j} V_i$, ($V_i = \exp(-\frac{u_{0,j}+u_{0,j-1}}{4\kappa} h_i)$) and $c_j^i(\mathbf{x})$, $i = 1, 2$, $j = 0, \cdots, m-1$ are functions in $\mathbf{x}$. In this step, the approximation error can be bounded by (Theorem 3.3, Deng et al., 2022), for any $\mathbf{x} = (x, t) \in (-\pi, \pi) \times (0, +\infty)$,

$$\sup_{u_0 \in \mathcal{S}} \left| \mathcal{G}(u_0)(\mathbf{x}) - \mathcal{G}_{m,p}(T_m u_0; \mathbf{x}) \right| \le Cp^{-1} + \sup_{u_0 \in \mathcal{S}} \left| \mathcal{G}(u_0)(\mathbf{x}) - \mathcal{G}_m(T_m u_0; \mathbf{x}) \right|$$
$$\le Cp^{-1} + 2\Big(\frac{M_0^2}{\kappa} + M_1\Big)h. \tag{8.14}$$

Third, we can view $\mathcal{G}_m(T_m u_0)(x)$ as a rational function in $(V_0, V_1, \cdots, V_{m-1})^\top$ and thus can be well approximated by a ReLu neural network as in Telgarsky (2017). Specifically, there exists a ReLU network $g^{\mathcal{N}}(T_m u_{0,m}; \Theta)$ of size $|\Theta| = O(m^2 \ln(m))$ and a constant $C = C(\kappa, M_0, M_1) > 0$, such that for any $\mathbf{x} \in [-\pi, \pi) \times (0, \infty)$, there exists a set of parameters $\Theta_{\mathbf{x}}$ such that

$$\sup_{u_0 \in \mathcal{S}} \left| \mathcal{G}_m(T_m u_0; \mathbf{x}) - g^{\mathcal{N}}(T_m u_0; \Theta_{\mathbf{x}}) \right| \le Ch.$$

Rewriting the basis L_k^p by a ReLu neural network, we then obtain the desired conclusion.

Extensions to 1D Burgers equation with Dirichlet boundary condition and/or forcing terms and 2D Burgers equation have been discussed in Deng et al. (2022).

Example 8.8 (2D diffusion-reaction equation). Consider the following 2D diffusion-reaction equation:

$$-\Delta u + a(x, y)u = f, \quad \text{in } \Omega \subset \mathbb{R}^2, \quad \mathcal{B}u = 0, \quad \text{on } \partial\Omega, \tag{8.15}$$

where Ω is a rectangular domain and $\mathcal{B}$ can be the Dirichlet, Neumann or Robin boundary operator.

Let $a(x, y) \in \Sigma$, where

$$\Sigma = \{a(x, y) \in W^{r-2,2}(\Omega) \cap L^\infty(\Omega) : 0 \le a(x, y) \le C_0\}, \quad C_0 > 0.$$

Here $1 < r \le 3$ and $W^{s,2}$ is the Sobolev space with square-integrable s-th order weak derivatives. The theorem below states the rate of DeepONets approximating the solution operator of 2D advection-reaction-diffusion equation (8.15).

Theorem 8.9 (Approximation rate of DeepONet for 2D diffusion-reaction equations). *Assume $f \in W^{r-2,2}(\Omega)$, $1 < r \le 3$. Let $\mathcal{G}^f(a)$ be the solution operator. There exist a branch network (blessed representation) $b^{\mathcal{N}}(\mathbf{a}_m; \Theta) \in \mathbb{R}^{p\times 1}$ of size $O(m^3 \ln(\varepsilon^{-1}))$ and ReLU trunk networks $t^{\mathcal{N}}(x; \theta^{(k)}) \in \mathbb{R}^{1\times 1}$ of size $|\theta^{(k)}| = O(1)$, $k = 1, \cdots, p$ $(p = m)$, such that*

$$\sup_{a\in\Sigma} \left\| \mathcal{G}^f(a) - \mathcal{G}^f_{\mathbb{N}}(\mathbf{a}_m) \right\|_{L^\infty} \le C\Big((\log m)^{\frac{1}{2}} m^{-\frac{r-1}{2}} + \left|N_{b^{\mathcal{N}}} L_{b^{\mathcal{N}}}\right|^{-\frac{1}{3}+\epsilon}\Big),$$

where $\epsilon > 0$ *is arbitrarily small and* $C > 0$ *is independent of* m, $N_{b^{\mathcal{N}}}$, $L_{b^{\mathcal{N}}}$ *and* $a(x, y)$. *Here* $\mathcal{G}^{f}_{\mathbb{N}}(\mathbf{a}_m) = (b^{\mathcal{N}}(\mathbf{a}_m; \Theta))^{\top}\vec{t}^{\mathcal{N}}$ *and* $\vec{t}^{\mathcal{N}} = (t^{\mathcal{N}}(x; \theta^{(1)}), \ldots, t^{\mathcal{N}}(x; \theta^{(p)}))$.

As we don't have analytical solutions, we emulate the finite difference method to establish the approximation rate of a DeepONet. The proof is established in three steps. First, we approximate the solution by the central finite difference scheme and obtain the approximation error $(\log(m))^{\frac{1}{2}} m^{-\frac{r-1}{2}}$ when m grid points are used. Second, we rewrite the solver for the linear system resulting from the central finite difference scheme by applying the Sherman-Morrison's formula. Third, we emulate the Sherman-Morrison's formula by a blessed representation (a tree structure) (Mhaskar and Poggio, 2016) for branch networks $b^{\mathcal{N}}$. Then we can obtain the approximation error estimates by the estimates of the blessed representation in Mhaskar and Poggio (2016). Details of the proof can be bounded in Section 4 of Deng et al. (2022).

8.5 Training DeepONets

To train DeepONets, we need the following:

- Sufficient amount of data, which represents diverse inputs and output;
- A suitable architecture for the underlying operator;
- A proper training method such as stochastic gradient descent methods.

High quality data generation is perhaps the first essential step for the success of DeepONets. In real applications, data may be captured by placing affordable and feasible number of sensors. When working with solution operators, we need sufficient amount of data from efficient solvers of the underlying problems with various representative inputs. The inputs are usually modelled with Gaussian random fields with covariance functions such as Gaussian $\exp(-|x-y|^2/l^2)$ in Lu et al. (2021a, 2022), Matern kernel (Li et al., 2020). It is crucial to tune the parameters in these covariance functions to have representative inputs. Over a bounded domain D, we can obtain via Mercer's theorem that the Gaussian process can be written as (known as Karhunen-Loeve expansion) $\sum_{k=1}^{\infty} \sqrt{\lambda_k} e_k(x) \xi_k$, where (λ_k, e_k) is an eigenpair of the covariance function, ξ_k's are independent and identically distributed standard Gaussian random variables. Also, $\{e_k\}_{k=1}^{\infty}$ forms an orthonormal basis in $L^2(D)$ over a bounded domain D and thus has the capacity to represent a large class of functions in a Hilbert space, which is called the reproduced kernel Hilbert space.

The architectures for DeepONets may be adjusted for different operators or even for the same operator for a different level of accuracy. We will not discuss this aspect further.

In the DeepONets, we often use the l_2 regression. Suppose that we are given the data pairs $(u^{(j)}, v^{(j)})_{j=1}^{J}$ where $v^{(j)} \approx \mathcal{G}(u^{(j)})$ (such as those obtained from certain numerical methods or from noisy observations). Here the functions $v^{(j)}$

are given at certain points y_i's.

$$\sum_{j=1}^{J}\sum_{i=1}^{I}\left\|\mathcal{G}_{\mathbb{N}}(T_m u^{(j)})(y_i) - v^{(j)}(y_i))\right\|^2. \tag{8.16}$$

Here the numbers I and J and the size of neural networks should be chosen carefully to have an efficient calculation. These numbers can be estimated according to error estimates, e.g. those in last subsections.

If we know the physics laws that the data obey, e.g. in the case of solution operators from $\mathcal{D}v = f$ with certain boundary or initial conditions, we may use the physics-informed DeepONets, e.g., in Wang et al. (2021) and Goswami et al. (2023) to improve the efficiency. The formulation may read as follows.

$$\sum_{j=1}^{J}\Big(\sum_{i=1}^{I}\left\|\mathcal{G}_{\mathbb{N}}(T_m u^{(j)})(y_i) - v^j(y_i))\right\|^2 + \sum_{l=1}^{L}\left|\mathcal{L}\mathcal{G}_{\mathbb{N}}(T_m u^{(j)}) - f\right|^2(z_l)\Big) \tag{8.17}$$

with possibly extra terms on boundary and initial conditions. Here the points $\{z_l\}$ can different from the points $\{y_i\}$. The accuracy of DeepONets may be enhanced by adding a norm of the gradients, e.g. Luo et al. (2023a) and in Section 8.7.

The training of the DeepONets can be performed with stochastic gradient descent methods. However, due to the structure of DeepONets, the training of the DeepONets can be split into two steps as in DeepONets with POD (Lu et al., 2022) or SVD (Venturi and Casey, 2023): one can first perform singular value decomposition and use the eigenvectors scaled by the positive eigenvalues as the basis and emulate this basis to obtain trunk networks and then one trains branch networks. In Lee and Shin (2023), a two-step training method is proposed based on the Gram-Schmidt orthonormalization. The two-step method of Lee and Shin (2023) is based on the matrix expression of the loss. Since the DeepONet can be written as

$$\mathcal{G}_{\mathbb{N}}(u)(y) = T(y)B(u) \quad \text{where} \quad T(y;\theta) = [t^N(y;\theta^{(1)}), \ldots, t^N(y;\theta^{(p)})],$$

$$B(u;\Theta) = \begin{bmatrix} b^N(T_m u;\Theta^{(1)}) \\ \vdots \\ b^N(T_m u;\Theta^{(p)}) \end{bmatrix},$$

where θ and Θ are the collections of all the parameters of the trunk and branch networks, respectively, it can be checked that the loss of (8.16) can be expressed as

$$\mathcal{L}(\theta,\Theta) = \|T(\theta)B(\Theta) - V\|_F^2,$$

where T is the matrix of size $I \times p$ whose i-th row is $T(y_i;\theta)$, B is the matrix of size $p \times J$ whose j-th column is $B(u^{(j)};\Theta)$, and V is the matrix of size $I \times J$

whose j-th column is $[v^{(j)}(y_1), \ldots, v^{(j)}(y_I)]^\top$. In the first step, the method solves

$$\min_{\theta, A} \|T(\theta)A - V\|_F^2,$$

where A is a trainable matrix of size $p \times J$ and let (θ^*, A^*) be an optimal solution. Then, the Gram-Schmidt orthonormalization is applied to the trunk network, which can be done by the QR-factorization (or equivalently SVD), which gives $T(\theta^*) = Q^* R^*$. The second step then trains the branch network to fit $R^* A^*$, i.e.,

$$\Theta^* = \underset{\Theta}{\operatorname{argmin}} \|B(\Theta) - R^* A^*\|.$$

Accordingly, the trunk network is given by $T(y; \theta^*)(R^*)^{-1}$ and the branch network is $B(u; \Theta^*)$, which gives the DeepONet trained by the two-step of Lee and Shin (2023) is $\mathcal{G}_{\mathbb{N}}(u)(y) = T(y; \theta^*)(R^*)^{-1} B(u; \Theta^*)$.

Remark 8.10. It is worth mentioning some differences between the above two-step training methods and the direct use of either SVD (Lu et al., 2022) or POD (Venturi and Casey, 2023) for the trunk networks. A major difference lies in the linear learning versus the nonlinear learning. The direct use of SVD or POD for the trunk network means that one should expect only a linear relation between data and the basis functions on a fixed grid. Yet, the use of the trunk network in the learning process creates a nontrivial nonlinear relation between data and the resulting basis functions. Furthermore, one can evaluate the basis (the trunk network) on any points aside from the given grid. Admittedly, if one is interested in problems where the standard SVD/POD approaches perform well, one may skip the first step of the training and use the SVD or POD basis instead of training the trunk network. However, if one is concerned with problems where the standard SVD/POD approaches suffer, e.g. advection dominated problems, the nonlinear learning approach provides a viable option. In particular, Theorem 3.5 of Lee and Shin (2023) shows that there exists a trunk network architecture having the smallest width of 4 that approximates any SVD basis. This indicates that one advantage expected from nonlinear learning may be an efficient dimension reduction capability. A similar discussion can be found in the context of the reduced order modeling (ROM) – linear manifold ROM versus nonlinear manifold ROM (Kim et al., 2022).

8.6 Extending DeepONets

In Chen and Chen (1995b), t^N and b^N are two-layer (shallow neural networks) while multilayer feedforward neural networks are used in Lu et al. (2021a). Various extensions in architectures are made such as using convolutions neural networks for branch neural networks in order to accelerate the computation as the input of the branch networks is very high-dimensional, adding Fourier features trunk networks or using cosine and sine functions as the inner layer of the trunk network, e.g. in Lu et al. (2022).

DeepONets with proper orthogonal decomposition (POD) has been developed in Lu et al. (2022), where the trunk networks are replaced by the empirical orthogonal basis from performing POD on the data. A similar idea is implemented in Venturi and Casey (2023), where singular value decomposition is applied.

Convolutional neural networks can be employed in branch networks to reduce the computational cost in Lu et al. (2022). Similarly, Fourier neural operators (Li et al., 2020; Kovachki et al., 2021a) can be applied as branch networks.

As mentioned in Section 8.1, the vanilla DeepONets depends on a fixed discretization and thus can be extended to take data from discretization of different mesh size in Zhang et al. (2023b) by adding one more layer as a function of y in branch networks. In Franco et al. (2023), the branch networks are designed to accommodate mesh-based data.

Architectures of DeepONets can also be established via emulating efficient numerical methods for the underlying operators. For example, the solution operators arising from partial differential equations. For example, in Deng et al. (2022), the above architecture of DeepONets for solution operators of 1D and 2D diffusion-reaction equations (on an interval and a square) can be established by emulating finite difference methods and a numerical solver for the resulting linear system. The approximation error estimates can be obtained accordingly.

8.7 Benchmark test: Burgers' equation

Consider the 1-D Burgers equation with periodic boundary condition

$$\begin{cases} u_t + uu_x = \kappa u_{xx}, \quad (x,t) \in \mathbb{R} \times (0,\infty), \quad \kappa > 0, \\ u(x-\pi,t) = u(x+\pi,t), \quad u(x,0) = u_0(x). \end{cases} \tag{8.18}$$

Then, by the Cole-Hopf transformation, the solution to (8.18) can be written as $u = \frac{-2\kappa v_x}{v}$, where $v_t = \kappa v_{xx}$, $v(x,0) = v_0(x) = \exp\Big(-\frac{1}{2\kappa}\int_{-\pi}^{x} u_0(s)ds\Big)$. With the heat kernel $\mathcal{K}(x,y,t) = \frac{1}{\sqrt{4\pi\kappa t}}\exp\Big(-\frac{(x-y)^2}{4\kappa t}\Big)$, we obtain

$$u(x,t) = -2\kappa \frac{\int_{\mathbb{R}} \partial_x \mathcal{K}(x,y,t) v_0(y)dy}{\int_{\mathbb{R}} \mathcal{K}(x,y,t) v_0(y)dy}. \tag{8.19}$$

Here we assume that the initial condition $u_0(x)$ has zero mean in a period $\bar{u}_0 := \int_{-\pi}^{\pi} u_0(s)ds = 0$ and thus $v_0(x)$ in (8.19) is 2π-periodic.

Remark 8.11. The condition we need in the Cole-Hopf transformation is $\int_{\mathbb{R}} \exp(-\epsilon x^2)\,|v_0(x)|\,dx < \infty$, which can be satisfied when $u_0(x)$ is piecewise linear as assumed. Here $\epsilon > 0$. If the initial condition $u_0(x)$ has the average

$\bar{u}_0 \neq 0$, we write the solution $u(x,t) := v(x - \bar{u}_0 t, t) + \bar{u}_0$, where v satisfies the Burgers equation (8.18) with the initial condition of zero average $u_0 - \bar{u}_0$.

In the following experiments, we test two different formulations for DeepONets for the solution operator $\mathcal{G}$ defined by $u = \mathcal{G}(u_0)$ over the interval [0, 1] (by scaling the domain from $[0, 2\pi]$ to [0, 1], e.g., by $x' = x/(2\pi)$, denoted also by x), when $\kappa = 0.001$.

Data generation. We use a Gaussian random field (GRF) to generate various initial conditions. Specifically, we let $u_0(x) = f(\sin^2(ax))$, where $f(x) \sim \mathcal{GP}(0, k_l(x_1, x_2))$, a is a constant s.t. u_0 satisfies the desired periodic condition. The covariance kernel $k_l(x_1, x_2)$ is taken to be the Gaussian kernel with a length scale parameter $l > 0$, i.e., $k_l(x_1, x_2) = \exp\left(-\|x_1 - x_2\|^2 / 2l^2\right)$. We use a high resolution Fourier spectral method to generate data u.

Architecture and Hyperparameters of DeepONets. We apply the plain DeepONet of the form (8.4) with the following setting:

- m (data length, #sensors) $= 641$; $\#u_{\text{data}}$ train $= 30{,}000$; $\#u_{\text{data}}$ test $= 3000$,
- (ADAM optimizer) learning rate $= 6 \times 10^{-4}$ with exponential decay; epochs $= 500{,}000$.

For the branch network, we use ReLU neural network with layers $[m] + [200] \times 10$; and for the trunk network, we use the swish neural network with layers $[1] + [200] \times 10$;

We use the following two Loss functions: the standard MSE:

$$\begin{aligned}\mathcal{L}_1 &= \frac{1}{N_u \times N_y} \|T_{\text{data}} - \mathcal{G}_T(u_{\text{data}})\|_{l^2}^2 \\ &= \frac{1}{N_u \times N_y} \sum_{i=1}^{N_u} \sum_{j=1}^{N_y} \left| T_j^{(i)} - \mathcal{G}_T\left(\mathbf{u}^{(i)}\right)(y_j) \right|^2 \end{aligned} \tag{8.20}$$

and MSE with regularization by first-order derivatives:

$$\mathcal{L}_2 = \frac{1}{N_u \times N_y} \|T_{\text{data}} - \mathcal{G}_T(u_{\text{data}})\|_{l^2}^2 + \frac{\lambda_g}{N_u \times N_y} |\partial_y T_{\text{data}} - \partial_y \mathcal{G}_T(u_{\text{data}})|_{l^2}^2 \tag{8.21}$$

Here we obtain $\partial_y T_{\text{data}}$ by finite difference or spectral differentiation. The DeepONet are implemented in Python3 and TensorFlow2.

Index	1	2
Loss function	$\mathcal{L}_1$	$\mathcal{L}_2$
test MSE	5.313×10^{-3}	5.963×10^{-4}
test relative l^2	8.146×10^{-2}	2.115×10^{-2}
test mean l^∞	3.065×10^{-1}	1.093×10^{-1}
test max l^∞	2.888×10^{0}	1.999×10^{0}

From the table, we observe that using H^1 regularization can enhance the accuracy.

More benchmark problems can be found in Lu et al. (2022) and Luo et al. (2023b) (fluid flows) and in many related works.

Acknowledgments

GEK acknowledges support by the DOE SEA-CROGS project (DE-SC0023191), the MURI-AFOSR FA9550-20-1-0358 project, and the ONR Vannevar Bush Faculty Fellowship (N00014-22-1-2795).

Appendix 6.A Approximation of elementary functions with ReLU NNs

Note that the ReLU activation function satisfies the positive-homogeneity of order 1, i.e., for any $a > 0$, we have

$$\phi(ax) = a\phi(x),$$

where $\phi(x) = \max\{x, 0\}$. Therefore, for any $\mathrm{C} > 0$ and any ReLU network emulated by $\{W^\ell, b^\ell\}_{\ell=1}^L$, we have

$$\mathcal{R}[\theta](x) = \mathrm{C}^L \mathcal{R}[\tilde{\theta}_\mathrm{C}](x) \quad \text{where} \quad \tilde{\theta}_\mathrm{C} = \{\mathrm{C}^{-1}W^\ell, \mathrm{C}^{-1}b^\ell\}_{\ell=1}^L.$$

Therefore, in principle, there is a simple way of controlling the magnitude of weights and biases by setting $\mathrm{C} \gg 1$ and by expressing a (possibly) large number C^L with additional layers of controlled weights. However, such a naive construction will not consider the potentially complex structure of $\mathcal{R}[\theta]$ and may lead to unnecessarily large NN architectures at the end.

Lemma 6.A.1 (Magnitude control by depth and width). *For any $C > 0$, suppose $s \in \mathbb{N}$ and $r \in \mathbb{N}_{\geq 2}$ satisfy $C \leq s^{r-1}\mathrm{M}^r$. Then, there exists a r-layer ReLU network $\mathcal{R}[\theta] : \mathbb{R} \to \mathbb{R}$ with the architecture of $\vec{\boldsymbol{n}}_\theta = (s^{\oplus(r-1)})$ and $\mathcal{M}[\theta] = 2s + (r-2)s^2$ such that*

$$\mathcal{R}[\theta](x) = C\phi(x), \qquad |\theta|_\infty \leq \mathrm{M}.$$

Furthermore, assuming $r \in \mathbb{N}_{\geq 2}$, there exists a r-layer ReLU network $\mathcal{R}[\theta'] : \mathbb{R} \to \mathbb{R}$ with the architecture of $\vec{\boldsymbol{n}}_{\theta'} = (2s^{\oplus(r-1)})$ and $\mathcal{M}[\theta'] = 4s + 4(r-2)s^2$ such that

$$\mathcal{R}[\theta'](x) = Cx, \qquad |\theta'|_\infty \leq \mathrm{M}.$$

Proof. Let $W^1 = [\mathrm{M}, \cdots, \mathrm{M}]^\top \in \mathbb{R}^{s\times 1}$ and $W^r = \frac{\mathrm{C}}{s^{r-1}\mathrm{M}^r}[\mathrm{M}, \cdots, \mathrm{M}] \in \mathbb{R}^{1\times s}$. For $2 \leq l < r$, let $[W^l]_{ij} = \mathrm{M} \in \mathbb{R}^{s\times s}$. By letting $b^l = 0$ for all $1 \leq l \leq r$, it then can be checked that $\mathcal{R}[\theta](x) = \mathrm{C}\phi(x)$ with $\theta = \{W^l, b^l\}_{l=1}^r$ and $|\theta|_\infty \leq \mathrm{M}$.

Similarly, let $W^1 = [\mathrm{M}, \cdots, \mathrm{M}, -\mathrm{M}, \cdots, -\mathrm{M}]^\top \in \mathbb{R}^{2s\times 1}$, and $W^r = \frac{\mathrm{C}}{s^{r-1}\mathrm{M}^r}(W^1)^\top \in \mathbb{R}^{1\times 2s}$. For $2 \le l < r$, let $W_i^l = (W^1)^\top \in \mathbb{R}^{1\times 2s}$ for all $1 \le i \le 2s$. By letting $b^l = 0$ for all $1 \le l \le r$, it then can be checked that $\mathcal{R}[\theta'](x) = \mathrm{C}x$ with $\theta' = \{W^l, b^l\}_{l=1}^r$ and $|\theta'|_\infty \le \mathrm{M}$. □

Lemma 6.A.2 (Duplication of Width). *Let $\theta = \{W^\ell, b^\ell\}_{\ell=1}^L$ be a ReLU NN whose architecture is $\vec{n}_\theta = (n_1, \ldots, n_{L-1})$, let $k \in \mathbb{N}_{\ge 1}$, and $j \in \{1, \ldots, L-1\}$. Let $\theta'_{[j,k]} := \{\boldsymbol{W}^\ell_{[j,k]}, \boldsymbol{b}^\ell_{[j,k]}\}_{\ell=1}^L$ where*

$$(\textit{if } j = 1): \quad \begin{cases} \boldsymbol{W}^1_{[j,k]} = 1_{k\times 1} \otimes W^1, \boldsymbol{b}^1_{[j,k]} = 1_{k\times 1} \otimes b^1 \\ \boldsymbol{W}^\ell_{[j,k]} = 1_{k\times k} \otimes W^\ell, \boldsymbol{b}^\ell_{[j,k]} = k^{\ell-1} 1_{k\times 1} \otimes b^\ell \quad \textit{if } 2 \le \ell \le L-1 \\ \boldsymbol{W}^L_{[j,k]} = 1_{1\times k} \otimes W^L, \boldsymbol{b}^L_{[j,k]} = k^{L-1} b^L, \end{cases}$$

$$(\textit{if } 2 \le j \le L-1): \quad \begin{cases} \boldsymbol{W}^\ell_{[j,k]} = W^\ell, \qquad \boldsymbol{b}^\ell_{[j,k]} = b^\ell & \textit{if } 1 \le \ell < j \\ \boldsymbol{W}^\ell_{[j,k]} = 1_{k\times k} \otimes W^\ell, \boldsymbol{b}^\ell_{[j,k]} = k^{\ell-j} 1_{k\times 1} \otimes b^\ell & \textit{if } j \le \ell \le L-1 \\ \boldsymbol{W}^L_{[j,k]} = 1_{1\times k} \otimes W^L, \boldsymbol{b}^L_{[j,k]} = k^{L-j} b^L. \end{cases}$$

Then, the ReLU network $\theta'_{[j,k]}$ satisfies

$$\mathcal{R}[\theta'_{[j,k]}](\cdot) = k^{L-j} \cdot \mathcal{R}[\theta](\cdot),$$

with the architecture of $\vec{n}_{\theta'_{[j,k]}} = (n_1, \ldots, n_{j-1}, kn_j, \ldots, kn_{L-1})$. Furthermore, $|\boldsymbol{W}^\ell_{[k]}|_\infty = |W^\ell|_\infty$ and $|\boldsymbol{b}^\ell_{[k]}|_\infty = k^{\max\{0,\ell-j\}}|b^\ell|_\infty$ for all $1 \le \ell \le L$. In particular, if $b^\ell = 0$ for all $\ell \ge j$, we have $|\theta'|_\infty = |\theta|_\infty$.

Lemma 6.A.3 (Concatenation of two networks). *Let $\theta_1 = \{W^l, b^l\}_{l=1}^{L_1}$ and $\theta_2 = \{A^l, c^l\}_{l=1}^{L_2}$ be two networks whose architectures are $\vec{n}_i$ and the output dimension of $\mathcal{R}[\theta_1]$ is equal to the input dimension of $\mathcal{R}[\theta_2]$. Then, there is a ReLU network θ that emulates the composition of the two networks, i.e., $\mathcal{R}[\theta](x) = \mathcal{R}[\theta_2] \circ \mathcal{R}[\theta_1](x)$ for all x with the architecture of $\vec{n} = (\vec{n}_1, \vec{n}_2)$ where*

$$\theta := \{(W^1, b^1), \cdots, (A^1 W^{L_1}, A^1 b^{L_1} + c^1), (A^2, c^2), \cdots, (A^{L_2}, c^{L_2})\}. \tag{6.A.1}$$

Proposition 6.A.4. *Let $\mathcal{R}[\theta_0] : \mathbb{R}^d \to \mathbb{R}$ be a L-layer ReLU NN with the architecture of $\vec{n}_0 = (n_1, \ldots, n_{L-1})$ and $|\theta_0|_\infty = \mathrm{B} > 0$. For any M such that $\mathrm{M} \in (0, \mathrm{B})$, $k \in \mathbb{N}$ and $j \in \{1, \ldots, L-1\}$, suppose $(s, r) \in \mathbb{N} \times \mathbb{N}$ satisfying $(\frac{\mathrm{B}}{\mathrm{M}})^L \le k^{L-j}(s\mathrm{M})^{r-1}$. Then, there exists a ReLU NN $\mathcal{R}[\theta]$ such that*

$$\mathcal{R}[\theta](x) = \mathcal{R}[\theta_0](x) \quad \forall x \in \mathbb{R}^d, \quad |\theta|_\infty = \mathrm{M},$$

with the architecture of $\vec{n}_\theta = (n_1, \ldots, n_{j-1}, kn_j, \ldots, kn_{L-1}, 2s^{\oplus(r-1)})$.

Proof. Let $\theta_0 = \{W^\ell, b^\ell\}_{\ell=1}^L$ and observe that

$$\mathcal{R}[\theta_0](x) = (\frac{\mathrm{B}}{\mathrm{M}})^L \mathcal{R}[\theta_{0,\mathrm{M}}](x), \quad \theta_{0,\mathrm{M}} = \{\frac{\mathrm{M}}{\mathrm{B}} W^\ell, \frac{\mathrm{M}}{\mathrm{B}} b^\ell\}_{\ell=1}^L,$$

which gives $|\theta_{0,\mathrm{M}}|_\infty \le \mathrm{M}$.

Suppose $(s, r) \in \mathbb{N} \times \mathbb{N}_{\ge 2}$ satisfying $(\frac{\mathrm{B}}{\mathrm{M}})^L \le (s\mathrm{M})^{r-1}$. Following the construction given in Lemma 6.A.1, but by setting the weight matrix of the 1st hidden layer as $[1, \ldots, 1, -1, \ldots, 1]^\top \in \mathbb{R}^{2s\times 1}$, we obtain a r-layer ReLU NN $\mathcal{R}[\theta']$ such that $\mathcal{R}[\theta'](x) = (\frac{\mathrm{B}}{\mathrm{M}})^L x$ for all x with the architecture of $\vec{n}_{\theta'} = (2s^{\oplus(r-1)})$. Let $\mathcal{R}[\theta]$ be the composition of $\mathcal{R}[\theta']$ and $\mathcal{R}[\theta_{0,\mathrm{M}}]$, which gives $\mathcal{R}[\theta](x) = \mathcal{R}[\theta'] \circ \mathcal{R}[\theta_0](x) = (\frac{\mathrm{B}}{\mathrm{M}})^L \mathcal{R}[\theta_{0,\mathrm{M}}](x)$ for all x. It then can be checked that the architecture of $\mathcal{R}[\theta]$ is $\vec{n}_\theta = (\vec{n}_{\theta_0}, 2s^{\oplus(r-1)})$. The layer that connects the last layer of θ_0 and the first layer of θ' (see (6.A.1)) is given by

$$[1, \cdots, 1, -1, \cdots, -1]^\top (\frac{\mathrm{M}}{\mathrm{B}}) W^L, \qquad [1, \cdots, 1, -1, \cdots, -1]^\top (\frac{\mathrm{M}}{\mathrm{B}}) b^L,$$

whose maximum norm is bounded by M. Therefore, $|\theta|_\infty = \mathrm{M}$. □

Lemma 6.A.5 (Heaviside). *For* $x = (x_1, \cdots, x_d) \in \mathbb{R}^d$, *let*

$$H(x) = \begin{cases} \mathbb{I}_{[0,\infty)}(x) & \text{if } d = 1, \\ \mathbb{I}_{[0,\infty)\times\mathbb{R}^{d-1}}(x) & \text{if } d \ge 2. \end{cases}$$

Suppose $(s, r, k) \in \mathbb{N}^3$ *such that* $\epsilon^{-1}\mathrm{M}^{-2} \le k(s\mathrm{M})^{r-1}$. *Then, there exists a ReLU network* θ *such that* $\vec{n}_\theta = (2k, 2s^{\oplus(r-1)})$ *and* $|\theta|_\infty = \mathrm{M}$ *and*

$$\|\mathcal{R}[\theta](\cdot) - H(\cdot)\|_{L^p(\mathbb{R}^d)} \le \epsilon^{-\frac{1}{p}}.$$

In particular, by letting $s = k$, *we have* $\vec{n}_\theta = (2s^{\oplus r})$.

Proof. First, we observe that $\forall (x_2, \ldots, x_d) \in \mathbb{R}^{d-1}$,

$$\tilde{H}_\epsilon(x) = \tilde{H}_\epsilon(x_1, \ldots, x_d) = \epsilon^{-1}[\phi(x_1) - \phi(x_1 - \epsilon)] = \begin{cases} 1 & \text{if } \epsilon \le x_1 \\ 0 & \text{if } x_1 \le 0 \\ \epsilon^{-1}x_1 & \text{if } 0 \le x_1 \le \epsilon \end{cases}$$

and $\|\tilde{H}_\epsilon(\cdot) - H(\cdot)\|^p_{L^p(\mathbb{R}^d)} \le \epsilon$. Assuming $\epsilon \in (0, 1]$, observe that $\tilde{H}_\epsilon(x) = \epsilon^{-1}\mathrm{M}^{-2}\mathcal{R}[\theta_0](x)$ where $\mathcal{R}[\theta_0](x) = \mathrm{M}(\phi([\mathrm{M}, \mathbf{0}]x) - \phi([\mathrm{M}, \mathbf{0}]x - \epsilon\mathrm{M}))$ with $\vec{n}_{\theta_0} = (2)$ $|\theta_0| = \mathrm{M}$. Suppose $(s, r) \in \mathbb{N} \times \mathbb{N}_{\ge 3}$ such that $\epsilon^{-1}\mathrm{M}^{-2} \le (s\mathrm{M})^{r-1}$. It then follows from Proposition 6.A.4 that there exists θ such that $\mathcal{R}[\theta] = \tilde{H}_\epsilon$ with $\vec{n}_\theta = (2, 2s^{\oplus(r-1)})$ and $|\theta| = \mathrm{M}$. □

Lemma 6.A.6 (Indicator 1D). *Let $(s,r,k) \in \mathbb{N}^3$ satisfying $\epsilon^{-1}\mathrm{M}^{-2} \le k(s\mathrm{M})^{r-1}$. For any $a, b \in [-1/2, 1/2]$ and $\epsilon \in (0, \frac{1}{2}]$, there exists a ReLU NN, θ such that*

$$\|\mathcal{R}[\theta](\cdot) - \mathbb{I}_{[a,b]}(\cdot)\|_{L_p(\mathbb{R})} = (\frac{2}{p+1}\epsilon)^{\frac{1}{p}},$$

with $\vec{n}_\theta = (4k, 2s^{\oplus(r-1)})$ and $|\theta|_\infty = \mathrm{M}$. In particular, if $s = 2k$, we have $\vec{n}_\theta = (2s^{\oplus r})$.

Proof. For any $a, b \in [-1/2, 1/2]$ such that $a < b$, let us consider a two-layer NN, θ_ϵ such that

$$\mathcal{R}[\theta_\epsilon](x) = \frac{1}{\epsilon}\left[\phi(x-a) - \phi(x-a-\epsilon) - \phi(x-b+\epsilon) + \phi(x-b)\right],$$

which satisfies $\|\mathcal{R}[\theta_0](\cdot) - \mathbb{I}_{[a,b]}(\cdot)\|_{L_p(\mathbb{R})} = (2\int_0^\epsilon (\frac{1}{\epsilon}x)^p dx)^{1/p} = (\frac{2}{p+1}\epsilon)^{1/p}$.

Assuming $\epsilon \le \frac{1}{2}$, observe that $\mathcal{R}[\theta_\epsilon](x) = \epsilon^{-1}\mathrm{M}^{-2} \cdot \mathcal{R}[\theta_0](x)$ where

$$\begin{aligned}\mathcal{R}[\theta_0](x) = \mathrm{M}\big[&\phi(\mathrm{M}x - \mathrm{M}a) - \phi(\mathrm{M}x - \mathrm{M}(a+\epsilon)) - \phi(\mathrm{M}x - \mathrm{M}(b-\epsilon)) \\ &+ \phi(\mathrm{M}x - \mathrm{M}b)\big],\end{aligned}$$

with $\vec{n}_{\theta_0} = (4)$ and $|\theta_0| = \mathrm{M}$.

Let $(s, r) \in \mathbb{N} \times \mathbb{N}_{\ge 3}$ satisfying $\epsilon^{-1}\mathrm{M}^{-2} \le (s\mathrm{M})^{r-1}$. It then follows from Proposition 6.A.4 that there is a ReLU NN, θ, such that $\mathcal{R}[\theta] = \mathcal{R}[\theta_\epsilon]$ with $\vec{n}_\theta = (4, 2s^{\oplus(r-1)})$ and $|\theta|_\infty = \mathrm{M}$. □

Lemma 6.A.7 (Parallelization of multiple NNs, Petersen and Voigtlaender, 2018). *For $j = 1, \ldots, N$, let θ^j be a L-layer NN whose architecture is $\vec{\boldsymbol{n}}_{\theta^j}$. Let us define the separate $\mathrm{P}_{\mathrm{sp}}(\{\theta^j\}_{j=1}^N)$ and the joint $\mathrm{P}_{\mathrm{jt}}(\{\theta^j\}_{j=1}^N)$ parallelizations of $\{\theta^j\}$ by*

$$\mathcal{R}[\mathrm{P}_{\mathrm{sp}}(\{\theta^j\}_{j=1}^N)](z_1, \ldots, z_N) = \begin{bmatrix} \mathcal{R}[\theta^1](z_1) \\ \vdots \\ \mathcal{R}[\theta^N](z_N) \end{bmatrix},$$

$$\mathcal{R}[\mathrm{P}_{\mathrm{jt}}(\{\theta^j\}_{j=1}^N)](x) = \begin{bmatrix} \mathcal{R}[\theta^1](x) \\ \vdots \\ \mathcal{R}[\theta^N](x) \end{bmatrix},$$

whose architectures are

$$\vec{n}_{\mathrm{P}_{\mathrm{sp}}(\{\theta^j\}_{j=1}^N)} = \vec{n}_{\mathrm{P}_{\mathrm{jt}}(\{\theta^j\}_{j=1}^N)} = \sum_{j=1}^N \vec{n}_{\theta^j}$$

and

$$|\mathrm{P}_{\mathrm{sp}}(\{\theta^j\}_{j=1}^N)|_\infty = |\mathrm{P}_{\mathrm{jt}}(\{\theta^j\}_{j=1}^N)|_\infty = \max_j |\theta^j|_\infty.$$

Remark: If the output dimension of $\mathcal{R}[\theta^j]$ is 1, the weight matrix of the last layer of $\mathrm{P}_{\mathrm{sp}}(\{\theta^j\}_{j=1}^N)$ is the form of the block diagonal matrix. This allows one to express $\sum_{j=1}^N \mathcal{R}[\theta^j](z_j)$ without affecting the magnitude of NN. For the notational simplicity, if there is no ambiguity, we denote the NN for such summation by $\mathrm{P}_{\mathrm{sp}}(\{\theta^j\}_{j=1}^N)$. A similar statement works for $\mathrm{P}_{\mathrm{jt}}(\{\theta^j\}_{j=1}^N)$ as well.

Lemma 6.A.8 (Indicator in d-dimension). *For $d \in \mathbb{N}$, let $a_i, b_i \in [-1/2, 1/2]$ with $a_i < b_i$ and $\frac{b_i - a_i}{2} > \epsilon$ for all $i = 1, \cdots, d$. Let $\mathrm{M} \geq 1$ be a given. For any $B > 1$ and $\epsilon \in (0, \frac{1}{2}]$, let $(s, r, r') \in \mathbb{N} \times \mathbb{N}^2_{\geq 2}$ be integers satisfying*

$$\epsilon^{-1}\mathrm{M}^{-1} \leq 0.5(s\mathrm{M})^{r-1},$$
$$B \leq ((ds+1)\mathrm{M})^{r'-1}$$

Then, there exists a ReLU network Φ such that for any $|f| \leq B$,

$$\|\mathcal{R}[\Phi](\cdot, f(\cdot)) - \mathbb{I}_{\prod_{i=1}^d [a_i, b_i]}(\cdot) f(\cdot)\|_{L_p(\mathbb{R}^d)} \leq 4dB\epsilon,$$

whose architecture is $\vec{n}_\Phi = (2ds+2)^{\oplus(r+r')}$ and $|\Phi|_\infty = \mathrm{M}$. Also, for any $|f|, |g| \leq B$, we have

$$\|\mathcal{R}[\Phi](\cdot, f(\cdot)) - \mathcal{R}[\Phi](\cdot, g(\cdot))\|_{L_p(\mathbb{R}^d)} \leq 2\|f - g\|_{L_p(\prod_{i=1}^d [a_i, b_i])}.$$

Proof. Let $\mathbf{x} = (x_1, \cdots, x_d) \in \mathbb{R}^d$ and $y \in \mathbb{R}$, and define

$$n(\mathbf{x}, y) = dB\phi\left(\frac{1}{d}\sum_{i=1}^d \mathcal{R}[\mathcal{I}_i](x_i) + \frac{1}{d}\phi(y/B) - 1\right)$$
$$- dB\phi\left(\frac{1}{d}\sum_{i=1}^d \mathcal{R}[\mathcal{I}_i](x_i) + \frac{1}{d}\phi(-y/B) - 1\right),$$

where $B \geq 1$ and $\mathcal{R}[\mathcal{I}_i](\cdot)$ is a neural network from Lemma 6.A.6 satisfying $\|\mathbb{I}_{[a_i, b_i]}(x) - \mathcal{R}[\theta_{\mathcal{I}_i}](x)\|_{L_p(\mathbb{R})} \leq \epsilon^{-\frac{1}{p}}$. It then can be checked that

$$n(\mathbf{x}, y) = \begin{cases} y & \text{if } \mathbf{x} \in \prod_{i=1}^d [a_i + \epsilon, b_i - \epsilon], \\ 0 & \text{if } \mathbf{x} \notin \prod_{i=1}^d [a_i, b_i]. \end{cases}$$

Since ϕ is 1-Lipschitz, we have $|n(\mathbf{x}, y)| \leq 2B$ whenever $|y| \leq B$. Thus, for any $|f| \leq B$, since the Lebesgue measure of $\prod_{i=1}^d [a_i, b_i] \backslash \prod_{i=1}^d [a_i + \epsilon, b_i - \epsilon]$ is bounded by $(2d)\epsilon$, we have

$$\|n(\mathbf{x}, f(\cdot)) - \mathbb{I}_{\prod_{i=1}^d [a_i,b_i]}(\cdot) f(\cdot)\|_{L_p(\mathbb{R}^d)} \leq (2B)(2d)\epsilon.$$

The rest of the proof constructs a ReLU network that exactly emulates $n(\mathbf{x}, y)$. Let the architecture of $\mathcal{I}_i$ be $\vec{n}_{\mathcal{I}} = (2s^{\oplus r})$ for all i.

Suppose $B > 1$ and $\mathrm{M} \geq 1$. Let $\theta^4 = \{W_4^\ell, b_4^\ell\}_{\ell=1}^{r+1}$ where $W_4^1 = [B^{-1}; -B^{-1}]$, $W_4^\ell = I_2$ for $2 \leq \ell$ and $b_4^\ell = 0$. Then, $\mathcal{R}[\theta^4](y) = \begin{bmatrix} \phi(y/B) \\ \phi(-y/B) \end{bmatrix}$ with $\vec{n}_{\theta^4} = (2^{\oplus r})$ and $|\theta^4|_\infty = \mathrm{M}$.

By modifying the bias vector of the last layer in $\mathcal{I}_i$, one can find $\mathcal{I}_i'$ such that $\mathcal{R}[\mathcal{I}_i'](x) = \mathcal{R}[\mathcal{I}_i](x) - 1$ while having the same architectures and magnitude. Let $\theta^1 = \mathrm{P}_{\mathrm{sp}}(\{\mathcal{I}_i'\}_{i=1}^d \cup \theta^4)$ such that

$$\mathcal{R}\left[\mathrm{P}_{\mathrm{sp}}(\{\mathcal{I}_i\}_{i=1}^d \cup \theta^4)\right](\mathbf{x}, y) = \begin{bmatrix} \mathcal{R}[\mathcal{I}_1](x_1) - 1 \\ \vdots \\ \mathcal{R}[\mathcal{I}_d](x_d) - 1 \\ \mathcal{R}[\theta^4](y) \end{bmatrix}$$

with $|\theta^1|_\infty = \mathrm{M}$ and $\vec{n}_{\theta^1} = (d2s + 2)^{\oplus r}$.

Let $\theta^5 = \{W_5^l, b_5^l\}_{l=1}^2$ where $b_5^l = 0$, $W_5^2 = 1_{1\times k} \otimes [1, -1]$, and

$$W_5^1 = (ds+1)^{-1} 1_{k\times 1} \begin{bmatrix} 1 & 1 & \cdots & 1 & 1 & 0 \\ 1 & 1 & \cdots & 1 & 0 & 1 \end{bmatrix} \in \mathbb{R}^{2k\times(d+2)}.$$

Then consider $\Theta := \theta^5 \bullet \theta^1$. Then, it can be checked that $\mathcal{R}[\Theta](\mathbf{x}, y) = \frac{k}{(ds+1)B} n(\mathbf{x}, y)$ with $|\Theta|_\infty = \mathrm{M}$. By letting $k = ds + 1$, we have $\vec{n}_\Theta = (2ds + 2)^{\oplus(r+1)}$.

Choose $r' \in \mathbb{N}$ such that $B \leq ((ds + 1)\mathrm{M})^{r'-1}$. It then follows from Proposition 6.A.4 that there exists a NN θ such that $\mathcal{R}[\theta] = B\mathcal{R}[\Theta] = n(\mathbf{x}, y)$ with $|\theta| = \mathrm{M}$ and $\vec{n}_\theta = (2ds + 2)^{\oplus(r+r'+1)}$. □

Lemma 6.A.9 (Sum of Indicators). *For $d \in \mathbb{N}$, let $a_{i,l}, b_{i,l} \in [-1/2, 1/2]$ with $a_{i,l} < b_{i,l}$ and $\frac{b_{i,l}-a_{i,l}}{2} > \epsilon$ for all $i = 1, \cdots, d$ and $l = 1, \cdots, N$. Let $\mathrm{M} \geq 1$ be a given. For any $B > 1$ and $\epsilon \in (0, \frac{1}{2}]$, let $(s, r, r') \in \mathbb{N} \times \mathbb{N}_{\geq 2}^2$ be integers satisfying*

$$\epsilon^{-1}\mathrm{M}^{-1} \leq 0.5(s\mathrm{M})^{r-1},$$
$$B \leq ((ds + 1)\mathrm{M})^{r'-1}$$

Let Ψ be a $\mathbb{R}^N$-valued ReLU NN whose architecture is $\vec{n}_\Psi$ with $|\Psi|_\infty \leq \mathrm{M}$ and $|\mathcal{R}[\Psi]_l(x)| \leq B$ for all l and for all $x \in \mathbb{R}^d$. Then, there exists a ReLU NN Φ

such that

$$\left\| \mathcal{R}[\Phi](\cdot) - \sum_{l=1}^{N} \mathbb{I}_{\prod_{i=1}^{d}[a_{i,l},b_{i,l}]} \mathcal{R}[\Psi]_l(\cdot) \right\|_{L_p([-1/2,1/2]^d)} \leq 4dBN\epsilon,$$

whose architecture is $\vec{n}_\Phi = (\vec{n}_\Psi, (2N(ds+1))^{\oplus(r+r')})$ *and* $|\Phi|_\infty \leq \mathrm{M}$.

Proof. Let Ψ be a given network having L_0 layers and Id_d be a L_0-layer identity network such that $\mathcal{R}[\mathrm{Id}_d](x) = x$ for all $x \in \mathbb{R}^d$ with $\vec{\boldsymbol{n}}_{\mathrm{Id}_d} = (2d, \cdots, 2d)$. Let $\theta_0 = \mathrm{P_{jt}}(\{\mathrm{Id}_d, \Psi\})$. Then, $\vec{\boldsymbol{n}}_{\theta_0} = (n + 2d)^{\oplus(L_0-1)}$ and $\mathcal{R}[\theta_0](x) = \begin{bmatrix} x \\ \mathcal{R}[\Psi](x) \end{bmatrix}$ with $|\theta_0| \leq \mathrm{M}$.

Let $\Theta_{0,k}$ be a $L = (r + r' + 1)$-layer network from Lemma 6.A.8 such that $\mathcal{R}[\Theta_{0,k}](x, y) \approx \mathbb{I}_{\prod_{i=1}^{d}[a_{i,k},b_{i,k}]}(x)y$ for all $|y| \leq B$ with $\vec{n}_{\Theta_{0,k}} = (2ds+2)^{\oplus(L-1)}$. Let $\tilde{\theta}_{1,k}$ be a NN such that $\mathcal{R}[\tilde{\theta}_{1,k}](x, v) = \mathcal{R}[\Theta_{0,k}](x, v_k)$ which can be obtained by modifying the first layer of $\Theta_{0,k}$. Then, $\vec{n}_{\tilde{\theta}_{1,k}} = \vec{n}_{\Theta_{0,k}}$ with $|\tilde{\theta}_{1,k}|_\infty = |\Theta_{0,k}|_\infty$. Let $\theta_1 = \mathrm{P_{jt}}(\{\tilde{\theta}_{1,j}\}_{j=1}^{N})$ such that $\mathcal{R}[\theta_1](v, x) = \sum_{k=1}^{N} \mathcal{R}[\Theta_{0,k}](x, v_k)$ with $\vec{n}_{\theta_1} = (2N(ds+1))^{\oplus(r+r')}$ and $|\theta_1| \leq \mathrm{M}$.

Finally, let $\Phi := \theta_1 \bullet \theta_0$. It can be checked that $\mathcal{R}[\Phi](x) = \sum_{l=1}^{N} \mathcal{R}[\Theta_{0,k}](x, \mathcal{R}[\Psi]_l(x))$ with $\vec{n}_\Phi = (\vec{n}_\Psi, (2N(ds+1))^{\oplus(L-1)})$ and $|\Phi|_\infty \leq \mathrm{M}$.

Therefore,

$$\begin{aligned} &\left\| \mathcal{R}[\Phi](x) - \sum_{l=1}^{N} \mathbb{I}_{\prod_{i=1}^{d}[a_{i,l},b_{i,l}]} \mathcal{R}[\Psi]_l(x) \right\|_{L_p([-1/2,1/2]^d)} \\ &= \left\| \sum_{l=1}^{N} n(x, \mathcal{R}[\Psi]_l(x)) - \sum_{l=1}^{N} \mathbb{I}_{\prod_{i=1}^{d}[a_{i,l},b_{i,l}]} \mathcal{R}[\Psi]_l(x) \right\|_{L_p([-1/2,1/2]^d)} \\ &\leq \sum_{l=1}^{N} \| n(x, \mathcal{R}[\Psi]_l(x)) - \mathbb{I}_{\prod_{i=1}^{d}[a_{i,l},b_{i,l}]} \mathcal{R}[\Psi]_l(x) \|_{L_p([-1/2,1/2]^d)} \\ &\leq 4dBN\epsilon, \end{aligned}$$

which completes the proof. □

Lemma 6.A.10 (Square x^2). *For any* $m \in \mathbb{N}$, *let* $L \geq 2$ *and* $k \geq 1$ *such that* $m = k(L-1)$. *Then, there exists a ReLU network* θ *such that*

$$\|\mathcal{R}[\theta](x) - x^2\|_{L_\infty[0,1]} \leq 2^{-2(m+1)},$$

with the architecture of $\vec{n}_\theta = (2^k + 2)^{\oplus(L-1)}$ *and* $|\theta|_\infty \leq 1$.

Proof. We mainly follow the proof of Lemma A.3 of Petersen and Voigtlaender (2018). Let us first define as in Yarotsky (2022) the function

$$g(x) = \begin{cases} 2x & \text{if } 0 \le x < 0.5 \\ 2(1-x) & \text{if } 0.5 \le x \ge 1 \\ 0 & \text{elsewhere.} \end{cases}$$

For $t \in \mathbb{N}$, let $g_t(x) = \underbrace{g \circ g \circ \cdots g}_{t\text{-times}}$ be the t-times composition of g. It then can be checked that

$$g_t(x) = \begin{cases} 2^t(x - k/2^{t-1}) & \text{if } \frac{2k}{2^t} \le x \le \frac{2k+1}{2^t} \text{ for some } k \in \{0, 1, \cdots, 2^{t-1}-1\} \\ -2^t(x - k/2^{t-1}) & \text{if } \frac{2k-1}{2^t} \le x \le \frac{2k}{2^t} \text{ for some } k \in \{1, \cdots, 2^{t-1}\}, \\ 0 & \text{elsewhere.} \end{cases}$$

Let $f_m(x) = x - \sum_{t=1}^{m} 4^{-t} g_t(x)$. It then follows from Proposition 2 of Yarotsky (2022) that

$$\|f_m(x) - x^2\|_{L_\infty[0,1]} \le 2^{-2-2m}.$$

With this known result in mind, we now construct a ReLU network that exactly represents $f_m(x)$, which already has been done in Petersen and Voigtlaender (2018). However, we require all the network parameters to be bounded in absolute values by M.

Let us define $g_t(x; k)$ as follows:

$$g_t(x; k) = 2^t \left[\phi(x) + 2 \sum_{s=1}^{2^t-1} (-1)^s \phi\left(x - 4^{-k} \frac{s}{2^t}\right) + \phi(x - 4^{-k}) \right].$$

We note that $g_t(x; 0) = g_t(x)$ and $4^{-t} g_t(4^{-s} g_s(x); s) = 4^{-t-s} g_{t+s}(x)$.

Let $\boldsymbol{g}_{t,k}, \boldsymbol{c}_{t,k}, \boldsymbol{d}_{t,k} \in \mathbb{R}^{2^t+1}$ defined by

$$[\boldsymbol{g}_{t,k}]_i = \begin{cases} 4^{-k} & \text{if } i = 1 \text{ or } 2^t + 1 \\ 2 \cdot 4^{-k} & \text{if } 1 < i < 2^t + 1 \end{cases},$$

$$[\boldsymbol{d}_{t,k}]_i = \begin{cases} 0 & \text{if } i = 1 \\ -4^{-k} & \text{if } i = 2^t + 1 \\ -4^{-k} \frac{i}{2^{t-1}} & \text{if } 1 < i < 2^t + 1 \end{cases},$$

and $[\boldsymbol{c}_t]_i = 2^{-t}(-1)^{i+1}$, $[\boldsymbol{s}_t]_i = \sum_{(s,p) \in \Omega_t^i} 2^{-s}(-1)^{p+1}$ where

$$\Omega_t^i = \{(s, p) \in \{1, \cdots, t\} \times \{1, \cdots, 2^s\} : i = p 2^{t-s}\}.$$

It then can be checked that

$$4^{-m}g_m(x) = \boldsymbol{c}_m^T\phi(\boldsymbol{g}_{m,0}x + \boldsymbol{d}_{m,0}), \qquad \sum_{s=1}^{m} 4^{-s}g_s(x) = \boldsymbol{s}_m^T\phi(\boldsymbol{g}_{m,0}x + \boldsymbol{d}_{m,0}).$$

Since $x = \phi(x)$ in $[0, 1]$ and $f_m(x)$ can be exactly represented by a two-layer network of width $2^m + 1$ whose weights and biases are all bounded by 1 (in absolute values) and the number of nonzero weights and biases is $3 \cdot 2^m + 2$.

For the deep ReLU construction, consider

$$C^1\phi(W^1x + b^1) = \begin{bmatrix} 4^{-t}g_t(x) \\ x - \sum_{s=1}^{t} 4^{-s}g_s(x) \end{bmatrix}$$

where

$$W^1 = \begin{bmatrix} \boldsymbol{g}_{t,0} \\ 1 \end{bmatrix} \in \mathbb{R}^{2^t+2}, \quad b^1 = \begin{bmatrix} \boldsymbol{d}_{t,0} \\ 0 \end{bmatrix} \in \mathbb{R}^{2^t+2},$$

$$C^1 = \begin{bmatrix} \boldsymbol{c}_t^T & 0 \\ -\boldsymbol{s}_t^T & 1 \end{bmatrix} \in \mathbb{R}^{2\times(2^t+2)}.$$

Next, we observe that

$$\begin{aligned} \boldsymbol{s}_t^T\phi(\boldsymbol{g}_{k,0}(4^{-t}g_t(x)) + \boldsymbol{d}_{k,t}) &= 4^{-t}\boldsymbol{s}_t^T\phi(\boldsymbol{g}_{k,0}(g_t(x)) + \boldsymbol{d}_{k,0}) \\ &= 4^{-t}\sum_{s=1}^{k} 4^{-s}g_s(g_t(x)) = \sum_{s=t+1}^{t+k} 4^{-s}g_s(x). \end{aligned}$$

Thus, we have

$$C\phi\left(V\begin{bmatrix} 4^{-t}g_t(x) \\ x - \sum_{s=1}^{t} 4^{-s}g_s(x) \end{bmatrix} + b\right) = \begin{bmatrix} 4^{-t-k}g_{t+k}(x) \\ x - \sum_{s=1}^{t+k} 4^{-s}g_s(x) \end{bmatrix},$$

where

$$V = \begin{bmatrix} \boldsymbol{g}_{k,0} & 0 \\ 0 & 1 \end{bmatrix} \in \mathbb{R}^{(2^k+2)\times 2}, \quad b = \begin{bmatrix} \boldsymbol{d}_{k,t} \\ 0 \end{bmatrix} \in \mathbb{R}^{2^k+2},$$

$$C = \begin{bmatrix} \boldsymbol{c}_k^T & 0 \\ -\boldsymbol{s}_k^T & 1 \end{bmatrix} \in \mathbb{R}^{2\times(2^k+2)}.$$

For any integers $L \geq 3$ and $k_j \geq 1$ for $j = 2, \cdots, L$, let $m = \sum_{j=2}^{L} k_j$ and set

$$W^j = \begin{bmatrix} \boldsymbol{g}_{k_j,0}\boldsymbol{c}_{k_{j-1}}^T & 0 \\ -\boldsymbol{s}_{k_{j-1}}^T & 1 \end{bmatrix}, \quad b^j = \begin{bmatrix} \boldsymbol{d}_{k_j, \sum_{s=2}^{j} k_s} \\ 0 \end{bmatrix}, \qquad 1 < j < L,$$

and $W^L = [-s_{k_L}^T, 1] \in \mathbb{R}^{1\times(2^{k_L}+2)}$ and $b^L = 0$. Then, $\theta = \{W^j, b^j\}_{j=1}^L$ is a L-layer ReLU network with the architecture of

$$\vec{n} = (2^{k_2}+2, 2^{k_3}+2, \cdots, 2^{k_L}+2),$$

such that $\mathcal{R}[\theta](x) = f_m(x)$. Also, since every elements of $g_{k,0}$ and c_k are less than one, all the parameters are bounded by 1 in absolute values. □

Lemma 6.A.11 (Multiplication). *For $B \geq \frac{1}{2}$ and $m \in \mathbb{N}$, let $(L, k, r) \in \mathbb{N}^3$ such that*

$$m = k(L-1),$$
$$B^2 \leq (3(2^k+2))^{r-1}.$$

Then, there exists a ReLU network $\tilde{\times}$ satisfying

- *for all $x, y \in [-B, B]$, we have $|xy - \mathcal{R}[\tilde{\times}](x, y)| \leq 2^{-2m}$;*
- *for all $x, y \in [-B, B]$ with $xy = 0$, we have $\mathcal{R}[\tilde{\times}](x, y) = 0$;*
- *$|\tilde{\times}|_\infty \leq 1$ and $\vec{n}_{\tilde{\times}} = (3(2^k+2))^{\oplus(L+r-1)}$*

Proof. As in the proof of Proposition 3 in Yarotsky (2022), it follows from $xy = \frac{(x+y)^2 - x^2 - y^2}{2}$ that we define

$$h_m(x, y) = \frac{B^2}{2}\left[f_m\left(\frac{|x+y|}{B}\right) - f_m\left(\frac{|x|}{B}\right) - f_m\left(\frac{|y|}{B}\right)\right].$$

Then, $|h_m(x, y) - xy| \leq \left(\frac{B}{2^m}\right)^2$.

Let $\theta^0 = \{W_0^l, b_0^l\}_{l=1}^2$ be a two-layer network of width 12 such that $\mathcal{R}[\theta^0](x, y) = (|x+y|/B, |x|/B, |y|/B)^\top$ and $|\theta^0|_\infty \leq 1$.

By Lemma 6.A.10, there exists a network θ^1 such that $\mathcal{R}[\theta^1](x) = f_m(x)$. Then $\mathrm{P_{sp}}(\theta^1)$ gives $\mathcal{R}[\mathrm{P_{sp}}(\theta^1)](x, y, z) = \frac{f_m(x) - f_m(y) - f_m(z)}{2}$ with $|\theta^1|_\infty \leq 1$ and $\vec{n}_{\theta^1} = (3(2^k+2))^{\oplus(L-1)}$.

Let $\theta = \mathrm{P_{sp}}(\theta^1) \bullet \theta^0$. It then can be checked that $\mathcal{R}[\theta](x, y) = B^{-2} h_m(x, y)$ with $|\theta|_\infty \leq 1$ and $\vec{n}_\theta = (12, (3(2^k+2))^{\oplus(L-1)})$. Lastly, let $(s, r) \in \mathbb{N}^2$ such that $B^2 \leq s^{r-1}$. It then follows from Proposition 6.A.4 that there exists θ' such that $\mathcal{R}[\theta'](x, y) = h_m(x, y)$ with $|\theta'| \leq 1$ and $\vec{n}_{\theta'} = (12, (3(2^k+2))^{\oplus(L-1)}, s^{\oplus(r-1)})$. □

Lemma 6.A.12 (Monomial). *Let $\boldsymbol{\alpha} = (\alpha_1, \cdots, \alpha_d)$ be a multiindex and let $\mathfrak{n} = \lceil \log_2 |\boldsymbol{\alpha}| \rceil$. For any integers $L \in \mathbb{N}$, let $k \geq 1$ such that $m = k(L-1) \geq \frac{1}{2}\log_2\left(\frac{\mathfrak{n}}{2}\right)$. Then, there exists a ReLU NN $\Theta_{\boldsymbol{\alpha}}$ satisfying*

- *$|x^{\boldsymbol{\alpha}} - \mathcal{R}[\Theta_{\boldsymbol{\alpha}}](x)| \leq \mathfrak{n}2^{-2(m+1)}$ for all $x \in [-1/2, 1/2]^d$,*
- *$|\Theta_{\boldsymbol{\alpha}}|_\infty \leq 1$, and $\vec{n}_{\Theta_{\boldsymbol{\alpha}}} = [2^{\mathfrak{n}-1}, \ldots, 2^0] \otimes \vec{n}_{\tilde{\times}}$ where $\vec{n}_{\tilde{\times}} = (3(2^k+2))^{\oplus L}$.*

Proof. Let $x = (x_1, \cdots, x_{n_0}) \in \mathbb{R}^{n_0}$. By letting $\theta^1 = \{W_1^1, b_1^1\}$ where $W_1^1 = e_j^T \in \mathbb{R}^{1\times n_0}$ and $b_1^1 = 0$, it can be checked that $\mathcal{R}[\theta^1](x) = x_j$. Also, by letting $\theta^0 = \{W_0^1 = 0_{1\times n_0}, b_0^1 = 1\}$, we have $\mathcal{R}[\theta^0](x) = 1$. Thus, any monomials of degree ≤ 1 can be represented by ReLU networks.

Let $\boldsymbol{\alpha} = (\alpha_1, \cdots, \alpha_d)$ be a multiindex such that

$$x^{\boldsymbol{\alpha}} = x_1^{\alpha_1} \cdots x_d^{\alpha_d}, \qquad |\boldsymbol{\alpha}| = \sum_{j=1}^{d} \alpha_j.$$

For any $\boldsymbol{\alpha}$, let us define $\pi_{\boldsymbol{\alpha}} : \{1, \cdots, |\boldsymbol{\alpha}|\} \to \{1, \cdots, d\}$ such that $\pi_{\boldsymbol{\alpha}}(i) = j$ if $\sum_{l=1}^{j-1} \alpha_l < i \le \sum_{l=1}^{j} \alpha_l$. Then, there is a 1-layer ReLU network $\bar{\theta}$ such that $\mathcal{R}[\bar{\theta}](x) \in \mathbb{R}^{|\boldsymbol{\alpha}|}$ where $[\mathcal{R}[\bar{\theta}](x)]_i = x_{\pi_{\boldsymbol{\alpha}}(i)}$ for $1 \le i \le |\boldsymbol{\alpha}|$. It then can be checked that $\mathcal{M}(\bar{\theta}) = |\boldsymbol{\alpha}|$.

First, let us assume that $|\boldsymbol{\alpha}| = 2^{\mathfrak{n}}$ for any $\mathfrak{n} \in \mathbb{N}$. Let $\tilde{\times}$ be the multiplication network from Lemma 6.A.11 with $B = 1/2$ such that $|\mathcal{R}[\tilde{\times}](x, y) - xy| \le 2^{-2(m+1)}$ for all $x, y \in [-0.5, 0.5]$. Then, let $\times_{\mathfrak{n}} := \mathrm{P}_{\mathrm{sp}}^{2^{\mathfrak{n}-1}}(\tilde{\times})$ if $\mathfrak{n} > 1$ and $\times_1 := \tilde{\times}$. By recursively applying similar procedures, we define

$$\times_{\mathrm{Full}}^{\boldsymbol{\alpha}} := \bar{\times}_1^{\mathfrak{n}}, \quad \text{where} \quad \bar{\times}_j^{\mathfrak{n}} := \times_j \bullet \cdots \bullet \times_{\mathfrak{n}-1} \bullet \times_{\mathfrak{n}} \bullet \bar{\theta}, \quad 1 \le j \le \mathfrak{n},$$

whose realization approximates $x^{\boldsymbol{\alpha}}$ within the error of $\mathfrak{n}2^{-2(m+1)}$, which can be verified as follows. First, note that the network architecture of $\times_{\mathrm{Full}}^{\boldsymbol{\alpha}}$ is given by

$$\vec{\boldsymbol{n}}_{\times_{\mathrm{Full}}^{\boldsymbol{\alpha}}} = [2^{\mathfrak{n}-1}, \ldots, 2^0] \otimes \vec{n}_{\tilde{\times}}.$$

We prove the error bound by induction on $\mathfrak{n}$. If $\mathfrak{n} = 1$, we have $|\boldsymbol{\alpha}| = 2$ and

$$\begin{aligned} |x_{\pi_{\boldsymbol{\alpha}}(1)} x_{\pi_{\boldsymbol{\alpha}}(2)} - \mathcal{R}[\times_{\mathrm{Full}}^{\boldsymbol{\alpha}}](x)| &= |x_{\pi_{\boldsymbol{\alpha}}(1)} x_{\pi_{\boldsymbol{\alpha}}(2)} - \mathcal{R}[\tilde{\times}](x_{\pi_{\boldsymbol{\alpha}}(1)}, x_{\pi_{\boldsymbol{\alpha}}(2)})| \\ &\le 2^{-2(m+1)}, \end{aligned}$$

which proves the claim for the case of $\mathfrak{n} = 1$.

Suppose the statement is true for $\tilde{\mathfrak{n}} < \mathfrak{n}$. For notational convenience, let us denote $\mathcal{R}[\theta]$ as θ. Observe that

$$\begin{aligned} &\left| \prod_{i=1}^{2^{\mathfrak{n}}} x_{\pi_{\boldsymbol{\alpha}}(i)} - \mathcal{R}[\times_{\mathrm{Full}}^{\boldsymbol{\alpha}}](x) \right| = \left| \prod_{i=1}^{2^{\mathfrak{n}}} x_{\pi_{\boldsymbol{\alpha}}(i)} - \mathcal{R}[\tilde{\times}]([\bar{\times}_2^{\mathfrak{n}}(x)]_1, [\bar{\times}_2^{\mathfrak{n}}(x)]_2) \right| \\ &\le \left| \prod_{i=1}^{2^{\mathfrak{n}}} x_{\pi_{\boldsymbol{\alpha}}(i)} - [\bar{\times}_2^{\mathfrak{n}}(x)]_1 [\bar{\times}_2^{\mathfrak{n}}(x)]_2 \right| \\ &\quad + \left| [\bar{\times}_2^{\mathfrak{n}}(x)]_1 [\bar{\times}_2^{\mathfrak{n}}(x)]_2 - \mathcal{R}[\tilde{\times}]([\bar{\times}_2^{\mathfrak{n}}(x)]_1, [\bar{\times}_2^{\mathfrak{n}}(x)]_2) \right| \\ &\le 2^{-2(m+1)} + \left| \prod_{i=1}^{2^{\mathfrak{n}-1}} x_{\pi_{\boldsymbol{\alpha}}(i)} - [\bar{\times}_2^{\mathfrak{n}}(x)]_1 \right| \cdot \left| [\bar{\times}_2^{\mathfrak{n}}(x)]_2 \right| \end{aligned}$$

$$
\begin{aligned}
&+\left|\prod_{i=2^{\mathfrak{n}-1}+1}^{2^{\mathfrak{n}}} x_{\pi_{\boldsymbol{\alpha}}(i)}-[\bar{\times}_2^{\mathfrak{n}}(x)]_2\right|\left|\prod_{i=1}^{2^{\mathfrak{n}-1}} x_{\pi_{\boldsymbol{\alpha}}(i)}\right| \\
&\leq 2^{-2(m+1)}+(\mathfrak{n}-1)2^{-2(m+1)}=\mathfrak{n}2^{-2(m+1)},
\end{aligned}
$$

where the last inequality uses the induction hypothesis and the facts that $\left|\prod_{i=1}^{2^{\mathfrak{n}-1}} x_{\pi_{\boldsymbol{\alpha}}(i)}\right| \leq 2^{-2}$ and $\left|[\times_2^{\mathfrak{n}}(x)]_2\right| \leq (\mathfrak{n}-1)2^{-2(m+1)}+2^{-2} \leq 1-2^{-2}$. Note that m is chosen to satisfy

$$
m \geq \frac{1}{2}\log_2\left(\frac{\mathfrak{n}-1}{2}\right).
$$

It now suffices to prove the case where $2^{\mathfrak{n}-1} < |\boldsymbol{\alpha}| < 2^{\mathfrak{n}}$. Let $\boldsymbol{\alpha}_1$, $\boldsymbol{\alpha}_2$ be multiindices such that $\boldsymbol{\alpha}_1+\boldsymbol{\alpha}_2=\boldsymbol{\alpha}$ where $|\boldsymbol{\alpha}_1|=2^{\mathfrak{n}-1}$ and $|\boldsymbol{\alpha}_2|=|\boldsymbol{\alpha}|-2^{\mathfrak{n}-1}$. Let us consider the following network. Given $\boldsymbol{\alpha}_2$, let $\hat{\theta}$ be a one-layer network with the architecture of $\vec{\boldsymbol{n}}_{\hat{\theta}}=(n_0, |\boldsymbol{\alpha}|)$ such that $[\mathcal{R}[\hat{\theta}](x)]_i = x_{\pi_{\boldsymbol{\alpha}}(i)}$. Let $\mathrm{H}(x) = \lceil \frac{x}{2} \rceil$, $\mathrm{H}^{(j)}(x) = \overbrace{\mathrm{H}\circ\cdots\circ\mathrm{H}}^{j \text{ times}}(x)$ and $\mathrm{H}^{(0)}(x)=x$. Let $h^j_{|\boldsymbol{\alpha}|} := \mathrm{H}^{(\mathfrak{n}-j)}(|\boldsymbol{\alpha}|)$. Since $2^{\mathfrak{n}-1} < |\boldsymbol{\alpha}| < 2^{\mathfrak{n}}$, $h^1_{|\boldsymbol{\alpha}|}=2$.

Let us recursively define

$$
\underline{\times}^{\boldsymbol{\alpha}}_{\mathrm{Full}} := \underline{\times}^{\mathfrak{n}}_1, \quad \text{where} \quad \underline{\times}^{\mathfrak{n}}_j := \hat{\times}_j \bullet \cdots \bullet \hat{\times}_{\mathfrak{n}-1} \bullet \hat{\times}_{\mathfrak{n}} \bullet \hat{\theta}, \quad 1 \leq j \leq \mathfrak{n},
$$

where $\hat{\times}_j$ is a network such that

$$
\hat{\times}_j = \begin{cases}
\mathrm{P}^{\lfloor \frac{h^j_{\boldsymbol{\alpha}}}{2} \rfloor}(\tilde{\times}), & \text{if } h^j_{\boldsymbol{\alpha}} \text{ is even,} \\
\mathrm{P}(\mathrm{P}^{\lfloor \frac{h^j_{\boldsymbol{\alpha}}}{2} \rfloor}(\tilde{\times}), \mathrm{Id}_{L+1,1}) & \text{if } h^j_{\boldsymbol{\alpha}} \text{ is odd,}
\end{cases}
$$

whose architecture is

$$
\vec{\boldsymbol{n}}_{\hat{\times}_j} = \begin{cases}
\lfloor \frac{h^j_{\boldsymbol{\alpha}}}{2} \rfloor \tilde{\times} & \text{if } h^j_{\boldsymbol{\alpha}} \text{ is even,} \\
\lfloor \frac{h^j_{\boldsymbol{\alpha}}}{2} \rfloor \tilde{\times} + 2^{\oplus L} & \text{if } h^j_{\boldsymbol{\alpha}} \text{ is odd}
\end{cases}
$$

that satisfies

$$
\begin{aligned}
&\left[\mathcal{R}[\underline{\times}^{\mathfrak{n}}_j](x)\right]_i \\
&= \begin{cases}
\mathcal{R}[\tilde{\times}]\Big([\mathcal{R}[\underline{\times}^{\mathfrak{n}}_{j-1}](x)]_{2i-1}, [\mathcal{R}[\underline{\times}^{\mathfrak{n}}_{j-1}](x)]_{2i}\Big), & \text{if } 1 \leq i \leq \lfloor \frac{h^j_{\boldsymbol{\alpha}}}{2} \rfloor, \\
[\mathcal{R}[\underline{\times}^{\mathfrak{n}}_{j-1}](x)]_{h^j_{\boldsymbol{\alpha}}}, & \text{if } h^j_{\boldsymbol{\alpha}} \text{ is odd}
\end{cases}
\end{aligned}
$$

It then can be checked that $\left|x^{\boldsymbol{\alpha}} - \mathcal{R}[\underline{\times}^{\boldsymbol{\alpha}}_{\text{Full}}](x)\right| \leq \mathfrak{n}2^{-2(m+1)}$, and the architecture of $\underline{\times}^{\boldsymbol{\alpha}}_{\text{Full}}$ is given by

$$\vec{\boldsymbol{n}}_{\underline{\times}^{\boldsymbol{\alpha}}_{\text{Full}}} = (d, \boldsymbol{v}^{\mathfrak{n}-1}_{\tilde{\times}}, \cdots, \boldsymbol{v}^{0}_{\tilde{\times}}, 1),$$
$$\text{where} \quad \boldsymbol{v}^{\mathfrak{n}-1-j}_{\tilde{\times}} = \lfloor \frac{h^j_{\boldsymbol{\alpha}}}{2} \rfloor \tilde{\times} + s_j 2^{\oplus L}, \quad s_j \equiv h^j_{\boldsymbol{\alpha}} \pmod 2),$$

for $0 \leq j < \mathfrak{n}$. By observing $\lfloor \frac{h^j_{\boldsymbol{\alpha}}}{2} \rfloor + s_j \leq 2^{\mathfrak{n}-1-j}$, the proof is completed. □

Lemma 6.A.13 (Polynomials). *Let $q \in \mathbb{N}$, $\{x_l\}_{l=1}^q \subset [-1/2, 1/2]^d$, $\mathfrak{n} = \lceil \log_2 n \rceil$. Let $\gamma_{\boldsymbol{j},\ell} = \sum_{\boldsymbol{j} \leq \boldsymbol{\alpha}, |\boldsymbol{\alpha}| \leq n} c_{\boldsymbol{\alpha},\ell} \binom{\boldsymbol{\alpha}}{\boldsymbol{j}} (-x_\ell)^{\boldsymbol{\alpha}-\boldsymbol{j}}$ and let $\overline{\gamma} := \max_{|\boldsymbol{j}| \leq n, 1 \leq \ell \leq q} |\gamma_{\boldsymbol{j},\ell}|$. For any $m \in \mathbb{N}$, let $(k, L) \in \mathbb{N}^2$ such that $m = k(L-1) \geq \frac{1}{2}\log_2\left(\frac{\log_2 n}{2}\right)$. For a given $\mathrm{M} \geq 1$, let $(s, r) \in \mathbb{N}^2$ satisfying $\overline{\gamma} \leq (s\mathrm{M})^{r-1}$. Then, there exists a $\mathbb{R}^q$-valued ReLU NN Φ_n such that $|\Phi_n|_\infty \leq \mathrm{M}$ satisfying*

$$\left| \mathcal{R}[\Phi_n]_l(x) - \sum_{|\boldsymbol{\alpha}| \leq n} c_{\boldsymbol{\alpha},l}(x - x_l)^{\boldsymbol{\alpha}} \right| \leq \overline{\gamma} \cdot \mathfrak{n} \cdot \binom{d+n}{n} \cdot 2^{-2(m+1)},$$

for all $x \in [-1/2, 1/2]^d$ and for all $l \in \{1, \cdots, q\}$. The architecture of Φ_n is given by

$$\vec{\boldsymbol{n}}_{\Phi_n} = \left(P_{d,n} \cdot [2^{\mathfrak{n}-1}, \ldots, 2^0] \otimes \vec{n}_{\tilde{\times}}, (2s)^{\oplus(r-1)}\right)$$
$$\textit{where} \quad \vec{n}_{\tilde{\times}} = (3(2^k + 2))^{\oplus L}.$$

Proof. Let $x, x_\ell \in \mathbb{R}^d$. By applying the binomial theorem, we have

$$(x - x_\ell)^{\boldsymbol{\alpha}} = \sum_{0 \leq \mathbf{j} \leq \boldsymbol{\alpha}} \binom{\boldsymbol{\alpha}}{\mathbf{j}} (-x_\ell)^{\boldsymbol{\alpha}-\mathbf{j}} (x)^{\mathbf{j}}, \qquad \binom{\boldsymbol{\alpha}}{\mathbf{j}} = \prod_{i=1}^d \binom{\alpha_i}{j_i}.$$

Thus, we have

$$\begin{aligned}
\sum_{|\boldsymbol{\alpha}| \leq n} c_{\boldsymbol{\alpha},\ell}(x - x_\ell)^{\boldsymbol{\alpha}} &= \sum_{|\boldsymbol{\alpha}| \leq n} c_{\boldsymbol{\alpha},\ell} \sum_{0 \leq \mathbf{j} \leq \boldsymbol{\alpha}} \binom{\boldsymbol{\alpha}}{\mathbf{j}} (-x_\ell)^{\boldsymbol{\alpha}-\mathbf{j}} (x)^{\mathbf{j}} \\
&= \sum_{|\mathbf{j}| \leq n} (x)^{\mathbf{j}} \left[\sum_{\mathbf{j} \leq \boldsymbol{\alpha}, |\boldsymbol{\alpha}| \leq n} c_{\boldsymbol{\alpha},\ell} \binom{\boldsymbol{\alpha}}{\mathbf{j}} (-x_\ell)^{\boldsymbol{\alpha}-\mathbf{j}} \right].
\end{aligned}$$

Let $\gamma_{\mathbf{j},\ell} = \sum_{\mathbf{j} \leq \boldsymbol{\alpha}, |\boldsymbol{\alpha}| \leq n} c_{\boldsymbol{\alpha},\ell} \binom{\boldsymbol{\alpha}}{\mathbf{j}} (-x_\ell)^{\boldsymbol{\alpha}-\mathbf{j}}$ and let $\overline{\gamma} := \max_{\mathbf{j} \leq n, 1 \leq \ell \leq q} |\gamma_{\mathbf{j},\ell}|$. Let $s, r \in \mathbb{N}$ satisfy $\overline{\gamma} \leq (s\mathrm{M})^{r-1}$.

Let $\{\boldsymbol{\alpha} : 1 \le |\boldsymbol{\alpha}| \le n\} = \{\boldsymbol{\alpha}_1, \cdots, \boldsymbol{\alpha}_{P'_{d,n}}\}$ where $P'_{d,n} = P_{d,n} - 1$ and $P_{d,n} = \binom{d+n}{n}$, and let $\mathfrak{n} = \lceil \log_2 n \rceil$. Let $\Theta_{\boldsymbol{\alpha}}$ be the network from Lemma 6.A.12 such that $\mathcal{R}[\Theta_{\boldsymbol{\alpha}}](x) \approx x^{\boldsymbol{\alpha}}$. If $1 \le |\boldsymbol{\alpha}| < n$, let us consider $\Theta_{\boldsymbol{\alpha}}^{\text{ext}} = \Theta_{\boldsymbol{\alpha}} \bullet \text{Id}$ where Id is the identity network such that $\mathcal{R}[\text{Id}_{L,d}](x) = x$ for all $x \in \mathbb{R}^d$ and that makes $\Theta_{\boldsymbol{\alpha}}^{\text{ext}}$ a $(\mathfrak{n}L + 1)$-layer network. Specifically,

$$\vec{\boldsymbol{n}}_{\Theta_{\boldsymbol{\alpha}}^{\text{ext}}} = (d, \vec{\boldsymbol{v}}_{\Theta_{\boldsymbol{\alpha}}^{\text{ext}}}, 1), \qquad \vec{\boldsymbol{v}}_{\Theta_{\boldsymbol{\alpha}}^{\text{ext}}} = (\overbrace{2d, \cdots, 2d}^{(\mathfrak{n} - \lceil \log_2 |\boldsymbol{\alpha}| \rceil) L \text{ times}}, \boldsymbol{v}_{\tilde{\times},\boldsymbol{\alpha}}^{\lceil \log_2 |\boldsymbol{\alpha}| \rceil - 1}, \cdots, \boldsymbol{v}_{\tilde{\times},\boldsymbol{\alpha}}^{0}),$$

where $\boldsymbol{v}_{\tilde{\times},\boldsymbol{\alpha}}^{j}$ is defined in Lemma 6.A.12.

Let $\Theta'_n := P_{jt}(\Theta_{\boldsymbol{\alpha}_j}^{\text{ext}} : 1 \le j \le P'_{d,n})$ be a $(\mathfrak{n}L + 1)$-layer network such that

$$\mathcal{R}[\Theta'_n]_j(x) = \mathcal{R}[\Theta_{\boldsymbol{\alpha}_j}^{\text{ext}}](x), \qquad 1 \le j \le P'_{d,n},$$

and whose architecture is $\vec{\boldsymbol{n}}_{\Theta'_n} = (d, \sum_{j=1}^{P'_{d,n}} \vec{\boldsymbol{v}}_{\Theta_{\boldsymbol{\alpha}_j}^{\text{ext}}}, P'_{d,n})$.

Let $\theta_1 = \{G_c, g_c\}$ be a layer such that

$$G_c \in \mathbb{R}^{q \times P'_{d,n}}, \quad g_c \in \mathbb{R}^q,$$
$$\text{where} \quad [G_c]_{\ell j} = \frac{1}{\gamma} \gamma_{\boldsymbol{\alpha}_j, \ell}, [g_c]_\ell = \frac{1}{\gamma} \gamma_{\boldsymbol{\alpha}_0, \ell}, \quad 1 \le \ell \le q, 1 \le j \le P'_{d,n}.$$

Let $\Theta_n = \theta_1 \bullet \Theta'_n$. Then, Θ_n is a network with the architecture of $\vec{\boldsymbol{n}}_{\Theta_n} = (d, \sum_{j=1}^{P'_{d,n}} \vec{\boldsymbol{v}}_{\Theta_{\boldsymbol{\alpha}_j}^{\text{ext}}}, q)$ that satisfies

$$|\mathcal{R}[\Theta_n]_l(x) - \sum_{|\boldsymbol{\alpha}| \le n} c_{\boldsymbol{\alpha},l} (x - x_l)^{\boldsymbol{\alpha}}| \le \overline{\gamma} \sum_{|\boldsymbol{\alpha}| \le n} |x^{\boldsymbol{\alpha}} - \mathcal{R}[\theta_{\boldsymbol{\alpha}}](x)|$$
$$\le \overline{\gamma} \left(\sum_{k=1}^{\mathfrak{n}} k |\{\boldsymbol{\alpha} : |\boldsymbol{\alpha}| = k\}| \right) \cdot 2^{-2(m+1)} \le \overline{\gamma} \mathfrak{n} \binom{d+n}{n} 2^{-2(m+1)}. \qquad \square$$

Lemma 6.A.14 (Lemma A.4 in Petersen and Voigtlaender, 2018). *Let $n \in \mathbb{N}_0$ and $\beta = n + \sigma$ where $\sigma \in (0, 1]$. For each $f \in \mathcal{F}_{\beta,d,B}$ and $x_0 \in (-1/2, 1/2)^d$, there exists a polynomial $p(x) = \sum_{|\boldsymbol{k}| \le n} c_{\boldsymbol{k}} (x - x_0)^{\boldsymbol{k}}$ with $|c_{\boldsymbol{k}}| \le \frac{B}{\boldsymbol{k}!}$ for all $|\boldsymbol{k}| \le n$, and such that*

$$|f(x) - p(x)| \le \frac{d^n}{n!} B \|x - x_0\|^\beta, \quad \forall x \in [-1/2, 1/2]^d.$$

Proof. See Petersen and Voigtlaender (2018). $\square$

Lemma 6.A.15 (Truncation). *For $B > 0$, there exists a two-layer network $\theta = \{W^l, b^l\}_{l=1}^2$ with $|\theta|_\infty \le 1$ and $\vec{\boldsymbol{n}}_\theta = (d, 2d, d)$ that exactly represents the truncation function $\tau_B(x) = sign(x) \min\{|x|, B\}$, i.e., $\max\{1, B\} \mathcal{R}_k[\theta](x) = \tau_B(x_k)$ for all $1 \le k \le d$.*

Proof. Let $A = [1, -1]$, $1_{2\times 1} = [1, 1]^T$ and consider the following network $\theta = \{W^l, b^l\}_{l=1}^2$ such that

$$W^1 = \min\{1, B^{-1}\} I_d \otimes 1_{2\times 1} \in \mathbb{R}^{2d\times d}, \quad b^1 = \min\{1, B\} 1_{d\times 1} \otimes A^T \in \mathbb{R}^{2d\times 1},$$
$$W^2 = I_d \otimes A \in \mathbb{R}^{d\times 2d}, \quad b^2 = -\min\{1, B\} 1_{d\times 1} \mathbb{R}^{d\times 1}.$$

It then can be checked that $\max\{1, B\}\mathcal{R}[\theta]_l(x, B) = \tau_B(x_l)$ for all $1 \le l \le d$ and $|\theta|_\infty \le 1$. □

Appendix 6.B Approximation of piecewise polynomials

Theorem 6.B.1. *Let $d \in \mathbb{N}$, $\beta = n + \sigma$ where $\sigma \in (0, 1]$. Let $\mathfrak{n} = \lceil \log_2 n \rceil$. Let $\mathrm{M} \ge 1$ be given. The following seven integers $(k, L, s, s', r, r', r'')$ decide the specific architecture that depends on the approximation accuracy (controlled by m, the magnitude M of weights and biases.*

- *For any $m \in \mathbb{N}$, choose $(k, L) \in \mathbb{N}^2$ such that $m = k(L-1) \ge \frac{1}{2}\log_2\left(\frac{\log_2 n}{2}\right)$.*
- *For any $(s, r) \in \mathbb{N}^2$, let $\tilde{\gamma}(d, n, B) \le (s\mathrm{M})^{r-1}$ where $\tilde{\gamma}$ is defined in (6.B.1).*
- *For any $(s', r', r'') \in \mathbb{N}^3$, let $2^{2(m+1)}\mathrm{M}^{-1} \le 0.5(s'\mathrm{M})^{r'-1}$ and $(1 + \frac{d^{n+\beta}}{n!})B \le ((ds+1)\mathrm{M})^{r''-1}$.*

Then, there exists a ReLU NN Φ such that $|\Psi|_\infty \le \mathrm{M}$ and for any $f \in \mathcal{F}_{\beta,d,B}$, there are network parameters that depend on f satisfying

$$\|f(x) - \mathcal{R}[\Psi](x)\|_{L_p([-1/2,1/2]^d)} \le C'(d, n, \beta, B) 2^{-\frac{\beta}{d+\beta}(2m+1)},$$

where $C'(d, n, \beta, B) = 2\max\{\max\{d, \mathfrak{n}\}\max\{(1 + \frac{d^n}{n!}d^\beta)B, \tilde{\gamma}(d, n, B)P_{d,n}\}, \frac{d^{n+\beta}}{n!}B\}$. The architecture of Φ is given by

$$\vec{n}_\Phi = (\vec{n}_{\Theta_0}, 2(N^d + d), (2N^d(ds'+1))^{\oplus(r'+r'')}), \qquad N^d = 2^{\frac{d}{d+\beta}(2m+1)},$$

where $P_{d,n} = \binom{d+n}{n}$ and

$$\vec{\boldsymbol{n}}_{\Phi_n} = \left(P_{d,n} \cdot [2^{\mathfrak{n}-1}, \ldots, 2^0] \otimes \vec{n}_{\tilde{\times}}, (2s)^{\oplus(r-1)}\right) + (2d)^{\oplus(\mathfrak{n}L+r-1)},$$
$$\text{where} \quad \vec{n}_{\tilde{\times}} = (3(2^k+2))^{\oplus L}.$$

Proof. Let $N \in \mathbb{N}$ which will be chosen later. For $\lambda \in \{1, \cdots, N\}^d =: [N]^d$, let us consider a partition of $[-1/2, 1/2]^d$ (with disjointness up to null sets):

$$I_\lambda := \prod_{i=1}^d \left[\frac{\lambda_i - 1}{N} - \frac{1}{2}, \frac{\lambda_i}{N} - \frac{1}{2}\right], \qquad \bigcup_{\lambda\in[N]^d} I_\lambda = [-1/2, 1/2]^d,$$
$$x_\lambda = [x_\lambda]_i = \left(\frac{\lambda_i - \frac{1}{2}}{N} - \frac{1}{2}\right).$$

Note that $I_\lambda \subset \bar{B}^{|\cdot|}_{d/N}(x)$ for all $x \in I_\lambda$. It then follows from Lemma 6.A.14 that for each λ, there exists a polynomial $p_{\lambda,n}$ of degree up to n such that

$$\|f - p_{\lambda,n}\|_{C^0(I_\lambda)} \le \frac{d^n}{n!} B \left(\frac{d}{N}\right)^\beta \implies$$
$$\|f - p_{\lambda,n}\|_{L_p(I_\lambda)} \le N^{-d/p} \frac{d^n}{n!} B \left(\frac{d}{N}\right)^\beta .$$

Here we use $\|f\|_{L_p(\Omega)} \le \mu(\Omega)^{1/p} \|f\|_{C^0}$, where $\mu(\Omega)$ is the Lebesgue measure of Ω. Then, we have

$$\left\| f(x) - \sum_{\lambda \in [N]^d} \mathbb{I}_{I_\lambda}(x) \cdot p_{\lambda,n}(x) \right\|_{L_p([-1/2,1/2]^d)}$$
$$= \left(\sum_{\lambda \in [N]^d} \|f - p_{\lambda,n}\|^p_{L_p(I_\lambda)} \right)^{1/p} \le N^{-\beta} \frac{d^{n+\beta}}{n!} B.$$

Also, since $f \in \mathcal{F}_{\beta,d,B}$, $\|f\|_{C^0} \le B$ and thus, it follows from Lemma 6.A.14 that

$$\|p_{\lambda,n}\|_{C^0(I_\lambda)} \le (1 + \frac{d^n}{n!} d^\beta) B := \tilde{B}(d, \beta, B), \qquad \forall \lambda \in [N]^d,$$

where $\tilde{B}$ depends only on d, β and B. Let $x_\lambda = [x_\lambda]_i = (\frac{\lambda_i - \frac{1}{2}}{N} - \frac{1}{2})$ for $\lambda \in [N]^d$. Here $p_{\lambda,n}$ is the Taylor polynomial of degree n centered at x_λ, which can be written as

$$p_{\lambda,n} = \sum_{|\boldsymbol{\alpha}| \le n} c_{\boldsymbol{\alpha},\lambda} (x - x_\lambda)^{\boldsymbol{\alpha}} = \sum_{|\mathbf{k}| \le n} \gamma_{\mathbf{k}}(\mathbf{C}_\lambda, x_\lambda) x^{\mathbf{k}}, \qquad c_{\boldsymbol{\alpha},\lambda} = \frac{\partial^{\boldsymbol{\alpha}} f(x_\lambda)}{\boldsymbol{\alpha}!},$$

where $\gamma_{\mathbf{k}}(\mathbf{C}_\lambda, x_\lambda) = \sum_{\mathbf{k} \le \boldsymbol{\alpha}, |\boldsymbol{\alpha}| \le n} \frac{c^f_{\boldsymbol{\alpha},\lambda}}{\boldsymbol{\alpha}!} \binom{\boldsymbol{\alpha}}{\mathbf{k}} (-x_\lambda)^{\boldsymbol{\alpha} - \mathbf{k}}$ and $\mathbf{C}_\lambda = \{c_{\boldsymbol{\alpha},\lambda}\}_{|\boldsymbol{\alpha}| \le n} \subset [-B, B]^{P_{d,n}}$. Note that for any $f \in \mathcal{F}_{\beta,d,B}$, $|c_{\boldsymbol{\alpha},\lambda}| \le B$ for any λ and $\boldsymbol{\alpha}$ such that $|\boldsymbol{\alpha}| \le n$. Let

$$\tilde{\gamma}(d, n, B) := \max_{\mathbf{C} \subset [-B,B]^{P_{d,n}}} \max_{|\mathbf{k}| \le n, x \in [-1/2,1/2]^d} |\gamma_{\mathbf{k}}(\mathbf{C}, x)|. \tag{6.B.1}$$

For any $m \in \mathbb{N}$, let $(k, L) \in \mathbb{N}^2$ such that $m = k(L-1) \ge \frac{1}{2} \log_2 \left(\frac{\log_2 n}{2} \right)$ where $\mathfrak{n} = \lceil \log_2 n \rceil$. Also, let $s, r \in \mathbb{N}^2$ satisfy $\tilde{\gamma}(d, n, B) \le (s\mathrm{M})^{r-1}$. By Lemma 6.A.13, we have a $\mathbb{R}^{N^d + d}$-valued ReLU NN Θ_n that approximates $p_{\lambda,n}$ for all $\lambda \in [N]^d$ such that

$$\|\mathcal{R}[\Theta_n]_\lambda(x) - p_{\lambda,n}(x)\|_{L^\infty(I_\lambda)} \le \tilde{\gamma}(d, n, B) \cdot P_{d,n} \cdot \mathfrak{n} \cdot 2^{-2(m+1)}, \quad \forall \lambda \in [N]^d,$$

and $\mathcal{R}[\Theta_n]_\lambda(x) = x_j$ for $\lambda = N^d + j$ and $1 \le j \le d$, whose architecture is

$$\vec{\boldsymbol{n}}_{\Phi_n} = \left(P_{d,n} \cdot [2^{\mathrm{n}-1}, \ldots, 2^0] \otimes \vec{n}_{\tilde{\times}}, (2s)^{\oplus(r-1)}\right) + (2d)^{\oplus(\mathrm{n}L+r-1)},$$
$$\text{where} \quad \vec{n}_{\tilde{\times}} = (3(2^k+2))^{\oplus L}.$$

Let τ be the separate concatenation of the truncation network from Lemma 6.A.15 and the identity network. That is, let $\Theta_n' := \tau_{\tilde{B}} \bullet \Theta_n$. Then, $\mathcal{R}[\Theta_n']_l(x) = \tau_{\tilde{B}}(\mathcal{R}[\Theta_n]_l(x))$ for $1 \le l \in N^d$, and $\mathcal{R}[\Theta_n']_l(x) = x_j$ for $l = N^d + j$ and $1 \le j \le d$. Its architecture is

$$\vec{\boldsymbol{n}}_{\Theta_n'} = (\vec{\boldsymbol{n}}_{\Phi_n}, 2(N^d + d)),$$

and $|\Theta_n'|_\infty \le \mathrm{M}$.

We now apply Lemma 6.A.9 with Θ_n'. Let $\epsilon < \frac{1}{2N}$. Let $(s', r', r'') \in \mathbb{N} \times \mathbb{N}^2_{\ge 2}$ such that

$$\epsilon^{-1}\mathrm{M}^{-1} \le 0.5(s'\mathrm{M})^{r'-1},$$
$$(1 + \frac{d^n}{n!} d^\beta)B = \tilde{B} \le ((ds+1)\mathrm{M})^{r''-1}.$$

Then, there exists a ReLU NN Φ such that $|\Phi|_\infty \le \mathrm{M}$, $\vec{n}_\Phi = (\vec{n}_{\Theta_0'}, (2N^d(ds' + 1))^{\oplus(r'+r'')})$ and

$$\left\| \mathcal{R}[\Phi](x) - \sum_{\lambda \in [N]^d} \mathbb{I}_{I_\lambda}(x) \mathcal{R}[\Theta_n']_\lambda(x) \right\|_{L_p([-1/2,1/2]^d)} \le 4d\tilde{B}N^d\epsilon.$$

It then can be checked that

$$\begin{aligned}
&\left\| \mathcal{R}[\Phi](x) - \sum_{\lambda \in [N]^d} \mathbb{I}_{I_\lambda}(x) \cdot p_{\lambda,n}(x) \right\|_{L_p([-1/2,1/2]^d)} \\
&\le \sum_{\lambda \in [N]^d} \left\| n_\epsilon(x, \tilde{p}_{\lambda,n}(x)) - \mathbb{I}_{I_\lambda}(x) \cdot p_{\lambda,n}(x) \right\|_{L_p(I_\lambda)} \\
&\le \sum_{\lambda \in [N]^d} \left\{ \left\| n_\epsilon(x, p_{\lambda,n}(x)) - \mathbb{I}_{I_\lambda}(x) \cdot p_{\lambda,n}(x) \right\|_{L_p(I_\lambda)} \right. \\
&\quad \left. + \left\| n_\epsilon(x, \tilde{p}_{\lambda,n}(x)) - n_\epsilon(x, p_{\lambda,n}(x)) \right\|_{L_p(I_\lambda)} \right\} \\
&\le N^d d\tilde{B}\epsilon + \sum_{\lambda \in [N]^d} \left\| \tilde{p}_{\lambda,n}(x) - p_{\lambda,n}(x) \right\|_{L_p(I_\lambda)} \\
&\le N^d d\tilde{B}\epsilon + N^d \tilde{\gamma}(d, n, B) \cdot P_{d,n} \cdot \mathrm{n} \cdot 2^{-2(m+1)} \\
&\le N^d C(d, n, B)\{d\epsilon + \mathrm{n}2^{-2(m+1)}\},
\end{aligned}$$

where $C(d,n,B) = \max\{\tilde{B}(d,n,B), \tilde{\gamma}(d,n,B)P_{d,n}\}$. Therefore, we have

$$\begin{aligned}
&\|f(x) - \mathcal{R}[\Phi](x)\|_{L_p([-1/2,1/2]^d)} \\
&\le \left\| \mathcal{R}[\Phi](x) - \sum_{\lambda \in [N]^d} \mathbb{I}_{I_\lambda}(x) \cdot p_{\lambda,n}(x) \right\|_{L_p([-1/2,1/2]^d)} \\
&\quad + \left\| f(x) - \sum_{\lambda \in [N]^d} \mathbb{I}_{I_\lambda}(x) \cdot p_{\lambda,n}(x) \right\|_{L_p([-1/2,1/2]^d)} \\
&\le N^d C(d,n,B)\{d\epsilon + \mathrm{n}2^{-2(m+1)}\} + N^{-\beta} \frac{d^{n+\beta}}{n!} B \\
&\le C'(d,n,\beta,B)((\epsilon + 2^{-2(m+1)})N^d + N^{-\beta}),
\end{aligned}$$

where $C'(d,n,\beta,B) = \max\{\max\{d,\mathrm{n}\}C(d,n,B), \frac{d^{n+\beta}}{n!}B\}$.

By letting $N = 2^{s''}$ where $s'' = \frac{2m+1}{d+\beta}$ and $\epsilon = 2^{-2(m+1)}$, we have

$$\begin{aligned}
&\|f(x) - \mathcal{R}[\Psi](x)\|_{L_p([-1/2,1/2]^d)} \\
&\le C'(d,n,\beta,B)((\epsilon + 2^{-2(m+1)})2^{ds'} + 2^{-\beta s'}) \\
&= 2C'(d,n,\beta,B)2^{-\frac{\beta}{d+\beta}(2m+1)}.
\end{aligned}$$

□

Appendix 6.C Approximation of horizon functions

Lemma 6.C.1. *Let* $\mathrm{M} \ge 1$ *be given. Let* $\gamma \in \mathcal{F}_{\beta,d-1,B}$ *and* θ_γ *be a ReLU network that approximates* γ *from Theorem 6.B.1 with* $|\theta_\gamma|_\infty \le \mathrm{M}$. *For any* $\epsilon \in (0,1]$, *let* $(s_h, r_h) \in \mathbb{N}^2$ *such that* $\epsilon^{-1}\mathrm{M}^{-1} \le (s_h\mathrm{M})^{r_h}$. *Then, there exists a network* Φ *such that*

$$\begin{aligned}
&\|\mathcal{R}[\Phi](x) - \mathbb{I}_{[0,\infty)}(x_1 + \gamma(x_{-1}))\|_{L^p([-1/2,1/2]^d)} \\
&\le 2^{\frac{1+p}{p}} \max\left\{ \|\gamma - \mathcal{R}[\theta_\gamma]\|_{L^1([-1/2,1/2]^{d-1})}^{\frac{1}{p}}, \epsilon^{\frac{1}{p}} \right\},
\end{aligned}$$

with $\vec{n}_\Phi = (\vec{n}_{\theta_\gamma} + 2^{\oplus L_{\theta_\gamma}}, (2s_h)^{\oplus r_h})$ *and* $|\Phi|_\infty \le \mathrm{M}$.

Proof. For $\gamma \in \mathcal{F}_{\beta,d-1,B}$, let θ_γ be a ReLU NN from Theorem 6.B.1 that approximates γ. Since the permutation matrix does not change the accuracy of the resulting networks, without loss of generality (up to a constant), we assume $T = I_d$.

Let $x = (x_1, \cdots, x_d)$ and $x_{-1} = (x_2, \cdots, x_d)$. After modifying the first hidden layer of θ_γ, we obtain a NN, θ'_γ such that $\mathcal{R}[\theta'_\gamma](x) = x_1 + \mathcal{R}[\theta_\gamma](x_{-1})$ with $|\theta'_\gamma|_\infty = |\theta_\gamma|_\infty$ and $\vec{n}_{\theta'_\gamma} = \vec{n}_{\theta_\gamma} + 2^{\oplus L_{\theta_\gamma}}$ where L_{θ_γ} is the size of $\vec{n}_{\theta_\gamma}$.

For any $\epsilon \in (0, 1]$, let $(s_h, r_h) \in \mathbb{N}^2$ such that $\epsilon^{-1}\mathrm{M}^{-1} \le (s\mathrm{M})^r$. Then, there is a NN θ_H such that $\|\mathcal{R}[\theta_H] - \mathbb{I}_{[0,\infty)}\|_{L_p} \le \epsilon^{-1/p}$.

Let $\Phi = \theta_H \bullet \theta'_\gamma$ with $\vec{n}_\Phi = (\vec{n}_{\theta'_\gamma}, \vec{n}_{\theta_H})$ and $|\Phi|_\infty \le \mathrm{M}$. It can be checked that

$$\|\mathcal{R}[\Phi](x) - \mathbb{I}_{[0,\infty)}(x_1 + \gamma(x_{-1}))\|_{L^p([-1/2,1/2]^d)} \le 2^q \max\{E_I, E_{II}\},$$

where $q = 1 + p^{-1}$ and

$$\begin{aligned} E_I &= \|\mathbb{I}_{[0,\infty)}(\mathcal{R}[\theta'_\gamma](x)) - \mathbb{I}_{[0,\infty)}(x_1 + \gamma(x_{-1}))\|_{L^p([-1/2,1/2]^d)}, \\ E_{II} &= \|\mathcal{R}[\theta_H](\mathcal{R}[\theta'_\gamma](x)) - \mathbb{I}_{[0,\infty)}(\mathcal{R}[\theta'_\gamma](x))\|_{L^p([-1/2,1/2]^d)}. \end{aligned}$$

First, we note that for fixed $x_{-1} \in [-1/2, 1/2]^{d-1}$,

$$\begin{aligned} & x_1 + \gamma(x_{-1}) \ge 0 \text{ and } x_1 + \mathcal{R}[\theta'_\gamma](\tilde{x}) < 0 \\ & \iff x_1 \in [-\gamma(x_{-1}), -\mathcal{R}[\theta'_\gamma](x_{-1})), \\ & x_1 + \gamma(x_{-1}) < 0 \text{ and } x_1 + \mathcal{R}[\theta'_\gamma](x_{-1}) \ge 0 \\ & \iff x_1 \in [-\mathcal{R}[\theta'_\gamma](x_{-1}), -\gamma(x_{-1})). \end{aligned}$$

Thus, it then can be seen that

$$(2^q E_I)^p \le 2^{1+p}\|\gamma - \mathcal{R}[\theta_\gamma]\|_{L^1([-1/2,1/2]^{d-1})}.$$

For E_{II}, since $|\mathbb{I}_{[0,\infty)}(x_1) - \mathcal{R}[\theta_H](x_1)| \le \mathbb{I}_{[0,\epsilon]}(x_1)$ for all $x_1 \in \mathbb{R}$, we have

$$\begin{aligned} (2^q E_{II})^p &\le 2^{1+p} \int_{[-1/2,1/2]^{d-1}} \int_{[-1/2,1/2]} \mathbb{I}_{0 \le x_1 + \mathcal{R}[\theta_\gamma](x_{-1}) \le \epsilon}(x)\, dx_1 dx_{-1} \\ &\le 2^{1+p} \int_{[-1/2,1/2]^{d-1}} \epsilon\, dx_{-1} = 2^{1+p}\epsilon. \end{aligned}$$

Therefore,

$$\begin{aligned} &\|\mathcal{R}[\Phi](x) - \mathbb{I}_{[0,\infty)}(x_1 + \gamma(\tilde{x}))\|_{L^p([-1/2,1/2]^d)} \\ &\le 2^{\frac{1+p}{p}} \max\left\{ \|\gamma - \mathcal{R}[\theta_\gamma]\|^{\frac{1}{p}}_{L^1([-1/2,1/2]^{d-1})}, \epsilon^{\frac{1}{p}} \right\}. \end{aligned}$$

□

Theorem 6.C.2. *Let $d \in \mathbb{N}$, $\beta = n + \sigma$ where $\sigma \in (0, 1]$, $r \in \mathbb{N}$ and $\mathrm{M} \ge 1$ be given. Let $\mathfrak{n} = \lceil \log_2 n \rceil$. Let $m \in \mathbb{N}$ such that $(2m + 1) \ge (r + 1)^{\frac{p(d-1+\beta)}{\beta}}$.*

- *Let $(k, L) \in \mathbb{N}^2$ such that $m = k(L - 1) \ge \frac{1}{2}\log_2\left(\frac{\log_2 n}{2}\right)$.*
- *Let $(s', r') \in \mathbb{N}^2$ such that $\tilde{\gamma}(d - 1, n, B) \le (s'\mathrm{M})^{r'-1}$ where $\tilde{\gamma}$ is defined in (6.B.1).*
- *Let $(s'', r'', r''') \in \mathbb{N}^3$ such that $2^{2(m+1)}\mathrm{M}^{-1} \le 0.5(s''\mathrm{M})^{r''-1}$ and $(1 + \frac{(d-1)^{n+\beta}}{n!})B \le (((d - 1)s' + 1)\mathrm{M})^{r'''-1}$.*

- *Let* $(s_h, r_h) \in \mathbb{N}^2$ *such that* $2^{\frac{\beta}{d-1+\beta}(2m+1)}\mathrm{M}^{-1} \leq 0.5(s_h\mathrm{M})^{r_h}$.
- *Let* $(s_I, r_I) \in \mathbb{N}^2$ *such that* $2^{\frac{\beta}{p(d-1+\beta)}(2m+1)}\mathrm{M}^{-1} \leq 0.5(s_I\mathrm{M})^{r_I-1}$.

For any $K \in \mathcal{K}_{r,\beta,d,B}$*, there exists a ReLU NN,* Φ *such that* $|\Phi|_\infty \leq \mathrm{M}$ *and*

$$\|\mathcal{R}[\Psi](x) - \mathbb{I}_K(x)\|_{L_p} \leq C(r,\beta,d,B,p)2^{-\frac{\beta}{p(d-1+\beta)}(2m+1)},$$

where $C(r,\beta,d,B,p) = 2^{1+p^{-1}+rd}(4d + 2^{\frac{1+p}{p}}C'(d-1,n,\beta,B)^{1/p})$ *and* C' *is defined in Theorem 6.B.1. The architecture of* Φ *is*

$$\vec{n}_\Phi = (2^{rd}\cdot(\vec{n}_{\theta_\gamma} + 2^{\oplus L_{\theta_\gamma}}, (2s_h)^{\oplus r_h}), (2^{rd+1}(ds_I+1))^{\oplus r_I}),$$

where L_{θ_γ} *is the length of* $\vec{n}_{\theta_\gamma}$*, and*

$$\vec{n}_{\theta_\gamma} = (\vec{n}_{\Theta_0}, 2(N^{d-1}+d-1), (2N^{d-1}((d-1)s'+1))^{\oplus(r'+r'')}),$$
$$N^{d-1} = 2^{\frac{d-1}{d-1+\beta}(2m+1)},$$

where $P_{d,n} = \binom{d+n}{n}$ *and*

$$\vec{n}_{\Theta_0} = \left(P_{d,n}\cdot[2^{\mathfrak{n}-1},\ldots,2^0]\otimes\vec{n}_{\tilde{\times}}, (2s)^{\oplus(r-1)}\right) + (2d)^{\oplus(\mathfrak{n}L+r-1)},$$
$$\textit{where}\quad \vec{n}_{\tilde{\times}} = (3(2^k+2))^{\oplus L}.$$

Proof. For $\lambda = (\lambda_1,\cdots,\lambda_d) \in \{1,\cdots,2^{r_0}\}^d =: [2^{r_0}]^d$, let us consider a partition of $\Omega = [-1/2, 1/2]^d$ (disjointness upto null sets):

$$I_\lambda = \prod_{i=1}^d \left[(\lambda_i - 1)2^{-r_0} - \frac{1}{2}, \lambda_i 2^{-2} - \frac{1}{2}\right]. \tag{6.C.1}$$

Note that for $x \in I_\lambda$, $I_\lambda \subset \overline{B}^{\|\cdot\|_{\ell^\infty}}_{2^{-r_0}}(x)$.

From the definition of $\mathcal{K}_{r_0,\beta,d,B}$, for each $\lambda \in \{1,\cdots,2^{r_0}\}^d$, there is a horizon function $f_\lambda \in \mathcal{HF}_{\beta,d,B}$ such that $\mathbb{I}_{I_\lambda}(x)\mathbb{I}_K(x) = \mathbb{I}_{I_\lambda}(x)f_\lambda(x)$. Thus, for any $K \in \mathcal{K}_{r_0,\beta,d,B}$, we have

$$\mathbb{I}_K(x) = \sum_{\lambda\in[2^{r_0}]^d}\mathbb{I}_{I_\lambda}(x)f_\lambda(x).$$

The goal is to find a deep ReLU network that approximates $\sum_{\lambda\in[2^{r_0}]^d}\mathbb{I}_{I_\lambda}(x)f_\lambda(x)$.

Let $d-1 \in \mathbb{N}$, $\beta = n + \sigma$ where $\sigma \in (0,1]$. Let $\mathfrak{n} = \lceil\log_2 n\rceil$. Let $\mathrm{M} \geq 1$ be given

- For any $m \in \mathbb{N}$, choose $(k, L) \in \mathbb{N}^2$ such that $m = k(L-1) \geq \frac{1}{2}\log_2\left(\frac{\log_2 n}{2}\right)$.

- For any $(s, r) \in \mathbb{N}^2$, let $\tilde{\gamma}(d-1, n, B) \le (s\mathrm{M})^{r-1}$ where $\tilde{\gamma}$ is defined in (6.B.1).
- For any $(s', r', r'') \in \mathbb{N}^3$, let $2^{2(m+1)}\mathrm{M}^{-1} \le 0.5(s'\mathrm{M})^{r'-1}$ and $(1 + \frac{(d-1)^{n+\beta}}{n!})B \le (((d-1)s+1)\mathrm{M})^{r''-1}$.

From Theorem 6.B.1, for any $\gamma_\lambda \in \mathcal{F}_{d-1,\beta,B}$, there exists a ReLU network θ_{γ_λ} such that

$$\|\gamma - \mathcal{R}[\theta_{\gamma_\lambda}]\|_{L_p} \le C'(d-1, n, \beta, B) 2^{-\frac{\beta}{d-1+\beta}(2m+1)},$$

whose architecture is given by

$$\vec{n}_{\theta_{\gamma_\lambda}} = (\vec{n}_{\Theta_0}, 2(N^{d-1}+d-1), (2N^{d-1}((d-1)s'+1))^{\oplus(r'+r'')}),$$
$$N^{d-1} = 2^{\frac{d-1}{d-1+\beta}(2m+1)},$$

where $P_{d,n} = \binom{d+n}{n}$ and

$$\vec{n}_{\Theta_0} = \left(P_{d,n} \cdot [2^{\mathrm{n}-1}, \ldots, 2^0] \otimes \vec{n}_{\tilde{\times}}, (2s)^{\oplus(r-1)}\right) + (2d)^{\oplus(\mathrm{n}L+r-1)},$$
$$\text{where} \quad \vec{n}_{\tilde{\times}} = (3(2^k+2))^{\oplus L},$$

while $|\theta_{\gamma_\lambda}|_\infty \le \mathrm{M}$. By Lemma 6.C.1, for any $\epsilon \in (0, 1]$, let $(s_h, r_h) \in \mathbb{N}^2$ such that $\epsilon^{-1}\mathrm{M}^{-1} \le (s_h\mathrm{M})^{r_h}$. Then, there exists a ReLU NN Θ_{γ_λ} such that $\mathcal{R}[\Theta_{\gamma_\lambda}](x) = \mathcal{R}[\theta_H](x_1 + \mathcal{R}[\theta_{\gamma_\lambda}](x_{-1}))$ whose architecture is given by $\vec{n}_{\Theta_\gamma} := \vec{n}_{\Theta_{\gamma_\lambda}} = (\vec{n}_{\theta_{\gamma_\lambda}} + 2^{\oplus L_{\theta_{\gamma_\lambda}}}, (2s_h)^{\oplus r_h})$ and $|\Phi|_\infty \le \mathrm{M}$.

By jointly concatenating $\{\Theta_{\gamma_\lambda}\}_{\lambda \in [2^r]^d}$, we have a ReLU NN θ_F whose architecture is $\vec{n}_{\theta_F} = 2^{r_0 d}\vec{n}_{\Theta_\gamma}$ satisfying $|\theta_F|_\infty \le \mathrm{M}$ such that $\mathcal{R}[\theta_F]_\lambda(x) = \mathcal{R}[\theta_H](x_1 + \mathcal{R}[\theta_{\gamma_\lambda}](\tilde{x}))$ for $\lambda \in [2^r]^d$.

For $\epsilon' \le \frac{1}{2^{r_0+1}}$, let $(s_I, r_I) \in \mathbb{N}^2$ such that $\epsilon'^{-1}\mathrm{M}^{-1} \le 0.5(s_I\mathrm{M})^{r_I-1}$. By applying Lemma 6.A.9, we have a ReLU network Φ such that

$$\left\| \mathcal{R}[\Phi](\cdot) - \sum_{\lambda \in [2^{r_0}]^d} \mathbb{I}_{I_\lambda} \mathcal{R}[\theta_F]_\lambda(\cdot) \right\|_{L_p([-1/2,1/2]^d)} \le 4 \cdot d \cdot 2^{r_0 d} \cdot \epsilon',$$

whose architecture is $\vec{n}_\Phi = (\vec{n}_{\theta_F}, (2^{r_0 d+1}(ds_I+1))^{\oplus r_I}$ and $|\Phi|_\infty \le \mathrm{M}$.

Lastly, observe that

$$\|\mathcal{R}[\Phi](x) - \mathbb{I}_K(x)\|_{L_p}$$
$$\le 2^q \left\| \mathcal{R}[\Phi](x) - \sum_{\lambda \in [2^r]^d} \mathbb{I}_{I_\lambda}(x) \mathcal{R}[\theta_F]_\lambda(x) \right\|_{L_p}$$

$$
\begin{aligned}
&+2^q \left\| \sum_{\lambda \in [2^r]^d} \mathbb{I}_{I_\lambda}(x)\,(\mathcal{R}[\theta_F]_\lambda(x) - f_\lambda(x)) \right\|_{L_p} \\
&\leq 2^{q+r_0 d} \left\{ 4d\epsilon' + \max_{\lambda \in [2^r]^d} \{ \| \mathcal{R}[\theta_F]_\lambda(x) - f_\lambda(x) \|_{L_p} \} \right\} \\
&\leq 2^{q+r_0 d} \left\{ 4d\epsilon' + 2^{\frac{1+p}{p}} \max\{ C'(d-1, n, \beta, B)^{1/p} 2^{-\frac{\beta}{p(d-1+\beta)}(2m+1)}, \epsilon^{1/p} \} \right\}.
\end{aligned}
$$

By letting $\epsilon^{1/p} = \epsilon' = 2^{-\frac{\beta}{p(d-1+\beta)}(2m+1)}$ we have

$$
\| \mathcal{R}[\Psi](x) - \mathbb{I}_K(x) \|_{L_p} \leq C(r_0, \beta, d, B, p) 2^{-\frac{\beta}{p(d-1+\beta)}(2m+1)},
$$

where $C(r_0, \beta, d, B, p) = 2^{q+r_0 d}(4d + 2^{\frac{1+p}{p}} C'(d-1, n, \beta, B)^{1/p})$ and the proof is completed. □

Appendix 6.D Proof of Theorem 6.1

Proof. For $K \in \mathcal{K}_{r,\beta,d,B}$, let θ_K be a ReLU network θ_K from Theorem 6.C.2 that approximates $\mathbb{I}_K(x)$. For $g \in \mathcal{F}_{\beta',d,B}$, let θ_g be a deep ReLU network from Theorem 6.B.1. Without loss of generality, let us assume that the number of layers of θ_g and θ_K are the same. Let $\tilde{\times}$ be the multiplication network from Lemma 6.A.11. We then define $\Phi = \tilde{\times} \bullet \mathrm{P_{jt}}(\theta_g, \theta_K)$. It then can be checked that

$$
\begin{aligned}
&\| \mathcal{R}[\Phi](x) - \mathbb{I}_K(x) g(x) \|_{L_p} \\
&\leq \| \mathcal{R}[\tilde{\times}](\mathcal{R}[\theta_g](x), \mathcal{R}[\theta_K](x)) - \mathcal{R}[\theta_K](x)\mathcal{R}[\theta_g](x) \|_{L_p} \\
&\qquad + \| \mathcal{R}[\theta_K](x)\mathcal{R}[\theta_g](x) - \mathbb{I}_K(x) g(x) \|_{L_p} \\
&\leq 2^{-2m} + \| \mathcal{R}[\theta_K](x) \left(\mathcal{R}[\theta_g](x) - g(x) \right) \|_{L_p} + \| g(x) (\mathcal{R}[\theta_K](x) - \mathbb{I}_K(x)) \|_{L_p}.
\end{aligned}
$$

Note that $\|g\|_{C^0} \leq B$ and $|\mathcal{R}[\theta_K](x)| \leq 1$. Thus, it follows from Theorems 6.B.1 and 6.C.2 that

$$
\begin{aligned}
\| \mathcal{R}[\theta_K](x) \left(\mathcal{R}[\theta_g](x) - g(x) \right) \|_{L_p} &\leq \| \mathcal{R}[\theta_g](x) - g(x) \|_{L_p} \\
&\leq C'(d, n, \beta', B) 2^{-\frac{\beta'}{d+\beta'}(2m+1)}, \\
\| g(x) (\mathcal{R}[\theta_K](x) - \mathbb{I}_K(x)) \|_{L_p} &\leq C(r, \beta, d, B, p) 2^{-\frac{\beta}{p(d-1+\beta)}(2m+1)},
\end{aligned}
$$

which shows that

$$
\begin{aligned}
&\| \mathcal{R}[\Phi](x) - \mathbb{I}_K(x) g(x) \|_{L_p} \\
&\leq 3^{-1} \tilde{C} \left[2^{-2m} + 2^{-\frac{\beta'}{d+\beta'}(2m+1)} + 2^{-\frac{\beta}{p(d-1+\beta)}(2m+1)} \right],
\end{aligned}
$$

where $\tilde{C}(r,\beta,d,B,p) = 3\max\{C(r,\beta,d,B,p), C'(d,n,\beta',B),1\}$. Since $\frac{\beta'}{d+\beta'} = \frac{\beta}{p(d-1+\beta)+\beta} \leq \frac{\beta}{p(d-1+\beta)} \leq 1$, we have

$$\|\mathcal{R}[\Phi](x) - \mathbb{I}_K(x)g(x)\|_{L_p} \leq \tilde{C} \cdot 2^{-\frac{\beta}{p(d-1+\beta)+\beta}(2m+1)}.$$

The architecture of Φ is $\vec{n}_\Phi = (\vec{n}_{\theta_K} + \vec{n}_{\theta_g}, \vec{n}_{\tilde{\times}})$. □

References

Ainsworth, M., Dong, J., 2021. Galerkin neural networks: a framework for approximating variational equations with error control. SIAM Journal on Scientific Computing 43 (4), A2474–A2501.

Berg, J., Nyström, K., 2018. A unified deep artificial neural network approach to partial differential equations in complex geometries. Neurocomputing 317, 28–41.

Bochev, P., Gunzburger, M., 1998. Finite element methods of least-squares type. SIAM Review 40 (4), 789–837.

Bramble, J.H., Schatz, A.H., 1970. Rayleigh-Ritz-Galerkin methods for Dirichlet's problem using subspaces without boundary conditions. Communications on Pure and Applied Mathematics 23, 653–675.

Burman, E., Oksanen, L., 2018. Weakly consistent regularisation methods for ill-posed problems. In: Numerical Methods for PDEs. In: SEMA SIMAI Springer Ser., vol. 15. Springer, Cham, pp. 171–202.

Chen, T., 1998. A unified approach for neural network-like approximation of non-linear functionals. Neural Networks 11 (6), 981–983.

Chen, T., Chen, H., 1993. Approximations of continuous functionals by neural networks with application to dynamic systems. IEEE Transactions on Neural Networks 4 (6), 910–918.

Chen, T., Chen, H., 1995a. Approximation capability to functions of several variables, nonlinear functionals, and operators by radial basis function neural networks. IEEE Transactions on Neural Networks 6 (4), 904–910.

Chen, T., Chen, H., 1995b. Universal approximation to nonlinear operators by neural networks with arbitrary activation functions and its application to dynamical systems. IEEE Transactions on Neural Networks 6 (4), 911–917.

Cho, J., Nam, S., Yang, H., Yun, S.-B., Hong, Y., Park, E., 2022. Separable PINN: mitigating the curse of dimensionality in physics-informed neural networks. arXiv preprint. arXiv: 2211.08761.

Dahmen, W., Monsuur, H., Stevenson, R., 2023. Least squares solvers for ill-posed PDEs that are conditionally stable. ESAIM: Mathematical Modelling and Numerical Analysis 57 (4), 2227–2255.

De Ryck, T., Jagtap, A.D., Mishra, S., 2023. Error estimates for physics-informed neural networks approximating the Navier–Stokes equations. IMA Journal of Numerical Analysis, drac085.

Deng, B., Shin, Y., Lu, L., Zhang, Z., Karniadakis, G.E., 2022. Approximation rates of DeepONets for learning operators arising from advection–diffusion equations. Neural Networks 153, 411–426.

E, W., Yu, B., 2018. The deep Ritz method: a deep learning-based numerical algorithm for solving variational problems. Communications in Mathematics and Statistics 6 (1), 1–12.

Franco, N.R., Manzoni, A., Zunino, P., 2023. Mesh-informed neural networks for operator learning in finite element spaces. Journal of Scientific Computing 97 (2), 35.

Glorot, X., Bengio, Y., 2010. Understanding the Difficulty of Training Deep Feedforward Neural Networks. Proceedings of the Thirteenth International Conference on Artificial Intelligence and Statistics, vol. 9. PMLR, pp. 249–256.

Goswami, S., Bora, A., Yu, Y., Karniadakis, G.E., 2023. Physics-informed deep neural operator networks. In: Machine Learning in Modeling and Simulation: Methods and Applications. Springer, pp. 219–254.

He, K., Zhang, X., Ren, S., Sun, J., 2015. Delving deep into rectifiers: surpassing human-level performance on imagenet classification. In: Proc. IEEE Int. Conf. Comput. Vis, pp. 1026–1034.

Hu, Z., Jagtap, A.D., Karniadakis, G.E., Kawaguchi, K., 2022. When do extended physics-informed neural networks (XPINNs) improve generalization? SIAM Journal on Scientific Computing 44 (5), A3158–A3182.

Hu, Z., Shukla, K., Karniadakis, G.E., Kawaguchi, K., 2023. Tackling the curse of dimensionality with physics-informed neural networks. arXiv preprint. arXiv:2307.12306.

Jagtap, A.D., Karniadakis, G.E., 2020. Extended physics-informed neural networks (XPINNs): a generalized space-time domain decomposition based deep learning framework for nonlinear partial differential equations. Communications in Computational Physics 28 (5), 2002–2041.

Jagtap, A.D., Kharazmi, E., Karniadakis, G.E., 2020. Conservative physics-informed neural networks on discrete domains for conservation laws: applications to forward and inverse problems. Computer Methods in Applied Mechanics and Engineering 365, 113028.

Jagtap, A.D., Mao, Z., Adams, N., Karniadakis, G.E., 2022a. Physics-informed neural networks for inverse problems in supersonic flows. Journal of Computational Physics 466, 111402.

Jagtap, A.D., Shin, Y., Kawaguchi, K., Karniadakis, G.E., 2022b. Deep Kronecker neural networks: a general framework for neural networks with adaptive activation functions. Neurocomputing 468, 165–180.

Kharazmi, E., Zhang, Z., Karniadakis, G.E., 2019. Variational physics-informed neural networks for solving partial differential equations. arXiv preprint. arXiv:1912.00873.

Kharazmi, E., Zhang, Z., Karniadakis, G.E., 2021. hp-vpinns: variational physics-informed neural networks with domain decomposition. Computer Methods in Applied Mechanics and Engineering 374, 113547.

Khodayi-Mehr, R., Zavlanos, M., 2020. Varnet: variational neural networks for the solution of partial differential equations. In: Learning for Dynamics and Control. PMLR, pp. 298–307.

Kim, H.H., Yang, H.J., 2023. Domain decomposition algorithms for physics-informed neural networks. In: Domain Decomposition Methods in Science and Engineering XXVI. Springer, pp. 697–704.

Kim, Y., Choi, Y., Widemann, D., Zohdi, T., 2022. A fast and accurate physics-informed neural network reduced order model with shallow masked autoencoder. Journal of Computational Physics 451, 110841.

Klibanov, M.V., Yamamoto, M., 2006. Lipschitz stability of an inverse problem for an acoustic equation. Applicable Analysis 85 (5), 515–538.

Kopaničáková, A., Kothari, H., Karniadakis, G.E., Krause, R., 2023. Enhancing training of physics-informed neural networks using domain-decomposition based preconditioning strategies. arXiv preprint. arXiv:2306.17648.

Kovachki, N., Lanthaler, S., Mishra, S., 2021a. On universal approximation and error bounds for Fourier neural operators. Journal of Machine Learning Research 22 (1), 13237–13312.

Kovachki, N., Li, Z., Liu, B., Azizzadenesheli, K., Bhattacharya, K., Stuart, A., Anandkumar, A., 2021b. Neural operator: learning maps between function spaces. arXiv preprint. arXiv:2108.08481.

Lagaris, I.E., Likas, A., Fotiadis, D.I., 1998. Artificial neural networks for solving ordinary and partial differential equations. IEEE Transactions on Neural Networks 9 (5), 987–1000.

Lanthaler, S., Mishra, S., Karniadakis, G.E., 2021. Error estimates for DeepOnets: a deep learning framework in infinite dimensions.

Lee, S., Shin, Y., 2023. On the training and generalization of deep operator networks. arXiv preprint. arXiv:2309.01020.

Li, Z., Kovachki, N., Azizzadenesheli, K., Liu, B., Bhattacharya, K., Stuart, A., Anandkumar, A., 2020. Fourier neural operator for parametric partial differential equations. arXiv:2010.08895.

Lu, L., Jin, P., Karniadakis, G.E., 2019. Deeponet: learning nonlinear operators for identifying differential equations based on the universal approximation theorem of operators. arXiv preprint. arXiv:1910.03193.
Lu, L., Jin, P., Pang, G., Zhang, Z., Karniadakis, G.E., 2021a. Learning nonlinear operators via deeponet based on the universal approximation theorem of operators. Nature Machine Intelligence 3 (3), 218–229.
Lu, L., Meng, X., Cai, S., Mao, Z., Goswami, S., Zhang, Z., Karniadakis, G.E., 2022. A comprehensive and fair comparison of two neural operators (with practical extensions) based on FAIR data. Computer Methods in Applied Mechanics and Engineering 393, 114778.
Lu, L., Pestourie, R., Yao, W., Wang, Z., Verdugo, F., Johnson, S.G., 2021b. Physics-informed neural networks with hard constraints for inverse design. SIAM Journal on Scientific Computing 43 (6), B1105–B1132.
Lu, L., Shin, Y., Su, Y., Karniadakis, G., 2020. Dying ReLU and initialization: theory and numerical examples. Communications in Computational Physics 28, 1671–1706.
Luo, D., O'Leary-Roseberry, T., Chen, P., Ghattas, O., 2023a. Efficient pde-constrained optimization under high-dimensional uncertainty using derivative-informed neural operators.
Luo, Y., Chen, Y., Zhang, Z., 2023b. Cfdbench: a comprehensive benchmark for machine learning methods in fluid dynamics. arXiv preprint. arXiv:2310.05963.
Marcati, C., Schwab, C., 2023. Exponential convergence of deep operator networks for elliptic partial differential equations. SIAM Journal on Numerical Analysis 61 (3), 1513–1545.
Meng, X., Li, Z., Zhang, D., Karniadakis, G.E., 2020. PPINN: parareal physics-informed neural network for time-dependent PDEs. Computer Methods in Applied Mechanics and Engineering 370, 113250.
Mhaskar, H.N., Poggio, T., 2016. Deep vs. shallow networks: an approximation theory perspective. Analysis and Applications 14 (06), 829–848.
Mishra, S., Molinaro, R., 2022. Estimates on the generalization error of physics-informed neural networks for approximating a class of inverse problems for PDEs. IMA Journal of Numerical Analysis 42 (2), 981–1022.
Mishra, S., Molinaro, R., 2023. Estimates on the generalization error of physics-informed neural networks for approximating PDEs. IMA Journal of Numerical Analysis 42 (1), 1–43.
Penwarden, M., Jagtap, A.D., Zhe, S., Karniadakis, G.E., Kirby, R.M., 2023. A unified scalable framework for causal sweeping strategies for physics-informed neural networks (PINNs) and their temporal decompositions. Journal of Computational Physics 493, 112464.
Petersen, P., Voigtlaender, F., 2018. Optimal approximation of piecewise smooth functions using deep ReLU neural networks. Neural Networks 108, 296–330.
Raissi, M., Perdikaris, P., Karniadakis, G.E., 2019. Physics-informed neural networks: a deep learning framework for solving forward and inverse problems involving nonlinear partial differential equations. Journal of Computational Physics 378, 686–707.
Richter-Powell, J., Lipman, Y., Chen, R.T., 2022. Neural conservation laws: a divergence-free perspective. Advances in Neural Information Processing Systems 35, 38075–38088.
Shang, Y., Wang, F., Sun, J., 2022. Deep Petrov-Galerkin method for solving partial differential equations. arXiv preprint. arXiv:2201.12995.
Sheng, H., Yang, C., 2022. PFNN-2: a domain decomposed penalty-free neural network method for solving partial differential equations. Communications in Computational Physics 5, 19–25.
Shin, Y., Karniadakis, G.E., 2020. Trainability of ReLU networks and data-dependent initialization. Journal of Machine Learning for Modeling and Computing 1 (1), 39–74.
Shin, Y., Zhang, Z., Karniadakis, G.E., 2023. Error estimates of residual minimization using neural networks for linear PDEs. Journal of Machine Learning for Modeling and Computing 4 (4).
Shukla, K., Jagtap, A.D., Karniadakis, G.E., 2021. Parallel physics-informed neural networks via domain decomposition. Journal of Computational Physics 447, 110683.
Sirignano, J., Spiliopoulos, K., 2018. DGM: a deep learning algorithm for solving partial differential equations. Journal of Computational Physics 375, 1339–1364.

Son, H., Cho, S.W., Hwang, H.J., 2023. Enhanced physics-informed neural networks with augmented Lagrangian relaxation method (AL-PINNs). Neurocomputing, 126424.

Son, H., Jang, J.W., Han, W.J., Hwang, H.J., 2021. Sobolev training for physics informed neural networks. arXiv preprint. arXiv:2101.08932.

Sukumar, N., Srivastava, A., 2022. Exact imposition of boundary conditions with distance functions in physics-informed deep neural networks. Computer Methods in Applied Mechanics and Engineering 389, 114333.

Sun, Q., Xu, X., Yi, H., 2023. Dirichlet-Neumann learning algorithm for solving elliptic interface problems. arXiv preprint. arXiv:2301.07361.

Telgarsky, M., 2017. Neural networks and rational functions. In: International Conference on Machine Learning. PMLR, pp. 3387–3393.

Venturi, S., Casey, T., 2023. Svd perspectives for augmenting deeponet flexibility and interpretability. Computer Methods in Applied Mechanics and Engineering 403, 115718.

Wang, S., Wang, H., Perdikaris, P., 2021. Learning the solution operator of parametric partial differential equations with physics-informed deeponets. Science Advances 7 (40), eabi8605.

Wang, Y., Jin, P., Xie, H., 2022a. Tensor neural network and its numerical integration. arXiv preprint. arXiv:2207.02754.

Wang, Y., Liao, Y., Xie, H., 2022b. Solving Schrödinger equation using tensor neural network. arXiv preprint. arXiv:2209.12572.

Yarotsky, D., 2022. Universal approximations of invariant maps by neural networks. Constructive Approximation 55 (1), 407–474.

Yu, J., Lu, L., Meng, X., Karniadakis, G.E., 2022. Gradient-enhanced physics-informed neural networks for forward and inverse pde problems. Computer Methods in Applied Mechanics and Engineering 393, 114823.

Zang, Y., Bao, G., Ye, X., Zhou, H., 2020. Weak adversarial networks for high-dimensional partial differential equations. Journal of Computational Physics 411, 109409.

Zhang, H., Xu, Y., Liu, Q., Li, Y., 2023a. Deep learning framework for solving Fokker-Planck equations with low-rank separation representation. Engineering Applications of Artificial Intelligence 121, 106036.

Zhang, Z., Wing Tat, L., Schaeffer, H., 2023b. Belnet: basis enhanced learning, a mesh-free neural operator. Proceedings of the Royal Society A 479 (2276), 20230043.

Chapter 7

Computability of optimizers for AI and data science

Yunseok Lee[a], **Holger Boche**[b,c,d,e,f], **and Gitta Kutyniok**[a,g,h,i,*]

[a]*Ludwig-Maximilians-Universität München, Munich, Germany,* [b]*Technical University of Munich, Munich, Germany,* [c]*Ruhr University Bochum, Bochum, Germany,* [d]*Munich Center for Quantum Science and Technology, Munich, Germany,* [e]*Munich Quantum Valley, Munich, Germany,* [f]*BMBF Research Hub 6G-life, Munich, Germany,* [g]*Munich Center for Machine Learning, Munich, Germany,* [h]*University of Tromsø, Tromsø, Norway,* [i]*German Aerospace Center, Oberpfaffenhofen, Germany*

**Corresponding author: e-mail address: kutyniok@math.lmu.de*

The fundamental question underlying all computing is 'What can be (efficiently) automated'?

–Computing as a discipline (Denning et al., 1989)

Contents

Handbook of Numerical Analysis, Volume 25, ISSN 1570-8659, https://doi.org/10.1016/bs.hna.2024.05.005

Abstract

Artificial Intelligence (AI) and Data Science stand as pivotal innovations that revolutionize methods and processes across a multitude of industries. In unison, they facilitate the management, storage, transmission, and analysis of vast data volumes. A significant portion of these challenges are articulated as optimization problems. The potency of AI and Data Science is deeply rooted in the successful resolution of these optimization problems, which are prevalent in areas such as machine learning model fine-tuning, operational research, and logistics.

However, it is crucial to acknowledge that solutions to these optimization problems do not always come with guarantees. This does not necessarily imply that research is lacking in this direction. Instead, it is often a manifestation of the constraints imposed by the nature of the digital hardware used for calculations.

Digital hardware is bound by physical limitations, including constraints on processing power, storage capacity, and time. A key limitation, however, is its inherent binary nature, which can handle only discrete values precisely and may struggle with mathematical functions on real variables. This chapter aims to summarize key results from the field of computability and highlight critical, yet lesser-known issues in optimization theory.

Keywords

Optimization, Artificial Intelligence, Computability, Turing machine, Digital computing, Information theory, Portfolio optimization, Cryptography

MSC Codes

65G50, 68Q05, 68T01, 49J99

1 Introduction

Artificial Intelligence (AI) is a branch of computer science that aims to create systems capable of performing tasks that would normally require human intelligence. These tasks include learning from data, processing natural language, recognizing patterns, and making decisions. Mathematically, AI often involves optimization problems. For example, in deep learning (a subset of AI), we often try to minimize a loss function that measures the difference between the model's predictions and the actual data. Often, an enormous amount of data is needed to achieve satisfactory results. The process of gathering, transmitting, and analyzing this data presents its own set of challenges, and is referred to as data science.

Data Science is an interdisciplinary field that uses scientific methods, processes, algorithms, and systems to extract knowledge and insights from structured and unstructured data. It involves various disciplines such as classical statistics, data mining, AI, and database systems. AI plays a crucial role in data science since it provides the means to automatically identify patterns in data and make data-driven predictions or decisions. By using AI, data scientists can create models that learn from data, and these models can be used to predict future outcomes or discover underlying patterns. But data science is not just about analyzing data, but also about its effective management. This involves the communication, transfer, and safety-critical sharing, storing, and collecting of data. Many of these challenges, including those related to data management, can be formulated as optimization problems. For instance, when transferring data, we might want to minimize the time it takes or the bandwidth used. When storing data, we might want to minimize the storage space or the cost. Even privacy is commonly formulated as an optimization problem with constraints.

Optimization plays a pivotal role in a multitude of domains, including but not limited to artificial intelligence, data science, mathematics, natural sciences, and engineering. The primary objective in these fields is to either minimize or maximize a specific function, all while adhering to certain constraints. Furthermore, optimization finds extensive applications in diverse areas such as artificial intelligence, medical imaging, and communication. In these fields, it serves as a powerful tool to tackle complex and high-dimensional problems. However, most optimization algorithms are designed and executed on digital devices, such as computers, which can be mathematically modeled as Turing machines. This means that they have inherent limitations in their ability to represent and manipulate real numbers, which often require infinite precision. Therefore, most real numbers have to be approximated by finite representations, such as floating-point numbers, which introduce errors and uncertainties in the computation. These errors can affect the performance and reliability of optimization algorithms, depending on how they handle the approximation of inputs, that cannot be exactly represented on any digital device.

In this chapter, we examine central questions of AI, like deep learning, and of data science, like effective communication and security/cryptography. More precisely, we survey these issues in the specific context of optimization. However, most optimization algorithms only focus on finding the optimal value of a function that measures the quality of a solution. In many cases and applications, the optimizer, which is the solution that achieves the optimal value instead of the optimal value itself, is more important. Furthermore, we survey some of the challenges of finding the optimizer on digital hardware and show examples of optimizers that cannot be computed.

1.1 Overview

In Section 2, we delve into the core concepts of computability and examine various hardware models. Section 3 is dedicated to the introduction of neural networks and the underlying principles of deep learning. The discussion extends into Section 4, where we explore a range of computability and noncomputability results, emphasizing the intractability of certain functions or properties. Section 5 establishes a general criterion for noncomputability. Subsequent sections provide an analysis of specific instances of noncomputable problems of practical relevance, encompassing areas such as neural networks, inverse problems, and information theory.

1.2 Numerical computations on digital hardware

Numerical computations are ubiquitous in modern science and engineering. One of the most remarkable achievements of the past decade has been the emergence of artificial intelligence and deep learning as powerful tools for solving various problems. However, these methods also face significant challenges and limitations that hinder their reliability, robustness, and applicability in critical domains such as medical imaging or autonomous driving. For instance, deep learning models are often unstable, untrustworthy, vulnerable to adversarial attacks, and lack theoretical guarantees. We hypothesize that one of the possible reasons for these difficulties is the inherent noncomputability of some optimization problems on digital hardware.

1.3 Computability and hardware

Algorithms are not just abstract entities, but also concrete implementations on particular hardware devices. The way numerical data is handled by the hardware, such as how π is represented and computed, can influence the behavior and results of an algorithm. To fully describe an algorithm, one needs to account for the details of how numbers are stored and processed on the hardware platform. We focus on the distinction between digital and analog hardware. Digital hardware makes use of discrete representation to store numbers, i.e., bits, while analog hardware can store and manipulate real quantities directly.

1.3.1 Turing machines

The most fundamental and widely-known hardware model in computer science is the notion of a (deterministic) Turing machine (Turing et al., 1936), which is an abstract device that can perform any computation that can be expressed by a finite set of instructions. A Turing machine operates on an unbounded tape that contains symbols from a finite alphabet, and it can manipulate the tape according to a transition function that depends on its current state and the symbol under the tape head. Turing machines capture the essential features of real-world computation and serve as a theoretical model for modern computers. Therefore, they are a useful tool for studying the computational capabilities and limitations of real machines. The finite alphabet of Turing machines—one might imagine them as bits—is a crucial aspect of their functioning. As a consequence, the number of states that Turing machines can attain is countable. Therefore they are unable to accurately represent uncountable sets, for example, the set of real numbers.

1.3.2 Blum-Shub-Smale machines

A Blum-Shub-Smale machine (Blum et al., 2000) is an analog extension of a Turing machine (Turing et al., 1936), which allows for storage and manipulation of real numbers. The stored numbers can be updated by applying functions from a fixed set of operations, in our case addition, subtraction, multiplication, division, and relations like “$<$”, “$>$”, and “$=$”. Blum-Shub-Smale machines can also output real numbers as a result of their computations. By choosing different sets of operations, one can obtain different classes of Blum-Shub-Smale machines with different computational power and complexity. We refer the reader to Blum et al. (2000) for a comprehensive explanation of BSS machines.

1.3.3 Quantum computers

Quantum computers (Benioff, 1980) are a paradigm of computation that uses the principles of quantum mechanics to perform operations. Quantum computers use quantum bits, or qubits, as the basic unit of information, which can, unlike classical bits, exist in superpositions of two states, 0 and 1. Quantum computers manipulate qubits by applying quantum logic gates, which are operations that change the state of one or more qubits. By designing a sequence of quantum gates, one constructs a Hamiltonian, which describes the evolution of all qubits according to the Schrödinger equation. Through the clever choice of those quantum gates, one can perform calculations more efficiently than a Turing machine would be able to. Quantum computers also fall in the realm of analog hardware models, since the “storage” of qubits is performed physically and does not have to be truncated in its representation.

1.3.4 Neuromorphic computing

Inspired by the architecture and operation of biological cells, neuromorphic hardware (Mead, 1990) is a new approach to computing that uses analog components to mimic the behavior of neurons and synapses. Neuromorphic hardware differs from traditional digital hardware, which relies on binary logic and discrete states, by using chemical and electrical phenomena to process information. The potential benefit of neuromorphic hardware might be its higher energy efficiency. Neuromorphic hardware is a type of analog hardware since it encodes all its internal values through the physical quantities that control the system, such as electric current or voltage.

2 Basic notions

This chapter investigates optimization problems from the perspective of computability theory and effective analysis, which are fields of mathematics that study the boundaries and capabilities of computation and algorithmic approximation. Optimization lies at the heart of AI and data science. Ultimately, the extent to which these fields can be reliably applied hinges largely on our ability to solve optimization problems reliably and effectively. The notation and concepts required are explained in this section.

2.1 Search for the optimal value

By optimization problems, we refer to the minimization or maximization of some functional $F : X \times Y \to \mathbb{R}$ over a solution space $X \subset \mathbb{R}^n$ and a parameter space $Y \subset \mathbb{R}^m$ for $n, m \in \mathbb{N}$, i.e.,

$$\min_{x \in X(y)} F(x, y) \text{ or } \max_{x \in X(y)} F(x, y), \tag{1}$$

where $y \in Y$ is a parameter in the parameter space and $X(y) \subset X$ is a subset of the solution space, depending on y. From now on, and without loss of generality, we only consider maximization problems. This general form allows us to treat most problems from applications (Boyd and Vandenberghe, 2004).

Two main problem settings can be identified in this context: The first asks for the optimal value or an approximation of it, i.e.,

Optimal Value

Construct or approximate a function $\varphi : Y \to \mathbb{R}$ such that

$$\forall_{y \in Y} : \varphi(y) = \max_{x \in X(y)} F(x, y). \tag{2}$$

The second problem setting aims to find an optimizer, i.e.,

Optimizer

Construct or approximate a function $G : Y \to X$ such that

$$\forall_{y \in Y} : \varphi(y) = F(G(y), y) \wedge G(y) \in X(y), \tag{3}$$

where φ is the function defined by Eq. (2).

It is evident that constructing a function G yields a construction of a function φ. However, in general, the opposite direction does not hold. It is in this sense that finding φ is "easier" than finding G.

Optimization problems suffer the same curse as many other problems of wide interest, namely, there does in general not exist a closed-form solution for either φ or G. Therefore, solutions usually have to be approximated by numerical algorithms. For a wide variety of optimization problems, established algorithms that aim to approximate the optimal solution do exist. A classical class of approaches are iterative solvers, which construct a sequence of approximators. In some cases, it has been proven that this sequence does indeed converge to the optimizer (Arimoto, 1972; Cover, 1984). Depending on the application, either the function G or the function φ is of greater interest. Examples for the former are portfolio optimization or compressed sensing and, for the latter, exemplary problems are computing the capacity of a channel or solving a deep learning problem. In practice, one often approximates G by an iterative scheme to obtain a sequence $G_n : Y \to X$ and, correspondingly, $\varphi_n : Y \to \mathbb{R}$ through $\varphi_n(y) := F(G_n(y), y)$. Depending on the applied algorithm, one might obtain one or a combination of the following guarantees:

- $\forall_{y \in Y} : \varphi_n(y) \to \varphi(y)$ with or without known convergence speed,
- $\forall_{y \in Y} : G_n(y) \to G(y)$ with or without known convergence speed.

Notice that in this case "known convergence speed" of a convergent Banach space sequence $a_n \to a$ refers to having an explicit description of a function $f : \mathbb{N} \to \mathbb{R}_+$ such that $\lim_{n\to\infty} f(n) = 0$ and $\|a_n - a\| \leq f(n)$ for all $n \in \mathbb{N}$.

Since finding φ is "easier" than finding G a significant amount of research in optimization has (successfully) focused on finding φ over finding G. There are numerous examples of iterative algorithms, which construct $\varphi_n \to \varphi$ with known convergence speed, as well as $G_n \to G$ without known convergence speed. A natural question arises if there is a way to bound the convergence speed of G_n since this would give a stop criterion, which ensures a small error of the computed optimizer to the true optimizer.

Usually, the existence of an optimizer is ensured through compactness of X or $X(y)$. Note that compactness only proves abstract existence, but does not provide a description or approximation of the optimizer itself. In this chapter, we use the following definition, if the optimization problem and its parameter space are evident,

$$Opt(y) := \left\{ x \in X(y) \,\middle|\, F(x, y) = \max_{x' \in X(y)} F(x', y) \right\}.$$

Using this notation, the objective of computing an optimizer can be interpreted as the process of identifying a function $G : Y \to X$ such that

$$\forall_{y \in Y} : G(y) \in Opt(y)$$

or an approximation of G, i.e., a function $G^* : Y \to X$ such that G and G^* are close. In our case, we define closeness by the supremum norm

$$\|G - G^*\|_\infty = \sup_{y \in Y} |G(y) - G^*(y)| < \alpha$$

for some $\alpha > 0$.

2.2 Computability

Computability theory studies the limitations and possibilities of various models of computation, in this case, deterministic Turing machines. One of the earliest and most influential results in this field was Turing's introduction of the first steps in effective analysis and his proof of the undecidability of the Entscheidungsproblem. This proof showed that no algorithm can decide whether every Turing machine with a given input terminates after a finite amount of steps or not. Building on this foundation, the theory of effective analysis, also called computable analysis, emerged. This area focuses on real number functions and, in contrast to other computability areas, is not entirely discrete. We follow the definitions and notation from Pour-El and Richards (2017), which provides a comprehensive introduction to computability theory and its applications.

2.2.1 Turing machines

Deterministic Turing machines are a theoretical model of computation used to classify algorithmic tasks (Turing et al., 1936; Turing, 1938). They consist of a finite set of states, a finite alphabet of symbols, a tape divided into cells that can store one symbol each, a tape head that can read and write symbols on the tape, and a transition function that determines the next state, symbol, and head movement based on the current state and symbol. Deterministic Turing machines are digital in nature because they only allow a finite alphabet of discrete symbols, e.g., bits, and not analog because they cannot represent continuous values with infinite precision.

Definition 1. A function $f : \mathbb{N} \to \mathbb{N}$ is called **recursive** or **computable**, if there exists a Turing machine, which, given the input $x \in \mathbb{N}$, leaves $f(x)$ on its tape after termination. With slight abuse of notation, we equate a recursive function with its corresponding Turing machine.

2.2.2 Computable numbers and sequences

One of the main characteristics of Turing machines is the fact that they operate on a finite alphabet. This means that only natural numbers, and by extension, rational numbers, can be represented exactly by Turing machines. However, many real numbers, such as π or $\sqrt{2}$, are irrational and cannot be expressed as the ratio of two integers. A possible approach is to use sequences of rational numbers that converge to a real number. We say that a sequence of rational numbers is computable if there exists a Turing machine that outputs this sequence.

Definition 2. A sequence of rational numbers $(r_n)_{n\in\mathbb{N}}$ is **computable**, if there exist recursive functions $s, p, q : \mathbb{N} \to \mathbb{N}$ such that

$$\forall_{k\in\mathbb{N}} : r_k = (-1)^{s(k)} \frac{p(k)}{q(k)}.$$

Also, we adopt a notion of convergence, which is more natural for Turing machines.

Definition 3. A sequence of real numbers $(x_n)_{n\in\mathbb{N}}$ does **converge effectively** to a limit $x \in \mathbb{R}$, if

$$\forall_{n\in\mathbb{N}} : |x_n - x| \leq 2^{-n}.$$

Using these definitions, one can say that a real number is computable if it is the effective limit of a computable sequence of rational numbers. It turns out that most commonly used numbers such as algebraic numbers, π and e are computable. For instance, Chaitin's constant, which is defined as the probability that a randomly chosen Turing machine halts, is noncomputable because approximating it would violate the noncomputability of the Entscheidungsproblem.

Definition 4. A real number $x \in \mathbb{R}$ is called **computable**, if there exists a rational computable sequence $(q_n)_{n\in\mathbb{N}}$, such that

$$q_n \to x,$$

where the convergence is effective. The sequence $(q_n)_{n\in\mathbb{N}}$ is called a **representation** of x. We refer to the set of computable real numbers $\mathbb{R}_c$.

We can prove that the computable real numbers $\mathbb{R}_c$ form an algebraic field by showing that each field operation can be computed by a Turing machine, given the Turing machines that compute the operands. As we have also mentioned, most of the commonly used numbers in mathematics and science, such as rational numbers, algebraic numbers, and transcendental numbers like π, and e are computable. Therefore, the computable real numbers encompass a large and rich class of relevant numbers. This explains the practical usefulness of Turing machines for numerical analysis and optimization problems.

2.2.3 Computable continuous functions

A computable function is a function whose values can be calculated by a Turing machine. We already established discrete computable functions on natural numbers. We extend this notion to the continuous setting of functions on computable real numbers. For this, we begin with a formal definition of a computable function that is based on Turing's original paper. The idea is straightforward: Given any representation of an input, we can obtain a representation of the corresponding output.

Definition 5. Let $N, M \in \mathbb{N}$. A function $f : \mathbb{R}_c^N \to \mathbb{R}_c^M$ is **Borel-Turing computable**, if there exists a Turing machine, which transforms all representations $(r_n)_{n\in\mathbb{N}}$ of a vector $x \in \mathbb{R}_c^N$ to representations of $f(x)$.

A more general notion of computability can be introduced by defining the concept of computable real sequences. A real sequence is computable if there exists a computable double sequence of rational numbers that converges effectively to the real sequence:

Definition 6. A sequence of real numbers $(x_n)_{n\in\mathbb{N}}$ is **computable**, if there exists a rational computable double sequence $(r_{n,k})_{n,k\in\mathbb{N}}$ such that

$$\forall_{k,n\in\mathbb{N}} : |q_{n,k} - x_n| \leq 2^{-k}.$$

Based on this definition, we can generalize the concept of computable functions to a more technical level. A function is Banach-Mazur computable if it maps computable real sequences to computable real sequences.

Definition 7. Let $N, M \in \mathbb{N}$. A function $f : \mathbb{R}_c^N \to \mathbb{R}_c^M$ is **Banach-Mazur computable**, if for every computable real vector-valued sequence $(a_n)_{n\in\mathbb{N}}$, the sequence $(f(a_n))_{n\in\mathbb{N}} \subset \mathbb{R}^M$ is computable.

Mazur (1963) established that every Borel-Turing computable function is also Banach-Mazur computable, but the converse is not true in general.

Another important concept is the effective uniform continuity of a function. This definition is a computable version of the widely used notion of uniform continuity in calculus.

Definition 8. A function $f : [a, b] \to \mathbb{R}_c$ with $a, b \in \mathbb{R}_c$ is **effectively uniformly continuous** if there exists a recursive function $d : \mathbb{N} \to \mathbb{N}$ s.t.

$$\forall_{x,y\in[a,b]} : \forall_{N\in\mathbb{N}} : |x - y| \leq d(N)^{-1} \Rightarrow |f(x) - f(y)| \leq 2^{-N}.$$

The all-quantor ranges over every element in the interval, which includes noncomputable numbers as well. Therefore, any function that is effectively uniformly continuous on this interval must also satisfy the standard definition of uniform continuity.

2.2.4 Decidable sets

We next present the concept of decidable sets. A set is said to be decidable if, for any given element, there exists a Turing machine that can determine whether it belongs to the set or not. This notion is most widely used for subsets of natural numbers, which we will start with.

Definition 9. A set $A \subset \mathbb{N}$ is called **decidable**, if the function $\mathbb{1}_A : \mathbb{N} \to \mathbb{N}$, defined by

$$\mathbb{1}_A(n) := \begin{cases} 1, & n \in A, \\ 0, & n \in A^c, \end{cases}$$

is recursive.

A relaxation of computable sets are semidecidable sets equipped with a partial decision procedure. This means that there exists a Turing machine, that can always confirm if a given element belongs to the set, but does not reject it if it is not an element. In other words, the algorithm will halt and output yes if the element is in the set, but does not terminate if the element is not in the set. Therefore, one can never be certain whether an element is outside the set or the algorithm is still computing.

Definition 10. A set $A \subset \mathbb{N}$ is called **semidecidable**, if there exists a Turing machine $\mathcal{TM}_A$ such that $\mathcal{TM}_A(n)$ outputs 1, if $n \in A$, and $\mathcal{TM}_A(n)$ does not terminate, if $n \in A^c$.

The concept of (semi-)decidability can be readily applied to subsets of real numbers, with a minor adjustment in the definition. Rather than testing if a given element is part of a set, we take for granted that the element is already contained in a bigger set, and we only have to determine if it is also included in a narrower subset of that set.

Definition 11. Given $B \subset \mathbb{R}_c^n$, a set $A \subset B$ is called **(semi-)decidable**, if there exists a Turing machine $\mathcal{TM}_A$ such that $\mathcal{TM}_A(x)$ outputs 1, if $x \in A$ and $\mathcal{TM}_A(x)$ outputs 0 (resp. does not terminate), if $x \in B \backslash A$. $\mathcal{TM}_A(x)$ can either output an arbitrary symbol or not terminate for $x \in B^c$.

One of the most famous noncomputable problems is the Entscheidungsproblem, which involves deciding whether a given Turing machine will ever stop on a given input or not. No algorithm can solve this problem for all Turing machines and inputs (Turing, 1938). A direct implication of this result is the existence of sets that are semidecidable but not decidable. This can be shown by identifying the set of Turing machines $\mathcal{T}$ and the set of possible inputs I with the set of natural numbers, which is feasible since both $\mathcal{T}$ and I are countably infinite sets. For this define

$$F := \{(t, i) \in \mathcal{T} \times I | t(i) \text{ terminates}\} \subset \mathcal{T} \times I \cong \mathbb{N},$$

the set of Turing machines and inputs, which do terminate. Then F is semidecidable using the Turing machine, which applies i on t for any $(t, i) \in \mathcal{T} \times I$ and outputs 1 if it terminates. By definition, this will output 1 if $(t, i) \in F$ and not terminate otherwise. But the noncomputability of the Entscheidungsproblem implies there is no Turing machine, which can decide if (i, t) is in F or not.

3 Deep learning as a key technique of artificial intelligence

Artificial intelligence is a tremendously successful field that aims to create programs which perform tasks that normally require human intelligence. A key component of AI is deep learning, which has achieved remarkable results in many challenging domains, by now already surpassing human performance in some cases. This section introduces the basic ideas of deep learning and discusses some of its drawbacks.

3.1 Essence of deep learning

Deep neural networks are composed of multiple layers of artificial neurons, which are mathematical functions consisting of adjustable linear-affine functions followed by a nonlinear activation function. By adjusting the parameters of each neuron's linear-affine function, the neural network is able to "learn" a function. This learning process is facilitated by employing a variant of gradient descent on a large data set.

Definition 12. A **(feed-forward) neural network** are functions $\Phi : \mathbb{R}_c^n \to \mathbb{R}_c^m$ of the form

$$\Phi(x) := (A_L \circ \rho \circ A_{L-1} \circ \ldots \circ \rho \circ A_1)(x)$$

where $L \in \mathbb{N}$ and

$$\forall_{l=1,\ldots,L} : A_l x = W_l x + b_l, \quad W_l \in \mathbb{R}_c^{n_l \times n_{l-1}}, b_l \in \mathbb{R}_c^{n_l}$$

with $n_0 = n$, $n_L = m$ and $\rho : \mathbb{R} \to \mathbb{R}$ is a (nonlinear) activation function, applied component-wise. The coefficients of W_l are called **weights** and b_l are called **biases**.

For a more general theory on neural networks, we refer to Berner et al. (2021).

The choice of activation functions plays an important role in the performance of deep learning models. Among the commonly used activation functions are ReLU, tanh, and sigmoid with the most popular being ReLU defined as $\rho(x) = \max(0, x)$. Therefore, in this chapter, we will focus on the analysis of neural networks using the ReLU activation functions. We define $\mathcal{NN}$ as the set of all neural networks. Neural networks are optimized using a large dataset $(x_i, y_i)_{i=1,\ldots,n} \subset \mathbb{R}_c^n \times \mathbb{R}_c$ by minimizing a loss function $\mathcal{L} : \mathcal{P}(\mathbb{R}_c^n \times \mathbb{R}_c) \times \mathcal{NN} \to \mathbb{R}_c$, which measures the "fit" of the neural network to the given data. For a simple regression task, a possible choice of loss function is the L^2-distance

$$\mathcal{L}((x_i, y_i)_{i=1,\ldots,n}, \Phi) = \frac{1}{n} \sum_{i=1}^{n} (\Phi(x_i) - y_i)^2.$$

Depending on the specific objective of the problem, such as classification or more complex forms of regression, other loss functions are commonly employed. Given a dataset and a loss function, the goal then is to find the optimal values of the weights and biases that minimize the loss. This is achieved by applying a variant of gradient descent, which is an iterative optimization algorithm. The basic idea of gradient descent is to update each parameter λ by subtracting a small fraction $\mu > 0$ of the gradient of the loss function with respect to that parameter

$$\lambda \leftarrow \lambda - \mu \frac{\partial \mathcal{L}}{\partial \lambda}((x_i, y_i)_{i=1,\ldots,n}, \Phi).$$

The efficient calculation of the gradient is done by backpropagation.

3.2 Drawbacks of deep learning

Despite its impressive achievements in various fields, such as computer vision, natural language processing, and speech recognition, deep learning also has serious drawbacks, especially when it comes to safety-critical applications like medical imaging and autonomous driving. Some of the current main challenges of deep learning are:

1. *Intransparency:* Neural networks are often viewed as black boxes, meaning that their internal mechanisms and reasoning processes are not clear or understandable. This poses a problem for explaining and validating the results of deep learning models, as well as for detecting and fixing potential mistakes or biases.
2. *Theory-to-Practice gap:* Deep learning lacks rigorous theoretical foundations that can guarantee its performance and robustness. It is not well understood why deep learning works so well for some problems but not for others, or under what conditions it will fail or degrade.
3. *Instability:* Neural networks are susceptible to showing inconsistent behavior or poor generalization in some situations, even when they perform well on average. For instance, they are vulnerable to adversarial attacks, which are malicious inputs that are designed to fool or mislead the models by introducing subtle perturbations that are imperceptible to humans.

In addition to the well-known challenges of deep learning, mentioned above, there is another aspect that is sometimes overlooked: The computational limits of running neural networks on digital hardware. As we explained in Section 2, real numbers cannot be represented precisely by digital devices, which introduces errors and uncertainties in the computations. This issue is not inherent to neural networks themselves, but rather a consequence of the interplay of neural networks with digital hardware. Thus, these issues could be reduced or avoided on analog hardware.

4 Computability of optimal values and existence of computable optimizers

Depending on the dimension and regularity of the functional we can make different statements about the computability of the maximum, the maximizer, and if you can calculate the maximum and maximizer by computable means. We present a collection of results, encompassing both computable and noncomputable optimal values and optimizers. Our survey begins with the simplest one-dimensional scenario, before progressing to the more complex multidimensional cases.

4.1 One-dimensional optimization

For the one-dimensional case, consider a computable function $f : [a, b] \to \mathbb{R}_c$ that is effectively uniformly continuous on a computable interval. Then by Pour-El and Richards (2017, Theorem 7) $\max_{x \in [a,b]} f(x)$ is computable. Furthermore, the proof of the theorem provides an extra insight. It is not only the case that the maximum of every computable effectively uniformly continuous function is computable, but also that there exists an algorithm for finding this maximum. We present the proof here for completeness:

Theorem 1. *Let $a, b \in \mathbb{R}_c$ with $a < b$ and define $\mathcal{F} := \{f : [a,b] \to \mathbb{R}_c | \ f$ Borel-Turing computable and effectively uniformly continuous$\}$. Then define $M : \mathcal{F} \to \mathbb{R}_c$ by*

$$M(f) = \max_{x\in[a,b]} f(x).$$

For a function $f \in \mathcal{F}$ define T_f as the Turing machine, which given a representation of $x \in \mathbb{R}_c$, outputs a representation of $f(x)$. Then there exists a Turing machine $T_{\max}$, which given a Turing machine T_f with $f \in \mathcal{F}$, outputs a representation of $M(f)$.

Remark. We write with slight abuse of notation,

$$T_{\max}(T_f) \text{ outputs a representation of } M(f).$$

Proof. Define the computable sequence

$$s_n = \max\left\{ f\left(a + \frac{j}{n}(b-a)\right) \middle| 1 \le j \le n \right\}.$$

By effective uniform continuity of f, there exists a recursive function $d : \mathbb{N} \to \mathbb{N}$, such that,

$$\forall_{x,y\in[a,b]} : \forall_{n\in N} : |x-y| \le d(n)^{-1} \Rightarrow |f(x) - f(y)| \le 2^{-n}. \tag{4}$$

Let $M \in \mathbb{N}$ with $M > b - a$ and $e(n) := M \cdot d(n)$. Next, we will show that

$$n \ge e(N) \Rightarrow |s_n - M(f)| \le 2^{-n}.$$

Given $n \ge e(N)$, this implies

$$\frac{b-a}{n} < \frac{M}{n} \le \frac{M}{e(N)} = \frac{1}{d(N)}.$$

Since each interval

$$I_j := \left[a + \frac{j}{n}(b-a), a + \frac{j+1}{n}(b-a)\right], \ j = 0, \ldots, n-1$$

has length $\frac{b-a}{n}$, we can deduce that

$$\forall_{j=1\ldots n-1} : \left| \max_{x\in I_j} f(x) - f\left(a + \frac{j}{n}(b-a)\right)\right| \le 2^{-n}.$$

This immediately implies

$$\forall_{n\ge e(N)} : |s_n - M(f)| \le 2^{-n},$$

hence (4) is proven. Since $(s_n)_{n\in\mathbb{N}}$ is a computable sequence, it is a representation of $M(f)$. Thus, one can use the Turing machine, which maps

$$f \mapsto (s_n)_{n\in\mathbb{N}}$$

to construct $M(f)$ for any $f \in \mathcal{F}$. □

This theorem is remarkable, as it guarantees the existence of a Turing machine that can calculate the maximal value for any computable and effectively uniform continuous function on a compact interval. However, when we focus on the maximizer instead, we encounter a surprising phenomenon. Specker showed in Specker (1959) that there does exist a function whose maximal value is computable, but whose maximizers are not. This circles back to the fact mentioned in Subsection 2.2.1 of finding a maximal value being "easier" to solve (see Problem (2)) than constructing a maximizer as in Problem (3). As shown, this is already observable in the one-dimensional case.

If a maximizer is at an isolated point though, the maximizer is always computable.

Theorem 2. *Let $a, b \in \mathbb{R}_c$ computable numbers with $a < b$ and $f : [a, b] \to \mathbb{R}_c$ be a computable and effectively uniformly continuous function with an isolated maximum,*

$$\exists_{\hat{x}\in[a,b]} : f(\hat{x}) = \max_{x\in[a,b]} f(x) \wedge \forall_{x\in[a,b]\setminus\{\hat{x}\}} : f(x) < f(\hat{x}).$$

Then $\hat{x} \in \mathbb{R}_c$ is computable.

Proof. Define

$$F_1(x) = \max_{\tau\in[a,x]} f(\tau).$$

Then F_1 is monotonously nondecreasing with $\forall_{x<\hat{x}} : F(x) < F(\hat{x})$ and $\forall_{x\geq\hat{x}} : F(x) = F(\hat{x})$. Also by Theorem 1, $F_1(x)$ is a computable and continuous function. Now for fixed $n \in \mathbb{N}$ and $k = 0, \ldots, 2^n$ consider the computable numbers

$$l_n^k := F_1\left(a + \frac{(b-a)k}{2^n}\right).$$

By comparing the $(n+2)$th element of the representations of l_n^k and $f(\hat{x}) - 2^{-n+1}$ one can compare l_n^k and $f(\hat{x}) - 2^{-n+1}$ up to an error of 2^{-n+1}, i.e., there exists a Turing machine which can check if

$$l_n^k > f(\hat{x}) - 2^{-n}$$

holds or not. Next, define the computable sequence

$$x_n := \min\left\{a + \frac{(b-a)k}{2^n} \,\middle|\, k \in \{0, \ldots, 2^n\} \wedge l_n^k > f(\hat{x}) - 2^{-n}\right\}.$$

Then x_n is a nondecreasing sequence with $x_n \to \hat{x}$.

Now by choosing a subsequence (n_k) one can construct a computable sequence $(x_{n_k})_{k\in\mathbb{N}}$ that converges to $\hat{x}$ effectively

$$\forall_{k\in\mathbb{N}} : |x_{n_k} - \hat{x}| < 2^{-k}.$$

So $\hat{x}$ is computable. □

For certain functions Theorem 2 implies the existence of a computable optimizer $\hat{x}$, which can be computed by a Turing machine $\mathcal{T}_{\hat{x}}$. For every $n \in \mathbb{N}$, this Turing machine outputs a rational number r_n with the property that

$$|\hat{x} - r_n| \leq 2^{-n}.$$

But this does not imply that there is a Turing machine that can generate $\mathcal{T}_{\hat{x}}$ given a Turing machine that defines the matching function. On the contrary, we will demonstrate later that this is not feasible in general. More examples of computable and noncomputable functions can be found in Pour-El and Richards (2017).

4.2 Computability of convex optimizers in higher dimensions

The previous sections focused on one-dimensional functions. However, many real-world applications require functions of more than one variable. Therefore, in this section, we extend the concepts and methods of optimization to multivariate functions and present the similarities and differences to the univariate case. Using the same proof as in 1 one can establish the multidimensional analogue of Theorem 1.

Theorem 3. *Let $I \subset \mathbb{R}_c^n$ be a computable rectangle and define $\mathcal{F} := \{f : I \to \mathbb{R}_c | \, f$ Borel-Turing computable and effectively uniformly continuous$\}$. Then define $M : \mathcal{F} \to \mathbb{R}_c$ by*

$$M(f) = \max_{x\in I} f(x).$$

For a function $f \in \mathcal{F}$ define T_f as the Turing machine, which given a representation of $x \in R_c^n$, outputs a representation of $f(x)$. Then there exists a Turing machine $T_{\max}$, which given a Turing machine T_f with $f \in \mathcal{F}$, outputs a representation of $M(f)$.

The following theorem can be considered a generalization of Theorem 2. It establishes the existence of a computable maximizer for convex functionals in the multidimensional case.

Theorem 4 (Wong, 1996). *Let $I \subset \mathbb{R}^n$ be a computable rectangle, and let $f : I \to \mathbb{R}$ be a convex, continuous, and Banach-Mazur computable. Then there exists a computable $x^* \in I \cap \mathbb{R}_c^n$, which minimizes f, i.e.,*

$$f(x^*) = \min_{x\in I} f(x).$$

We, again, remark that this theorem does not imply that the maximizer x^* can be calculated by computable means given f and I.

4.3 Example of multidimensional optimization in information theory

One way to generalize Theorem 4 is by relaxing the domain I to other geometries than computable rectangles or parameterizing I. For illustration, we present the well-known convex optimization problem of channel capacity from information theory.

Information theory plays a crucial role in AI. The sparse and compressed representation of data is not just useful in storing and transmitting data but also in designing efficient AI algorithms and neural network architectures. In this subsection, we consider the case of optimal transmission capacity of data through a discrete channel. The channel capacity is again described through a maximization problem. The goal is to maximize the mutual information between the input and output of a discrete memoryless channel, which measures the amount of information transmitted through the channel.

4.3.1 Communication model

In information theory, a point-to-point channel with one receiver and one transmitter is modeled by two discrete random variables X and Y over the probability spaces $\mathcal{X}$ and $\mathcal{Y}$. If we choose $\mathcal{X}$ and $\mathcal{Y}$ to be finite, we are describing a discrete memoryless channel (DMC). The channel itself is then given by a stochastic matrix $W \in \mathbb{R}^{m\times n}$, where $|\mathcal{X}| = n$ and $|\mathcal{Y}| = m$. X, Y and W are related by

$$W(x) = P(Y|X = x).$$

We define the mutual information of two discrete random variables X, Y over $\mathcal{X}$, $\mathcal{Y}$ as

$$I(X,Y) = \sum_{x\in\mathcal{X}}\sum_{y\in\mathcal{Y}} P_{(X,Y)}(x,y)\log\left(\frac{P_{(X,Y)}(x,y)}{P_X(x)P_Y(y)}\right),$$

where $P_{(X,Y)}$ is the probability mass function of (X,Y), and P_X and P_Y are the probability mass functions of X and Y. Now the capacity $C(W)$ of a DMC W is the maximal mutual information over all possible distributions over $\mathcal{X}$

$$C(W) := \max_{X\in\mathcal{P}(\mathcal{X})} I(X,Y).$$

In this case $\mathcal{P}(\mathcal{X})$ can be modeled as a convex set

$$\mathcal{P}(\mathcal{X}) = \left\{(P_1,\ldots P_n) \in \mathbb{R}^n_+ \,\middle|\, \sum_i P_i = 1\right\}.$$

The capacity of a DMC is well established and goes back to Shannon (1948).

4.3.2 Blahut-Arimoto

The Blahut-Arimoto algorithm (Blahut, 1972; Arimoto, 1972) establishes a computable way to compute the capacity of any DMC. It works by iteratively approximating an estimate of the input distribution. So given a channel W, Blahut-Arimoto outputs a sequence of probability distributions $X_n \in \mathcal{P}(\mathcal{X})$. It is proven that the corresponding mutual information converges to the capacity

$$\lim_{n\to\infty} I(X_n, Y) = C(W).$$

Also, X_n do converge to a probability distribution X^*, which maximizes the mutual information

$$\lim_{n\to\infty} X_n = X^* \;\wedge\; I(X^*, Y) = C(W).$$

However, there is a significant difference in the effectiveness of the convergence of both quantities. On one hand, there is a stopping criterion for the capacity, which guarantees arbitrarily small error, i.e., for any $N \in N$ one can computably choose $n \in \mathbb{N}$, such that,

$$\|I(X_n, Y) - C(W)\| \leq 2^{-N}.$$

On the other hand, there is no computable stopping criterion known, that guarantees arbitrarily small approximation error for X^*. The question if such a stopping criterion exists has been answered lately in Boche et al. (2022) with a clear no. More precisely, the authors proved that no Banach-Mazur computable function exists at all, which maps all DMCs to an approximation of the optimal input distribution X^* with an arbitrarily small error. We refer to Theorem 9 for more details.

This result is unexpected, as both the set we optimize over and the objective function of the optimization problem are convex. Convexity is usually considered to be a desirable property that simplifies the analysis and solution of optimization problems. However, we show that there exist convex optimization problems that are not computable. On the contrary, we can use convexity to prove the existence of computable optimizers, even though finding those might not be computable. Define for the channel W the convex set of optimizers $\mathcal{P}_{opt}(W) := \{X \in \mathcal{P}(\mathcal{X}) | I(X, Y) = C(W)\}$ where Y is determined through W and X.

Theorem 5 (Hinted at in Boche et al. (2022)). *Given a computable stochastic matrix W, there exists a $X^* \in \mathcal{P}_{opt}(W)$, s.t. X^* is computable, i.e., $X_i^* \in \mathbb{R}_c$ for all $i = 1, \ldots, n$.*

Proof. If $|\mathcal{P}_{opt}(W)| = 1$, we can apply Theorem 3 on every dimension separately, which finishes the proof.

If $|\mathcal{P}_{opt}(W)| > 1$, let $X^0, X^1 \in \mathcal{P}_{opt}(W)$ with $X^0 \neq X^1$. Since the mutual information is convex, $\mathcal{P}_{opt}(W)$ is convex, too. So for any $\lambda \in [0, 1]$ the following distribution is also an optimizer $X^\lambda := (1-\lambda)X^0 + \lambda X^1 \in \mathcal{P}_{opt}(W)$. Now let $i_1 \in \{1, \ldots, n\}$ be an index s.t. $X^0_{i_1} \neq X^1_{i_1}$. So $X^\lambda_{i_1} = (1-\lambda)X^0_{i_1} + \lambda X^1_{i_1} \in [\min(X^0_{i_1}, X^1_{i_1}), \max(X^0_{i_1}, X^1_{i_1})]$. Then there exists a $\lambda^* \in [0, 1]$ s.t. $X^{\lambda^*}_{i_1} \in \mathbb{R}_c$ since $\mathbb{R}_c \subset \mathbb{R}$ is dense.

Now consider the set $\mathcal{P}^1_{opt}(W) := \{X \in \mathcal{P}_{opt}(W) | X_{i_1} = X^{\lambda^*}_{i_1}\}$ and the optimizer $X^*_1 := (X^0_1, \ldots X^0_{i_1-1}, X^{\lambda^*}_{i_1}, X^0_{i_1+1}, \ldots X^0_n) \in \mathcal{P}^1_{opt}(W)$. Now we can repeat the argument as before. If $|\mathcal{P}^1_{opt}(W)| = 1$, one can conclude that the only optimizer is computable by applying Theorem 3 on every dimension separately.

If $|\mathcal{P}^1_{opt}(W)| > 1$, we can repeat the procedure by fixing $X^1, X^2 \in \mathcal{P}^1_{opt}(W)$ and choosing an index $i_2 \in \{1, \ldots, n\}$ with $i_1 \neq i_2$, s.t. $X^1_{i_2} \neq X^1_{i_2}$. Then we can construct an optimizer X^{λ^*}, s.t., $X^{\lambda^*}_{i_2} \in \mathbb{R}_c$. After repeating this process $k < n$ times, we end up with a set of optimizer $\mathcal{P}^k_{opt}(W)$ with just one computable element. □

We note that this proof is an adaptation of the proof of Theorem 4 and again does not provide any computable procedure to identify a computable optimizer, just its existence. As can be seen from our adaptation, the assumption in Theorem 4 of I of being a rectangle can be relaxed.

4.4 Other computable and noncomputable functions

Most commonly used functions are computable over any computable rectangle as domain without a singularity. This includes e^x, $\sin x$, $\cos x$, $\sqrt{x}$, $\log(x)$, and $\Gamma(x)$ as well as its compositions. To prove this it suffices to consider the approximation formulas of these functions commonly used in numerical mathematics and confirming they have a computable stopping criterion, which ensures arbitrarily small approximation error. As an example we can prove computability of $\sqrt{x}$ using Heron's formula (Schwarz and Köckler, 2013) for $x^* \in \mathbb{R}_c$

$$x_{n+1} := \frac{1}{2}\left(x_n + \frac{x_0}{x_n}\right)$$

with any suitable initial guess $x_0 \in \mathbb{R}_c$. It is well-known that the relative error

$$r_n := \frac{x_n}{x^*} - 1,$$

is always positive (assuming an appropriate initial guess) and decreases at least exponentially, i.e., $r_n \leq \frac{r_{n-1}}{2}$. Using this convergence result one can calculate for any $N \in \mathbb{N}$ an index $N_0 \in \mathbb{N}$ s.t.

$$\forall_{m > N_0} : |\sqrt{x^*} - x_m| \leq 2^{-N}.$$

Also, definite integrals of computable, uniformly effective continuous functions are computable. The derivative of a computable function is not computable in general, even if the derivative exists. However computable functions in C^2 do have a computable derivative. For a comprehensive theory in computable analysis, we refer the reader to Pour-El and Richards (2017) and Weihrauch (2000).

5 Finding the optimizer is not effectively solvable

This section begins by addressing a key question: Does there exist a Turing machine capable of outputting the optimizer when provided with a description of the functional and the domain? Previously, we focused on determining which optimal values and optimizers are computable. However, a subtly different yet arguably more important question is whether a computable optimizer can be discovered through computable methods. This section concentrates on results addressing this question.

5.1 General noncomputability theorem

In order to determine whether optimizers can be computed, we examine general optimization problems in multiple dimensions, as referenced in Section 2. If the solving function, denoted as G, which is defined in Problem (3), exhibits a certain type of computable discontinuity along with other technical properties, it becomes impossible to efficiently calculate the optimizer. Interestingly, this can occur even if the function F is both continuous and convex. More importantly, it has been established that not only is the calculation of the optimizer noncomputable, but also its approximation. This implies that there will always be at least one instance where any computable machine will fail to approximate the optimizer by a fixed positive constant.

Theorem 6 (Main Theorem (Lee et al., 2023)). *Let X, Y, $X(y)$ and F be as described in Section 2. Let $G : Y \to X$ such that, for all $y \in Y$, we have $G(y) \in Opt(y)$. Now let $Y_1, Y_2 \subset Y$, $y_1^* \in Y_1$, $y_2^* \in Y_2$ and $y_* \in Y$, and $\gamma : [-1, 1] \to Y$, a Turing computable, continuous path such that:*

- **(i)** $Y_1 \cap Y_2 = \varnothing$ *and* $G(Y_1) \cap G(Y_2) = \varnothing$,
- **(ii)** $\inf_{y_1 \in Y_1 y_2 \in Y_2} \|y_1 - y_2\| = 0$,
- **(iii)** $\inf_{y_1 \in Y_1 y_2 \in Y_2} \|G(y_1) - G(y_2)\| = \kappa > 0$,
- **(iv)** $\gamma(-1) = y_1^*$, $\gamma(1) = y_2^*$, $\gamma(t_0) = y_*$, *for some* $t_0 \in (-1, 1)$,
- **(v)** $\gamma([-1, t_0)) \subset Y_1$ *and* $\gamma((t_0, 1]) \subset Y_2$,
- **(vi)** $G(Y_1) \subset G(Y_1) \cup G(Y_2)$ *is decidable.*

Then G cannot be Borel-Turing computable. In fact, there does not even exist a Borel-Turing computable function, which can approximate G by up to an absolute error of $\alpha < \frac{\kappa}{2}$, i.e. there does not exist a Borel-Turing computable function $G^* : Y \to X$ such that $\|G - G^*\|_\infty \le \alpha$. If we replace condition (vi) by

(vii) *$Y_1 \subset Y_1 \cup Y_2$ is decidable,*

then G can even not be Banach-Mazur computable. In fact, there does not even exist a Banach-Mazur computable function, which can approximate G by up to an absolute error of $\alpha < \frac{\kappa}{2}$.

This theorem is noteworthy as it suggests the impossibility of computing an optimizer, even when the optimizer is computable. This holds true even under favorable conditions, such as the convexity of the domain and the function. As we will discuss, this scenario occurs frequently in many practically relevant cases.

5.2 Noncomputability of neural networks and other optimizers

The previous theorem is a powerful tool that can be applied to various optimization problems. In this section, we demonstrate its effectiveness and generality by presenting some selected examples from different fields of study.

5.2.1 Neural networks

Smale's 1998 discussion on the limitations of AI (Smale et al., 1998) can be seen as a precursor to the specific case of neural networks, which can be viewed as an extension of these initial concepts (Colbrook et al., 2022). A remarkable example of noncomputability in neural networks is the impossibility of finding or approximating the optimal weights for a given loss function. Although other noncomputability results exist for neural networks, these typically focus on a particular application (Colbrook et al., 2022; Boche et al., 2023), or they diverge from the domain of effective analysis by examining general functions that are not tied to Turing machines (Grohs and Voigtlaender, 2023). For more on the works by Colbrook et al. (2022); Boche et al. (2023) we refer to Subsection 2.6.5. The noncomputability of neural networks were demonstrated in Lee et al. (2023) for a simple shallow neural network model with ReLU activation function without biases. Of course more sophisticated and deeper architectures exist, like convolutional neural networks, transformers, recursive neural networks, U-nets, and a whole zoo of combinations of these. Still, the most simple case of a simple shallow neural network is "complicated enough" to lead to noncomputability.

Theorem 7 (Neural Network (Lee et al., 2023)). *Given $d \geq 14$ and a data set $\mathcal{D} = \prod_{i=1}^{d}(x_i, y_i) \in (\mathbb{R}_c^3 \times \mathbb{R}_c)^d$, consider the minimization problem*

$$\min_{A \in \mathbb{R}_c^{3\times 3}} \sum_{i=1}^{d} (\|\rho(Ax_i)\|_1 - y_i)^2.$$

Let $G : (\mathbb{R}_c^3 \times \mathbb{R}_c)^d \to \mathbb{R}_c^{3\times 3}$ such that, for all $\mathcal{D} \in (\mathbb{R}_c^3 \times \mathbb{R}_c)^d$, we have: $G(\mathcal{D}) \in Opt(\mathcal{D})$. Then G is not Banach-Mazur computable.

All functions $G^* : (\mathbb{R}_c^3 \times \mathbb{R}_c)^d \to \mathbb{R}_c^{3\times 3}$ *satisfying*

$$\|G - G^*\|_\infty \le \alpha < 4$$

are also not Banach-Mazur computable.

This theorem establishes that perfect loss-minimizing algorithms for neural networks cannot exist in the special case of a shallow neural network with the architecture described above. While this conclusion does not necessarily extend to wider and deeper neural networks, it is reasonable to expect similar results in more complex scenarios. Intuitively, as the number of parameters increases, finding loss-minimizing neural networks becomes more challenging. Consequently, we anticipate that analogous results hold true for general neural networks. It is essential to note that this theorem does not imply nonapproximability of neural networks when interpreted as functions; rather, it pertains specifically to the nonapproximability of their weights. Although some might view this limitation as a drawback of the theorem—since researchers are often more interested in the neural network's function itself rather than its precise weights—it serves as a caution against methods that directly manipulate or utilize trained neural network weights. Examples include techniques like Dropout (Hinton et al., 2012) and Layer-Wise Relevance Propagation (Montavon et al., 2019).

Furthermore, the nonapproximability of weights highlights the delicate nature of neural networks. It underscores the importance of ensuring that the corresponding neural network function indeed approximates the desired function accurately. Interestingly, if there were a computable method to recover all possible weight configurations from a given neural network function consistently, it would imply nonapproximability of the neural network function itself. We believe that additional noncomputability results for deep learning exist, and inherent issues such as instability (Gottschling et al., 2020) may represent fundamental challenges for neural networks on digital hardware—one that cannot be entirely overcome.

5.2.2 Financial mathematics - information theory

In portfolio optimization, the stock market can be modeled by a random vector $X \in \mathbb{R}_+^m$, where each component describes a separate stock. A portfolio $b \in \mathbb{R}_+^m$ is a vector such that $\sum_i b_i = 1$, which describes the allocation of the available funds. The vector $b^t X$ describes the evolution of the portfolio after one time step. Due to the multiplicative nature of investments, it is natural to maximize the convex functional (Cover, 1984; Latane, 1959)

$$\max_b \mathbb{E}[\log(b^t X)],$$

which is the expected return after one time step. A well-known approach to this optimization problem is an iterative algorithm found by (Cover, 1984). Cover's algorithm uses similar ideas as the Blahut-Arimoto algorithm (Arimoto, 1972; Blahut, 1972) from information theory.

We present the result of Lee et al. (2023) that even though such an effective algorithm exists, finding a maximizing portfolio is noncomputable in general. This implies no approximation guarantee for optimal portfolios in Cover's algorithm—or any other algorithm—can be made.

We consider the case of a discrete random vector $X(\cdot) = \sum_{i=1}^{n}\sum_{j=1}^{m} p_{i,j}\delta_{x_{i,j}}(\cdot)e_j$, where $p_{i,j} > 0$ are probabilities, i.e., $\sum_{i,j} p_{i,j} = 1$, and $x_{i,j} \in \mathbb{R}_+$ are the possible outcomes. Also $e_j \in \mathbb{R}^m$ are the standard basis vectors and we assume $\forall_{j,i_1 \neq i_2} : x_{i_1,j} \neq x_{i_2,j}$. We define this set of discrete random vectors as $\mathcal{D}_{n,m}$, which is identified with the Euclidean space $\mathbb{R}_c^{2nm}$ here by equating $p_{i,j}$ and $x_{i,j}$ to vector entries.

Theorem 8 (Log-Optimal Portfolio). *Let $X \in \mathcal{D}_{n,m}$. Define $W : \{b \in \mathbb{R}_+^m | \sum_i b_i = 1\} \to \mathbb{R}$ by*

$$W(b) := \mathbb{E}[\log(b^t X)],$$

and consider the corresponding maximization problem. For all $n, m \in \mathbb{N}_+$ with $m > 1$ define a function $G : \mathcal{D}_{n,m} \to \{b \in \mathbb{R}_+^m | \sum_i b_i = 1\}$ such that for all $y \in \mathcal{D}_{n,m}$, it holds $G(y) \in Opt(y)$. Then G is not Banach-Mazur computable.

All functions $G^* : \mathcal{D}_{n,m} \to \{b \in \mathbb{R}_+^m | \sum_i b_i = 1\}$ satisfying

$$\|G - G^*\|_\infty \leq \alpha < 1$$

are also not Banach-Mazur computable.

5.2.3 Optimal input distribution - information theory

In information theory, a point-to-point channel with one receiver and one transmitter is modeled by two discrete random variables X and Y over the probability spaces $\mathcal{X}$ and $\mathcal{Y}$. If we choose $\mathcal{X}$ and $\mathcal{Y}$ to be finite, we are describing a discrete memoryless channel (DMC). The channel itself is then given by a stochastic matrix $W \in \mathbb{R}^{m \times n}$, where $|\mathcal{X}| = n$ and $|\mathcal{Y}| = m$. X, Y and W are related by

$$W(x) = P(Y|X = x).$$

We define the mutual information of two discrete random variables X, Y over $\mathcal{X}$, $\mathcal{Y}$ as

$$I(X, Y) = \sum_{x \in \mathcal{X}} \sum_{y \in \mathcal{Y}} P_{(X,Y)}(x, y) \log\left(\frac{P_{(X,Y)}(x, y)}{P_X(x)P_Y(y)}\right),$$

where $P_{(X,Y)}$ is the probability mass function of (X, Y), and P_X and P_Y are the probability mass functions of X and Y. Now the capacity $C(W)$ of a DMC W is the maximal mutual information over all possible distributions over $\mathcal{X}$

$$C(W) := \max_{X \in \mathcal{P}(\mathcal{X})} I(X, Y).$$

The capacity of a DMC is well established and goes back to Shannon (1948). Also, more recently, the capacity has been considered in more complicated settings (Boche et al., 2019b,a). Trying to find the capacity of a DMC is a classical optimization problem for which a well-known approach using an iterative algorithm with convergence guarantee exists (Arimoto, 1972; Blahut, 1972).

We will show that even though such an effective algorithm exists, it is still impossible to compute a maximizing distribution in general or give an approximation guarantee. This was first proven in Boche et al. (2022) and again proven as a special case of Theorem 6 in Boche et al. (2022).

Theorem 9 (Channel Capacity). *Let $\mathcal{X}$ and $\mathcal{Y}$ be finite sets, such that $|\mathcal{X}| = n \geq 3$ and $|\mathcal{Y}| = m \geq 2$ and W is a stochastic matrix. We define $\mathcal{P}(\mathcal{X}) := \{\text{random variables in } \mathcal{X}\}$ and $\mathcal{P}(\mathcal{Y}) := \{\text{random variables in } \mathcal{Y}\}$. Since discrete random vectors can be identified by their probabilities for each event, i.e., we uniquely describe $Y \in \mathcal{P}(\mathcal{Y})$ by the vector $(P(Y = y_1), \ldots, P(Y = y_m)) \in \mathbb{R}_c^m$, with slight abuse of notation we equate those two objects by*

$$\mathcal{P}(\mathcal{Y}) \hat{=} \{(v_1, \ldots, v_m) \in \mathbb{R}_c^m \mid \sum_{i=1}^{m} v_i = 1,\ \forall_i : v_i \geq 0\}$$

and analogously for $\mathcal{P}(\mathcal{X})$. Also, define $\mathcal{W}$ as the set of all stochastic matrices in $\mathbb{R}_c^{m \times n}$. Let $G : \mathcal{W} \to \mathcal{P}(\mathcal{X})$ be a function, such that, regarding the maximization problem

$$\max_{X \in \mathcal{P}(\mathcal{X})} I(X, Y),$$

for all $W \in \mathcal{W}$ we have $G(W) \in Opt(W)$. Then G is not Banach-Mazur computable.

All functions $G^* : \mathcal{W} \to \mathcal{P}(\mathcal{X})$ satisfying

$$\|G - G^*\|_\infty \leq \alpha < 1$$

are also not Banach-Mazur computable.

5.3 Wasserstein distance

The Wasserstein-1 distance, originally formulated by Kantorovich (1960) and Vaserstein (1969) to tackle optimal transport problems, is a metric defined on the set of real probability distributions with finite first moment, i.e.,

$$\mathcal{P}_1 := \left\{ \pi \text{ probability distributions} \,\middle|\, \inf_{c \in \mathbb{R}} \int |x - c| d\pi(x) < \infty \right\}.$$

One way to define the Wasserstein-1 metric $\mathbb{W}_1$ is by

$$\mathbb{W}_1(\pi_1, \pi_2) := \sup_{f \in Lip_1} \left| \mathbb{E}_{x \in \pi_1}[f(x)] - \mathbb{E}_{x \in \pi_2}[f(x)] \right|,$$

where $\pi_1, \pi_2 \in \mathcal{P}_1$ and

$$Lip_1 := \{f : \mathbb{R} \to \mathbb{R} | \forall_{x,y \in \mathbb{R}} : |f(x) - f(y)| \leq |x - y|\}.$$

This is the Kantorovich-Rubenstein duality formulation of the Wasserstein-1 distance. Recently this formulation of the Wasserstein distance came to particular interest in the context of Wasserstein-GANs (Arjovsky et al., 2017). The basic idea is to train a neural network, which is able to discriminate between the distribution of "nice" objects and the distribution of "adversarial" objects. This is done by maximizing over $\left|\mathbb{E}_{x\in\pi_1}[f(x)] - \mathbb{E}_{x\in\pi_2}[f(x)]\right|$, where f is the neural network to be trained and adding some regularizer to ensure that the Lipschitz constant is close to 1. We consider the following relaxed setting. First, we only consider probability distribution with computable density functions, supported in $[-\frac{1}{2}, \frac{1}{2}]$,

$$P([-1/2, 1/2]) := \left\{ f : [-1/2, 1/2] \to \mathbb{R}_+ \middle| \int f = 1,\ f \text{ Borel-Turing computable} \right\}.$$

Second, we restrict ourselves to a function space $\mathbb{F} \subset Lip_1$, which is made of Borel-Turing computable functions. The only additional assumption on $\mathbb{F}$ is:

$$\exists_{f_1, f_2 \in \mathbb{F}} : \exists_{c_1, c_2 \in \mathbb{R}} : \forall_{x \in [-1/2, 1/2]} : f_1(x) = x + c_1 \text{ and } f_2(x) = |x| + c_2.$$

This assumption holds in the example case of normalized neural networks.

It was shown in Lee et al. (2023) that calculating such a Wasserstein maximizer, or even approximating it is not possible with a Turing machine in these settings.

Theorem 10 (Wasserstein distance). *We define*

$$\mathbb{W}_1'(\pi_1, \pi_2) := \sup_{f \in \mathbb{F}} \left|\mathbb{E}_{x\in\pi_1}[f(x)] - \mathbb{E}_{x\in\pi_2}[f(x)]\right|.$$

Let $p_1, p_2 \in P[-\frac{1}{2}, \frac{1}{2}]$ be two computable probability densities. Then the problem of finding a function $f \in \mathbb{F}$, such that $\mathbb{W}_1'(\pi_1, \pi_2) = \mathbb{E}_{x\in\pi_1}[f(x)] - \mathbb{E}_{x\in\pi_2}[f(x)]$ or $|\mathbb{W}_1'(\pi_1, \pi_2) - \mathbb{E}_{x\in\pi_1}[f(x)] + \mathbb{E}_{x\in\pi_2}[f(x)]| \leq \alpha < \frac{\sqrt{5}}{8\sqrt{3}}$ is not Banach-Mazur computable.

It might very well happen that such a maximizing function does not exist at all. In this case, finding such a function is trivially noncomputable. However, it was proven that finding such a maximizing function might not be computable even in the case a computable maximizing function does exist.

5.4 Lattice problem for cryptographic applications

The basic idea of encryption is to apply a function F to a message m together with a (secret) key k to obtain an encrypted message $F(k, m) = e$. Ideally, it is hard to recover m from e without knowledge of k and easy to do with knowledge of k. Usually, F is motivated by using problems that are hard or suspected to be hard for all Turing machines to solve (Diffie and Hellman, 2022; Rivest et al., 1978; Daemen and Rijmen, 2002). So complexity and computability questions on Turing machines are central for well working encryption schemes. We focus on the question of computability of one particular problem from the family of lattice problems.

Lattice problems have become a topic of interest with the rising feasibility of quantum computers. Since quantum computers are able to crack conventional encryptions efficiently (Hallgren, 2007; Shor, 1999), lattice problems are seen as a new viable source for encryptions. It is wildly believed, but not proven, that lattice problems are hard to solve not only for Turing machines (Blömer and Seifert, 1999) but also for quantum computers.

Different optimization problems are highly relevant candidates for post-quantum cryptography, among others (Regev, 2009) are the shortest vector problem, the shortest independent vector problem, the closest vector problem, and the short generator principal ideal problem. Since Regev's discoveries (Regev, 2009), tremendous efforts have been made to solve the mentioned problems (Eisenträger et al., 2014; Biasse and Song, 2016). We consider the shortest independent vectors problem (SIVP), for which a randomized algorithm with exponential runtime exists if the complexity of the input is bounded (Ajtai et al., 2001).

To formulate the SIVP, we first have to define lattices over a field. Commonly, finite fields such as $\mathbb{Z}/p\mathbb{Z}$, where $p \in \mathbb{N}$ is a large prime number, are considered for the message space. In the following, we consider lattices over the field of real computable numbers $\mathbb{R}_c$. As it was shown in Lee et al. (2023), the transition from large p to the "continuous" field $\mathbb{R}_c$ is problematic, and the corresponding optimization problem becomes noncomputable on Turing machines, while for finite fields there have been recent successes (Eisenträger et al., 2014; Biasse and Song, 2016). Given $n \in \mathbb{N}$ and a basis $B = \{b_1, \dots, b_n\} \subset \mathbb{R}_c^n$, we define the corresponding lattice as

$$\Lambda(B) := \left\{ \sum_{i=1}^{n} \lambda_i b_i \in \mathbb{R}_c^n | \forall_{i=1,\dots,n} : \lambda_i \in \mathbb{Z} \right\}.$$

Now define $\mathcal{B}(B)$ to be the set of bases in $\Lambda(B)$:

$$\mathcal{B}(B) = \{\beta \subset \Lambda(B) | \beta \text{ is basis of } \mathbb{R}_c^n\}.$$

The SIVP is described by the minimization problem

$$\min_{\beta \in \mathcal{B}(B)} \sum_{b \in \beta} \|b\|_2.$$

Theorem 11 (SIVP). *Let $n \in \mathbb{N}$ and define V as the set of all bases in $\mathbb{R}^n$. Let $G : V \to V$ such that regarding SIVP and all bases $B \in V$ we have $G(B) \in Opt(B)$. Then G is not Banach-Mazur computable.*

All functions $G^* : V \to V$ satisfying

$$\|G - G^*\|_\infty \leq \alpha < \frac{\sqrt{2}}{2}$$

are also not Banach-Mazur computable.

5.5 Inverse problems

Let $A : \mathbb{R}_c^n \to \mathbb{R}_c^m$, $x \in \mathbb{R}^n$, $e \in \mathbb{R}^m$, and $y = Ax + e$. Usually A is called the forward operator, x the ground truth image, e the noise, and y the measurement. An inverse problem is the task of recovering x only knowing A, y, and having insufficient knowledge about e. Common examples of A are the identity operator for denoising, the convolution operator for deblurring, the Radon transformation for computer-tomography, and the (subsampled) Fourier transformation for magnet-resonance-tomography.

There are different approaches to solving inverse problems. Three of the most popular ones are to formulate this as a minimization problem. In the spirit of signal processing theory sparse solutions for x are often preferred, i.e. solutions where $\|x\|_0$ is small. The idea of recovering sparse solutions is called sparse recovery. We introduce three popular sparse recovery approaches.

Define the Basis Pursuit (BP) problem as

$$\min_x \|x\|_{l^1} \ s.t. \ \|Ax - y\|_{l^2} \leq \epsilon.$$

For some $\lambda > 0$ define the Lasso problem as

$$\min_x \lambda \|x\|_{l^1} + \|Ax - y\|_{l^2}.$$

For some $\lambda > 0$ define the quadratic Lasso problem as

$$\min_x \lambda \|x\|_{l^1} + \|Ax - y\|_{l^2}^2.$$

These formulations are widely used in sparse recovery, but they have limitations that were exposed by the impossibility results in Boche et al. (2023).

Theorem 12 (Boche et al., 2023). *Fix the parameters $n \geq 2$ and $m < n$ as well as $\epsilon \in (0, \frac{1}{4}) \cap \mathbb{Q}$ and $\lambda \in (0, \frac{5}{4}) \cap \mathbb{Q}$ respectively. Consider the Basis Pursuit* *problem and let* $K \geq C_{bp} > 0$ *be sufficiently large. Let* $G^* : \mathbb{C}^{m \times n} \times \mathbb{C}^m \to X$ *be an arbitrary function with*

$$\sup_{\substack{(A,y)\mathbb{C}^{m\times n}\times\mathbb{C}^m \\ \|A\|\leq K, \|y\|\leq 1}} \|G^* - G\|_{l_2} < \frac{1}{4}.$$

Then G^* *is not Banach-Mazur computable.*

Now consider the Lasso problem and let $K \geq C_l > 0$*. Let* $G^* : \mathbb{C}^{m \times n} \times \mathbb{C}^m \to X$ *be an arbitrary function with*

$$\sup_{\substack{(A,y)\mathbb{C}^{m\times n}\times\mathbb{C}^m \\ \|A\|\leq K, \|y\|\leq 1}} \|G^* - G\|_{l_2} < \frac{1}{8}.$$

Then G^* *is not Banach-Mazur computable.*

A different branch of noncomputability research in inverse problems was established in Colbrook et al. (2022), where a different notion of algorithm was used. The authors proved that for all three instances of sparse recovery, i.e., BP, Lasso, and quadratic Lasso, neural networks cannot achieve exact recovery. They proved the existence of training sets that allow neural networks to approximate the sparse solution up to $N - 1$ digits of precision, but not up to N digits, regardless of the network architecture or parameters.

Acknowledgments

The work of Yunseok Lee was supported by the German Research Foundation under Grant DFG-SPP-2298, KU 1446/32-1. The work of Holger Boche and Gitta Kutyniok was supported in part by the ONE Munich Strategy Forum (LMU Munich, TU Munich, and the Bavarian Ministery for Science and Art). The work of Holger Boche was also supported in part by the German Federal Ministry of Education and Research (BMBF) in the project Hardware Platforms and Computing Models for Neuromorphic Computing (NeuroCM) under Grant 16ME0442 and within the national initiative on 6G Communication Systems through the research hub 6G-life under Grant 16KISK002. The work of Gitta Kutyniok was also supported in part by the Konrad Zuse School of Excellence in Reliable AI (DAAD), the Munich Center for Machine Learning (BMBF) as well as the German Research Foundation under Grants DFG-SPP-2298, KU 1446/31-1 and KU 1446/32-1 and under Grant DFG-SFB/TR 109, Project C09 and the Federal Ministry of Education and Research under Grant MaGriDo.

References

Ajtai, Miklós, Kumar, Ravi, Sivakumar, Dandapani, 2001. A sieve algorithm for the shortest lattice vector problem. In: Proceedings of the Thirty-Third Annual ACM Symposium on Theory of Computing, pp. 601–610.

Arimoto, S., 1972. An algorithm for computing the capacity of arbitrary discrete memoryless channels. IEEE Transactions on Information Theory 18 (1), 14–20. https://doi.org/10.1109/TIT.1972.1054753.

Arjovsky, Martin, Chintala, Soumith, Bottou, Léon, 2017. Wasserstein GAN. https://arxiv.org/abs/1701.07875. https://dx.doi.org/10.48550/ARXIV.1701.07875.

Benioff, Paul, 1980. The computer as a physical system: a microscopic quantum mechanical Hamiltonian model of computers as represented by Turing machines. Journal of Statistical Physics 22 (5), 563–591.

Berner, Julius, et al., 2021. The modern mathematics of deep learning. https://doi.org/10.48550/ARXIV.2105.04026. https://arxiv.org/abs/2105.04026.

Biasse, Jean-François, Song, Fang, 2016. Efficient quantum algorithms for computing class groups and solving the principal ideal problem in arbitrary degree number fields. In: Proceedings of the Twenty-Seventh Annual ACM-SIAM Symposium on Discrete Algorithms. SIAM, pp. 893–902.

Blahut, R., 1972. Computation of channel capacity and rate-distortion functions. IEEE Transactions on Information Theory 18 (4), 460–473. https://doi.org/10.1109/TIT.1972.1054855.

Blömer, Johannes, Seifert, Jean-Pierre, 1999. On the complexity of computing short linearly independent vectors and short bases in a lattice. In: Proceedings of the Thirty-First Annual ACM Symposium on Theory of Computing, pp. 711–720.

Blum, Lenore, Shub, Mike, Smale, Steve, 2000. On a theory of computation and complexity over the real numbers: NP-completeness, recursive functions and universal machines. In: The Collected Papers of Stephen Smale, vol. 3. World Scientific, pp. 1293–1338.

Boche, Holger, Fono, Adalbert, Kutyniok, Gitta, 2023. Limitations of deep learning for inverse problems on digital hardware. https://dx.doi.org/10.48550/ARXIV.2202.13490. http://arxiv.org/abs/1207.0580.

Boche, Holger, Schaefer, Rafael F., Poor, H. Vincent, 2019a. Coding for non-iid sources and channels: entropic approximations and a question of Ahlswede. In: 2019 IEEE Information Theory Workshop (ITW). IEEE, pp. 1–5.

Boche, Holger, Schaefer, Rafael F., Poor, H. Vincent, 2019b. Secure communication and identification systems—effective performance evaluation on Turing machines. IEEE Transactions on Information Forensics and Security 15, 1013–1025.

Boche, Holger, Schaefer, Rafael F., Poor, H. Vincent, 2022. Algorithmic computability and approximability of capacity-achieving input distributions. https://arxiv.org/abs/2202.12617. https://dx.doi.org/10.48550/ARXIV.2202.12617.

Boyd, Stephen, Vandenberghe, Lieven, 2004. Convex Optimization. Cambridge University Press.

Colbrook, Matthew J., Antun, Vegard, Hansen, Anders C., 2022. The difficulty of computing stable and accurate neural networks: on the barriers of deep learning and Smale's 18th problem. Proceedings of the National Academy of Sciences 119, 12. https://doi.org/10.1073/pnas.2107151119.

Cover, T.M., 1984. An algorithm for maximizing expected log investment return. In: IEEE Transactions on Information Theory IT-30.2.

Daemen, Joan, Rijmen, Vincent, 2002. The Design of Rijndael, vol. 2. Springer.

Denning, Peter J., et al., 1989. Computing as a discipline. Communications of the ACM 32 (1), 9–23.

Diffie, Whitfield, Hellman, Martin E., 2022. New directions in cryptography. In: Democratizing Cryptography: The Work of Whitfield Diffie and Martin Hellman, pp. 365–390.

Eisenträger, Kirsten, et al., 2014. A quantum algorithm for computing the unit group of an arbitrary degree number field. In: Proceedings of the Forty-Sixth Annual ACM Symposium on Theory of Computing, pp. 293–302.

Gottschling, Nina M., et al., 2020. The troublesome kernel: why deep learning for inverse problems is typically unstable. CoRR. arXiv:2001.01258. arXiv:2001.01258. http://arxiv.org/abs/2001.01258.

Grohs, Philipp, Voigtlaender, Felix, 2023. Proof of the theory-to-practice gap in deep learning via sampling complexity bounds for neural network approximation spaces. Foundations of Computational Mathematics, 1–59.

Hallgren, Sean, 2007. Polynomial-time quantum algorithms for Pell's equation and the principal ideal problem. Journal of the ACM 54 (1), 1–19.

Hinton, Geoffrey E., et al., 2012. Improving neural networks by preventing co-adaptation of feature detectors. CoRR. arXiv:1207.0580. arXiv:1207.0580. http://arxiv.org/abs/1207.0580.

Kantorovich, Leonid V., 1960. Mathematical methods of organizing and planning production. Management Science 6 (4), 366–422.

Latane, Henry Allen, 1959. Criteria for choice among risky ventures. Journal of Political Economy 67 (2), 144–155.

Lee, Yunseok, Boche, Holger, Kutyniok, Gitta, 2023. Computability of optimizers. arXiv:2301.06148 [math.OC].

Mazur, Stanisław, 1963. Computable analysis.

Mead, Carver, 1990. Neuromorphic electronic systems. Proceedings of the IEEE 78 (10), 1629–1636.

Montavon, Grégoire, et al., 2019. Layer-wise relevance propagation: an overview. In: Explainable AI: Interpreting, Explaining and Visualizing Deep Learning, pp. 193–209.

Pour-El, Marian B., Richards, J. Ian, 2017. Computability in analysis and physics. In: Perspectives in Logic. Cambridge University Press.

Regev, Oded, 2009. On lattices, learning with errors, random linear codes, and cryptography. Journal of the ACM 56 (6), 1–40.

Rivest, Ronald L., Shamir, Adi, Adleman, Leonard, 1978. A method for obtaining digital signatures and public-key cryptosystems. Communications of the ACM 21 (2), 120–126.

Schwarz, Hans-Rudolf, Köckler, Norbert, 2013. Numerische mathematik. Springer-Verlag.

Shannon, Claude Elwood, 1948. A mathematical theory of communication. The Bell System Technical Journal 27 (3), 379–423.

Shor, Peter W., 1999. Polynomial-time algorithms for prime factorization and discrete logarithms on a quantum computer. SIAM Review 41 (2), 303–332.

Smale, Steve, et al., 1998. Mathematical problems for the next century. The Mathematical Intelligencer 20 (2), 7–15.

Specker, E., 1959. Der Satz vom Maximum in der rekursiven Analysis. In: Heyting, A. (Ed.), Constructivity in Mathematics: Proceedings of the Colloquium Held at Amsterdam, 1957.

Turing, Alan Mathison, 1938. On computable numbers, with an application to the Entscheidungsproblem. A correction. Proceedings of the London Mathematical Society 2 (1), 544–546.

Turing, Alan Mathison, et al., 1936. On computable numbers, with an application to the Entscheidungsproblem. Journal of Mathematics 58 (345–363), 5.

Vaserstein, Leonid Nisonovich, 1969. Markov processes over denumerable products of spaces, describing large systems of automata. Problemy Peredači Informacii 5 (3), 64–72.

Weihrauch, Klaus, 2000. Computable Analysis: an Introduction. Springer Science & Business Media.

Wong, Kam-Chau, 1996. Computability of minimizers and separating hyperplanes. Mathematical Logic Quarterly 42 (1), 564–568.

Chapter 8

Neural Galerkin schemes for sequential-in-time solving of partial differential equations with deep networks

Jules Berman, Paul Schwerdtner, and Benjamin Peherstorfer*
Courant Institute of Mathematical Sciences, New York University, New York, NY, United States
**Corresponding author: e-mail address: pehersto@cims.nyu.edu*

Contents

Handbook of Numerical Analysis, Volume 25, ISSN 1570-8659, https://doi.org/10.1016/bs.hna.2024.05.006

Abstract

This survey discusses Neural Galerkin schemes that leverage nonlinear parametrizations such as deep networks to numerically solve time-dependent partial differential equations (PDEs) in a variational sense. Neural Galerkin schemes build on the Dirac-Frenkel variational principle to train networks by minimizing the residual sequentially over time, which is in contrast to many other methods that approximate PDE solution fields with deep networks globally in time. Because of the sequential-in-time training, Neural Galerkin solutions inherently respect causality and approximate solution fields locally in time so that often fewer parameters are required than by global-in-time methods. Additionally, the sequential-in-time training enables adaptively sampling the spatial domain to efficiently evaluate the residual objectives over time, which is key for numerically realizing the expressive power of deep networks and other nonlinear parametrizations in high dimensions and when solution features are local such as wave fronts.

Keywords

Deep networks, Dirac-Frenkel variational principle, Time-dependent partial differential equations, Kolmogorov n-width, Curse of dimensionality, Model reduction

MSC Codes

65M20, 65M60, 65M22, 65F55

1 Introduction

Partial differential equations (PDEs) are broadly used in science and engineering to model systems of interest. Because analytic solutions of PDEs are available only in limited settings, one often has to resort to numerical computations to obtain approximate solutions.

1.1 Linear parametrizations in numerical analysis

A typical approach to numerically solving PDEs in numerical analysis is to first parametrize the PDE solution field with a finite number of parameters and then to derive a system of algebraic equations to solve for the parameters such that the parametrization approximates the PDE solution in some numerical sense. Common parametrizations are linear combinations of basis functions with local supports centered on grid points (Ern and Guermond, 2004). However, there are classes of PDEs for which linear parametrizations are inefficient in the sense that the best-approximation error decays slowly with increasing numbers of parameters. One important class of PDEs for which linear approximations on grid points become inefficient is given by PDEs that are formulated over high-dimensional spatial domains, which are often affected by the curse of dimensionality so that an exponential increase in the number of grid points—and thus typically computational costs—with the dimension is required to maintain the same accuracy. Another class is given by PDEs with slowly decaying Kolmogorov n-widths (Ohlberger and Rave, 2016; Greif and Urban, 2019; Arbes et al., 2023), which, e.g., provides a lower bound on the error that can be achieved with linear approximations in model reduction (Antoulas, 2005; Rozza et al., 2008; Benner et al., 2015; Antoulas et al., 2021; Kramer et al., 2024). In general, examples of PDEs that lead to slowly decaying n-widths are often found when modeling transport-dominated problems (Peherstorfer, 2022).

1.2 Nonlinear parametrizations for discretizing PDE solution fields

One approach to overcome the limitations of linear approximations is to use parametrizations that have a nonlinear dependence on the parameters. Nonlinear parametrizations are given by, for example, deep neural networks (LeCun et al., 2015), tensor networks (Orús, 2019), and Gaussian wave packets (Lubich, 2008). While nonlinear parametrizations can achieve faster error decays than linear ones under certain assumptions from an approximation-theoretic perspective, these results are mostly existence results that fall short of providing methods for numerically computing the parameters (DeVore et al., 1989, 1993; Cohen et al., 2022; DeVore et al., 2021; Daubechies et al., 2022).

1.3 Neural Galerkin schemes

The purpose of this survey is to discuss Neural Galerkin schemes (Bruna et al., 2024; Berman and Peherstorfer, 2023; Wen et al., 2024) that leverage nonlinear parametrizations to numerically solve time-dependent PDEs in a variational sense. The focus of this survey is on computational aspects rather than approximation theory. Neural Galerkin schemes build on the Dirac-Frenkel variational principle (Dirac, 1930; Frenkel, 1934; Lubich, 2008) to train networks by minimizing the residual sequentially over time, which is in contrast to many other methods that leverage deep networks that are global in time (Dissanayake and Phan-Thien, 1994; Sirignano and Spiliopoulos, 2018; Raissi et al., 2019; Berg and Nyström, 2018); see also Du and Zaki (2021); Anderson and Farazmand (2022); Kast and Hesthaven (2023); Zhang et al. (2024) for other, related sequential-in-time training methods. Because of the sequential-in-time training, Neural Galerkin solutions inherently respect causality. Furthermore, because solution fields are approximated only locally in time, often fewer parameters (e.g., networks with lower number of weights) are sufficient to provide accurate approximations when compared to global-in-time methods (Berman and Peherstorfer, 2023). In particular, locally in time, the network weights typically are of low rank, which can be leveraged via randomized sparse updates (Berman and Peherstorfer, 2023) and pretraining schemes (Berman and Peherstorfer, 2024). Additionally, the sequential-in-time training enables adaptively sampling the spatial domain to efficiently evaluate the residual objectives over time. The adaptive sampling is guided by the dynamics described by the PDE, which means that samples are placed where they are needed to efficiently evaluate the residual objective function as the solution field evolves. In fact, the numerical results in Bruna et al. (2024); Wen et al. (2024), which are also reported in this survey, show that the adaptive sampling of Neural Galerkin schemes is key for numerically realizing the expressive power of deep networks and other nonlinear parametrizations.

1.4 Literature overview

There are several other numerical methods that can leverage nonlinear parametrizations in PDE settings. Besides the large body of work on global-in-time methods such as Dissanayake and Phan-Thien (1994); Sirignano and Spiliopoulos (2018); Raissi et al. (2019); Berg and Nyström (2018), there are methods that leverage specific properties of classes of PDEs (E et al., 2017; Han et al., 2018) and focus on other, related approximation tasks such as learning committor functions (Khoo et al., 2018; Li et al., 2019; Rotskoff et al., 2022), closure modeling (Bar-Sinai et al., 2019; Kochkov et al., 2021; Wang et al., 2020), and de-noising (Rudy et al., 2019).

One major motivation for Neural Galerkin schemes is to circumvent the Kolmogorov barrier (Berman and Peherstorfer, 2024), which is a challenge in model reduction (Peherstorfer, 2022). There are other model reduction methods that can overcome the Kolmogorov barrier, and we survey several of them now. First, there are localized model reduction methods that learn a dictionary of candidate basis functions and then adaptively select a subset of basis functions (Jens et al., 2011; Dihlmann et al., 2011; Amsallem et al., 2012; Eftang and Stamm, 2012; Maday and Stamm, 2013; Peherstorfer et al., 2014; Kaulmann et al., 2015; Geelen and Willcox, 2022). By relying on a fixed dictionary, such localized model reduction methods restrict their flexibility in terms of what functions to approximate. In particular, all dynamics must have been seen in the pretraining (offline) phase. Another class of methods uses nonlinear maps to transform the solution fields such that the solution manifolds induced by the solution fields are not affected by a slowly decaying Kolmogorov n-width anymore (Rowley and Marsden, 2000; Ohlberger and Rave, 2013; Reiss et al., 2018; Ehrlacher et al., 2020; Qian et al., 2020; Papapicco et al., 2022; Issan and Kramer, 2023; Taddei et al., 2015; Cagniart et al., 2019). Related to transformations are methods based on nonlinear embeddings such as autoencoders (Lee and Carlberg, 2020; Kim et al., 2022; Romor et al., 2023). In the works (Geelen et al., 2023; Barnett and Farhat, 2022; Geelen et al., 2024; Sharma et al., 2023; Schwerdtner and Peherstorfer, 2024), approximations on quadratic manifolds have been proposed, which rely on a polynomial feature map to achieve a nonlinear parametrization.

Closer to Neural Galerkin schemes are online adaptive methods that aim to adapt basis functions over time (Koch and Lubich, 2007; Sapsis and Lermusiaux, 2009; Iollo and Lombardi, 2014; Gerbeau and Lombardi, 2014; Carlberg, 2015; Peherstorfer and Willcox, 2015; Zahr and Farhat, 2015; Peherstorfer, 2020; Black et al., 2020; Billaud-Friess and Nouy, 2017; Ramezanian et al., 2021), where a particular challenge is to achieve online updates of the basis functions efficiently (Peherstorfer and Willcox, 2015; Zimmermann et al., 2018; Peherstorfer, 2020; Huang and Duraisamy, 2023; Singh et al., 2023). Especially dynamic low-rank approximations (Koch and Lubich, 2007; Musharbash et al., 2015; Einkemmer and Lubich, 2019; Einkemmer et al., 2021; Musharbash and Nobile, 2017, 2018; Hesthaven et al., 2022) have been widely used, of which many build on the Dirac-Frenkel variational principle (Dirac, 1930; Frenkel, 1934; Lubich, 2008), just as Neural Galerkin schemes.

1.5 Outline

This manuscript is organized as follows. We first discuss the need for nonlinear parametrizations in Section 2. Neural Galerkin schemes are described in Section 3. Adaptive sampling (active data acquisition) is an important part of Neural Galerkin schemes that is discussed in Section 4. A randomized sparse extension of Neural Galerkin schemes is presented in Section 5. It numerically achieves faster runtimes and stabler approximations. Conclusions are drawn in Section 6.

2 The need for nonlinear parametrizations in approximating solution fields of PDEs

2.1 Setup

Consider a time-dependent PDE of the form

$$\partial_t q(t, \boldsymbol{x}) = f(t, \boldsymbol{x}, q), \qquad \boldsymbol{x} \in \Omega,\ t \in (0, T], \tag{1}$$

with the spatial domain $\Omega \subseteq \mathbb{R}^d$ and the solution field $q : [0, T] \times \Omega \to \mathbb{R}$. The right-hand side function f depends on time $t \in [0, T]$, the spatial coordinate $\boldsymbol{x} \in \Omega$, and the field q. The function f can depend on partial derivatives of q as well. For example, we obtain an advection-diffusion-reaction equation by setting

$$f(t, \boldsymbol{x}, q) = b(t, \boldsymbol{x}) \cdot \nabla q(t, \boldsymbol{x}) + \operatorname{div}(a(t, \boldsymbol{x})\nabla q(t, \boldsymbol{x})) + g(t, \boldsymbol{x}, q(t, \boldsymbol{x}))$$

for coefficient functions $b : [0, \infty) \times \Omega \to \mathbb{R}^d$ and $a : [0, \infty) \times \Omega \to \mathbb{R}^d \times \mathbb{R}^d$ and a source term $g : [0, \infty) \times \Omega \times \mathbb{R} \to \mathbb{R}$. In the following, we only consider situations where the function $q(t, \cdot) : \Omega \to \mathbb{R}$ over the spatial domain Ω is in an appropriate Hilbert space $\mathcal{V}$ with inner product $\langle \cdot, \cdot \rangle_{\mathcal{V}}$ at all times $t \in [0, T]$.

We focus exclusively on Dirichlet and periodic boundary conditions, which can be imposed by restricting the space $\mathcal{V}$ so that all functions in $\mathcal{V}$ satisfy the boundary conditions. The initial conditions of (1) are denoted as $q_0 : \Omega \to \mathbb{R}$ and are elements of the set $\mathcal{V}^{(0)} \subseteq \mathcal{V}$.

Let now $\mathcal{M}$ be the set of solutions of (1), which formally is just a set of functions defined as

$$\mathcal{M} = \{q(t, \cdot) | t \in [0, T], q_0 \in \mathcal{V}^{(0)}\} \subset \mathcal{V}. \tag{2}$$

The set (2) does not necessarily have to have a manifold structure but the convention is to call it the solution manifold and we stick to this convention in the following. We also note that typically solutions of PDEs are considered in a variational sense rather than in a classical sense as in (2); however, the formulation via the classical solutions in (2) will be sufficient to demonstrate the limitations of linear parametrizations in the following.

2.2 Linear parametrizations of solution fields

Let us now consider linear parametrizations with a finite number of parameters of the solution field q, which we can write as

$$\tilde{q}(\boldsymbol{\theta}(t), \boldsymbol{x}) = \sum_{i=1}^{n} \theta_i(t)\varphi_i(\boldsymbol{x}). \tag{3}$$

There are n parameters $\theta_1(t), \ldots, \theta_n(t) \in \mathbb{R}$, which depend on time t. We collect the parameters in the vector $\boldsymbol{\theta}(t) = [\theta_1(t), \ldots, \theta_n(t)]^T \in \mathbb{R}^n$. The functions $\varphi_1, \ldots, \varphi_n$ span a subspace of $\mathcal{V}$ of dimension at most n, in which the solution field q is approximated. Linear combinations of basis functions such as (3) are widely used in scientific computing. For example, finite-element methods (Ern and Guermond, 2004) typically discretize the spatial domain Ω with a grid and place the basis functions $\varphi_1, \ldots, \varphi_n$ at the grid points. The basis functions are often chosen with a local support in the context of finite-element methods. Another example is given by model reduction methods that build on linear combinations (3) with basis functions $\varphi_1, \ldots, \varphi_n$ that have global support (Antoulas, 2005; Rozza et al., 2008; Benner et al., 2015; Antoulas et al., 2021; Kramer et al., 2024).

The best-approximation error that can be achieved with parametrizations of the linear type (3) can be described with the concept of the Kolmogorov n-width (Pinkus, 1985). We consider the following version of the Kolmogorov n-width:

$$d_n(\mathcal{M}) = \inf_{\substack{\mathcal{V}_n \subset \mathcal{V} \\ \dim(\mathcal{V}_n) \leq n}} \sup_{q(t,\cdot) \in \mathcal{M}} \inf_{\tilde{q}^* \in \mathcal{V}_n} \|q(t, \cdot) - \tilde{q}^*\|. \tag{4}$$

The Kolmogorov n-width as given in (4) is the lowest worst-case error that any n-dimensional subspace $\mathcal{V}_n$ of $\mathcal{V}$ can achieve over the elements of $\mathcal{M}$. Note that (4) gives no indication how to construct a sequence of subspaces $(\mathcal{V}_n)_n$ that achieves $d_n(\mathcal{M})$.

The decay rate of $d_n(\mathcal{M})$ has been studied for solution manifolds induced by certain classes of PDEs. The work by Maday et al. (2002) shows an exponential decay of $d_n(\mathcal{M})$ for $\mathcal{M}$ induced by specific elliptic coercive PDEs over a single parameter. Additional results are given in Cohen and DeVore (2016). The works by Binev et al. (2011); Buffa et al. (2012); Cohen et al. (2020) show how to construct sequences of subspaces that achieve an exponential decay rate. The class of equations for which the Kolmogorov n-width decays exponentially fast are well suited for classical model reduction with linear parametrizations (Antoulas, 2005; Rozza et al., 2008; Benner et al., 2015; Antoulas et al., 2021; Kramer et al., 2024).

2.3 The Kolmogorov barrier

Let us now move from elliptic PDEs to hyperbolic ones and more generally to models that describe transport phenomena. It has been shown in Ohlberger and Rave (2016) that the linear advection equation can lead to a solution manifold that has a slowly decaying n-width. To see this, consider the equation

$$\partial_t q(t,x) + \partial_x q(t,x) = 0\,, \qquad x \in (0,1), t \in (0,1]\,, \tag{5}$$

with initial condition

$$q_0(x) = \begin{cases} 1\,, & x \le 0 \\ 0\,, & \text{else}\,. \end{cases}$$

The solution to (5) is given by $q(t,x) = q_0(x-t)$ and thus the solution manifold $\mathcal{M}$ consists of the step functions that have a discontinuity in the spatial domain $[0,1]$. Ohlberger and Rave (2016) show the lower bound

$$d_n(\mathcal{M}) \ge c\frac{1}{\sqrt{n}}\,, \tag{6}$$

for a constant $c > 0$ independent of n. The bound (6) means that there cannot exist a sequence of subspaces $(\mathcal{V}_n)_n$ of $\mathcal{V}$ that achieves a faster error decay in the sense of (4) than $1/\sqrt{n}$, which is a slow rate compared to the exponential rate achieved for some PDEs of the elliptic type. Lower bounds with similarly slow decay rates have been shown in Arbes et al. (2023) for the linear advection equation with smoother initial conditions, where the smoothness of the initial condition determines the decay rate. A slow decay of $1/\sqrt{n}$ has also been shown for instances of the wave equation (Greif and Urban, 2019). Additionally, it is empirical observed that models that describe transport phenomena can lead to matrices of coefficient vectors with slowly decaying singular values, which is insufficient to draw conclusions about the decay of the Kolmogorov n-width but it provides indication that linear parametrizations are inefficient for such transport-dominated problems (Rowley and Marsden, 2000; Ohlberger and Rave, 2013; Reiss et al., 2018; Huang and Duraisamy, 2023; Peherstorfer, 2020; Uy et al., 2024). The slow decay of the Kolmogorov n-width is often referred to as the Kolmogorov barrier of linear parametrizations; see also the survey by Peherstorfer (2022).

2.4 Numerical illustrations of limitations of linear parametrizations

Let us consider a numerical experiment to illustrate the limitations of linear parametrizations. The following results are taken from Peherstorfer (2022). As a prototypical model of a diffusion-dominated problem, let us consider the heat equation

$$\partial_t q(t,x) - \partial_x^2 q(t,x) = 1\,, \qquad x \in \Omega\,, \tag{7}$$

with spatial domain $\Omega = (0, 1) \subset \mathbb{R}$. The boundary conditions are of homogeneous Dirichlet type and the initial condition is the zero function. We set end time to $T = 0.4$ and discretize on $N = 1024$ linear finite elements in space and with the implicit Euler method with time-step size 10^{-3} in time. Denote with $\boldsymbol{q}(t_k) \in \mathbb{R}^{1024}$ the coefficient vector of the finite-element approximation at time step $t_k = \delta t k$ for $k = 0, \ldots, 400 = K$. In Fig. 1a we show the numerical solution over the time-space domain.

We now collect snapshots $\boldsymbol{q}(t_1), \ldots, \boldsymbol{q}(t_K)$ over time and assemble the snapshot matrix $\boldsymbol{Q} = [\boldsymbol{q}(t_0), \boldsymbol{q}(t_1), \ldots, \boldsymbol{q}(t_K)] \in \mathbb{R}^{1024\times 401}$. Recall that the decay of the singular values of the snapshot matrix $\boldsymbol{Q}$ indicates how well the snapshots can be approximated in the space spanned by the first few left-singular vectors. Fig. 1b shows the singular values, which we normalized so that the largest singular value is one. Only about 20 singular vectors are sufficient to approximate the snapshots up to double precision. A space of dimension 20 is therefore sufficient for achieving double precision even though the dimension of the snapshot vectors is $N = 1024$. It is important to note that the decay of the singular values does not allow to draw conclusions about the Kolmogorov n-width decay of the corresponding solution manifold; however, the singular values can serve as a numerical indication if classical methods with linear approximations in subspaces can be efficient.

Let us now consider a transport-dominated problem modeled by the linear advection equation (5) with spatial domain $\Omega = (0, 1)$, time domain $(0, 0.4]$ and periodic boundary conditions. We choose the Gaussian probability density function with mean 0.1 and standard deviation 1.5×10^{-2} as initial condition. The solution is plotted over time and space in Fig. 1c. Analogously to the example with the heat equation, we collect snapshot data and plot the decay of the normalized singular values in Fig. 1d. The singular values decay orders of magnitude slower, which provides further indication that solution fields of transport-dominated problems can be challenging to approximate with linear approximations in subspaces. Even though this is just a numerical illustration,

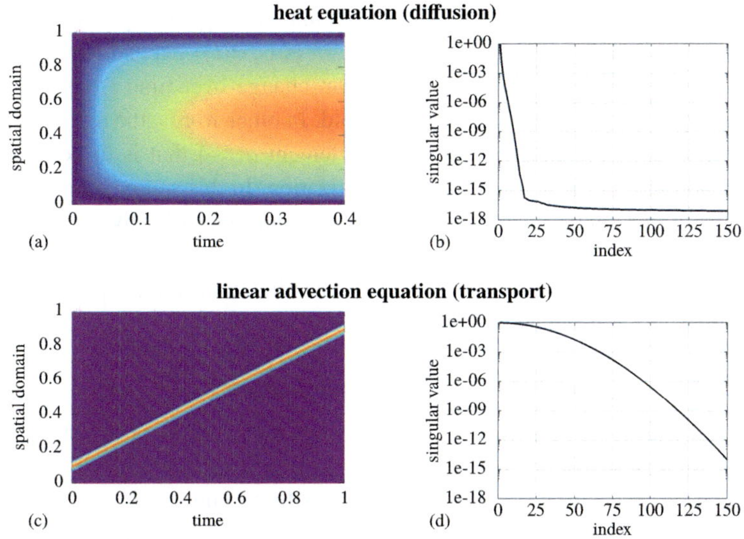

FIGURE 1 For diffusion-dominated problems such as the heat equation in this example, the singular values of snapshot matrices typically decay exponentially fast. In contrast, if dynamics are dominated by transport such as with models given by the linear advection equation, the singular values of snapshot matrices can decay orders of magnitude slower, which motivates the use of nonlinear parametrizations in model reduction for transport-dominated problems (Peherstorfer, 2022). (First published in Notices of the American Mathematical Society in 69, Number 5 (2022), published by American Mathematical Society. © 2022 American Mathematical Society.)

it is representative of the challenges of linear approximations for transport-dominated problems; see the survey by Peherstorfer (2022).

2.5 Nonlinear parametrizations

The Kolmogorov barrier can be circumvented with nonlinear parametrizations (DeVore et al., 1989, 1993; Cohen et al., 2022). We call a parametrization nonlinear if the representation can depend on the element of $\mathcal{M}$ that one wants to approximate, which is in contrast to the linear parametrizations discussed in Section 2.2, where the representation $\varphi_1, \ldots, \varphi_n$ is fixed for all elements in $\mathcal{M}$.

2.5.1 Generic nonlinear parametrizations

Consider the generic nonlinear parametrization

$$\tilde{q}(\boldsymbol{\theta}(t), \boldsymbol{x}) = \sum_{i=1}^{n} \beta_i(t)\varphi_i(\boldsymbol{x}; \boldsymbol{\alpha}(t)), \tag{8}$$

with the parameter vector $\boldsymbol{\theta}(t) = [\boldsymbol{\alpha}(t); \boldsymbol{\beta}(t)] \in \mathbb{R}^p$ that consists of the feature vector $\boldsymbol{\alpha}(t) = [\alpha_1(t), \ldots, \alpha_{n'}(t)]^T$ with n' components and the vector

$\boldsymbol{\beta}(t) = [\beta_1(t), \ldots, \beta_n(t)]^T$ of n coefficients that enter $\tilde{q}$ linearly. The representation given by the functions $\varphi_1(\cdot; \boldsymbol{\alpha}(t)), \ldots, \varphi_n(\cdot; \boldsymbol{\alpha}(t))$ depends on the time-dependent feature vector $\boldsymbol{\alpha}(t)$ and thus can change over time t based on the element $q(t, \cdot) \in \mathcal{M}$ of the solution manifold. In other words, the representation $\varphi_1, \ldots, \varphi_n$ can be adapted based on the element $q(t, \cdot)$ that is to be approximated and thus (8) is a nonlinear parametrization. In this sense, adaptive mesh refinement in scientific computing, which was introduced for hyperbolic problems in Berger and Colella (1989); Berger and LeVeque (1998), is also a form of nonlinear parametrization because the basis functions are adapted based on how the solution fields evolve over time.

2.5.2 Examples of nonlinear parametrizations

The nonlinear parametrization given in (8) is generic and we now consider specific instances of it. Let us first consider dictionary approaches, which we can write as (8) by restricting us to have only a finite number $L \in \mathbb{N}$ of feature vectors $\boldsymbol{\alpha}^{(1)}, \ldots, \boldsymbol{\alpha}^{(L)} \in \mathbb{R}^{n'}$ that are independent of time t. Each feature vector $\boldsymbol{\alpha}^{(i)}$ corresponds to a representation $\varphi_1^{(i)} = \varphi_1(\cdot; \boldsymbol{\alpha}^{(i)}), \ldots, \varphi_n^{(i)} = \varphi_n(\cdot; \boldsymbol{\alpha}^{(i)})$ and a corresponding subspace $\mathcal{V}_n^{(i)} \subset \mathcal{V}$ for $i = 1, \ldots, L$. Based on a classification function such as $I : \mathcal{V} \to \{1, \ldots, L\}$, one of the representations is selected based on the element of $\mathcal{M}$ that is to be approximated. Because the representation is chosen based on the element that is to be approximated, it provides a nonlinear parametrization. Such nonlinear parametrizations with a finite number of feature vectors have been studied extensively in the context of model reduction under the umbrella term of localized model reduction (Jens et al., 2011; Dihlmann et al., 2011; Amsallem et al., 2012; Eftang and Stamm, 2012; Maday and Stamm, 2013; Peherstorfer et al., 2014; Kaulmann et al., 2015; Geelen and Willcox, 2022). Localized model reduction is closely related to dictionary approaches, because the combination of all functions $\varphi_1^{(1)}, \ldots, \varphi_n^{(1)}, \varphi_1^{(2)}, \ldots, \varphi_n^{(L)}$ can be considered as the dictionary from which an indicator function selects n elements.

Another type of nonlinear parametrizations that is widely used in model reduction is building on nonlinear transformations. For example, if the manifold $\mathcal{M}$ describes solution fields with moving coherent structures in the spatial domain, then the functions $\varphi_1, \ldots, \varphi_n$ in (8) can account for this transport via the feature vector $\boldsymbol{\alpha}(t)$. A frequently given example is the linear advection equation (5), which has as solution $q(t, x) = q_0(x - t)$ and thus setting $n = 1$ and $n' = 1$ with $\alpha_1(t) = -t$ so that $\varphi_1(x, \boldsymbol{\alpha}(t)) = q_0(x - t)$ provides an exact representation of the solution field; see also Peherstorfer (2022). More sophisticated transformations can be constructed either analytically (Rowley et al., 2004; Ehrlacher et al., 2020) or via optimization (Reiss et al., 2018; Taddei, 2020; Taddei and Zhang, 2021).

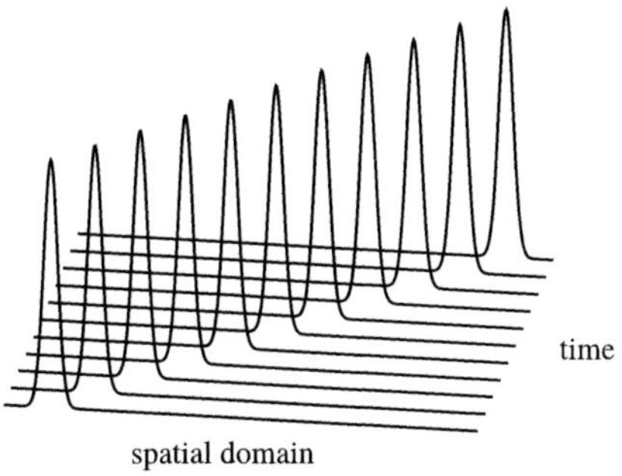

FIGURE 2 The classical method of lines first discretizes the spatial domain of time-dependent partial differential equations to obtain a system of ODEs, which is then integrated in time to obtain approximate coefficient vectors for linear combinations of basis functions that represent the solution field. Neural Galerkin schemes are also sequentially in time approximating solution fields but allow to use nonlinear parametrizations such as deep networks.

2.5.3 Nonlinear parametrizations via time-dependent parameters

In this survey, we consider nonlinear parametrizations with time-dependent parameters $\boldsymbol{\theta}(t)$. In the following, there is no need to distinguish between features $\boldsymbol{\alpha}(t)$ and coefficients $\boldsymbol{\beta}(t)$ and thus we can consider the nonlinear parametrization as just depending on a p-dimensional vector $\boldsymbol{\theta}(t)$ as

$$\tilde{q}(\boldsymbol{\theta}(t), \cdot) : \Omega \to \mathbb{R}. \tag{9}$$

Examples of such parametrizations are given by dynamic low-rank approximations (Koch and Lubich, 2007; Sapsis and Lermusiaux, 2009), deep neural networks with time-dependent weights (Bruna et al., 2024; Du and Zaki, 2021; Finzi et al., 2023), tensor networks (Orús, 2019), Gaussian wave packets (Lubich, 2008), and other nonlinear parametrizations (Black et al., 2020).

3 Neural Galerkin schemes based on the Dirac-Frenkel variational principle and deep networks

We now discuss Neural Galerkin schemes (Bruna et al., 2024) that provide dynamical systems for the time-dependent parameters $\boldsymbol{\theta}(t)$ so that the nonlinear parametrizations $\tilde{q}(\boldsymbol{\theta}(t), \cdot)$ solve the PDEs of interest in a variational sense; see Fig. 2. The two key building blocks of Neural Galerkin schemes are the Dirac-Frenkel variational principle (Lasser and Lubich, 2020, Section 3.8) for deriving the dynamical systems and deep networks with time-dependent weights for providing the nonlinear parametrizations.

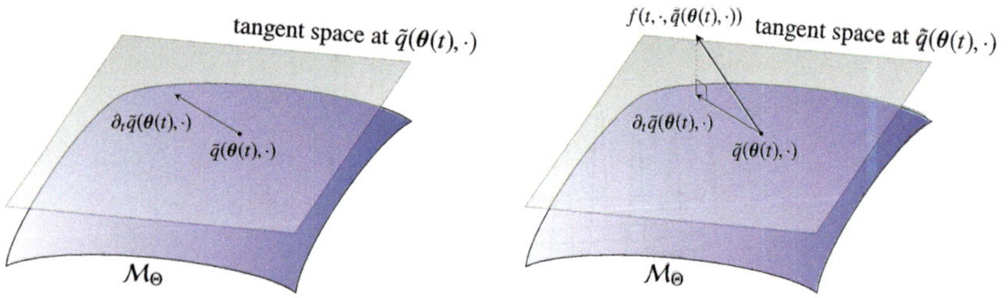

FIGURE 3 The time derivative $\dot{\boldsymbol{\theta}}(t)$ of the parameter $\boldsymbol{\theta}(t)$ is given by approximating the right-hand side function $f(t, \cdot, \tilde{q}(\boldsymbol{\theta}(t), \cdot))$ in the tangent space of the parametrization manifold $\mathcal{M}_\Theta$ at the current solution $\tilde{q}(\boldsymbol{\theta}(t), \cdot)$ in a least-squares sense, which is the Dirac-Frenkel variational principle (Lasser and Lubich, 2020, Section 3.8).

3.1 The Dirac-Frenkel variational principle

Recall the nonlinear parametrizations with p parameters described in (9). Let us drop the time dependence of $\boldsymbol{\theta}$ for now to obtain

$$\tilde{q}(\boldsymbol{\theta}, \cdot) : \Omega \to \mathbb{R}, \tag{10}$$

with $\boldsymbol{\theta} \in \Theta \subseteq \mathbb{R}^p$. The parametrization (10) induces the manifold

$$\mathcal{M}_\Theta = \{\tilde{q}(\boldsymbol{\theta}, \cdot) \,|\, \boldsymbol{\theta} \in \Theta\},$$

which is different from the manifold $\mathcal{M}$ induced by the PDE solutions; see Section 2.1. The manifold $\mathcal{M}_\Theta$ is depicted in Fig. 3a. The tangent space of $\mathcal{M}_\Theta$ at a point $\tilde{q}(\boldsymbol{\theta}, \cdot) \in \mathcal{M}_\Theta$ is spanned by the component functions of the gradient $\nabla_{\boldsymbol{\theta}} \tilde{q}$,

$$\mathcal{T}_{\tilde{q}(\boldsymbol{\theta}, \cdot)} \mathcal{M}_\Theta = \operatorname{span}\left\{\partial_{\theta_1} \tilde{q}, \ldots, \partial_{\theta_p} \tilde{q}\right\}. \tag{11}$$

Let us now make the parameter $\boldsymbol{\theta}(t)$ depend on time t again so that we can consider the time derivative of $\tilde{q}(\boldsymbol{\theta}(t), \cdot)$. Notice that time t enters the parametrization $\tilde{q}$ only via the parameter vector $\boldsymbol{\theta}(t)$. We thus apply the chain rule to obtain

$$\partial_t \tilde{q}(\boldsymbol{\theta}(t), \cdot) = \nabla_{\boldsymbol{\theta}} \tilde{q}(\boldsymbol{\theta}(t), \cdot) \cdot \dot{\boldsymbol{\theta}}(t), \tag{12}$$

where $\dot{\boldsymbol{\theta}}(t)$ is a vector in $\mathbb{R}^p$, which we interpret as the time derivative of the parameter vector $\boldsymbol{\theta}(t)$. Eq. (12) shows that the time derivative $\partial_t \tilde{q}(\boldsymbol{\theta}(t), \cdot)$ is an element of the tangent space $\mathcal{T}_{\tilde{q}(\boldsymbol{\theta}(t), \cdot)} \mathcal{M}_\Theta$ of $\mathcal{M}_\Theta$ at $\tilde{q}(\boldsymbol{\theta}(t), \cdot)$, which is spanned by the component functions of the gradient $\nabla_{\boldsymbol{\theta}} \tilde{q}(\boldsymbol{\theta}(t), \cdot)$.

Let us now consider the PDE given in (1) again so that we can define the residual function at time t as

$$r_t(\boldsymbol{\theta}(t), \dot{\boldsymbol{\theta}}(t), \cdot) = \underbrace{\partial_t \tilde{q}(\boldsymbol{\theta}(t), \cdot)}_{\nabla_{\boldsymbol{\theta}} \tilde{q}(\boldsymbol{\theta}(t), \cdot) \cdot \dot{\boldsymbol{\theta}}(t)} - f(t, \cdot, \tilde{q}(\boldsymbol{\theta}(t), \cdot)), \tag{13}$$

where we plugged $\tilde{q}(\boldsymbol{\theta}(t), \cdot)$ into (1). Notice that the residual function r_t has the parameter vector $\boldsymbol{\theta}(t)$ as well as the time derivative $\dot{\boldsymbol{\theta}}(t)$ as arguments. We now discuss multiple options to seek $\dot{\boldsymbol{\theta}}(t)$, which all will result in the same dynamics and thus be equivalent. One option is to project the right-hand side function f onto the tangent space and then to seek a $\dot{\boldsymbol{\theta}}(t)$ that sets the corresponding residual to zero. Because we know that the time derivative $\partial_t \tilde{q}(\boldsymbol{\theta}(t), \cdot)$ is in the tangent space (11), such a $\dot{\boldsymbol{\theta}}(t)$ can be found, which closes the equation,

$$\partial_t \tilde{q}(\boldsymbol{\theta}(t), \cdot) = \mathrm{P}_{\boldsymbol{\theta}(t)} f(t, \cdot, \tilde{q}(\boldsymbol{\theta}(t), \cdot)) \tag{14}$$

with the projection with respect to an L^2 inner product $\langle \cdot, \cdot \rangle_\nu$ with measure ν

$$\mathrm{P}_{\boldsymbol{\theta}} g = \nabla_{\boldsymbol{\theta}} \tilde{q} \cdot \langle \nabla_{\boldsymbol{\theta}} \tilde{q}, g \rangle_\nu ,$$

for $g \in \mathcal{V}$. Eq. (14) is found in a wide range of literature that builds on the Dirac-Frenkel variational principle (Dirac, 1930; Frenkel, 1934; Lubich, 2008; Koch and Lubich, 2007; Sapsis and Lermusiaux, 2009; Hesthaven et al., 2022; Du and Zaki, 2021; Anderson and Farazmand, 2022). Another option, that will result in the same equation (14), can be derived by imposing Galerkin conditions for finding $\dot{\boldsymbol{\theta}}(t)$ so that the residual r_t is orthogonal to the tangent space with respect to the inner product $\langle \cdot, \cdot \rangle_\nu$,

$$\langle \partial_{\theta_i} \tilde{q}(\boldsymbol{\theta}(t), \cdot), r_t(\boldsymbol{\theta}(t), \dot{\boldsymbol{\theta}}(t), \cdot) \rangle_\nu = 0, \quad i = 1, \ldots, p . \tag{15}$$

Yet another option is to minimize the squared residual norm locally over time, which once more will lead to the same equation dynamics; see Section 3.2.

Transformations show that (14) and equivalently (15) provide the dynamics of the parameter vector $\boldsymbol{\theta}(t)$ as

$$M(t, \boldsymbol{\theta}(t)) \dot{\boldsymbol{\theta}}(t) = F(t, \boldsymbol{\theta}(t)) , \qquad \boldsymbol{\theta}(0) = \boldsymbol{\theta}^{(0)} , \tag{16}$$

with

$$M(t, \boldsymbol{\theta}) = \int_\Omega \nabla_{\boldsymbol{\theta}} \tilde{q}(\boldsymbol{\theta}, \boldsymbol{x}) \otimes \nabla_{\boldsymbol{\theta}} \tilde{q}(\boldsymbol{\theta}, \boldsymbol{x}) \mathrm{d}\nu(\boldsymbol{x}) ,$$
$$F(t, \boldsymbol{\theta}) = \int_\Omega \nabla_{\boldsymbol{\theta}} \tilde{q}(\boldsymbol{\theta}, \boldsymbol{x}) f(t, \boldsymbol{x}, \tilde{q}(\boldsymbol{\theta}, \cdot)) \mathrm{d}\nu(\boldsymbol{x}) ,$$

and an initial condition $\boldsymbol{\theta}^{(0)} \in \Theta$. The symbol $\otimes$ means the outer product of the gradients. We refer to (16) as the Neural Galerkin equations because they use deep networks as a parametrization and correspond to the Galerkin conditions (15). But we stress that analogous equations have been derived for various other parametrizations as well as deep networks under various names (Lubich, 2008; Koch and Lubich, 2007; Sapsis and Lermusiaux, 2009; Hesthaven et al., 2022; Du and Zaki, 2021; Anderson and Farazmand, 2022). Because the matrix $M(t, \boldsymbol{\theta})$ in (16) can become singular, the dynamical system can have multiple solutions $\boldsymbol{\theta}(t)$; see Section 5 for a more detailed discussion.

3.2 An optimization perspective of the Dirac-Frenkel variational principle

Many methods in machine learning are motivated via an optimization perspective rather than the Galerkin projection perspective often found in scientific computing. Let us now derive the dynamical system given by the Dirac-Frenkel variational principle by starting with an objective function

$$H_t(\boldsymbol{\theta}, \eta) = \mathbb{E}_{\boldsymbol{x}\sim\nu}\left[|r_t(\boldsymbol{\theta}, \eta, \boldsymbol{x})|^2\right], \tag{17}$$

as in Du and Zaki (2021). We then seek $\dot{\boldsymbol{\theta}}(t)$ that minimizes objective H_t at time t with respect to η,

$$\min_{\eta\in\Theta} H_t(\boldsymbol{\theta}(t), \eta)$$

Problem (17) does not necessarily have a unique global optimum. We only ask for first-order optimality so that it is sufficient to set the gradient of H_t with respect to η to zero,

$$\nabla_\eta J_t(\boldsymbol{\theta}(t), \dot{\boldsymbol{\theta}}(t)) = 0, \tag{18}$$

which leads to the dynamical system for $\boldsymbol{\theta}(t)$ given in (16).

A remark is in order. The optimization perspective given by introducing the dynamical system (16) via the objective function (17) clearly contrasts the present approach to widely used time-space discretizations, which parametrize the whole time-space domain with a neural network (Raissi et al., 2019). In this sense, Neural Galerkin schemes can be interpreted as being nonlinear extensions of the classical method of lines (Zafarullah, 1970; Verwer and Sanz-Serna, 1984) for solving time-dependent PDEs, where first the spatial domain is discretized to obtain a semidiscrete system of ordinary differential equations (ODEs). The system of ODEs is then discretized in time and numerically integrated. Similarly, Neural Galerkin schemes first parametrize the spatial domain of the PDEs with time-dependent parametrizations, which are then numerically computed by integrating a dynamical system in time. Again, this is in contrast to time-space approximations with deep networks as Raissi et al. (2019).

3.3 Least-squares formulation and discretization in time

The dynamical system (16) for $\boldsymbol{\theta}(t)$ describes the normal equations of the least-squares problem

$$\min_{\dot{\boldsymbol{\theta}}(t)\in\Theta} \|\nabla_{\boldsymbol{\theta}}\tilde{q}(\boldsymbol{\theta}(t), \cdot)\cdot\dot{\boldsymbol{\theta}}(t) - f(t, \cdot, \tilde{q}(\boldsymbol{\theta}(t), \cdot))\|_\nu . \tag{19}$$

Numerically it thus is often beneficial with respect to the condition of the problem to work with the least-squares problem (19) rather than the dynamical system (16). An important insight is that the unknown $\dot{\boldsymbol{\theta}}(t)$ enters linearly in the least-squares problem (19), even though the parametrization $\tilde{q}$ depends nonlinearly on the parameter vector $\boldsymbol{\theta}(t)$.

We can now discretize (19) in time. Consider therefore $K \in \mathbb{N}$ time steps with $0 = t_0 < t_1 < \cdots < t_K = T$. At each time step $k = 0, \ldots, K-1$, we obtain a parameter vector $\boldsymbol{\theta}_k \in \mathbb{R}^p$, which is an approximation of the time-continuous parameter $\boldsymbol{\theta}(t_k)$ at time step t_k. If we take an explicit time integration scheme such as forward Euler, we obtain the least-squares problems

$$\min_{\boldsymbol{\theta}_{k+1} \in \Theta} \|\nabla_{\boldsymbol{\theta}} \tilde{q}(\boldsymbol{\theta}_k, \cdot) \cdot (\boldsymbol{\theta}_{k+1} - \boldsymbol{\theta}_k) - \delta t f(t_k, \cdot, \tilde{q}(\boldsymbol{\theta}_k, \cdot))\|_\nu \,, k = 0, \ldots, K-1 \,. \tag{20}$$

Because we used an explicit time integration scheme, the linearity of the time-continuous least-squares problem (19) is preserved in (20): At each time step $k = 0, \ldots, K-1$, a linear least-squares problem has to be solved. In contrast, if we take an implicit time integration scheme such as backward Euler, then we obtain

$$\begin{aligned} &\min_{\boldsymbol{\theta}_{k+1} \in \Theta} \|\nabla_{\boldsymbol{\theta}} \tilde{q}(\boldsymbol{\theta}_{k+1}, \cdot) \cdot (\boldsymbol{\theta}_{k+1} - \boldsymbol{\theta}_k) - \delta t f(t_{k+1}, \cdot, \tilde{q}(\boldsymbol{\theta}_{k+1}, \cdot))\|_\nu \,, \\ &\qquad k = 0, \ldots, K-1 \,, \end{aligned} \tag{21}$$

where now the unknown $\boldsymbol{\theta}_{k+1}$ enters nonlinearly via the parametrizations $\tilde{q}$. Thus, at each time step, a nonlinear optimization problem has to be solved, which can be computationally more expensive than solving the linear least-squares problems corresponding to the explicit schemes. The observation that explicit time integration schemes lead to computationally cheaper time steps than implicit ones is often the case in numerical analysis and scientific computing and reflects that Neural Galerkin schemes are rooted in numerical analysis. As a side remark, we state that constraints can be added to (19) and their discrete counterparts to conserve mass, momentum, Hamiltonians (Schwerdtner et al., 2023); see also Anderson and Farazmand (2022) for a method based on nonlinear parametrizations with conserving quantities.

4 Adaptive sampling in Neural Galerkin schemes

To numerically solve for the parameter $\boldsymbol{\theta}(t)$ in Neural Galerkin schemes, the objective of the least-squares problem (19) has to be numerically evaluated, which is the topic of this section.

4.1 The sampling challenge

The least-squares problem (19) and its discrete counterparts (20) and (21) are formulated with objectives that depend on the norm $\|\cdot\|_\nu$ with measure ν over the spatial domain Ω. To numerically optimize for the parameter $\boldsymbol{\theta}(t)$, it is necessary to numerically estimate the objectives and thus to evaluate the norm $\|\cdot\|_\nu$ via the inner product $\langle\cdot,\cdot\rangle_\nu$. If we can draw samples $\boldsymbol{x}_1, \ldots, \boldsymbol{x}_m \sim \nu$ from the distribution corresponding to the measure ν, then we can estimate the objective of (19) via a Monte Carlo estimator as

$$\frac{1}{m}\sum_{i=1}^{m}|\nabla_{\boldsymbol{\theta}}\tilde{q}(\boldsymbol{\theta}(t),\boldsymbol{x}_i)\cdot\dot{\boldsymbol{\theta}}(t)-f(t,\boldsymbol{x}_i,\tilde{q}(\boldsymbol{\theta}(t),\cdot))|^2\,. \tag{22}$$

Efficiently evaluating objective functions with Monte Carlo estimators such as (22) is a pervasive challenge in machine learning, where it is typically referred to as estimating the population loss with the empirical loss (Vapnik, 1991).

Analogously, in scientific computing, inner products have to be evaluated in, e.g., finite-element methods to assemble systems of equations corresponding to Galerkin conditions. Linear parametrizations such as linear combinations of basis functions centered on grid points allow in many cases to precompute inner products of local basis elements, which then can be re-used to efficiently evaluate inner products globally (Ern and Guermond, 2004). In case of nonlinear parametrizations, the superposition principle of linear approximations is lost and thus the inner product needs to be evaluated explicitly at each time step. The evaluation of the objective can be a major numerical runtime bottleneck when working with nonlinear parametrizations (Wen et al., 2024).

In certain limited cases, objectives can be evaluated analytically (Lubich, 2008; Lasser and Lubich, 2020). In other settings, it has been proposed to build on quadrature rules (Du and Zaki, 2021; Schwerdtner et al., 2023), which can be sufficient for problems with low-dimensional spatial domains. But even in low dimensions, local features such as wave fronts can make nonadaptive quadrature rules inefficient. Another common option therefore is falling back to Monte Carlo estimators that simply sample from the measure ν in the spatial domain to estimate the objectives. However, in particular for transport-dominated problems where local features such as wave fronts move through the spatial domain, a static sampling from a measure that is fixed over all times t means that these local features that change over time t need to be discovered without guidance; see Rotskoff et al. (2022); Bruna et al. (2024); Wen et al. (2024). For example, consider the linear advection equation with a Gaussian bump as initial condition, for which we plotted the solution field over the time-space domain in Fig. 1. A uniform sampling of the domain quickly reveals that many samples are required to resolve the local Gaussian bump as it is transported through the spatial domain. Thus, even though the nonlinear parametrization can be expressive, evaluating the objective function to numerically find parameters that realize the expressiveness can still be challenging.

4.2 Objectives with time-dependent measures

The works by Bruna et al. (2024); Wen et al. (2024) propose to consider time-dependent inner products $\langle\cdot,\cdot\rangle_{\mu_t}$ for formulating the least-squares problem (19) to obtain

$$\min_{\dot{\boldsymbol{\theta}}(t)\in\Theta} \|\nabla_{\boldsymbol{\theta}}\tilde{q}(\boldsymbol{\theta}(t),\cdot)\cdot\dot{\boldsymbol{\theta}}(t) - f(t,\cdot,\tilde{q}(\boldsymbol{\theta}(t),\cdot))\|_{\mu_t} \tag{23}$$

in continuous time. The inner product depends on time via the time-dependent measure μ_t, which is in contrast to the formulation (19) that builds on a measure ν that is fixed in time. As shown in Wen et al. (2024), if the parametrization is so rich that the residual is zero, then the optima of the objective of (19) are invariant to the measure as long as the measure has full support in the spatial domain. If a zero residual is not reached, then this insight serves as a heuristic when the norm of residual is small. Notice that we are indeed interested in cases where the norm of the residual is small because otherwise it would mean we incur large errors during time integration. Thus, in this sense, we are free to choose measure μ_t as long as it is fully supported on the spatial domain Ω.

The goal is now to choose a measure μ_t such that the objective of (23) can be estimated accurately with few samples. We set $\mu_t \propto \exp(-V_{\boldsymbol{\theta}(t),\dot{\boldsymbol{\theta}}(t)})$ with the potential

$$V_{\boldsymbol{\theta}(t),\dot{\boldsymbol{\theta}}(t)}(\boldsymbol{x}) = |r_t(\boldsymbol{\theta}(t),\dot{\boldsymbol{\theta}}(t),\boldsymbol{x})|^2, \tag{24}$$

where r_t is the residual function defined in (13). The measure μ_t defined via (24) distributes mass proportional to the magnitude of the residual, which agrees with the intuition that one should sample where the residual is high.

The time-dependent measure μ_t is then coupled to the dynamics of the parameter vector $\boldsymbol{\theta}(t)$ because the potential $V_{\boldsymbol{\theta}(t),\dot{\boldsymbol{\theta}}(t)}$ of μ_t depends on $\boldsymbol{\theta}(t)$ and vice versa. Thus, the parameter vector $\boldsymbol{\theta}(t)$ and the measure μ_t are updated together over time via the coupled dynamical system

$$\begin{aligned} M(t,\boldsymbol{\theta}(t))\dot{\boldsymbol{\theta}}(t) &= F(t,\boldsymbol{\theta}(t)), \\ \partial_t\mu_t &= \gamma\nabla\cdot(\nabla\mu_t + \mu_t\nabla V_{\boldsymbol{\theta}(t),\dot{\boldsymbol{\theta}}(t)}), \end{aligned} \tag{25}$$

where γ is a scaling parameter that controls how much faster the measure μ_t moves forward in time versus the parameter vector $\boldsymbol{\theta}(t)$; details can be found in Wen et al. (2024).

Note that other options than sampling proportional to the squared residual are possible such as sampling proportional to the magnitude of the solution field $\tilde{q}$, which can be useful if the solution field itself is a density as in the case when the PDE (1) is a Fokker-Planck equation.

4.3 A computational procedure for adaptive sampling based on particles and Stein variational gradient descent

Let us discretize the time-dependent measure μ_t introduced in the previous section with an empirical measure $\tilde{\mu}_t$ that depends on a set of particles $\{\boldsymbol{x}_i(t)\}_{i=1}^m$

at time t

$$\tilde{\mu}_t = \frac{1}{m}\sum\nolimits_{i=1}^{m} \delta_{\boldsymbol{x}_i(t)}\,.$$

We can evaluate gradients of the potential V, which means we can evaluate the score of the density of μ_t,

$$\nabla \log \mu_t\,.$$

Building on Stein variational gradient descent (SVGD) with a kernel $\mathcal{K}$ (Liu and Wang, 2016), we obtain a system of ODEs for the particles,

$$\frac{\mathrm{d}}{\mathrm{d}\tau}\boldsymbol{x}_i^{(\tau)}(t_k) = \mathbb{E}_{\boldsymbol{x}' \sim \tilde{\mu}_{t_k}^{(\tau)}}\left[\mathcal{K}(\boldsymbol{x}', \boldsymbol{x}_i^{(\tau)}(t_k))\nabla \log \mu_{t_k}(\boldsymbol{x}') + \nabla_1 \mathcal{K}(\boldsymbol{x}', \boldsymbol{x}_i^{(\tau)}(t_k))\right],$$

where τ is an artificial time in which the particles move, which is different from the physical time t. The relation between the artificial time τ and the physical time t is controlled by the parameter γ in the dynamics (25); see Wen et al. (2024) for details. Other sampling techniques than SVGD can be used such as Langevin and Markov chain Monte Carlo methods. In discrete time, it is beneficial to initialize the empirical measure at time t_{k+1} with the empirical measure from time t_k and thus particle methods are particularly well suited here.

4.4 Using adaptive samples in least-squares formulations of Neural Galerkin schemes

Recall the time-dependent sampling points, which we called particles in the previous section,

$$\boldsymbol{x}_1(t), \ldots, \boldsymbol{x}_m(t) \sim \mu_t\,.$$

We use the samples to form the batch gradient $\boldsymbol{J}_t(\boldsymbol{\theta}(t)) \in \mathbb{R}^{m \times p}$ as

$$\boldsymbol{J}_t(\boldsymbol{\theta}(t)) = \begin{bmatrix} — & \nabla_{\boldsymbol{\theta}} \tilde{q}(\boldsymbol{\theta}(t), \boldsymbol{x}_1(t))^T & — \\ & \vdots & \\ — & \nabla_{\boldsymbol{\theta}} \tilde{q}(\boldsymbol{\theta}(t), \boldsymbol{x}_m(t))^T & — \end{bmatrix}, \tag{26}$$

and the batch right-hand side $\boldsymbol{f}_t(\boldsymbol{\theta}(t)) \in \mathbb{R}^m$ as

$$\boldsymbol{f}_t(\boldsymbol{\theta}(t)) = \begin{bmatrix} f(t, \boldsymbol{x}_1(t), \tilde{q}(\boldsymbol{\theta}(t), \cdot)) \\ \vdots \\ f(t, \boldsymbol{x}_m(t), \tilde{q}(\boldsymbol{\theta}(t), \cdot)) \end{bmatrix}. \tag{27}$$

If we now discretize with the explicit Euler method in time, we obtain the regression problems

$$\min_{\delta\boldsymbol{\theta}_k \in \Theta} \|\hat{\boldsymbol{J}}_{t_k}(\boldsymbol{\theta}_k)\delta\boldsymbol{\theta}_k - \hat{\boldsymbol{f}}_{t_k}(\boldsymbol{\theta}_k)\|_2^2$$

over time steps $k = 0, \ldots, K-1$, where $\hat{\boldsymbol{J}}_{t_k}$ and $\hat{\boldsymbol{f}}_{t_k}$ are the time-discrete counterparts of $\boldsymbol{J}_t$ and $\boldsymbol{f}_t$, respectively. At each time step $k = 0, \ldots, K-1$, we update $\boldsymbol{\theta}_{k+1} = \boldsymbol{\theta}_k + \delta t_k \delta\boldsymbol{\theta}_k$. Notice that the time-step size $\delta t_k > 0$ can depend on the time step, which allows adaptively choosing the time-step size. We alternate between taking a time step to compute $\boldsymbol{\theta}_{k+1}$ from $\boldsymbol{\theta}_k$ and updating the set of particles from $\{\boldsymbol{x}(t_k)\}_{i=1}^m$ to $\{\boldsymbol{x}(t_{k+1})\}_{i=1}^m$; see Wen et al. (2024) for details.

4.5 Example: Fokker-Planck equations in moderately high dimensions

We report the experiment of Bruna et al. (2024) here. Consider a system of $d = 8$ particles that are attracted by a time-varying trap. The positions $X_1(t) \ldots, X_d(t) \in \mathbb{R}$ of the particles are governed by the stochastic differential equation

$$\mathrm{d}X_i = -(X_i - a(t))\mathrm{d}t - \frac{\alpha}{d}\sum_{j=1}^{d}(X_i - X_j)\mathrm{d}t + \sqrt{2\beta^{-1}}\mathrm{d}W_i\,, \quad i = 1, \ldots, d\,,$$

with the Wiener processes W_i, $\alpha = 1/4$, $\beta = 10^2$, and transport coefficient

$$a(t) = 5/4(\sin(\pi t) + 3/2)\,.$$

The density of the particle positions is governed by the Fokker-Planck equation over the $d = 8$ dimensional spatial domain. As the particles get trapped, the density concentrates, which leads to local features in high dimensions. We solve for the density function with Neural Galerkin schemes, with a shallow network with Gaussian units and 30 nodes per layer. Time is discretized with Runge-Kutta 4 and time-step size $\delta t = 10^{-3}$. The adaptive sampling is based on $m = 1000$ particles. The particles are adapted by sampling proportional to the magnitude of the current solution function. Details of the numerical setup are provided in Bruna et al. (2024).

Fig. 4 shows the positions of particles X_1, X_4, X_8 as well as of particles X_6, X_7, X_8. A benchmark solution is computed, which is indicated via black dots. The Neural Galerkin solution with adaptive sampling closely matches the benchmark, whereas the static sampling over the spatial domain leads too poor approximations of the particle positions. A quantitive comparison is shown in Fig. 5. The relative error in the mean particle position is on the order of 10^{-3} with adaptive sampling, whereas a static sampling leads to a relative error larger than one. The covariance of the particle distribution is approximated to a relative error of about 10^{-2} in this example.

Because we compute the density function, we can compute quantities of interest that require more than just the moments, as provided by Monte Carlo methods. We thus can compute the entropy of the system, which we show in Fig. 6a. Comparing to a Monte Carlo sampling with subsequent density estimation shows that the Neural Galerkin approximation leads to more accurate

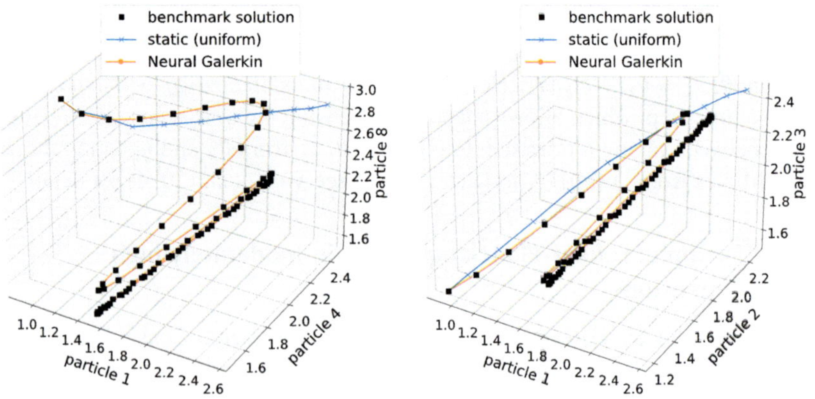

FIGURE 4 Neural Galerkin schemes with adaptive sampling accurately predict the positions of the physics particles, whereas using the same network as in the Neural Galerkin solution but with a static, uniform sampling fails to provide accurate predictions of the particle positions. (Figure from Bruna et al. (2024).)

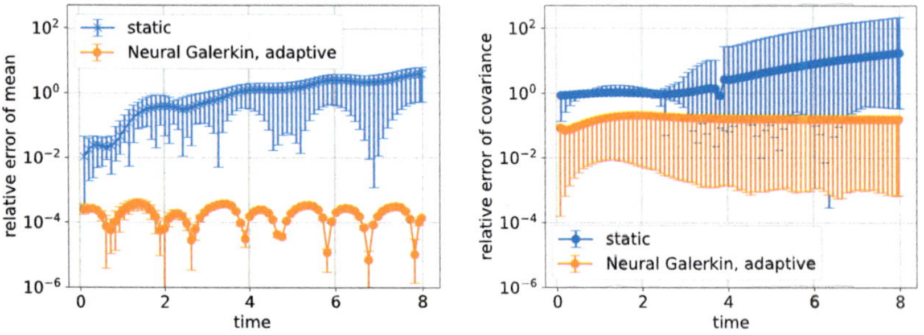

(a) mean of particle density

(b) covariance of particle density

FIGURE 5 Neural Galerkin schemes with adaptive sampling achieve accurate approximations of the mean and covariance of the particle density, whereas a static, uniform sampling leads to large relative errors. (Figure from Bruna et al., 2024.)

entropy estimates. We also show in Fig. 6b the entropy computed for an aharmonic trap; details in Bruna et al. (2024). It is known that such a system has an oscillating entropy, which agrees with the prediction of Neural Galerkin in this case.

5 Randomized sparse Neural Galerkin schemes

The work by Berman and Peherstorfer (2023) introduces randomized sparse Neural Galerkin (RSNG) schemes that update only sparse subsets of the components of $\boldsymbol{\theta}(t)$ and randomize which components of $\boldsymbol{\theta}(t)$ are updated. Updating only a sparse subset is sufficient because many nonlinear parametrizations lead to batch gradients $\boldsymbol{J}_t(\boldsymbol{\theta})$ that are of low rank, which indicates that components of

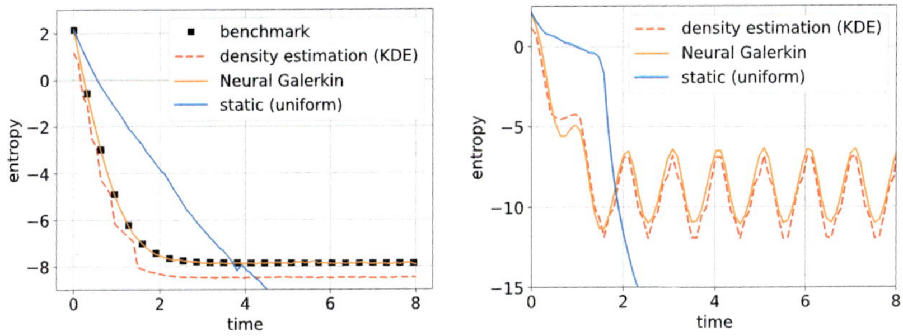

(a) entropy, harmonic trap (b) entropy, aharmonic trap

FIGURE 6 With Neural Galerkin schemes in this example, we approximate the particle density function rather than only moments. We thus can compute quantities such as the entropy. In contrast, to estimate the entropy with Monte Carlo-based methods, the density function has to be estimated, which is challenging and leads to less accurate results in this example than the Neural Galerkin schemes. (Figure from Bruna et al., 2024.)

the time derivative $\dot{\boldsymbol{\theta}}(t)$ are redundant and can be ignored for updating $\boldsymbol{\theta}(t)$ without losing expressiveness. Additionally, the randomization can be interpreted analogously to dropout (Srivastava et al., 2014; Sung et al., 2021; Zaken et al., 2022), which helps preventing overfitting locally in time.

5.1 The importance of the tangent spaces of the parametrization manifold

The tangent spaces play a critical role in the error of Neural Galerkin schemes (Zhang et al., 2024). To see this, let us assume there exists an $\epsilon : [0, T] \to [0, \infty)$ that bounds the projection error of the right-hand side function f onto the tangent space $\mathcal{T}_{\tilde{q}(\boldsymbol{\theta}(t),\cdot)}\mathcal{M}_\Theta$ at the current field $\tilde{q}(\boldsymbol{\theta}(t), \cdot)$,

$$\|f(t, \cdot, \tilde{q}(\boldsymbol{\theta}(t), \cdot)) - \mathrm{P}_{\boldsymbol{\theta}(t)} f(t, \cdot, \tilde{q}(\boldsymbol{\theta}(t), \cdot))\|_\nu \leq \epsilon(t)\,.$$

Under standard assumptions (Lubich, 2005, 2008; Lasser and Lubich, 2020; Zhang et al., 2024) on f and $\tilde{q}$, the error in the Neural Galerkin solution is bounded as

$$\|q(t, \cdot) - \tilde{q}(\boldsymbol{\theta}(t), \cdot)\|_\nu \leq \mathrm{e}^{Ct}\left(e_0 + \int_0^t \mathrm{e}^{-Cs}\epsilon(s)\mathrm{d}s\right), \tag{28}$$

where $C > 0$ is a constant independent of the time t and e_0 is the error in representing the initial condition in the parametrization,

$$e_0 = \|q(0, \cdot) - \tilde{q}(\boldsymbol{\theta}(0), \cdot)\|_\nu\,.$$

The bound (28) shows that the error is driven by the bound on the projection error $\epsilon(t)$ of projecting the right-hand side function onto the tangent space. The bound (28) underlines the importance of the tangent spaces.

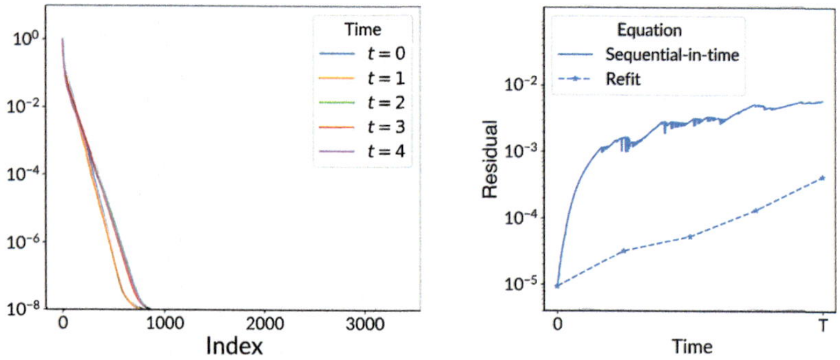

(a) low-rank batch gradients (b) local approximations versus re-fitting from scratch

FIGURE 7 The low-rankness of the batch gradient $\boldsymbol{J}_t$ in Neural Galerkin schemes motivates a randomized sparse version of Neural Galerkin (Berman and Peherstorfer, 2023) that updates only a subset of $s \ll p$ of the total number of p parameters in the deep network at each time step, which leads to speedups and has a regularization effect. (Figure from Berman and Peherstorfer, 2023.)

5.2 Tangent space collapse

Nonlinear parametrizations tend to lead to low-rank batch gradients $\boldsymbol{J}_t \in \mathbb{R}^{m \times p}$ as defined in (26). The space spanned by the columns of $\boldsymbol{J}_t$ therefore is of lower rank than $\min\{m, p\}$, which is a phenomenon that is referred to as tangent space collapse in Zhang et al. (2024); see Fig. 7a. Consequently, if the right-hand side function (or the batch right-hand side function (27)) cannot be represented well anymore in the spanned space, then this leads to a quick deterioration of the accuracy of Neural Galerkin solutions because of the bound given in (28).

The low-rankness of the batch gradient additionally means that there are multiple trajectories $\boldsymbol{\theta}(t)$ that solve the dynamical system (16) because the matrix $M(t, \boldsymbol{\theta}(t))$ can become singular. Adding a regularization term can enforce a unique solution; however, it is unclear if such a regularization can prevent the collapsing tangent space phenomenon and thus avoid the loss of expressiveness. In particular, only local moves in the parameter domain are made via the time stepping rather than global jumps, which means the scheme can get stuck in poor parameter regions despite regularization. Detailed discussions about the collapsing tangent phenomena and local moves are provided in Zhang et al. (2024); Berman and Peherstorfer (2023).

5.3 Leveraging low-rankness of batch gradients with subsampling

Let us recall the least-squares problem from (23), the batch gradient $\boldsymbol{J}_t(\boldsymbol{\theta}(t))$ defined in (26), and the batch right-hand side $\boldsymbol{f}_t(\boldsymbol{\theta}(t))$ given in (27). In continuous time, we obtain the least-squares problem

$$\min_{\dot{\boldsymbol{\theta}}(t) \in \mathbb{R}^p} \|\boldsymbol{J}_t(\boldsymbol{\theta}(t))\dot{\boldsymbol{\theta}}(t) - \boldsymbol{f}_t(\boldsymbol{\theta}(t))\|_2, \tag{29}$$

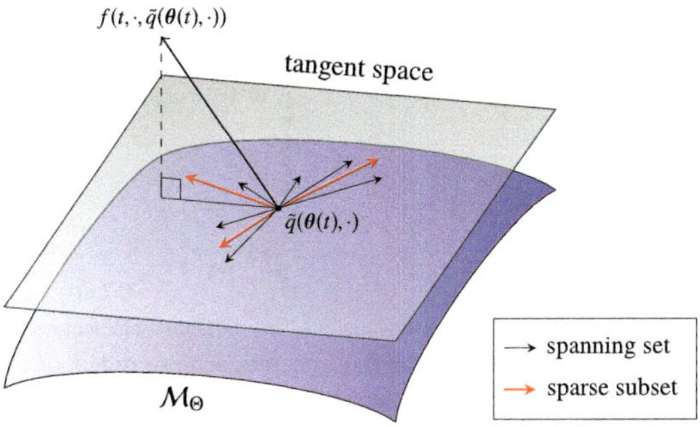

FIGURE 8 Randomized sparse Neural Galerkin schemes update only a subset of $s \ll p$ parameters of the deep network at each time step, which is motivated by the low-rankness of the batch gradient. (Figure from Berman and Peherstorfer, 2023.)

where the batch gradient and the batch right-hand side are obtained by sampling from the measure ν. The low-rankness of the batch gradient $\boldsymbol{J}_t$ means that considering a subset of the columns of $\boldsymbol{J}_t$ is sufficient in the least-squares problem (29); see Fig. 8. Motivated by this, the work by Berman and Peherstorfer (2023) proposes to update only $s \ll p$ components of $\boldsymbol{\theta}(t)$ over time t.

Consider therefore the subsampled parameter vector

$$\boldsymbol{\theta}_s(t) = \boldsymbol{S}_t^T \boldsymbol{\theta}(t),$$

where $\boldsymbol{S}_t \in \mathbb{R}^{p \times s}$ subselects s components out of the p components of $\boldsymbol{\theta}(t)$. The subsampled parameter vector is called $\boldsymbol{\theta}_s(t) \in \mathbb{R}^s$. It is then proposed in Berman and Peherstorfer (2023) to solve the sketched least-squares problem

$$\min_{\dot{\boldsymbol{\theta}}_s(t) \in \mathbb{R}^s} \|\boldsymbol{J}_t(\boldsymbol{\theta}(t))\boldsymbol{S}_t\dot{\boldsymbol{\theta}}_s(t) - \boldsymbol{f}_t(t,\cdot,\tilde{q}(\boldsymbol{\theta}(t),\cdot))\|_\nu, \tag{30}$$

where the sketching happens over the parameter vector $\boldsymbol{\theta}(t)$. Notice that the columns of the batch gradient that correspond to indices of the parameter vector $\boldsymbol{\theta}(t)$ that are not selected can be ignored because they are multiplied with zeros in the objective of (30). It is found in Berman and Peherstorfer (2023) that one way to prevent the tangent spaces from collapsing is to randomly select the s components that are updated over time t. Thus, the matrix $\boldsymbol{S}_t$ becomes a random matrix that uniformly draws s out of p components of $\boldsymbol{\theta}(t)$.

Updating only $s \ll p$ components per time step leads to lower runtimes because the number of unknowns in the randomly sketched least-squares problem (30) is lower than in the least-squares problem (29) that densely updates all components of $\boldsymbol{\theta}(t)$. In particular, if a direct, dense numerical linear algebra method is used to solve the least-squares problems such as based on the singular value decomposition or the QR decomposition, then the speedup scales quadratically in $1/s$.

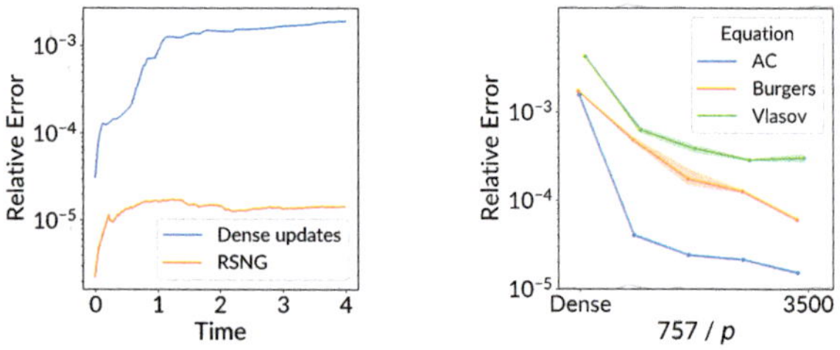

(a) comparison to dense updates (b) larger ambient networks

FIGURE 9 Randomized sparse Neural Galerkin schemes achieve orders of magnitude lower errors because the randomization of the parameters that are updated has a regularization effect, as plot (a) shows. Even though the number of parameters $s \ll p$ that is updated at each time step is low, randomized sparse Neural Galerkin schemes leverage the increasing expressiveness as the total number of parameters p of the network is increased, as plot (b) shows. (Figure from Berman and Peherstorfer, 2023.)

5.4 Numerical experiments with randomized sparse Neural Galerkin schemes

We report two numerical experiments from Berman and Peherstorfer (2023). Let us first consider the Allen-Cahn equation with a quadratic potential over a one-dimensional spatial domain as well as a fully connected feedforward network with rational activation functions; the details of the setup are given in Berman and Peherstorfer (2023). Fig. 9a compares the error obtained with a three-layer network for which all parameters are updated at each time step ("dense") versus a seven-layer network where only a sparse subset of the parameters are updated so that the size of the sparse subset is the same as the total number of parameters in the three-layer network. The plot in Fig. 9a shows that two orders of magnitude improvement in the relative error is achieved with the sparse updates compared to dense updates. Fig. 9b keeps the number of parameters s that are updated per time step fixed at $s = 757$ but increases the number of layers in the network from which the s parameters are taken. The plot shows that the error decays and thus the increasing expressiveness of the larger, ambient network can be leveraged even though the same number of parameters s are updated at each time step.

Besides reducing the error with sparse updates, we also obtain speedups compared to updating all parameters in the network; see Fig. 10. The bars with label "direct" correspond to dense updates with direct least-squares solvers based on the singular value decomposition. The bars with "iterative" correspond to iterative methods as described in Berman and Peherstorfer (2023). And the bars with "RSNG" show sparse updates with a direct least-squares solver, which achieves the lowest runtime by orders of magnitude.

6 Conclusions

Nonlinear parametrizations have been shown to achieve faster best-approximation error decays than linear parametrizations, which can be interpreted as them

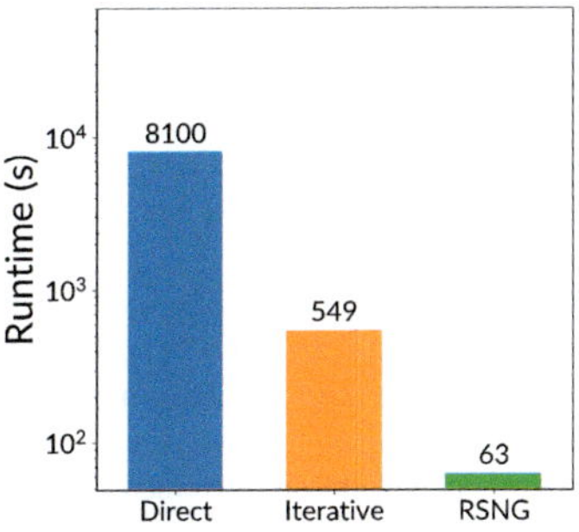

FIGURE 10 Because randomized sparse Neural Galerkin schemes update only $s \ll p$ out of the total of p parameters of the deep network, speedups of almost two orders of magnitude can be achieved compared to dense updates with direct least-squares solvers and about one order of magnitude compared to iterative least-squares solvers. (Figure from Berman and Peherstorfer, 2023.)

being more expressiveness with the same number of parameters than linear parametrizations. However, from a computational mathematics and numerical analysis perspective, major challenges remain regarding the development and analysis of numerical methods that can leverage that increased expressiveness and realize it numerically. In this survey, we discussed Neural Galerkin schemes that are motivated by the success of the method of lines from numerical analysis with linear parametrizations. The key building block is the Dirac-Frenkel variational principle to derive dynamical systems for the time-dependent, finite-dimensional parameter vectors of the nonlinear parametrizations. Another key ingredient is adaptive sampling to efficiently evaluate the residual objective. The numerical results in this survey and in the original publications where they have been presented first indicate that sequential-in-time approaches can be beneficial in terms of inherently providing causal numerical solutions, enforcing physics constraints, and requiring fewer parameters to well approximate solution fields compared to global-in-time methods. However, major challenges lie ahead to develop numerical methods for nonlinear parametrizations that are as rigorously analyzable, stable, robust, and efficient as today's numerical methods for linear parametrizations.

References

Amsallem, D., Zahr, M.J., Farhat, C., 2012. Nonlinear model order reduction based on local reduced-order bases. International Journal for Numerical Methods in Engineering 92 (10), 891–916.

Anderson, W., Farazmand, M., 2022. Evolution of nonlinear reduced-order solutions for PDEs with conserved quantities. SIAM Journal on Scientific Computing 44 (1), A176–A197.

Antoulas, A.C., 2005. Approximation of Large-Scale Dynamical Systems. SIAM.

Antoulas, A.C., Beattie, C.A., Gugercin, S., 2021. Interpolatory Methods for Model Reduction. SIAM.

Arbes, F., Greif, C., Urban, K., 2023. The Kolmogorov N-width for linear transport: exact representation and the influence of the data. arXiv:2305.00066.

Bar-Sinai, Y., Hoyer, S., Hickey, J., Brenner, M.P., 2019. Learning data-driven discretizations for partial differential equations. Proceedings of the National Academy of Sciences 116 (31), 15344–15349.

Barnett, J., Farhat, C., 2022. Quadratic approximation manifold for mitigating the Kolmogorov barrier in nonlinear projection-based model order reduction. Journal of Computational Physics 464, 111348.

Benner, P., Gugercin, S., Willcox, K., 2015. A survey of projection-based model reduction methods for parametric dynamical systems. SIAM Review 57 (4), 483–531.

Berg, J., Nyström, K., 2018. A unified deep artificial neural network approach to partial differential equations in complex geometries. Neurocomputing 317, 28–41.

Berger, M., Colella, P., 1989. Local adaptive mesh refinement for shock hydrodynamics. Journal of Computational Physics 82 (1), 64–84.

Berger, M.J., LeVeque, R.J., 1998. Adaptive mesh refinement using wave-propagation algorithms for hyperbolic systems. SIAM Journal on Numerical Analysis 35, 2298–2316.

Berman, J., Peherstorfer, B., 2023. Randomized sparse Neural Galerkin schemes for solving evolution equations with deep networks. In: Thirty-Seventh Conference on Neural Information Processing Systems.

Berman, J., Peherstorfer, B., 2024. CoLoRA: continuous low-rank adaptation for reduced implicit neural modeling of parameterized partial differential equations. arXiv:2402.14646.

Billaud-Friess, M., Nouy, A., 2017. Dynamical model reduction method for solving parameter-dependent dynamical systems. SIAM Journal on Scientific Computing 39 (4), A1766–A1792.

Binev, P., Cohen, A., Dahmen, W., DeVore, R., Petrova, G., Wojtaszczyk, P., 2011. Convergence rates for greedy algorithms in reduced basis methods. SIAM Journal on Mathematical Analysis 43 (3), 1457–1472.

Black, Felix, Schulze, Philipp, Unger, Benjamin, 2020. Projection-based model reduction with dynamically transformed modes. ESAIM: M2AN 54 (6), 2011–2043.

Bruna, J., Peherstorfer, B., Vanden-Eijnden, E., 2024. Neural Galerkin schemes with active learning for high-dimensional evolution equations. Journal of Computational Physics 496, 112588.

Buffa, A., Maday, Y., Patera, A.T., Prud'homme, C., Turinici, G., 2012. A priori convergence of the greedy algorithm for the parametrized reduced basis method. ESAIM: M2AN 46 (3), 595–603.

Cagniart, N., Maday, Y., Stamm, B., 2019. Model order reduction for problems with large convection effects. In: Chetverushkin, B.N., Fitzgibbon, W., Kuznetsov, Y., Neittaanmäki, P., Periaux, J., Pironneau, O. (Eds.), Contributions to Partial Differential Equations and Applications. Springer International Publishing, Cham, pp. 131–150.

Carlberg, K., 2015. Adaptive h-refinement for reduced-order models. International Journal for Numerical Methods in Engineering 102 (5), 1192–1210.

Cohen, A., DeVore, R., 2016. Kolmogorov widths under holomorphic mappings. IMA Journal of Numerical Analysis 36 (1), 1–12.

Cohen, A., DeVore, R., Petrova, G., Wojtaszczyk, P., 2022. Optimal stable nonlinear approximation. Foundations of Computational Mathematics 22 (3), 607–648.

Cohen, Albert, Dahmen, Wolfgang, DeVore, Ronald, Nichols, James, 2020. Reduced basis greedy selection using random training sets. ESAIM: M2AN 54 (5), 1509–1524.

Daubechies, I., DeVore, R., Foucart, S., Hanin, B., Petrova, G., 2022. Nonlinear approximation and (deep) ReLU networks. Constructive Approximation 55 (1), 127–172.

DeVore, R., Hanin, B., Petrova, G., 2021. Neural network approximation. Acta Numerica 30, 327–444.

DeVore, R.A., Howard, R., Micchelli, C., 1989. Optimal nonlinear approximation. Manuscripta Mathematica 63 (4), 469–478.

DeVore, R.A., Kyriazis, G., Leviatan, D., Tikhomirov, V.M., 1993. Wavelet compression and nonlinear n-widths. Advances in Computational Mathematics 1 (2), 197–214.

Dihlmann, M., Drohmann, M., Haasdonk, B., 2011. Model reduction of parametrized evolution problems using the reduced basis method with adaptive time-partitioning. In: Proc. of ADMOS 2011.

Dirac, P.A.M., 1930. Note on exchange phenomena in the Thomas atom. Mathematical Proceedings of the Cambridge Philosophical Society 26 (3), 376–385.
Dissanayake, M.W.M.G., Phan-Thien, N., 1994. Neural-network-based approximations for solving partial differential equations. Communications in Numerical Methods in Engineering 10 (3), 195–201.
Du, Y., Zaki, T.A., 2021. Evolutional deep neural network. Physical Review E 104 (4).
E, W., Han, J., Jentzen, A., 2017. Deep learning-based numerical methods for high-dimensional parabolic partial differential equations and backward stochastic differential equations. Communications in Mathematics and Statistics 5 (4), 349–380.
Eftang, J.L., Stamm, B., 2012. Parameter multi-domain 'hp' empirical interpolation. International Journal for Numerical Methods in Engineering 90 (4), 412–428.
Ehrlacher, V., Lombardi, D., Mula, O., Vialard, F.-X., 2020. Nonlinear model reduction on metric spaces. Application to one-dimensional conservative PDEs in Wasserstein spaces. ESAIM: M2AN 54 (6), 2159–2197.
Einkemmer, L., Hu, J., Wang, Y., 2021. An asymptotic-preserving dynamical low-rank method for the multi-scale multi-dimensional linear transport equation. Journal of Computational Physics 439, 110353.
Einkemmer, L., Lubich, C., 2019. A quasi-conservative dynamical low-rank algorithm for the Vlasov equation. SIAM Journal on Scientific Computing 41 (5), B1061–B1081.
Ern, A., Guermond, J.-L., 2004. Theory and Practice of Finite Elements. Springer.
Finzi, M.A., Potapczynski, A., Choptuik, M., Wilson, A.G., 2023. A stable and scalable method for solving initial value PDEs with neural networks. In: The Eleventh International Conference on Learning Representations.
Frenkel, J., 1934. Wave Mechanics, Advanced General Theory. Clarendon Press, Oxford.
Geelen, R., Balzano, L., Wright, S., Willcox, K., 2024. Learning physics-based reduced-order models from data using nonlinear manifolds. Chaos 34 (3), 033122.
Geelen, R., Willcox, K., 2022. Localized non-intrusive reduced-order modelling in the operator inference framework. Philosophical Transactions - Royal Society. Mathematical, Physical and Engineering Sciences 380 (2229), 20210206.
Geelen, R., Wright, S., Willcox, K., 2023. Operator inference for non-intrusive model reduction with quadratic manifolds. Computer Methods in Applied Mechanics and Engineering 403, 115717.
Gerbeau, J.-F., Lombardi, D., 2014. Approximated Lax pairs for the reduced order integration of nonlinear evolution equations. Journal of Computational Physics 265, 246–269.
Greif, C., Urban, K., 2019. Decay of the Kolmogorov N-width for wave problems. Applied Mathematics Letters 96, 216–222.
Han, J., Jentzen, A., E, W., 2018. Solving high-dimensional partial differential equations using deep learning. Proceedings of the National Academy of Sciences 115 (34), 8505–8510.
Hesthaven, J.S., Pagliantini, C., Rozza, G., 2022. Reduced basis methods for time-dependent problems. Acta Numerica 31, 265–345.
Huang, C., Duraisamy, K., 2023. Predictive reduced order modeling of chaotic multi-scale problems using adaptively sampled projections. Journal of Computational Physics 491, 112356.
Iollo, A., Lombardi, D., 2014. Advection modes by optimal mass transfer. Physical Review E 89, 022923.
Issan, O., Kramer, B., 2023. Predicting solar wind streams from the inner-heliosphere to earth via shifted operator inference. Journal of Computational Physics 473, 111689.
Jens, D.J.K., Eftang, L., Patera, A.T., 2011. An hp certified reduced basis method for parametrized parabolic partial differential equations. Mathematical and Computer Modelling of Dynamical Systems 17 (4), 395–422.
Kast, M., Hesthaven, J.S., 2023. Positional embeddings for solving PDEs with evolutional deep neural networks. arXiv:2308.03461.
Kaulmann, S., Flemisch, B., Haasdonk, B., Lie, K.A., Ohlberger, M., 2015. The localized reduced basis multiscale method for two-phase flows in porous media. International Journal for Numerical Methods in Engineering 102 (5), 1018–1040.

Khoo, Y., Lu, J., Ying, L., 2018. Solving for high-dimensional committor functions using artificial neural networks. Research in the Mathematical Sciences 6 (1), 1.

Kim, Y., Choi, Y., Widemann, D., Zohdi, T., 2022. A fast and accurate physics-informed neural network reduced order model with shallow masked autoencoder. Journal of Computational Physics 451, 110841.

Koch, O., Lubich, C., 2007. Dynamical low-rank approximation. SIAM Journal on Matrix Analysis and Applications 29 (2), 434–454.

Kochkov, D., Smith, J.A., Alieva, A., Wang, Q., Brenner, M.P., Hoyer, S., 2021. Machine learning–accelerated computational fluid dynamics. Proceedings of the National Academy of Sciences 118 (21).

Kramer, B., Peherstorfer, B., Willcox, K.E., 2024. Learning nonlinear reduced models from data with operator inference. Annual Review of Fluid Mechanics 56 (1), 521–548.

Lasser, C., Lubich, C., 2020. Computing quantum dynamics in the semiclassical regime. Acta Numerica 29, 229–401.

LeCun, Y., Bengio, Y., Hinton, G., 2015. Deep learning. Nature 521 (7553), 436–444.

Lee, K., Carlberg, K.T., 2020. Model reduction of dynamical systems on nonlinear manifolds using deep convolutional autoencoders. Journal of Computational Physics 404, 108973.

Li, Q., Lin, B., Ren, W., 2019. Computing committor functions for the study of rare events using deep learning. Journal of Chemical Physics 151 (5), 054112.

Liu, Q., Wang, D., 2016. Stein variational gradient descent: a general purpose Bayesian inference algorithm. In: Lee, D., Sugiyama, M., Luxburg, U., Guyon, I., Garnett, R. (Eds.), Advances in Neural Information Processing Systems, vol. 29. Curran Associates, Inc., pp. 2378–2386.

Lubich, C., 2005. On variational approximations in quantum molecular dynamics. Mathematics of Computation 74 (250), 765–779.

Lubich, C., 2008. From Quantum to Classical Molecular Dynamics: Reduced Models and Numerical Analysis, vol. 12. European Mathematical Society.

Maday, Y., Patera, A.T., Turinici, G., 2002. Global a priori convergence theory for reduced-basis approximations of single-parameter symmetric coercive elliptic partial differential equations. Comptes Rendus. Mathématique 335 (3), 289–294.

Maday, Y., Stamm, B., 2013. Locally adaptive greedy approximations for anisotropic parameter reduced basis spaces. SIAM Journal on Scientific Computing 35 (6), A2417–A2441.

Musharbash, E., Nobile, F., 2017. Symplectic dynamical low rank approximation of wave equations with random parameters. Mathicse Technical Report nr 18.2017.

Musharbash, E., Nobile, F., 2018. Dual dynamically orthogonal approximation of incompressible Navier Stokes equations with random boundary conditions. Journal of Computational Physics 354, 135–162.

Musharbash, E., Nobile, F., Zhou, T., 2015. Error analysis of the dynamically orthogonal approximation of time dependent random pdes. SIAM Journal on Scientific Computing 37 (2), A776–A810.

Ohlberger, M., Rave, S., 2013. Nonlinear reduced basis approximation of parameterized evolution equations via the method of freezing. Comptes Rendus. Mathématique 351 (23), 901–906.

Ohlberger, M., Rave, S., 2016. Reduced basis methods: success, limitations and future challenges. In: Proceedings of the Conference Algoritmy, pp. 1–12.

Orús, R., 2019. Tensor networks for complex quantum systems. Nature Reviews Physics 1 (9), 538–550.

Papapicco, D., Demo, N., Girfoglio, M., Stabile, G., Rozza, G., 2022. The neural network shifted-proper orthogonal decomposition: a machine learning approach for non-linear reduction of hyperbolic equations. Computer Methods in Applied Mechanics and Engineering 392, 114687.

Peherstorfer, B., 2020. Model reduction for transport-dominated problems via online adaptive bases and adaptive sampling. SIAM Journal on Scientific Computing 42, A2803–A2836.

Peherstorfer, B., 2022. Breaking the Kolmogorov barrier with nonlinear model reduction. Notices of the American Mathematical Society 69, 725–733.

Peherstorfer, B., Butnaru, D., Willcox, K., Bungartz, H.-J., 2014. Localized discrete empirical interpolation method. SIAM Journal on Scientific Computing 36 (1), A168–A192.

Peherstorfer, B., Willcox, K., 2015. Online adaptive model reduction for nonlinear systems via low-rank updates. SIAM Journal on Scientific Computing 37 (4), A2123–A2150.

Pinkus, A., 1985. n-Widths in Approximation Theory. Springer, Berlin, Heidelberg.

Qian, E., Kramer, B., Peherstorfer, B., Willcox, K., 2020. Lift & learn: physics-informed machine learning for large-scale nonlinear dynamical systems. Physica D. Nonlinear Phenomena 406, 132401.

Raissi, M., Perdikaris, P., Karniadakis, G., 2019. Physics-informed neural networks: a deep learning framework for solving forward and inverse problems involving nonlinear partial differential equations. Journal of Computational Physics 378, 686–707.

Ramezanian, D., Nouri, A.G., Babaee, H., 2021. On-the-fly reduced order modeling of passive and reactive species via time-dependent manifolds. Computer Methods in Applied Mechanics and Engineering 382, 113882.

Reiss, J., Schulze, P., Sesterhenn, J., Mehrmann, V., 2018. The shifted proper orthogonal decomposition: a mode decomposition for multiple transport phenomena. SIAM Journal on Scientific Computing 40 (3), A1322–A1344.

Romor, F., Stabile, G., Rozza, G., 2023. Non-linear manifold reduced-order models with convolutional autoencoders and reduced over-collocation method. Journal of Scientific Computing 94 (3), 74.

Rotskoff, G.M., Mitchell, A.R., Vanden-Eijnden, E., 2022. Active importance sampling for variational objectives dominated by rare events: consequences for optimization and generalization. In: Bruna, J., Hesthaven, J., Zdeborova, L. (Eds.), Proceedings of the 2nd Mathematical and Scientific Machine Learning Conference. In: Proceedings of Machine Learning Research, vol. 145. PMLR, pp. 757–780.

Rowley, C.W., Colonius, T., Murray, R.M., 2004. Model reduction for compressible flows using POD and Galerkin projection. Physica D. Nonlinear Phenomena 189 (1), 115–129.

Rowley, C.W., Marsden, J.E., 2000. Reconstruction equations and the Karhunen–Loève expansion for systems with symmetry. Physica D. Nonlinear Phenomena 142 (1), 1–19.

Rozza, G., Huynh, D.B.P., Patera, A.T., 2008. Reduced basis approximation and a posteriori error estimation for affinely parametrized elliptic coercive partial differential equations. Archives of Computational Methods in Engineering 15 (3), 229–275.

Rudy, S.H., Nathan Kutz, J., Brunton, S.L., 2019. Deep learning of dynamics and signal-noise decomposition with time-stepping constraints. Journal of Computational Physics 396, 483–506.

Sapsis, T.P., Lermusiaux, P.F., 2009. Dynamically orthogonal field equations for continuous stochastic dynamical systems. Physica D. Nonlinear Phenomena 238 (23), 2347–2360.

Schwerdtner, P., Peherstorfer, B., 2024. Greedy construction of quadratic manifolds for nonlinear dimensionality reduction and nonlinear model reduction. arXiv:2403.06732.

Schwerdtner, P., Schulze, P., Berman, J., Peherstorfer, B., 2023. Nonlinear embeddings for conserving Hamiltonians and other quantities with Neural Galerkin schemes. arXiv:2310.07485.

Sharma, H., Mu, H., Buchfink, P., Geelen, R., Glas, S., Kramer, B., 2023. Symplectic model reduction of Hamiltonian systems using data-driven quadratic manifolds. Computer Methods in Applied Mechanics and Engineering 417, 116402.

Singh, R., Uy, W., Peherstorfer, B., 2023. Lookahead data-gathering strategies for online adaptive model reduction of transport-dominated problems. Chaos.

Sirignano, J., Spiliopoulos, K., 2018. DGM: a deep learning algorithm for solving partial differential equations. Journal of Computational Physics 375, 1339–1364.

Srivastava, N., Hinton, G., Krizhevsky, A., Sutskever, I., Salakhutdinov, R., 2014. Dropout: a simple way to prevent neural networks from overfitting. Journal of Machine Learning Research 15 (56), 1929–1958.

Sung, Y.-L., Nair, V., Raffel, C., 2021. Training neural networks with fixed sparse masks. arXiv: 2111.09839.

Taddei, T., 2020. A registration method for model order reduction: data compression and geometry reduction. SIAM Journal on Scientific Computing 42 (2), A997–A1027.

Taddei, T., Perotto, S., Quarteroni, A., 2015. Reduced basis techniques for nonlinear conservation laws. ESAIM: M2AN 49 (3), 787–814.

Taddei, Tommaso, Zhang, Lei, 2021. Space-time registration-based model reduction of parameterized one-dimensional hyperbolic pdes. ESAIM: M2AN 55 (1), 99–130.

Uy, W., Wentland, C., Huang, C., Peherstorfer, B., 2024. Reduced models with nonlinear approximations of latent dynamics for model premixed flame problems. In: Rozza, G., Stabile, G., Gunzburger, M., D'Elia, M. (Eds.), Reduction, Approximation, Machine Learning, Surrogates, Emulators and Simulators. In: Lecture Notes in Computational Science and Engineering, vol. 151. Springer. In press.

Vapnik, V., 1991. Principles of risk minimization for learning theory. In: Moody, J., Hanson, S., Lippmann, R. (Eds.), Advances in Neural Information Processing Systems, vol. 4. Morgan-Kaufmann.

Verwer, J.G., Sanz-Serna, J.M., 1984. Convergence of method of lines approximations to partial differential equations. Computing 33 (3), 297–313.

Wang, Q., Ripamonti, N., Hesthaven, J.S., 2020. Recurrent neural network closure of parametric POD-Galerkin reduced-order models based on the Mori-Zwanzig formalism. Journal of Computational Physics 410, 109402.

Wen, Y., Vanden-Eijnden, E., Peherstorfer, B., 2024. Coupling parameter and particle dynamics for adaptive sampling in Neural Galerkin schemes. Physica D.

Zafarullah, A., 1970. Application of the method of lines to parabolic partial differential equations with error estimates. Journal of the ACM 17 (2), 294–302.

Zahr, M.J., Farhat, C., 2015. Progressive construction of a parametric reduced-order model for PDE-constrained optimization. International Journal for Numerical Methods in Engineering 102 (5), 1111–1135.

Zaken, E.B., Ravfogel, S., Goldberg, Y., 2022. BitFit: simple parameter-efficient fine-tuning for transformer-based masked language-models. arXiv:2106.10199.

Zhang, H., Chen, Y., Vanden-Eijnden, E., Peherstorfer, B., 2024. Sequential-in-time training of nonlinear parametrizations for solving time-dependent partial differential equations. arXiv: 2404.01145.

Zimmermann, R., Peherstorfer, B., Willcox, K., 2018. Geometric subspace updates with applications to online adaptive nonlinear model reduction. SIAM Journal on Matrix Analysis and Applications 39 (1), 234–261.

Chapter 9

Operator learning

Algorithms and analysis

Nikola B. Kovachki[a], Samuel Lanthaler[b,*], and Andrew M. Stuart[b]
[a]*NVIDIA, Santa Clara, CA, United States,* [b]*Computing and Mathematical Sciences, California Institute of Technology, Pasadena, CA, United States*
**Corresponding author: e-mail address: slanth@caltech.edu*

Contents

Abstract

Operator learning refers to the application of ideas from machine learning to approximate (typically nonlinear) operators mapping between Banach spaces of functions. Such operators often arise from physical models expressed in terms of partial differential equations (PDEs). In this context, such approximate operators hold great potential as efficient surrogate models to complement traditional numerical methods in many-query tasks. Being

Handbook of Numerical Analysis, Volume 25, ISSN 1570-8659, https://doi.org/10.1016/bs.hna.2024.05.009

data-driven, they also enable model discovery when a mathematical description in terms of a PDE is not available. This review focuses primarily on neural operators, built on the success of deep neural networks in the approximation of functions defined on finite dimensional Euclidean spaces. Empirically, neural operators have shown success in a variety of applications, but our theoretical understanding remains incomplete. This review article summarizes recent progress and the current state of our theoretical understanding of neural operators, focusing on an approximation theoretic point of view.

Keywords
Neural operator, Approximation theory

MSC Codes
68T01, 68Q17, 41A46

1 Introduction

This paper overviews algorithms and analysis related to the subject of operator learning: finding approximations of maps between Banach spaces, from data. Our focus is primarily on neural operators, which leverage the success of neural networks in finite dimensions; but we also cover related literature in the work specific to learning linear operators, and the use of Gaussian processes and random features. In Subsection 1.1 we discuss the motivation for our specific perspective on operator learning. Subsection 1.2 contains a literature review. Subsection 1.3 overviews the remainder of the paper.

1.1 High dimensional vectors versus functions

Many tasks in machine learning require operations on high dimensional tensors[1] arising, for example, from pixellation of images or from discretization of a real-valued mapping defined over a subset of $\mathbb{R}^d$. The main idea underlying the work that we overview in this paper is that it can be beneficial, when designing and analyzing algorithms in this context, to view these high dimensional vectors as functions $u : \mathfrak{D} \to \mathbb{R}^c$ defined on a domain $\mathfrak{D} \subset \mathbb{R}^d$. For example $(c, d) = (3, 2)$ for RGB images and $(c, d) = (1, 3)$ for a scalar field such as temperature in a room. Pixellation, or discretization, of $\mathfrak{D}$ will lead to a tensor with size scaling like N, the number of pixels or discretization points; N will be large and hence the tensor will be of high dimension. Working with data-driven algorithms designed to act on function u, rather than the high dimensional tensor, captures intrinsic properties of the problem, and not details related to specific pixellation or discretization; as a consequence models learned from data can be transfered from one pixellation or discretization level to another.

Consider the image shown in Fig. 1a, at four levels of resolution. As an RGB image it may be viewed as a vector in $\mathbb{R}^{3N}$ where N is the number of pixels. However by the time we reach the highest resolution (bottom right) it is

[1] "Tensor" here may be a vector, matrix or object with more than two indices.

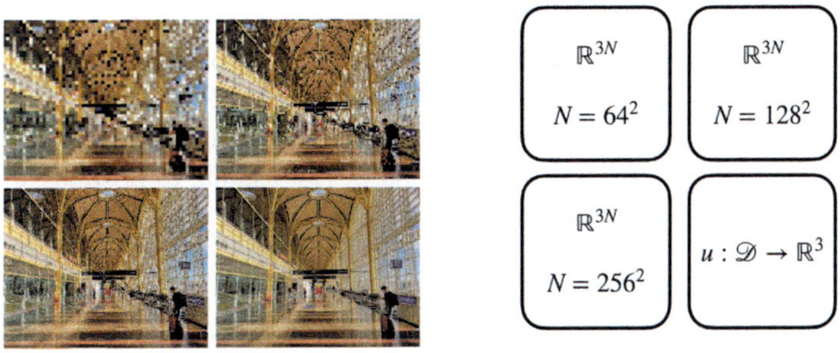

(a) Same images at different resolutions

(b) Different resolutions as vectors and (bottom right) as a function

FIGURE 1 High-dimensional vectors versus functions.

more instructive to view it as a function mapping $\mathfrak{D} := [0, 1]^2 \subset \mathbb{R}^2$ into $\mathbb{R}^3$. This idea is summarized in Fig. 1b. Even if the original machine learning task presents as acting on high dimensional tensor of dimension proportional to N, it is worth considering whether it may be formulated in the continuum limit $N \to \infty$, conceiving of algorithms in this setting, and only then approximating to finite dimension again to obtain practical algorithms. These ideas are illustrated in Figs. 2a and 2b.

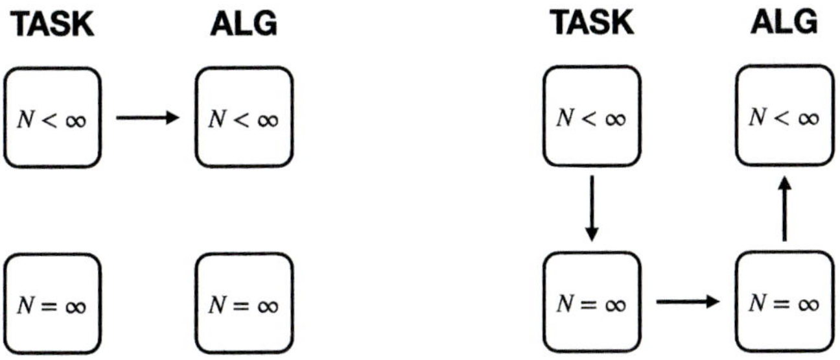

(a) Directly design algorithm at fixed resolution N

(b) Design algorithm at limit of infinite resolution

FIGURE 2 Discretize-then-approximate versus approximate-then-discretize.

1.2 Literature review

We give a brief overview of the literature in this field; greater depth, and more citations, are given in subsequent sections.

1.2.0.1 Algorithms on function space

The idea of conceiving algorithms in the continuum, and only then discretizing, is prevalent in numerous areas of computational science and engineering. For example in the field of PDE constrained optimization the relative merits of optimize-then-discretize, in comparison with discretize-then-optimize, are frequently highlighted (Hinze et al., 2008). In the field of Bayesian inverse problems (Kaipio and Somersalo, 2006) formulation on Banach space (Stuart, 2010) leads to new perspectives on algorithms, for example in MAP estimation (Klebanov and Sullivan, 2023). And sampling probability measures via MCMC can be formulated on Banach spaces (Cotter et al., 2012), leading to provably dimension independent convergence rates (Hairer et al., 2014).

1.2.0.2 Supervised learning on function space

Supervised learning (Goodfellow et al., 2016) rose to prominence in the context of the use of deep neural network (DNN) methods for classifying digits and then images. In such contexts the task is formulated as learning a mapping from a Euclidean space (pixellated image) to a set of finite cardinality. Such methods are readily generalized to regression in which the output space is also a Euclidean space. However, applications in science and engineering, such as surrogate modeling (Sacks et al., 1989) and scientific discovery (Raghu and Schmidt, 2020), often suggest supervised learning tasks in which input and/or output spaces are infinite dimensional; in particular they comprise spaces of functions defined over subsets of Euclidean space. We refer to the resulting methods, conceived to solve supervised learning tasks where the inputs and outputs are functions, as neural operators. Whilst there is earlier work on regression in function space (Ramsay and Dalzell, 1991), perhaps the earliest paper to conceive of neural network-based supervised learning between spaces of functions is by Chen and Chen (1995). This work was generalized in the seminal DeepONet paper (Lu et al., 2021). Concurrently with the development of DeepONet other methods were being developed, including methods based on model reduction (Bhattacharya et al., 2021) (PCA-Net) and on random features (Nelsen and Stuart, 2021). The random feature approach in Nelsen and Stuart (2021) included the use of manipulations in the Fourier domain, to learn the solution operator for viscous Burgers' equation, whose properties are well-understood in Fourier space. We also mention a related Fourier-based approach in Patel and Desjardins (2018); Patel et al. (2021). The idea of using the Fourier transform was exploited more systematically through development of the Fourier Neural Operator (FNO) (Li et al., 2021). The framework introduced in this paper has subsequently been generalized to work with sets of functions other than Fourier, such as wavelets (Tripura and Chakraborty, 2023), spherical harmonics (Bonev et al., 2023) and more general sets of functions (Benitez et al., 2023; Lanthaler et al., 2023a). The FNO architecture is related to convolutional neural networks, which have also been explored for operator learning, see e.g. Raonic et al. (2023a,b); Franco et al. (2023); Lippe et al. (2024), and Rahman et al. (2023); Gupta and Brandstetter (2022) for relevant work. We mention also similar developments in computer graphics where, in Ovsjanikov et al. (2012), a method is proposed based on projections onto the eigenfunctions of the Laplace-Beltrami operator and is subsequently extended in Yi et al. (2017); Litany et al. (2017); Sharp et al. (2022). At a more foundational level, recent work (Bartolucci et al., 2023) develops a theoretical framework to study neural operators, aiming to pinpoint theoretical distinctions between these infinite-dimensional architectures from conventional finite-dimensional approaches, based on a frame-theoretic notion of representation equivalence.

1.2.0.3 Approximation theory

The starting point for approximation theory is universal approximation. Such theory is overviewed in the finite dimensional setting in Pinkus (1999). It is developed systematically for neural operators in Kovachki (2022), work that also appeared in Kovachki et al. (2023). However, the first paper to study universal approximation, in the context of mappings between spaces of scalar-valued functions, is Chen and Chen (1995). This was followed by work extending their analysis to DeepONet (Lanthaler et al., 2022), analysis for the FNO (Kovachki et al., 2021) and analysis for PCA-Net in Bhattacharya et al. (2021), and a number of more recent contributions, e.g. Zhang et al. (2023); Hua and Lu (2023); Jin et al. (2022b); Castro (2023); Castro et al. (2022); Huang et al. (2024).

The paper by Lu et al. (2021) first introduced DeepONets and studied their practical application on a number of prototypical problems involving differential equations. The empirical paper by De Hoop et al. (2022) studies various neural operators from the perspective of the cost-accuracy trade-off, studying how many parameters, or how much data, is needed to achieve a given error. Such complexity issues are studied theoretically for DeepONet, in the context of learning the solution operator for the incompressible Navier-Stokes equation and several other PDEs, in Lanthaler et al. (2022), with analogous analysis for PCA-Net in Lanthaler (2023). In Marcati and Schwab (2023) the coefficient to solution map is studied for divergence form elliptic PDEs, and analyticity of the coefficient and the solution is exploited to study complexity of the resulting neural operators. The paper by Herrmann et al. (2022) studies complexity for the same problem, but exploits operator holomorphy. In Lanthaler and Stuart (2023) complexity is studied for Hamilton-Jacobi equations, using approximation of the underlying characteristic flow. The work by Furuya et al. (2024) discusses conditions under which neural operator layers are injective and surjective. The sample complexity of operator learning with DeepONet and related architectures is discussed in Liu et al. (2024). Out-of-distribution bounds are discussed in Benitez et al. (2023).

The paper by de Hoop et al. (2023) studies the learning of linear operators from data. This subject is developed for elliptic and parabolic equations, and in particular for the learning of Greens functions, in Gin et al. (2020); Boullé et al. (2022a,b); Boullé and Townsend (2023); Wang and Townsend (2023); Stepaniants (2023) and, for spectral properties of the Koopman operator, the solution operator for advection equations, in Kostic et al. (2022); Colbrook and Townsend (2024).

1.3 Overview of paper

In Section 2 we introduce operator learning as a supervised learning problem on Banach space; we formulate testing and training in this context, and provide an example from porous medium flow. Section 3 is devoted to definitions of the supervised learning architectures that we focus on in this paper: PCA-Net, DeepONet, the Fourier Neural Operator (FNO) and random features methods. Section 4 describes various aspects of universal approximation theories in the context of operator learning. In Section 5 we study complexity of these approximations, including discussion of linear operator learning; specifically we study questions such as how many parameters, or how much data, is required to achieve an operator approximation with a specified level of accuracy; and what properties of the operator can be exploited to reduce complexity? We summarize and conclude in Section 6.

2 Operator learning

In Subsection 2.1 we define supervised learning, followed in subsection 2.2 by discussion of the topic in the specific case of operator learning. Subsection 2.3 is devoted to explaining how the approximate operator is found from data (training) and how it is evaluated (testing). Subsection 2.4 describes how latent structure can be built into operator approximation, and learned from data. Subsection 2.5 contains an example from parametric partial differential equations (PDEs) describing flow in a porous medium.

2.1 Supervised learning

The objective of supervised learning is to determine an underlying mapping $\Psi^\dagger : \mathcal{U} \to \mathcal{V}$ from samples[2]

$$\{u_n, \Psi^\dagger(u_n)\}_{n=1}^N, \quad u_n \sim \mu. \tag{2.1}$$

Here the probability measure μ is supported on $\mathcal{U}$. Often supervised learning is formulated by use of the data model

$$\{u_n, v_n\}_{n=1}^N, \quad (u_n, v_n) \sim \pi, \tag{2.2}$$

where the probability measure π is supported on $\mathcal{U} \times \mathcal{V}$. The data model (2.1) is a special case which is sufficient for the exposition in this article.

In the original applications of supervised learning $\mathcal{U} = \mathbb{R}^{d_x}$ and $\mathcal{V} = \mathbb{R}^{d_y}$ (regression) or $\mathcal{V} = \{1, \ldots, K\}$ (classification). We now go beyond this setting.

---

[2] Note that from now on N denotes the data volume (and not the size of a finite dimensional problem as in Subsection 1.1).

2.2 Supervised learning of operators

In many applications arising in science and engineering it is desirable to consider a generalization of the finite-dimensional setting to separable Banach spaces $\mathcal{U}$, $\mathcal{V}$ of vector-valued functions:

$$\mathcal{U} = \{u : \mathfrak{D}_x \to \mathbb{R}^{d_i}\}, \quad \mathfrak{D}_x \subseteq \mathbb{R}^{d_x}$$
$$\mathcal{V} = \{v : \mathfrak{D}_y \to \mathbb{R}^{d_o}\}, \quad \mathfrak{D}_y \subseteq \mathbb{R}^{d_y}.$$

Given data (2.1) we seek to determine an approximation to $\Psi^\dagger : \mathcal{U} \to \mathcal{V}$ from within a family of parameterized functions

$$\Psi : \mathcal{U} \times \Theta \mapsto \mathcal{V}.$$

Here $\Theta \subseteq \mathbb{R}^p$ denotes the parameter space from which we seek the optimal choice of parameter, denoted $\theta^\star$. Parameter $\theta^\star$ may be chosen in a data-driven fashion to optimize the approximation of $\Psi^\dagger$ by $\Psi(\cdot\,;\theta^\star)$; see the next subsection. In Section 4 we will discuss the choice of $\theta^\star$ from the perspective of approximation theory.

2.3 Training and testing

The data (2.1) is used to train the model Ψ; that is, to determine a choice of θ. To this end we introduce an error, or relative error, measure $r : \mathcal{V}' \times \mathcal{V}' \to \mathbb{R}^+$. Here $\mathcal{V}'$ is another Banach space containing the range of $\Psi^\dagger$ and $\Psi(\cdot\,;\theta)$. Typical choices for r include the error

$$r(v_1, v_2) = \|v_1 - v_2\|_{\mathcal{V}'} \tag{2.3}$$

and, for $\varepsilon \in (0, \infty)$, one of the relative errors

$$r(v_1, v_2) = \frac{\|v_1 - v_2\|_{\mathcal{V}'}}{\varepsilon + \|v_1\|_{\mathcal{V}'}}, \quad \text{or} \quad r(v_1, v_2) = \frac{\|v_1 - v_2\|_{\mathcal{V}'}}{\max\{\varepsilon, \|v_1\|_{\mathcal{V}'}\}}. \tag{2.4}$$

Now let μ_N be the empirical measure

$$\mu_N = \frac{1}{N}\sum_{n=1}^{N} \delta_{u_n}.$$

Then the parameter $\theta^\star$ is determined from

$$\theta^* = \text{argmin}_\theta\, \mathcal{R}_N(\theta), \quad \mathcal{R}_N(\theta) := \mathbb{E}^{u\sim\mu_N}\left[r\big(\Psi^\dagger(u), \Psi(u;\theta)\big)^q\right],$$

for some positive q, typically $q = 1$. Function $\mathcal{R}_N(\theta)$ is known as the empirical risk; also of interest is the expected (or population) risk

$$\mathcal{R}_\infty(\theta) := \mathbb{E}^{u\sim\mu}\left[r\big(\Psi^\dagger(u), \Psi(u;\theta)\big)^q\right].$$

Note that $\mathcal{R}_\infty(\theta)$ requires knowledge of data in the form of the entire probability measure μ.

Once trained, models are typically tested by evaluating the error

$$\text{error} := \mathbb{E}^{u \sim \mu'} \left[r\left(\Psi^\dagger(u), \Psi(u;\theta)\right)^q \right].$$

Here μ' is defined on the support of μ. For computational purposes the measure μ' is often chosen as another empirical approximation of μ, independently of μ_N; other empirical measures may also be used. For theoretical analyses μ' may be chosen equal to μ, but other choices may also be of interest in determining the robustness of the learned model; see, for example, Benitez et al. (2023).

2.4 Finding latent structure

Behind many neural operators is the extraction of latent finite dimensional structure, as illustrated in Fig. 3. Here we have two encoder/decoder pairs on $\mathcal{U}$ and $\mathcal{V}$, namely

$$G_{\mathcal{U}} \circ F_{\mathcal{U}} \approx I_{\mathcal{U}}, \quad G_{\mathcal{V}} \circ F_{\mathcal{V}} \approx I_{\mathcal{V}}$$

where $I_{\mathcal{U}}, I_{\mathcal{V}}$ are the identity maps on $\mathcal{U}$ and $\mathcal{V}$ respectively. Then φ is chosen so that

$$G_{\mathcal{V}} \circ \varphi \circ F_{\mathcal{U}} \approx \Psi^\dagger.$$

The map $F_{\mathcal{U}}$ extracts a finite dimensional latent space from the input Banach space while the map $G_{\mathcal{V}}$ returns from a second finite dimensional latent space to the output Banach space. These encoder-decoder pairs can be learned, reducing the operator approximation to a finite dimensional problem.

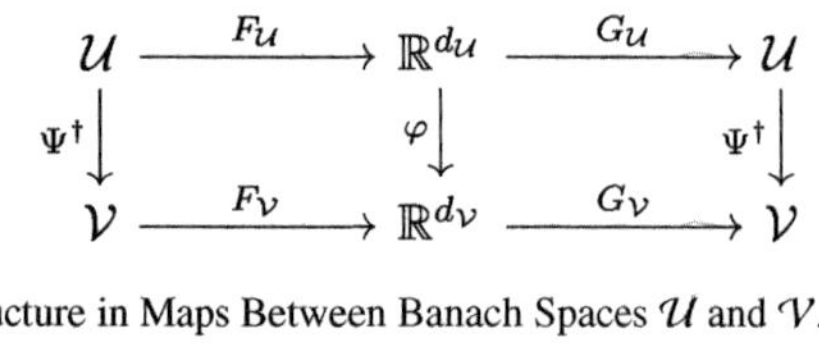

FIGURE 3 Latent Structure in Maps Between Banach Spaces $\mathcal{U}$ and $\mathcal{V}$.

2.5 Example (fluid flow in a porous medium)

We consider the problem of finding the piezometric head v from permeability a in a porous medium assumed to be governed by the Darcy Law in domain $\mathfrak{D} \subset \mathbb{R}^2$. This results in the need to solve the PDE

$$\begin{cases} -\nabla \cdot (a \nabla v) = f, & z \in \mathfrak{D} \\ v = 0, & z \in \partial\mathfrak{D}. \end{cases} \tag{2.5}$$

Here we consider $f \in H^{-1}(\mathfrak{D})$ to be given and fixed. The operator of interest[3] $\Psi^\dagger : a \mapsto v$ then maps from a subset of the Banach space $L^\infty(\mathfrak{D})$ into $H_0^1(\mathfrak{D})$. An example of a typical input-output pair is shown in Figs. 4a, 4b. Because the equation requires strictly positive a, in order to be well-defined mathematically and to be physically meaningful, the probability measure μ must be chosen carefully. Furthermore, from the point of view of approximation theory, it is desirable that the space $\mathcal{U}$ is separable; for this reason it is often chosen to be $L^2(\mathfrak{D})$ and the measure supported on functions a satisfying a positive lower bound and a finite upper bound. Draws from such a measure are in $L^\infty(\mathfrak{D})$ and satisfy the necessary positivity and boundedness inequalities required for a solution to the Darcy problem to exist (Evans, 2010).

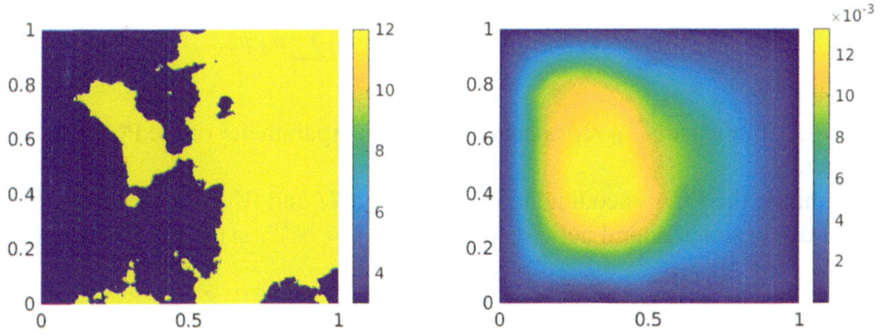

(a) Input a

(b) Output v

FIGURE 4 Illustrative input-output function pair for the Darcy problem.

3 Specific supervised learning architectures

Having reviewed the general philosophy behind operator learning, we next aim to illustrate how this methodology is practically implemented. To this end, we review several representative proposals for neural operator architectures, below.

3.1 PCA-Net

The PCA-Net architecture was proposed as an operator learning framework in Bhattacharya et al. (2021), anticipated in Hesthaven and Ubbiali (2018); Swischuk et al. (2019). In the setting of PCA-Net, $\Psi^\dagger : \mathcal{U} \to \mathcal{V}$ is an operator mapping between Hilbert spaces $\mathcal{U}$ and $\mathcal{V}$, and inputs are drawn from a probability measure μ on $\mathcal{U}$. Principal component analysis (PCA) is employed to obtain data-driven encoders and decoders, which are combined with a neural network mapping to give rise to the PCA-Net architecture.

[3] We use the notation a for input functions here, because it is a commonly adopted notation in applications to porous medium flow.

The encoder is defined from PCA basis functions on the input space $\mathcal{U}$, computed from the covariance under μ: the encoder $F_{\mathcal{U}}$ is determined by projection onto the first $d_{\mathcal{U}}$ PCA basis functions $\{\phi_j\}_{j=1}^{d_{\mathcal{U}}}$. The encoding dimension $d_{\mathcal{U}}$ represents a hyperparameter of the architecture. The resulting encoder is given by a linear mapping to the PCA coefficients,

$$F_{\mathcal{U}} : \mathcal{U} \to \mathbb{R}^{d_{\mathcal{U}}}, \quad F_{\mathcal{U}}(u) = Lu := \{\langle \phi_j, u \rangle\}_{j=1}^{d_{\mathcal{U}}}.$$

The decoder $G_{\mathcal{V}}$ on $\mathcal{V}$ is similarly obtained from PCA under the push-forward measure $\Psi^{\dagger}_{\#}\mu$. Denoting by $\{\psi_j\}_{j=1}^{d_{\mathcal{V}}}$ the first $d_{\mathcal{V}}$ basis functions under this push-forward, the PCA-Net decoder is defined by an expansion in this basis, i.e.

$$G_{\mathcal{U}} : \mathbb{R}^{d_{\mathcal{V}}} \to \mathcal{V}, \quad G_{\mathcal{U}}(\alpha) = \sum_{j=1}^{d_{\mathcal{V}}} \alpha_j \psi_j.$$

The PCA dimension $d_{\mathcal{V}}$ represents another hyperparameter of the PCA-Net architecture.

Finally, the PCA encoding and decoding on $\mathcal{U}$ and $\mathcal{V}$ are combined with a finite-dimensional neural network $\alpha : \mathbb{R}^{d_{\mathcal{U}}} \times \Theta \to \mathbb{R}^{d_{\mathcal{V}}}$, $w \mapsto \alpha(w;\theta)$ where

$$\alpha(w;\theta) := (\alpha_1(w;\theta), \ldots, \alpha_{d_{\mathcal{V}}}(w;\theta)),$$

parametrized by $\theta \in \Theta$. This results in an operator $\Psi_{PCA} : \mathcal{U} \to \mathcal{V}$, of the form

$$\Psi_{PCA}(u;\theta)(y) = \sum_{j=1}^{d_{\mathcal{V}}} \alpha_j(Lu;\theta)\psi_j(y), \quad \forall u \in \mathcal{U}, \qquad y \in \mathfrak{D}_y$$

which defines the PCA-Net architecture. Hyperparameters include the dimensions of PCA $d_{\mathcal{U}}$, $d_{\mathcal{V}}$, and additional hyperparameters determining the neural network architecture of α. In practice we are given samples of input-/output-function pairs with u_j sampled i.i.d. from μ:

$$\{(u_1, v_1), \ldots, (u_N, v_N)\},$$

where $v_j := \Psi^{\dagger}(u_j)$. Then the PCA basis functions are determined from the covariance under an empirical approximation μ_N of μ, and its pushforward under $\Psi^{\dagger}$. The same data is then used to train neural network parameter $\theta \in \Theta$ defining $\alpha(w;\theta)$.

3.2 DeepONet

The DeepONet architecture was first proposed as a practical operator learning framework in Lu et al. (2021), building on early work by Chen and Chen (1995). Similar to PCA-Net, the DeepONet architecture is also defined in terms of an encoder $F_{\mathcal{U}}$ on the input space, a finite-dimensional neural network α between the latent finite-dimensional spaces, and a decoder $G_{\mathcal{V}}$ on the output space. To simplify notation, we only summarize this architecture for real-valued input and output functions. Extension to operators mapping between vector-valued functions is straightforward.

The encoder in the DeepONet architecture is given by a general linear map L,

$$F_{\mathcal{U}} : \mathcal{U} \to \mathbb{R}^{d_{\mathcal{U}}}, \quad F_{\mathcal{U}}(u) = Lu.$$

Popular choices for the encoding include a mapping to PCA coefficients, or could comprise pointwise observations $\{u(x_\ell)\}_{\ell=1}^{d_{\mathcal{U}}}$ at a predetermined set of so-called sensor points x_ℓ. Another alternative of note is active subspaces (Zahm et al., 2020), which combine information about the input distribution and forward mapping through its gradient. This approach has been explored for function encoding in scientific ML in O'Leary-Roseberry et al. (2024); Luo et al. (2023).

The decoder in the DeepONet architecture is given by expansion with respect to a neural network basis. Given a neural network $\psi : \mathfrak{D}_y \times \Theta_\psi \to \mathbb{R}^{d_{\mathcal{V}}}$, which defines a parametrized function from the domain $\mathfrak{D}_y$ of the output functions to $\mathbb{R}^{d_{\mathcal{V}}}$, the DeepONet decoder is defined as

$$G_{\mathcal{V}} : \mathbb{R}^{d_{\mathcal{V}}} \to \mathcal{V}, \quad G_{\mathcal{V}}(\alpha) = \sum_{j=1}^{d_{\mathcal{V}}} \alpha_j \psi_j.$$

The above encoder and decoders on $\mathcal{U}$ and $\mathcal{V}$ are combined with a finite-dimensional neural network $\alpha : \mathbb{R}^{d_{\mathcal{U}}} \times \Theta_\alpha \to \mathbb{R}^{d_{\mathcal{V}}}$, to define the parametrized DeepONet,

$$\Psi_{DEEP}(u;\theta)(y) = \sum_{j=1}^{d_{\mathcal{V}}} \alpha_j(Lu;\theta_\alpha)\psi_j(y;\theta_\psi), \quad \forall u \in \mathcal{U}, \qquad y \in \mathfrak{D}_y.$$

This architecture is specified by choice of the linear encoding L, and choice of neural network architectures for α and ψ. Following Lu et al. (2021) the network α is conventionally referred to as the "branch-net" (often denoted b or β), while ψ is referred to as the "trunk-net" (often denoted t or τ). The combined parameters $\theta = \{\theta_\alpha, \theta_\psi\}$ of these neural networks are learned from data of input- and output-functions.

3.3 FNO

The Fourier Neural Operator (FNO) architecture was introduced in Li et al. (2021); Kovachki et al. (2023). In contrast to the PCA-Net and DeepONet architectures above, FNO is not based on an approach that combines an encoding/decoding to a finite-dimensional latent space with a finite-dimensional neural network. Instead, neural operators such as the FNO generalize the structure of finite-dimensional neural networks to a function space setting, as summarized below. We will assume that the domain of the input and output functions $\mathfrak{D}_x = [0, 2\pi]^d$ can be identified with the d-dimensional periodic torus.

FNO is an operator architecture of the form,

$$\Psi_{FNO}(u; \theta) = \mathcal{Q} \circ \mathcal{L}_L \circ \cdots \mathcal{L}_2 \circ \mathcal{L}_1 \circ \mathcal{R}(u), \ \forall u \in \mathcal{U},$$
$$\mathcal{L}_\ell(v)(x; \theta) = \sigma\big(W_\ell v(x) + b_\ell + \mathcal{K}(v)(x; \gamma_\ell)\big).$$

It comprises of input and output layers $\mathcal{Q}$, $\mathcal{R}$, given by a pointwise composition with either a shallow neural network or a linear transformation, and several hidden layers $\mathcal{L}_\ell$.

Upon specification of a "channel width" d_c, the ℓ-the hidden layer takes as input a vector-valued function $v : x \mapsto v(x) \in \mathbb{R}^{d_c}$ and outputs another vector-valued function $\mathcal{L}_\ell(v) : x \mapsto \mathcal{L}_\ell(v)(x) \in \mathbb{R}^{d_c}$.[4] Each hidden layer involves an affine transformation

$$v(x) \mapsto w(x) := W_\ell v(x) + b_\ell + \mathcal{K}(v)(x; \gamma_\ell),$$

and a pointwise composition with a standard activation function

$$w(x) \mapsto \sigma(w(x)),$$

where σ could e.g. be the rectified linear unit or a smooth variant thereof.

In the affine transformation, the matrix-vector pair (W_ℓ, b_ℓ) defines a pointwise affine transformation of the input $v(x)$, i.e. multiplication by matrix W_ℓ and adding a bias b_ℓ. $\mathcal{K}$ is a convolutional integral operator, parameterized by γ_ℓ,

$$\mathcal{K}(v)(x; \gamma_\ell) = \int_{\mathfrak{D}_x} \kappa(x - y; \gamma_\ell) v(y)\, dy,$$

with $\kappa(\,\cdot\,; \gamma_\ell)$ a matrix-valued integral kernel. The convolutional operator can be conveniently evaluated via the Fourier transform $\mathcal{F}$, giving rise to a matrix-valued Fourier multiplier,

$$\mathcal{K}(v)(x; \gamma_\ell) = \mathcal{F}^{-1}(\mathcal{F}(\kappa(\,\cdot\,; \gamma_\ell))\mathcal{F}(v)),$$

where the Fourier transform is computed componentwise, and given by $\mathcal{F}(v)(k) = \int_{\mathfrak{D}_x} v(x) e^{-ik\cdot x}\, dx$. To be more specific, if we write $\kappa(x) = (\kappa_{ij}(x))_{ij=1}^{d_c}$ in terms of its components, and if $\widehat{\kappa}_{k,ij}$ denotes the k-th Fourier coefficient of $\kappa_{ij}(x)$, then the i-th component $\mathcal{K}(v)_i$ of the vector-valued output function $\mathcal{K}(v)$ is given by

$$[\mathcal{K}(v)_i](x; \gamma_\ell) = \frac{1}{(2\pi)^d} \sum_{k \in \mathbb{Z}^d} \left(\sum_{j=1}^{d_c} \widehat{\kappa}_{k,ij} \mathcal{F}(v_j)(k) \right) e^{ik\cdot x}.$$

[4] Channel width can change from layer to layer; we simplify the exposition by fixing it.

Here, the inner sum represents the action of $\widehat{\kappa} = \mathcal{F}(\kappa)$ on $\mathcal{F}(v)$, and the outer sum is the inverse Fourier transform $\mathcal{F}^{-1}$. The Fourier coefficients $\widehat{\kappa}_{k,ij}$ represent the tunable parameters of the convolutional operator. In a practical implementation, a Fourier cut-off $k_{\max}$ is introduced and the sum over k is restricted to Fourier wavenumbers $|k|_{\ell^\infty} \le k_{\max}$, with $|\cdot|_{\ell^\infty}$ the ℓ^∞-norm, resulting in a finite number of tunable parameters $\gamma_\ell = \{\widehat{\kappa}_{k,ij} : |k|_{\ell^\infty} \le k_{\max},\ i, j = 1, \dots, d_c\}$.

To summarize, the FNO architecture is defined by

1. an input layer $\mathcal{R}$, given by pointwise composition of the input function with a shallow neural network or a linear transformation,
2. hidden layers $\mathcal{L}_1, \dots, \mathcal{L}_L$ involving, for each $\ell = 1, \dots, L$, matrix W_ℓ, bias b_ℓ and convolutional operator $\mathcal{K}(\cdot\,; \gamma_\ell)$ with parameters γ_ℓ identified with the corresponding Fourier multipliers $\widehat{\kappa}_{k,ij}$,
3. an output layer $\mathcal{Q}$, given by pointwise composition with a shallow neural network or a linear transformation.

The composition of these layers defines a parametrized operator $u \mapsto \Psi_{FNO}(u; \theta)$, where θ collects parameters from (1), (2) and (3). The parameters contained in θ are to be trained from data. The hyperparameters of FNO include the channel width d_c, the Fourier cut-off $k_{\max}$, the depth L and additional hyperparameters specifying the input and output layers $\mathcal{R}$ and $\mathcal{Q}$.

In theory, the FNO is formulated directly on function space and does not involve a reduction to a finite-dimensional latent space. In a practical implementation, it is usually discretized by identifying the input and output functions with their point-values on an equidistant grid. In this case, the discrete Fourier transform can be conveniently evaluated using the fast Fourier transform algorithm (FFT), and straightforward implementation in popular deep learning libraries is possible.

3.4 Random features method

The operator learning architectures above are usually trained from data using stochastic gradient descent. Whilst this shows great empirical success, the inability to analyze the optimization algorithms used by practitioners makes it difficult to make definitive statements about the networks that are trained in practice. Nelsen and Stuart (2021) have proposed a randomized alternative, by extending the random features methodology (Rahimi and Recht, 2007) to a function space setting; this methodology has the advantage of being trainable through solution of a quadratic optimization problem.

The random feature model (RFM) requires specification of a parametrized operator $\psi : \mathcal{U} \times \Gamma \to \mathcal{V}$ with parameter set Γ, and a probability measure ν on the parameters Γ. Each draw $\gamma \sim \nu$ specifies a random feature $\psi(\cdot\,; \gamma) : \mathcal{U} \to \mathcal{V}$, i.e. a random operator. Given iid samples $\gamma_1, \dots, \gamma_M \sim \nu$, the RFM operator is then defined as

$$\Psi_{RFM}(u;\theta)(y) = \sum_{j=1}^{M} \theta_j \psi(u;\gamma_j)(y) \quad \forall u \in \mathcal{U},\ y \in \mathfrak{D}_y; \quad \gamma_j \text{ i.i.d.}.$$

Here $\theta_1, \ldots, \theta_M$ are scalar parameters. In contrast to the methodologies outlined above, the random feature model keeps the randomly drawn parameters $\gamma_1, \ldots, \gamma_M$ fixed, and only optimizes over the coefficient vector $\theta = (\theta_1, \ldots, \theta_M)$. With conventional loss functions, the resulting optimization over θ is convex, allowing for efficient and accurate optimization and a unique minimizer to be determined.

A suitable choice of random features is likely problem-dependent. Among others, DeepONet and FNO with randomly initialized weights are possible options. In the original work by Nelsen and Stuart (2021), the authors employ Fourier space random features (RF) for their numerical experiments, resembling a single-layer FNO. These Fourier space RF are specified by $\psi(u;\gamma) = \sigma\big(\mathcal{F}^{-1}(\chi \mathcal{F} \gamma \mathcal{F} u)\big)$, where $\mathcal{F}$ denotes Fourier transform, σ an activation function, and χ is a Fourier space reshuffle, and γ is drawn from a Gaussian random field.

3.5 Discussion

The architectures above can be roughly divided into two categories, depending on how the underlying ideas from deep learning are leveraged to define a parametrized class of mappings on function space.

3.5.0.1 Encoder-decoder network structure

The first approach, which we refer to as encoder-decoder-net and which includes the PCA-Net and DeepONet architectures, involves three steps: first, the input function is encoded by a finite-dimensional vector; second, an ordinary neural network, such as a fully connected or convolutional neural network, is employed to map the encoded input to a finite-dimensional output; third, a decoder maps the finite-dimensional output to an output function in the infinite-dimensional function space.

This approach is very natural from a numerical analysis point of view, sharing the basic structure of many numerical schemes, such as finite element methods (FEM), finite volume methods (FVM), and finite difference methods (FDM), as illustrated in Table 1. From this point of view, encoder-decoder-nets mainly differ from standard numerical schemes by replacing the hand-crafted algorithm and choice of numerical discretization by a data-driven algorithm encoded in the weights and biases of a neural network, and the possibility for a data-driven encoding and reconstruction. While appealing, such structure yields approximations within a fixed, finite-dimensional, linear subspace of $\mathcal{V}$. In particular, each output function from the approximate operator belongs to this linear subspace independently of the input function. Therefore these methods fall within the category of linear approximation, while methods for which outputs lie on a nonlinear manifold in $\mathcal{V}$ lead to what is known as *nonlinear approximation*. The benefits of nonlinear approximation are well understood in the context of functions (DeVore, 1998), however, for the case of operators, results are still sparse but benefits for some specific cases have been observed (Lee and Carlberg, 2020; Cohen et al., 2022; Lanthaler et al., 2023b; Kramer et al., 2024). The FNO (Li et al., 2021) and random features (Nelsen and Stuart, 2021) are concrete examples of operator learning methodologies for which the outputs lie on a nonlinear manifold in $\mathcal{V}$.

TABLE 1 Numerical schemes vs. Encoder-Decoder-Net.

Method	Encoding	Finite-dim. Mapping	Reconstruction
FEM	Galerkin projection	Numerical scheme	Finite element basis
FVM	Cell averages	Numerical scheme	Piecewise polynomial
FD	Point values	Numerical scheme	Interpolation
PCA-Net	PCA projection	Neural network	PCA basis
DeepONet	Linear encoder	Neural network	Neural network basis

3.5.0.2 Neural operators generalizing neural network structure

A second approach to defining a parametrized class of operators on function space, distinct from encoder-decoder-nets, is illustrated by FNO. Following this approach, the structure of neural networks, which consist of an alternating composition of affine and nonlinear layers, is retained and generalized to function space. Nonlinearity is introduced via composition with a standard activation function, such as rectified linear unit or smooth variants thereof. The affine layers are obtained by integrating the input function against an integral kernel; this introduces nonlocality which is clearly needed if the architecture is to benefit from universal approximation.

3.5.0.3 Optimization and randomization

The random features method (Nelsen and Stuart, 2021) can in principle be combined with any operator learning architecture. The random features approach opens up a less explored direction of combining optimization with randomization in operator learning. In contrast to optimization of all parameters (by gradient descent) within a neural operator approach, the RFM allows for in-depth analysis resulting in error and convergence guarantees that take into account the finite number of samples, the finite number of parameters and the optimization (Lanthaler and Nelsen, 2023). One interesting, and largely unresolved question is how to design efficient random features for the operator learning setting.

The RFM is closely related to kernel methods which have a long pedigree in machine learning. In this context, we mention a related kernel-based approach proposed in Batlle et al. (2024), which employs kernel methods for operator learning within the encoder-decoder-net paradigm. This approach has shown to be competitive on several benchmark operator learning problems, and has been analyzed in Batlle et al. (2024).

3.5.0.4 Other approaches

This review is mostly focused on methods that fall into one of the neural network-based approaches above, but it should be emphasized that other approaches are being actively pursued with success. Without aiming to present an exhaustive list, we mention nonlinear reduced-order modeling (Qian et al., 2020, 2022; Swischuk et al., 2019; Ling et al., 2016; Lee and Carlberg, 2020), approaches based on the theory of Koopman operators (Yeung et al., 2019; Peherstorfer and Willcox, 2016; Li et al., 2017; Morton et al., 2018), work aiming to augment and speed up numerical solvers (Stanziola et al., 2021), or work on data-driven closure modeling (Huang et al., 2020; Xu et al., 2021; Liu et al., 2021; Wang et al., 2017; Wu et al., 2018), to name just a few examples. For a broader overview of other approaches to machine learning for PDEs, we refer to the recent review by Brunton and Kutz (2023). While most "operator learning" is focused on operators mapping between functions with spatial or spatio-temporal dependence and often arising in connection with PDEs, we note that problems involving time-series represent another important avenue of machine learning research, which can also be viewed from, and may benefit from, the continuous viewpoint by Lanthaler et al. (2023c).

4 Universal approximation

The goal of the methodologies summarized in the last section is to approximate operators mapping between infinite-dimensional Banach spaces of functions. The first theoretical question to be addressed is whether these methods can achieve this task, even in principle? The goal of this line of research is to identify classes of operators for which operator learning methodologies possess a universal approximation property, i.e. the ability to approximate a wide class of operators to any given accuracy in the absence of any constraints on the model size, the number of data samples and without any constraints on the optimization.

Universal approximation theorems are well-known for finite dimensional neural networks mapping between Euclidean spaces (Hornik et al., 1989; Cybenko, 1989), providing a theoretical underpinning for their use in diverse applications. These results show that neural networks with nonpolynomial activation can approximate very general classes of continuous (and even measurable) functions to any degree of accuracy. Universal approximation theorems for operator learning architectures provide similar guarantees in the infinite-dimensional context.

4.1 Encoder-decoder-nets

As pointed out in the last section, a popular type of architecture follows the encoder-decoder-net paradigm. Examples include PCA-Net and DeepONet. The theoretical basis for operator learning broadly, and within this paradigm more specifically, was laid out in a paper by Chen and Chen (1995), only a few years after the above cited results on the universality of neural networks. In that work, the authors introduce a generalization of neural networks, called operator networks, and prove that the proposed architecture possesses a universal property: it is shown that (shallow) operator networks can approximate, to arbitrary accuracy, continuous operators mapping between spaces of continuous functions. This architecture and analysis forms the basis of DeepONet, where the shallow neural networks of the original architecture of Chen and Chen (1995) are replaced by their deep counterparts. We present first a general, abstract version of an encoder-decoder-net and give a criterion on the spaces $\mathcal{U}$, $\mathcal{V}$ for which such architectures satisfy universal approximation. We then summarize specific results for DeepONet and PCA-Net architectures.

We call an encoder-decoder-net a mapping $\Psi_{ED} : \mathcal{U} \to \mathcal{V}$ which has the form

$$\Psi_{ED} = F_{\mathcal{U}} \circ \alpha \circ G_{\mathcal{V}}$$

where $F_{\mathcal{U}} : \mathcal{U} \to \mathbb{R}^{d_{\mathcal{U}}}$, $G_{\mathcal{V}} : \mathbb{R}^{d_{\mathcal{V}}} \to \mathcal{V}$ are bounded, linear maps and $\alpha : \mathbb{R}^{d_{\mathcal{U}}} \to \mathbb{R}^{d_{\mathcal{V}}}$ is a continuous function. The following theorem (Kovachki et al., 2023, Lemma 22) asserts that encoder-decoder-nets satisfy universal approximation over a large class of spaces $\mathcal{U}$ and $\mathcal{V}$.

Theorem 4.1. *Suppose that $\mathcal{U}$, $\mathcal{V}$ are separable Banach spaces with the approximation property. Let $\Psi^\dagger : \mathcal{U} \to \mathcal{V}$ be a continuous operator. Fix a compact set $K \subset \mathcal{U}$. Then for any $\epsilon > 0$, there exist positive integers $d_{\mathcal{U}}$, $d_{\mathcal{V}}$, bounded linear maps $F_{\mathcal{U}} : \mathcal{U} \to \mathbb{R}^{d_{\mathcal{U}}}$, $G_{\mathcal{V}} : \mathbb{R}^{d_{\mathcal{V}}} \to \mathcal{V}$, and a function $\alpha \in C(\mathbb{R}^{d_{\mathcal{U}}}; \mathbb{R}^{d_{\mathcal{V}}})$, such that*

$$\sup_{u \in K} \|\Psi^\dagger(u) - (F_{\mathcal{U}} \circ \alpha \circ G_{\mathcal{V}})(u)\|_{\mathcal{V}} \leq \epsilon.$$

◊

A Banach space is said to have the *approximation property* if, over any compact set, the identity map can be resolved as the limit of finite rank operators (Lindenstrauss and Tzafriri, 2013). Although it is a fundamental property useful in many areas in approximation theory, it is not satisfied by all separable Banach spaces (Enflo, 1973). However, many of the Banach spaces used in PDE theory and numerical analysis such as Lebesgue spaces, Sobolev spaces, Besov spaces, and spaces of continuously differentiable functions all posses the approximation property (Kovachki et al., 2023, Lemma 26). The above therefore covers a large range of scenarios in which encoder-decoder-nets can be used. We now give examples of encoder-decoder-nets where we fix the exact functional form of $F_{\mathcal{U}}$ and $G_{\mathcal{V}}$ and show that universal approximation continues to hold.

4.1.1 Operator network and DeepONet

The specific form of operator networks as introduced and analyzed by Chen and Chen (1995) focuses on scalar-valued input and output functions. We will state the main result of Chen and Chen (1995) in this setting for notational simplicity. Extension to vector-valued functions is straightforward. In simplified form, Chen and Chen (1995, cf. Theorem 5) obtain the following result:

Theorem 4.2. *Suppose that $\sigma \in C(\mathbb{R})$ is a nonpolynomial activation function. Let $\mathfrak{D} \subset \mathbb{R}^d$ be a compact domain with Lipschitz boundary. Let $\Psi^\dagger : C(\mathfrak{D}) \to C(\mathfrak{D})$ be a continuous operator. Fix a compact set $K \subset C(\mathfrak{D})$. Then for any $\varepsilon > 0$, there are positive integers $d_{\mathcal{U}}$, $d_{\mathcal{V}}$, N, sensor points $x_1, \ldots, x_{d_{\mathcal{U}}} \in \mathfrak{D}$, and coefficients c_i^k, ξ_{ij}^k, b_i, ω_k, ζ_k with $i = 1, \ldots, N$, $j = 1, \ldots, d_{\mathcal{U}}$, $k = 1, \ldots, d_{\mathcal{V}}$, such that*

$$\sup_{u \in K} \sup_{x \in \mathfrak{D}} \left| \Psi^\dagger(u)(x) - \sum_{k=1}^{d_{\mathcal{V}}} \sum_{i=1}^{N} c_i^k \sigma \left(\sum_{j=1}^{d_{\mathcal{U}}} \xi_{ij}^k u(x_j) + b_i \right) \sigma(\omega_k x + \zeta_k) \right| \leq \varepsilon. \tag{4.1}$$

◊

Here, we can identify the linear encoder $Lu = (u(x_1), \ldots, u(x_{d_{\mathcal{U}}}))$, the shallow branch-net α, with components

$$\alpha_k(Lu) = \sum_{i=1}^{N} c_i^k \sigma \left(\sum_{j=1}^{d_{\mathcal{U}}} \xi_{ij}^k u(x_j) + b_i \right),$$

and trunk-net ψ, with components

$$\psi_k(y) = \sigma(\omega_k x + \zeta_k).$$

With these definitions, (4.1) can be written in the equivalent form,

$$\sup_{u \in K} \left\| \Psi^\dagger(u) - \sum_{k=1}^{d_{\mathcal{V}}} \alpha_k(Lu) \psi_k \right\|_{C(\mathfrak{D})} \leq \varepsilon.$$

Remark 4.3. Theorem 4.2 holds in much greater generality. In particular, it is not necessary to consider operator mapping input functions to output functions on the same domain $\mathfrak{D}$. In fact, the same result can be obtained for operators $\Psi^\dagger : C(V) \to C(\mathfrak{D})$, where the input "functions" $u \in C(V)$ can have domain a compact subset V of a general, potentially infinite-dimensional, Banach space. ◊

Theorem 4.2 provides the motivation and theoretical underpinning for DeepONet, extended to deep branch- and trunk-nets in Lu et al. (2021). These results demonstrate the universality of DeepONet for a very wide range of operators, with approximation error measured in the supremum-norm over a compact set of input functions.

4.1.1.1 Related work

Several extensions and variants of DeepONets have been proposed after the initial work by Lu et al. (2021), including extensions of the universal approximation analysis. We provide a short overview of relevant works that include a theoretical component below.

4.1.1.2 Input functions drawn from a probability measure

In Lanthaler et al. (2022), Theorem 4.2 has been generalized to input functions drawn from a general input measure μ, including the case of unbounded support, such as a Gaussian measure. The error is correspondingly measured in the Bochner $L^2(\mu)$-norm (cp. discussion of PCA-Net universality below), and it is demonstrated that DeepONet can approximate general Borel measurable operators in such a setting.

4.1.1.3 Alternative encoders

There is work addressing the discretization-invariance of the encoding in DeepONet, resulting in architectures that allow for encoding of the input function at arbitrary sensor locations include Bel(Basis enhanced learning)-Net (Zhang et al., 2023) and VIDON (Variable-input deep operator networks) (Prasthofer et al., 2022).

Hua and Lu (2023) introduce Basis Operator Network, a variant of DeepONet, where encoding is achieved by projection onto a neural network basis. Universal approximation results are obtained, including encoding error estimates for this approach.

Jin et al. (2022b) address the issue of multiple input functions, and propose MIO-Net (Multiple Input/Output Net), based on tensor-product representations. The authors prove a universal approximation property for the resulting architecture, and demonstrate its viability in numerical experiments.

4.1.1.4 DeepONets on abstract Hilbert spaces

DeepONets mapping between abstract Hilbert spaces have been considered in Castro (2023); Castro et al. (2022), including a discussion of their universality in that context.

4.1.2 PCA-Net

At a theoretical level, PCA-Net shares several similarities with DeepONet and much of the analysis can be carried out along similar lines. In addition to proposing the PCA-Net architecture and demonstrating its viability on numerical test problems including the Darcy flow and viscous Burgers equations, Bhattacharya et al. (2021) also prove that PCA-Net is universal for operators mapping between infinite-dimensional Hilbert spaces, with approximation error measured in the Bochner $L^2(\mu)$-norm with respect to the input measure μ. This initial analysis was developed and sharpened considerably in Lanthaler (2023); as an example of this we quote Lanthaler (2023, Proposition 31).

Theorem 4.4 (PCA-Net universality). *Let $\mathcal{U}$ and $\mathcal{V}$ be separable Hilbert spaces and let $\mu \in \mathcal{P}(\mathcal{U})$ be a probability measure on $\mathcal{U}$. Let $\Psi^\dagger : \mathcal{U} \to \mathcal{V}$ be a μ-measurable mapping. Assume the following moment conditions,*

$$\mathbb{E}_{u\sim\mu}[\|u\|_{\mathcal{U}}^2], \ \mathbb{E}_{u\sim\mu}[\|\Psi^\dagger(u)\|_{\mathcal{V}}^2] < \infty.$$

Then for any $\varepsilon > 0$, there are dimensions $d_{\mathcal{U}}$, $d_{\mathcal{V}}$, a requisite amount of data N, a neural network ψ depending on this data, such that the PCA-Net, $\Psi = G_{\mathcal{V}} \circ \psi \circ F_{\mathcal{U}}$, satisfies

$$\mathbb{E}_{\{u_j\}\sim\mu^{\otimes N}}\left[\mathbb{E}_{u\sim\mu}\left[\|\Psi^\dagger(u) - \Psi(u; \{u_j\}_{j=1}^N)\|_{\mathcal{V}}^2\right]\right] < \varepsilon,$$

where the outer expectation is with respect to the iid data samples $u_1, \dots, u_N \sim \mu$, which determine the empirical PCA encoder and reconstruction. ◇

4.2 Neural operators

The Fourier neural operator (FNO) is a specific instance of a more general notion of neural operators (Kovachki et al., 2023). The general structure of such neural operators is identical to that of the FNO, i.e. a composition

$$\begin{aligned} \Psi_{NO}(u;\theta) &= \mathcal{Q} \circ \mathcal{L}_L \circ \cdots \circ \mathcal{L}_2 \circ \mathcal{L}_1 \circ \mathcal{R}(u), \quad \forall u \in \mathcal{U}, \\ \mathcal{L}_\ell(v)(x;\theta) &= \sigma\left(W_\ell v(x) + b_\ell + \mathcal{K}(v)(x;\gamma_\ell)\right), \end{aligned} \tag{4.2}$$

except that the convolutional operator in each layer is replaced by a more general integral operator,

$$\mathcal{K}(v)(x;\gamma) = \int_{\mathfrak{D}} \kappa(x, y; \gamma) v(y)\, dy.$$

Here, the integral kernel $\kappa(x, y; \gamma)$ is a matrix-valued function of x and y, parametrized by γ. Additional nonlinear dependency on the input function is possible yielding, for example, $\kappa = \kappa(x, y, u(x), u(y); \gamma)$; this structure is present in transformer models (Vaswani et al., 2017). Different concrete implementations of such neural operators mostly differ in their choice of the parametrized integral kernel. For example, Li et al. (2021) use Fourier basis, Tripura and Chakraborty (2023) use wavelet basis, and Bonev et al. (2023) use spherical harmonics. Other approaches restrict the support of κ (Li et al., 2020a) or assume it decays quickly away from its diagonal (Li et al., 2020b; Lam et al., 2023). For a more thorough review of this methodology, we refer to Kovachki et al. (2023).

The universality of the FNO has first been established in Kovachki et al. (2021), using ideas from Fourier analysis and, in particular, building on the density of Fourier series to show that FNO can approximate a wide variety of operators. Given the great variety of possible alternative neural operator architectures, which differ from the FNO essentially only in their choice of the parametrized kernel, a proof of universality that does not explicitly rely on Fourier analysis, and which applies to a wide range of choice for the integral kernel is desirable. This has been accomplished in Lanthaler et al. (2023a), where the authors remove from the FNO all nonessential components, from the perspective of universal approximation, yielding a minimal architecture termed the "averaging neural operator" (ANO).

4.2.1 Averaging neural operator

Up to nonessential details, the ANO introduced in Lanthaler et al. (2023a) is a composition of nonlinear layers of the form,

$$\mathcal{L}(v; \gamma_\ell)(x) = \sigma\left(W_\ell v(x) + b_\ell(x) + V_\ell \int_{\mathfrak{D}} v(y)\, dy\right), \quad (\ell = 1, \dots, L),$$

where $W_\ell, V_\ell \in \mathbb{R}^{d_c \times d_c}$ are matrices, and $b_\ell(x)$ is a bias function. In the present work, to parametrize the bias functions, we consider bias of the form $b_\ell(x) = A_\ell x + c_\ell$ for matrix $A_\ell \in \mathbb{R}^{d_c \times d}$ and bias vector $c_\ell \in \mathbb{R}^{d_c}$. With this choice, the nonlinear layer $v \mapsto \mathcal{L}(v)$ takes the form,

$$\mathcal{L}(v; \gamma_\ell)(x) = \sigma\left(W_\ell v(x) + A_\ell x + c_\ell + V_\ell \int_{\mathfrak{D}} v(y)\, dy\right).$$

The parameter $\gamma_\ell = \{W_\ell, V_\ell, A_\ell, c_\ell\}$ collects the tunable parameters of the ℓ-th layer. To define an operator $\Psi : \mathcal{U}(D; \mathbb{R}^{d_i}) \to \mathcal{V}(D; \mathbb{R}^{d_o})$, we combine these nonlinear layers with linear input and output layers $\mathcal{R} : u(x) \mapsto Ru(x)$ and $Q : v(x) \mapsto Qv(x)$, obtained by multiplication with matrices $R \in \mathbb{R}^{d_c \times d_i}$ and $Q \in \mathbb{R}^{d_o \times d_c}$, respectively. The resulting ANO is an operator of the form,

$$\Psi(u; \theta) = Q \circ \mathcal{L}_L \circ \cdots \circ \mathcal{L}_1 \circ \mathcal{R}(u),$$

with θ collecting the tunable parameters in each hidden layer, and the input and output layers.

The ANO can be though of as a special case of FNO, where the convolutional integral kernel is constant, leading to the last term in each hidden layer being an integral or "average" of the input function. Similarly, the ANO is a special case of many other parametrizations of the integral kernel in neural operator architectures. Despite its simplicity, the ANO can nevertheless be shown to have a universal approximation property. We here cite a special case for operator mapping between continuous functions, and refer to Lanthaler et al. (2023a) for more general results:

Theorem 4.5. *Suppose that $\sigma \in C(\mathbb{R})$ is a nonpolynomial activation function. Let $\mathfrak{D} \subset \mathbb{R}^d$ be a compact domain with Lipschitz boundary. Let $\Psi^\dagger : C(\mathfrak{D}) \to C(\mathfrak{D})$ be a continuous operator. Fix a compact set $K \subset C(\mathfrak{D})$. Then for any $\varepsilon > 0$, there exists an ANO $\Psi : C(\mathfrak{D}) \to C(\mathfrak{D})$ such that*

$$\sup_{u \in K} \|\Psi^\dagger(u) - \Psi(u)\|_{C(\mathfrak{D})} < \varepsilon.$$

◊

As an immediate consequence, we have the following corollary which implies universality of neural operators for a wide range of choices:

Corollary 4.6. *Consider any neural operator architecture of the form* (4.2) *with parametrized integral kernel $\kappa(x, y; \gamma)$. If for any channel dimension d_c and matrix $V \in \mathbb{R}^{d_c \times d_c}$, there exists a parameter γ_V such that $\kappa(x, y; \gamma_V) \equiv V$, then the neural operator architecture is universal in the sense of Theorem 4.5.* ◊

The last corollary applies in particular to the FNO. Universality of the FNO was first established in Kovachki et al. (2021), there restricting attention to a periodic setting but allowing for operators mapping between general Sobolev spaces. The idea of the averaging neural operator can be found in Lanthaler et al. (2023a), where it was used to prove universality for a wide range of neural operator architectures, and for a class of operators mapping between general Sobolev spaces, or spaces of continuously differentiable functions.

4.2.1.1 Intuition

We would like to provide further intuition for the universality of ANO. A simple special case of the ANO is the two-layer ANO obtained as follows. We consider neural operators which can be written as a composition of two shallow neural networks $\phi : \mathbb{R}^{d_i} \times \mathfrak{D} \to \mathbb{R}^{d_c}$, $\psi : \mathbb{R}^{d_c} \times \mathfrak{D} \to \mathbb{R}^{d_o}$ and an additional integral (average):

$$\begin{cases} \alpha(u) = \displaystyle\int_{\mathfrak{D}} \phi(u(y), y)\, dy, \\ \Psi(u)(x) = \psi\left(\alpha(u), x\right), \end{cases}$$

where

$$\phi(w, x) = C_1 \sigma(A_1 w + B_1 x + d_1),$$

and

$$\psi(z, x) = C_2 \sigma(A_2 z + B_2 x + d_2).$$

Note that the above composition, i.e.

$$\Psi(u)(x) = \psi\left(\int_{\mathfrak{D}} \phi(u(y), y)\, dy, x\right),$$

indeed defines a special case of ANO that can be written as a composition of a trivial input layer, two hidden layers and an output layer:

$$\begin{aligned}
\mathcal{L}_1(u)(x) &= \sigma\left(W_1 u(x) + b_1(x)\right), & b_1(x) &= B_1 x + d_1,\, W_1 = A_1,\\
\mathcal{L}_2(v)(x) &= \sigma\left(b_2(x) + V_2 \int v(y)\, dy\right), & b_2(x) &= B_2 x + d_2,\, V_2 = A_2 C_1\\
\mathcal{Q}(v)(x) &= Qv(x), & Q &= C_2.
\end{aligned}$$

As depicted in Fig. 5, such shallow ANO can be interpreted as a composition of an nonlinear encoder $\alpha : \mathcal{U} \to \mathbb{R}^{d_c}$, $u \mapsto \alpha(u)$ defined via spatial averaging of ϕ, and mapping the input function to a finite-dimensional latent space, and a nonlinear decoder $\psi : \mathbb{R}^{d_c} \to \mathcal{V}$, $\alpha \mapsto \psi(\alpha, \cdot)$. This interpretation opens up a path for analysis, based on which universality can be established for the ANO, and any neural operator architecture that contains such ANO as a special case, such as the Fourier neural operator.

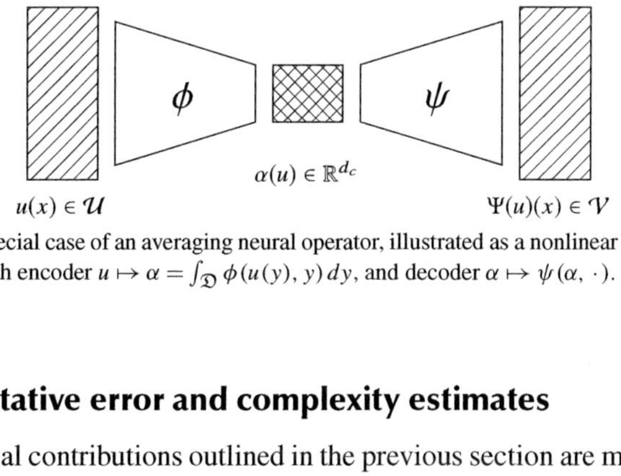

FIGURE 5 Special case of an averaging neural operator, illustrated as a nonlinear encoder-decoder architecture; with encoder $u \mapsto \alpha = \int_{\mathfrak{D}} \phi(u(y), y)\, dy$, and decoder $\alpha \mapsto \psi(\alpha, \cdot)$.

5 Quantitative error and complexity estimates

The theoretical contributions outlined in the previous section are mostly focused on methodological advances and a discussion of the universality of the resulting architectures. Universality of neural operator architectures, i.e. the ability to approximate a wide class of operators, is arguably a necessary condition for their success. But since universality is inherently qualitative, it cannot explain the efficiency of these methods in practice. Improving our understanding of the efficiency of neural operators in practice requires a quantitative theory of operator learning, providing explicit error and complexity estimates: given a desired accuracy $\varepsilon > 0$, what is the model size or the number of samples that is required to achieve such accuracy?

5.1 Linear operators

Learning a linear operator can be formulated as solving a linear inverse problem with a noncompact forward operator (Mollenhauer et al., 2022; de Hoop et al., 2023). We take this point of view and broadly describe the results from de Hoop et al. (2023). We consider the problem of learning $\Psi^\dagger = L^\dagger$, a linear operator, when $\mathcal{U} = \mathcal{V}$ is a separable Hilbert space. Studying the linear problem enables a thorough analysis of the complexity of operator learning and hence sheds light on the problem in the more general setting. The work proceeds by assuming that $L^\dagger$ can be diagonalized in a known Schauder basis of $\mathcal{U}$ denoted $\{\varphi_j\}_{j=1}^\infty$. Given input data $\{u_n\}_{n=1}^N \overset{\text{i.i.d.}}{\sim} \mu$, the noisy observations are assumed to be of the form

$$v_n = L^\dagger u_n + \gamma \xi_n$$

where $\gamma > 0$ and the sequence $\{\xi_n\}_{n=1}^N \overset{\text{i.i.d.}}{\sim} \mathcal{N}(0, I_{\mathcal{U}})$ comes from a Gaussian white noise process that is independent of the input data $\{u_n\}$. In what follows here, we assume that $\mu = \mathcal{N}(0, \Gamma)$, the measure on the input data, is Gaussian with a strictly positive covariance Γ and that this covariance is also diagonalizable by $\{\varphi_j\}$; note, however, that de Hoop et al. (2023) treat the more general case without assuming simultaneous diagonalizability of $L^\dagger$ and Γ.

Note that $\{l_j^\dagger, \varphi_j\}$ uniquely determines $L^\dagger$. Thus the problem as formulated here can be stated as learning the eigenvalue sequence $\{l_j^\dagger\}_{j=1}^\infty$ of $L^\dagger$ from the noisy observations

$$v_{jn} = \langle \varphi_j, u_n \rangle_{\mathcal{U}} l_j^\dagger + \xi_{jn}, \quad j \in \mathbb{N}, \quad n = 1, \dots, N$$

where $\{\xi_{jn}\} \overset{\text{i.i.d.}}{\sim} \mathcal{N}(0, \gamma^2)$. In this problem statement, the noise is crucial for obtaining meaningful estimates on the amount of data needed for learning. Indeed, without noise, the eigenvalues can simply be recovered as

$$l_j^\dagger = \frac{v_{j1}}{\langle \varphi_j, u_1 \rangle_{\mathcal{U}}}$$

for all $j \in \mathbb{N}$ from a single data point u_1 since the basis $\{\varphi_j\}$ is assumed to be known. While, in practice, the observations might not be noisy, the noise process can be used to model round-off or discretization errors which occur in computation.

Assuming a Gaussian prior on the sequence $\{l_j^\dagger\} \sim \otimes_{j=1}^\infty \mathcal{N}(0, \sigma_j^2)$, one may obtain a Bayesian estimator of $L^\dagger$, given the data $\big(\{v_{jn}\}, \{u_n\}\big)$. The Bayesian posterior is characterized as an infinite Gaussian product for the sequence of eigenvalues. We take as an estimator the mean of this Gaussian which is given as

$$l_j = \frac{\gamma^{-2}\sigma_j^2 \sum_{n=1}^N v_{jn} \langle \varphi_j, u_n \rangle_{\mathcal{U}}}{1 + \gamma^{-2}\sigma_j^2 \sum_{n=1}^N \left|\langle \varphi_j, u_n \rangle_{\mathcal{U}}\right|^2}.$$

Then our estimator is the operator Ψ, diagonalized in basis $\{\varphi_j\}$ with eigenvalues $\{l_j\}$. To quantify the smoothness of $L^\dagger$, we assume that $\{l_j^\dagger\}$ lives in an appropriately weighted ℓ^2 space, in particular,

$$\sum_{j=1}^{\infty} j^{2s} |l_j^\dagger|^2 < \infty$$

for some $s \in \mathbb{R}$. Then the following theorem holds (de Hoop et al., 2023, Theorem 1.3).

Theorem 5.1. *Suppose that for some $\alpha > 1/2$ and $p \in \mathbb{R}$, we have $\langle \varphi_j, \Gamma \varphi_j \rangle_{\mathcal{U}} \asymp j^{-2\alpha}$ and $\sigma_j^2 \asymp j^{-2p}$. Let $\alpha' \in [0, \alpha + 1/2)$ and assume that* $\min\{\alpha, \alpha'\} + \min\{p - 1/2, s\} > 0$. *Then, as $N \to \infty$, we have*

$$\mathbb{E} \sum_{j=1}^{\infty} j^{-2\alpha'} |l_j - l_j^\dagger|^2 \lesssim N^{-\frac{\alpha' + \min\{p-1/2, s\}}{\alpha + p}}$$

where the expectation is taken over the product measure defining the input data and noise. $\Diamond$

Theorem 5.1 quantifies the amount of data needed, on average, for the estimator $\{l_j\}$ to achieve ϵ-error in approximating $\{l_j^\dagger\}$ measured in a squared weighted ℓ^2 norm. In particular, we have

$$N \sim \epsilon^{-\frac{\alpha + p}{\alpha' + \min\{p-1/2, s\}}}.$$

The exact dependence of this rate on the parameters defining the smoothness of the truth, the input, the prior, and the error metric elucidates the optimal design choices for the estimator and sheds light on which pieces are most important for the learning process. We refer to de Hoop et al. (2023) for an in-depth discussion.

Within machine learning and functional data analysis, many works have focused on learning integral kernel operators (Rosasco et al., 2010; Abernethy et al., 2009; Crambes and Mas, 2013; Wang et al., 2022; Jin et al., 2022a). The operator learning problem can then be reduced to approximation of the kernel function and is typically studied in a Reproducing Kernel Hilbert Space setting. In numerical PDEs, some recent works have studied the problem of learning the Green's function arising from some elliptic, parabolic and hyperbolic problems (Boullé et al., 2022a,b; Boullé and Townsend, 2023; Wang and Townsend, 2023).

5.2 Holomorphic operators

Going beyond the linear case, holomorphic operators represent a very general class of operators for which *efficient* quantitative error and complexity estimates can be established. We mention the influential work by Cohen et al. (2010, 2011), as well as further developments (Chkifa et al., 2015, 2013; Schwab and Zech, 2019; Opschoor et al., 2022; Schwab and Zech, 2023; Herrmann et al., 2022; Adcock et al., 2022c,a,b; Marcati and Schwab, 2023; Adcock et al., 2023a,b). A detailed review, from 2015, can be found in Cohen and DeVore (2015). We will only describe a few main ideas. We mention in passing also the works by Kutyniok et al. (2022); Lei et al. (2022), which study the neural network approximation of parametric operators in a related setting.

Holomorphic operators have mostly been studied in a parametrized setting, where the input functions can be identified with the coefficients in a suitable basis (frame) expansion. A prototypical example is the elliptic Darcy flow equation (2.5), where the coefficient field $a = a(\cdot\,; \boldsymbol{y})$ is parametrized by a sequence $\boldsymbol{y} = (y_1, y_2, \dots) \in [-1, 1]^{\mathbb{N}}$, e.g. in the form of a linear expansion,

$$a(x; \boldsymbol{y}) = \overline{a}(x) + \sum_{j=1}^{\infty} y_j \varphi_j(x), \quad x \in \mathfrak{D},$$

where $\overline{a} \in \mathcal{U}$ and $\varphi_1, \varphi_2, \dots \in \mathcal{U}$ are fixed. The operator $\Psi^\dagger : a \mapsto u$ can then (loosely) be identified with the parametrized mapping,

$$\mathsf{F} : [-1, 1]^{\mathbb{N}} \to \mathcal{V}, \quad \mathsf{F}(\boldsymbol{y}) := \Psi^\dagger(a(\cdot\,; \boldsymbol{y})).$$

In the above prototypical setting, and assuming that the sequence $\boldsymbol{b}$ with coefficients $b_j = \|\varphi_j\|_{\mathcal{V}}$ decays sufficiently fast, it can be shown in Cohen and DeVore (2015) that F possesses a holomorphic extension to a subset of the infinite product $\mathbb{C}^{\mathbb{N}} = \prod_{j=1}^{\infty} \mathbb{C} = \mathbb{C} \times \mathbb{C} \times \dots$ of the complex plane $\mathbb{C}$. More precisely, there exists a holomorphic extension

$$\mathsf{F} : \prod_{j=1}^{\infty} \mathcal{E}_{\rho_j} \to \mathcal{V}_{\mathbb{C}}, \quad \boldsymbol{z} \mapsto \mathsf{F}(\boldsymbol{z}). \tag{5.1}$$

Here, for $\rho_j > 1$, the set $\mathcal{E}_{\rho_j} = \left\{ \frac{1}{2}\left(z + z^{-1}\right) \,\middle|\, z \in \mathbb{C},\ 1 \le |z| \le \rho_j \right\} \subset \mathbb{C}$ denotes the Bernstein ellipse with focal points ± 1 and major and minor semiaxes lengths $\frac{1}{2}(\rho_j \pm \rho_j^{-1})$, and $\mathcal{V}_{\mathbb{C}}$ is the complexification of the Banach space $\mathcal{V}$.

In general, given a nonnegative sequence $\boldsymbol{b} \in [0, \infty)^{\mathbb{N}}$ and $\varepsilon > 0$, a parametrized operator $\mathsf{F} : [-1, 1]^{\mathbb{N}} \to \mathcal{V}$ is called *$(\boldsymbol{b}, \varepsilon)$-holomorphic*, if it possesses a holomorphic extension (5.1) for any $\boldsymbol{\rho} = (\rho_1, \rho_2, \dots) \in [1, \infty)^{\mathbb{N}}$, s.t.

$$\sum_{j=1}^{\infty} \left(\frac{\rho_j + \rho_j^{-1}}{2} - 1 \right) b_j \le \varepsilon. \tag{5.2}$$

As reviewed in Adcock et al. (2022c, Chapter 4), a number of parametric differential equations of practical interest give rise to $(\boldsymbol{b}, \varepsilon)$-holomorphic operators.

The approximation theory of this class of operators is well-developed (Cohen and DeVore, 2015; Adcock et al., 2022c). The underlying reason why efficient approximation of such operators is possible is that holomorphic operators possess convergent expansions in multivariate polynomial bases, where each polynomial basis element only depends on a finite number of the components of the complex input $z = (z_1, z_2, \dots)$.

A standard setting considers $\boldsymbol{b} \in \ell^p(\mathbb{N})$, for some $0 < p < 1$. For example, if $b_j \sim j^{-s}$ decays algebraically, then $\boldsymbol{b} \in \ell^p(\mathbb{N})$ for $s > p^{-1} > 1$. Assuming that $\boldsymbol{b} \in \ell^p(\mathbb{N})$, it can be shown (e.g. Cohen and DeVore, 2015, Corollary 3.11) that the best n-term polynomial approximation to F converges at rate $n^{1-1/p}$ in a sup-norm setting, and rate $n^{1/2-1/p}$ in a Bochner $L^2(\mu)$-setting, with specific input measure μ on $[-1, 1]^{\mathbb{N}}$. Importantly, these convergence rates are polynomial in the number of degrees of freedom n, even in this infinite-dimensional parametric setting. When restricting to a finite-dimensional input space with d input components, i.e. considering only inputs of the form $z = (z_1, \dots, z_d, 0, 0, \dots)$, this fact implies that convergence rates independent of the dimension d can be obtained, and thus such approximation of $(\boldsymbol{b}, \varepsilon)$-holomorphic operators can provably overcome the *curse of dimensionality* (Cohen and DeVore, 2015).

The above mentioned results in the parametrized setting can also be used to prove efficient approximation of holomorphic operators by operator learning frameworks in a nonparametric setting (Lanthaler et al., 2022; Schwab and Zech, 2019; Opschoor et al., 2022; Schwab and Zech, 2023; Herrmann et al., 2022). For example, Herrmann et al. (2022) consider the DeepONet approximation of holomorphic operators with general Riesz frame encoders and decoders, demonstrating algebraic error and complexity estimates; Under suitable conditions, the authors prove that ReLU deep operator networks (DeepONet) approximating holomorphic operators can achieve convergence rates arbitrarily close to n^{1-s} in a worst-error setting (supremum norm) and at rate $n^{1/2-s}$ in a Bochner $L^2(\mu)$-norm setting. Here n denotes the number of tunable parameters of the considered DeepONet, and the parameter s determines the decay of the coefficients in the frame expansion of the considered input functions. Under the (loose) identification $s \sim 1/p$, these rates for DeepONet recover the rates discussed above. These results show that there exist operator surrogates which essentially achieve the approximation rates afforded by best n-term approximation schemes mentioned above.

The complementary question of the sample complexity of operator learning for holomorphic operators has been studied in Bachmayr and Cohen (2017); Adcock et al. (2023a,b). Building on the theory of N-widths, Adcock et al. (2023a,b) show that on the class $(\boldsymbol{b}, \varepsilon)$-holomorphic operators and in a Bochner L^2-setting, data-driven methods relying on N samples cannot achieve convergence rates better than $N^{1/2-1/p}$. In addition, it is shown that the optimal rate can be achieved up to logarithmic terms. We refer to Adcock et al. (2023a,b) for further details.

To summarize: holomorphic operators represent a class of operators of practical interest for which approximation theory by neural operator learning frameworks can be developed. The approximation theory of this class of operators is well-developed, especially in the parametrized setting. In a parametrized setting, these operators allow for efficient approximation by multivariate (sparse) polynomials. This fact can be leveraged to show that efficient approximation by neural network-based methods is possible, and such results can be extended beyond a parametric setting, e.g. via frame expansions. Optimal approximation rates, and methods achieving these optimal rates, up to logarithmic terms, are known under specific assumptions.

5.3 General (Lipschitz) operators

The last two sections provide an overview of theoretical results on the approximation of linear and holomorphic operators. While these classes of operators include several operators of practical interest and allow for the development of general approximation theory, not all operators of relevance are holomorphic (or indeed linear). Examples of nonholomorphic operators include the solution operator associated with nonlinear hyperbolic conservation laws such as the compressible Euler equations. Solutions of such equations can develop shocks in finite time, and it can be shown that the associated solution operators themselves are not holomorphic. It is therefore of interest to extend the approximation theory of operator learning frameworks beyond the restrictive class of holomorphic operators.

A general and natural class of nonlinear operators are general Lipschitz continuous operators, the approximation theory of which has been considered from an operator learning perspective e.g. in Bhattacharya et al. (2021); Liu et al. (2024); Franco et al. (2023); Galimberti et al. (2022); Schwab et al. (2023). We present a brief outline of the general approach and mention relevant work on model complexity estimates in the following subsections 5.3.1–5.3.3. Relevant results on the data complexity of operator learning in this setting are summarized in Subsection 5.3.4.

5.3.1 Error decomposition

Encoder-decoder-net architectures arguably follow the basic mathematical intuition of how to approach the operator approximation problem most closely, and most theoretical work has focused on this approach. We recall that within this framework, the infinite-dimensional input and output spaces $\mathcal{U}$, $\mathcal{V}$ are first encoded through suitable finite-dimensional latent spaces. This involves an encoder/decoder pair $(F_{\mathcal{U}}, G_{\mathcal{U}})$ on $\mathcal{U}$,

$$F_{\mathcal{U}} : \mathcal{U} \to \mathbb{R}^{d_{\mathcal{U}}}, \quad G_{\mathcal{U}} : \mathbb{R}^{d_{\mathcal{U}}} \to \mathcal{U},$$

and another encoder/decoder pair $(F_{\mathcal{V}}, g_{\mathcal{V}})$ on $\mathcal{V}$,

$$F_{\mathcal{V}} : \mathcal{V} \to \mathbb{R}^{d_{\mathcal{V}}}, \quad G_{\mathcal{V}} : \mathbb{R}^{d_{\mathcal{V}}} \to \mathcal{V}.$$

We recall that the composition $G_{\mathcal{U}} \circ F_{\mathcal{U}}$, $G_{\mathcal{V}} \circ F_{\mathcal{V}}$ are interpreted as approximations to the identity on $\mathcal{U}$ and $\mathcal{V}$, respectively. These encode/decoder pairs in turn imply an encoding of the underlying infinite-dimensional operator $\Psi^\dagger : \mathcal{U} \to \mathcal{V}$, resulting in a finite-dimensional function

$$\varphi : \mathbb{R}^{d_{\mathcal{U}}} \to \mathbb{R}^{d_{\mathcal{V}}}, \quad \varphi(\alpha) = F_{\mathcal{V}} \circ \Psi^\dagger \circ G_{\mathcal{U}}(\alpha),$$

as depicted earlier, in Fig. 3.

While the encoder and decoder of these architectures perform dimension reduction, the neural network $\psi : \mathbb{R}^{d_{\mathcal{U}}} \to \mathbb{R}^{d_{\mathcal{V}}}$ at the core of encoder-decoder-net architectures is interpreted as approximating this resulting finite-dimensional function $\varphi : \mathbb{R}^{d_{\mathcal{U}}} \to \mathbb{R}^{d_{\mathcal{V}}}$. To summarize, an encoder-decoder-net can conceptually be interpreted as involving three steps:

1. Dimension reduction on the input space $\mathcal{U} \approx \mathbb{R}^{d_{\mathcal{U}}}$,
2. Dimension reduction on the output space $\mathcal{V} \approx \mathbb{R}^{d_{\mathcal{V}}}$,
3. Encoding of the operator $\Psi^\dagger$ yielding $\varphi : \mathbb{R}^{d_{\mathcal{U}}} \to \mathbb{R}^{d_{\mathcal{V}}}$, approximated by neural network $\psi : \mathbb{R}^{d_{\mathcal{U}}} \to \mathbb{R}^{d_{\mathcal{V}}}$.

Each part of this conceptual decomposition, the encoding of $\mathcal{U} \approx \mathbb{R}^{d_{\mathcal{U}}}$, the decoding $\mathbb{R}^{d_{\mathcal{V}}} \approx \mathcal{V}$ and the approximation $\psi \approx \varphi$, represents a source of error, and the total encoder-decoder-net approximation error $\mathscr{E}$ is bounded by three contributions $\mathscr{E} \lesssim \mathscr{E}_{\mathcal{U}} + \mathscr{E}_{\psi} + \mathscr{E}_{\mathcal{V}}$, where

$$\mathscr{E}_{\mathcal{U}} = \sup_{u} \| u - G_{\mathcal{U}} \circ F_{\mathcal{U}}(u) \|_{\mathcal{U}},$$

quantifies the encoding error, with supremum taken over the relevant set of input functions u,

$$\mathscr{E}_{\mathcal{V}} = \sup_{v} \| v - G_{\mathcal{V}} \circ F_{\mathcal{V}}(v) \|_{\mathcal{V}},$$

quantifies the decoding error, and

$$\mathscr{E}_{\psi} = \sup_{\alpha} \| F_{\mathcal{V}} \circ \Psi^\dagger \circ G_{\mathcal{U}}(\alpha) - \psi(\alpha) \|, \tag{5.3}$$

is the neural network approximation error.

Given this decomposition, the derivation of error and complexity estimates for encoder-decoder-net architectures thus boils down to the estimation of encoding error $\mathscr{E}_{\mathcal{U}}$, neural network approximation error $\mathscr{E}_{\psi}$ and reconstruction error $\mathscr{E}_{\mathcal{V}}$, respectively.

5.3.1.1 Encoding and reconstruction errors

Encoding and reconstruction errors are relatively well understood on classical function spaces such as Sobolev and Besov spaces, by various linear and non-linear methods of approximation (DeVore and Lorentz, 1993).

For linear encoder/decoder pairs, the analysis of encoding and reconstruction errors amounts to principal component analysis (PCA) when measuring the error in the Bochner norm L^2_μ, or to Kolmogorov n-widths when measuring the error in the sup-norm over a compact set. Relevant discussion of PCA in the context of operator learning is given in Bhattacharya et al. (2021) (see also Lanthaler et al., 2022, and Lanthaler, 2023).

In certain settings, such as for PDEs with discontinuous output functions, it has been shown by Lanthaler et al. (2023b) that relying on linear reconstruction imposes fundamental limitations on the approximation accuracy of operator methodologies, which can be overcome by methods with nonlinear reconstruction; specifically, it was shown both theoretically and experimentally in Lanthaler et al. (2023b) that FNO and shift-DeepONet, a variant of DeepONet with nonlinear reconstruction, achieve higher accuracy than vanilla DeepONet for prototypical PDEs with discontinuous solutions. We also mention closely related work on the nonlinear manifold decoder architecture of Seidman et al. (2022).

5.3.1.2 Neural network approximation error

At their core, encoder-decoder-net architectures employ a neural network to approximate the encoded version $F_\mathcal{V} \circ \Psi^\dagger \circ G_\mathcal{U}$ of the underlying operator $\Psi^\dagger$ (cp. (5.3)). The practical success of these frameworks thus hinges on the ability of ordinary neural networks to approximate the relevant class of high-dimensional functions in the latent-dimensional spaces, which are obtained through the encoding of such operators. While the empirical success of neural networks in high-dimensional approximation tasks is undeniable, our theoretical understanding and the mathematical foundation underpinning this empirical success remains incomplete.

General approximation theoretic results on the neural network approximation of functions have been obtained, and some available quantitative bounds in operator learning (Franco et al., 2023; Galimberti et al., 2022) build on these results to estimate the neural network approximation error $\mathscr{E}_\psi$. Notably, the seminal work by Yarotsky (2017) of D. Yarotsky presents general error and complexity estimates for functions with Lipschitz continuous derivatives:

Theorem 5.2. *A function $f \in W^{k,\infty}([0,1]^d)$ can be approximated to uniform accuracy $\varepsilon > 0$,*

$$\sup_{x\in[0,1]^d} |f(x) - \psi(x)| \le \varepsilon,$$

by a ReLU neural network ψ with at most $O(\varepsilon^{-d/k}\log(\varepsilon^{-1}))$ tunable parameters. ◊

Remark 5.3. Note that the relevant dimension in the operator learning context is the latent dimension $d = d_{\mathcal{U}}$. Neglecting logarithmic terms, we note that each component of the function $G : \mathbb{R}^{d_{\mathcal{U}}} \to \mathbb{R}^{d_{\mathcal{V}}}$ can be approximated individually by a neural network of size at most $O(\varepsilon^{-d_{\mathcal{U}}/k})$, and hence, we expect that G can be approximated to accuracy ε by a neural network ψ of size at most $O(d_{\mathcal{V}}\varepsilon^{-d_{\mathcal{U}}/k})$. $\Diamond$

Without aiming to provide a comprehensive overview of this very active research direction on neural network approximation theory, adjacent to operator learning theory, we mention that similar error and complexity estimates can also be obtained on more general Sobolev spaces, e.g. Yarotsky and Zhevnerchuk (2020); Siegel (2023). Lower bounds illuminating the limitations of neural networks on model classes are for example discussed in Achour et al. (2022); Yarotsky (2017); Bolcskei et al. (2019). Approximation rates leveraging additional structure beyond smoothness have also been considered, e.g. compositional structure is explored in Mhaskar and Poggio (2020); Shen et al. (2019); Schmidt-Hieber (2020).

5.3.1.3 Nonstandard architectures and hyperexpressive activations

While the general research area of neural network approximation theory is too broad to adequately summarize here, we mention relevant work on hyperexpressive activations, which can formally break the curse of dimensionality observed Theorem 5.2; it has been shown that neural networks employing nonstandard activations can formally achieve arbitrary convergence on model function classes (Pinkus, 1999; Yarotsky, 2021; Shen et al., 2021; Liang et al., 2021), when the complexity is measured in terms of number of tunable parameters. This is not true for the ReLU activation (Yarotsky, 2017). Another way to break the curse of dimensionality is via architectures with nonstandard "three-dimensional" structure (Zhang et al., 2022).

While nonstandard architectures can overcome the curse of dimensionality in the sense that the number of parameters does not grow exponentially with d (or is independent of d), it should be pointed out that this necessarily comes at the expense of the number of bits that are required to represent each parameter in a practical implementation. Indeed, from work on quantized neural networks (Bolcskei et al., 2019) (with arbitrary activation function), it can be inferred that the total number of bits required to store all parameters in such architectures is lower bounded by the Kolmogorov ε-entropy of the underlying model class; For the specific model class $W^{k,\infty}([0, 1]^d)$, this entropy scales as $\varepsilon^{-d/k}$. Hence, architectures which achieve error ε with a number of parameters that scales strictly slower than $\varepsilon^{-d/k}$ must do so at the expense of the precision that is required to represent each individual parameter in a practical implementation, keeping the total number of bits above the entropy limit. For related discussion, we e.g. refer to Yarotsky and Zhevnerchuk (2020, section 7) or Siegel (2023, discussion on page 5). Another implication of this fact is that the constructed nonstandard architectures are necessarily very sensitive to minute changes in the network parameters.

5.3.2 Upper complexity bounds

Quantitative error estimates for operator learning based on the general approach outlined in the last Subsection 5.3.1 have been derived in a number of recent works (Liu et al., 2024; Franco et al., 2023; Galimberti et al., 2022; Bhattacharya et al., 2021; Hua and Lu, 2023; Mhaskar, 2023).

5.3.2.1 Relevant work

The two papers by Bhattacharya et al. (2021); Lanthaler et al. (2022), analyzing PCA-Net and DeepONet respectively, both introduce a splitting of the error into encoder, neural network approximation and reconstruction errors. A similar error analysis is employed in Hua and Lu (2023) for so-called "basis operator network", a variant of DeepONet. An in-depth analysis of DeepONets with various encoder/decoder pairs, including generalization error estimates, is given in Liu et al. (2024). Quantitative approximation error estimates for convolution neural networks applied to operator learning are derived in Franco et al. (2023). General error estimates motivated by infinite-dimensional dynamical systems in stochastic analysis can be found in Galimberti et al. (2022). An alternative approach to operator learning with explicit algorithms for all weights is proposed in Mhaskar (2023), including error estimates for this approach.

5.3.2.2 Alternative decompositions

Finally, we point out that while the error decomposition in encoding, neural network approximation and reconstruction errors is natural, alternative error decompositions, potentially more fine-grained, are possible. We mention the work by Patel et al. (2024) which proposes a mimetic neural operator architecture inspired by the weak variational form of elliptic PDEs, discretized by the finite-element method; starting from this idea, the authors arrive at an architecture that can be viewed as a variant of DeepONet, including a specific mixed nonlinear and linear branch network structure and a nonlinear trunk net. In this work, an a priori error analysis is conducted resulting in a splitting of the overall approximation in numerical approximation, stability, training and quadrature errors depending on the data-generation with a numerical scheme (no access to the actual operator), the Lipschitz stability of the underlying operator, the finite number of training samples and quadrature errors to approximate integrals, respectively.

5.3.3 Lower complexity bounds

Operator learning frameworks are based on neural networks and provide highly nonlinear approximation (DeVore, 1998). Despite their astonishing approximation capabilities, even highly nonlinear approximation methods have intrinsic limitations.

5.3.3.1 Combined error analysis for encoder-decoder-nets

To illustrate some of these intrinsic limitations, we first outline the combined error analysis that results from the decomposition summarized in the last section. To this end, we combine the encoding, reconstruction and neural network analysis to derive quantitative error and complexity estimates within the encoder-decoder-net paradigm.

Firstly, under reasonable assumptions on the input functions, the encoding error can often be shown to decay at an algebraic rate in the $d_{\mathcal{U}}$, e.g.

$$\mathscr{E}_{\mathcal{U}} \lesssim d_{\mathcal{U}}^{-\alpha}. \tag{5.4}$$

For example, if we assume that the input functions are defined on a bounded domain $\mathfrak{D} \subset \mathbb{R}^d$ and subject to a smoothness constraint such as a uniform bound on their k-th derivative, then a decay rate $\alpha = k/d$ can be achieved (depending on the precise setting); For dimension reduction by principal component analysis, the exponent α instead relates to the decay rate of the eigenvalues of the covariance operator of the input distribution.

Under similar assumptions on the set of output functions, depending on the properties of the underlying operator $\Psi^\dagger$, the reconstruction error on the output space often also decays algebraically,

$$\mathscr{E}_{\mathcal{V}} \lesssim d_{\mathcal{V}}^{-\beta}, \tag{5.5}$$

where the decay rate β can e.g. be estimated in terms of the smoothness of the output functions under $\Psi^\dagger$, or could be related to the decay of the PCA eigenvalues of the output distribution (push-forward under $\Psi^\dagger$).

Finally, given latent dimensions $d_{\mathcal{U}}$ and $d_{\mathcal{V}}$, the size of the neural network ψ that is required to approximate the encoded operator mapping $G : \mathbb{R}^{d_{\mathcal{U}}} \to \mathbb{R}^{d_{\mathcal{V}}}$, with NN approximation error bound,

$$\mathscr{E}_{\psi} \leq \varepsilon,$$

roughly scales as (cp. Remark 5.3),

$$\text{size}(\psi) \sim d_{\mathcal{V}} \varepsilon^{-d_{\mathcal{U}}/k}, \tag{5.6}$$

when the only information on the underlying operator is captured by its degree of smoothness k ($k = 1$ corresponding to Lipschitz continuity). Note that this is the scaling consistent with Kolmogorov entropy bounds.

Given the error decomposition $\mathscr{E} \lesssim \mathscr{E}_{\mathcal{U}} + \mathscr{E}_{\psi} + \mathscr{E}_{\mathcal{V}}$, we require each error contribution individually to be bounded by ε. In view of (5.4) and (5.5), this can be achieved provided that $d_{\mathcal{U}} \sim \varepsilon^{-1/\alpha}$, $d_{\mathcal{V}} \sim \varepsilon^{-1/\beta}$. Inserting such choice of $d_{\mathcal{U}}, d_{\mathcal{V}}$ in (5.6), we arrive at a neural network size of roughly the form,

$$\operatorname{size}(\psi) \sim \varepsilon^{-1/\beta} \varepsilon^{-c\varepsilon^{-1/\alpha}/k}.$$

In particular, we note the exponential dependence on ε^{-1}, resulting in a size estimate,

$$\operatorname{size}(\psi) \gtrsim \exp\left(\frac{c\varepsilon^{-1/\alpha}}{k}\right). \tag{5.7}$$

As pointed out after (5.4), when the set of input functions consists of functions defined on a d-dimensional domain with uniformly bounded s-th derivatives (in a suitable norm), then we expect a rate $\alpha = s/d$, in which case we obtain,

$$\operatorname{size}(\psi) \gtrsim \exp\left(\frac{c\varepsilon^{-d/s}}{k}\right). \tag{5.8}$$

For operator learning frameworks, this superalgebraic (even exponential) dependence of the complexity on ε^{-1} has been termed the "curse of dimensionality" in Kovachki et al. (2021); Lanthaler et al. (2022) or more recently "curse of parametric complexity" in Lanthaler (2023). The latter term was introduced to avoid confusion, which may arise because in these operator learning problems, there is no fixed dimension to speak of. The curse of parametric complexity can be viewed as an infinite-dimensional scaling limit of the finite-dimensional curse of dimensionality, represented by the $d_{\mathcal{U}}$-dependency of the bound $\varepsilon^{-d_{\mathcal{U}}/k}$, and arises from the finite-dimensional CoD by observing that the required latent dimension $d_{\mathcal{U}}$ itself depends on ε, with scaling $d_{\mathcal{U}} \sim \varepsilon^{-1/\alpha}$. We note in passing that even if $d_{\mathcal{U}} \sim \log(\varepsilon^{-1})$ were to scale only logarithmically in ε^{-1}, the complexity bound implied by (5.6) would still be superalgebraic, consistent with the main result of Lanthaler (2023).

5.3.3.2 Nonlinear n-width estimates

The rather pessimistic complexity bound outlined in (5.7) is based on an upper bound on the operator approximation error $\mathscr{E}$, and is not necessarily tight. One may therefore wonder if more careful estimates could yield complexity bounds that do not scale exponentially in ε^{-1}.

In this context, we would like to highlight the early work on operator approximation by Mhaskar and Hahm (1997) which presents first quantitative bounds for the approximation of nonlinear functionals; most notably, this work identifies the continuous nonlinear n-widths of spaces of Hölder continuous functionals defined on L^2-spaces; it is shown that the relevant n-widths decay only (poly-)logarithmically in n, including both upper and lower bounds.

We will presently state a simplified version of the main result of Mhaskar and Hahm (1997), and refer to the original work for the general version. To this end, we recall that the continuous nonlinear n-width $d_{\mathcal{N}}(\mathcal{K}; \|\cdot\|_{\mathcal{X}})$ (DeVore et al., 1989) of a subset $\mathcal{K} \subset \mathcal{X}$, with $(\mathcal{X}, \|\cdot\|_{\mathcal{X}})$ Banach, is defined as the optimal reconstruction error,

$$d_{\mathcal{N}}(\mathcal{K}; \|\cdot\|_{\mathcal{X}}) = \inf_{(a,M)} \sup_{f \in \mathcal{K}} \|f - M(a(f))\|_{\mathcal{X}},$$

where the infimum is over all encoder/decoder pairs (a, M), consisting of a continuous map $a : \mathcal{K} \to \mathbb{R}^n$ and general map $M : \mathbb{R}^n \to \mathcal{X}$.

To derive lower n-width bounds, we consider spaces of nonlinear Lipschitz functionals $\Psi^\dagger \in \mathfrak{F}_d$, where d denotes the spatial dimension of the input functions. More, precisely define

$$\mathfrak{F}_d = \left\{ \Psi^\dagger : L^2([-1,1]^d) \to \mathbb{R} \,\middle|\, \|\Psi^\dagger(u)\|_{\mathrm{Lip}} \le 1 \right\},$$

with

$$\|\Psi^\dagger\|_{\mathrm{Lip}} := \sup_{u \in L^2} |\Psi^\dagger(u)| + \sup_{u,v \in L^2} \frac{|\Psi^\dagger(u) - \Psi^\dagger(v)|}{\|u - v\|_{L^2}}.$$

Given a smoothness parameter $s > 0$, we consider approximation of $\Psi^\dagger \in \mathfrak{F}_d$, uniformly over a compact set of input functions $K^s \subset L^2([-1,1]^d)$, obtained as follows: we expand input functions $f \in L^2([-1,1]^d)$ in a Legendre expansion, $f(x) = \sum_{k \in \mathbb{N}^d} \widehat{f}_k P_k(x)$, and consider functionals defined on the "Sobolev" ball,

$$K^s := \left\{ f \in L^2([-1,1]^d) \,\middle|\, \sum_{k \in \mathbb{N}^d} |k|^{2s} |\widehat{f}_k|^2 \le 1 \right\}.$$

We measure the approximation error between $\Psi^\dagger, \Psi : K^s \subset L^2 \to \mathbb{R}$ with respect to the supremum norm over K^s,

$$\|\Psi^\dagger - \Psi\|_{C(K^s)} = \sup_{u \in K^s} |\Psi^\dagger(u) - \Psi(u)|.$$

It follows from the main results of Mhaskar and Hahm (1997) that the continuous nonlinear n-widths of the set of functionals $\mathfrak{F}_d$ decay only (poly-)logarithmically, as

$$d_n(\mathfrak{F}_d)_{C(K^s)} \sim \log(n)^{-s/d}.$$

In particular, to achieve uniform approximation accuracy $\varepsilon > 0$ with a *continuous* encoder/decoder pair (a, M), requires at least

$$n \gtrsim \exp\left(c\varepsilon^{-d/s}\right), \tag{5.9}$$

parameters. This last lower bound should be compared with (5.8) (for $k = 1$); the lower n-width bound (5.9) would imply (5.8) under the assumption that the architecture of $\psi = \psi(\cdot\,;\theta)$ was fixed and assuming that the weight assignment $\Psi^\dagger \mapsto \theta_{\Psi^\dagger}$ from the functional $\Psi^\dagger$ and the optimal tuning of neural network parameters $\theta_{\Psi^\dagger}$ was continuous. The latter assumption may not be satisfied if parameters are optimized using gradient descent.

5.3.3.3 Curse of parametric complexity

Given the result outlined in the previous paragraph, one may wonder if the pessimistic bound (5.9) and (5.7), i.e. the "curse of parametric complexity", can be broken by (a) a dis-continuous weight assignment, and (b) an adaptive choice of architecture optimized for specific $\Psi^\dagger$. This question has been raised in Lanthaler (2023); Lanthaler and Stuart (2023).

It turns out that, with operator learning frameworks such as DeepONet, FNO or PCA-Net, and relying on standard neural network architectures, it is not possible to overcome the curse of parametric complexity when considering approximation on the full class of Lipschitz continuous or Fréchet differentiable operators. We mention the following illustrative result for DeepONet, which follows from Lanthaler and Stuart (2023, Example 2.17):

Proposition 5.4 (Curse of Parametric Complexity). *Let $\mathfrak{D} \subset \mathbb{R}^d$ be a domain. Let $k \in \mathbb{N}$ be given, and consider the compact set of input functions,*

$$K = \left\{ u \in C^k(\mathfrak{D}) \,\middle|\, \|u\|_{C^k} \le 1 \right\} \subset \mathcal{U} := C(\mathfrak{D}).$$

Fix $\alpha > 2 + \frac{k}{d}$. Then for any $r \in \mathbb{N}$, there exists a r-times Fréchet differentiable functional $\Psi^\dagger : \mathcal{U} \to \mathbb{R}$ and constant $c, \overline{\varepsilon} > 0$, such that approximation to accuracy $\varepsilon \le \overline{\varepsilon}$ by a DeepONet $\Psi : \mathcal{U} \to \mathbb{R}$ with ReLU activation,

$$\sup_{u \in K} |\Psi^\dagger(u) - \Psi(u)| \le \varepsilon, \tag{5.10}$$

with linear encoder $\mathcal{E}$ and neural network ψ, implies complexity bound

$$\operatorname{size}(\psi) \ge \exp(c\varepsilon^{-1/\alpha r}). \tag{5.11}$$

Here $c, \overline{\varepsilon} > 0$ are constants depending only on k, α and r. $\Diamond$

As mentioned above, analogous lower complexity bounds can be obtained for PCA-Net, Fourier neural operator and many other architectures, and the more general version of this lower complexity bound applies to Sobolev input functions and beyond. We refer to Lanthaler and Stuart (2023) for a detailed discussion.

Remark 5.5. In Lanthaler and Stuart (2023), no attempt was made to optimize the exponent of ε^{-1} in (5.11). It would be interesting to know whether the appearance of the degree of operator Féchet differentiability, i.e. the parameter r, in the lower bound is merely an artifact of the proof in Lanthaler and Stuart

(2023). The back-of-the-envelope calculation leading to (5.7) indicates that r should not appear in the exponent, and that the factor $\alpha = k/d + 2 + \delta$ could be replaced by $k/d + \delta$. ◇

5.3.3.4 Breaking the curse of parametric complexity with nonstandard architectures

As mentioned in a previous section, there exist nonstandard neural network architectures which either employ nonstandard activations (Pinkus, 1999; Yarotsky, 2021; Shen et al., 2021), or impose a nonstandard "three-dimensional" connectivity (Zhang et al., 2022) which can overcome the curse of dimensionality in finite dimensions. In particular, encoder-decoder-nets based on such nonstandard architectures can achieve neural network approximation error $\mathscr{E}_\psi \leq \varepsilon$ with a complexity (as measured by the number of tunable degrees of freedom) that grows much slower than the rough scaling we considered in (5.6). Based on such architectures, it has recently been shown by Schwab et al. (2023) that DeepONets can achieve approximation rates for general Lipschitz and Hölder continuous operators which break the curse of parametric complexity implied by (5.11). In fact, such architectures achieve algebraic expression rate bounds for general Lipschitz and Hölder continuous operators.

5.3.4 *Sample complexity results*

There is a rapidly growing body of work on the approximation theory of operator learning with focus on parametric complexity. The complementary question of the sample complexity of operator learning, i.e. how many samples are needed to achieve a given approximation accuracy, has not received as much attention. The work described in Subsection 5.1 addresses this question in the setting of learning linear operators, and the question is also addressed in Subsection 5.2 for holomorphic operators. We now develop this subject further. Of particular note in the general Lipschitz setting of this subsection is the paper by Liu et al. (2024), as well as related recent work in Chen et al. (2023), which studies the nonparametric error estimation of Lipschitz operators for general encoder-decoder-net architectures. In Liu et al. (2024), nonasymptotic upper bounds for the generalization error of empirical risk minimizers on suitable classes of operator networks are derived. The results are stated in a Bochner $L^2(\mu)$ setting with input functions drawn from a probability measure μ, and variants are derived for general (Lipschitz) encoder/decoder pairs, for fixed basis encoder/decoder pairs, and for PCA encoder/decoder pairs. The analysis underlying the approximation error estimates in Liu et al. (2024) is based on a combination of best-approximation error estimates (parametric complexity) which are combined with statistical learning theory to derive sample complexity bounds.

For detailed results applicable to more general settings, we refer the reader to Liu et al. (2024). Here, we restrict attention to a representative result for fixed basis encoder/decoder pair (Liu et al., 2024, Corollary 3), obtained by projection onto a trigonometric basis.

To state this simplified result, consider a Lipschitz operator $\Psi^\dagger : \mathcal{U} \to \mathcal{V}$, mapping between spaces $\mathcal{U}, \mathcal{V} = L^2([-1,1]^d)$. We assume that there exists a constant $C > 0$, such that the probability measure $\mu \in \mathcal{P}(\mathcal{U})$ and its push-forward $\Psi^\dagger_\# \mu \in \mathcal{P}(\mathcal{V})$ are supported on periodic, continuously differentiable functions belonging to the set,

$$K := \left\{ u \in L^2([-1,1]^d) \,\middle|\, u \text{ is periodic}, \; \|u\|_{C^{k,\alpha}} \le C \right\}.$$

Then the squared approximation error

$$\mathcal{E}^2 := \mathbb{E}_{\text{data}} \mathbb{E}_{u \sim \mu} \| \mathcal{D}_{\mathcal{Y}} \circ \psi \circ \mathcal{E}_{\mathcal{X}}(u) - \Psi^\dagger(u) \|^2_{L^2}$$

satisfies the upper bound,

$$\mathcal{E}^2 \lesssim d_{\mathcal{V}}^{\frac{4+d_{\mathcal{U}}}{2+d_{\mathcal{U}}}} N^{-\frac{2}{2+d_{\mathcal{U}}}} \log^6(n) + d_{\mathcal{U}}^{-\frac{2s}{d}} + d_{\mathcal{V}}^{-\frac{2s}{d}}, \tag{5.12}$$

where N is the number of samples and $s = k + \alpha$ is the smoothness on the input and output spaces. The neural network ψ is a ReLU network of depth L and width p, satisfying (up to logarithmic terms),

$$Lp \sim d_{\mathcal{V}}^{\frac{d_{\mathcal{U}}}{4+2d_{\mathcal{U}}}} N^{\frac{d_{\mathcal{U}}}{4+2d_{\mathcal{U}}}}.$$

Comparing with (5.4) and (5.5), we can identify the last two terms in (5.12) as the encoding and reconstruction errors. The first term corresponds to a combination of neural network approximation and generalization errors.

To ensure that the total error $\mathcal{E} \le \varepsilon$ is below accuracy threshold ε, we first choose $d_{\mathcal{U}}, d_{\mathcal{V}} \sim \varepsilon^{-d/s}$, consistent with our discussion in Subsection 5.3.3. And according to the above estimate, we choose a number of samples of roughly the size $N \sim \varepsilon^{-(2+d_{\mathcal{U}})/2}$. Note that, once more, the additional ε-dependency of $d_{\mathcal{U}}$ implies that

$$N \gtrsim \exp\left(c\varepsilon^{-d/s}\right),$$

exhibits an exponential curse of complexity. This time, the curse is reflected by an exponential number of samples N that are required to achieve accuracy ε, rather than the parametric complexity. In turn, this implies that the size of the product Lp of the depth L and width p of the neural network ψ satisfies the lower bound,

$$Lp \gtrsim \exp\left(c\varepsilon^{-d/s}\right),$$

consistent with the expected curse of parametric complexity, (5.7). It is likely that the results of Liu et al. (2024) cannot be substantially improved in the considered setting of Lipschitz operators. Extending their work to a slightly different setting, Liu et al. (2024) also raise the question of low dimensional structure in operator learning, and derive error bounds decaying with a fast rate under suitable conditions, relying on low-dimensional latent structure of the input space.

In a related direction, Kim and Kang (2022) provide estimates on the Rademacher complexity of FNO. Generalization error estimates are derived based on these Rademacher complexity estimates, and the theoretical insights are compared with the empirical generalization error and the proposed capacity of FNO, in numerical experiments. Out-of-distribution risk bounds for neural operators with focus on the Helmholtz equation are discussed in depth in Benitez et al. (2023).

We finally mention the recent work by Mukherjee and Roy (2023), where a connection is made between the number of available samples n and the required size of the DeepONet reconstruction dimension $d_{\mathcal{V}}$. It is shown that when only noisy measurements are available, a scaling of the number of trunk basis functions $d_{\mathcal{V}} \gtrsim \sqrt{n}$ is required to achieve accurate approximation.

5.4 Structure beyond smoothness

The results summarized in the previous sections indicate that, when relying on standard neural network architectures, *efficient* operator learning on general spaces of Lipschitz continuous, or Fréchet differentiable, operators may not be possible: the class of all such operators on infinite-dimensional Banach spaces is arguably too rich, and operator learning on this class suffers from a curse of parametric complexity, requiring exponential model sizes of the form $\gtrsim \exp(c\varepsilon^{-\gamma})$.

This is in contrast to operator learning for $(\boldsymbol{b}, \varepsilon)$-holomorphic operators, for which approximation to accuracy ε is possible with a parametric complexity $O(\varepsilon^{-\gamma})$ scaling only algebraically in ε^{-1}. In this case, the curse of parametric complexity is broken by the extraordinary amount of smoothness of the underlying operators, going far beyond Lipschitz continuity or Fréchet differentiability.

These contrasting results rely only on the smoothness of the approximated operator: Is such smoothness the deciding factor for the practical success of operator learning methodologies? While we currently cannot provide a theoretical answer to this important question, we finally would like to mention several approximation theoretic results addressing how operator learning frameworks can break the curse of parametric complexity by leveraging structure beyond holomorphy.

5.4.0.1 Operator Barron spaces

A celebrated result in the study of shallow neural networks on finite-dimensional spaces is Barron's discovery (Barron, 1993) of a function space on which *dimension-independent* Monte-Carlo approximation rates $O(1/\sqrt{n})$ can be obtained. In particular, the approximation, by shallow neural networks, of functions belonging to this Barron class does not suffer from the well-known curse of dimensionality.

In the recent paper by Korolev (2022), a suitable generalization of the Barron spaces is introduced, and it is shown that Monte-Carlo approximation rates $O(1/\sqrt{n})$ can be obtained even in this infinite-dimensional setting, under precisely specified conditions. Quantitative error estimates (convergence rates) for the approximation of nonlinear operators are obtained by extending earlier results (Bach, 2017; E et al., 2022; E and Wojtowytsch, 2022) from the finite-dimensional setting $f : \mathbb{R}^d \to \mathbb{R}$ to the vector-valued and infinite-dimensional case $f : \mathcal{U} \to \mathcal{V}$, where $\mathcal{U}$ and $\mathcal{V}$ are Banach spaces.

The operator Barron spaces identified in Korolev (2022) represent a general class of operators, distinct from the holomorphic operators discussed in a previous section, which allow for efficient approximation by a class of "shallow neural operators". Unfortunately, a priori, it is unclear which operators of practical interest belong to this class, leaving the connection between these theoretical results and the practically observed efficiency of neural operator somewhat tenuous. In passing, we also mention the operator reproducing kernel Hilbert spaces (RKHS) considered in the context of the random feature model in Nelsen and Stuart (2021), for which similar Monte-Carlo convergence rates have been derived in Lanthaler and Nelsen (2023).

5.4.0.2 Representation formulae and emulation of numerical methods

To conclude our discussion of complexity and error bounds, we mention work focused on additional structure, separate from smoothness and the above-mentioned idea of Barron spaces, which can be leveraged by operator learning frameworks to achieve efficient approximation: these include operators with explicit representation formulae, and operators for which efficient approximation by traditional numerical schemes is possible. Such representations by classical methods can often be efficiently emulated by operator learning methodologies, resulting in error and complexity estimates that beat the curse of parametric complexity.

The complexity estimates for DeepONets in Deng et al. (2021) are mostly based on explicit representation of the solution; most prominently, this is achieved via the Cole-Hopf transformation for the viscous Burgers equation.

Results employing emulation of numerical methods to prove that operator learning frameworks such as DeepONet, FNO and PCA-Net can overcome the curse of parametric complexity for specific operators of interest can be found in Kovachki et al. (2021); Lanthaler et al. (2022); Lanthaler (2023); Marcati and Schwab (2023); specifically, such results have e.g. been obtained for the Darcy flow equation, the Navier-Stokes equations, reaction-diffusion equations and the inviscid Burgers equation. For the solution operators associated with these PDEs, it has been shown that operator learning frameworks can achieve approximation accuracy ε with a total number of tunable degrees of freedom which either only scales algebraically in ε, i.e. with $\text{size}(\psi) = O(\varepsilon^{-\gamma})$, or scales only logarithmically, size$(\psi) = O(|\log \varepsilon|^\gamma)$ in certain settings (Marcati and Schwab, 2023). This should be contrasted with the general curse of dimensionality (5.7). It is expected that the underlying ideas apply to many other PDEs.

Results in this direction are currently only available for very specific operators, and an abstract characterization of the relevant structure that can be exploited by operator learning frameworks is not available. First steps towards a more general theory have been proposed in Ryck and Mishra (2022), where generic bounds for operator learning are derived, relating the approximation error for physics-informed neural networks (PINNs) and operator learning architectures such as DeepONets and FNOs.

5.4.1 Discussion

Ultimately, the overarching theme behind many of the above cited results is that neural operators, or neural networks more generally, can efficiently emulate numerical algorithms, which either result from bespoke numerical methods or are a consequence of explicit representation formulae. The total complexity of a neural network emulator, and reflected by its size, is composed of the complexity of the emulated numerical algorithm and an additional overhead cost of emulating this algorithm by a neural network (translation to neural network weights). From an approximation-theoretic point of view, it could be conjectured that, for a suitable definition of "numerical algorithm", neural networks can efficiently approximate *all numerical algorithms*, hence implying efficient approximation of a great variety of operators, excluding only those operators for which no efficient numerical algorithms exist. Formalizing a suitable notion of numerical algorithm and proving that neural networks can efficiently emulate any such algorithm would be of interest and could provide a general way for proving algebraic expression rate bound for a general class of operators that can be approximated by a numerical method with algebraic memory and run-time complexity (i.e. any "reasonable" approximation method).

6 Conclusions

Neural operator architectures employ neural networks to approximate nonlinear operators mapping between Banach spaces of functions. Such operators often arise from physical models which are expressed as partial differential equations (PDEs). Despite their empirical success in a variety of applications, our theoretical understanding of neural operators remains incomplete. This review article summarizes recent progress and the current state of our theoretical understanding of neural operators, focusing on an approximation theoretic point of view.

The starting point of the theoretical analysis is universal approximation. Very general universal approximation results are now available for many of the proposed neural operator architectures. These results demonstrate that, given a sufficient number of parameters, neural operators can approximate a very wide variety of infinite-dimensional operators, providing a theoretical underpinning for diverse applications. Such universal approximation is arguably a necessary but not sufficient condition for the success of these architectures. In particular, universal approximation is inherently qualitative and does not guarantee that approximation to a desired accuracy is feasible at a practically realistic model size.

A number of more recent works thus aim to provide quantitative bounds on the required model size and the required number of input-/output-samples to achieve a desired accuracy ϵ. Most such results consider one of three settings: general Lipschitz (or Fréchet differentiable) operators, holomorphic operators, or specific PDE operators. While Lipschitz operators are a natural and general class to consider, it turns out that approximation to error ϵ with standard architectures requires an exponential (in ϵ^{-1}) number of tunable parameters, bringing into question whether operator learning at this level of generality is possible. In contrast, the class of holomorphic operators allows for complexity bounds that scale only algebraically in ϵ^{-1}, both in terms of models size as well as sample complexity. Holomorphic operators represent a general class of operators of practical interest, for which rigorous approximation theory has been developed building on convergent (generalized) polynomial expansions.

Going beyond notions of operator smoothness, it has been shown that operator learning frameworks can leverage intrinsic structure of (PDE-) operators to achieve algebraic convergence rates in theory; this intrinsic structure is distinct from holomorphy. Available results in this direction currently rely on a case-by-case analysis and often leverage emulation of traditional numerical methods. The authors of the present article view the development of a general approximation theory, including a characterization of the relevant structure that can be leveraged by neural operators, as one of the great challenges of this field.

Acknowledgments

The authors are grateful to Dima Burov and Edo Calvello for creating Figs. 1a and 1b, 2a, 2b respectively. They are also grateful to Nick Nelsen for reading, and commenting on, a draft of the paper. NBK is grateful to the NVIDIA Corporation for full time employment. SL is grateful for support from the Swiss National Science Foundation, Postdoc.Mobility grant P500PT-206737. AMS is grateful for support from a Department of Defense Vannevar Bush Faculty Fellowship.

References

Abernethy, J., Bach, F., Evgeniou, T., Vert, J.-P., 2009. A new approach to collaborative filtering: operator estimation with spectral regularization. Journal of Machine Learning Research 10, 803–826.

Achour, E.M., Foucault, A., Gerchinovitz, S., Malgouyres, F., 2022. A general approximation lower bound in L^p norm, with applications to feed-forward neural networks. In: Advances in Neural Information Processing Systems.

Adcock, B., Brugiapaglia, S., Dexter, N., Moraga, S., 2022a. On efficient algorithms for computing near-best polynomial approximations to high-dimensional, Hilbert-valued functions from limited samples. arXiv preprint. arXiv:2203.13908.

Adcock, B., Brugiapaglia, S., Dexter, N., Moraga, S., 2022b. Near-optimal learning of Banach-valued, high-dimensional functions via deep neural networks. arXiv preprint. arXiv:2211.12633.
Adcock, B., Brugiapaglia, S., Webster, C.G., 2022c. Sparse Polynomial Approximation of High-Dimensional Functions, vol. 25. SIAM.
Adcock, B., Dexter, N., Moraga, S., 2023a. Optimal approximation of infinite-dimensional holomorphic functions. arXiv preprint. arXiv:2305.18642.
Adcock, B., Dexter, N., Moraga, S., 2023b. Optimal approximation of infinite-dimensional holomorphic functions ii: recovery from iid pointwise samples. arXiv preprint. arXiv:2310.16940.
Bach, F., 2017. Breaking the curse of dimensionality with convex neural networks. Journal of Machine Learning Research 18 (1), 629–681.
Bachmayr, M., Cohen, A., 2017. Kolmogorov widths and low-rank approximations of parametric elliptic PDEs. Mathematics of Computation 86 (304), 701–724.
Barron, A.R., 1993. Universal approximation bounds for superpositions of a sigmoidal function. IEEE Transactions on Information Theory 39 (3), 930–945.
Bartolucci, F., de Bézenac, E., Raonić, B., Molinaro, R., Mishra, S., Alaifari, R., 2023. Are neural operators really neural operators? Frame theory meets operator learning. arXiv preprint. arXiv:2305.19913.
Batlle, P., Darcy, M., Hosseini, B., Owhadi, H., 2024. Kernel methods are competitive for operator learning. Journal of Computational Physics 496, 112549.
Benitez, J.A.L., Furuya, T., Faucher, F., Kratsios, A., Tricoche, X., de Hoop, M.V., 2023. Out-of-distributional risk bounds for neural operators with applications to the Helmholtz equation. arXiv preprint. arXiv:2301.11509.
Bhattacharya, K., Hosseini, B., Kovachki, N.B., Stuart, A.M., 2021. Model reduction and neural networks for parametric PDEs. The SMAI Journal of Computational Mathematics 7, 121–157.
Bolcskei, H., Grohs, P., Kutyniok, G., Petersen, P., 2019. Optimal approximation with sparsely connected deep neural networks. SIAM Journal on Mathematics of Data Science 1 (1), 8–45.
Bonev, B., Kurth, T., Hundt, C., Pathak, J., Baust, M., Kashinath, K., Anandkumar, A., 2023. Spherical Fourier neural operators: learning stable dynamics on the sphere. In: Proceedings of the 40th International Conference on Machine Learning.
Boullé, N., Earls, C.J., Townsend, A., 2022a. Data-driven discovery of Green's functions with human-understandable deep learning. Scientific Reports 12 (1), 4824.
Boullé, N., Kim, S., Shi, T., Townsend, A., 2022b. Learning Green's functions associated with time-dependent partial differential equations. Journal of Machine Learning Research 23 (1), 9797–9830.
Boullé, N., Townsend, A., 2023. Learning elliptic partial differential equations with randomized linear algebra. Foundations of Computational Mathematics 23 (2), 709–739.
Brunton, S.L., Kutz, J.N., 2023. Machine learning for partial differential equations. arXiv preprint. arXiv:2303.17078.
Castro, J., 2023. The Kolmogorov infinite dimensional equation in a Hilbert space via deep learning methods. Journal of Mathematical Analysis and Applications 527 (2), 127413.
Castro, J., Muñoz, C., Valenzuela, N., 2022. The Calderón's problem via DeepONets. arXiv preprint. arXiv:2212.08941.
Chen, K., Wang, C., Yang, H., 2023. Deep operator learning lessens the curse of dimensionality for pdes. arXiv preprint. arXiv:2301.12227.
Chen, T., Chen, H., 1995. Universal approximation to nonlinear operators by neural networks with arbitrary activation functions and its application to dynamical systems. IEEE Transactions on Neural Networks 6 (4), 911–917.
Chkifa, A., Cohen, A., DeVore, R., Schwab, C., 2013. Sparse adaptive Taylor approximation algorithms for parametric and stochastic elliptic PDEs. ESAIM: Mathematical Modelling and Numerical Analysis 47 (1), 253–280.
Chkifa, A., Cohen, A., Schwab, C., 2015. Breaking the curse of dimensionality in sparse polynomial approximation of parametric PDEs. Journal de Mathématiques Pures et Appliquées 103 (2), 400–428.

Cohen, A., Dahmen, W., Mula, O., Nichols, J., 2022. Nonlinear reduced models for state and parameter estimation. SIAM/ASA Journal on Uncertainty Quantification 10 (1), 227–267.

Cohen, A., DeVore, R., 2015. Approximation of high-dimensional parametric PDEs. Acta Numerica 24, 1–159.

Cohen, A., DeVore, R., Schwab, C., 2010. Convergence rates of best n-term Galerkin approximations for a class of elliptic SPDEs. Foundations of Computational Mathematics 10 (6), 615–646.

Cohen, A., Devore, R., Schwab, C., 2011. Analytic regularity and polynomial approximation of parametric and stochastic elliptic PDEs. Analysis and Applications 9 (01), 11–47.

Colbrook, M.J., Townsend, A., 2024. Rigorous data-driven computation of spectral properties of Koopman operators for dynamical systems. Communications on Pure and Applied Mathematics 77 (1), 221–283.

Cotter, S., Roberts, G., Stuart, A., White, D., 2012. MCMC methods for functions: modifying old algorithms to make them faster. Statistical Science 28, 02.

Crambes, C., Mas, A., 2013. Asymptotics of prediction in functional linear regression with functional outputs. Bernoulli 19 (5B), 2627–2651.

Cybenko, G., 1989. Approximation by superpositions of a sigmoidal function. MCSS. Mathematics of Control, Signals and Systems 2 (4), 303–314.

De Hoop, M., Huang, D.Z., Qian, E., Stuart, A.M., 2022. The cost-accuracy trade-off in operator learning with neural networks. Journal of Machine Learning 1 (3), 299–341.

de Hoop, M.V., Kovachki, N.B., Nelsen, N.H., Stuart, A.M., 2023. Convergence rates for learning linear operators from noisy data. SIAM/ASA Journal on Uncertainty Quantification 11 (2), 480–513.

Deng, B., Shin, Y., Lu, L., Zhang, Z., Karniadakis, G.E., 2021. Convergence rate of DeepONets for learning operators arising from advection-diffusion equations. arXiv preprint. arXiv:2102.10621.

DeVore, R.A., 1998. Nonlinear approximation. Acta Numerica 7, 51–150.

DeVore, R.A., Howard, R., Micchelli, C., 1989. Optimal nonlinear approximation. Manuscripta Mathematica 63, 469–478.

DeVore, R.A., Lorentz, G.G., 1993. Constructive Approximation, vol. 303. Springer Science & Business Media.

E, W., Ma, C., Wu, L., 2022. The Barron space and the flow-induced function spaces for neural network models. Constructive Approximation 55 (1), 369–406.

E, W., Wojtowytsch, S., 2022. Representation formulas and pointwise properties for Barron functions. Calculus of Variations and Partial Differential Equations 61 (2), 1–37.

Enflo, P., 1973. A counterexample to the approximation problem in Banach spaces. Acta Mathematica 130, 309–316.

Evans, L.C., 2010. Partial Differential Equations, vol. 19. American Mathematical Soc.

Franco, N.R., Fresca, S., Manzoni, A., Zunino, P., 2023. Approximation bounds for convolutional neural networks in operator learning. Neural Networks 161, 129–141.

Furuya, T., Puthawala, M., Lassas, M., de Hoop, M.V., 2024. Globally injective and bijective neural operators. In: Thirty-Eight Conference on Neural Information Processing Systems.

Galimberti, L., Kratsios, A., Livieri, G., 2022. Designing universal causal deep learning models: the case of infinite-dimensional dynamical systems from stochastic analysis. arXiv preprint. arXiv:2210.13300.

Gin, C.R., Shea, D.E., Brunton, S.L., Kutz, J.N., 2020. Deepgreen: deep learning of Green's functions for nonlinear boundary value problems. arXiv preprint. arXiv:2101.07206.

Goodfellow, I., Bengio, Y., Courville, A., 2016. Deep Learning. MIT Press.

Gupta, J.K., Brandstetter, J., 2022. Towards multi-spatiotemporal-scale generalized PDE modeling. arXiv preprint. arXiv:2209.15616.

Hairer, M., Stuart, A.M., Vollmer, S.J., 2014. Spectral gaps for a Metropolis–Hastings algorithm in infinite dimensions. The Annals of Applied Probability 24 (6), 2455–2490.

Herrmann, L., Schwab, C., Zech, J., 2022. Neural and GPC operator surrogates: construction and expression rate bounds. arXiv preprint. arXiv:2207.04950.

Hesthaven, J.S., Ubbiali, S., 2018. Non-intrusive reduced order modeling of nonlinear problems using neural networks. Journal of Computational Physics 363, 55–78.
Hinze, M., Pinnau, R., Ulbrich, M., Ulbrich, S., 2008. Optimization with PDE Constraints, vol. 23. Springer Science & Business Media.
Hornik, K., Stinchcombe, M., White, H., 1989. Multilayer feedforward networks are universal approximators. Neural Networks 2 (5), 359–366.
Hua, N., Lu, W., 2023. Basis operator network: a neural network-based model for learning nonlinear operators via neural basis. Neural Networks 164, 21–37.
Huang, D.Z., Nelsen, N.H., Trautner, M., 2024. An operator learning perspective on parameter-to-observable maps. arXiv preprint. arXiv:2402.06031.
Huang, D.Z., Xu, K., Farhat, C., Darve, E., 2020. Learning constitutive relations from indirect observations using deep neural networks. Journal of Computational Physics 416, 109491.
Jin, J., Lu, Y., Blanchet, J., Ying, L., 2022a. Minimax optimal kernel operator learning via multilevel training. In: Eleventh International Conference on Learning Representations.
Jin, P., Meng, S., Lu, L., 2022b. Mionet: learning multiple-input operators via tensor product. SIAM Journal on Scientific Computing 44 (6), A3490–A3514.
Kaipio, J., Somersalo, E., 2006. Statistical and Computational Inverse Problems, vol. 160. Springer Science & Business Media.
Kim, T., Kang, M., 2022. Bounding the Rademacher complexity of Fourier neural operator. arXiv preprint. arXiv:2209.05150.
Klebanov, I., Sullivan, T., 2023. Aperiodic table of modes and maximum a posteriori estimators. arXiv preprint. arXiv:2306.16278.
Korolev, Y., 2022. Two-layer neural networks with values in a Banach space. SIAM Journal on Mathematical Analysis 54 (6), 6358–6389.
Kostic, V., Novelli, P., Maurer, A., Ciliberto, C., Rosasco, L., Pontil, M., 2022. Learning dynamical systems via Koopman operator regression in reproducing kernel Hilbert spaces. In: Thirty-Sixth Annual Conference on Neural Information Processing Systems.
Kovachki, N.B., 2022. Machine Learning and Scientific Computing. PhD thesis. California Institute of Technology.
Kovachki, N.B., Lanthaler, S., Mishra, S., 2021. On universal approximation and error bounds for Fourier neural operators. Journal of Machine Learning Research 22 (1).
Kovachki, N.B., Li, Z., Liu, B., Azizzadenesheli, K., Bhattacharya, K., Stuart, A., Anandkumar, A., 2023. Neural operator: learning maps between function spaces with applications to PDEs. Journal of Machine Learning Research 24 (89).
Kramer, B., Peherstorfer, B., Willcox, K.E., 2024. Learning nonlinear reduced models from data with operator inference. Annual Review of Fluid Mechanics 56 (1), 521–548.
Kutyniok, G., Petersen, P., Raslan, M., Schneider, R., 2022. A theoretical analysis of deep neural networks and parametric PDEs. Constructive Approximation 55 (1), 73–125.
Lam, R., Sanchez-Gonzalez, A., Willson, M., Wirnsberger, P., Fortunato, M., Alet, F., Ravuri, S., Ewalds, T., Eaton-Rosen, Z., Hu, W., Merose, A., Hoyer, S., Holland, G., Vinyals, O., Stott, J., Pritzel, A., Mohamed, S., Battaglia, P., 2023. Learning skillful medium-range global weather forecasting. Science 382 (6677), 1416–1421.
Lanthaler, S., 2023. Operator learning with PCA-Net: upper and lower complexity bounds. Journal of Machine Learning Research 24 (318).
Lanthaler, S., Li, Z., Stuart, A.M., 2023a. The nonlocal neural operator: universal approximation. arXiv preprint. arXiv:2304.13221.
Lanthaler, S., Mishra, S., Karniadakis, G.E., 2022. Error estimates for DeepONets: a deep learning framework in infinite dimensions. Transactions of Mathematics and Its Applications 6 (1).
Lanthaler, S., Molinaro, R., Hadorn, P., Mishra, S., 2023b. Nonlinear reconstruction for operator learning of PDEs with discontinuities. In: Eleventh International Conference on Learning Representations.
Lanthaler, S., Nelsen, N.H., 2023. Error bounds for learning with vector-valued random features. In: Thirty-Seventh Conference on Neural Information Processing Systems.

Lanthaler, S., Rusch, T.K., Mishra, S., 2023c. Neural oscillators are universal. In: Thirty-Seventh Conference on Neural Information Processing Systems.

Lanthaler, S., Stuart, A.M., 2023. The curse of dimensionality in operator learning. arXiv preprint. arXiv:2306.15924.

Lee, K., Carlberg, K.T., 2020. Model reduction of dynamical systems on nonlinear manifolds using deep convolutional autoencoders. Journal of Computational Physics 404, 108973.

Lei, Z., Shi, L., Zeng, C., 2022. Solving parametric partial differential equations with deep rectified quadratic unit neural networks. Journal of Scientific Computing 93 (3), 80.

Li, Q., Dietrich, F., Bollt, E.M., Kevrekidis, I.G., 2017. Extended dynamic mode decomposition with dictionary learning: a data-driven adaptive spectral decomposition of the Koopman operator. Chaos 27 (10), 103111.

Li, Z., Kovachki, N.B., Azizzadenesheli, K., Liu, B., Bhattacharya, K., Stuart, A.M., Anandkumar, A., 2020a. Neural operator: graph kernel network for partial differential equations. arXiv preprint. arXiv:2003.03485.

Li, Z., Kovachki, N.B., Azizzadenesheli, K., Liu, B., Bhattacharya, K., Stuart, A.M., Anandkumar, A., 2021. Fourier neural operator for parametric partial differential equations. In: Ninth International Conference on Learning Representations.

Li, Z., Kovachki, N.B., Azizzadenesheli, K., Liu, B., Stuart, A.M., Bhattacharya, K., Anandkumar, A., 2020b. Multipole graph neural operator for parametric partial differential equations. In: Thirty-Fourth Annual Conference on Neural Information Processing Systems.

Liang, S., Lyu, L., Wang, C., Yang, H., 2021. Reproducing activation function for deep learning. arXiv preprint. arXiv:2101.04844.

Lindenstrauss, J., Tzafriri, L., 2013. Classical Banach Spaces I: Sequence Spaces. Ergebnisse der Mathematik und ihrer Grenzgebiete. 2. Folge. Springer, Berlin Heidelberg.

Ling, J., Kurzawski, A., Templeton, J., 2016. Reynolds averaged turbulence modelling using deep neural networks with embedded invariance. Journal of Fluid Mechanics 807, 155–166.

Lippe, P., Veeling, B., Perdikaris, P., Turner, R., Brandstetter, J., 2024. PDE-refiner: achieving accurate long rollouts with neural PDE solvers. In: Thirty-Seventh Annual Conference on Neural Information Processing Systems.

Litany, O., Remez, T., Rodola, E., Bronstein, A., Bronstein, M., 2017. Deep functional maps: structured prediction for dense shape correspondence. In: Proceedings of the IEEE International Conference on Computer Vision, pp. 5659–5667.

Liu, H., Yang, H., Chen, M., Zhao, T., Liao, W., 2024. Deep nonparametric estimation of operators between infinite dimensional spaces. Journal of Machine Learning Research 25 (24), 1–67.

Liu, X., Tian, S., Tao, F., Yu, W., 2021. A review of artificial neural networks in the constitutive modeling of composite materials. Composites. Part B, Engineering 224, 109152.

Lu, L., Jin, P., Pang, G., Zhang, Z., Karniadakis, G.E., 2021. Learning nonlinear operators via DeepONet based on the universal approximation theorem of operators. Nature Machine Intelligence 3 (3), 218–229.

Luo, D., O'Leary-Roseberry, T., Chen, P., Ghattas, O., 2023. Efficient PDE-constrained optimization under high-dimensional uncertainty using derivative-informed neural operators. arXiv preprint. arXiv:2305.20053.

Marcati, C., Schwab, C., 2023. Exponential convergence of deep operator networks for elliptic partial differential equations. SIAM Journal on Numerical Analysis 61 (3), 1513–1545.

Mhaskar, H., 2023. Local approximation of operators. Applied and Computational Harmonic Analysis 64, 194–228.

Mhaskar, H., Poggio, T., 2020. Function approximation by deep networks. Communications on Pure and Applied Analysis 19 (8), 4085–4095.

Mhaskar, H.N., Hahm, N., 1997. Neural networks for functional approximation and system identification. Neural Computation 9 (1), 143–159.

Mollenhauer, M., Mücke, N., Sullivan, T., 2022. Learning linear operators: infinite-dimensional regression as a well-behaved non-compact inverse problem. arXiv preprint. arXiv:2211.08875.

Morton, J., Witherden, F.D., Jameson, A., Kochenderfer, M.J., 2018. Deep dynamical modeling and control of unsteady fluid flows. In: Proceedings of the 32nd International Conference on Neural Information Processing Systems.
Mukherjee, A., Roy, A., 2023. Size lowerbounds for deep operator networks. arXiv preprint. arXiv: 2308.06338.
Nelsen, N.H., Stuart, A.M., 2021. The random feature model for input-output maps between Banach spaces. SIAM Journal on Scientific Computing 43 (5), A3212–A3243.
O'Leary-Roseberry, T., Chen, P., Villa, U., Ghattas, O., 2024. Derivative-informed neural operator: an efficient framework for high-dimensional parametric derivative learning. Journal of Computational Physics 496, 112555.
Opschoor, J.A., Schwab, C., Zech, J., 2022. Exponential ReLU DNN expression of holomorphic maps in high dimension. Constructive Approximation 55 (1), 537–582.
Ovsjanikov, M., Ben-Chen, M., Solomon, J., Butscher, A., Guibas, L., 2012. Functional maps: a flexible representation of maps between shapes. ACM Transactions on Graphics 31 (4).
Patel, D., Ray, D., Abdelmalik, M.R., Hughes, T.J., Oberai, A.A., 2024. Variationally mimetic operator networks. Computer Methods in Applied Mechanics and Engineering 419, 116536.
Patel, R.G., Desjardins, O., 2018. Nonlinear integro-differential operator regression with neural networks. arXiv preprint. arXiv:1810.08552.
Patel, R.G., Trask, N.A., Wood, M.A., Cyr, E.C., 2021. A physics-informed operator regression framework for extracting data-driven continuum models. Computer Methods in Applied Mechanics and Engineering 373, 113500.
Peherstorfer, B., Willcox, K., 2016. Data-driven operator inference for nonintrusive projection-based model reduction. Computer Methods in Applied Mechanics and Engineering 306, 196–215.
Pinkus, A., 1999. Approximation theory of the mlp model in neural networks. Acta Numerica 8, 143–195.
Prasthofer, M., De Ryck, T., Mishra, S., 2022. Variable-input deep operator networks. arXiv preprint. arXiv:2205.11404.
Qian, E., Farcas, I.-G., Willcox, K., 2022. Reduced operator inference for nonlinear partial differential equations. SIAM Journal on Scientific Computing 44 (4), A1934–A1959.
Qian, E., Kramer, B., Peherstorfer, B., Willcox, K., 2020. Lift & Learn: physics-informed machine learning for large-scale nonlinear dynamical systems. Physica D. Nonlinear Phenomena 406, 132401.
Raghu, M., Schmidt, E., 2020. A survey of deep learning for scientific discovery. arXiv preprint. arXiv:2003.11755.
Rahimi, A., Recht, B., 2007. Random features for large-scale kernel machines. In: Twenty-First Annual Conference on Neural Information Processing Systems.
Rahman, M.A., Ross, Z.E., Azizzadenesheli, K., 2023. U-NO: U-shaped neural operators. In: Transactions on Machine Learning Research.
Ramsay, J.O., Dalzell, C., 1991. Some tools for functional data analysis. Journal of the Royal Statistical Society, Series B, Statistical Methodology 53 (3), 539–561.
Raonic, B., Molinaro, R., Rohner, T., Mishra, S., de Bezenac, E., 2023a. Convolutional neural operators. In: ICLR 2023 Workshop on Physics for Machine Learning.
Raonic, B., Molinaro, R., Ryck, T.D., Rohner, T., Bartolucci, F., Alaifari, R., Mishra, S., de Bezenac, E., 2023b. Convolutional neural operators for robust and accurate learning of PDEs. In: Thirty-Seventh Conference on Neural Information Processing Systems.
Rosasco, L., Belkin, M., Vito, E.D., 2010. On learning with integral operators. Journal of Machine Learning Research 11, 905–934.
Ryck, T.D., Mishra, S., 2022. Generic bounds on the approximation error for physics-informed (and) operator learning. In: Thirty-Sixth Annual Conference on Neural Information Processing Systems.
Sacks, J., Welch, W.J., Mitchell, T.J., Wynn, H.P., 1989. Design and analysis of computer experiments. Statistical Science 4 (4), 409–423.

Schmidt-Hieber, J., 2020. Nonparametric regression using deep neural networks with ReLU activation function. The Annals of Statistics 48 (4), 1875–1897. https://doi.org/10.1214/19-AOS1875.

Schwab, C., Stein, A., Zech, J., 2023. Deep operator network approximation rates for Lipschitz operators. arXiv preprint. arXiv:2307.09835.

Schwab, C., Zech, J., 2019. Deep learning in high dimension: neural network expression rates for generalized polynomial chaos expansions in UQ. Analysis and Applications 17 (01), 19–55.

Schwab, C., Zech, J., 2023. Deep learning in high dimension: neural network expression rates for analytic functions in $L^2(\mathbb{R}^d, \gamma_d)$. SIAM/ASA Journal on Uncertainty Quantification 11 (1), 199–234.

Seidman, J., Kissas, G., Perdikaris, P., Pappas, G.J., 2022. NOMAD: nonlinear manifold decoders for operator learning. In: Thirty-Sixth Annual Conference on Neural Information Processing Systems.

Sharp, N., Attaiki, S., Crane, K., Ovsjanikov, M., 2022. Diffusionnet: discretization agnostic learning on surfaces. ACM Transactions on Graphics.

Shen, Z., Yang, H., Zhang, S., 2019. Nonlinear approximation via compositions. Neural Networks 119, 74–84.

Shen, Z., Yang, H., Zhang, S., 2021. Neural network approximation: three hidden layers are enough. Neural Networks 141, 160–173.

Siegel, J.W., 2023. Optimal approximation rates for deep ReLU neural networks on Sobolev and Besov spaces. Journal of Machine Learning Research 24 (357), 1–52.

Stanziola, A., Arridge, S.R., Cox, B.T., Treeby, B.E., 2021. A Helmholtz equation solver using unsupervised learning: application to transcranial ultrasound. Journal of Computational Physics 441, 110430.

Stepaniants, G., 2023. Learning partial differential equations in reproducing kernel Hilbert spaces. Journal of Machine Learning Research 24 (86), 1–72.

Stuart, A.M., 2010. Inverse problems: a Bayesian perspective. Acta Numerica 19, 451–559.

Swischuk, R., Mainini, L., Peherstorfer, B., Willcox, K., 2019. Projection-based model reduction: formulations for physics-based machine learning. Computers & Fluids 179, 704–717.

Tripura, T., Chakraborty, S., 2023. Wavelet neural operator for solving parametric partial differential equations in computational mechanics problems. Computer Methods in Applied Mechanics and Engineering 404, 115783.

Vaswani, A., Shazeer, N., Parmar, N., Uszkoreit, J., Jones, L., Gomez, A.N., Kaiser, Ł., Polosukhin, I., 2017. Attention is all you need. In: Thirty-First Annual Conference on Neural Information Processing Systems.

Wang, C., Townsend, A., 2023. Operator learning for hyperbolic partial differential equations. arXiv preprint. arXiv:2312.17489.

Wang, D., Zhao, Z., Yu, Y., Willett, R., 2022. Functional linear regression with mixed predictors. Journal of Machine Learning Research 23 (1), 12181–12274.

Wang, J.-X., Wu, J.-L., Xiao, H., 2017. Physics-informed machine learning approach for reconstructing Reynolds stress modeling discrepancies based on DNS data. Physical Review Fluids 2, 034603.

Wu, J.-L., Xiao, H., Paterson, E., 2018. Physics-informed machine learning approach for augmenting turbulence models: a comprehensive framework. Physical Review Fluids 3, 074602.

Xu, K., Huang, D.Z., Darve, E., 2021. Learning constitutive relations using symmetric positive definite neural networks. Journal of Computational Physics 428, 110072.

Yarotsky, D., 2017. Error bounds for approximations with deep ReLU networks. Neural Networks 94, 103–114.

Yarotsky, D., 2021. Elementary superexpressive activations. In: Thirty-Eighth International Conference on Machine Learning.

Yarotsky, D., Zhevnerchuk, A., 2020. The phase diagram of approximation rates for deep neural networks. In: Thirty-Fourth Annual Conference on Neural Information Processing Systems.

Yeung, E., Kundu, S., Hodas, N., 2019. Learning deep neural network representations for Koopman operators of nonlinear dynamical systems. In: 2019 American Control Conference (ACC).

Yi, L., Su, H., Guo, X., Guibas, L.J., 2017. Syncspeccnn: synchronized spectral CNN for 3D shape segmentation. In: Proceedings of the IEEE Conference on Computer Vision and Pattern Recognition.

Zahm, O., Constantine, P.G., Prieur, C., Marzouk, Y.M., 2020. Gradient-based dimension reduction of multivariate vector-valued functions. SIAM Journal on Scientific Computing 42 (1), A534–A558.

Zhang, S., Shen, Z., Yang, H., 2022. Neural network architecture beyond width and depth. In: Thirty-Sixth Annual Conference on Neural Information Processing Systems.

Zhang, Z., Tat, L., Schaeffer, H., 2023. BelNet: basis enhanced learning, a mesh-free neural operator. Proceedings of the Royal Society A 479.

Chapter 10

A structure-preserving domain decomposition method for data-driven modeling☆

Shuai Jiang[a], Jonas Actor[a], Scott Roberts[b], and Nathaniel Trask[c,*]
[a]*Center for Computing Research, Sandia National Laboratories, Albuquerque, NM, United States,*
[b]*Engineering Sciences Center, Sandia National Laboratories, Albuquerque, NM, United States,*
[c]*Department of Mechanical Engineering and Applied Mechanics, University of Pennsylvania, Philadelphia, PA, United States*
**Corresponding author: e-mail address: ntrask@seas.upenn.edu*

Contents

☆ **Funding:** This article has been authored by an employee of National Technology & Engineering Solutions of Sandia, LLC under Contract No. DE-NA0003525 with the U.S. Department of Energy (DOE). The employee owns all right, title and interest in and to the article and is solely responsible for its contents. The United States Government retains and the publisher, by accepting the article for publication, acknowledges that the United States Government retains a nonexclusive, paid-up, irrevocable, world-wide license to publish or reproduce the published form of this article or allow others to do so, for United States Government purposes. The DOE will provide public access to these results of federally sponsored research in accordance with the DOE Public Access Plan https://www.energy.gov/downloads/doe-public-access-plan.

Abstract

We present a domain decomposition strategy for developing structure-preserving finite element discretizations from data when exact governing equations are unknown. On subdomains, trainable Whitney form elements are used to identify structure-preserving models from data, providing a Dirichlet-to-Neumann map which may be used to globally construct a mortar method. The reduced-order local elements may be trained offline to reproduce high-fidelity Dirichlet data in cases where first principles model derivation is either intractable, unknown, or computationally prohibitive. In such cases, particular care must be taken to preserve structure on both local and mortar levels without knowledge of the governing equations, as well as to ensure well-posedness and stability of the resulting monolithic data-driven system. This strategy provides a flexible means of both scaling to large systems and treating complex geometries, and is particularly attractive for multi-scale problems with complex microstructure geometry. While consistency is traditionally obtained in finite element methods via quasioptimality results and the Bramble-Hilbert lemma as the local element diameter $h \to 0$, our analysis establishes notions of accuracy and stability for finite h with accuracy coming from matching data. Numerical experiments and analysis establish properties for $H(\text{div})$ problems in small data limits ($\mathcal{O}(1)$ reference solutions).

Keywords

Structure preservation, Mortar method, Domain decomposition, Whitney forms, Model reduction, Data-driven modelling, Scientific machine learning

MSC Codes

68T01, 65N30, 65N55

1 Introduction

We consider the problem of identifying a model from data when the governing equations are unknown, but the conservation structure is known. Namely, one may know that fluxes associated with mass, momentum, or energy are conserved, but be unable to derive specific expressions for those fluxes.

We assume a class of models of the form

$$\begin{aligned} \nabla \cdot \boldsymbol{u} &= -f && \text{on } \Omega, \\ \boldsymbol{u} &= h(p;\theta) && \text{on } \Omega, \\ p &= g && \text{on } \partial\Omega \end{aligned} \tag{1.1}$$

where $\Omega \in \mathbb{R}^d$ is a Lipschitz domain, $f \in L^2(\Omega)$ forcing term, g Dirichlet data, and h a closure for the flux of unknown functional form approximated by a family of nonparametric regressors parameterized by θ. We demonstrate on $\Omega \in \mathbb{R}^2$ exclusively, but the techniques shown here generalize to higher dimensions and arbitrary manifolds. For this class of problems, data is provided in the form

$\mathcal{D} = \{(\boldsymbol{u}_k, f_k, h_k, g_k)\}_{k=1}^N$ and one identifies parameters θ which minimize error in a suitable norm, providing a model which may generalize by solving for choices of f and h outside the training set.

By casting data-driven modeling in such a *structure-preserving* framework, one aims to identify a model which balances a trade-off between rigorous preservation of physical/algebraic/stability structure while maintaining "black-box" approximation of as large a class of models as possible. This lies on a spectrum of methods in the literature spanning a trade-off between expressivity and exploitable structure. For example, operator regression methods aim to directly identify a solution map $(f, h) \to \boldsymbol{u}$ via interpolation in unconstrained Hilbert spaces (high expressivity), while PDE-constrained optimization (Biegler et al., 2003; Hinze et al., 2008) assumes a known functional form for h which requires only estimation of material parameters (highly structured with simplified analysis).

For the purposes of this work we consider elliptic systems of $H(\text{div})$-type where structure-preservation amounts to preserving notions of flux continuity. In the literature, preservation of other types of structure is a key challenge for data-driven models: gauge invariances associated with nontrivial null-spaces (Trask et al., 2022), geometric structure associated with bracket dynamics (Gruber et al., 2023; Greydanus et al., 2019; Desai et al., 2021; Hernández et al., 2021), group equivariance (Bergomi et al., 2019; Villar et al., 2021) and other structures (Celledoni et al., 2021). Many of these approaches aim to enforce the invariances by construction rather than rely on data or training to "learn" them, allowing better performance in small-data limits and improved theoretical properties.

In our previous works (Actor et al., 2024; Trask et al., 2022), we have developed structure-preserving machine learning frameworks generalizing the discrete exterior calculus (DEC) and finite element exterior calculus (FEEC) (see Subsection 2.1). Both frameworks pose the learning of physics as identifying maps between cochains associated with a de Rham complex, and provide a number of desirable theoretical guarantees: preservation of exact sequence structure (e.g. $\nabla \cdot (\nabla \times) = 0$), exact local conservation of generalized fluxes, an exact Hodge decomposition, a Lax-Milgram stability theory for Hodge Laplacians, well-posedness theory for nonlinear problems, and a framework for treating problems with nontrivial null-spaces (e.g. electromagnetism). In the FEEC setting, a Dirichlet-to-Neumann map prescribing the exchange of generalized fluxes between subdomains is expressed in terms of parameterized Whitney forms, allowing the machine learning of geometric control volumes which optimally admit integral balance laws. While effective for providing rigorous structure-preservation, the scheme provides poor computational scaling whereby the number of degrees of freedom scale as $\mathcal{O}(N^k)$, where N is the number of partitions and k is the order of the Whitney form.

The current work applies a divide-and-conquer strategy to mitigate this by partitioning the domain into disjoint, nonoverlapping subdomains $\Omega = \cup_i \Omega_i$, whose exact specifications will be discussed later, and seeks local models restricted to each Ω_i of the form

$$\begin{aligned} \nabla \cdot \boldsymbol{u}_i &= -f_i, \\ \boldsymbol{u}_i &= h(p_i;\theta_i), \\ p_i &= g_i \qquad\qquad \text{on } \partial\Omega_i, \end{aligned} \tag{1.2}$$

with the subscript $\cdot_i$ denoting appropriate restrictions of fields to Ω_i. The framework for regressing local models is introduced in Section 3. To train subdomain models, we can perform offline training over data $\mathcal{D}_i = \left\{(u_{i,k}, f_{i,k}, h_{i,k}, g_{i,k})\right\}_{k=1}^{N_i}$. This can be obtained either by taking the restriction of global data onto the subdomain ($g_i = p|_{\partial\Omega}$), or by performing simulations directly on each subdomain to identify the local response to a representative mortar space (e.g. $g_i \in \mathbb{P}_m(\partial\Omega_i)$ the space of mth-order polynomials). After obtaining local models, a mortar method is presented in Section 4 which is used to assemble local models into a global model on Ω.

For this data-driven mortar strategy, we impose two desired requirements:

1. **R1: Preservation of structure across both scales:** For the $H(\text{div})$ problems under consideration, the Whitney form construction admits interpretation as an integral balance law where fluxes are discretely treated as equal and opposite, providing a local conservation principle on each subdomain Ω_i. We require that the mortar formulation be compatible with this, so that when local elements are stitched together through the mortar we preserve conservation globally on Ω.
2. **R2: Stability of error at global scale:** If, during pretraining, local models may be obtained to a given optimization error, we would like to quantify the error induced at a global level by the coupling process. Ideally this would be bound by a constant independent of the number of subdomains, so that the global error remains comparable to that of the locally trained models as many elements are coupled together and performance does not degenerate in the limit of many data-driven elements.

We demonstrate both requirements either in analytical proofs in Section 4, or via numerical example in Section 5. Finally, the technical proofs and more details regarding training are shown in Appendix 10.A.

2 Relation to previous work

The proposed strategy exploits a connection to structure-preserving PDE discretization to ensure that physics are enforced by construction, rather than via the penalty formulation typically pursued in the physics-informed machine learning literature. We summarize the relationship between this approach and the literature, as well as how our strategy relates to classical domain decomposition methods.

2.1 Data-driven DEC/FEEC and Dirichlet-to-Neumann maps

In traditional numerical analysis the discrete exterior calculus (DEC) is a framework for constructing and analyzing staggered finite volume schemes (Hirani, 2003; Nicolaides, 1992). The generalized Stokes theorem is used to define discrete vector calculus operators (e.g. grad/curl/div) which map between differential forms on a pair of primal/dual computational meshes. The finite element exterior calculus (FEEC) generalizes DEC by constructing finite element spaces which interpolate differential forms and provides variational extensions (Arnold, 2018).

In the data-driven exterior calculus (DDEC) (Trask et al., 2022), DEC operators are parameterized in a manner allowing the learning of well-posed models on graphs, where data is used to identify the inner-product associated with codifferential operators. In Actor et al. (2024), it was shown that a family of data-driven Whitney forms may be constructed from parameterized partitions-of-unity (POUs). The Whitney forms admit a de Rham complex which encodes POU geometry as differentiable control volumes and their higher order boundaries (faces/edges/etc) without reference to a traditional mesh. An inner-product is induced by the geometry of the control volumes, supporting the discovery of models in terms of control volume balances. This allows a data-driven FEEC extension of DDEC which we use extensively in this work. Furthermore, by posing integral balances as relationships between domains and fluxes on their boundaries, we work with degrees of freedom which naturally conform to the trace spaces necessary for a mortar strategy.

2.2 Structure-preserving ML vs. physics-informed ML

In the recent scientific machine learning literature, physics-informed methods broadly encompass frameworks where physical constraints are incorporated by adding (typically collocation) residuals to a loss function as a Tikhonov regularization with a penalty parameter (Cai et al., 2021). This technique is simple to implement and, when used together with automatic differentiation, admits a simple treatment of inverse problems, discovery of "missing physics" or closures (Karniadakis et al., 2021; Patel et al., 2022), and uncertainty quantification (Yang and Perdikaris, 2019; Zhang et al., 2019).

The flexibility of the framework comes at the expense of solving a multiobjective optimization problem whereby the physics residual must be empirically weighted against the data loss, and can only be enforced to within optimization error (Wang et al., 2021). For certain classes of problems it is necessary to enforce physics to machine precision to obtain qualitatively correct answers; e.g. subsurface transport and lubrication flows depend crucially on exact conservation of mass (Trask et al., 2018), while electromagnetic problems which fail to provide an exactly divergence-free magnetic field predict qualitatively incorrect spectra (Arnold, 2018). In the context of physics-informed learning, some works have pursued a penalty-based domain decomposition strategy with the goal of efficient distributed computation and more flexibility in neural network approximation (Jagtap and Karniadakis, 2021). While effective, the collocation scheme and penalty formulation complicate analysis and preclude exact conservation, respectively. Because the desired conservation structure only holds to within optimization error, penalization may be insufficient for certain classes of applications.

2.3 Choice of mortar scheme

Domain decomposition is a mature field, with many established options for how to couple solutions across arbitrary finite element subdomains (Toselli and Widlund, 2004; Smith, 1997). Representative rigorous methods range from (e.g. finite element tearing and interconnecting (FETI) Farhat et al. (2001), mortar methods Bernardi et al. (1993), and hybridizable discontinuous Galerkin methods Cockburn et al. (2009)) impose continuity of fluxes and state at subdomain interfaces either strongly via Lagrange multipliers or weakly by using Nietsche's trick to introduce a variational penalty.

For the div-grad problem, there is also a choice of working in either H^1- or $H(\text{div})$-conforming spaces (e.g. $\mathbb{P}_1$/Nedelec or Raviart-Thomas/$\mathbb{P}_0$ mixed spaces), and whether one chooses to apply a mortar on the state or flux variables. Working in $H(\text{div})$ is perhaps most natural, as the mortar space admits interpretation as a conservative flux that trivially preserves conservation structure (Arbogast et al., 2007). However, this requires working with d- and $(d-1)$-dimensional Whitney forms. Our Whitney form construction scales with computational complexity $\mathcal{O}(N^k)$, where N is the dimension of 0th-order Whitney forms and k is the maximal order Whitney form. It is therefore preferable to exploit primal/dual structure and work in H^1, meaning only 0th- and 1st-order Whitney forms are used. This forces us to adopt an H^1 domain decomposition strategy similar to that developed by Glowinski and Wheeler (1987, §7). Further extensions are needed to easily incorporate and analyze the case where data-driven FEEC elements are used as the local solvers.

3 Local learning of Whitney form elements

For brevity, we discuss only the fundamental aspects of data-driven DEC/FEEC necessary to describe the local element construction. For a complete exposition we direct readers to references for: data-driven exterior calculus (Trask et al., 2022), data-driven finite element exterior calculus (Actor et al., 2024), classical finite element exterior calculus for forward simulation (Arnold, 2018), and Whitney forms (Gillette et al., 2016).

Given a compact domain $\omega \in \mathbb{R}^2$ with finite open cover $\{U_i\}_{i=1}^N$, a *partition of unity* (POU) is a collection of functions $\phi_i : \omega \to [0, 1]$ such that $\phi_i(\boldsymbol{x}) \geq 0$, $\text{supp}(\phi_i) \subseteq U_i$, $\phi_i < \infty$ and $\sum_i \phi_i(\boldsymbol{x}) = 1$ for all $\boldsymbol{x} \in \omega$. We assume access to

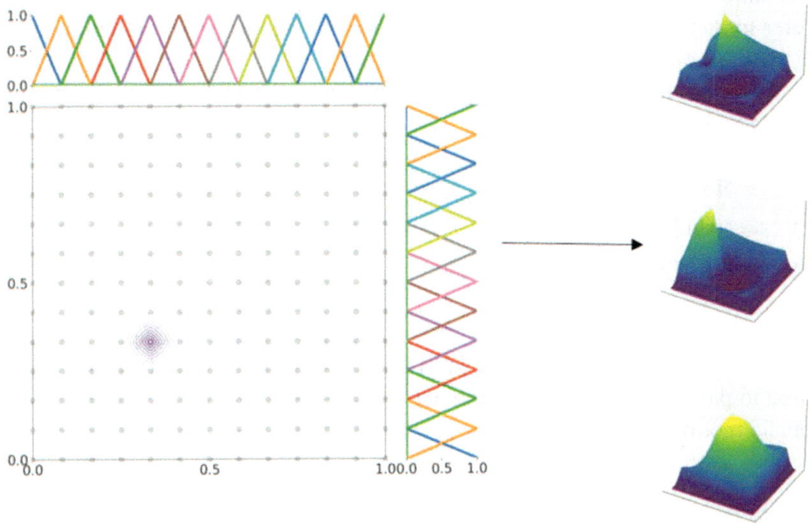

FIGURE 1 To construct a PPOU, we first consider an underlying tensor product grid of B-splines with trainable vertex locations. By taking a trainable convex combination of these shape functions, we arrive at more complex geometries. Noting that B-splines form a partition of unity, and that partitions of unity are closed under convex combination, this process provides a trainable partition of unity which may be integrated exactly via a pull-back onto the fine grid. For purposes of illustration, the underlying tensor product is shown to be uniform in the figure, but they are allowed to shift in the general case.

a *parameterized POU* (PPOU) $\left\{\phi_i(\boldsymbol{x};\theta)\right\}_{i=1}^{N}$, which is continuous with respect to a parameter θ.

To construct Whitney forms, any trainable PPOU may be used, although in this work we adopt the same used in Actor et al. (2024). Starting with tensor-product B-splines on the unit domain, we refer to trainable vertex locations as *fine-scale nodes/knots*. To approximate complex geometries, we consider a coarsening via convex combinations of the knots into our ultimate PPOU, $\left\{\phi_i(\boldsymbol{x};\theta)\right\}_{i=1}^{N}$ where θ denotes parameters corresponding to both knot locations and trainable entries of the convex combination tensor; see Fig. 1 for an illustrative figure of this process.

In Actor et al. (2024), the tensor-product grid points are parameterized using the distances between the grid points to avoid inversion of elements. In particular, we can define the grid points in one dimension $\{t_i\}_{i=0}^{n}$ with $t_0 = 0$ and $t_1 = 1$ by parameterizing

$$\sigma(\delta)_i = t_{i+1} - t_i, \qquad i \in \{0, \ldots, n-1\} \tag{3.1}$$

where δ_i is a trainable parameter, and σ is a sigmoid activation enforcing positivity. To parameterize a map of convex combination of knots, we consider a trainable two-tensor with softmax activation applied to each row; for details we refer to Actor et al. (2024). In what follows we adopt the simplified notation $\phi_i(\boldsymbol{x};\theta)=\phi_i$.

We construct finite element spaces consisting of the 0th-, 1st- and 2nd-order Whitney forms from ϕ_i:

$$
\begin{aligned}
V^0 &:= \operatorname{span}\left\{\phi_i \mid 1\le i\le N\right\},\\
V^1 &:= \operatorname{span}\left\{\phi_i\nabla\phi_j-\phi_j\nabla\phi_i \mid 1\le i,j\le N\right\},\\
V^2 &:= \operatorname{span}\left\{\phi_i\nabla\phi_j\times\nabla\phi_k-\phi_j\nabla\phi_i\times\nabla\phi_k-\phi_k\nabla\phi_j\times\nabla\phi_i \mid 1\le i,j,k\le N\right\},
\end{aligned}
\tag{3.2}
$$

adopting the notation $\psi_{j_1,\dots,j_K}\in V^{k-1}$, to identify elements of spaces by their constituent 0-forms (e.g. $\psi_{ij}\in V^1$). As shown in Actor et al. (2024), the tensor used to parameterize convex combinations of B-splines may be manipulated to obtain modifications of these spaces with zero trace

$$
V_0^k := \left\{u\in V^k \,\middle|\, u|_{\partial\omega}=0\right\}. \tag{3.3}
$$

Consider now the variational form of divergence $(q,\nabla\cdot\boldsymbol{u})$ and curl $(\boldsymbol{v},\nabla\times\boldsymbol{w})$, where $q\in V^0$, $\boldsymbol{u}\in V_0^1$, $\boldsymbol{v}\in V^1$, and $\boldsymbol{w}\in V_0^2$. After integration by parts, Whitney forms induce the following discrete vector calculus operators (Actor et al., 2024, §3)

$$
\begin{aligned}
(\mathrm{DIV})_{i,(ab)} &:= (\psi_{ab},-\nabla\psi_i)=\sum_{j\neq i}(\psi_{ab},\psi_{ij}),\\
(\mathrm{CURL})_{(ij),(abc)} &:= (\psi_{abc},\nabla\times\psi_{ij})=2\sum_{k\neq i,j}(\psi_{abc}.\psi_{ijk}).
\end{aligned}
\tag{3.4}
$$

These discrete exterior derivatives maintain a powerful connection to the graph exterior calculus from combinatorial Hodge theory. Consider a complete graph $\mathcal{G}=(\mathcal{V},\mathcal{E})$ with the vertex set $\mathcal{V}$, edge set $\mathcal{E}$, and higher-order k-cliques denoted by the oriented tuples $(i_1,\dots,i_k)$. The standard k^{th}-order coboundary operator δ_k is simply associated with the oriented incidence matrix between $k+1$- and k-cliques. Specifically, the graph gradient δ_0 and graph curl δ_1 are defined by

$$
\begin{aligned}
(\delta_0 u)_{ij} &= u_j-u_i\\
(\delta_1 u)_{ijk} &= u_{ij}+u_{jk}+u_{ki},
\end{aligned}
$$

where u_i denote a scalar value associated with the node i, $u_{ij}=-u_{ji}$ denotes a scalar associated with the edge $(i,j)\in\mathcal{E}$, and u_{ijk} a value associated with the 3-cliques (e.g. faces) which is antisymmetric with respect to the index ordering

$$
u_{ijk}=-u_{ikj}=-u_{jik}=-u_{kji}=u_{kij}=u_{jki}.
$$

The adjoint of coboundary operators induces the so-called codifferential operators, which in this setting provide definitions of graph divergence and curl:

$$\begin{aligned}(DIV\,u)_i &:= (\delta_0^T u)_i = \sum_{j \neq i} u_{ij}, \\ (CURL\,u)_{ij} &:= (\delta_1^T u)_i = \sum_{k \neq i,j} u_{ijk}.\end{aligned} \tag{3.5}$$

These graph operators have a number of properties mimicking the familiar vector calculus, but follow only from the topological properties of graphs. For example, the *exact sequence* property $DIV \circ CURL = 0$ discretely parallels $\nabla \cdot \nabla \times = 0$, and conservation structure is reflected in DIV calculating the sum of antisymmetric generalized fluxes.

The connection between the parameterized Whitney form space and the combinatorial Hodge theory follows by rewriting (3.4) as

$$\mathsf{DIV} = DIV\,\mathbf{M}_1, \quad \mathsf{CURL} = CURL\,\mathbf{M}_2$$

where $(\mathbf{M}_1)_{(ij),(ab)} = (\psi_{ab}, \psi_{ij})$ and $(\mathbf{M}_2)_{(ijk),(abc)} = (\psi_{abc}, \psi_{ijk})$ are mass matrices associated with the finite element spaces V^1 and V^2, respectively. Therefore, we see that the geometry of the PPOUs implicitly induces a weighting on the graph exterior calculus, with the boundaries of learned partitions inducing a topology associated with conservation structure.

We may finally revisit the original task of identifying a model of the form (1.2). Let the Whitney forms associated with subdomain Ω_i be $V^0(\Omega_i)$ and $V^1(\Omega_i)$ by taking $\omega = \Omega_i$. Mirroring (1.2), the model on each individual subdomain is equivalent to the following variational problem: find $(p_i, \boldsymbol{u}_i) \in V^0(\Omega_i) \times V^1(\Omega_i)$ such that for all $(w_i, \boldsymbol{v}_i) \in V_0^0(\Omega_i) \times V^1(\Omega_i)$,

$$\begin{aligned}(\boldsymbol{u}_i, \boldsymbol{v}_i) - (h(p_i; \theta_i), \boldsymbol{v}_i) = 0 \\ (\boldsymbol{u}_i, \nabla w_i) = (f_i, w_i)\end{aligned}$$

with Dirichlet boundary condition $p_i = g_i$ on $\partial\Omega$, which is enforced by using a standard lift.

Following the theory laid out in Trask et al. (2022), we could assume the unknown fluxes take the form of a nonlinear perturbation of a diffusive flux while maintaining a tractable stability analysis, e.g.

$$h(p_i; \theta_i) = \nabla p_i + N[p_i; \theta_i]$$

However in the current work, we will consider only the linear case ($N[p_i; \theta_i] = 0$). In this setting the Whitney forms will identify the geometry and properties associated with material heterogeneities under an assumed diffusion process, providing the following variational problem on each element.

$$\begin{aligned} (\boldsymbol{u}_i, \boldsymbol{v}_i) - (\nabla p_i, \boldsymbol{v}_i) &= 0 \\ (\boldsymbol{u}_i, \nabla w_i) &= (f_i, w_i). \end{aligned} \tag{3.6}$$

Finally we substitute in the discrete exterior derivatives associated with the PPOUs to obtain a discrete parametric model, posing the following equality constrained optimization problem to calibrate the POU geometry to data,

$$\begin{aligned} &\min_{W, \boldsymbol{B}_0, \boldsymbol{B}_1, \boldsymbol{D}_0 \boldsymbol{D}_1} \left\| p_{\text{data}} - \sum_i \hat{p}_i \psi_i \right\|_2^2 + \alpha^2 \left\| F_{\text{data}} - \sum_{ij} \hat{F}_{ij} \psi_{ij} \right\|_2^2, \\ &\text{such that} \begin{bmatrix} \mathbf{M}_1 & -\mathbf{M}_1 \mathbf{D}_1^{-1} \delta_0 \mathbf{D}_0 \\ -\mathbf{B}_0^{-1} \delta_0^T \mathbf{B}_1 \mathbf{M}_1 & \mathbf{0} \end{bmatrix} \begin{bmatrix} \widehat{\mathbf{F}} \\ \widehat{\mathbf{p}} \end{bmatrix} = \begin{bmatrix} \mathbf{b}_D \\ -\mathbf{b}_f \end{bmatrix} \end{aligned} \tag{3.7}$$

where $\mathbf{B_k}$ and $\mathbf{D}_k$ are diagonal matrices with trainable positive coefficients, $\mathbf{b}_D$, $\mathbf{b}_f$ the terms arising from the Dirichlet boundary condition and forcing term respectively, α a normalization parameter, and W the remaining weights associated with the POUs such as the location of knots and the convex combination tensor. As shown in Actor et al. (2024), $\mathbf{B_k}$ and $\mathbf{D}_k$ infer metric information from data without impacting the topological structure of the model. For further details regarding the specific construction of POUs we refer to Actor et al. (2024).

Remark 3.1. The Whitney form construction supports a number of theoretical constructions: a Hodge decomposition, Poincare inequality, a corresponding Lax-Milgram theory, a well-posedness theory for certain nonlinear elliptic problems, and discrete preservation of exact sequence properties which exactly preserve conservation structure. When we use the Whitney form elements to construct the subdomain spaces V_i in the mortar method in the following section, we aim to carefully construct the mortar space so that this structure is not lost at the global level.

4 Mortar method

After the local models are trained, we seek to construct a mortar method which is flexible enough to couple FEEC elements on the different subdomains together. Note that since the fine-scale knots are able to move during pretraining, the mortar is necessarily nonconforming, with possible "hanging" mortar nodes which do not coincide with the neighboring local element nodes; this necessitates an analysis of stability associated with projecting between local and mortar spaces. Furthermore, we would like the mortar method to preserve the conservation and stability properties outlined in the introduction (**R1**, **R2**).

As discussed in Section 3, we assume that our data $\{(\boldsymbol{u}(x_k), p(x_k)), g_k\}_{k=0}^N$ (with $x_k \in \Omega$ sampled randomly) satisfy the following variational equation: seek

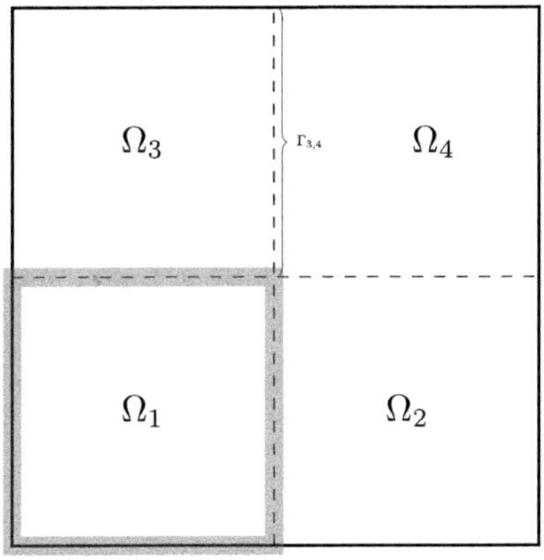

FIGURE 2 Figure of a square domain Ω divided into four subdomains. The edge $\Gamma_{3,4}$ is denoted explicitly and the highlighted boundary is Γ_1.

solution $(\boldsymbol{u}, p) \in (L^2(\Omega)^2, H^1_g(\Omega))$ such that

$$\begin{aligned} (\boldsymbol{u}, \boldsymbol{v}) - (K\nabla p, \boldsymbol{v}) &= 0, & \forall \boldsymbol{v} \in L^2(\Omega)^2 \\ (\boldsymbol{u}, \nabla w) &= (f, w), & \forall w \in H^1_0(\Omega) \end{aligned} \tag{4.1}$$

where H^1 is the standard Sobolev space and $H^1_g(\Omega) = \{u \in H^1(\Omega) \mid u|_{\partial\Omega} = g_k\}$ (Braess, 2007), and the tensor $K \in L^\infty$ is a positive-definite matrix. Finally, we assume the problem is of at least $p \in H^{3/2}(\Omega)$ regularity, which arises naturally if, for example, $f \in L^2(\Omega)$, $g \in H^{3/2}(\partial\Omega)$ with Lipschitz coefficients K and Ω is convex (Grisvard, 2011). We will see in our numerical results that the above regularity result is a sufficient condition for the error analysis, and not a necessary one.

Let Ω be divided into n nonoverlapping, polygonal subdomain blocks Ω_i of similar aspect ratios. Let Γ_i be the edges of Ω_i, $\Gamma = \cup_i \Gamma_i$ the set of all boundaries of the subdomains (including those intersecting $\partial\Omega$), and let $\Gamma_{ij} = \Gamma_i \cap \Gamma_j$ for all i, j be the boundary between two adjacent subdomains. See Fig. 2 for an illustrative figure.

Define

$$\Lambda := \{v \in L^2(\Gamma) \mid \exists u \in H^1(\Omega), u|_\Gamma = v\} \tag{4.2}$$

as the space of L^2 functions on the interfaces which are the traces of H^1 functions, and the subspaces

$$\Lambda_0 := \{\lambda \in \Lambda \mid \lambda|_{\partial\Omega} = 0\},$$
$$\Lambda_g := \{\lambda \in \Lambda \mid \lambda|_{\partial\Omega} = g\}.$$

Note that since Λ consists of the trace of H^1 functions, we may endow Λ with the $H^{1/2}$ norm on Γ.

4.1 Stability analysis for continuous case

Before proceeding to the model discovery problem and the discrete, we first consider smooth solutions coming from solutions from diffusion problem to guide the design of a suitable mortar method. It is straightforward to decompose (4.1) into problems on the subdomains $\{\Omega_i\}_{i=1}^n$ by introducing a mortar representing the pressure on the space Λ:

Lemma 4.1. *For $1 \le i \le n$, let $(\boldsymbol{u}_i, p_i, \lambda) \in (L^2(\Omega_i)^2, H^1(\Omega_i), \Lambda_g)$ such that*

$$\begin{aligned} (\boldsymbol{u}_i, \boldsymbol{v}_i)_{\Omega_i} - (K\nabla p_i, \boldsymbol{v}_i)_{\Omega_i} &= 0, & \forall \boldsymbol{v}_i \in L^2(\Omega_i)^2 \\ (\boldsymbol{u}_i, \nabla w_i)_{\Omega_i} &= (f, w_i)_{\Omega_i}, & \forall w_i \in H_0^1(\Omega_i) \end{aligned} \tag{4.3}$$

with continuity of state and flux enforced via the boundary condition $p_i|_{\Gamma_i} = \lambda|_{\Gamma_i}$ and weak flux continuity condition

$$\sum_{i=1}^{n} (\boldsymbol{u}_i, \nabla w)_{\Omega_i} = (f, w), \qquad \forall w \in H_0^1(\Omega). \tag{4.4}$$

Then $\boldsymbol{u} = \sum_{i=1}^n \boldsymbol{u}_i \in L^2(\Omega)^2$, $p = \sum_{i=1}^n p_i \in H_g^1(\Omega)$ solves (4.1).

Proof. The existence of functions $(\boldsymbol{u}_i, p_i)$ and λ comes trivially by restricting the solution from (4.1) to the individual subdomains and mortar space.

To see that (4.3) and (4.4) implies (4.1), we note that $L^2(\Omega)^2 = \bigoplus_{i=1}^n L^2(\Omega_i)^2$, and thus by summing the first equation of (4.3) and choosing $\boldsymbol{v}_i = \boldsymbol{v}|_{\Omega_i}$ as test functions, we have

$$\left(\sum_{i=1}^{n} \boldsymbol{u}_i, \boldsymbol{v}\right) - \left(K\nabla \sum_{i=1}^{n} p_i, \boldsymbol{v}\right) = 0, \qquad \forall \boldsymbol{v} \in L^2(\Omega)^2$$

with $\sum_{i=1}^n p_i \in H_g^1(\Omega)$ since continuity is enforced with λ. As for the test functions arising in $w \in H_0^1(\Omega)$, we simply decompose w into $\sum_{i=1}^n w_i + w_0$ where $w_i \in H_0^1(\Omega_i)$ for $1 \le i \le n$ and $w_0 := w - \sum_{i=1}^n w_i$, so that the summation of the second equation of (4.3) and (4.4) gives us the desired result. □

The condition (4.4) can be simplified. Consider the space $H^\gamma(\Omega)$ satisfying the decomposition

$$H^1(\Omega) = H_0^1(\Omega_1) \oplus \cdots \oplus H_0^1(\Omega_n) \oplus H^\gamma(\Omega) \tag{4.5}$$

with $H^\gamma \perp H_0^1(\Omega_i)$ relative to the H^1 norm for each i. Then, using to (4.3) and (4.5), (4.4) can be rewritten as

$$\sum_{i=1}^{n}(\boldsymbol{u}_i, \nabla w) = (f, w), \qquad \forall w \in H_0^\gamma(\Omega) \tag{4.6}$$

where $H_0^\gamma := \{u \in H^\gamma(\Omega) \mid u|_{\partial\Omega} = 0\}$. We also define the subset $H_g^\gamma := \{u \in H^\gamma(\Omega) \mid u|_{\partial\Omega} = g\}$. The space H^γ corresponds to a minimal energy extension (Toselli and Widlund, 2004) as the following lemma shows:

Lemma 4.2. *For all $u \in H^1(\Omega)$, there exists a unique decomposition $u = u_\gamma + \sum_{i=1}^n u_i$ such that $u_\gamma \in H^\gamma(\Omega)$, $u_i \in H_0^1(\Omega_i)$. Furthermore, one has*

$$\|u_\gamma\|_{H^1(\Omega)} = \inf_{v \in H^1(\Omega), v|_\Gamma = u} \|v\|_{H^1(\Omega)} \simeq \sum_{i=1}^{n} \|u_\gamma\|_{H^{1/2}(\Gamma_i)}.$$

Proof. Given u, consider $u_I \in H_0^1(\Omega_1) \oplus \cdots \oplus H_0^1(\Omega_n)$ such that for $1 \le i \le n$,

$$(u_I, v_i)_{H^1(\Omega_i)} = (u, v_i)_{H^1(\Omega_i)}, \qquad \forall v_i \in H_0^1(\Omega_i).$$

Then the decomposition is simply $u = \sum_{i=1}^n u_I|_{\Omega_i} + u_\gamma$ where $u_\gamma = u - u_I$. The orthogonality is enforced since, for all w_i in $H_0^1(\Omega_i)$ and $1 \le i \le n$,

$$(u_\gamma, w_i)_{H^1(\Omega)} = (u - u_I, w_i)_{H^1(\Omega_i)} = (u, w_i)_{H^1(\Omega_i)} - (u_I, w_i)_{H^1(\Omega_i)} = 0.$$

As for the minimal condition, let $v = u_\gamma + \sum_{i=1}^n v_i$ with $v_i \in H_0^1(\Omega_i)$ arbitrary, then by orthogonality

$$\|v\|_{H^1(\Omega)}^2 = \|u_\gamma\|_{H^1(\Omega)}^2 + \left\|\sum_{i=1}^{n} v_i\right\|_{H^1(\Omega)}^2 \ge \|u_\gamma\|_{H^1(\Omega)}^2$$

and the $H^{1/2}$ equivalence is well known (Bertoluzza and Kunoth, 2000; Cowsar et al., 1995). □

With the above decomposition, we can further reduce (4.1) to be a variational problem only on H^γ and Λ. Let $\lambda, \mu \in H^\gamma$, define the bilinear form and linear functional

$$b(\lambda, \mu) = \sum_{i=1}^{n}(\boldsymbol{u}^*(\lambda), \nabla\mu)_{\Omega_i} \tag{4.7}$$

and

$$L(\mu) = (f, \mu)_\Omega - \sum_{i=1}^{n}(\bar{\boldsymbol{u}}, \nabla\mu)_{\Omega_i} \tag{4.8}$$

where $(\boldsymbol{u}^*(\lambda), p^*(\lambda)) \in (L^2(\Omega)^2, H^1(\Omega))$ solves the local problems, for $1 \leq i \leq n$,

$$\begin{aligned}(\boldsymbol{u}^*(\lambda), \boldsymbol{v}) - (K\nabla p^*(\lambda), \boldsymbol{v}) = 0, &\quad \forall \boldsymbol{v} \in L^2(\Omega_i)^2 \\ (\boldsymbol{u}^*(\lambda), \nabla w) = 0, &\quad \forall w \in H_0^1(\Omega_i)\end{aligned} \tag{4.9}$$

with boundary condition $p^*(\lambda)|_{\Gamma_i} = \lambda|_{\Gamma_i}$, and where $(\bar{\boldsymbol{u}}, \bar{p}) \in (L(\Omega)^2, H_0^1(\Omega))$ solves, for $1 \leq i \leq n$,

$$\begin{aligned}(\bar{\boldsymbol{u}}, \boldsymbol{v}) - (K\nabla \bar{p}, \boldsymbol{v}) = 0, &\qquad \forall \boldsymbol{v} \in L^2(\Omega_i)^2 \\ (\bar{\boldsymbol{u}}, \nabla w) = (f, w), &\quad \forall w \in H_0^1(\Omega_i)\end{aligned} \tag{4.10}$$

with boundary condition $\bar{p}|_{\Gamma_i} = 0$. The bilinear form and linear functional closely resemble those of the $H(\text{div})$ case from Arbogast et al. (2007, 2000). Note that the problems (4.9) and (4.10) above are local in nature and can be solved in parallel.

The following lemma shows that one can recover the original variational equations by working with the above bilinear form:

Lemma 4.3. *Let $\lambda \in H_g^\gamma$ be the solution to the variational equation,*

$$b(\lambda, \mu) = L(\mu), \qquad \forall \mu \in H_0^\gamma \tag{4.11}$$

then $\boldsymbol{u} := \boldsymbol{u}^*(\lambda) + \bar{\boldsymbol{u}}$, $p := p^*(\lambda) + \bar{p}$ is the solution to (4.1).

Proof. Summing (4.9) and (4.10) results in

$$\begin{aligned}(\boldsymbol{u}, \boldsymbol{v}) - (K\nabla p, \boldsymbol{v}) = 0, &\qquad \forall \boldsymbol{v} \in L^2(\Omega_i)^2 \\ (\boldsymbol{u}, \nabla w) = (f, w), &\qquad \forall w \in H_0^1(\Omega_i)\end{aligned}$$

with $p|_\Gamma = \lambda$ for each $1 \leq i \leq n$.

It remains to check (4.6), but this is simply because if (4.11) holds, then

$$\sum_{i=1}^{n} (\boldsymbol{u}, \nabla\mu)_{\Omega_i} = (f, \mu)$$

for all $\mu \in H_0^\gamma$ and the results follows from Lemma 4.1 and (4.6). □

Finally, we note that the variational equation (4.11) is well-defined as the bilinear form is coercive as shown in the following lemma, whose proof is delayed until the appendix:

Lemma 4.4. *The bilinear form* (4.7) *is symmetric and coercive on Λ_0.*

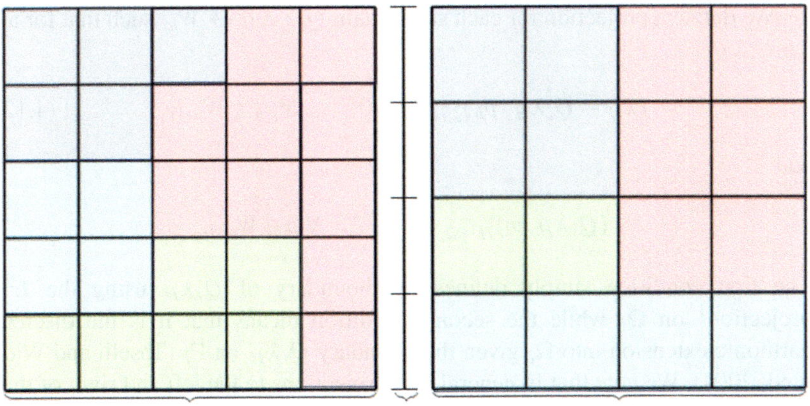

FIGURE 3 Sketch of a 4 element mortar Λ_H and its two adjacent subdomains. The colors on the subdomains represent the PPOUs constructed as convex combinations of a fine-scale B-splines. We note that while the mortar matches the fine-scale nodes on Ω_2, it is disjoint from Ω_1, requiring analysis of a remap/projection between the two meshes. Because the FEEC fine-scale nodes on Ω_i evolve during training, they will generally not coincide with mortar nodes.

4.2 Discretized case

The discrete case is more technical, since both the spaces and the bilinear form are discretized as (4.9) and (4.10) cannot be solved exactly. Further, care must be taken to treat the nonconforming grids that emerge naturally as nodes between adjacent subdomains evolve.

In what follows, the subscripts h, H will denote a discretized version of a continuous space. On each subdomain Ω_i, let W_{hi}, $\boldsymbol{V}_{hi}$ be the discretized versions of $H^1(\Omega_i)$, $L^2(\Omega_i)^2$ respectively. We require the standard inf-sup compatibility between W_{hi}, $\boldsymbol{V}_{hi}$, which in this case is simply the condition $\nabla W_{hi} \subseteq \boldsymbol{V}_{hi}$ (Arnold, 2018; Braess, 2007). In particular, we can choose W_{hi} and $\boldsymbol{V}_{hi}$ to be the spaces V^0 and V^1 defined in (3.2) in the case of FEEC elements; by construction then we have $\nabla V^0 = V^1$. We will also use the case where local elements are taken to be traditional finite elements to show convergence; in this case we will consider V^0 and V^1 as continuous $\mathbb{Q}_1$ and lowest-order Nedelec elements, respectively. Finally, we let $W_{hi,0}$ be the subspace with homogeneous Dirichlet boundary condition (e.g. (3.3)).

On the interfaces, we choose $\Lambda_H \subset \Lambda$ to be the space of continuous, piecewise linear functions. Let $\Lambda_{H,0} := \{\mu_H \in \Lambda_H \mid \mu_H|_{\partial\Omega} = 0\} \subset \Lambda_0$ and similarly let $\Lambda_{H,g} \subset \Lambda_g$ be the subset whereby the boundary is equal to g. We allow the discretization between two subdomains to be different and also allow for the mortars to be nonmatching. See Fig. 3 for a simplified figure where there are nonmatching tensor-product grids.

We define a projection for each subdomain $Q_i : \Lambda_H \to W_{hi}$ such that for all $\lambda_H \in \Lambda_H$

$$(\lambda_H - Q_i\lambda_H, p_h)_{L^2(\Gamma_i)} = 0, \qquad \forall p_h \in W_{hi}|_{\Gamma_i} \tag{4.12}$$

and

$$(Q_i\lambda_H, w)_{H^1(\Omega_i)} = 0, \qquad \forall w \in W_{hi,0}.$$

The first condition simply defines the boundary of $Q_i\lambda_H$ using the L^2-projection[1] on Ω_i while the second condition means that it is the discrete harmonic extension into Ω_i given the boundary $Q_i\lambda_H$ on Γ_i (Toselli and Widlund, 2004). We note that in general, the projections to the left and right of that interface are different since the discretization can be different on either sides as can be seen in Fig. 3.

With the above in hand, we can define the discretized bilinear operator and linear functional similar to (4.7) and (4.8). For $\lambda_H, \mu_H \in \Lambda_H$, let

$$b_h(\lambda_H, \mu_H) := \sum_{i=1}^{n} (\boldsymbol{u}_h^*(Q_i\lambda_H), \nabla(Q_i\mu_H))_{\Omega_i} \tag{4.13}$$

and

$$L_h(\mu_H) := \sum_{i=1}^{n} (f, Q_i\mu_H)_{\Omega_i} - (\bar{\boldsymbol{u}}_h, \nabla(Q_i\mu_H))_{\Omega_i} \tag{4.14}$$

where $p_h^*(Q_i\lambda_H) \in \oplus_{i=1}^n W_{hi}, \boldsymbol{u}_h^*(Q_i\lambda_H) \in \oplus_{i=1}^n \boldsymbol{V}_{hi}$ satisfies, for $1 \le i \le n$,

$$(\boldsymbol{u}_h^*(Q_i\lambda_H), \boldsymbol{v}_h) - (K\nabla p_h^*(Q_i\lambda_H), \boldsymbol{v}_h) = 0, \qquad \forall \boldsymbol{v}_h \in \boldsymbol{V}_{hi} \tag{4.15}$$

$$(\boldsymbol{u}_h^*(Q_i\lambda_H), \nabla w_h) = 0, \qquad \forall w_h \in W_{hi,0} \tag{4.16}$$

with $p_h^*(Q_i\lambda_H) = Q_i\lambda_H$ on Γ_i, and $\bar{p}_h \in \oplus_{i=1}^n W_{hi}, \bar{\boldsymbol{u}}_h \in \oplus_{i=1}^n V_{hi}$ satisfying

$$(\bar{\boldsymbol{u}}_h, \boldsymbol{v}_h) - (K\nabla \bar{p}_h, \boldsymbol{v}_h) = 0, \qquad \forall \boldsymbol{v}_h \in \boldsymbol{V}_{hi} \tag{4.17}$$

$$(\bar{\boldsymbol{u}}_h, \nabla w_h) = (f, w_h), \qquad \forall w_h \in W_{hi,0} \tag{4.18}$$

with $\bar{p}_h = 0$ on Γ_i. As before, the above problems are defined locally and can be solved in parallel.

We state the discrete variational equation as follows. Find $\lambda_H \in \Lambda_{H,g}$ such that

$$b_h(\lambda_H, \mu_H) = L_h(\mu_H), \qquad \forall \mu_H \in \Lambda_{H,0}. \tag{4.19}$$

[1] We found in our numerical examples that using the interpolant suffices, however we will carry out the analysis using the projection.

The well-posedness of the variational form can be deduced from Lax-Milgram if the coercivity condition

$$b_h(\lambda_H, \lambda_H) \geq \alpha \sum_{i=1}^{n} \|\lambda_H\|^2_{H^{1/2}(\Gamma_i)} \tag{4.20}$$

is true. The coercivity condition (4.20) will require two assumptions which excludes pathological discretizations:

1. Assumption 1 (injectivity): for all $\lambda_H \in \Lambda_H$, there exists a constant C such that

$$\sum_{i=1}^{n} \|Q_i\lambda_H\|_{H^{1/2}(\Gamma_i)} \geq C \sum_{i=1}^{n} \|\lambda_H\|_{H^{1/2}(\Gamma_i)} \tag{4.21}$$

meaning we have unisolvency when projecting from the mortar space onto the local subdomains.

2. Assumption 2 (strengthened triangle inequality): for each shared edge Γ_{ij} and for all $\lambda_H \in \Lambda_H$, that

$$\frac{C_p}{|\Gamma_{ij}|} \left\| Q_i\lambda_H - Q_j\lambda_H \right\|^2_{L^2(\Gamma_{ij})} \leq \frac{1}{2}(\| Q_i\lambda_H \|^2_{H^{1/2}(\Gamma_{ij})} + \left\| Q_j\lambda_H \right\|^2_{H^{1/2}(\Gamma_{ij})}) \tag{4.22}$$

where C_p is the Poincare constant arising in Brenner (2003, (1.3)) and $|\Gamma_{ij}|$ is the length of the shared edge. The condition means two adjacent subdomains cannot have too large of a difference in their discretization parameter. In particular, if two adjacent subdomains have the same, symmetric discretization parameters, then the left side of (4.22) is trivially zero.

In the case of data-driven elements, extra care must be paid to Assumption 1 since a training procedure might move the fine-scale nodes such that unisolvency is lost. However, this can be circumvented by either placing restrictions on the movement of the nodes, or, as in some of our numerical examples, using a very coarse mortar space.

With the above assumptions, we can now state the stability result:

Lemma 4.5. *With the above two assumptions, the discretized bilinear form* (4.13) *is coercive (e.g.* (4.20)*) over* $\Lambda_{H,0}$.

The proof of the above lemma is technical and is delayed to the appendix.

Lemma 4.5 means that one is allowed to apply Strang's second lemma to obtain error estimates. We assume that an a priori estimate exists: let δ be a constant such that the discrete approximations on each subdomain $1 \leq i \leq n$ satisfy

$$\left\| p^*(\lambda) - p_h^*(Q_i\lambda) \right\|_{\Omega_i} \leq \delta, \qquad \left\| \boldsymbol{u}^*(\lambda) - \boldsymbol{u}_h^*(Q_i\lambda) \right\|_{\Omega_i} \leq \delta \tag{4.23}$$

$$\left\|\bar{p}-\bar{p}_h\right\|_{\Omega_i} \leq \delta \qquad \|\bar{\boldsymbol{u}}-\bar{\boldsymbol{u}}_h\|_{\Omega_i} \leq \delta \tag{4.24}$$

for all $\lambda \in \Lambda$. The constant δ corresponds to the ability of the local solvers to solve for p^*, $\boldsymbol{u}^*$ accurately for an arbitrary mortar. In the case where standard FEM is used on the subdomain, then δ can be replaced with the respective a priori estimate whereas for the DDEC methods, this corresponds to an optimization threshold.

We can now state a simple convergence guarantee **R2** on the mortar space, whose proof is delayed until the appendix:

Theorem 4.6. *Suppose the solution to the* (4.1) *is such that* $p \in H^2(\Omega)$ *with homogeneous Dirichlet boundary condition. Then there exists a constant* C *independent of* λ^* *such that*

$$\sum_{i=1}^{n}\left\|\lambda^*-\lambda_H^*\right\|_{H^{1/2}(\Gamma_i)} \leq Cn\left|p\right|_{H^2(\Omega)}(H+h+\delta) \tag{4.25}$$

where λ^* *is the true solution to* (4.11), *and* λ_H^* *is the solution to* (4.19), *and* H, h *are the maximal mesh sizes on* Λ_H *and the boundary of the subdomains* Ω_i, *respectively.*

Remark 4.7. As mentioned, the constant δ associated with (4.23) corresponds to the accuracy of the local solvers while the h term relates to the accuracy of projecting the mortar to the local subdomains using (4.12), though in general we can assume that $h < H$. We also note that (4.25) implies that a combination of refinement of both the local solvers and the mortar space is needed to obtain convergence.

Finally, we can easily bound the error on the pressure and velocity explicitly.

Lemma 4.8. *With the same assumptions and constants as in Theorem 4.6, there exists a constant* C *independent of* $\boldsymbol{u}$ *and* p *such that*

$$\left\|p-p_h\right\|_{\Omega}+\|\boldsymbol{u}-\boldsymbol{u}_h\|_{\Omega} \leq Cn\left|p\right|_{H^2(\Omega)}(H+h+\delta)$$

where $\boldsymbol{u}$, p *are the true solutions arising from* (4.3) *and* $\boldsymbol{u}_h=\sum_{i=1}^{n}\boldsymbol{u}_h^*(Q_i\lambda_H^*)+\bar{\boldsymbol{u}}_h$, $p_h=\sum_{i=1}^{n}p_h^*(Q_i\lambda_H^*)+\bar{p}_h$.

Proof. By Lemma 4.3, we have

$$\|\boldsymbol{u}-\boldsymbol{u}_h\|_{\Omega} \leq \sum_{i=1}^{n}\left\|\boldsymbol{u}^*(\lambda^*)-\boldsymbol{u}_h^*(Q_i\lambda_H^*)\right\|_{\Omega_i}+\|\bar{\boldsymbol{u}}-\bar{\boldsymbol{u}}_h\|_{\Omega_i}.$$

The latter term on the right hand side is bounded by δ by assumption. Thus the result follows by

$$\sum_{i=1}^{n}\left\|\boldsymbol{u}^*(\lambda^*)-\boldsymbol{u}_h^*(Q_i\lambda_H^*)\right\|_{\Omega_i} \leq \sum_{i=1}^{n}\left\|\boldsymbol{u}^*(\lambda^*)-\boldsymbol{u}^*(\lambda_H^*)\right\|_{\Omega_i}$$

$$
\begin{aligned}
&+\left\|\boldsymbol{u}^*(\lambda_H^*)-\boldsymbol{u}_h^*(Q_i\lambda_H^*)\right\|_{\Omega_i}\\
&\le n\delta+\sum_{i=1}^{n}\left\|\boldsymbol{u}^*(\lambda^*-\lambda_H^*)\right\|_{\Omega_i}\\
&\le n\delta+\sum_{i=1}^{n}\left\|\lambda^*-\lambda_H^*\right\|_{H^{1/2}(\Gamma_i)}
\end{aligned}
$$

where we used standard regularity estimates at the last step. The same estimates also follow for the pressure and the result follows from applying Theorem 4.6. □

The above error analysis partially shows that requirement **R2** from the introduction is met, as the total error is indeed controlled by a combination of the local optimization error, and coupling error from the mortars. However, due to the use of crude bounds on the sum, it is not independent with the number of subdomains, though we will later observe it holds numerically (cf. Subsection 5.4).

4.3 Data-driven elements with mortar method

Classical finite elements such as Nedelec elements can be used for the local solvers in (4.15) and (4.17) on the subdomains Ω_i in a straightforward manner (see Subsection 5.1 for an example). However, the true strength of the above mortar method is its ability to interface with the data-driven structure-preserving models discussed in Section 3. We briefly discuss combining the usage of the Whitney form elements with the mortar method.

As before, we assume the data is of the form $\{(\boldsymbol{u}(x_k), p(x_k)), g_k\}_{k=0}^{N}$ with x_k sampled randomly on Ω. This can either be supplied via physical data or high-fidelity PDE solvers. Let M be the total number of unique boundary conditions g_k (e.g. $M = 1$ if all data points originate from the same boundary value problem). We assume Ω is divided into subdomains Ω_i. As with most data-driven applications, a large number of data points N is needed, however, only one boundary condition M is needed (see Subsection 5.4.3 for an example with $M = 1$), though more is always better.

The iterative solving process for the mortar (4.19) involves different Dirichlet boundary conditions λ_H being passed into (4.15), meaning that the ability for the data-driven Whitney form solvers to be able to correctly respond to different Dirichlet data is important. Ideally M is large so that a good sampling of Dirichlet conditions around each Ω_i is achieved.

In cases where simulations on each Ω_i are possible, one should perform simulations to obtain responses to a possible mortar boundary conditions. In particular, in our numerical examples, we choose to use either nodal functions $\{(1-x)(1-y), x(1-y), (1-x)y, xy\}$ or edge Bernstein polynomials. The Bernstein polynomials are chosen as they provide a complete basis on $\partial\Omega_i$

and their gradients are very smooth, however other boundary conditions can be chosen. We note that these data are usually cheaper to generate since the subdomains are smaller than Ω, and they can be performed in parallel.

However, the ability to perform these simulations on each subdomain is not always possible. In this case, a simple approach consisting of taking $g_k := p|_{\Omega_i}$ and the corresponding data points $(\boldsymbol{u}(x_k), p(x_k))$ restricted to each Ω_i can be done. While easier, this does lead to higher errors due to undersampling from certain mortar modes. Nevertheless, the structure-preserving nature of the data-driven elements ensures adherence to the underlying invariance.

With the data on each Ω_i chosen, we then solve the minimization problem (3.7) giving us fine-scale nodes, and a coarsening to POUs. *These data-driven elements are then used as the local solvers for* (4.15) *and* (4.17). Some care must be exercised to ensure that Assumption 1 is satisfied; the projection from the mortar space onto the local solvers must be unique. One can mix and match the local solvers, and only use the data-driven elements where the fluxes are unknown and use traditional finite elements elsewhere; see Subsection 5.3 for an example. Specific details regarding the training process for the numerical examples are given in Appendix 10.A.

4.4 Neumann boundary conditions and conservation

We briefly discuss modifications needed to solve the pure Neumann problem $\boldsymbol{u} \cdot \vec{n} = g$ on $\partial\Omega$, and show that the critical conservation and compatibility property of

$$\int_{\Omega} f + \int_{\partial\Omega} g = 0 \tag{4.26}$$

is satisfied by the discrete mortar method. Such conservation is exhibited in the FEEC elements also Actor et al. (2024), and thus by showing the mortar method exhibits this behavior as well, requirement **R1** is satisfied.

The assumed global model is now to find $(\boldsymbol{u}, p) \in (L^2(\Omega)^2, H^1(\Omega))$ satisfying

$$\begin{aligned} (\boldsymbol{u}, \boldsymbol{v}) - (K\nabla p, \boldsymbol{v}) = 0, \qquad & \forall \boldsymbol{v} \in L^2(\Omega)^2 \\ (\boldsymbol{u}, \nabla w) = (f, w) + (g, w)_{\partial\Omega}, \quad & \forall w \in H^1(\Omega) \end{aligned} \tag{4.27}$$

with the condition that $(p, 1) = 0$ for uniqueness.

Due to the differences in boundary conditions, a slightly different choice of spaces and decomposition akin to (4.5) is needed. Define

$$H^{\gamma}_{B}(\Omega) := \{u \in H^{\gamma} \mid u|_{\Gamma_i \setminus \partial\Omega} = 0, \forall 1 \le i \le n\} \tag{4.28}$$

and let H^{γ}_{D} be such that

$$H^{\gamma} = H^{\gamma}_{D} \oplus H^{\gamma}_{B}. \tag{4.29}$$

The space H_B^γ is simply the subspace which vanishes on the interior mortar spaces, while H_D^γ is its complement. Finally, for each $1 \le i \le n$, let

$$H_D^1(\Omega_i) := \{u \in H^1(\Omega_i) \mid u|_{\Gamma_i \setminus \partial\Omega} = 0\}, \tag{4.30}$$

the set of H^1 functions vanishing only on the interior boundary. Note that all functions in H_B^γ can be written as a sum of functions in $H_D^1(\Omega_i)$. Hence, a new decomposition can be written

$$H^1(\Omega) = H_D^1(\Omega_1) \oplus \cdots \oplus H_D^1(\Omega_n) \oplus H_D^\gamma(\Omega). \tag{4.31}$$

With the spaces above, we can introduce a mortar that is equivalent to (4.27), up to a constant: for $1 \le i \le n$, let $(\boldsymbol{u}_i, p_i, \lambda) \in (L^2(\Omega_i)^2, H^1(\Omega_i), H_D^\gamma(\Omega))$ satisfy

$$\begin{aligned} (\boldsymbol{u}_i, \boldsymbol{v}_i) - (K\nabla p_i, \boldsymbol{v}_i) &= 0, & \forall \boldsymbol{v}_i \in L^2(\Omega_i)^2 \\ (\boldsymbol{u}_i, \nabla w_i) &= (f, w_i) + (g, w_i)_{\Gamma_i \cap \partial\Omega}, & \forall w_i \in H_D^1(\Omega_i) \end{aligned} \tag{4.32}$$

with the boundary condition that $p|_{\Gamma_i \setminus \partial\Omega} = \lambda|_{\Gamma_i \setminus \partial\Omega}$, and

$$\sum_{i=1}^{n} (\boldsymbol{u}_i, \nabla w)_{\Omega_i} = (f, w) + (g, w_i)_{\partial\Omega}, \qquad \forall w \in H_D^\gamma(\Omega). \tag{4.33}$$

Finally, we can impose $\int_\Gamma \lambda = 0$ for uniqueness. The proof is similar to that of Lemma 4.1 and is omitted.

With the above, it's easy to define the variational problem as before. Small changes are needed in the bilinear form (4.7) and linear functional (4.8): the definition of $(\boldsymbol{u}^*(\lambda), p^*(\lambda))$, $(\bar{\boldsymbol{u}}, \bar{p})$ should be changed to

$$\begin{aligned} (\boldsymbol{u}^*(\lambda), \boldsymbol{v}) - (K\nabla p^*(\lambda), \boldsymbol{v}) &= 0, \quad \forall \boldsymbol{v} \in L^2(\Omega_i)^2 \\ (\boldsymbol{u}^*(\lambda), \nabla w) &= 0, \quad \forall w \in H_D^1(\Omega_i) \end{aligned} \tag{4.34}$$

with boundary conditions $p^*(\lambda)|_{\Gamma_i \setminus \partial\Omega} = \lambda|_{\Gamma_i \setminus \partial\Omega}$, and

$$\begin{aligned} (\bar{\boldsymbol{u}}, \boldsymbol{v}) - (K\nabla \bar{p}, \boldsymbol{v}) &= 0, & \forall \boldsymbol{v} \in L^2(\Omega_i)^2 \\ (\bar{\boldsymbol{u}}, \nabla w) &= (f, w) + (g, w)_{\Gamma_i \cap \partial\Omega}, & \forall w \in H_D^1(\Omega_i) \end{aligned} \tag{4.35}$$

with boundary condition $\bar{p}|_{\Gamma_i \setminus \partial\Omega} = 0$. We note that (4.34) and (4.35) are both well-defined for all subdomains due to the Dirichlet boundary conditions on the mortar space, except for the degenerate case where there is only one subdomain. Finally, the variational form is similar, where we seek $\lambda \in H_D^\gamma$

$$\sum_{i=1}^{n} (\boldsymbol{u}^*(\lambda), \nabla \mu)_{\Omega_i} = (f, \mu)_\Omega + (g, \mu)_{\partial\Omega} - \sum_{i=1}^{n} (\bar{\boldsymbol{u}}, \nabla \mu)_{\Omega_i}, \qquad \forall \mu \in H_D^\gamma.$$

Turning to the discrete case, let W_{hi}, $\boldsymbol{V}_{hi}$ be as before and let $W_{hi,D}$ be the discretization of $H^1_D(\Omega_i)$. Let $\Lambda_{H,D} \subset \Lambda_H$ be the discretized mortar space consisting of continuous, piecewise linear functions that vanish where H^γ_D is zero. The projection Q_i should be changed to $Q_i : H^\gamma_D \to W_{hi}$ with the same alteration to (4.12).

Thus, the discretized bilinear form and linear functional is similar to before, with the exception that $p^*_h(Q_i\lambda_H) \in \oplus^n_{i=1} W_{hi}$, $\boldsymbol{u}^*_h(Q_i\lambda_H) \in \oplus^n_{i=1} \boldsymbol{V}_{hi}$ satisfies, for $1 \le i \le n$,

$$(\boldsymbol{u}^*_h(Q_i\lambda_H), \boldsymbol{v}_h) - (K\nabla p^*_h(Q_i\lambda_H), \boldsymbol{v}_h) = 0, \qquad \forall \boldsymbol{v}_h \in \boldsymbol{V}_{hi} \tag{4.36}$$

$$(\boldsymbol{u}^*_h(Q_i\lambda_H), \nabla w_h) = 0, \qquad \forall w_h \in W_{hi,D} \tag{4.37}$$

with $p^*_h(Q_i\lambda_H) = Q_i\lambda_H$ on $\Gamma_i \setminus \partial\Omega$, and $\bar{p}_h \in \oplus^n_{i=1} W_{hi}$, $\bar{\boldsymbol{u}}_h \in \oplus^n_{i=1} V_{hi}$ satisfying

$$(\bar{\boldsymbol{u}}_h, \boldsymbol{v}_h) - (K\nabla \bar{p}_h, \boldsymbol{v}_h) = 0, \qquad \forall \boldsymbol{v}_h \in \boldsymbol{V}_{hi} \tag{4.38}$$

$$(\bar{\boldsymbol{u}}_h, \nabla w_h) = (f, w_h) + (g, w_h)_{\partial\Omega}, \qquad \forall w_h \in W_{hi,D} \tag{4.39}$$

with $\bar{p}_h = 0$ on $\Gamma_i \setminus \partial\Omega_i$. The variational form (written explicitly) is to find $\lambda_H \in \Lambda_{H,D}$, with mean zero, such that

$$\sum_{i=1}^{n} (\boldsymbol{u}^*_h(Q_i\lambda_H), \nabla(Q_i\mu_H))_{\Omega_i} = \sum_{i=1}^{n} (f, Q_i\mu_H)_{\Omega_i} + (g, Q_i\mu_H)_{\Omega_i \cap \partial\Omega} - (\bar{\boldsymbol{u}}_h, \nabla(Q_i\mu_H))_{\Omega_i}, \qquad \forall \mu_H \in \Lambda_{H,D}. \tag{4.40}$$

Turning to (4.26) and **R1**, it's easy to see that for a function $\mu_H \in \Lambda_{H,D}$, the projection is exact $Q_i\mu_H|_{\Gamma_i\setminus\partial\Omega} = \mu_H|_{\Gamma_i\setminus\partial\Omega}$ if μ_H is constant on the interior edges (e.g. $\mu_H = C$ on $\Gamma_i \setminus \partial\Omega$). Without loss of generality, let $\Lambda_{H,D} \ni \mu_H = 1$ on all $\Gamma_i \setminus \partial\Omega$, then we can choose w_h such that $w_h + Q_i\mu_H = 1$ on $W_{hi,D}$ for each $1 \le i \le n$. Thus, adding (4.37) and (4.39) for $1 \le i \le n$ to (4.40) and rearranging, we obtain

$$\begin{aligned}
&\sum_{i=1}^{n} (\boldsymbol{u}^*_h(Q_i\lambda_H) + \bar{\boldsymbol{u}}_h, \nabla(w_h + Q_i\mu_H))_{\Omega_i} \\
&\quad = \sum_{i=1}^{n} (\boldsymbol{u}^*_h(Q_i\lambda_H) + \bar{\boldsymbol{u}}_h, \nabla 1)_{\Omega_i} \\
&\quad = \sum_{i=1}^{n} (f, w_h + Q_i\mu_H)_{\Omega_i} + (g, w_h + Q_i\mu_H)_{\partial\Omega} \\
&\quad = \sum_{i=1}^{n} (f, 1)_{\Omega_i} + (g, 1)_{\partial\Omega} = 0,
\end{aligned}$$

meaning (4.26) is valid even in the discrete case with nonmatching mortars.

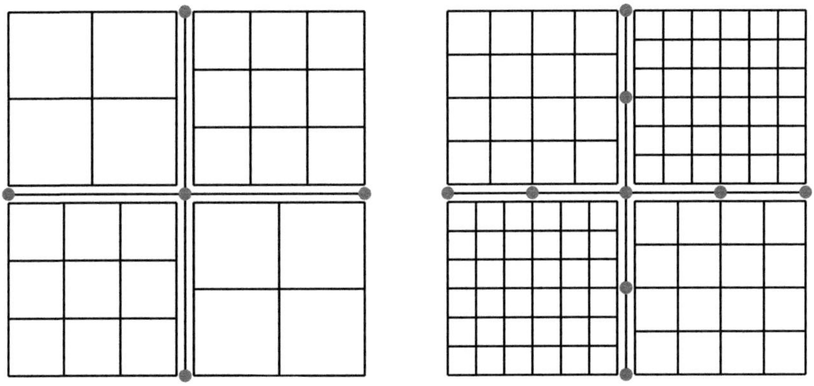

FIGURE 4 Figure illustrating the initial mesh, the corresponding mortar space, and their first refinement for the example in Subsection 5.1.

5 Numerical results

In this section, we present numerical results obtained from applying the above mortar method to several representative examples. All except the first example will involve using pretrained FEEC elements as local subdomain solvers as discussed in Section 3.

5.1 Example 1: pure finite elements

We start by validating the accuracy and well-posedness of the mortar method in the classical setting by using only finite element solvers on each subdomain. In particular, no model training is used for this particular example and we only seek to show that the above mortar method converges in the forward problem. We examine the problem (4.1) with true solution $p(x, y) = xy + y^2$ on the domain $\Omega = [0, 2]^2$ with

$$K = \begin{pmatrix} (x+1)^2 & 0.5 \\ 0.5 & y^2+1 \end{pmatrix}. \tag{5.1}$$

The domain is subdivided into four equal squares. Our initial mesh is depicted in Fig. 4 with only one degree of freedom on the mortar (with the remaining four fixed due to the homogeneous Dirichlet boundary condition). For refinement, we divide each subdomain diameter and the mortar diameter by half; see the right hand side of Fig. 4 for a figure of the first refinement.

On each subdomain, we will use the standard $\mathbb{Q}_1$ space for pressure with Nedelec elements of the lowest order for the velocity. The quantities in Lemma 4.8 can be replaced with results from standard FEM a priori estimates

(Roberts and Thomas, 1991). As a result, we obtain a convergence result of

$$\|\boldsymbol{u} - \boldsymbol{u}_h\|_\Omega + \left\| p - p_h \right\|_\Omega \le CH \tag{5.2}$$

where H is the size of the mortar. The $\mathcal{O}(H)$ convergence in the velocity is clearly illustrated in Table 1 while we obtain $\mathcal{O}(H^2)$ superconvergence in the pressure, which was observed in smooth solutions using mortar methods (Arbogast et al., 2007, 2000). Furthermore, since the estimate in Lemma 4.8 is in the $H^{1/2}$ norm, we expect convergence of $\mathcal{O}(H^{3/2})$ as we are measuring the L^2 norm but we also observe a level of superconvergence.

TABLE 1 Table illustrating the absolute errors, and the convergence rates for Example 1. The rates are in agreement with Lemma 4.8.

H	$\|p - p_h\|_\Omega$	$\|u - u_h\|_\Omega$	$\|\lambda - \lambda_h\|_{L^2(\Gamma)}$
1	2.73E-01	4.66E+00	2.44E-01
1/2	6.23E-02	2.16E+00	5.75E-02
1/4	1.49E-02	1.04E+00	1.43E-02
1/8	3.66E-03	5.12E-01	3.56E-03
1/16	9.07E-04	2.54E-01	8.91E-04
1/32	2.31E-04	1.26E-01	2.41E-04
Rate	$\mathcal{O}(H^{2.04})$	$\mathcal{O}(H^{1.04})$	$\mathcal{O}(H^{2.00})$

5.2 Example 2: pure FEEC and pure FEM elements comparison

We now consider the data arising from the problem (4.1) with $\Omega = [0,3] \times [0,3]$,

$$f := 2\pi^2 \cos(\pi x)\sin(\pi y), \qquad K = \boldsymbol{I} \tag{5.3}$$

with boundary condition determined by the true solution $p(x, y) = \cos(\pi x)\sin(\pi y)$.

The domain Ω is split into 9 uniform squares whereby either a $\mathbb{Q}_1$ FEM or a pretrained FEEC element is used in each subdomain. The FEEC element is trained on 20,480 uniformly drawn points from $[0,1]^2$ with 16 POUs on the interior and 16 on the boundary with varying number of fine-scale knots. As discussed in Actor et al. (2024), increasing the number of fine-scale grids is akin to h-refinement in the FEM sense.

To train the FEEC elements, we use data arising from different boundary conditions and forcing terms which corresponds to approximating (4.17) and (4.15):

1. a problem with the same forcing term as in (5.3) but homogeneous zero Dirichlet boundary condition. This corresponds to (4.17).

2. Sixteen different boundary conditions consisting of the Bernstein polynomials of fourth order on the boundary (e.g. x^4y^4, $\binom{4}{1}x^4y(1-y)^3$ etc) and forcing term of $f=0$. This is needed so that (4.15) can be approximated accurately on the FEEC elements when different boundary conditions are passed in from the mortar.

The solutions to the above boundary value problems were calculated by a low-order finite element solver. For more details regarding the training, we refer the reader to Subsection 10.A.2.

The mortar refinement level was chosen to be $H=4h$ in for the FEM case. For the FEEC local solvers, we note that the fine-scale nodes can move, resulting in nonuniform meshes; nevertheless, we still choose the same H as the FEM case for comparison's sake.

In Table 2 and Table 3, we show the error resulting from using purely FEEC elements or purely FEM elements on all the subdomain respectively. The convergence rates among the two different solvers are similar, and reflect superconvergence due to the smoothness of the problem. In Fig. 5, we plot the true solution and its fluxes, and the approximate solution and its fluxes on the whole $[0,3]^2$ domain solved using FEEC elements, while Fig. 6 plots the quantities on the diagonal line from $(0,0)$ through $(3,3)$. In both cases, the true solution is well-approximated.

TABLE 2 Table of absolute error for the sine-cosine problem Subsection 5.2 using trained FEEC elements as the subdomain solver. The convergence rates are similar to the method using pure FEM elements. We note that the fine-scale grid roughly corresponds to h-scaling in a standard FEM method (Actor et al., 2024).

FEEC fine-scale grid and mortar size	$\|p-p_h\|_\Omega$	$\|u-u_h\|_\Omega$	$\|\lambda-\lambda_H\|_{L^2(\Gamma)}$
8×8, $H=1/2$	1.47E-01	1.48E+00	2.20E-01
12×12, $H=1/3$	6.66E-02	8.97E-01	8.47E-02
16×16, $H=1/4$	4.07E-02	5.41E-01	4.41E-02
20×20, $H=1/5$	2.76E-02	4.40E-01	2.71E-02
24×24, $H=1/6$	2.10E-02	3.97E-01	1.97E-02
	$\mathcal{O}(h^{1.78})$	$\mathcal{O}(h^{1.25})$	$\mathcal{O}(h^{2.22})$

5.3 Example 3: hybrid methods

We next showcase the ability to use a hybrid approach whereby standard finite elements are interfaced to FEEC elements allowing for areas with unknown features to be learned using FEEC elements, and smooth areas using classical FEM methods.

TABLE 3 Table of absolute error for the sine-cosine problem Subsection 5.2 with FEM elements as the local solvers.

FEM fine-scale grid and mortar size	$\|p-p_h\|_\Omega$	$\|u-u_h\|_\Omega$	$\|\lambda-\lambda_H\|_{L^2(\Gamma)}$
8×8, $H=1/2$	1.67E-01	1.54E+00	2.16E-01
12×12, $H=1/3$	7.04E-02	8.54E-01	8.16E-02
16×16, $H=1/4$	3.91E-02	5.76E-01	4.31E-02
20×20, $H=1/5$	2.49E-02	4.30E-01	2.68E-02
24×24, $H=1/6$	1.72E-02	3.41E-01	1.83E-02
28×28, $H=1/7$	1.26E-02	2.82E-01	1.33E-02
32×32, $H=1/8$	9.64E-03	2.40E-01	1.01E-02
36×36, $H=1/9$	7.61E-03	2.08E-01	7.98E-03
40×40, $H=1/10$	6.15E-03	1.84E-01	6.45E-03
	$\mathcal{O}(h^{2.04})$	$\mathcal{O}(h^{1.31})$	$\mathcal{O}(h^{2.16})$

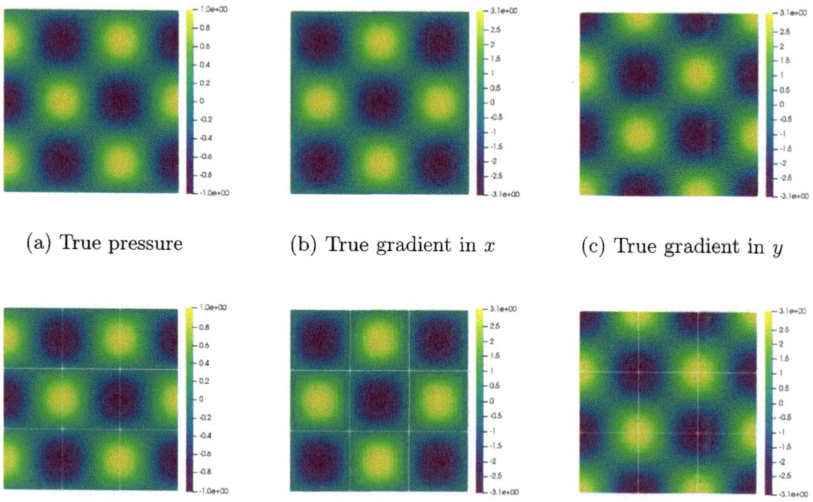

(a) True pressure (b) True gradient in x (c) True gradient in y

(d) Estimated pressure (e) Estimated gradient in x (f) Estimated gradient in y

FIGURE 5 Plot of the true (first row) and estimated (second row) solution for the sine-cosine problem Subsection 5.2 with pure FEEC elements consisting of 24×24 fine scale nodes, and $H=1/6$. As expected, the solution is well-approximated by the FEEC elements.

We assume data is obtained from the problem (4.1) on $\Omega=[-1.5,1.5]^2$ with the parameters

$$f:=0, \qquad K(\boldsymbol{x})=\begin{cases}\begin{pmatrix}k & 0\\ 0 & k\end{pmatrix}, & \|\boldsymbol{x}\|\le b\\[2ex] \begin{pmatrix}1 & 0\\ 0 & 1\end{pmatrix}, & \|\boldsymbol{x}\|> b\end{cases} \tag{5.4}$$

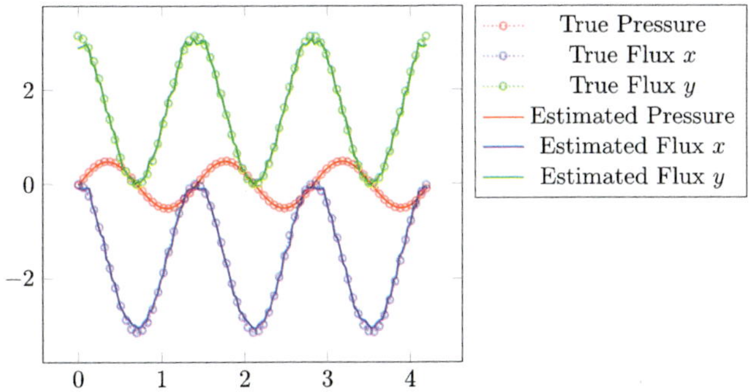

FIGURE 6 Profile of the true and estimate solutions for the sine-cosine problem Subsection 5.2 on the line (0, 0) to (3, 3) using pure FEEC elements with 24×24 fine scale nodes and $H = 1/6$. While there are small fluctuations in the FEEC approximation, it is clear that both the pressure and fluxes are captured.

with $b = .2$, $k = 10$. The Dirichlet boundary imposed such that the true solution is

$$u := \begin{cases} x\left(1 - \frac{b^2(k-1)}{(k+1)(x^2+y^2)}\right) & \|\boldsymbol{x}\| > b \\ \frac{2}{k+1}x & \|\boldsymbol{x}\| \leq b \end{cases}$$

This particular equation arises in electrostatics when examining the case where a conducting cylinder with radius b and capacitance k is placed within a uniform field of strength 1 (Smythe, 1989, §4.03). Note that outside of a radius around the origin, the diffusion problem is easy to solve.

We split the domain into 9 congruent squares with the center square $[-0.5, 0.5]^2$ consisting of a FEEC element to capture the change in material coefficients while the remaining eight subdomains utilizing a simple, low-order FEM space with 8×8 quads. The FEEC element is trained on 12 different boundary conditions corresponding to the 12 third-order Bernstein polynomials on the boundary as in the previous example. We note that in training, only the solution and its fluxes are provided, meaning the material coefficient (5.4) is not fully exposed to the FEEC element. A total of 16 POUs are used on the interior and the boundary. We choose to use a mortar of $H = 1/4$.

We show the error over the whole domain in Table 4 from only refining the fine-scale grid of the FEEC element in $[-.5, .5]^2$. A full rate of convergence is not expected since Theorem 4.6 requires both the mortar space and the local subdomain solvers to be refined in tandem. We do not consider refinement with the mortar here as Assumption 1 might be violated from either the movement of fine-scale knots of the FEEC elements, or the fact that the mesh size of the FEM solvers are fixed to be very coarse.

In Fig. 7, we plot the true and estimated solution to the problem. Note that the trained FEEC element managed to resolve the circular inclusion and the sub-

TABLE 4 Table of absolute and relative errors for the cylinder problem Subsection 5.3 using a hybrid approach. While we do not expect a full convergence as we are only refining the singular FEEC element on $[-.5, .5]^2$ while keeping the mortar spaces and FEM spaces constant, we do observe that using the finest FEEC element gives significantly better results.

FEEC fine-scale grid	$\|p - p_h\|_\Omega$	$\|u - u_h\|_\Omega$	$\|\lambda - \lambda_H\|_{L^2(\Gamma)}$
8×8	5.69E-03 (0.224%)	1.64E-01 (5.48%)	6.41E-3
16×16	3.07E-03 (0.121%)	1.16E-01 (3.87%)	4.66E-3
24×24	2.01E-03 (0.079%)	8.29E-02 (2.77%)	2.50E-3
	$\mathcal{O}(h^{.939})$	$\mathcal{O}(h^{.607})$	$\mathcal{O}(h^{.814})$

tleties in the fluxes when the true solution is not explicitly given in the training data. Furthermore, we plot the true and estimated solution profiles in Fig. 8. From the plots, it is clear that while there are small spurious fluctuations in the estimated solutions, that the error decreases as we refine the FEEC model. In Fig. 9, we compare the FEEC profiles to the profile obtained using a 24×24 FEM on $[-.5, .5]^2$ instead. Note that the oscillations are greatly reduced by using the FEEC elements due to the adaptivity of the fine-scale mesh.

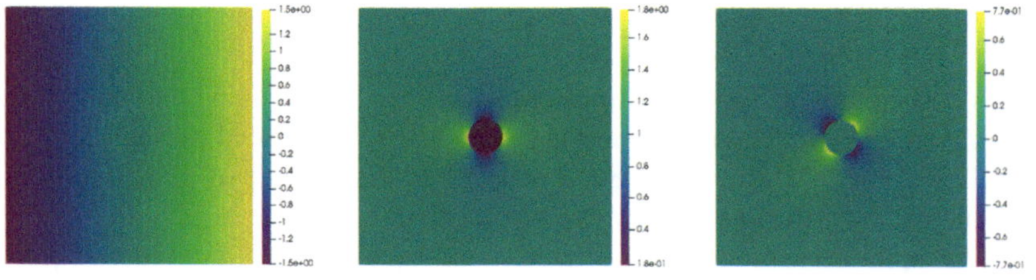

(a) True pressure (b) True gradient in x (c) True gradient in y

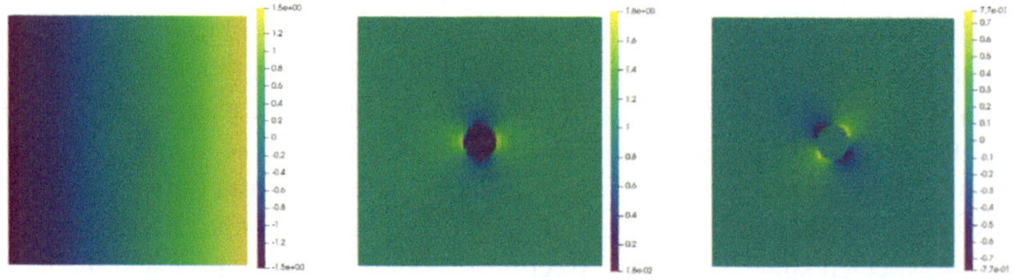

(d) Estimated pressure (e) Estimated gradient in x (f) Estimated gradient in y

FIGURE 7 Figure of the true solution and estimated value for the cylinder problem Subsection 5.3. The estimated solution uses a single FEEC element with 24×24 fine-scale knots in the center-most subdomain with the remaining subdomains using FEM of just 8×8 elements.

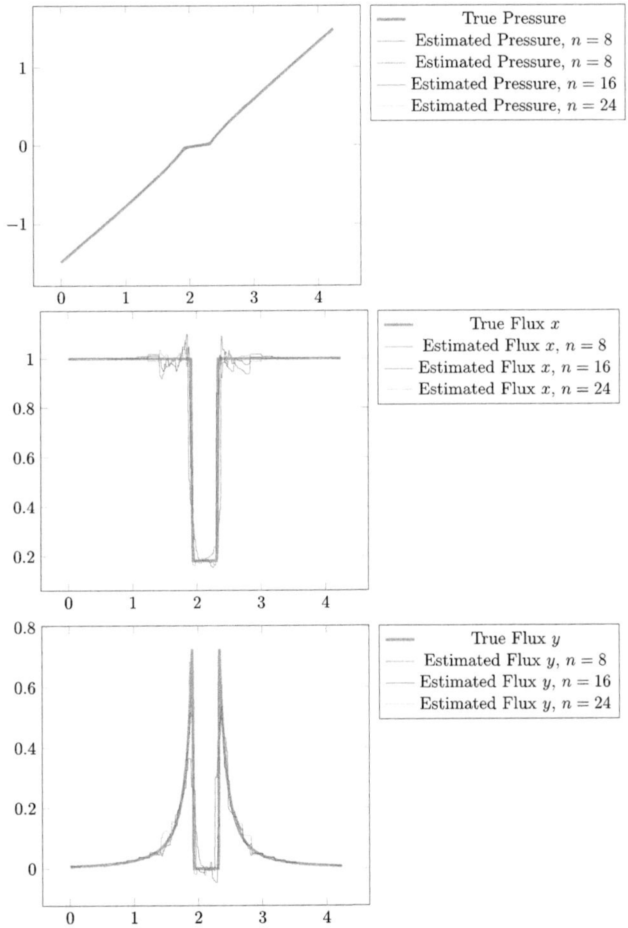

FIGURE 8 Trace plot from $[-1.5, -1.5]$ to $[1.5, 1.5]$ of the cylinder problem Subsection 5.3 for FEEC models with different fine-scale nodes n in the subdomain $[-.5, .5]^2$. As we refine the number of fine-scale nodes we use, the jumps in the fluxes are increasingly more well-resolved with less fluctuations.

5.4 Example 4: subdomain refinement with FEEC

In this next class of examples, we will consider three separate problems whereby the number of subdomains is increased with no further refinement in either the subdomain-level solver, or the number of mortar degrees of freedom per subdomain. This is a nonstandard example case in the context of domain decomposition methods, but is extremely useful in the case where machine-learned elements are used.

We hypothesize that smaller subdomains means that there are fewer features for each FEEC element to learn, meaning that the optimization procedure will

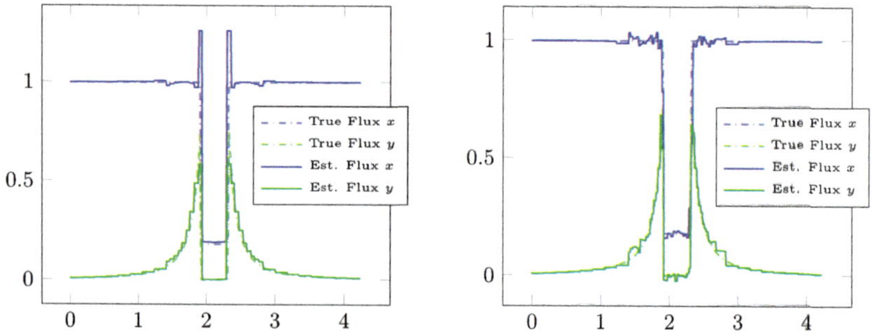

(a) Line from $(-1.5, -1.5) \to (1.5, 1.5)$ with 24×24 FEM in center.

(b) Line from $(-1.5, -1.5) \to (1.5, 1.5)$ with 24×24 fine scale FEEC in center

FIGURE 9 Comparison between using FEM (left) and FEEC (right) solvers in the material discontinuity region $[-.5, .5]^2$ for the cylinder problem Subsection 5.3. Note the overshoot in the discontinuity in the x component of the flux for the pure finite elements case, resulting in a relative error of over 25% near the discontinuity. On the other hand, the FEEC element is able to reduce that fluctuation near the discontinuity to less than 5% using the same number of fine-scale knots due to the adaptivity.

usually result in smaller local losses. The smaller number of features to capture also means that we can use FEEC elements without as many fine-scale nodes, decreasing computational costs in training. Furthermore, in the case with large amount of data points, smaller subdomains means that one can speed up the training tremendously as all the training points can now fit on a single GPU.

In the first two examples, we perform a similar training procedure as before where on each subdomain, a suite of boundary conditions are used to train the local Whitney elements. The last example is more representative of a possible usage case where only a single reference solution is provided with realistic multiscale features.

5.4.1 *Stripe problem*

Consider data arising from the problem (4.1) with $\Omega = [0, n] \times [0, n]$ for n a positive integer,

$$f := 0, \quad g := x, \qquad K = \kappa_i \mathbf{I} \tag{5.5}$$

where $\mathbf{I}$ is the $\mathbb{R}^{2\times 2}$ identity matrix, and where if $\lfloor y \rfloor$ is even, then

$$\kappa_i = \begin{cases} 1 & 0 \le y < .4 \\ .4 & .4 \le y < .8 \\ .8 & .8 \le y < 1 \end{cases}$$

otherwise,

$$\kappa_i = \begin{cases} .8 & 0 \le y < .2 \\ .3 & .4 \le y < .6 \\ .9 & .8 \le y < 1 \end{cases}$$

While the true solution for the pressure is trivially $p(x) = x$ for all n, the difficulty lies in the ability of the discrete solution to capture the discontinuous velocities

$$\boldsymbol{u}(x) := \begin{pmatrix} \kappa_i \\ 0 \end{pmatrix}$$

which arises.

Two FEEC elements of size $[0, 1]^2$ are trained: one to capture the case where $\lfloor y \rfloor$ is even, and another for the odd case. For both FEEC elements, a total of 20×20 fine scale nodes were used, which was subsequently compressed down to 14 POUs on the interior and 14 on the boundary. To train the two FEEC systems, we minimize the MSE against only four PDEs corresponding to the Laplace equation $f = 0$ with the boundary conditions xy, $x(1-y)$, $(1-x)y$, $(1-x)(1-y)$ on 20,480 randomly sampled points on $[0, 1]^2$. As for the mortar space, the lowest order space $H = 1$ is used. Note that in this case, Assumption 1 is trivially satisfied.

In Fig. 10, we show the solutions of the pressure for $n = 2, 3, 5$. We see that we recover the true pressure easily as it is just a simple linear function. We note that the notion of convergence is not applicable in this case since the domain and problem itself are actually changing as we increase n.

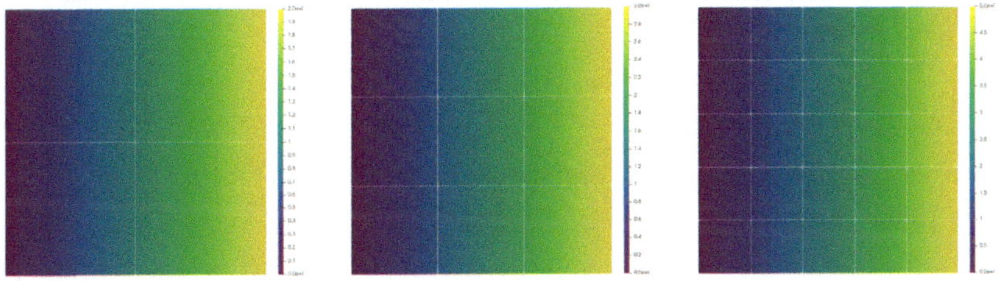

FIGURE 10 Figure of the pressure solutions obtained for the stripes problem Subsection 5.4.1 for on increasingly larger domain $[0, 2]^2$, $[0, 3]^2$, $[0, 5]^2$ using FEEC elements of 14×14 fine-scale knots and a very coarse mortar of $H = 1$. Importantly, we note that, from left to right, the domain Ω of the problem is being increased and we are not depicting a refinement process.

In Fig. 11, we show the x-component of the gradient; it is clear that the stripes structure is well-preserved even as we introduce more subdomains into the mortar space. While the error estimates Lemma 4.8 cannot support this statement due to the usage of crude L^∞ norms, this is indication that, at least numerically, requirement **R2** is satisfied. We also plot the estimate solution profile on the line (2.5, 0) to (2.5, 5) in Fig. 12 for the case of $n = 5$. From this view, it's clear that the actual numerical values are in good agreement with the true solution.

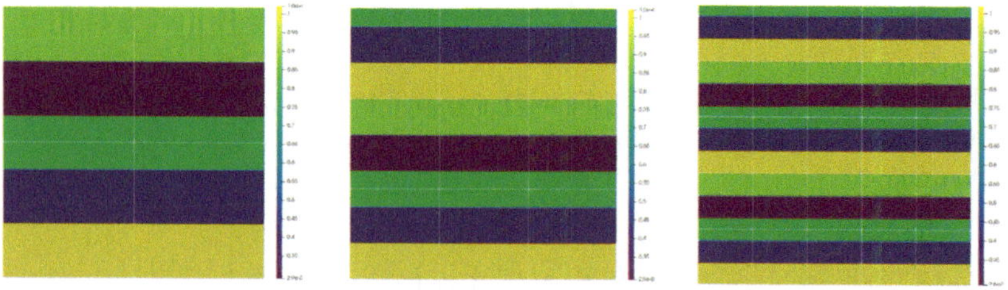

FIGURE 11 Figure of the flux in x of the solutions obtained for the stripes problem Subsection 5.4.1 on $[0, 2]^2$, $[0, 3]^3$, $[0, 5]^2$ with FEEC elements of 14×14 fine-scale knots and $H = 1$. We remark that the discontinuities are well-preserved using the FEEC elements even as the domain of the problem is increased.

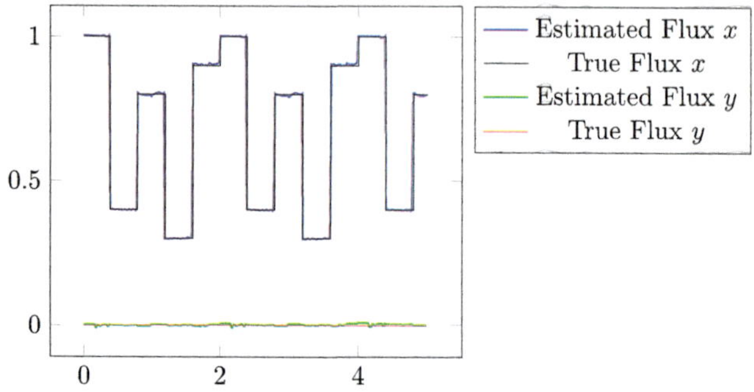

FIGURE 12 Profile on the line $(2.5, 0)$ to $(2.5, n)$ for $n = 5$ of the fluxes of the estimated and true stripes problem Subsection 5.4.1 obtained from the FEEC elements with 14×14 fine-scale knots and $H = 1$ mortar space.

5.4.2 Path problem

Consider data arising from the problem (4.1) on $\Omega = [0, 1] \times [0, 1]$ with $f = 0$, $g = x$ and

$$K = \begin{cases} \frac{1}{5}\mathbf{I} & x \in \Omega_{\text{path}} \\ \frac{1}{2}\mathbf{I} & x \in \Omega_{\text{circ}} \\ \mathbf{I} & x \in \text{elsewhere} \end{cases} \tag{5.6}$$

where Ω_{path} is defined as the region lying in

$$R((0, .625), (.375, .875)) \cup R((.375, .125), (.625, .875)) \cup R((.625, .125), (1, .375))$$

with $R(p_1, p_2)$ is the rectangle with lower left point p_1 and upper right corner p_2, and Ω_{circ} are two circles centered at $(.125, .25)$ and $(.875, .75)$ with radius $.075$. See the first column of Fig. 13 for figures of the true solution.

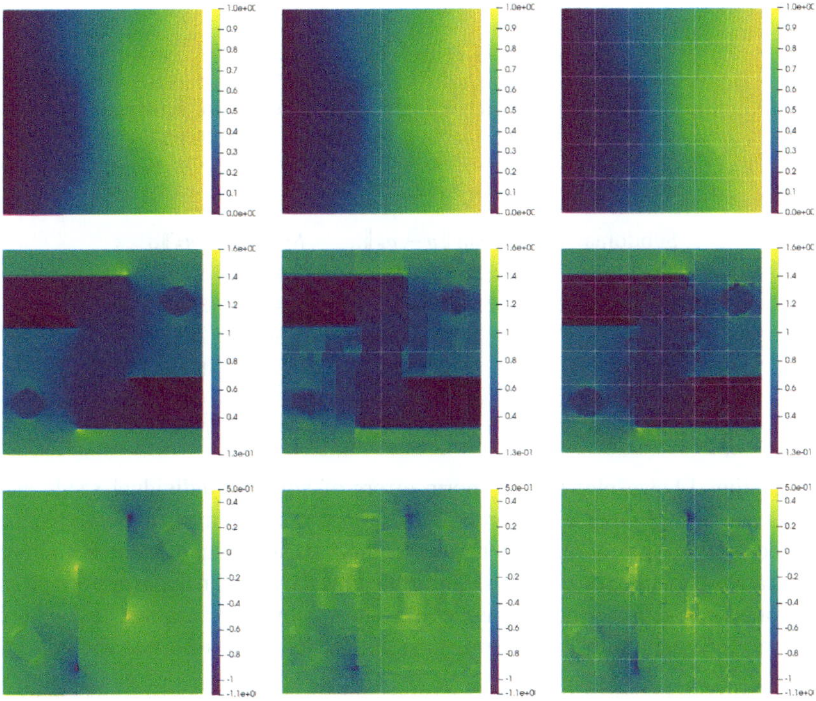

FIGURE 13 Plot of the true solution (first column), and subdomain with 2 (second column) and 6 (third column) refinements for the path problem Subsection 5.4.2. Note that the features are increasingly more refined and match the true solutions as the number of subdomains are increased.

Let our domain Ω be subdivided into n^2 equal squares as our subdomains, and let $H = \frac{1}{4n}$ meaning each subdomain has a total of 16 mortar degrees of freedom. On each of the subdomains, we train a FEEC element on 20,480 uniformly sampled points from the subdomain with 10 fine scale nodes and 14 POUs on the interior and boundary. As before, the FEEC elements are trained on 12 total boundary conditions corresponding to the third order Bernstein polynomials on squares. Rather than refining the mortar discretization relative to the number of subdomains, or increasing the fine-scale nodes on the local solvers, we *strictly increase the number of subdomains in this study.* We reiterate the fact that as the number of subdomains increases, the number of mortar degrees of freedom per subdomain remains the same at 16 and each FEEC element has the same number of parameters (e.g. 10 fine scale nodes and 14 POUs on the interior).

In Table 5, we show the average error resulting from increasing the number of subdomains over five different random seeds for training. We note that while the error in the pressure is already captured quite accurately by a single FEEC element owing to its almost linear nature on the whole domain, the error in the gradient decreases much more dramatically, due to the higher resolution by increasing the number of subdomains.

TABLE 5 Table of average absolute and relative errors for the path problem Subsection 5.4.2 whereby the domain is increasingly subdivided into finer pieces. While the pressure does not exhibit convergence, the flux converges at a rate of $\mathcal{O}(h)$ and so does the full H^1 norm (see Fig. 14).

Subdomains	Mean $\|p - p_h\|_\Omega$	Mean $\|u - u_h\|_\Omega$
2×2	3.53E-03 (0.596%)	3.47E-02 (5.21%)
3×3	3.25E-03 (0.549%)	2.49E-02 (3.74%)
4×4	3.13E-03 (0.528%)	1.56E-02 (2.34%)
5×5	2.97E-03 (0.501%)	1.33E-02 (1.99%)
6×6	2.70E-03 (0.456%)	9.80E-03 (1.47%)
8×8	3.28E-03 (0.554%)	6.50E-03 (0.97%)

In Fig. 14, we plot the H^1 norm errors of both the individual seeds and the mean. We observe a first-order convergence in the number of subdomains, supporting the notion that our mortar method satisfies requirement **R2** as we increase the number of elements. Unfortunately, the error analysis performed in the previous section is not fine enough to show convergence in this case where we increase the number of subdomains due to the usage of crude triangle inequalities.

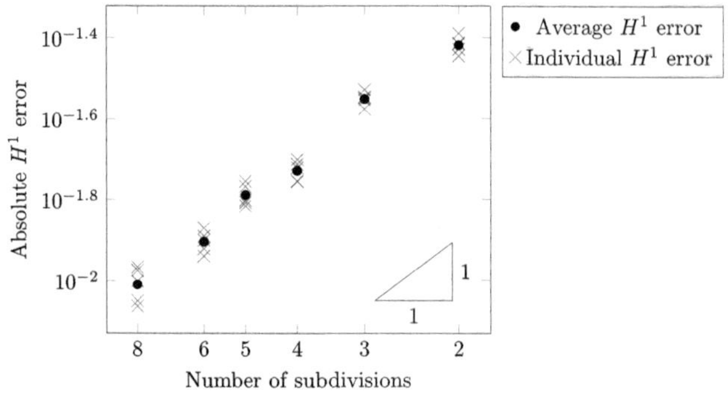

FIGURE 14 Plot of the absolute H^1 error resulting from refinement for the path problem Subsection 5.4.2. We observe a linear convergence rate by dividing the domain into increasingly smaller domains for the FEEC problem.

5.4.3 Battery problem: single solution training

We now consider data from the problem (4.1) on $\Omega = [0, 1] \times [0, 1]$ with $f = 0$ and a nontrivial material data and boundary condition corresponding to a voltage difference across a lithium-ion battery. The true pressure and fluxes, which are sampled at 5.89 million points, are provided via a high-fidelity solver SIERRA/ARIA (Notz et al., 2016) and will be treated as the *only* source of provided data with no additional methods of augmentation. In other words, we assume a full simulation of the response for the subdomains to arbitrary mortars is not available, meaning the local FEEC elements will have to extrapolate the correct Dirichlet-to-Neumann maps. For a figure of the true pressure and flux, see Fig. 15. More details regarding the data can be found in Appendix B of Actor et al. (2024); note that for simplicity, we consider the problem as a purely Dirichlet boundary condition problem whilst (Actor et al., 2024) included Neumann boundary conditions.

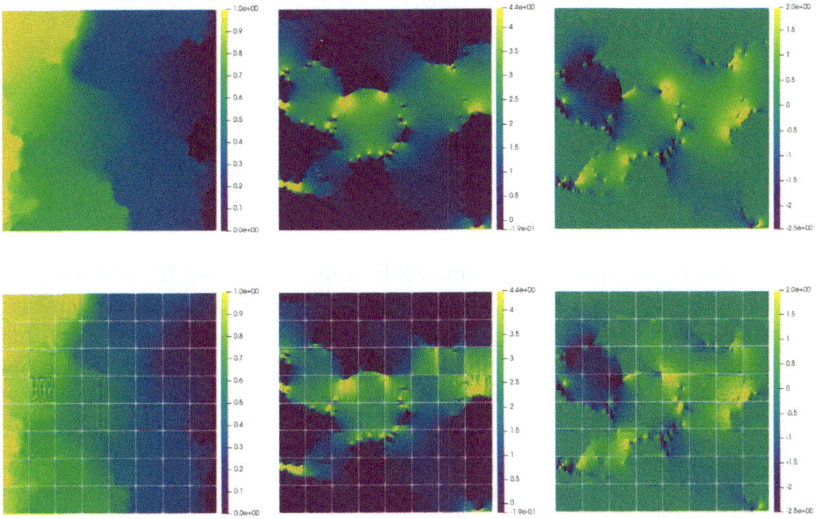

FIGURE 15 Figures of the given pressure/fluxes for the battery problem Subsection 5.4.3 in the first row, and the approximation obtained from solving the Darcy's flow problem in the second row for 8×8 subdivisions. Overall, the estimated solution matches the data fairly accurately with many most small details captured.

We again split the domain $[0, 1]^2$ into n^2 uniform squares, but only employ four mortar degrees of freedom per subdomain with $H = \frac{1}{n}$ (i.e. the mortar degrees of freedom lie on the corners of the subdomain).[2] A FEEC element with 12 fine scale nodes in both the x and y direction, and 12 POUs on the interior and boundary are used on each subdomain.

Since only a single reference solution is provided, we train the FEEC element with boundary condition obtained from interpolating the given solution and the data given (e.g. $g_i = p|_{\partial\Omega_i}$). For example, suppose $n = 2$, then the FEEC element on subdomain corresponding to $\Omega' = [0, .5]^2$ will have $\frac{5.89}{4} \approx 1.5$ mil-

[2] The coarsest mortar mesh is chosen since the fine scale nodes may move substantially, due to only one training set, and violate assumption (4.21).

lion data points, and boundary conditions corresponding to the nearest neighbor interpolation of those points on $\partial\Omega'$. This is unlike Subsection 5.4.2 or even Subsection 5.4.1 where each FEEC element was provided with multiple examples to train on. Note that the number of training data points per FEEC element decrease as we increase the number of subdomains, we have found that it can lead to some instability in pretraining.

TABLE 6 Table of absolute and relative errors for the battery problem Subsection 5.4.3. The right "true mortar" (TM) columns essentially capture the training error by simply fixing the mortar space to the true values, while the left columns result from actually solving the Darcy's flow equations. Similar to Subsection 5.4.2, the error in pressure only decreases slightly with most of the benefits arising from the convergence in the H^1-seminorm.

Subdomains	$L^2(\Omega)$	H^1-seminorm	TM $L^2(\Omega)$	TM H^1-seminorm
2×2	7.15E-03 (1.27%)	1.22E+00 (67.6%)	5.37E-03	1.39E+00
3×3	3.01E-03 (0.54%)	6.45E-01 (35.7%)	2.76E-03	6.03E-01
4×4	2.71E-03 (0.48%)	4.57E-01 (25.3%)	2.42E-03	3.27E-01
6×6	2.46E-03 (0.44%)	2.44E-01 (13.5%)	1.67E-03	1.37E-01
8×8	2.40E-03 (0.43%)	1.43E-01 (7.92%)	1.41E-03	1.19E-01

In Table 6, we show the absolute MSE of the L^2 and the H^1 seminorm resulting from solving the Darcy's flow equation with the trained FEEC elements. In the case of 2×2 refinement, the error is quite large since the mortar only has one degree of freedom in the interior (cf. Fig. 4) and the boundary conditions are not even well-resolved; however, it's clear that as additional refinements are made that the relative error decreases. In addition, we also show the absolute MSE of the "true mortar" (TM) which is obtained by setting the mortar degrees of freedom to be the interpolant from the data set. This "true mortar" indicates how much of the error is due to the training procedure as no actual solves of the bilinear form is performed and allows us to see how much error arises from the actual mortar coupling. Since this true mortar errors are similar to the errors obtained from solving the bilinear form, this suggests that very little error arises due to the mortar coupling. In Fig. 16, we observe that the error obtained from solving the Darcy flow equation decreases as we increase the number of subdomains, with the finest level obtaining a better H^1 error than the errors obtained in Actor et al. (2024).

Appendix 10.A Technical details

10.A.1 Technical proofs

Proof of Lemma 4.4. By (4.5), for any $\mu \in H_0^\gamma(\Omega)$, we can decompose it as

$$\mu = p^*(\mu) + \sum_{i=1}^{n} p_{i0}$$

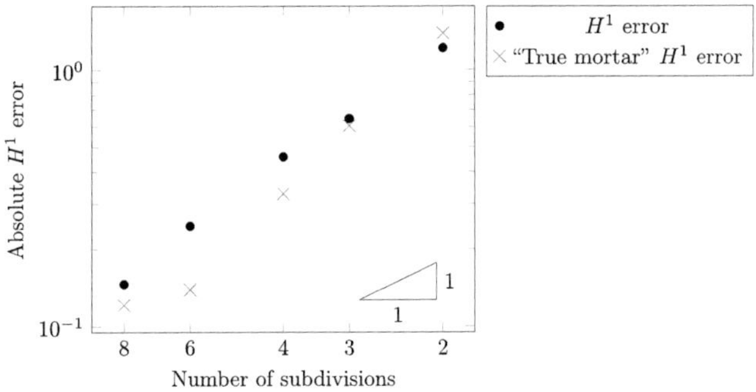

FIGURE 16 Plot of the H^1 error and true mortar error resulting from refinement for the battery problem Subsection 5.4.3. For an explanation of what the true mortar error is, we refer the reader to the corresponding discussion Subsection 5.4.3. We note that the error is quite close to the true mortar error, meaning that the coarse mortar space does not negatively effect the convergence that much.

where $p^*(\mu) \in H_0^1(\Omega)$ satisfying (4.9) (hence $p^*(\mu)|_\Gamma = \mu|_\Gamma$) and $p_{i0} \in H_0^1(\Omega_i)$ are bubble functions.

Thus,

$$\begin{aligned} b(\lambda, \mu) &= \sum_{i=1}^{n} \left(\boldsymbol{u}^*(\lambda), \nabla p^*(\mu) + \sum_{i=1}^{n} \nabla p_{i0} \right)_{\Omega_i} \\ &= \sum_{i=1}^{n} \left(\boldsymbol{u}^*(\lambda), \nabla p^*(\mu) \right)_{\Omega_i} = \sum_{i=1}^{n} \left(K \nabla p^*(\lambda), \nabla p^*(\mu) \right)_{\Omega_i} \end{aligned} \tag{10.A.1}$$

since (4.9) implies the inner products of $\boldsymbol{u}^*(\lambda)$ with the gradient of bubble functions are zero. From the above, the bilinear form is clearly symmetric and positive definite.

For coercivity, using (10.A.1), Poincare inequality and trace inequality (Evans, 2022),

$$\begin{aligned} b(\lambda, \lambda) &\geq \left\| \nabla p^*(\lambda) \right\|^2 \geq \frac{1}{C} \left\| p^*(\lambda) \right\|^2_{H^1(\Omega)} = \frac{1}{C} \sum_{i=1}^{n} \left\| p^*(\lambda) \right\|^2_{H^1(\Omega_i)} \\ &\geq \frac{1}{C} \sum_{i=1}^{n} \|\lambda\|^2_{H^{1/2}(\Gamma_i)}, \end{aligned}$$

meaning $b(\lambda, \lambda) \geq \alpha \sum_{i=1}^{n} \|\lambda\|^2_{H^{1/2}(\Gamma_i)} \sim \alpha \|\lambda\|^2_{H^\gamma}$ for some constant α independent of λ. □

The remaining proofs are for the coercivity and the error estimate for the discrete mortar. We introduce the shorthand notation $p_h^*(Q\lambda_H) :=$

$\sum_{i=1}^{n} p_h^*(Q_i\lambda_H)$, and let $\|\cdot\|_\Omega$ denote the L^2 norm over the domain Ω unless otherwise stated. Before proceeding, we define the inclusion map $P_i : \Lambda_H \subset \Lambda \to H^\gamma|_{\Omega_i}$ through the isomorphism. We need a preparatory lemma:

Lemma 10.A.1. *Let* $\delta := \frac{2}{C_p+1}$, *where* C_p *is the constant arising from Corollary 6.3 of Brenner (2003), then*

$$\frac{\delta}{2}\left\|p_h^*(Q\lambda_H)\right\|_{H^1(\Omega)}^2 \leq \left[\left\|\nabla p_h^*(Q\lambda_H)\right\|_\Omega^2 + \frac{C_p}{C_p+1}\sum_{\Gamma_{ij}}\frac{1}{|\Gamma_{ij}|}\left\|Q_i\lambda_H - Q_j\lambda_H\right\|_{L^2(\Gamma_{ij})}^2\right].$$

Proof. By a simple application of Corollary 6.3 of Brenner (2003):

$$\begin{aligned}\left\|p_h^*(Q\lambda_H)\right\|_{H^1(\Omega)}^2 &\leq C_p\left[(1+\frac{1}{C_p})\left\|\nabla p_h^*(Q\lambda_H)\right\|_\Omega^2 + \sum_{\Gamma_{ij}}\frac{1}{|\Gamma_{ij}|^2}\left(\int_{\Gamma_{ij}} Q_i\lambda_H - Q_j\lambda_H\, ds\right)^2\right]\\ &\leq C_p\left[(1+\frac{1}{C_p})\left\|\nabla p_h^*(Q\lambda_H)\right\|_\Omega^2 + \sum_{\Gamma_{ij}}\frac{1}{|\Gamma_{ij}|}\left\|Q_i\lambda_H - Q_j\lambda_H\right\|_{L^2(\Gamma_{ij})}^2\right]\end{aligned}$$

where we used Cauchy-Schwarz inequality on $(\int_\sigma f)^2 \leq |\sigma|\,\|f\|^2$. □

Proof of Lemma 4.5. An identity like (10.A.1) can also be verified for the discrete version as well since on any subdomain i and $\mu_H \in \Lambda_H$, $Q_i\mu_H = p_h^*(Q_i\mu_H) + p_{hi}$ where $p_{hi} \in W_{hi,0}$ bubble functions, one has

$$\begin{aligned}b_h(\lambda_H,\mu_H) &= \sum_{i=1}^{n}\left(\boldsymbol{u}_h^*(Q_i\lambda_H), \nabla p_h^*(Q_i\mu_H) + \nabla p_{hi}\right)_{\Omega_i}\\ &= \sum_{i=1}^{n}\left(\boldsymbol{u}_h^*(Q_i\lambda_H), \nabla p_h^*(Q_i\mu_H)\right)_{\Omega_i}\\ &= \sum_{i=1}^{n}\left(K\nabla p_h^*(Q_i\lambda_H), \nabla p_h^*(Q_i\mu_H)\right)_{\Omega_i}.\end{aligned} \tag{10.A.2}$$

Thus, the bilinear form b_h is symmetric, and, at least, positive semidefinite. Coercivity requires a bit more work.

Since for each subdomain i, $p_h^*(Q_i\lambda_H)|_{\Gamma_i} = Q_i\lambda_H|_{\Gamma_i}$, we add by zero and expand to obtain

$$\begin{aligned}
b_h(\lambda_H,\lambda_H) &= \sum_{i=1}^{n} \big(K\nabla p_h^*(Q_i\lambda_H), \nabla p_h^*(Q_i\lambda_H)\big)_{\Omega_i} \\
&\quad + \delta\langle Q_i\lambda_H - p_h^*(Q_i\lambda_H), Q_i\lambda_H\rangle_{H^{1/2}(\Gamma_i)} \\
&\geq \sum_{i=1}^{n} \|\nabla p_h^*(Q_i\lambda_H)\|_{\Omega_i}^2 + \delta\|Q_i\lambda_H\|_{H^{1/2}(\Gamma_i)}^2 \\
&\quad - \frac{\delta}{2}\|p_h^*(Q_i\lambda_H)\|_{H^{1/2}(\Gamma_i)}^2 - \frac{\delta}{2}\|Q_i\lambda_H\|_{H^{1/2}(\Gamma_i)}^2 \\
&\geq \sum_{i=1}^{n} \|\nabla p_h^*(Q_i\lambda_H)\|_{\Omega_i}^2 + \frac{\delta}{2}\|Q_i\lambda_H\|_{H^{1/2}(\Gamma_i)}^2 \\
&\quad - \frac{\delta}{2}\|p_h^*(Q_i\lambda_H)\|_{H^1(\Omega_i)}^2 \\
&= \|\nabla p_h^*(Q\lambda_H)\|_{\Omega}^2 - \frac{\delta}{2}\|p_h^*(Q\lambda_H)\|_{H^1(\Omega)}^2 \\
&\quad + \sum_{\Gamma_{ij}} \frac{\delta}{2}(\|Q_i\lambda_H\|_{H^{1/2}(\Gamma_{ij})}^2 + \|Q_j\lambda_H\|_{H^{1/2}(\Gamma_{ij})}^2)
\end{aligned}$$

by using Cauchy-Schwarz, the trace inequality, and the trivial inequality $ab \leq 2a^2 + 2b^2$.

Now, using the assumption (4.22)

$$\begin{aligned}
b_h(\lambda_H,\lambda_H) &\geq \|\nabla p_h^*(Q\lambda_H)\|^2 - \frac{\delta}{2}\|p_h^*(Q\lambda_H)\|_{H^1(\Omega)}^2 \\
&\quad + \sum_{\Gamma_{ij}} \frac{\delta C_p}{2|\Gamma_{ij}|}\|Q_i\lambda_H - Q_j\lambda_H\|_{L^2(\Gamma_{ij})}^2 \\
&\quad + \sum_{\Gamma_{ij}} \frac{\delta}{4}(\|Q_i\lambda_H\|_{H^{1/2}(\Gamma_{ij})}^2 + \|Q_j\lambda_H\|_{H^{1/2}(\Gamma_{ij})}^2) \\
&= \|\nabla p_h^*(Q\lambda_H)\|^2 - \frac{\delta}{2}\|p_h^*(Q\lambda_H)\|_{H^1(\Omega)}^2 \\
&\quad + \sum_{\Gamma_{ij}} \frac{C_p}{(C_p+1)|\Gamma_{ij}|}\|Q_i\lambda_H - Q_j\lambda_H\|_{L^2(\Gamma_{ij})}^2 \\
&\quad + \sum_{\Gamma_{ij}} \frac{\delta}{4}(\|Q_i\lambda_H\|_{H^{1/2}(\Gamma_{ij})}^2 + \|Q_j\lambda_H\|_{H^{1/2}(\Gamma_{ij})}^2)
\end{aligned}$$

Then, by Lemma 10.A.1 and the assumption (4.21)

$$b_h(\lambda_H,\lambda_H) \geq \sum_{\Gamma_{ij}} \frac{\delta}{4}(\|Q_i\lambda_H\|^2_{H^{1/2}(\Gamma_{ij})} + \|Q_j\lambda_H\|^2_{H^{1/2}(\Gamma_{ij})})$$

$$\geq \sum_{i=1}^{n} \frac{\delta}{4}\|Q_i\lambda_H\|^2_{H^{1/2}(\Gamma_i)} \geq \frac{\delta}{4}\sum_{i=1}^{n}\|\lambda_H\|^2_{H^{1/2}(\Gamma_i)}. \qquad \square$$

For the sake of notation, we assume that $\|\cdot\|_{H^{1/2}}$ denote the sum of the $H^{1/2}$ norms over all the Γ_i unless otherwise denoted:

Proof of Theorem 4.6. By Strang's second lemma, there exists a constant C such that

$$\|\lambda^* - \lambda_h^*\|_{H^{1/2}}$$
$$\leq C\left(\inf_{\mu_H\in\Lambda_0}\|\lambda^* - \mu_H\|_{H^{1/2}} + \sup_{\mu_H\in\Lambda_0}\frac{|b_h(\lambda^*,\mu_H) - L_h(\mu_H)|}{\|\mu_H\|_{H^{1/2}}}\right).$$

The first term, otherwise known as the approximation error, is bounded by our assumption that $|p|_{H^2} < \infty$, meaning that the traces on the interior are at least in $H^{3/2}(\Gamma_i)$ for all $1 \leq i \leq n$, hence

$$\inf_{\mu_H\in\Lambda_0}\|\lambda^* - \mu_H\|_{H^{1/2}} \leq H\sum_{i=1}^{n}\|\lambda^*\|_{H^{3/2}(\Gamma_i)} \leq H|p|_{H^2(\Omega)}$$

by standard approximation results.

For the consistency error, we substitute the definition into the definition of our bilinear form and linear functional in, and noting that $-\nabla K\nabla p = f$ by definition of our problem, we have for all $\mu_H \in \Lambda_0$

$$\frac{|b_h(\lambda^*,\mu_H) - L_h(\mu_H)|}{\|\mu_H\|_{H^{1/2}}}$$
$$= \frac{|\sum_{i=1}^{n}(\boldsymbol{u}_h^*(Q_i\lambda^*) + \bar{\boldsymbol{u}}_h, \nabla(Q_i\mu_H)) - (f, Q_i\mu_H)|}{\|\mu_H\|_{H^{1/2}}}$$
$$= \frac{|\sum_{i=1}^{n}(K\nabla(p_h(\lambda^*) - p), \nabla Q_i\mu_H) + (K\nabla p, \nabla Q_i\mu_H) - (-\nabla K\nabla p, Q_i\mu_H)|}{\|\mu_H\|_{H^{1/2}}}$$
$$= \frac{|\sum_{i=1}^{n}(K\nabla(p_h(\lambda^*) - p), \nabla Q_i\mu_H) - \int_{\Gamma_i} K\nabla p Q_i\mu_H\cdot\boldsymbol{n}_i\, ds|}{\|\mu_H\|_{H^{1/2}}}$$

where $\boldsymbol{n}_i$ is the outward normal to the subdomain Ω_i, and $p_h(\lambda^*) := p^*(Q_i\lambda^*) + \bar{p}$. The first term can be estimate using Cauchy-Schwarz inequality,

$$
\begin{aligned}
&\frac{\left|\sum_{i=1}^{n}(K\nabla(p_h(\lambda^*) - p), \nabla Q_i \mu_H)\right|}{\|\mu_H\|_{H^{1/2}}} \\
&\leq \frac{\sum_{i=1}^{n}\|K\nabla(p_h(\lambda^*) - p)\|_{\Omega_i}\|\nabla Q_i\mu_H\|_{\Omega_i}}{\|\mu_H\|_{H^{1/2}}} \\
&\leq n\max_i\|K\nabla(p_h(\lambda^*) - p)\|_{\Omega_i}\frac{\sum_{i=1}^{n}\|\nabla Q_i\mu_H\|_{\Omega_i}}{\|\mu_H\|_{H^{1/2}}} \\
&\leq Cn\max_i\|K\nabla(p_h(\lambda^*) - p)\|_{\Omega_i} \leq Cn\delta,
\end{aligned}
$$

where we use the fact that

$$
\|\nabla Q_i\mu_H\|_{\Omega_i} \leq \|Q_i\mu_H\|_{H^1(\Omega_i)} \leq C\|Q_i\mu_H\|_{H^{1/2}(\Gamma_i)} \leq C\|\mu_H\|_{H^{1/2}(\Gamma_i)}
$$

where we used the properties of discrete harmonic extensions (Toselli and Widlund, 2004), and the fact that L^2 projection is stable in $H^{1/2}$ due to interpolation (Bramble and Xu, 1991).

As for the second term, we note that if two subdomains Ω_i, Ω_j are adjacent, then $\boldsymbol{n}_i = -\boldsymbol{n}_j$ meaning

$$
\begin{aligned}
&\frac{\left|\sum_{i=1}^{n}\int_{\Gamma_i} K\nabla p Q_i\mu_H \cdot \boldsymbol{n}_i\, ds\right|}{\|\mu_H\|_{H^{1/2}}} \\
&\leq \frac{\sum_{\Gamma_{ij}}\left|\int_{\Gamma_{ij}} K\nabla p(Q_i\mu_H - Q_j\mu_H)\cdot \boldsymbol{n}\right|_i}{\|\mu_H\|_{H^{1/2}}} \\
&\leq \frac{\sum_{\Gamma_{ij}}\|K\nabla p\cdot \boldsymbol{n}_i\|_{H^{1/2}(\Gamma_{ij})}\|Q_i\mu_H - Q_j\mu_H\|_{H^{-1/2}(\Gamma_{ij})}}{\|\mu_H\|_{H^{1/2}}} \\
&\leq Cn|p|_{H^2(\Omega)}\max_i\frac{\|(I - Q_i)\mu_H\|_{H^{-1/2}(\Gamma_{ij})}}{\|\mu_H\|_{H^{1/2}}} \\
&\leq Cn|p|_{H^2(\Omega)}\max_i h_i
\end{aligned}
$$

where we used the inequality $\|Q_i - Q_j\| \leq \|I - Q_i\| + \|I - Q_j\|$, the trace inequality on normal derivatives (Grisvard, 2011, Thm. 1.5.1.2), L^2 projection approximation properties (Arbogast et al., 2007, (3.5)), and where h_i denotes the maximal mesh-size on each subdomain Ω_i. □

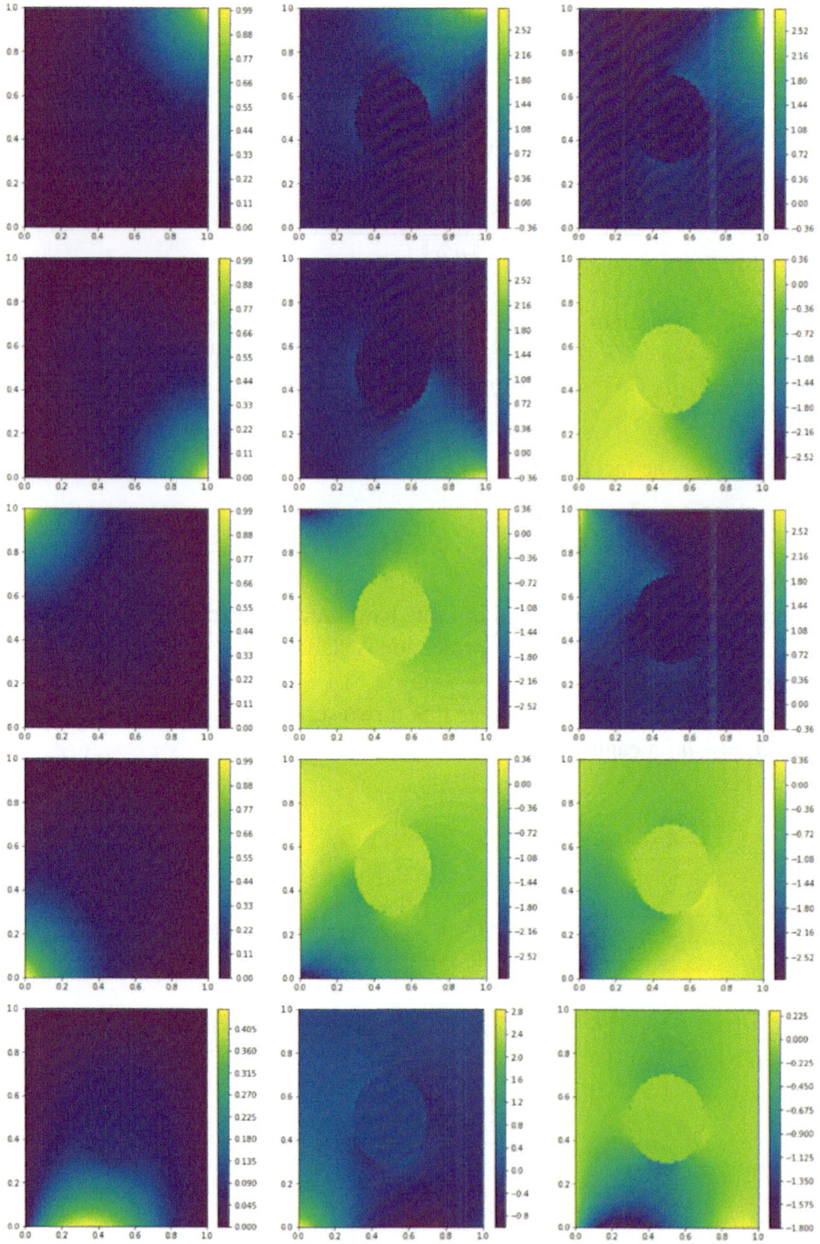

⋮

FIGURE 17 Plots illustrating some of the training data used for Subsection 5.3 with pressure, flux of x and flux of y in the columns respectively. The data is generated from a low order FEM method with $h = 1/100$. The key differences between each data set are that the boundary conditions are varied so that the element can respond to the different mortars.

10.A.2 FEEC element training

For each of the FEEC elements used in Examples 2 through 4 with the exception of the battery example (discussed below), a "monolithic" approach is used. For concreteness, we will exposit the details fully for Example 3 as the other examples only differ by model hyperparameters described in the relevant section and the training data.

The data used to train the FEEC elements are generated from 20,480 randomly sampled points from $[0, 1]^2$ evaluated by interpolating the solution of an elementary finite element solver. In the case of the FEEC element in Example 3, a grand total of 12 different solutions each with different boundary conditions, corresponding to the third-order Bernstein polynomials on the boundary (e.g. x^3y^3, $\binom{3}{1}x^3y^2(1-y)$, $\binom{3}{2}x^3y(1-y)^2$ etc), are used alongside the forcing term of $f = 0$. The Bernstein polynomials were used instead of simple hat functions as we found the additional smoothness meant pretraining of the FEEC element was more stable. In Fig. 17, we plot the first five, out of twelve, of the training data we generated for Subsection 5.3.

Let ξ correspond to all the hyperparameters in the FEEC model (e.g. knot location, POU coefficients, scaling coefficients). The loss function we use is

$$\min_{\xi} \sum_{k=1}^{12} \frac{\left\| p_{\xi,k} - p_{\text{data},k} \right\|_{MSE}}{\left\| p_{\text{data},k} \right\|_{\ell_2}} + \frac{\left\| \boldsymbol{u}_{\xi,k} - \boldsymbol{u}_{\text{data},k} \right\|_{MSE}}{\left\| \boldsymbol{u}_{\text{data},k} \right\|_{\ell_2} + 0.001} \tag{10.A.3}$$

where $p_{\xi,k}$, $\boldsymbol{u}_{\xi,k}$ are the FEEC solutions with the kth boundary condition, and $p_{\text{data},k}$, $\boldsymbol{u}_{\text{data},k}$ are the data for the kth boundary condition subject to the constraint. This is exactly (3.7), except we summed over all the different boundary conditions and minimized against all the boundary conditions in a single epoch (e.g. a monolithic approach). The computation of the loss is efficient since $p_{\xi,k}$ for $k = 1, \ldots, 12$ can be solved with a single linear solver step because their systems only differ in their right hand sides from the boundary conditions. Thus, the expensive stiffness matrix generation only has to be performed once at each optimization step. The standard Adams optimizer were used in each case as discussed in Actor et al. (2024).

As a result of the monolithic training and the basis generation of FEEC, the FEEC element will be able to accurately solve for the flux and pressure even when faced with Dirichlet boundary conditions which it has not seen before. For example, in Fig. 18, we plot the true and predicted solution of (5.4) with a boundary condition of y. Note that, while the boundary condition was never explicitly given in the training data, that the FEEC element was able to reproduce the behavior around the material discontinuity quite accurately.

As noted in Subsection 5.4.3, the battery example assumes only a single data set is available, with no additional data generation with varying boundary conditions as above. The data for each subdomain are simply obtained via a restriction operator, and the loss is exactly (3.7).

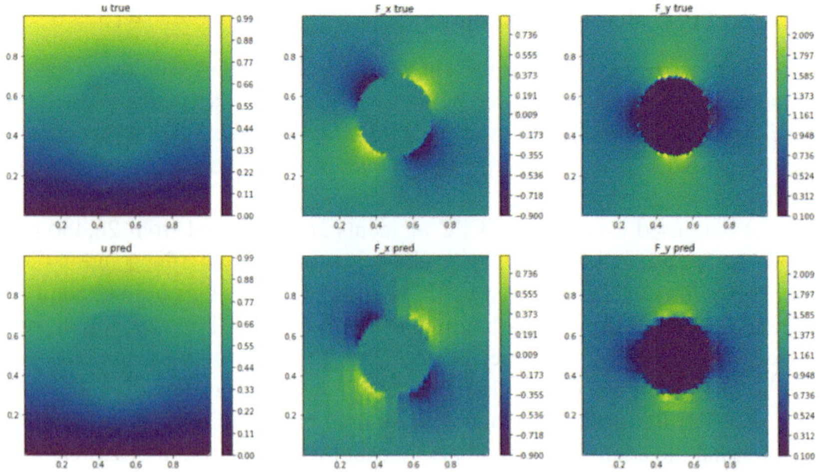

FIGURE 18 Plot of the true (first row) and predicted using a 24 × 24 fine-scale FEEC element (second row) solution to (5.4) with the boundary condition y on the domain $[-.5, .5]^2$. Note that while the boundary condition is not explicitly included in the training data, but rather a linear combination, we are able to reproduce the true solution accurately due to training against a large suite of boundary conditions.

References

Actor, J.A., Hu, X., Huang, A., Roberts, S.A., Trask, N., 2024. Data-driven Whitney forms for structure-preserving control volume analysis. Journal of Computational Physics 496, 112520.

Arbogast, T., Cowsar, L.C., Wheeler, M.F., Yotov, I., 2000. Mixed finite element methods on non-matching multiblock grids. SIAM Journal on Numerical Analysis 37, 1295–1315.

Arbogast, T., Pencheva, G., Wheeler, M.F., Yotov, I., 2007. A multiscale mortar mixed finite element method. Multiscale Modeling & Simulation 6, 319–346.

Arnold, D.N., 2018. Finite Element Exterior Calculus. SIAM.

Bergomi, M.G., Frosini, P., Giorgi, D., Quercioli, N., 2019. Towards a topological–geometrical theory of group equivariant non-expansive operators for data analysis and machine learning. Nature Machine Intelligence 1, 423–433.

Bernardi, C., Maday, Y., Patera, A.T., 1993. Domain decomposition by the mortar element method. In: Asymptotic and Numerical Methods for Partial Differential Equations with Critical Parameters. Springer, pp. 269–286.

Bertoluzza, S., Kunoth, A., 2000. Wavelet stabilization and preconditioning for domain decomposition. IMA Journal of Numerical Analysis 20, 533–559.

Biegler, L.T., Ghattas, O., Heinkenschloss, M., van Bloemen Waanders, B., 2003. Large-scale pde-constrained optimization: an introduction. In: Large-Scale PDE-Constrained Optimization. Springer, pp. 3–13.

Braess, D., 2007. Finite Elements: Theory, Fast Solvers, and Applications in Solid Mechanics. Cambridge University Press.

Bramble, J.H., Xu, J., 1991. Some estimates for a weighted l^2 projection. Mathematics of Computation 56, 463–476.

Brenner, S.C., 2003. Poincaré–Friedrichs inequalities for piecewise h 1 functions. SIAM Journal on Numerical Analysis 41, 306–324.

Cai, S., Mao, Z., Wang, Z., Yin, M., Karniadakis, G.E., 2021. Physics-informed neural networks (pinns) for fluid mechanics: a review. Acta Mechanica Sinica 37, 1727–1738.

Celledoni, E., Ehrhardt, M.J., Etmann, C., McLachlan, R.I., Owren, B., Schonlieb, C.-B., Sherry, F., 2021. Structure-preserving deep learning. European Journal of Applied Mathematics 32, 888–936.

Cockburn, B., Gopalakrishnan, J., Lazarov, R., 2009. Unified hybridization of discontinuous Galerkin, mixed, and continuous Galerkin methods for second order elliptic problems. SIAM Journal on Numerical Analysis 47, 1319–1365.

Cowsar, L.C., Mandel, J., Wheeler, M.F., 1995. Balancing domain decomposition for mixed finite elements. Mathematics of Computation 64, 989–1015.

Desai, S.A., Mattheakis, M., Sondak, D., Protopapas, P., Roberts, S.J., 2021. Port-Hamiltonian neural networks for learning explicit time-dependent dynamical systems. Physical Review E 104, 034312.

Evans, L.C., 2022. Partial Differential Equations, vol. 19. American Mathematical Society.

Farhat, C., Lesoinne, M., LeTallec, P., Pierson, K., Rixen, D., 2001. Feti-dp: a dual–primal unified feti method—part I: a faster alternative to the two-level feti method. International Journal for Numerical Methods in Engineering 50, 1523–1544.

Gillette, A., Rand, A., Bajaj, C., 2016. Construction of scalar and vector finite element families on polygonal and polyhedral meshes. Computational Methods in Applied Mathematics 16, 667–683.

Glowinski, R., Wheeler, M.F., 1987. Domain decomposition and mixed finite element methods for elliptic problems. Tech. Report.

Greydanus, S., Dzamba, M., Yosinski, J., 2019. Hamiltonian neural networks. Advances in Neural Information Processing Systems 32.

Grisvard, P., 2011. Elliptic Problems in Nonsmooth Domains. SIAM.

Gruber, A., Lee, K., Trask, N., 2023. Reversible and irreversible bracket-based dynamics for deep graph neural networks. arXiv preprint. arXiv:2305.15616.

Hernández, Q., Badías, A., González, D., Chinesta, F., Cueto, E., 2021. Structure-preserving neural networks. Journal of Computational Physics 426, 109950.

Hinze, M., Pinnau, R., Ulbrich, M., Ulbrich, S., 2008. Optimization with PDE Constraints, vol. 23. Springer Science & Business Media.

Hirani, A.N., 2003. Discrete Exterior Calculus. California Institute of Technology.

Jagtap, A.D., Karniadakis, G.E., 2021. Extended physics-informed neural networks (xpinns): a generalized space-time domain decomposition based deep learning framework for nonlinear partial differential equations. In: AAAI Spring Symposium: MLPS, vol. 10.

Karniadakis, G.E., Kevrekidis, I.G., Lu, L., Perdikaris, P., Wang, S., Yang, L., 2021. Physics-informed machine learning. Nature Reviews Physics 3, 422–440.

Nicolaides, R.A., 1992. Direct discretization of planar div-curl problems. SIAM Journal on Numerical Analysis 29, 32–56.

Notz, P.K., Subia, S.R., Hopkins, M.M., Moffat, H.K., Noble, D.R., Okusanya, T.O., 2016. Sierra multimechanics module: aria user manual–version 4.40. Tech. Report. Sandia National Lab. (SNL-NM), Albuquerque, NM (United States).

Patel, R.G., Manickam, I., Trask, N.A., Wood, M.A., Lee, M., Tomas, I., Cyr, E.C., 2022. Thermodynamically consistent physics-informed neural networks for hyperbolic systems. Journal of Computational Physics 449, 110754.

Roberts, J.E., Thomas, J.-M., 1991. Mixed and hybrid methods.

Smith, B.F., 1997. Domain decomposition methods for partial differential equations. In: Parallel Numerical Algorithms. Springer, pp. 225–243.

Smythe, W.R.w.R., 1989. Static and Dynamic Electricity. International Series in Pure and Applied Physics. Taylor & Francis, Philadelphia, PA.

Toselli, A., Widlund, O., 2004. Domain Decomposition Methods-Algorithms and Theory, vol. 34. Springer Science & Business Media.

Trask, N., Huang, A., Hu, X., 2022. Enforcing exact physics in scientific machine learning: a data-driven exterior calculus on graphs. Journal of Computational Physics 456, 110969.

Trask, N., Maxey, M., Hu, X., 2018. A compatible high-order meshless method for the Stokes equations with applications to suspension flows. Journal of Computational Physics 355, 310–326.

Villar, S., Hogg, D.W., Storey-Fisher, K., Yao, W., Blum-Smith, B., 2021. Scalars are universal: equivariant machine learning, structured like classical physics. Advances in Neural Information Processing Systems 34, 28848–28863.

Wang, S., Teng, Y., Perdikaris, P., 2021. Understanding and mitigating gradient flow pathologies in physics-informed neural networks. SIAM Journal on Scientific Computing 43, A3055–A3081.

Yang, Y., Perdikaris, P., 2019. Adversarial uncertainty quantification in physics-informed neural networks. Journal of Computational Physics 394, 136–152.

Zhang, D., Lu, L., Guo, L., Karniadakis, G.E., 2019. Quantifying total uncertainty in physics-informed neural networks for solving forward and inverse stochastic problems. Journal of Computational Physics 397, 108850.

Chapter 11

Two-layer neural networks for partial differential equations: optimization and generalization theory

Tao Luo[a,c] and Haizhao Yang[b,*,c]

[a]*School of Mathematical Sciences, Institute of Natural Sciences, MOE-LSC, and Qing Yuan Research Institute, Shanghai Jiao Tong University, Shanghai, PR China,* [b]*Department of Mathematics and Department of Computer Science, University of Maryland, College Park, MD, United States*

**Corresponding author: e-mail address: hzyang@umd.edu*

Contents

Abstract

The problem of solving partial differential equations (PDEs) can be formulated into a least-squares minimization problem, where neural networks are used to parametrize PDE solutions. A global minimizer corresponds to a neural network that solves the given PDE. In this paper, we show that the gradient descent method can identify a global minimizer of the least-squares optimization for solving second-order linear PDEs with two-layer neural networks under the assumption of overparametrization. We also analyze the generalization error of the least-squares optimization for second-order linear PDEs

[c] The work was finished when both authors were at Purdue University.

Handbook of Numerical Analysis, Volume 25, ISSN 1570-8659, https://doi.org/10.1016/bs.hna.2024.05.007

and two-layer neural networks, when the right-hand-side function of the PDE is in a Barron-type space and the least-squares optimization is regularized with a Barron-type norm, without the overparametrization assumption.

Keywords
Deep learning, Overparametrization, Partial differential equations, Optimization convergence, Generalization error

MSC Codes
68U99, 65N30, 65N25

1 Introduction

Deep learning, originated in computer science, has revolutionized many fields of science and engineering recently. This revolution also includes broad applications of deep learning in computational and applied mathematics, e.g., many breakthroughs in solving partial differential equations (PDEs) (Dissanayake and Phan-Thien, 1994; Lagaris et al., 1998; Rudd and Ferrari, 2015; Carleo and Troyer, 2017; Han et al., 2018; E et al., 2017; Berg and Nyström, 2018; Khoo et al., 2019; Raissi et al., 2019; Sirignano and Spiliopoulos, 2018; Huang et al., 2020; Gu et al., 2020b). The key idea of these approaches is to reformulate the PDE solution into a global minimizer of an expectation minimization problem, where deep neural networks (DNNs) are applied for discretization and the stochastic gradient descent (SGD) is adopted to solve the minimization problem. These methods probably date back to the 1990s (e.g., see Dissanayake and Phan-Thien, 1994; Lagaris et al., 1998) and were revisited recently (Rudd and Ferrari, 2015; Han et al., 2018; E et al., 2017; Berg and Nyström, 2018; Khoo et al., 2019; Sirignano and Spiliopoulos, 2018; Raissi et al., 2019) due to the significant development of GPU computing that accelerates DNN computation. Though these approaches have remarkable empirical successes, their theoretical justification remains vastly open.

For simplicity, let us use a PDE defined on a domain Ω in a compact form with equality constrains to illustrate the main idea, e.g.,

$$\begin{cases} \mathcal{L}u = f & \text{in } \Omega, \\ \mathcal{B}u = g & \text{on } \partial\Omega, \end{cases} \tag{1.1}$$

where $\mathcal{L}$ is a differential operator and $\mathcal{B}$ is the operator for specifying an appropriate boundary condition. In the least squares-type methods, DNNs, denoted as $\phi(\boldsymbol{x};\boldsymbol{\theta})$ with a parameter set $\boldsymbol{\theta}$, are applied to parametrize the solution space of the PDE and a best parameter set $\boldsymbol{\theta}_{\mathcal{D}}$ is identified via minimizing an expectation called the population risk (also known as the population loss):

$$\begin{aligned} \boldsymbol{\theta}_{\mathcal{D}} &= \arg\min_{\boldsymbol{\theta}} R_{\mathcal{D}}(\boldsymbol{\theta}) \\ &:= \mathbb{E}_{\boldsymbol{x}\sim U(\Omega)}\left[\ell(\mathcal{L}\phi(\boldsymbol{x};\boldsymbol{\theta}), f(\boldsymbol{x}))\right] + \gamma \mathbb{E}_{\boldsymbol{x}\sim U(\partial\Omega)}\left[\ell(\mathcal{B}\phi(\boldsymbol{x};\boldsymbol{\theta}), g(\boldsymbol{x}))\right], \end{aligned} \tag{1.2}$$

with a positive parameter γ and a loss function typically taken as $\ell(y, y') = \frac{1}{2}|y - y'|^2$, where the expectations are taken with uniform distributions $U(\Omega)$ and $U(\partial\Omega)$ over Ω and $\partial\Omega$, respectively. To implement the expectation minimization above using the gradient descent method (GD), a discrete set of samples are randomly drawn to obtain an empirical risk (or empirical loss) function

$$R_S(\boldsymbol{\theta}) := \frac{1}{n} \sum_{\{\boldsymbol{x}_i\}_{i=1}^n \subset \Omega} \ell(\mathcal{L}\phi(\boldsymbol{x}_i; \boldsymbol{\theta}), f(\boldsymbol{x}_i)) + \gamma \frac{1}{n} \sum_{\{\boldsymbol{x}_i\}_{i=1}^n \subset \partial\Omega} \ell(\mathcal{B}\phi(\boldsymbol{x}_i; \boldsymbol{\theta}), g(\boldsymbol{x}_i)) \tag{1.3}$$

used in each GD iteration to update $\boldsymbol{\theta}$. The set of random samples is usually renewed per iteration resulting in the SGD algorithm for minimizing (1.2). In this paper, we will focus on the case when these samples are fixed in all iterations. There are mainly three theoretical point of view to study the above deep learning-based PDE solver:

1. **Approximation theory:** given a budget of the size of DNNs, e.g. width m and depth L, or a budget of the total number of parameters N_{para}, what is the accuracy of $\phi(\boldsymbol{x}; \boldsymbol{\theta}_{\mathcal{D}})$ approximating the solution of the PDE?
2. **Optimization convergence:** under what condition can gradient descent converges to a global minimizer of (1.2) and (1.3)?
3. **Generalization analysis:** if only finitely many samples are available, how good is the global minimizer of (1.3) compared to the global minimizer of (1.2)?

Deep network approximation theory has shown that DNNs admit powerful approximation capacity. First, DNNs can approximate high-dimensional functions with an appealing approximation rate, e.g., Barron spaces (Barron, 1993; E et al., 2019b,a), Korobov spaces (Montanelli and Du, 2019), band-limited functions (Chen and Wu, 2019; Montanelli et al., 2019), compositional functions (Poggio et al., 2017; E et al., 2019c), smooth functions (Yarotsky and Zhevnerchuk, 2019; Lu et al., 2020a; Montanelli and Yang, 2020), solution spaces of certain PDEs (Hutzenthaler et al., 2019), and even general continuous functions (Shen et al., 2021b,a). Second, DNNs can achieve exponential approximation rates, i.e., the approximation error exponentially decays when the number of parameters increases, for target functions in the polynomial spaces (Yarotsky, 2017; Montanelli et al., 2019; Lu et al., 2020a), the smooth function spaces (Montanelli et al., 2019; Liang and Srikant, 2016), the analytic function space (E and Wang, 2018), the function space admitting a holomorphic extension to a Bernstein polyellipse (Opschoor et al., 2019), and even general continuous functions (Shen et al., 2021b). Theories in deep network approximation have provided attractive upper bounds of the accuracy of $\phi(\boldsymbol{x}; \boldsymbol{\theta}_{\mathcal{D}})$ approximating the solution of the PDE in various function spaces. In realistic applications, it might be more interesting to characterize deep network approximation in terms of m and L simultaneously than the characterization in terms of N_{para}. We refer reader to Shen et al. (2019, 2020); Lu et al. (2020a); Shen et al. (2021b); Yang and Wang (2020) for examples in terms of m and L.

Though DNNs are powerful in terms of approximation theory, obtaining the best DNN $\phi(\boldsymbol{x};\boldsymbol{\theta}_{\mathcal{D}})$ in (1.2) to approximate the PDE solution is still challenging. It is conjectured that, under certain conditions, SGD is able to identify an approximate global minimizer of (1.2) with accuracy depending on N_{para} and the sample size n. Though deep learning-based PDE solvers have been proposed since the 1990s, there might be no existing literature to investigate this conjecture, to the best of our knowledge. In this paper, assuming that the same set of random samples are used in minimizing (1.3), it is shown that GD can converge to a global minimizer of (1.3), denoted as $\boldsymbol{\theta}_{S}$, for second-order linear PDEs and two-layer neural networks, as long as N_{para} is sufficiently large depending on n, i.e., in the overparametrization regime. Furthermore, we will quantify how good the global minimizer $\boldsymbol{\theta}_{S}$ of the empirical loss in (1.3) is compared to the global minimizer $\boldsymbol{\theta}_{\mathcal{D}}$ of the population loss in (1.2), when the empirical loss is regularized with a penalty term using the path norm of $\boldsymbol{\theta}$ and the PDE solution is in a Barron-type space, a variant of the Barron-type space in Barron (1993); E et al. (2019b). Our analysis is an extension of the seminal work of neural tangent kernels (Jacot et al., 2018; Du et al., 2018, 2019) and the generalization analysis in Barron (1993); E et al. (2019b) for function regression problems to the case of PDE solvers.

Though the convergence of deep learning-based regression under the overparametrization assumption has been proposed recently (Jacot et al., 2018; Du et al., 2018; Mei et al., 2018; Du et al., 2019; Lu et al., 2020b), we would like to emphasize that the minimization of solving a PDE via (1.2) is more difficult and technical. In the case of solving PDEs, differential operators have changed the optimization objective function considered in the literature. Balancing between the differential operator and the boundary operator makes it more challenging to solve the optimization problem. For example, we consider a second order elliptic equation with variable coefficients, i.e., $\mathcal{L}u = f$ where $\mathcal{L}u = \sum_{\alpha,\beta=1}^{d} A_{\alpha\beta}(\boldsymbol{x})u_{x_\alpha x_\beta}$. Given a two-layer neural network $\phi(\boldsymbol{x};\boldsymbol{\theta}) = \sum_{k=1}^{m} a_k\sigma(\boldsymbol{w}_k^{\intercal}\boldsymbol{x})$ with an activation function $\sigma(z) = \max\{0, \frac{1}{6}z^3\}$ to parametrize the PDE solution, solving the original PDE via deep learning is equivalent to solving a regression problem with another type of neural network $f(\boldsymbol{x};\boldsymbol{\theta}) := \mathcal{L}\phi(\boldsymbol{x};\boldsymbol{\theta}) = \sum_{k=1}^{m} a_k\boldsymbol{w}_k^{\intercal}A(\boldsymbol{x})\boldsymbol{w}_k\sigma''(\boldsymbol{w}_k^{\intercal}\boldsymbol{x})$ to fit $f(\boldsymbol{x})$. Note that $\sigma''(z) = \text{ReLU}(z) = \max\{0, z\}$. Thus, the dependence of $f(\boldsymbol{x};\boldsymbol{\theta})$ on $\boldsymbol{w}_k$ is essentially cubic rather than linear (more precisely, positive homogeneous).

The generalization analysis of deep learning-based regression under the overparametrization assumption was studied recently in Jacot et al. (2018); Cao and Gu (2019); Chen et al. (2019). The generalization analysis with a regularization term based on the path norm without the overparametrization assumption was proposed in E et al. (2019b,a, 2020). In the case of PDE solvers, differential operators have enhanced the nonlinearity of the generalization analysis and hence make it more difficult to analyze. In the case of Linear Kolmogorov Equations and parabolic PDEs, examples of generalization analysis of PDE solvers were presented in Berner et al. (2018); Han and Long (2018). In the case of linear second-order elliptic and parabolic type PDEs, the generalization error of the physics-informed neural network was analyzed in Shin et al. (2020). However, the generalization analysis for generic PDEs is vastly open. Our attempt is for second-order linear PDEs with variable coefficients. Let us consider the second order elliptic equation with variable coefficients in the above paragraph again. The variable coefficients $A_{\alpha\beta}(\boldsymbol{x})$ lead to highly nonlinearity in the network $f(\boldsymbol{x};\boldsymbol{\theta})$ depending on $\boldsymbol{x}$, since we do not make any assumption on the smoothness of $\boldsymbol{A}(\boldsymbol{x})$. We develop new analysis of the Rademacher complexity to overcome these difficulties. Unlike existing work, our a priori estimates do not require any truncation on $f(\boldsymbol{x};\boldsymbol{\theta})$ (or $\phi(\boldsymbol{x};\boldsymbol{\theta})$). This is important because a common truncation trick does not lead to the boundedness of $f(\boldsymbol{x};\boldsymbol{\theta})$ in our PDE solver. In fact, if one considers the standard truncation on $\phi(\boldsymbol{x};\boldsymbol{\theta})$, e.g., $\mathcal{T}_{[0,1]}\phi(\boldsymbol{x};\boldsymbol{\theta}) := \min\{\max\{\phi(\boldsymbol{x};\boldsymbol{\theta}),0\},1\}$, then $\mathcal{L}[\mathcal{T}_{[0,1]}\phi(\boldsymbol{x};\boldsymbol{\theta})]$ might still be unbounded because $\mathcal{L}$ is a second order differential operator. Another naive trick is to truncate $f(\boldsymbol{x};\boldsymbol{\theta})$, i.e., $\mathcal{T}_{[0,1]}f(\boldsymbol{x};\boldsymbol{\theta}) := \min\{\max\{f(\boldsymbol{x};\boldsymbol{\theta}),0\},1\}$. But this does not make sense since we want to find a solution satisfying $\mathcal{L}\phi(\boldsymbol{x};\boldsymbol{\theta}) \approx f(\boldsymbol{x})$ instead of $\mathcal{T}_{[0,1]}\mathcal{L}\phi(\boldsymbol{x};\boldsymbol{\theta}) \approx f(\boldsymbol{x})$.

This paper will be organized as follows. In Section 2, deep learning-based PDE solvers will be introduced in detail. In Section 3, our main theorems for the convergence and generalization analysis of GD for minimizing (1.3) will be presented. In Section 4, the proof of the GD convergence theorems will be shown. In Section 5, the proof of the generalization bound will be given. Finally, we conclude our paper in Section 6.

2 Deep learning-based PDE solvers

We will introduce deep learning-based PDE solvers with necessary notations in this paper in preparation for our main theorems in Section 3.

2.1 Notations, definitions, and basic lemmas

The main notations of this paper are listed as follows.

- Vectors and matrices are denoted in bold font. All vectors are column vectors.
- For a parameter set Θ, $\mathrm{vec}\{\Theta\}$ denotes the vector consists of all the elements of Θ.
- $[n]$ denotes $\{1, 2, \ldots, n\}$.
- $\|\cdot\|_1$ and $\|\cdot\|_\infty$ represent the ℓ_1 and ℓ_∞ norms of a vector, respectively.
- Big "O" notation: for any functions $g_1, g_2 : \mathbb{R} \to \mathbb{R}^+$, $g_1(z) = O(g_2(z))$ as $z \to +\infty$ means that $g_1(z) \le C g_2(z)$ for some constants C, z_0 and any $z \ge z_0$.
- Small "o" notation: for any functions $g_1, g_2 : \mathbb{R} \to \mathbb{R}^+$, $g_1(z) = o(g_2(z))$ as $z \to +\infty$ means that $\lim_{z\to\infty} \frac{f(z)}{g(z)} = 0$.
- Let $\sigma : \mathbb{R} \to \mathbb{R}$ denote the activation function, e.g., $\sigma(x) = \max\{0, \frac{1}{6}x^3\}$ is the activation function used in this paper. With the abuse of notations, we define $\sigma : \mathbb{R}^d \to \mathbb{R}^d$ as $\sigma(\boldsymbol{x}) = (\max\{0, x_1\}, \ldots, \max\{0, x_d\})^\intercal$ for any $\boldsymbol{x} =$

$(x_1, \dots, x_d)^\top \in \mathbb{R}^d$, where $\top$ denotes the transpose of a matrix. Similarly, for any function f defined on $\mathbb{R}$ and vector $\boldsymbol{x} \in \mathbb{R}^d$, $f(\boldsymbol{x}) = [f(x_1), \dots, f(x_d)]^\top$.

Mathematically, DNNs are a form of function parametrization via the compositions of simple nonlinear functions (Goodfellow et al., 2016). Let us focus on the so-called fully connected feed-forward neural network (FNN) defined below. The FNN is a general DNN structure that includes other advanced structures as its special cases, e.g., convolutional neural network (Goodfellow et al., 2016), ResNet (He et al., 2015), and DenseNet (Huang et al., 2016).

Definition 2.1 (Fully connected feed-forward neural network (FNN)). An FNN of depth L defined on $\mathbb{R}^d$ is the composition of L simple nonlinear functions as follows:

$$\phi(\boldsymbol{x}; \boldsymbol{\theta}) := \boldsymbol{a}^\top \boldsymbol{h}^{[L]} \circ \boldsymbol{h}^{[L-1]} \circ \cdots \circ \boldsymbol{h}^{[1]}(\boldsymbol{x}),$$

where $\boldsymbol{h}^{[l]}(\boldsymbol{x}) = \sigma\left(\boldsymbol{W}^{[l]}\boldsymbol{x} + \boldsymbol{b}^{[l]}\right)$ with $\boldsymbol{W}^{[l]} \in \mathbb{R}^{m_l \times m_{l-1}}$, $\boldsymbol{b}_l \in \mathbb{R}^{m_l}$ for $l = 1, \dots, L$, $\boldsymbol{a} \in \mathbb{R}^{m_L}$, $m_0 = d$, and σ is a nonlinear activation function. Each $\boldsymbol{h}^{[l]}$ is referred as a hidden layer, m_l is the width of the l-th layer, and L is called the depth of the FNN. $\boldsymbol{\theta} := \mathrm{vec}\{\boldsymbol{a}, \{\boldsymbol{W}^{[l]}, \boldsymbol{b}^{[l]}\}_{l=1}^{L}\}$ denotes the set of all parameters in ϕ.

Without loss of generality, we consider FNNs omitting $\boldsymbol{b}^{[l]}$'s. In fact, for a network with $\boldsymbol{b}^{[l]}$'s, one can simply set $\tilde{\boldsymbol{x}} = (\boldsymbol{x}^\top, 1)^\top$ and $\tilde{\boldsymbol{W}}^{[l]} = (\boldsymbol{W}^{[l]}, \boldsymbol{b}^{[l]})$ for each $l \in [L]$, and work on $\boldsymbol{\theta} = \mathrm{vec}\{\boldsymbol{a}, \{\tilde{\boldsymbol{W}}^{[l]}\}_{l=1}^{L}\}$ by noting that $\tilde{\boldsymbol{W}}^{[l]}\tilde{\boldsymbol{x}} = \boldsymbol{W}^{[l]}\boldsymbol{x} + \boldsymbol{b}^{[l]}$. In this paper, we will focus on networks with $L = 1$.

To analyze PDE solvers, we introduce a new kind of Barron functions with their associated Barron norm, and a path norm defined below.

Definition 2.2 (Path norm). The path norm of a two-layer neural network

$$\phi(\boldsymbol{x}; \boldsymbol{\theta}) = \sum_{k=1}^{m} a_k \sigma(\boldsymbol{w}_k^\top \boldsymbol{x}),$$

with an activation function σ and a parameter set $\boldsymbol{\theta}$ is defined as

$$\|\boldsymbol{\theta}\|_{\mathcal{P}} := \sum_{j=1}^{m} |a_j| \|\boldsymbol{w}_j\|_1^3.$$

Definition 2.3. A function $f : \Omega \to \mathbb{R}$ is called a Barron-type function if f has an integral representation

$$f(\boldsymbol{x}) = \mathbb{E}_{(a,\boldsymbol{w})\sim\rho} a[\boldsymbol{w}^\top \boldsymbol{A}(\boldsymbol{x})\boldsymbol{w}\sigma''(\boldsymbol{w}^\top \boldsymbol{x}) + \boldsymbol{b}^\top(\boldsymbol{x})\boldsymbol{w}\sigma'(\boldsymbol{w}^\top \boldsymbol{x}) + c(\boldsymbol{x})\sigma(\boldsymbol{w}^\top \boldsymbol{x})]$$
$$\text{for all} \quad \boldsymbol{x} \in \Omega,$$

where ρ is a probability distribution over $\mathbb{R}^{d+1}$. The associated Barron norm of a Barron-type function is defined as

$$\|f\|_{\mathcal{B}} := \inf_{\rho \in \mathcal{P}_f} \left(\mathbb{E}_{(a,\boldsymbol{w})\sim\rho} |a|^2 \|\boldsymbol{w}\|_1^6 \right)^{1/2},$$

where $\mathcal{P}_f = \{\rho \mid f(\boldsymbol{x}) = \mathbb{E}_{(a,\boldsymbol{w})\sim\rho} a[\boldsymbol{w}^\top \boldsymbol{A}(\boldsymbol{x})\boldsymbol{w}\sigma''(\boldsymbol{w}^\top\boldsymbol{x}) + \boldsymbol{b}^\top(\boldsymbol{x})\boldsymbol{w}\sigma'(\boldsymbol{w}^\top\boldsymbol{x}) + c(\boldsymbol{x})\sigma(\boldsymbol{w}^\top\boldsymbol{x})], \boldsymbol{x} \in \Omega\}$. The Barron-type space is defined as $\mathcal{B}(\Omega) = \{f : \Omega \to \mathbb{R} \mid \|f\|_{\mathcal{B}} < \infty\}$.

Since $R_{\mathcal{D}}(\boldsymbol{\theta})$ cannot be realized in realistic applications due to the fact that the empirical loss $R_S(\boldsymbol{\theta})$ of finitely many samples is actually used in the computation, an immediate question is: how well $\phi(\boldsymbol{x};\boldsymbol{\theta}_S) \approx \phi(\boldsymbol{x};\boldsymbol{\theta}_{\mathcal{D}})$? Here $\boldsymbol{\theta}_S$ is a global minimizer when we minimize the empirical loss of $R_S(\boldsymbol{\theta})$. This is the generalization error analysis of deep learning-based PDE solvers and we will use the Rademacher complexity below to estimate the generalization error in terms of $|R_{\mathcal{D}}(\boldsymbol{\theta}_S) - R_S(\boldsymbol{\theta}_S)|$.

Definition 2.4 (The Rademacher complexity of a function class $\mathcal{F}$). Given a sample set $S = \{z_1, \ldots, z_n\}$ on a domain $\mathcal{Z}$, and a class $\mathcal{F}$ of real-valued functions defined on $\mathcal{Z}$, the empirical Rademacher complexity of $\mathcal{F}$ on S is defined as

$$\mathrm{Rad}_S(\mathcal{F}) = \frac{1}{n}\mathbb{E}_{\boldsymbol{\tau}}\left[\sup_{f\in\mathcal{F}} \sum_{i=1}^{n} \tau_i f(z_i)\right],$$

where $\tau_1, \ldots, \tau_n$ are independent random variables drawn from the Rademacher distribution, i.e., $\mathbb{P}(\tau_i = +1) = \mathbb{P}(\tau_i = -1) = \frac{1}{2}$ for $i = 1, \ldots, n$.

The Rademacher complexity is a basic tool for generalization analysis. In our analysis, we will use several important lemmas and theorems related to it. For the purpose of being self-contained, they are listed as follows.

First, we recall a well-known contraction lemma for the Rademacher complexity.

Lemma 2.1 (Contraction lemma (Shalev-Shwartz and Ben-David, 2014)). *Suppose that $\psi_i : \mathbb{R} \to \mathbb{R}$ is a C_{L}-Lipschitz function for each $i \in [n]$. For any $\boldsymbol{y} \in \mathbb{R}^n$, let $\boldsymbol{\psi}(\boldsymbol{y}) = (\psi_1(y_1), \cdots, \psi_n(y_n))^\top$. For an arbitrary set of functions $\mathcal{F}$ on an arbitrary domain $\mathcal{Z}$ and an arbitrary choice of samples $S = \{z_1, \ldots, z_n\} \subset \mathcal{Z}$, we have*

$$\mathrm{Rad}_S(\boldsymbol{\psi} \circ \mathcal{F}) \le C_{\mathrm{L}} \mathrm{Rad}_S(\mathcal{F}).$$

Second, the Rademacher complexity of linear predictors can be characterized by the lemma below.

Lemma 2.2 (Rademacher complexity for linear predictors (Shalev-Shwartz and Ben-David, 2014)). *Let $\Theta = \{\boldsymbol{w}_1, \cdots, \boldsymbol{w}_m\} \in \mathbb{R}^d$. Let $\mathcal{G} = \{g(\boldsymbol{w}) = \boldsymbol{w}^\top\boldsymbol{x} :$*

$\|\boldsymbol{x}\|_1 \le 1\}$ *be the linear function class with parameter* $\boldsymbol{x}$ *whose* ℓ^1 *norm is bounded by* 1. *Then*

$$\mathrm{Rad}_\Theta(\mathcal{G}) \le \max_{1\le k\le m} \|\boldsymbol{w}_k\|_\infty \sqrt{\frac{2\log(2d)}{m}}.$$

Finally, let us state a general theorem concerning the Rademacher complexity and generalization gap of an arbitrary set of functions $\mathcal{F}$ on an arbitrary domain $\mathcal{Z}$, which is essentially given in Shalev-Shwartz and Ben-David (2014).

Theorem 2.1 (Rademacher complexity and generalization gap (Shalev-Shwartz and Ben-David, 2014)). *Suppose that* f*'s in* $\mathcal{F}$ *are nonnegative and uniformly bounded, i.e., for any* $f \in \mathcal{F}$ *and any* $z \in \mathcal{Z}$, $0 \le f(z) \le B$. *Then for any* $\delta \in (0, 1)$, *with probability at least* $1 - \delta$ *over the choice of* n *i.i.d. random samples* $S = \{z_1, \ldots, z_n\} \subset \mathcal{Z}$, *we have*

$$\sup_{f\in\mathcal{F}} \left| \frac{1}{n}\sum_{i=1}^{n} f(z_i) - \mathbb{E}_z f(z) \right| \le 2\mathbb{E}_S \mathrm{Rad}_S(\mathcal{F}) + B\sqrt{\frac{\log(2/\delta)}{2n}},$$

$$\sup_{f\in\mathcal{F}} \left| \frac{1}{n}\sum_{i=1}^{n} f(z_i) - \mathbb{E}_z f(z) \right| \le 2\mathrm{Rad}_S(\mathcal{F}) + 3B\sqrt{\frac{\log(4/\delta)}{2n}}.$$

2.2 Expectation minimization

We will focus on the least-squares method in (1.2) for the boundary value problem (BVP) in (1.1) to discuss the expectation minimization, though the expectation minimization can either be formulated from the least-squares method (Berg and Nyström, 2018; Sirignano and Spiliopoulos, 2018; Raissi et al., 2019) or the variational formulation (E and Yu, 2018; Liao and Ming, 2019). As we shall see in the next subsection, an initial value problem (IVP) can also be formulated into a BVP and solved by the expectation minimization in this subsection.

The objective function in (1.2) consists of two parts: one part for the PDE operator in the domain interior and another part for the boundary condition at the boundary. Therefore, GD has to balance between these two parts and its performance heavily relies on the choice of the parameter γ in (1.2). To remove the hyperparameter γ and solve the balancing issue, we will introduce special DNNs in Gu et al. (2020b,a) satisfying various boundary conditions by design, i.e., $\mathcal{B}\phi(\boldsymbol{x};\boldsymbol{\theta}) = g(\boldsymbol{x})$ is always fulfilled on $\partial\Omega$. Then the expectation minimization in (1.2) is reduced to

$$\boldsymbol{\theta}_{\mathcal{D}} = \arg\min_{\boldsymbol{\theta}} R_{\mathcal{D}}(\boldsymbol{\theta}) := \mathbb{E}_{\boldsymbol{x}\in\Omega}\left[\ell(\mathcal{L}\phi(\boldsymbol{x};\boldsymbol{\theta}), f(\boldsymbol{x}))\right]. \tag{2.1}$$

Special neural networks for three types of boundary conditions will be introduced. Without loss of generality, we will take the example of one-dimensional problems on the domain $\Omega = [a, b]$. Networks for more complicated boundary conditions in high-dimensional domains can be constructed similarly.

Case 1. Dirichlet Boundary Conditions: $u(a) = a_0$, $u(b) = b_0$.

In this case, two special functions $h_1(x)$ and $h_2(x)$ are used to augment a neural network $\tilde{\phi}(x; \boldsymbol{\theta})$ to construct the final neural network $\phi(x; \boldsymbol{\theta})$ as the solution network:

$$\phi(x; \boldsymbol{\theta}) = h_1(x)\tilde{\phi}(x; \boldsymbol{\theta}) + h_2(x).$$

$h_1(x)$ and $h_2(x)$ are chosen such that $\phi(x; \boldsymbol{\theta})$ automatically satisfies the Dirichlet boundary conditions no matter what $\boldsymbol{\theta}$ is. Then $\phi(x; \boldsymbol{\theta})$ is trained to satisfy the differential operator in the interior of the domain Ω by solving (2.1).

To achieve this goal, $h_1(x)$ and $h_2(x)$ are constructed for two purposes: 1) construct $h_1(x)$ such that $h_1(x)\tilde{\phi}(x; \boldsymbol{\theta})$ satisfies the homogeneous Dirichlet boundary condition; 2) construct $h_2(x)$ such that $h_2(x)$ satisfies the given inhomogeneous Dirichlet boundary conditions. Therefore, $h_1(x)$ can be set as

$$h_1(x) = (x - a)^{p_a}(x - b)^{p_b},$$

where $0 < p_a, p_b \leq 1$, and $h_2(x)$ can be chosen as

$$h_2(x) = (b_0 - a_0)(x - a)/(b - a) + a_0.$$

Note that p_a and p_b should be chosen appropriately to avoid introducing a singular function that $\tilde{\phi}(x; \boldsymbol{\theta})$ needs to approximate. For instance, if the exact PDE solution is $u(x) = (x - a)^s(x - b)^s v(x) + h_1(x)$ with $v(x)$ as a smooth function and $s > 0$, $p_a = p_b > s$ results in $\tilde{\phi}(x; \boldsymbol{\theta}) \approx (x - a)^{s - p_a}(x - b)^{s - p_b} v(x)$, which makes the approximation very challenging.

Case 2. Mixed Boundary Conditions: $u'(a) = a_0$, $u(b) = b_0$.

Similar to Case 1, two special functions $h_1(x)$ and $h_2(x)$ are used to augment a neural network $\tilde{\phi}(x; \boldsymbol{\theta})$ to construct the final neural network $\phi(x; \boldsymbol{\theta})$ as the solution network:

$$\phi(x; \boldsymbol{\theta}) = h_1(x)\tilde{\phi}(x; \boldsymbol{\theta}) + h_2(x).$$

$h_1(x)$ and $h_2(x)$ are chosen such that $\phi(x; \boldsymbol{\theta})$ automatically satisfies the mixed boundary conditions no matter what $\boldsymbol{\theta}$ is. Then $\phi(x; \boldsymbol{\theta})$ is trained to satisfy the differential operator in the interior of the domain Ω by solving (2.1).

To achieve this goal, $h_1(x)$ and $h_2(x)$ are constructed as

$$h_1(x) = (x - a)^{p_a}$$

with $1 < p_a \leq 2$ and $h_2(x)$ can be chosen as

$$h_2(x) = -(b - a)^{p_a}\tilde{\phi}(b; \boldsymbol{\theta}) + a_0 x + b_0 - a_0 b.$$

Case 3. Neumann Boundary Conditions: $u'(a) = a_0$, $u'(b) = b_0$.

Similar to Cases 1 and 2, we augment a neural network $\tilde{\phi}(x;\boldsymbol{\theta})$ to construct the final neural network $\phi(x;\boldsymbol{\theta},c_1,c_2)$ as the solution network:

$$
\begin{aligned}
\phi(x;\boldsymbol{\theta},c_1,c_2) = &\exp(\frac{p_a x}{a-b})(x-a)^{p_a}\left((x-b)^{p_b}\tilde{\phi}(x;\boldsymbol{\theta}) + c_2\right) + c_1 \\
&+ \frac{(b_0 - a_0)}{2(b-a)}(x-a)^2 + a_0 x,
\end{aligned}
$$

where $1 < p_a, p_b \leq 2$, c_1 and c_2 are two parameters to be trained together with $\boldsymbol{\theta}$. Then $\phi(x;\boldsymbol{\theta},c_1,c_2)$ automatically satisfies the Neumann boundary conditions no matter what parameters are and $\phi(x;\boldsymbol{\theta},c_1,c_2)$ is trained to satisfy the differential operator in the interior of the domain Ω by solving (2.1).

2.3 Scope of analysis and applications

In Section 2.2, we have simplified the optimization problem from (1.2) to (2.1) for BVP in (1.1). Now we will show that various initial/boundary value problems can be formulated as a BVP in the form of (1.1). This helps us to simplify the optimization convergence and generalization analysis of deep learning-based PDE solvers to the case of BVP in (1.1) solved by (2.1). The analysis of a larger scope of applications has been naturally included in the analysis of BVPs.

Let us assume that the domain $\Omega \subset \mathbb{R}^d$ is bounded. Typical PDE problems of interest can be summarized as:

- Elliptic equation:

$$
\begin{aligned}
\mathcal{L}u(\boldsymbol{x}) &= f(\boldsymbol{x}) \text{ in } \Omega, \\
\mathcal{B}u(\boldsymbol{x}) &= g_0(\boldsymbol{x}) \text{ on } \partial\Omega.
\end{aligned}
\tag{2.2}
$$

- Parabolic equation:

$$
\begin{aligned}
&\frac{\partial u(\boldsymbol{x},t)}{\partial t} - \mathcal{L}u(\boldsymbol{x},t) = f(\boldsymbol{x},t) \text{ in } \Omega \times (0,T), \\
&\mathcal{B}u(\boldsymbol{x},t) = g_0(\boldsymbol{x},t) \text{ on } \partial\Omega \times (0,T), \\
&u(\boldsymbol{x},0) = h_0(\boldsymbol{x}) \text{ in } \Omega.
\end{aligned}
\tag{2.3}
$$

- Hyperbolic equation:

$$
\begin{aligned}
&\frac{\partial^2 u(\boldsymbol{x},t)}{\partial t^2} - \mathcal{L}u(\boldsymbol{x},t) = f(\boldsymbol{x},t) \text{ in } \Omega \times (0,T), \\
&\mathcal{B}u(\boldsymbol{x},t) = g_0(\boldsymbol{x},t) \text{ on } \partial\Omega \times (0,T), \\
&u(\boldsymbol{x},0) = h_0(\boldsymbol{x}), \quad \frac{\partial u(\boldsymbol{x},0)}{\partial t} = h_1(\boldsymbol{x}) \text{ in } \Omega.
\end{aligned}
\tag{2.4}
$$

In the above equations, u is the unknown solution function; f, g_0, h_0, h_1 are given data functions; $\mathcal{L}$ is a spatial differential operator with respect to x; $\mathcal{B}$ is a boundary operator specifying a certain type of boundary conditions.

As discussed in Gu et al. (2020b), when the temporal variable t is treated as an extra spatial coordinate, we can unify the above initial/boundary value problems in (2.2)-(2.4) in the following form

$$\begin{aligned} \mathcal{L}u(\boldsymbol{y}) &= f(\boldsymbol{y}) \text{ in } Q, \\ \mathcal{B}u(\boldsymbol{y}) &= g(\boldsymbol{y}) \text{ in } \Gamma, \end{aligned} \tag{2.5}$$

where $\boldsymbol{y}$ includes the spatial variable $\boldsymbol{x}$ and possibly the temporal variable t; $\mathcal{L}u = f$ represents a generic time-independent PDE; $\mathcal{B}u = g$ specifies the original boundary condition on $\boldsymbol{x}$ and possibly the initial condition of t; Q and Γ are the corresponding new domains of the equations. For the purpose of convenience, we will still use the BVP in (1.1) instead of (2.5) afterwards.

Though deep learning-based PDE solvers work for high-order differential equations in general domains, we consider second order differential equations with variable coefficients in $\Omega = [0, 1]^d$ in our analysis. The generalization to high-order differential equations and other domains follows straightforwardly and we leave it as future work. We will use the second order differential operator $\mathcal{L}$ in a nondivergence form

$$\mathcal{L}u = \sum_{\alpha,\beta=1}^{d} A_{\alpha\beta}(\boldsymbol{x})u_{x_\alpha x_\beta} + \sum_{\alpha=1}^{d} b_\alpha(\boldsymbol{x})u_{x_\alpha} + c(\boldsymbol{x})u. \tag{2.6}$$

If $\mathcal{L}$ is in a divergence form, e.g.,

$$\mathcal{L}u = \sum_{\alpha,\beta=1}^{d} \left(A_{\alpha\beta}(\boldsymbol{x})u_{x_\alpha}\right)_{x_\beta} + \sum_{\alpha=1}^{d} b_\alpha(\boldsymbol{x})u_{x_\alpha} + c(\boldsymbol{x})u,$$

then we can represent it in a nondivergence form as

$$\mathcal{L}u = \sum_{\alpha,\beta=1}^{d} A_{\alpha\beta}(\boldsymbol{x})u_{x_\alpha x_\beta} + \sum_{\alpha=1}^{d} \hat{b}_\alpha(\boldsymbol{x})u_{x_\alpha} + c(\boldsymbol{x})u$$

with

$$\hat{b}_\alpha = b_\alpha + \sum_{\beta=1}^{d} \frac{\partial A_{\alpha\beta}}{\partial x_\beta}.$$

Recall that we introduce two functions $h_1(\boldsymbol{x})$ and $h_2(\boldsymbol{x})$ to augment a neural network $\tilde{\phi}(x; \boldsymbol{\theta})$ to construct the final neural network

$$\phi(\boldsymbol{x}; \boldsymbol{\theta}) = h_1(\boldsymbol{x})\tilde{\phi}(\boldsymbol{x}; \boldsymbol{\theta}) + h_2(\boldsymbol{x})$$

as the solution network that automatically satisfies given Dirichlet boundary conditions, which makes it sufficient to solve the optimization problem in (2.1) to get the desired neural network. In this case, $\mathcal{L}\phi(\boldsymbol{x};\boldsymbol{\theta}) = f(\boldsymbol{x})$ is equivalent to $\tilde{\mathcal{L}}\tilde{\phi}(\boldsymbol{x};\boldsymbol{\theta}) = \tilde{f}(\boldsymbol{x})$, where

$$\tilde{\mathcal{L}} = \sum_{\alpha,\beta=1}^{d} \tilde{A}_{\alpha\beta}(\boldsymbol{x}) u_{x_\alpha x_\beta} + \sum_{\alpha=1}^{d} \tilde{b}_\alpha(\boldsymbol{x}) u_{x_\alpha} + \tilde{c}(\boldsymbol{x}),$$

$$\tilde{A}_{\alpha\beta}(\boldsymbol{x}) = A_{\alpha\beta}(\boldsymbol{x}) h_1(\boldsymbol{x}),$$

$$\tilde{b}_\alpha(\boldsymbol{x}) = b_\alpha(\boldsymbol{x}) h_1(\boldsymbol{x}) + \sum_{\beta=1}^{d} \left(A_{\alpha\beta}(\boldsymbol{x}) + A_{\beta\alpha}(\boldsymbol{x})\right) \partial_{x_\beta} h_1(\boldsymbol{x}),$$

$$\tilde{c}(\boldsymbol{x}) = \sum_{\alpha,\beta=1}^{d} A_{\alpha\beta}(\boldsymbol{x}) \partial_{x_\alpha} \partial_{x_\beta} h_1(\boldsymbol{x}) + \sum_{\alpha=1}^{d} b_\alpha(\boldsymbol{x}) \partial_{x_\alpha} h_1(\boldsymbol{x}) + c(\boldsymbol{x}) h_1(\boldsymbol{x}),$$

and

$$\tilde{f}(\boldsymbol{x}) = f(\boldsymbol{x}) - \mathcal{L}(h_2(\boldsymbol{x})).$$

Therefore, the optimization convergence and generalization analysis of (2.1) is equivalent to

$$\boldsymbol{\theta}_{\mathcal{D}} = \arg\min_{\boldsymbol{\theta}} R_{\mathcal{D}}(\boldsymbol{\theta}) := \mathbb{E}_{\boldsymbol{x}\in\Omega}\left[\ell(\tilde{\mathcal{L}}\tilde{\phi}(\boldsymbol{x};\boldsymbol{\theta}), \tilde{f}(\boldsymbol{x}))\right], \tag{2.7}$$

which gives

$$\phi(\boldsymbol{x};\boldsymbol{\theta}_{\mathcal{D}}) = h_1(\boldsymbol{x})\tilde{\phi}(\boldsymbol{x};\boldsymbol{\theta}_{\mathcal{D}}) + h_2(\boldsymbol{x})$$

as a best solution to the PDE in (1.1) parametrized by DNNs. The corresponding empirical risk is

$$R_S(\boldsymbol{\theta}) := \frac{1}{n} \sum_{\{\boldsymbol{x}_i\}_{i=1}^n \subset \Omega} \ell(\tilde{\mathcal{L}}\tilde{\phi}(\boldsymbol{x}_i;\boldsymbol{\theta}), \tilde{f}(\boldsymbol{x}_i)), \tag{2.8}$$

which gives $\boldsymbol{\theta}_S = \arg\min_{\boldsymbol{\theta}} R_S(\boldsymbol{\theta})$ and

$$\phi(\boldsymbol{x};\boldsymbol{\theta}_S) = h_1(\boldsymbol{x})\tilde{\phi}(\boldsymbol{x};\boldsymbol{\theta}_S) + h_2(\boldsymbol{x}).$$

Similarly, in the case of other two types of boundary conditions, the corresponding optimization problem in (1.2) can also be transformed to (2.7) and its discretization in (2.8) with an appropriate differential operator $\tilde{\mathcal{L}}$ and a right-hand-side function $\tilde{f}$.

In sum, the discussion in Section 2.2 and here indicates that the optimization and generalization analysis of deep learning-based PDE solvers for various IVPs and BVPs with different boundary conditions can be reduced to the analysis of (2.7) and (2.8) with $\tilde{\mathcal{L}}$ in a nondivergence form. In the next section, we will present our main theorems for this analysis. For simplicity, we will still use the notation of $\mathcal{L}$ and f instead of $\tilde{\mathcal{L}}$ and $\tilde{f}$ in our analysis afterwards.

3 Main results

In this section, we introduce our main results on the convergence of GD and the generalization error of neural network-based least-squares solvers for PDEs using two-layer neural networks on $\Omega = [0, 1]^d$. Throughout our analysis, we assume $|f| \leq 1$ and focus on second-order differential operators $\mathcal{L}$ given in (2.6) satisfying the assumption below.

Assumption 3.1 (Symmetry and boundedness of $\mathcal{L}$). Throughout the analysis of this paper, we assume $\mathcal{L}$ in (2.6) satisfies the condition: there exists $M \geq 1$[1] such that for all $\boldsymbol{x} \in \Omega = [0, 1]^d$, $\alpha, \beta \in [d]$, we have $A_{\alpha\beta} = A_{\beta\alpha}$

$$|A_{\alpha\beta}(\boldsymbol{x})| \leq M, \quad |b_\alpha(\boldsymbol{x})| \leq M, \quad \text{and} \quad |c(\boldsymbol{x})| \leq M. \tag{3.1}$$

First, we show that, under suitable assumptions, the empirical risk $R_S(\boldsymbol{\theta})$ of the PDE solution represented by an overparametrized two-layer neural network converges to zero, i.e., achieving a global minimizer, with a linear convergence rate by GD. In particular, as discussed in Section 2, it is sufficient to prove the convergence for minimizing the empirical loss

$$\boldsymbol{\theta}_S = \arg\min_{\boldsymbol{\theta}} R_S(\boldsymbol{\theta}) := \frac{1}{n} \sum_{S=\{\boldsymbol{x}_i\}_{i=1}^n \subset \Omega} \ell(\mathcal{L}\phi(\boldsymbol{x}_i; \boldsymbol{\theta}), f(\boldsymbol{x}_i)), \tag{3.2}$$

where $S := \{\boldsymbol{x}_i\}_{i=1}^n$ is a given set of i.i.d. samples with the uniform distribution $\mathcal{D}$ over $\Omega = [0, 1]^d$, and the two-layer neural network used here is constructed as

$$\phi(\boldsymbol{x}; \boldsymbol{\theta}) = \sum_{k=1}^{m} a_k \sigma(\boldsymbol{w}_k^{\mathsf{T}} \boldsymbol{x}), \tag{3.3}$$

where for $k \in [m]$, $a_k \in \mathbb{R}$, $\boldsymbol{w}_k \in \mathbb{R}^d$, $\boldsymbol{\theta} = \operatorname{vec}\{a_k, \boldsymbol{w}_k\}_{k=1}^m$, and $\sigma(x) = \max\{\frac{1}{6}x^3, 0\}$. Our main result of the linear convergence rate is summarized in Theorem 3.1 below.

Theorem 3.1 (Linear convergence rate). *Let* $\boldsymbol{\theta}^0 := \operatorname{vec}\{a_k^0, \boldsymbol{w}_k^0\}_{k=1}^m$ *at the GD initialization for solving* (3.2), *where* $a_k^0 \sim \mathcal{N}(0, \gamma^2)$ *and* $\boldsymbol{w}_k^0 \sim \mathcal{N}(\boldsymbol{0}, \boldsymbol{I}_d)$ *with*

[1] The upper bound M is not necessarily greater than 1. We set this for simplicity.

any $\gamma \in (0,1)$. *Let* $C_d := \mathbb{E}\|\boldsymbol{w}\|_1^{12} < +\infty$ *with* $\boldsymbol{w} \sim \mathcal{N}(\mathbf{0}, \boldsymbol{I}_d)$ *and* λ_S *be a positive constant in Assumption 4.1. For any* $\delta \in (0,1)$, *if*

$$m \ge \max\left\{\frac{512 n^4 M^4 C_d}{\lambda_S^2 \delta}, \frac{200\sqrt{2} M d^3 n \log(4m(d+1)/\delta)\sqrt{R_S(\boldsymbol{\theta}^0)}}{\lambda_S}, \right. \tag{3.4}$$

$$\left. \frac{2^{23} M^3 d^9 n^2 (\log(4m(d+1)/\delta))^4 \sqrt{R_S(\boldsymbol{\theta}^0)}}{\lambda_S^2}\right\}, \tag{3.5}$$

then with probability at least $1-\delta$ *over the random initialization* $\boldsymbol{\theta}^0$, *we have, for all* $t \ge 0$,

$$R_S(\boldsymbol{\theta}(t)) \le \exp\left(-\frac{m\lambda_S t}{n}\right) R_S(\boldsymbol{\theta}^0).$$

Remark 3.1. For the estimate of $R_S(\boldsymbol{\theta}^0)$, see Lemma 4.2. In particular, if $\gamma = O(\frac{1}{\sqrt{m}(\log m)^2})$, then $R_S(\boldsymbol{\theta}^0) = O(1)$. One may also use the Anti-Symmetrical Initialization (ASI) (Zhang et al., 2019), a general but simple trick that ensures $R_S(\boldsymbol{\theta}^0) \le \frac{1}{2}$.

Second, we prove that the a posteriori generalization error $|R_{\mathcal{D}}(\boldsymbol{\theta}) - R_S(\boldsymbol{\theta})|$ is bounded by $O\left(\frac{\|\boldsymbol{\theta}\|_{\mathcal{P}}^2 \log\|\boldsymbol{\theta}\|_{\mathcal{P}}}{\sqrt{n}}\right)$, where $\|\boldsymbol{\theta}\|_{\mathcal{P}}$ is the path norm introduced in Definition 2.2, and the a priori generalization error $R_{\mathcal{D}}(\boldsymbol{\theta}_{S,\lambda})$ is bounded by $O\left(\frac{\|f\|_{\mathcal{B}}^2}{m}\right) + O\left(\frac{\|f\|_{\mathcal{B}}^2 \log\|f\|_{\mathcal{B}}}{\sqrt{n}}\right)$, where $\|f\|_{\mathcal{B}}$ is the Barron norm for Barron-type functions $f(\boldsymbol{x})$ introduced in Definition 2.3, and $\boldsymbol{\theta}_{S,\lambda}$ is a global minimizer of a regularized empirical loss using the path norm. Our results of the generalization errors can be summarized in Theorems 3.2 and 3.3 below.

Theorem 3.2 (A posteriori generalization bound). *For any* $\delta \in (0,1)$, *with probability at least* $1-\delta$ *over the choice of random samples* $S := \{\boldsymbol{x}_i\}_{i=1}^n$ *in* (3.2), *for any two-layer neural network* $\phi(\boldsymbol{x};\boldsymbol{\theta})$ *in* (3.3), *we have*

$$|R_{\mathcal{D}}(\boldsymbol{\theta}) - R_S(\boldsymbol{\theta})| \le \frac{(\|\boldsymbol{\theta}\|_{\mathcal{P}}+1)^2}{\sqrt{n}} 2M^2 (14 d^2 \sqrt{2\log(2d)} + \log[\pi(\|\boldsymbol{\theta}\|_{\mathcal{P}}+1)] + \sqrt{2\log(1/3\delta)}).$$

Theorem 3.3 (A priori generalization bound). *Suppose that* $f(\boldsymbol{x})$ *is in the Barron-type space* $\mathcal{B}([0,1]^d)$ *and* $\lambda \ge 4M^2[2 + 14d^2\sqrt{2\log(2d)} + \sqrt{2\log(2/3\delta)}]$. *Let*

$$\boldsymbol{\theta}_{S,\lambda} = \arg\min_{\boldsymbol{\theta}} J_{S,\lambda}(\boldsymbol{\theta}) := R_S(\boldsymbol{\theta}) + \frac{\lambda}{\sqrt{n}} \|\boldsymbol{\theta}\|_{\mathcal{P}}^2 \log[\pi(\|\boldsymbol{\theta}\|_{\mathcal{P}}+1)].$$

Then for any $\delta \in (0, 1)$, *with probability at least* $1 - \delta$ *over the choice of random samples* $S := \{\boldsymbol{x}_i\}_{i=1}^{n}$ *in* (3.2), *we have*

$$\begin{aligned} R_{\mathcal{D}}(\boldsymbol{\theta}_{S,\lambda}) &:= \mathbb{E}_{\boldsymbol{x}\sim\mathcal{D}}\tfrac{1}{2}(\mathcal{L}\phi(\boldsymbol{x};\boldsymbol{\theta}_{S,\lambda}) - f(\boldsymbol{x}))^2 \\ &\leq \frac{6M^2\|f\|_{\mathcal{B}}^2}{m} + \frac{\|f\|_{\mathcal{B}}^2+1}{\sqrt{n}}(4\lambda + 16M^2) \\ &\quad \times \left\{\log[\pi(2\|f\|_{\mathcal{B}}+1)] + 14d^2\sqrt{\log(2d)} + \sqrt{\log(2/3\delta)}\right\}. \end{aligned} \tag{3.6}$$

The proof of Theorem 3.1 will be given in Section 4 and the proofs of Theorems 3.2 and 3.3 will be presented in Section 5.

4 Global convergence of gradient descent

In this section, we will prove the global convergence of GD with a linear convergence rate for deep learning-based PDE solvers as stated in Theorem 3.1. We will first summarize the notations and assumptions for the proof of Theorem 3.1 in Section 4.1. Several important lemmas will be proved in Section 4.2. Finally, Theorem 3.1 is proved in Section 4.3.

4.1 Notations and main ideas

Let us first summarize the notations and assumptions used in the proof of Theorem 3.1.

Recall that we use the two-layer neural network $\phi(\boldsymbol{x};\boldsymbol{\theta})$ in (3.3) with $\boldsymbol{\theta} = \mathrm{vec}\{a_k, \boldsymbol{w}_k\}_{k=1}^{m}$. In the GD iteration, we use t to denote the iteration or the artificial time variable in the gradient flow. Hence, we define the following notations for the evolution of parameters at time t:

$$a_k^t := a_k(t), \quad \boldsymbol{w}_k^t := \boldsymbol{w}_k(t), \quad \boldsymbol{\theta}^t := \boldsymbol{\theta}(t) := \mathrm{vec}\{a_k^t, \boldsymbol{w}_k^t\}_{k=1}^{m}.$$

In the analysis, we also use $\bar{a}_k^t := \bar{a}_k(t) := \gamma^{-1}a_k(t)$ with $0 < \gamma < 1$, e.g., $\gamma = \frac{1}{\sqrt{m}}$ or $\gamma = \frac{1}{m}$. $\bar{\boldsymbol{\theta}}(t)$ means $\mathrm{vec}\{\bar{a}_k^t, \boldsymbol{w}_k^t\}_{k=1}^{m}$. Similarly, we can introduce t to other functions or variables depending on $\boldsymbol{\theta}(t)$. When the dependency of t is clear, we will drop the index t. In the initialization of GD, we set

$$\begin{aligned} a_k^0 := a_k(0) \sim \mathcal{N}(0, \gamma^2), \quad \boldsymbol{w}_k^0 := \boldsymbol{w}_k(0) \sim \mathcal{N}(\boldsymbol{0}, \boldsymbol{I}_d), \\ \boldsymbol{\theta}^0 := \boldsymbol{\theta}(0) := \mathrm{vec}\{a_k^0, \boldsymbol{w}_k^0\}_{k=1}^{m}. \end{aligned} \tag{4.1}$$

Note that we use $\sigma(x) = \max\{\frac{1}{6}x^3, 0\}$ as the activation of our two-layer neural network. Therefore, $\sigma'(x) = \max\{\frac{1}{2}x^2, 0\}$, and $\sigma''(x) = \mathrm{ReLU}(x) = \max\{x, 0\}$. For simplicity, we define

$$f_{\boldsymbol{\theta}}(\boldsymbol{x}) := f(\boldsymbol{x};\boldsymbol{\theta}) := \mathcal{L}\phi(\boldsymbol{x};\boldsymbol{\theta})$$

$$= \sum_{k=1}^{m} a_k[\boldsymbol{w}_k^\top \boldsymbol{A}(\boldsymbol{x})\boldsymbol{w}_k \sigma''(\boldsymbol{w}_k^\top \boldsymbol{x}) + \boldsymbol{b}^\top(\boldsymbol{x})\boldsymbol{w}_k \sigma'(\boldsymbol{w}_k^\top \boldsymbol{x}) + c(\boldsymbol{x})\sigma(\boldsymbol{w}_k^\top \boldsymbol{x})], \tag{4.2}$$

which can be treated as a special two-layer neural network for a regression problem $f_{\boldsymbol{\theta}}(\boldsymbol{x}) \approx f(\boldsymbol{x})$.

For simplicity, we denote $e_i = f_{\boldsymbol{\theta}}(\boldsymbol{x}_i) - f(\boldsymbol{x}_i)$ for $i \in [n]$ and $\boldsymbol{e} = (e_1, e_2, \ldots, e_n)^\top$. Then the empirical risk can be written as

$$R_S(\boldsymbol{\theta}) = \frac{1}{2n} \sum_{i=1}^{n} (f_{\boldsymbol{\theta}}(\boldsymbol{x}_i) - f(\boldsymbol{x}_i))^2 = \frac{1}{2n} \boldsymbol{e}^\top \boldsymbol{e}.$$

Hence, the GD dynamics is

$$\dot{\boldsymbol{\theta}} = -\nabla_{\boldsymbol{\theta}} R_S(\boldsymbol{\theta}), \tag{4.3}$$

or equivalently in terms of a_k and $\boldsymbol{w}_k$ as follows:

$$\begin{aligned}
\dot{a}_k &= -\nabla_{a_k} R_S(\boldsymbol{\theta}) \\
&= -\frac{1}{n} \sum_{i=1}^{n} e_i \Big[\boldsymbol{w}_k^\top \boldsymbol{A}(\boldsymbol{x}_i)\boldsymbol{w}_k \sigma''(\boldsymbol{w}_k^\top \boldsymbol{x}_i) + \boldsymbol{b}^\top(\boldsymbol{x}_i)\boldsymbol{w}_k \sigma'(\boldsymbol{w}_k^\top \boldsymbol{x}_i) \\
&\quad + c(\boldsymbol{x}_i)\sigma(\boldsymbol{w}_k^\top \boldsymbol{x}_i) \Big], \\
\dot{\boldsymbol{w}}_k &= -\nabla_{\boldsymbol{w}_k} R_S(\boldsymbol{\theta}) \\
&= -\frac{1}{n} \sum_{i=1}^{n} e_i a_k \Big[2\boldsymbol{A}(\boldsymbol{x}_i)\boldsymbol{w}_k \sigma''(\boldsymbol{w}_k^\top \boldsymbol{x}_i) + \boldsymbol{w}_k^\top \boldsymbol{A}(\boldsymbol{x}_i)\boldsymbol{w}_k \sigma^{(3)}(\boldsymbol{w}_k^\top \boldsymbol{x}_i)\boldsymbol{x}_i \\
&\quad + \sigma'(\boldsymbol{w}_k^\top \boldsymbol{x}_i)\boldsymbol{b}(\boldsymbol{x}_i) + \boldsymbol{b}^\top(\boldsymbol{x}_i)\boldsymbol{w}_k \sigma''(\boldsymbol{w}_k^\top \boldsymbol{x}_i)\boldsymbol{x}_i + c(\boldsymbol{x}_i)\sigma'(\boldsymbol{w}_k^\top \boldsymbol{x}_i)\boldsymbol{x}_i \Big].
\end{aligned}$$

Adopting the neuron tangent kernel point of view (Jacot et al., 2018), in the case of a two-layer neural network with an infinite width, the corresponding kernels $k^{(a)}$ for parameters in the last linear transform and $k^{(w)}$ for parameters in the first layer are functions from $\Omega \times \Omega$ to $\mathbb{R}$ defined by

$$\begin{aligned}
k^{(a)}(\boldsymbol{x}, \boldsymbol{x}') &:= \mathbb{E}_{\boldsymbol{w} \sim \mathcal{N}(\boldsymbol{0}, \boldsymbol{I}_d)} g^{(a)}(\boldsymbol{w}; \boldsymbol{x}, \boldsymbol{x}'), \\
k^{(w)}(\boldsymbol{x}, \boldsymbol{x}') &:= \mathbb{E}_{(a, \boldsymbol{w}) \sim \mathcal{N}(\boldsymbol{0}, \boldsymbol{I}_{d+1})} g^{(w)}(a, \boldsymbol{w}; \boldsymbol{x}, \boldsymbol{x}'),
\end{aligned}$$

where

$$\begin{aligned}
& g^{(a)}(\boldsymbol{w}; \boldsymbol{x}, \boldsymbol{x}') \\
&:= \left[\boldsymbol{w}^\top \boldsymbol{A}(\boldsymbol{x})\boldsymbol{w}\sigma''(\boldsymbol{w}^\top \boldsymbol{x}) + \boldsymbol{b}^\top(\boldsymbol{x})\boldsymbol{w}\sigma'(\boldsymbol{w}^\top \boldsymbol{x}) + c(\boldsymbol{x})\sigma(\boldsymbol{w}^\top \boldsymbol{x}) \right] \\
&\quad \cdot \left[\boldsymbol{w}^\top \boldsymbol{A}(\boldsymbol{x}')\boldsymbol{w}\sigma''(\boldsymbol{w}^\top \boldsymbol{x}') + \boldsymbol{b}^\top(\boldsymbol{x}')\boldsymbol{w}\sigma'(\boldsymbol{w}^\top \boldsymbol{x}') + c(\boldsymbol{x}')\sigma(\boldsymbol{w}^\top \boldsymbol{x}') \right],
\end{aligned}$$

$$
\begin{aligned}
& g^{(w)}(a, \boldsymbol{w}; \boldsymbol{x}, \boldsymbol{x}') \\
& := a^2\big[2\boldsymbol{A}(\boldsymbol{x})\boldsymbol{w}\sigma''(\boldsymbol{w}^\top\boldsymbol{x}) + \boldsymbol{w}^\top\boldsymbol{A}(\boldsymbol{x})\boldsymbol{w}\sigma^{(3)}(\boldsymbol{w}^\top\boldsymbol{x})\boldsymbol{x} + \sigma'(\boldsymbol{w}^\top\boldsymbol{x})\boldsymbol{b}(\boldsymbol{x}) \\
& \quad + \boldsymbol{b}^\top(\boldsymbol{x})\boldsymbol{w}\sigma''(\boldsymbol{w}^\top\boldsymbol{x})\boldsymbol{x} + c(\boldsymbol{x})\sigma'(\boldsymbol{w}^\top\boldsymbol{x})\boldsymbol{x}\big] \cdot \big[2A(\boldsymbol{x}')\boldsymbol{w}\sigma''(\boldsymbol{w}^\top\boldsymbol{x}') \\
& \quad + \boldsymbol{w}^\top\boldsymbol{A}(\boldsymbol{x}')\boldsymbol{w}\sigma^{(3)}(\boldsymbol{w}^\top\boldsymbol{x}')\boldsymbol{x}' + \sigma'(\boldsymbol{w}^\top\boldsymbol{x}')\boldsymbol{b}(\boldsymbol{x}') \\
& \quad + \boldsymbol{b}^\top(\boldsymbol{x}')\boldsymbol{w}\sigma''(\boldsymbol{w}^\top\boldsymbol{x}')\boldsymbol{x}' + c(\boldsymbol{x})\sigma'(\boldsymbol{w}^\top\boldsymbol{x}')\boldsymbol{x}'\big].
\end{aligned}
$$

These kernels evaluated at $n \times n$ pairs of samples lead to $n \times n$ Gram matrices $\boldsymbol{K}^{(a)}$ and $\boldsymbol{K}^{(\boldsymbol{w})}$ with $K_{ij}^{(a)} = k^{(a)}(\boldsymbol{x}_i, \boldsymbol{x}_j)$ and $K_{ij}^{(w)} = k^{(w)}(\boldsymbol{x}_i, \boldsymbol{x}_j)$, respectively. Our analysis requires the matrix $\boldsymbol{K}^{(a)}$ to be positive definite, which has been verified for regression problems under mild conditions on random training data $S = \{\boldsymbol{x}_i\}_{i=1}^n$ and can be generalized to our case. Hence, we assume this as follows for simplicity.

Assumption 4.1. We assume that

$$\lambda_S := \lambda_{\min}\left(\boldsymbol{K}^{(a)}\right) > 0.$$

For a two-layer neural network with m neurons, the $n \times n$ Gram matrix $\boldsymbol{G}(\boldsymbol{\theta}) = \boldsymbol{G}^{(a)}(\boldsymbol{\theta}) + \boldsymbol{G}^{(w)}(\boldsymbol{\theta})$ is given by the following expressions for the (i, j)-th entry

$$
\begin{aligned}
\boldsymbol{G}_{ij}^{(a)}(\boldsymbol{\theta}) &:= \frac{1}{m}\sum_{k=1}^{m} g^{(a)}(\boldsymbol{w}_k; \boldsymbol{x}_i, \boldsymbol{x}_j), \\
\boldsymbol{G}_{ij}^{(w)}(\boldsymbol{\theta}) &:= \frac{1}{m}\sum_{k=1}^{m} g^{(w)}(a_k, \boldsymbol{w}_k; \boldsymbol{x}_i, \boldsymbol{x}_j).
\end{aligned}
$$

Clearly, $\boldsymbol{G}^{(a)}(\boldsymbol{\theta})$ and $\boldsymbol{G}^{(w)}(\boldsymbol{\theta})$ are both positive semi-definite for any $\boldsymbol{\theta}$. By using the Gram matrix $\boldsymbol{G}(\boldsymbol{\theta})$, we have the following evolution equations to understand the dynamics of GD:

$$\frac{\mathrm{d}}{\mathrm{d}t} f_{\boldsymbol{\theta}}(\boldsymbol{x}_i) = -\frac{1}{n}\sum_{j=1}^{n} \boldsymbol{G}_{ij}(\boldsymbol{\theta})(f_{\boldsymbol{\theta}}(\boldsymbol{x}_j) - f(\boldsymbol{x}_j))$$

and

$$\frac{\mathrm{d}}{\mathrm{d}t} R_S(\boldsymbol{\theta}) = -\|\nabla_{\boldsymbol{\theta}} R_S(\boldsymbol{\theta})\|_2^2 = -\frac{m}{n^2}\boldsymbol{e}^\top\boldsymbol{G}(\boldsymbol{\theta})\boldsymbol{e} \le -\frac{m}{n^2}\boldsymbol{e}^\top\boldsymbol{G}^{(a)}(\boldsymbol{\theta})\boldsymbol{e}. \tag{4.4}$$

Our goal is to show that the above evolution equation has a solution $f_{\boldsymbol{\theta}}(\boldsymbol{x}_i)$ converging to $f(\boldsymbol{x}_i)$ for all training samples $\boldsymbol{x}_i$, or equivalently, to show that $R_S(\boldsymbol{\theta})$ converges to zero. These goals are true if the smallest eigenvalue $\lambda_{\min}\left(\boldsymbol{G}^{(a)}(\boldsymbol{\theta})\right)$ of $\boldsymbol{G}^{(a)}(\boldsymbol{\theta})$ has a positive lower bound uniformly in t, since in this case we can solve (4.4) and bound $R_S(\boldsymbol{\theta})$ with a function in t converging to zero when $t \to \infty$ as shown in Lemma 4.4. In fact, a uniform lower bound of $\lambda_{\min}\left(\boldsymbol{G}^{(a)}(\boldsymbol{\theta})\right)$ can be $\frac{1}{2}\lambda_S$, which can be proved in the following three steps:

- **(Initial phase)** By Assumption 4.1 of $\boldsymbol{K}^{(a)}$, we can show $\lambda_{\min}\left(\boldsymbol{G}^{(a)}(\boldsymbol{\theta}(0))\right) \approx \lambda_S$ in Lemma 4.3 using the observation that $\boldsymbol{K}_{ij}^{(a)}$ is the mean of $g(\boldsymbol{w}; \boldsymbol{x}_i, \boldsymbol{x}_j)$ over the normal random variable $\boldsymbol{w}$, while $\boldsymbol{G}_{ij}^{(a)}(\boldsymbol{\theta}(0))$ is the mean of $g(\boldsymbol{w}; \boldsymbol{x}_i, \boldsymbol{x}_j)$ with m independent realizations.
- **(Evolution phase)** The GD dynamics results in $\boldsymbol{\theta}(t) \approx \boldsymbol{\theta}(0)$ under the assumption of overparametrization as shown in Lemma 4.5, which indicates that

$$\lambda_{\min}\left(\boldsymbol{G}^{(a)}(\boldsymbol{\theta}(0))\right) \approx \lambda_{\min}\left(\boldsymbol{G}^{(a)}(\boldsymbol{\theta}(t))\right).$$

- **(Final phase)** To show the uniform bound $\lambda_{\min}\left(\boldsymbol{G}^{(a)}(\boldsymbol{\theta}(t))\right) \geq \frac{1}{2}\lambda_S$ for all $t \geq 0$, we introduce a stopping time t^* via

$$t^* = \inf\{t \mid \boldsymbol{\theta}(t) \notin \mathcal{M}(\boldsymbol{\theta}^0)\}, \tag{4.5}$$

where

$$\mathcal{M}(\boldsymbol{\theta}^0) := \left\{\boldsymbol{\theta} \mid \|\boldsymbol{G}^{(a)}(\boldsymbol{\theta}) - \boldsymbol{G}^{(a)}(\boldsymbol{\theta}^0)\|_{\mathrm{F}} \leq \frac{1}{4}\lambda_S\right\}, \tag{4.6}$$

and show that t^* is in fact equal to infinity in the final proof of Theorem 3.1 in Section 4.3.

4.2 Proofs of lemmas for Theorem 3.1

In this subsection, we will prove several lemmas in preparation for the proof of Theorem 3.1.

Lemma 4.1. *For any $\delta \in (0, 1)$ with probability at least $1 - \delta$ over the random initialization in* (4.1), *we have*

$$\begin{aligned}\max_{k\in[m]}\left\{|\bar{a}_k^0|,\ \|\boldsymbol{w}_k^0\|_\infty\right\} &\leq \sqrt{2\log\frac{2m(d+1)}{\delta}},\\ \max_{k\in[m]}\left\{|a_k^0|\right\} &\leq \gamma\sqrt{2\log\frac{2m(d+1)}{\delta}}.\end{aligned} \tag{4.7}$$

Proof. If $\mathrm{X} \sim \mathcal{N}(0, 1)$, then $\mathbb{P}(|\mathrm{X}| > \varepsilon) \leq 2\mathrm{e}^{-\frac{1}{2}\varepsilon^2}$ for all $\varepsilon > 0$. Since $\bar{a}_k^0 \sim \mathcal{N}(0, 1)$, $(\boldsymbol{w}_k^0)_\alpha \sim \mathcal{N}(0, 1)$ for $k \in [m]$, $\alpha \in [d]$, and they are all independent, by setting

$$\varepsilon = \sqrt{2\log\frac{2m(d+1)}{\delta}},$$

one can obtain

$$
\begin{aligned}
&\mathbb{P}\left(\max_{k\in[m]}\left\{|\bar{a}_k^0|, \|\boldsymbol{w}_k^0\|_\infty\right\} > \varepsilon\right)\\
&= \mathbb{P}\left(\left(\bigcup_{k\in[m]}\left\{|\bar{a}_k^0| > \varepsilon\right\}\right)\bigcup\left(\bigcup_{k\in[m],\alpha\in[d]}\left\{|(\boldsymbol{w}_k^0)_\alpha| > \varepsilon\right\}\right)\right)\\
&\le \sum_{k=1}^{m}\mathbb{P}\left(|\bar{a}_k^0| > \varepsilon\right) + \sum_{k=1}^{m}\sum_{\alpha=1}^{d}\mathbb{P}\left(|(\boldsymbol{w}_k^0)_\alpha| > \varepsilon\right)\\
&\le 2m\mathrm{e}^{-\frac{1}{2}\varepsilon^2} + 2md\mathrm{e}^{-\frac{1}{2}\varepsilon^2}\\
&= 2m(d+1)\mathrm{e}^{-\frac{1}{2}\varepsilon^2}\\
&= \delta,
\end{aligned}
$$

which implies the conclusions of this lemma. □

Lemma 4.2. *For any $\delta \in (0, 1)$ with probability at least $1-\delta$ over the random initialization in* (4.1)*, we have*

$$
\begin{aligned}
R_S(\boldsymbol{\theta}^0) \le \frac{1}{2}\Bigg(1 + 32\gamma\sqrt{m}Md^3\left(\log\frac{4m(d+1)}{\delta}\right)^2\\
\times\left(\sqrt{2\log(2d)} + \sqrt{2\log(8/\delta)}\right)\Bigg)^2,
\end{aligned}
$$

Proof. From Lemma 4.1 we know that with probability at least $1-\delta/2$,

$$
|\bar{a}_k^0| \le \sqrt{2\log\frac{4m(d+1)}{\delta}} \quad \text{and} \quad \|\boldsymbol{w}_k^0\|_1 \le d\sqrt{2\log\frac{4m(d+1)}{\delta}}.
$$

Let

$$
\begin{aligned}
\mathcal{H} = \{h(\bar{a}, \boldsymbol{w}; \boldsymbol{x}) \mid h(\bar{a}, \boldsymbol{w}; \boldsymbol{x}) =& \bar{a}\Big[\boldsymbol{w}^\intercal \boldsymbol{A}(\boldsymbol{x})\boldsymbol{w}\sigma''(\boldsymbol{w}^\intercal\boldsymbol{x}) + \boldsymbol{b}^\intercal(\boldsymbol{x})\boldsymbol{w}\sigma'(\boldsymbol{w}^\intercal\boldsymbol{x})\\
&+ c(\boldsymbol{x})\sigma(\boldsymbol{w}^\intercal\boldsymbol{x})\Big], \boldsymbol{x}\in\Omega\}.
\end{aligned}
$$

Note that $\boldsymbol{A}$, $\boldsymbol{b}$, and c are known functions of $\boldsymbol{x}$. Each element in the above set is a function of $\bar{a}$ and $\boldsymbol{w}$ while $\boldsymbol{x}\in\Omega = [0,1]^d$ is a parameter. Since $\|\boldsymbol{x}\|_\infty \le 1$, we have

$$
\begin{aligned}
|h(\bar{a}_k^0, \boldsymbol{w}_k^0; \boldsymbol{x})| &\le |\bar{a}_k^0|\left[M\|\boldsymbol{w}_k^0\|_1^3 + \frac{1}{2}M\|\boldsymbol{w}_k^0\|_1^3 + \frac{1}{6}M\|\boldsymbol{w}_k^0\|_1^3\right]\\
&\le 2M|\bar{a}_k^0|\|\boldsymbol{w}_k^0\|_1^3\\
&\le 8Md^3\left(\log\frac{4m(d+1)}{\delta}\right)^2.
\end{aligned}
$$

Then with probability at least $1-\delta/2$, by the Rademacher-based uniform convergence theorem, we have

$$\frac{1}{\gamma m}\sup_{x\in\Omega}|f_{\theta^0}(x)| = \sup_{x\in\Omega}\left|\frac{1}{m}\sum_{k=1}^{m} h(\bar{a}_k^0, \boldsymbol{w}_k^0; \boldsymbol{x}) - \mathbb{E}_{(\bar{a},\boldsymbol{w})\sim\mathcal{N}(0,\boldsymbol{I}_{d+1})} h(\bar{a}, \boldsymbol{w}; \boldsymbol{x})\right|$$
$$\leq 2\mathrm{Rad}_{\bar{\boldsymbol{\theta}}^0}(\mathcal{H}) + 24Md^3\left(\log\frac{4m(d+1)}{\delta}\right)^2\sqrt{\frac{2\log(8/\delta)}{m}},$$

where

$$\mathrm{Rad}_{\bar{\boldsymbol{\theta}}^0}(\mathcal{H}) := \frac{1}{m}\mathbb{E}_{\boldsymbol{\tau}}\left[\sup_{\boldsymbol{x}\in\Omega}\sum_{k=1}^{m}\tau_k h(\bar{a}_k^0, \boldsymbol{w}_k^0; \boldsymbol{x})\right] \leq I_1 + I_2 + I_3,$$
$$I_1 = \frac{1}{m}\mathbb{E}_{\boldsymbol{\tau}}\left[\sup_{\boldsymbol{x}\in\Omega}\sum_{k=1}^{m}\tau_k\bar{a}_k^0\boldsymbol{w}_k^{0\mathsf{T}}\boldsymbol{A}(\boldsymbol{x})\boldsymbol{w}_k^0\sigma''(\boldsymbol{w}_k^{0\mathsf{T}}\boldsymbol{x})\right],$$
$$I_2 = \frac{1}{m}\mathbb{E}_{\boldsymbol{\tau}}\left[\sup_{\boldsymbol{x}\in\Omega}\sum_{k=1}^{m}\tau_k\bar{a}_k^0\boldsymbol{b}^{\mathsf{T}}(\boldsymbol{x})\boldsymbol{w}_k^0\sigma'(\boldsymbol{w}_k^{0\mathsf{T}}\boldsymbol{x})\right],$$
$$I_3 = \frac{1}{m}\mathbb{E}_{\boldsymbol{\tau}}\left[\sup_{\boldsymbol{x}\in\Omega}\sum_{k=1}^{m}\tau_k\bar{a}_k^0 c(\boldsymbol{x})\sigma(\boldsymbol{w}_k^{0\mathsf{T}}\boldsymbol{x})\right],$$

where $\boldsymbol{\tau}$ is a random vector in $\mathbb{N}^m$ with i.i.d. entries $\{\tau_k\}_{k=1}^m$ following the Rademacher distribution.

We only prove for I_1. It can be straightforwardly extended to I_2 and I_3.

$$\begin{aligned}
I_1 &= \frac{1}{m}\mathbb{E}_{\boldsymbol{\tau}}\left[\sup_{\boldsymbol{x}\in\Omega}\sum_{k=1}^{m}\tau_k\bar{a}_k^0\boldsymbol{w}_k^{0\mathsf{T}}\boldsymbol{A}(\boldsymbol{x})\boldsymbol{w}_k^0\sigma''(\boldsymbol{w}_k^{0\mathsf{T}}\boldsymbol{x})\right] \\
&\leq \frac{1}{m}\mathbb{E}_{\boldsymbol{\tau}}\left[\sup_{\boldsymbol{x},\boldsymbol{y}\in\Omega}\sum_{k=1}^{m}\tau_k\bar{a}_k^0\boldsymbol{w}_k^{0\mathsf{T}}\boldsymbol{A}(\boldsymbol{y})\boldsymbol{w}_k^0\sigma''(\boldsymbol{w}_k^{0\mathsf{T}}\boldsymbol{x})\right] \\
&= \frac{1}{m}\mathbb{E}_{\boldsymbol{\tau}}\left[\sup_{\boldsymbol{x},\boldsymbol{y}\in\Omega}\sum_{k=1}^{m}\sum_{\alpha,\beta=1}^{d}\tau_k\bar{a}_k^0(\boldsymbol{w}_k^{0\mathsf{T}})_\alpha A_{\alpha\beta}(\boldsymbol{y})(\boldsymbol{w}_k^0)_\beta\sigma''(\boldsymbol{w}_k^{0\mathsf{T}}\boldsymbol{x})\right] \\
&\leq \sum_{\alpha,\beta=1}^{d}\frac{1}{m}\mathbb{E}_{\boldsymbol{\tau}}\left[\sup_{\boldsymbol{x},\boldsymbol{y}\in\Omega}\sum_{k=1}^{m}\tau_k\bar{a}_k^0(\boldsymbol{w}_k^{0\mathsf{T}})_\alpha A_{\alpha\beta}(\boldsymbol{y})(\boldsymbol{w}_k^0)_\beta\sigma''(\boldsymbol{w}_k^{0\mathsf{T}}\boldsymbol{x})\right].
\end{aligned} \tag{4.8}$$

For any $\alpha, \beta \in [d]$, we have

$$\mathbb{E}_{\boldsymbol{\tau}}\left[\sup_{\boldsymbol{x},\boldsymbol{y}\in\Omega}\sum_{k=1}^{m}\tau_k\bar{a}_k^0(\boldsymbol{w}_k^{0\mathsf{T}})_\alpha A_{\alpha\beta}(\boldsymbol{y})(\boldsymbol{w}_k^0)_\beta\sigma''(\boldsymbol{w}_k^{0\mathsf{T}}\boldsymbol{x})\right]$$

$$
\begin{aligned}
&\leq \mathbb{E}_{\boldsymbol{\tau}}\left[\sup_{\boldsymbol{x},\boldsymbol{y}\in\Omega}\left|A_{\alpha\beta}(\boldsymbol{y})\right|\left|\sum_{k=1}^{m}\tau_k\bar{a}_k^0(\boldsymbol{w}_k^{0\mathsf{T}})_\alpha(\boldsymbol{w}_k^0)_\beta\sigma''(\boldsymbol{w}_k^{0\mathsf{T}}\boldsymbol{x})\right|\right]\\
&\leq M\mathbb{E}_{\boldsymbol{\tau}}\left[\sup_{\boldsymbol{x}\in\Omega}\left|\sum_{k=1}^{m}\tau_k\bar{a}_k^0(\boldsymbol{w}_k^{0\mathsf{T}})_\alpha(\boldsymbol{w}_k^0)_\beta\sigma''(\boldsymbol{w}_k^{0\mathsf{T}}\boldsymbol{x})\right|\right]\\
&\leq M\mathbb{E}_{\boldsymbol{\tau}}\left[\sup_{\boldsymbol{x}\in\Omega}\sum_{k=1}^{m}\tau_k\bar{a}_k^0(\boldsymbol{w}_k^{0\mathsf{T}})_\alpha(\boldsymbol{w}_k^0)_\beta\sigma''(\boldsymbol{w}_k^{0\mathsf{T}}\boldsymbol{x})\right]\\
&\quad+M\mathbb{E}_{\boldsymbol{\tau}}\left[\sup_{\boldsymbol{x}\in\Omega}\sum_{k=1}^{m}-\tau_k\bar{a}_k^0(\boldsymbol{w}_k^{0\mathsf{T}})_\alpha(\boldsymbol{w}_k^0)_\beta\sigma''(\boldsymbol{w}_k^{0\mathsf{T}}\boldsymbol{x})\right]\\
&=2M\mathbb{E}_{\boldsymbol{\tau}}\left[\sup_{\boldsymbol{x}\in\Omega}\sum_{k=1}^{m}\tau_k\bar{a}_k^0(\boldsymbol{w}_k^{0\mathsf{T}})_\alpha(\boldsymbol{w}_k^0)_\beta\sigma''(\boldsymbol{w}_k^{0\mathsf{T}}\boldsymbol{x})\right],
\end{aligned}
\tag{4.9}
$$

where in the third inequality, we have used the fact that $\sigma''(\boldsymbol{w}_k^{0\mathsf{T}}\boldsymbol{x})=0$ for $\boldsymbol{x}=0$ and for any $\boldsymbol{w}_k^0$. Applying Lemma 2.1 with $\psi_k(y_k)=\bar{a}_k(\boldsymbol{w}_k^{0\mathsf{T}})_\alpha(\boldsymbol{w}_k^0)_\beta\sigma''(y_k)$ for $k\in[m]$, whose Lipschitz constant is $\left(\sqrt{2\log\frac{4m(d+1)}{\delta}}\right)^3$, we have for all $\alpha,\beta\in[d]$

$$
\begin{aligned}
&\mathbb{E}_{\boldsymbol{\tau}}\left[\sup_{\boldsymbol{x}\in\Omega}\sum_{k=1}^{m}\tau_k\bar{a}_k^0(\boldsymbol{w}_k^{0\mathsf{T}})_\alpha(\boldsymbol{w}_k^0)_\beta\sigma''(\boldsymbol{w}_k^{0\mathsf{T}}\boldsymbol{x})\right]\\
&\quad\leq\left(\sqrt{2\log\frac{4m(d+1)}{\delta}}\right)^3\mathbb{E}_{\boldsymbol{\tau}}\left[\sup_{\boldsymbol{x}\in\Omega}\sum_{k=1}^{m}\tau_k\boldsymbol{w}_k^{0\mathsf{T}}\boldsymbol{x}\right].
\end{aligned}
\tag{4.10}
$$

Therefore, combining (4.8), (4.9), and (4.10), we obtain

$$
\begin{aligned}
I_1&\leq\frac{2Md^2}{m}\left(\sqrt{2\log\frac{4m(d+1)}{\delta}}\right)^3\mathbb{E}_{\boldsymbol{\tau}}\left[\sup_{\boldsymbol{x}\in\Omega}\sum_{k=1}^{m}\tau_k\boldsymbol{w}_k^{0\mathsf{T}}\boldsymbol{x}\right]\\
&\leq\frac{2Md^3}{\sqrt{m}}\left(\sqrt{2\log\frac{4m(d+1)}{\delta}}\right)^4\sqrt{2\log(2d)}\\
&\leq\frac{8Md^3\sqrt{2\log(2d)}}{\sqrt{m}}\left(\log\frac{4m(d+1)}{\delta}\right)^2,
\end{aligned}
$$

where the second inequality is by the Rademacher bound for linear predictors in Lemma 2.2. For I_2 and I_3, we note that $\sigma(z)=\frac{1}{6}z^2\sigma''(z)$ and $\sigma'(z)=\frac{1}{2}z\sigma''(z)$. Then by a similar argument, we have

$$
I_2\leq\frac{4Md^2\sqrt{2\log(2d)}}{\sqrt{m}}\left(\log\frac{4m(d+1)}{\delta}\right)^2,
$$

$$I_3 \leq \frac{4Md\sqrt{2\log(2d)}}{3\sqrt{m}}\left(\log\frac{4m(d+1)}{\delta}\right)^2,$$

$$\mathrm{Rad}_{\bar{\boldsymbol{\theta}}^0}(\mathcal{H}) \leq \frac{16Md^3\sqrt{2\log(2d)}}{\sqrt{m}}\left(\log\frac{4m(d+1)}{\delta}\right)^2.$$

So one can get

$$\begin{aligned}\sup_{\boldsymbol{x}\in\Omega}|f_{\boldsymbol{\theta}^0}(\boldsymbol{x})| &\leq 32\gamma Md^3\sqrt{m}\sqrt{2\log(2d)}\left(\log\frac{4m(d+1)}{\delta}\right)^2\\ &\quad + 24\gamma\sqrt{m}Md^3\left(\log\frac{4m(d+1)}{\delta}\right)^2\sqrt{2\log(8/\delta)}\\ &\leq 32\gamma\sqrt{m}Md^3\left(\log\frac{4m(d+1)}{\delta}\right)^2\left(\sqrt{2\log(2d)}+\sqrt{2\log(8/\delta)}\right).\end{aligned}$$

Then

$$\begin{aligned}R_S(\boldsymbol{\theta}^0) &\leq \frac{1}{2n}\sum_{i=1}^n\left(1+|f_{\boldsymbol{\theta}^0}(\boldsymbol{x}_i)|\right)^2\\ &\leq \frac{1}{2}\left(1+32\gamma\sqrt{m}Md^3\left(\log\frac{4m(d+1)}{\delta}\right)^2\right.\\ &\quad\left.\times\left(\sqrt{2\log(2d)}+\sqrt{2\log(8/\delta)}\right)\right)^2,\end{aligned}$$

where the first inequality comes from the fact that $|f|\leq 1$ by our assumption of the PDE. □

The following lemma shows the positive definiteness of $\boldsymbol{G}^{(a)}$ at initialization.

Lemma 4.3. *For any $\delta\in(0,1)$, if $m\geq\frac{256n^4M^4C_d}{\lambda_S^2\delta}$, then with probability at least $1-\delta$ over the random initialization in* (4.1), *we have*

$$\lambda_{\min}\left(\boldsymbol{G}^{(a)}(\boldsymbol{\theta}^0)\right)\geq\frac{3}{4}\lambda_S,$$

where $C_d:=\mathbb{E}\|\boldsymbol{w}\|_1^{12}<+\infty$ with $\boldsymbol{w}\sim\mathcal{N}(\boldsymbol{0},\boldsymbol{I}_d)$.

Proof. We define $\Omega_{ij}:=\{\boldsymbol{\theta}^0 \mid |\boldsymbol{G}_{ij}^{(a)}(\boldsymbol{\theta}^0)-\boldsymbol{K}_{ij}^{(a)}|\leq\frac{\lambda_S}{4n}\}$. Note that

$$|g^{(a)}(\boldsymbol{w}_k^0;\boldsymbol{x}_i,\boldsymbol{x}_j)|\leq\left(M\|\boldsymbol{w}_k^0\|_1^3+\frac{1}{2}M\|\boldsymbol{w}_k^0\|_1^3+\frac{1}{6}M\|\boldsymbol{w}_k^0\|_1^3\right)^2\leq 4M^2\|\boldsymbol{w}_k^0\|_1^6.$$

So

$$\mathrm{Var}\left(g^{(a)}(\boldsymbol{w}_k^0;\boldsymbol{x}_i,\boldsymbol{x}_j)\right)\leq\mathbb{E}\left(g^{(a)}(\boldsymbol{w}_k^0;\boldsymbol{x}_i,\boldsymbol{x}_j)\right)^2\leq 16M^4\mathbb{E}\|\boldsymbol{w}_k^0\|_1^{12}=16M^4C_d,$$

and

$$\operatorname{Var}\left(G_{ij}^{(a)}(\boldsymbol{\theta}^0)\right) = \frac{1}{m^2}\sum_{k=1}^{m}\operatorname{Var}\left(g^{(a)}(\boldsymbol{w}_k^0;\boldsymbol{x}_i,\boldsymbol{x}_j)\right) \le \frac{16M^4C_d}{m}.$$

Then the probability of the event Ω_{ij} has the lower bound:

$$\mathbb{P}(\Omega_{ij}) \ge 1 - \frac{\operatorname{Var}\left(G_{ij}^{(a)}(\boldsymbol{\theta}^0)\right)}{[\lambda_S/(4n)]^2} \ge 1 - \frac{256M^4n^2C_d}{\lambda_S^2 m}.$$

Thus, with probability at least $\left(1 - \frac{256M^4n^2C_d}{\lambda_S^2 m}\right)^{n^2} \ge 1 - \frac{256M^4n^4C_d}{\lambda_S^2 m}$, we have all events Ω_{ij} for $i, j \in [n]$ happen. This implies that with probability at least $1 - \frac{256M^4n^4C_d}{\lambda_S^2 m}$, we have

$$\|\boldsymbol{G}^{(a)}(\boldsymbol{\theta}^0) - \boldsymbol{K}^{(a)}\|_{\mathrm{F}} \le \frac{\lambda_S}{4}$$

and

$$\lambda_{\min}\left(\boldsymbol{G}^{(a)}(\boldsymbol{\theta}^0)\right) \ge \lambda_S - \|\boldsymbol{G}^{(a)}(\theta^0) - \boldsymbol{K}^{(a)}\|_{\mathrm{F}} \ge \frac{3}{4}\lambda_S.$$

For any $\delta \in (0,1)$, if $m \ge \frac{256n^4M^4C_d}{\lambda_S^2\delta}$, then with probability at least $1 - \frac{256M^4n^4C_d}{\lambda_S^2 m} \ge 1 - \delta$ over the initialization $\boldsymbol{\theta}^0$, we have $\lambda_{\min}\left(\boldsymbol{G}^{(a)}(\boldsymbol{\theta}^0)\right) \ge \frac{3}{4}\lambda_S$. □

The following lemma estimates the empirical loss dynamics before the stopping time t^* in (4.5).

Lemma 4.4. *For any $\delta \in (0,1)$, if $m \ge \frac{256n^4M^4C_d}{\lambda_S^2\delta}$, then with probability at least $1-\delta$ over the random initialization in* (4.1)*, we have for any $t \in [0, t^*)$*

$$R_S(\boldsymbol{\theta}(t)) \le \exp\left(-\frac{m\lambda_S t}{n}\right) R_S(\boldsymbol{\theta}^0).$$

Proof. From Lemma 4.3, for any $\delta \in (0,1)$ with probability at least $1-\delta$ over initialization $\boldsymbol{\theta}^0$ and for any $t \in [0, t^*)$ with t^* defined in (4.5), we have $\boldsymbol{\theta}(t) \in \mathcal{M}(\boldsymbol{\theta}^0)$ defined in (4.6) and

$$\begin{aligned}\lambda_{\min}\left(\boldsymbol{G}^{(a)}(\boldsymbol{\theta})\right) &\ge \lambda_{\min}\left(\boldsymbol{G}^{(a)}(\boldsymbol{\theta}^0)\right) - \|\boldsymbol{G}^{(a)}(\boldsymbol{\theta}) - \boldsymbol{G}^{(a)}(\boldsymbol{\theta}^0)\|_{\mathrm{F}}\\ &\ge \frac{3}{4}\lambda_S - \frac{1}{4}\lambda_S\\ &= \frac{1}{2}\lambda_S.\end{aligned}$$

Note that $\boldsymbol{G}_{ij} = \frac{1}{m}\nabla_{\boldsymbol{\theta}} f_{\boldsymbol{\theta}}(\boldsymbol{x}_i) \cdot \nabla_{\boldsymbol{\theta}} f_{\boldsymbol{\theta}}(\boldsymbol{x}_j)$ and $\nabla_{\boldsymbol{\theta}} R_S = \frac{1}{n}\sum_{i=1}^{n} e_i \nabla_{\boldsymbol{\theta}} f_{\boldsymbol{\theta}}(\boldsymbol{x}_i)$, so

$$\|\nabla_{\boldsymbol{\theta}} R_S(\boldsymbol{\theta}(t))\|_2^2 = \frac{m}{n^2}\boldsymbol{e}^\intercal \boldsymbol{G}(\boldsymbol{\theta}(t))\boldsymbol{e} \geq \frac{m}{n^2}\boldsymbol{e}^\intercal \boldsymbol{G}^{(a)}(\boldsymbol{\theta}(t))\boldsymbol{e},$$

where the last equation is true by the fact that $G^{(w)}(\boldsymbol{\theta}(t))$ is a Gram matrix and hence positive semi-definite. Together with

$$\frac{m}{n^2}\boldsymbol{e}^\intercal \boldsymbol{G}^{(a)}(\boldsymbol{\theta}(t))\boldsymbol{e} \geq \frac{2m}{n}\lambda_{\min}\left(\boldsymbol{G}^{(a)}(\boldsymbol{\theta}(t))\right) R_S(\boldsymbol{\theta}(t)) \geq \frac{m}{n}\lambda_S R_S(\boldsymbol{\theta}(t)),$$

then finally we get

$$\frac{\mathrm{d}}{\mathrm{d}t} R_S(\boldsymbol{\theta}(t)) = -\|\nabla_{\boldsymbol{\theta}} R_S(\boldsymbol{\theta}(t))\|_2^2 \leq -\frac{m}{n}\lambda_S R_S(\boldsymbol{\theta}(t)).$$

Integrating the above equation yields the conclusion in this lemma. □

The following lemma shows that the parameters in the two-layer neural network are uniformly bounded in time during the training before time t^*.

Lemma 4.5. *For any $\delta \in (0, 1)$, if*

$$m \geq \max\left\{\frac{512n^4M^4C_d}{\lambda_S^2\delta}, \frac{200\sqrt{2}Md^3n\log(4m(d+1)/\delta)\sqrt{R_S(\boldsymbol{\theta}^0)}}{\lambda_S}\right\},$$

then with probability at least $1-\delta$ over the random initialization in (4.1), *for any $t \in [0, t^*)$ and any $k \in [m]$,*

$$\begin{aligned} |a_k(t) - a_k(0)| &\leq q, \quad & \|\boldsymbol{w}_k(t) - \boldsymbol{w}_k(0)\|_\infty &\leq q, \\ |a_k(0)| &\leq \gamma\eta, & \|\boldsymbol{w}_k(0)\|_\infty &\leq \eta, \end{aligned}$$

where

$$q := \frac{320Md^3(\log\frac{4m(d+1)}{\delta})^{3/2}n\sqrt{R_S(\boldsymbol{\theta}^0)}}{m\lambda_S}$$

and

$$\eta := \sqrt{2\log\frac{4m(d+1)}{\delta}}.$$

Proof. Let $\xi(t) = \max_{k\in[m],s\in[0,t]}\{|a_k(s)|, \|\boldsymbol{w}_k(s)\|_\infty\}$. Note that

$$\begin{aligned} |\nabla_{a_k} R_S(\boldsymbol{\theta})|^2 = \Bigg\{\frac{1}{n}\sum_{i=1}^{n} e_i\Big[&\boldsymbol{w}_k^\intercal \boldsymbol{A}(\boldsymbol{x}_i)\boldsymbol{w}_k\sigma''(\boldsymbol{w}_k^\intercal \boldsymbol{x}_i) + \boldsymbol{b}^\intercal(\boldsymbol{x}_i)\boldsymbol{w}_k\sigma'(\boldsymbol{w}_k^\intercal \boldsymbol{x}_i) \\ &+ c(\boldsymbol{x}_i)\sigma(\boldsymbol{w}_k^\intercal \boldsymbol{x}_i)\Big]\Bigg\}^2 \end{aligned}$$

$$
\begin{aligned}
&\leq 8M^2\|\boldsymbol{w}_k\|_1^6 R_S(\boldsymbol{\theta}) \\
&\leq 8M^2 d^6 (\xi(t))^6 R_S(\boldsymbol{\theta}),
\end{aligned}
$$

and

$$
\begin{aligned}
&\|\nabla_{\boldsymbol{w}_k} R_S(\boldsymbol{\theta})\|_\infty^2 \\
&= \Bigg\| \frac{1}{n}\sum_{i=1}^n e_i a_k \Big[2\boldsymbol{A}(\boldsymbol{x}_i)\boldsymbol{w}_k \sigma''(\boldsymbol{w}_k^\mathsf{T}\boldsymbol{x}_i) + \boldsymbol{w}_k^\mathsf{T}\boldsymbol{A}(\boldsymbol{x}_i)\boldsymbol{w}_k \sigma^{(3)}(\boldsymbol{w}_k^\mathsf{T}\boldsymbol{x}_i)\boldsymbol{x}_i \\
&\quad + \sigma'(\boldsymbol{w}_k^\mathsf{T}\boldsymbol{x}_i)\boldsymbol{b}(\boldsymbol{x}_i) + \boldsymbol{b}^\mathsf{T}(\boldsymbol{x}_i)\boldsymbol{w}_k\sigma''(\boldsymbol{w}_k^\mathsf{T}\boldsymbol{x}_i)\boldsymbol{x}_i + c(\boldsymbol{x}_i)\sigma'(\boldsymbol{w}_k^\mathsf{T}\boldsymbol{x}_i)\boldsymbol{x}_i \Big] \Bigg\|_\infty^2 \\
&\leq |a_k|^2 2R_S(\boldsymbol{\theta}) \Big(2M\|\boldsymbol{w}_k\|_1^2 + M\|\boldsymbol{w}_k\|_1^2 + \frac{1}{2}M\|\boldsymbol{w}_k\|_1^2 \\
&\quad + M\|\boldsymbol{w}_k\|_1^2 + M\frac{1}{2}\|\boldsymbol{w}_k\|_1^2 \Big)^2 \\
&\leq 50M^2\|\boldsymbol{w}_k\|_1^4 |a_k|^2 R_S(\boldsymbol{\theta}) \\
&\leq 50M^2 d^4 (\xi(t))^6 R_S(\boldsymbol{\theta}).
\end{aligned}
$$

From Lemma 4.4, if $m \geq \frac{512M^4n^4C_d}{\lambda_S^2\delta}$, then with probability at least $1-\delta/2$ over initialization

$$
\begin{aligned}
|a_k(t) - a_k(0)| &\leq \int_0^t |\nabla_{a_k} R_S(\boldsymbol{\theta}(s))|\,\mathrm{d}s \\
&\leq 2\sqrt{2}Md^3 \int_0^t \xi^3(t)\sqrt{R_S(\boldsymbol{\theta}(s))}\,\mathrm{d}s \\
&\leq 2\sqrt{2}Md^3\xi^3(t) \int_0^t \sqrt{R_S(\boldsymbol{\theta}^0)}\exp\left(-\frac{m\lambda_S s}{2n}\right)\mathrm{d}s \\
&\leq \frac{4\sqrt{2}Md^3 n\sqrt{R_S(\boldsymbol{\theta}^0)}}{m\lambda_S}\xi^3(t) \\
&\leq p\xi^3(t),
\end{aligned}
$$

where $p := \frac{10\sqrt{2}d^3Mn\sqrt{R_S(\boldsymbol{\theta}^0)}}{m\lambda_S}$. Similarly,

$$
\begin{aligned}
\|\boldsymbol{w}_k(t) - \boldsymbol{w}_k(0)\|_\infty &\leq \int_0^t \|\nabla_{\boldsymbol{w}_k} R_S(\boldsymbol{\theta}(s))\|_\infty\,\mathrm{d}s \\
&\leq 5\sqrt{2}Md^2 \int_0^t \xi^3(t)\sqrt{R_S(\boldsymbol{\theta}(s))}\,\mathrm{d}s \\
&\leq 5\sqrt{2}Md^2\xi^3(t) \int_0^t \sqrt{R_S(\boldsymbol{\theta}^0)}\exp\left(-\frac{m\lambda_S s}{2n}\right)\mathrm{d}s
\end{aligned}
$$

$$
\begin{aligned}
&\leq \frac{10\sqrt{2}Md^2n\sqrt{R_S(\boldsymbol{\theta}^0)}}{m\lambda_S}\xi^3(t)\\
&\leq p\xi^3(t).
\end{aligned}
$$

So

$$
\xi(t) \leq \xi(0) + p\xi^3(t). \tag{4.11}
$$

From Lemma 4.1 with probability at least $1-\delta/2$,

$$
\begin{aligned}
\xi(0) &= \max_{k\in[m]}\{|a_k(0)|, \|\boldsymbol{w}_k(0)\|_\infty\}\\
&\leq \max\left\{\gamma\sqrt{2\log\frac{4m(d+1)}{\delta}}, \sqrt{2\log\frac{4m(d+1)}{\delta}}\right\}\\
&\leq \sqrt{2\log\frac{4m(d+1)}{\delta}} = \eta.
\end{aligned} \tag{4.12}
$$

Since

$$
m \geq \frac{200\sqrt{2}Md^3n\log(4m(d+1)/\delta)\sqrt{R_S(\boldsymbol{\theta}^0)}}{\lambda_S} = 10mp\eta^2,
$$

then $p \leq \frac{1}{10}\left(2\log\frac{4m(d+1)}{\delta}\right)^{-1} = \frac{1}{10}\eta^{-2}$ and $p(2\eta)^2 \leq \frac{2}{5}$. Let

$$
t_0 := \inf\{t \mid \xi(t) > 2\eta\}.
$$

We will prove $t_0 \geq t^*$ by contradiction. Suppose that $t_0 < t^*$. For $t \in [0, t_0)$, by (4.11), (4.12), and $\xi(t) \leq 2\eta$, we have

$$
\xi(t) \leq \eta + p(2\eta)^2\xi(t) \leq \eta + \frac{2}{5}\xi(t),
$$

then

$$
\xi(t) \leq \frac{5}{3}\eta.
$$

After letting $t \to t_0$, the inequality just above contradicts with the definition of t_0. So $t_0 \geq t^*$ and then $\xi(t) \leq 2\eta$ for all $t \in [0, t^*)$. Thus

$$
\begin{aligned}
|a_k(t) - a_k(0)| &\leq 8\eta^3 p\\
\|\boldsymbol{w}_k(t) - \boldsymbol{w}_k(0)\|_\infty &\leq 8\eta^3 p.
\end{aligned}
$$

Finally, notice that

$$
\begin{aligned}
8\eta^3 p &= 8\sqrt{8}\left(\log\frac{4m(d+1)}{\delta}\right)^{3/2}\frac{10\sqrt{2}Md^3 n\sqrt{R_S(\boldsymbol{\theta}^0)}}{m\lambda_S} \\
&= \frac{320Md^3\left(\log\frac{4m(d+1)}{\delta}\right)^{3/2} n\sqrt{R_S(\boldsymbol{\theta}^0)}}{m\lambda_S} \\
&= q,
\end{aligned}
\tag{4.13}
$$

which ends the proof. □

4.3 Proof of Theorem 3.1

Proof of Theorem 3.1. From Lemma 4.4, it is sufficient to prove that the stopping time t^* in Lemma 4.4 is equal to $+\infty$. We will prove this by contradiction.

Suppose $t^* < +\infty$. Note that

$$
\begin{aligned}
&|\boldsymbol{G}_{ij}^{(a)}(\boldsymbol{\theta}(t^*)) - \boldsymbol{G}_{ij}^{(a)}(\boldsymbol{\theta}(0))| \\
&\le \frac{1}{m}\sum_{k=1}^{m}|g^{(a)}(\boldsymbol{w}_k(t^*); \boldsymbol{x}_i, \boldsymbol{x}_j) - g^{(a)}(\boldsymbol{w}_k(0); \boldsymbol{x}_i, \boldsymbol{x}_j)|.
\end{aligned}
\tag{4.14}
$$

By the mean value theorem,

$$
\begin{aligned}
&|g^{(a)}(\boldsymbol{w}_k(t^*); \boldsymbol{x}_i, \boldsymbol{x}_j) - g^{(a)}(\boldsymbol{w}_k(0); \boldsymbol{x}_i, \boldsymbol{x}_j)| \\
&\le \|\nabla g^{(a)}\left(c\boldsymbol{w}_k(t^*) + (1-c)\boldsymbol{w}_k(0); \boldsymbol{x}_i, \boldsymbol{x}_j\right)\|_\infty \|\boldsymbol{w}_k(t^*) - \boldsymbol{w}_k(0)\|_1
\end{aligned}
$$

for some $c \in (0, 1)$. Further computation yields

$$
\begin{aligned}
&\nabla g^{(a)}(\boldsymbol{w}; \boldsymbol{x}_i, \boldsymbol{x}_j) \\
=&\Big[2\boldsymbol{A}(\boldsymbol{x}_i)\boldsymbol{w}\sigma''(\boldsymbol{w}^\top\boldsymbol{x}_i) + \boldsymbol{w}^\top\boldsymbol{A}(\boldsymbol{x}_i)\boldsymbol{w}\sigma^{(3)}(\boldsymbol{w}^\top\boldsymbol{x}_i)\boldsymbol{x}_i + \sigma'(\boldsymbol{w}^\top\boldsymbol{x}_i)\boldsymbol{b}(\boldsymbol{x}_i) \\
&\quad + \boldsymbol{b}^\top(\boldsymbol{x}_i)\boldsymbol{w}\sigma''(\boldsymbol{w}^\top\boldsymbol{x}_i)\boldsymbol{x}_i + c(\boldsymbol{x}_i)\sigma'(\boldsymbol{w}^\top\boldsymbol{x}_i)\boldsymbol{x}_i\Big] \\
&\times\Big[\boldsymbol{w}^\top\boldsymbol{A}(\boldsymbol{x}_j)\boldsymbol{w}\sigma''(\boldsymbol{w}^\top\boldsymbol{x}_j) + \boldsymbol{b}^\top(\boldsymbol{x}_j)\boldsymbol{w}\sigma'(\boldsymbol{w}^\top\boldsymbol{x}_j) + c(\boldsymbol{x}_j)\sigma(\boldsymbol{w}^\top\boldsymbol{x}_j)\Big] \\
&+\Big[2\boldsymbol{A}(\boldsymbol{x}_j)\boldsymbol{w}\sigma''(\boldsymbol{w}^\top\boldsymbol{x}_j) + \boldsymbol{w}^\top\boldsymbol{A}(\boldsymbol{x}_j)\boldsymbol{w}\sigma^{(3)}(\boldsymbol{w}^\top\boldsymbol{x}_j)\boldsymbol{x}_i + \sigma'(\boldsymbol{w}^\top\boldsymbol{x}_i)\boldsymbol{b}(\boldsymbol{x}_i) \\
&\quad + \boldsymbol{b}^\top(\boldsymbol{x}_j)\boldsymbol{w}\sigma''(\boldsymbol{w}^\top\boldsymbol{x}_j)\boldsymbol{x}_j + c(\boldsymbol{x}_j)\sigma'(\boldsymbol{w}^\top\boldsymbol{x}_j)\boldsymbol{x}_j\Big] \\
&\times\Big[\boldsymbol{w}^\top\boldsymbol{A}(\boldsymbol{x}_i)\boldsymbol{w}\sigma''(\boldsymbol{w}^\top\boldsymbol{x}_i) + \boldsymbol{b}^\top(\boldsymbol{x}_i)\boldsymbol{w}\sigma'(\boldsymbol{w}^\top\boldsymbol{x}_i) + c(\boldsymbol{x}_i)\sigma(\boldsymbol{w}^\top\boldsymbol{x}_i)\Big]
\end{aligned}
$$

for all $\boldsymbol{w}$. Hence, it holds for all $\boldsymbol{w}$ that

$$\begin{aligned}
&\|\nabla g^{(a)}(\boldsymbol{w};\boldsymbol{x}_i,\boldsymbol{x}_j)\|_\infty \\
&\le 2\Big[2M\|\boldsymbol{w}\|_1^2 + M\|\boldsymbol{w}\|_1^2 + \frac{1}{2}M\|\boldsymbol{w}\|_1^2 + M\|\boldsymbol{w}\|_1^2 + \frac{1}{2}M\|\boldsymbol{w}\|_1^2\Big] \\
&\quad \times \Big[M\|\boldsymbol{w}\|_1^3 + \frac{1}{2}M\|\boldsymbol{w}\|_1^3 + \frac{1}{6}M\|\boldsymbol{w}\|_1^3\Big] \\
&\le 2(5M\|\boldsymbol{w}\|_1^2)(2M\|\boldsymbol{w}\|_1^3) \\
&= 20M^2\|\boldsymbol{w}\|_1^5.
\end{aligned}$$

Therefore, the bound in (4.14) becomes

$$\begin{aligned}
&|\boldsymbol{G}_{ij}^{(a)}(\boldsymbol{\theta}(t^*)) - \boldsymbol{G}_{ij}^{(a)}(\boldsymbol{\theta}(0))| \\
&\le \frac{20M^2}{m}\sum_{k=1}^{m}\|c\boldsymbol{w}_k(t^*) + (1-c)\boldsymbol{w}_k(0)\|_1^5\|\boldsymbol{w}_k(t^*) - \boldsymbol{w}_k(0)\|_1.
\end{aligned} \tag{4.15}$$

By Lemma 4.5,

$$\begin{aligned}
\|c\boldsymbol{w}_k(t^*) + (1-c)\boldsymbol{w}_k(0)\|_1 &\le \|\boldsymbol{w}_k(0)\|_1 + \|\boldsymbol{w}_k(t^*) - \boldsymbol{w}_k(0)\|_1 \\
&\le d(\eta + q) \le 2d\eta,
\end{aligned}$$

where η and q are defined in Lemma 4.5. So, (4.15) and the above inequalities indicate

$$|\boldsymbol{G}_{ij}^{(a)}(\boldsymbol{\theta}(t^*)) - \boldsymbol{G}_{ij}^{(a)}(\boldsymbol{\theta}(0))| \le 20M^2(2d\eta)^5 dq = 640M^2d^6\eta^5 q,$$

and

$$\begin{aligned}
\|\boldsymbol{G}^{(a)}(\boldsymbol{\theta}(t^*)) - \boldsymbol{G}^{(a)}(\boldsymbol{\theta}(0))\|_{\mathrm{F}} &\le 640M^2d^6 n\eta^5 q \\
&< \frac{2^{21}M^3d^9n^2(\log\frac{4m(d+1)}{\delta})^4\sqrt{R_S(\boldsymbol{\theta}^0)}}{m\lambda_S} \\
&\le \frac{1}{4}\lambda_S,
\end{aligned}$$

if we choose

$$m \ge \frac{2^{23}M^3d^9n^2(\log(4m(d+1)/\delta))^4\sqrt{R_S(\boldsymbol{\theta}^0)}}{\lambda_S^2}.$$

The fact that $\|\boldsymbol{G}^{(a)}(\boldsymbol{\theta}(t^*)) - \boldsymbol{G}^{(a)}(\boldsymbol{\theta}(0))\|_{\mathrm{F}} \le \frac{1}{4}\lambda_S$ above contradicts with the definition of t^* in (4.5). Hence, we have completed the proof. □

5 A priori estimates of generalization error for two-layer neural networks

To obtain good generalization, instead of minimizing R_S, we minimize the regularized risk of $R_S(\boldsymbol{\theta})$:

$$J_{S,\lambda}(\boldsymbol{\theta}) := R_S(\boldsymbol{\theta}) + \frac{\lambda}{\sqrt{n}}\|\boldsymbol{\theta}\|_{\mathcal{P}}^3 \tag{5.1}$$

to obtain

$$\boldsymbol{\theta}_{S,\lambda} = \arg\min_{\boldsymbol{\theta}} J_{S,\lambda}(\boldsymbol{\theta}). \tag{5.2}$$

Our work is inspired by the seminal work in E et al. (2019b,a) and the proof is a variant of the proof therein. But as we shall see, the differential operator increases the technical difficulty in the analysis: extra nonlinearity in the parameters, which makes existing mean field analysis (Mei et al., 2018) not applicable. We will use the path norm defined in Definition 2.2 adaptive to the PDE problem, instead of using the path norm in E et al. (2019b,a) for regression problems. We will show that the PDE solution network $\phi(\boldsymbol{x};\boldsymbol{\theta}_{S,\lambda})$ generalizes well if the true solution is in the Barron-type space defined in Definition 2.3, which is also a variance of the Barron-type space in E et al. (2019b,a). The generalization error is measured in terms of how well $f(\boldsymbol{x};\boldsymbol{\theta}_{S,\lambda}) := \mathcal{L}\phi(\boldsymbol{x};\boldsymbol{\theta}_{S,\lambda}) \approx f(\boldsymbol{x})$ generalizes from the random training samples $S = \{\boldsymbol{x}_i\}_{i=1}^n \subset \Omega$ to arbitrary samples in Ω.

Recall that $f(\boldsymbol{x};\boldsymbol{\theta})$, also denoted as $f_{\boldsymbol{\theta}}(\boldsymbol{x})$, is the result of the differential operator $\mathcal{L}$ acting on a two-layer neural network $\phi(\boldsymbol{x};\boldsymbol{\theta})$ in the domain Ω. In fact, $f(\boldsymbol{x};\boldsymbol{\theta})$ is also a two-layer neural network as explained in (4.2). Hence, the generalization error analysis of deep learning-based PDE solvers is reduced to the generalization analysis of the special two-layer neural network $f(\boldsymbol{x};\boldsymbol{\theta})$ fitting $f(\boldsymbol{x})$. The special structure of $f(\boldsymbol{x};\boldsymbol{\theta})$ leads to significant difficulty in analyzing the generalization error compared to traditional two-layer neural networks in the literature.

We will first summarize and prove several lemmas related to Rademacher complexity in Section 5.1. The proofs of our main theorems for the generalization bound in Theorems 3.2 and 3.3 are presented in Section 5.2.

5.1 Preliminary lemmas of Rademacher complexity

First, we define the set of functions

$$\mathcal{F}_Q = \Bigg\{ f(\boldsymbol{x};\boldsymbol{\theta}) = \sum_{k=1}^m a_k\big[\boldsymbol{w}_k^\intercal \boldsymbol{A}(\boldsymbol{x})\boldsymbol{w}_k\sigma''(\boldsymbol{w}_k^\intercal\boldsymbol{x}) + \boldsymbol{b}^\intercal(\boldsymbol{x})\boldsymbol{w}_k\sigma'(\boldsymbol{w}_k^\intercal\boldsymbol{x}) + c(\boldsymbol{x})\sigma(\boldsymbol{w}_k^\intercal\boldsymbol{x})\big] \mid \|\boldsymbol{\theta}\|_{\mathcal{P}} \le Q \Bigg\}.$$

Second, we estimate the Rademacher complexity of the class of special two-layer neural networks $\mathcal{F}_Q$.

Lemma 5.1 (Rademacher complexity of two-layer neural networks). *The Rademacher complexity of $\mathcal{F}_Q$ over a set of n uniform distributed random samples of Ω, denoted as $S = \{\boldsymbol{x}_1, \ldots, \boldsymbol{x}_n\}$, has an upper bound*

$$\mathrm{Rad}_S(\mathcal{F}_Q) \leq \frac{4MQd^2\sqrt{2\log(2d)}}{\sqrt{n}},$$

where M is the upper bound of the differential operator $\mathcal{L}$ introduced in (3.1).

Proof. Let $\hat{\boldsymbol{w}}_k = \boldsymbol{w}_k/\|\boldsymbol{w}_k\|_1$ for $k = 1, \cdots, m$ and $\boldsymbol{\tau}$ be a random vector in $\mathbb{N}^d$ with i.i.d. entries following the Rademacher distribution. Then

$$\begin{aligned}
& n\mathrm{Rad}_S(\mathcal{F}_Q) \\
&= \mathbb{E}_{\boldsymbol{\tau}}\Bigg\{\sup_{\|\boldsymbol{\theta}\|_{\mathcal{P}}\leq Q} \sum_{i=1}^{n} \tau_i \sum_{k=1}^{m} a_k[\boldsymbol{w}_k^\top \boldsymbol{A}(\boldsymbol{x}_i)\boldsymbol{w}_k\sigma''(\boldsymbol{w}_k^\top \boldsymbol{x}_i) + \boldsymbol{b}^\top(\boldsymbol{x}_i)\boldsymbol{w}_k\sigma'(\boldsymbol{w}_k^\top \boldsymbol{x}_i) \\
&\quad + c(\boldsymbol{x}_i)\sigma(\boldsymbol{w}_k^\top \boldsymbol{x}_i)]\Bigg\} \\
&\leq \mathbb{E}_{\boldsymbol{\tau}}\left[\sup_{\|\boldsymbol{\theta}\|_{\mathcal{P}}\leq Q} \sum_{i=1}^{n} \tau_i \sum_{k=1}^{m} a_k \boldsymbol{w}_k^\top \boldsymbol{A}(\boldsymbol{x}_i)\boldsymbol{w}_k\sigma''(\boldsymbol{w}_k^\top \boldsymbol{x}_i)\right] \\
&\quad + \mathbb{E}_{\boldsymbol{\tau}}\left[\sup_{\|\boldsymbol{\theta}\|_{\mathcal{P}}\leq Q} \sum_{i=1}^{n} \tau_i \sum_{k=1}^{m} a_k \boldsymbol{b}^\top(\boldsymbol{x}_i)\boldsymbol{w}_k\sigma'(\boldsymbol{w}_k^\top \boldsymbol{x}_i)\right] \\
&\quad + \mathbb{E}_{\boldsymbol{\tau}}\left[\sup_{\|\boldsymbol{\theta}\|_{\mathcal{P}}\leq Q} \sum_{i=1}^{n} \tau_i \sum_{k=1}^{m} a_k c(\boldsymbol{x}_i)\sigma(\boldsymbol{w}_k^\top \boldsymbol{x}_i)\right] \\
&=: I_1 + I_2 + I_3.
\end{aligned} \tag{5.3}$$

We first estimate I_1 as follows

$$\begin{aligned}
I_1 &= \mathbb{E}_{\boldsymbol{\tau}}\left[\sup_{\|\boldsymbol{\theta}\|_{\mathcal{P}}\leq Q} \sum_{i=1}^{n} \tau_i \sum_{k=1}^{m} a_k \|\boldsymbol{w}_k\|_1^3 \hat{\boldsymbol{w}}_k^\top \boldsymbol{A}(\boldsymbol{x}_i)\hat{\boldsymbol{w}}_k\sigma''(\hat{\boldsymbol{w}}_k^\top \boldsymbol{x}_i)\right] \\
&\leq \mathbb{E}_{\boldsymbol{\tau}}\left[\sup_{\|\boldsymbol{\theta}\|_{\mathcal{P}}\leq Q, \|\boldsymbol{u}_k\|_1=1, \forall k} \sum_{i=1}^{n} \tau_i \sum_{k=1}^{m} a_k \|\boldsymbol{w}_k\|_1^3 \boldsymbol{u}_k^\top \boldsymbol{A}(\boldsymbol{x}_i)\boldsymbol{u}_k\sigma''(\boldsymbol{u}_k^\top \boldsymbol{x}_i)\right] \\
&\leq \mathbb{E}_{\boldsymbol{\tau}}\left[\sup_{\|\boldsymbol{\theta}\|_{\mathcal{P}}\leq Q, \|\boldsymbol{u}_k\|_1=1, \forall k} \sum_{k=1}^{m} \left|a_k \|\boldsymbol{w}_k\|_1^3\right| \left|\sum_{i=1}^{n} \tau_i \boldsymbol{u}_k^\top \boldsymbol{A}(\boldsymbol{x}_i)\boldsymbol{u}_k\sigma''(\boldsymbol{u}_k^\top \boldsymbol{x}_i)\right|\right] \\
&= \mathbb{E}_{\boldsymbol{\tau}}\left[\sup_{\|\boldsymbol{\theta}\|_{\mathcal{P}}\leq Q, \|\boldsymbol{u}\|_1=1} \sum_{k=1}^{m} |a_k| \|\boldsymbol{w}_k\|_1^3 \left|\sum_{i=1}^{n} \tau_i \boldsymbol{u}^\top \boldsymbol{A}(\boldsymbol{x}_i)\boldsymbol{u}\sigma''(\boldsymbol{u}^\top \boldsymbol{x}_i)\right|\right]
\end{aligned}$$

$$
\begin{aligned}
&\le Q\mathbb{E}_{\boldsymbol{\tau}}\left[\sup_{\|\boldsymbol{u}\|_1\le 1,\|\boldsymbol{p}\|_1\le 1,\|\boldsymbol{q}\|_1\le 1}\left|\sum_{i=1}^{n}\tau_i\boldsymbol{p}^\top A(\boldsymbol{x}_i)\boldsymbol{q}\sigma''(\boldsymbol{u}^\top\boldsymbol{x}_i)\right|\right]\\
&= Q\mathbb{E}_{\boldsymbol{\tau}}\left[\sup_{\|\boldsymbol{u}\|_1\le 1,\|\boldsymbol{p}\|_1\le 1,\|\boldsymbol{q}\|_1\le 1}\left|\boldsymbol{p}^\top\left(\sum_{i=1}^{n}\tau_i A(\boldsymbol{x}_i)\sigma''(\boldsymbol{u}^\top\boldsymbol{x}_i)\right)\boldsymbol{q}\right|\right]\\
&= Q\mathbb{E}_{\boldsymbol{\tau}}\left[\sup_{\|\boldsymbol{u}\|_1\le 1,\|\boldsymbol{p}\|_1\le 1,\|\boldsymbol{q}\|_1\le 1}\sum_{\alpha,\beta=1}^{d}|p_\alpha||q_\beta|\left|\sum_{i=1}^{n}\tau_i A_{\alpha\beta}(\boldsymbol{x}_i)\sigma''(\boldsymbol{u}^\top\boldsymbol{x}_i)\right|\right]\\
&\le Q\mathbb{E}_{\boldsymbol{\tau}}\left[\sup_{\|\boldsymbol{u}\|_1\le 1}\max_{\alpha,\beta\in[d]}\left|\sum_{i=1}^{n}\tau_i A_{\alpha\beta}(\boldsymbol{x}_i)\sigma''(\boldsymbol{u}^\top\boldsymbol{x}_i)\right|\right]\\
&\le Q\mathbb{E}_{\boldsymbol{\tau}}\left[\sup_{\|\boldsymbol{u}\|_1\le 1}\sum_{\alpha,\beta=1}^{d}\left|\sum_{i=1}^{n}\tau_i A_{\alpha\beta}(\boldsymbol{x}_i)\sigma''(\boldsymbol{u}^\top\boldsymbol{x}_i)\right|\right]\\
&\le Q\mathbb{E}_{\boldsymbol{\tau}}\left[\sum_{\alpha,\beta=1}^{d}\sup_{\|\boldsymbol{u}\|_1\le 1}\left|\sum_{i=1}^{n}\tau_i A_{\alpha\beta}(\boldsymbol{x}_i)\sigma''(\boldsymbol{u}^\top\boldsymbol{x}_i)\right|\right]\\
&= Q\sum_{\alpha,\beta=1}^{d}\mathbb{E}_{\boldsymbol{\tau}}\left[\sup_{\|\boldsymbol{u}\|_1\le 1}\left|\sum_{i=1}^{n}\tau_i A_{\alpha\beta}(\boldsymbol{x}_i)\sigma''(\boldsymbol{u}^\top\boldsymbol{x}_i)\right|\right].
\end{aligned}
\tag{5.4}
$$

Note that $\sigma''(\boldsymbol{u}^\top\boldsymbol{x}_i)=0$ for $\boldsymbol{u}=0$ and for any $\boldsymbol{x}_i$. For any $\alpha,\beta\in[d]$, we have

$$
\begin{aligned}
&\mathbb{E}_{\boldsymbol{\tau}}\left[\sup_{\|\boldsymbol{u}\|_1\le 1}\left|\sum_{i=1}^{n}\tau_i A_{\alpha\beta}(\boldsymbol{x}_i)\sigma''(\boldsymbol{u}^\top\boldsymbol{x}_i)\right|\right]\\
&\le \mathbb{E}_{\boldsymbol{\tau}}\left[\sup_{\|\boldsymbol{u}\|_1\le 1}\sum_{i=1}^{n}\tau_i A_{\alpha\beta}(\boldsymbol{x}_i)\sigma''(\boldsymbol{u}^\top\boldsymbol{x}_i)\right]\\
&\quad+\mathbb{E}_{\boldsymbol{\tau}}\left[\sup_{\|\boldsymbol{u}\|_1\le 1}\sum_{i=1}^{n}-\tau_i A_{\alpha\beta}(\boldsymbol{x}_i)\sigma''(\boldsymbol{u}^\top\boldsymbol{x}_i)\right]\\
&=2\mathbb{E}_{\boldsymbol{\tau}}\left[\sup_{\|\boldsymbol{u}\|_1\le 1}\sum_{i=1}^{n}\tau_i A_{\alpha\beta}(\boldsymbol{x}_i)\sigma''(\boldsymbol{u}^\top\boldsymbol{x}_i)\right].
\end{aligned}
\tag{5.5}
$$

Applying Lemma 2.1 with $\psi_i(y_i)=A_{\alpha\beta}(\boldsymbol{x}_i)\sigma''(y_i)$ for $i\in[n]$, whose Lipschitz constant is M, we have for all $\alpha,\beta\in[d]$

$$
\mathbb{E}_{\boldsymbol{\tau}}\left[\sup_{\|\boldsymbol{u}\|_1\le 1}\sum_{i=1}^{n}\tau_i A_{\alpha\beta}(\boldsymbol{x}_i)\sigma''(\boldsymbol{u}^\top\boldsymbol{x}_i)\right]\le M\mathbb{E}_{\boldsymbol{\tau}}\left[\sup_{\|\boldsymbol{u}\|_1\le 1}\sum_{i=1}^{n}\tau_i\boldsymbol{u}^\top\boldsymbol{x}_i\right]. \tag{5.6}
$$

Therefore, combining (5.4), (5.5), and (5.6), we obtain

$$\begin{aligned} I_1 &\le 2MQd^2\mathbb{E}_{\tau}\left[\sup_{\|\boldsymbol{u}\|_1\le 1}\sum_{i=1}^{n}\tau_i\boldsymbol{u}^{\top}\boldsymbol{x}_i\right] \\ &\le 2MQd^2\sqrt{n}\sqrt{2\log(2d)}, \end{aligned}$$

where the last inequality comes from the Rademacher bound for linear predictors in Lemma 2.2.

For I_2 and I_3, we note that $\sigma(z) = \frac{1}{6}z^2\sigma''(z)$ and $\sigma'(z) = \frac{1}{2}z\sigma''(z)$. Then by similar arguments, we have

$$\begin{aligned} I_2 &\le MQd\sqrt{n}\sqrt{2\log(2d)}, \\ I_3 &\le \frac{1}{3}MQ\sqrt{n}\sqrt{2\log(2d)}. \end{aligned}$$

These estimates for I_1, I_2, I_3 combined with (5.3) complete the proof. □

5.2 Proofs of generalization bounds

In the proofs of this section, we will first show in Proposition 5.1 that two-layer neural networks $f(\boldsymbol{x};\boldsymbol{\theta})$ in (4.2) can approximate Barron-type functions with an approximation error $O\left(\frac{\|f\|_{\mathcal{B}}^2}{m}\right)$. Second, for an arbitrary $f(\boldsymbol{x};\boldsymbol{\theta}) = \mathcal{L}\phi(\boldsymbol{x};\boldsymbol{\theta})$, we show its a posteriori generalization bound $|R_{\mathcal{D}}(\boldsymbol{\theta}) - R_S(\boldsymbol{\theta})| \le O\left(\frac{\|\boldsymbol{\theta}\|_{\mathcal{P}}^2\log\|\boldsymbol{\theta}\|_{\mathcal{P}}}{\sqrt{n}}\right)$ in Theorem 3.2. Finally, the a priori generalization bound $R_{\mathcal{D}}(\boldsymbol{\theta}_{S,\lambda}) \le O\left(\frac{\|f\|_{\mathcal{B}}^2}{m} + \frac{\|f\|_{\mathcal{B}}^2\log\|f\|_{\mathcal{B}}}{\sqrt{n}}\right)$ is proved in Theorem 3.3, where the first and second terms come from the approximation error bound and the a posteriori generalization bound.

First, the approximation capacity of two-layer neural networks $f(\boldsymbol{x};\boldsymbol{\theta})$ can be characterized by Proposition 5.1 below.

Proposition 5.1 (Approximation Error). *For any $f \in \mathcal{B}(\Omega)$, there exists a two-layer neural network $f(\boldsymbol{x};\tilde{\boldsymbol{\theta}})$ of width m with $\|\tilde{\boldsymbol{\theta}}\|_{\mathcal{P}} \le 2\|f\|_{\mathcal{B}}$,*

$$R_{\mathcal{D}}(\tilde{\boldsymbol{\theta}}) := \mathbb{E}_{\boldsymbol{x}\sim\mathcal{D}}\tfrac{1}{2}(f(\boldsymbol{x},\tilde{\boldsymbol{\theta}}) - f(\boldsymbol{x}))^2 \le \frac{6M^2\|f\|_{\mathcal{B}}^2}{m},$$

where M introduced in (3.1) *controls the upper bound of the differential operator and m is the width of the neural network.*

Proof. Without loss of generality, let ρ be the best representation, i.e., $\|f\|_{\mathcal{B}}^2 = \mathbb{E}_{(a,\boldsymbol{w})\sim\rho}|a|^2\|\boldsymbol{w}\|_1^6$. We set $\bar{\boldsymbol{\theta}} = \{\frac{1}{m}a_k, \boldsymbol{w}_k\}_{k=1}^m$, where $(a_k, \boldsymbol{w}_k)$, $k = 1, \cdots, m$ are independent sampled from ρ. Let

$$f_{\bar{\theta}}(x) = \frac{1}{m}\sum_{k=1}^{m} a_k[w_k^\top A(x)w_k\sigma''(w_k^\top x) + b^\top(x)w_k\sigma'(w_k^\top x) + c(x)\sigma(w_k^\top x)].$$

Recall the definition $R_{\mathcal{D}}(\bar{\theta}) = \mathbb{E}_{x\sim\mathcal{D}}\frac{1}{2}|f_{\bar{\theta}}(x) - f(x)|^2$. Then

$$\begin{aligned}
&2\mathbb{E}_{\bar{\theta}}R_{\mathcal{D}}(\bar{\theta})\\
&= \mathbb{E}_{x\sim\mathcal{D}}\mathbb{E}_{\bar{\theta}}|f_{\bar{\theta}}(x) - f(x)|^2\\
&= \mathbb{E}_{x\sim\mathcal{D}}\mathrm{Var}_{\{(a_k,w_k)\}\text{i.i.d.}\sim\rho}\left(\frac{1}{m}\sum_{k=1}^{m} a_k[w_k^\top A(x)w_k\sigma''(w_k^\top x)\right.\\
&\qquad\left. + b^\top(x)w_k\sigma'(w_k^\top x) + c(x)\sigma(w_k^\top x)]\right)\\
&= \mathbb{E}_{x\sim\mathcal{D}}\frac{1}{m}\mathrm{Var}_{(a,w)\sim\rho}\big(a[w^\top A(x)w\sigma''(w^\top x) + b^\top(x)w\sigma'(w^\top x)\\
&\qquad + c(x)\sigma(w^\top x)]\big)\\
&\le \frac{1}{m}\mathbb{E}_{x\sim\mathcal{D}}\mathbb{E}_{(a,w)\sim\rho}\big(a[w^\top A(x)w\sigma''(w^\top x) + b^\top(x)w\sigma'(w^\top x)\\
&\qquad + c(x)\sigma(w^\top x)]\big)^2\\
&\le \frac{1}{m}\mathbb{E}_{x\sim\mathcal{D}}\mathbb{E}_{(a,w)\sim\rho}|a|^2\left(M\|w\|_1^3 + \tfrac{1}{2}M\|w\|_1^3 + \tfrac{1}{6}M\|w\|_1^3\right)^2\\
&\le \frac{4M^2}{m}\mathbb{E}_{(a,w)\sim\rho}|a|^2\|w\|_1^6\\
&= \frac{4M^2\|f\|_{\mathcal{B}}^2}{m}.
\end{aligned}$$

Also, we have

$$\begin{aligned}
\mathbb{E}_{\bar{\theta}}\|\bar{\theta}\|_{\mathcal{P}} &= \mathbb{E}_{\{(a_k,w_k)\}\text{i.i.d.}\sim\rho}\frac{1}{m}\sum_{k=1}^{m}|a_k|\|w_k\|_1^3\\
&= \mathbb{E}_{(a,w)\sim\rho}|a|\|w\|_1^3\\
&\le \|f\|_{\mathcal{B}}.
\end{aligned}$$

Define two events $E_1 := \{R_{\mathcal{D}}(\bar{\theta}) < \frac{6M^2\|f\|_{\mathcal{B}}^2}{m}\}$ and $E_2 := \{\|\bar{\theta}\|_{\mathcal{P}} < 2\|f\|_{\mathcal{B}}\}$. By Markov inequality, we have

$$\mathbb{P}(E_1) = 1 - \mathbb{P}\left(R_{\mathcal{D}}(\bar{\theta}) \ge \frac{6M^2\|f\|_{\mathcal{B}}^2}{m}\right) \ge 1 - \frac{\mathbb{E}_{\bar{\theta}}R_{\mathcal{D}}(\bar{\theta})}{6M^2\|f\|_{\mathcal{B}}^2/m} \ge \frac{2}{3},$$

$$\mathbb{P}(E_2) = 1 - \mathbb{P}(\|\bar{\theta}\|_{\mathcal{P}} \ge 2\|f\|_{\mathcal{B}}) \ge 1 - \frac{\mathbb{E}_{\bar{\theta}}\|\bar{\theta}\|_{\mathcal{P}}}{2\|f\|_{\mathcal{B}}} \ge \frac{1}{2}.$$

Thus

$$\mathbb{P}(E_1 \cap E_2) \geq \mathbb{P}(E_1) + \mathbb{P}(E_2) - 1 \geq \frac{2}{3} + \frac{1}{2} - 1 > 0. \qquad \square$$

Second, we use Theorem 2.1 with $\mathcal{F} = \mathcal{H}_Q := \{\ell(f(\boldsymbol{x}), f_{\boldsymbol{\theta}}(\boldsymbol{x})) \mid \|\boldsymbol{\theta}\|_{\mathcal{P}} \leq Q\}$ and $\mathcal{Z} = \Omega$ to show the a posteriori generalization bound in Theorem 3.2.

Proof of Theorem 3.2. Let $\mathcal{H}_Q := \{\ell(f(\boldsymbol{x}), f_{\boldsymbol{\theta}}(\boldsymbol{x})) \mid \|\boldsymbol{\theta}\|_{\mathcal{P}} \leq Q\}$, then $\mathcal{H} = \cup_{Q=1}^{\infty} \mathcal{H}_Q$. Note that

$$\begin{aligned}
&\sup_{\boldsymbol{x}\in\Omega} |f_{\boldsymbol{\theta}}(\boldsymbol{x})| \\
&= \sup_{\boldsymbol{x}\in\Omega} \left| \sum_{k=1}^{m} a_k [\boldsymbol{w}_k^{\mathsf{T}} \boldsymbol{A}(\boldsymbol{x}) \boldsymbol{w}_k \sigma''(\boldsymbol{w}_k^{\mathsf{T}} \boldsymbol{x}) + \boldsymbol{b}^{\mathsf{T}}(\boldsymbol{x}) \boldsymbol{w}_k \sigma'(\boldsymbol{w}_k^{\mathsf{T}} \boldsymbol{x}) + c(\boldsymbol{x}) \sigma(\boldsymbol{w}_k^{\mathsf{T}} \boldsymbol{x})] \right| \\
&\leq \sum_{k=1}^{m} |a_k| \|\boldsymbol{w}_k\|_1^3 \left[M + \frac{1}{2} M + \frac{1}{6} M \right] \\
&\leq \frac{5}{3} M \|\boldsymbol{\theta}\|_{\mathcal{P}}.
\end{aligned}$$

Therefore, for functions in $\mathcal{H}_Q$, since $|f(x)| \leq 1$ by assumption, we have

$$\begin{aligned}
0 \leq \ell(f(\boldsymbol{x}), f_{\boldsymbol{\theta}}(\boldsymbol{x})) &\leq \frac{1}{2}(1 + |f_{\boldsymbol{\theta}}(\boldsymbol{x})|)^2 \\
&\leq \frac{1}{2}\left(1 + \frac{5}{3} M \|\boldsymbol{\theta}\|_{\mathcal{P}}\right)^2 \\
&\leq \frac{32}{9} M^2 Q^2 \leq 4 M^2 Q^2
\end{aligned}$$

for all $\boldsymbol{x} \in \Omega$ and all $Q \geq 1$. For $\|\boldsymbol{\theta}\|_{\mathcal{P}} \leq Q$, we note that $\ell(y, \cdot)$ is a Lipschitz function with a Lipschitz constant which is no larger than $\sup_{\boldsymbol{x}\in\Omega} |f_{\boldsymbol{\theta}}(\boldsymbol{x})| \leq \frac{5}{3} M \|\boldsymbol{\theta}\|_{\mathcal{P}} + 1$. Let S' be an arbitrary set of n samples of Ω, then

$$\mathrm{Rad}_{S'}(\mathcal{H}_Q) \leq (\frac{5}{3} M \|\boldsymbol{\theta}\|_{\mathcal{P}} + 1) \mathrm{Rad}_{S'}(\mathcal{F}_Q) \leq (\frac{5}{3} M Q + 1) \mathrm{Rad}_{S'}(\mathcal{F}_Q).$$

Let us assume $MQ \geq \frac{3}{5}$ without loss of generality. By Lemma 5.1 and Theorem 2.1, for any δ given in Theorem 3.2 and any positive integer Q with probability at least $1 - \delta_Q$ over S with $\delta_Q = \frac{6\delta}{\pi^2 Q^2}$, we have

$$\begin{aligned}
&\sup_{\|\boldsymbol{\theta}\|_{\mathcal{P}} \leq Q} |R_{\mathcal{D}}(\boldsymbol{\theta}) - R_S(\boldsymbol{\theta})| \\
&\leq (\frac{5}{3} M Q + 1) 2 \mathbb{E}_{S'} \mathrm{Rad}_{S'}(\mathcal{F}_Q) + 4 M^2 Q^2 \sqrt{\frac{\log(2/\delta_Q)}{2n}}
\end{aligned}$$

$$\leq 27M^2Q^2d^2\sqrt{\frac{2\log(2d)}{n}} + 4M^2Q^2\sqrt{\frac{\log(\pi^2Q^2/3\delta)}{2n}}.$$

For any $\boldsymbol{\theta} \in \mathbb{R}^{m(d+1)}$ given in Theorem 3.2, choose the integer Q such that $\|\boldsymbol{\theta}\|_{\mathcal{P}} \leq Q \leq \|\boldsymbol{\theta}\|_{\mathcal{P}} + 1$. Then we have

$$\begin{aligned}
|R_{\mathcal{D}}(\boldsymbol{\theta}) - R_S(\boldsymbol{\theta})| &\leq 27M^2Q^2d^2\sqrt{\frac{2\log(2d)}{n}} + 4M^2Q^2\sqrt{\frac{\log(\pi^2Q^2/3\delta)}{2n}} \\
&\leq 27M^2(\|\boldsymbol{\theta}\|_{\mathcal{P}}+1)^2d^2\sqrt{\frac{2\log(2d)}{n}} \\
&\quad + 4M^2(\|\boldsymbol{\theta}\|_{\mathcal{P}}+1)^2\sqrt{\frac{\log\pi(\|\boldsymbol{\theta}\|_{\mathcal{P}}+1)}{n} + \frac{\log(1/3\delta)}{2n}} \\
&\leq 27M^2(\|\boldsymbol{\theta}\|_{\mathcal{P}}+1)^2d^2\sqrt{\frac{2\log(2d)}{n}} \\
&\quad + 4M^2(\|\boldsymbol{\theta}\|_{\mathcal{P}}+1)^2\left\{\frac{\log[\pi(\|\boldsymbol{\theta}\|_{\mathcal{P}}+1)]}{\sqrt{n}} + \sqrt{\frac{\log(1/3\delta)}{2n}}\right\} \\
&\leq \frac{(\|\boldsymbol{\theta}\|_{\mathcal{P}}+1)^2}{\sqrt{n}}2M^2(14d^2\sqrt{2\log(2d)} \\
&\quad + \log[\pi(\|\boldsymbol{\theta}\|_{\mathcal{P}}+1)] + \sqrt{2\log(1/3\delta)}),
\end{aligned}$$

where we have used the facts that $\sqrt{a+b} \leq \sqrt{a} + \sqrt{b}$ for $a, b > 0$ and that $\sqrt{a} \leq a$ for $a \geq 1$.

The bound just above holds with probability $1 - \delta_Q$ for any pair $(\boldsymbol{\theta}, Q)$ as long as $\|\boldsymbol{\theta}\|_{\mathcal{P}} \leq Q$. By the definition $\delta_Q = \frac{6\delta}{\pi^2Q^2}$, we have $\sum_{Q=1}^{\infty}\delta_Q = \delta$. Therefore, for any $\boldsymbol{\theta} \in \mathbb{R}^{m(d+1)}$ given in Theorem 3.2, the above bound holds with probability $1 - \delta$, which finishes the proof of Theorem 3.2. □

Finally, based on the approximation bound in Proposition 5.1 and the a posteriori generalization bound in Theorem 3.2, we show the a priori generalization bound in Theorem 3.3.

Proof of Theorem 3.3. Note that

$$\begin{aligned}
R_{\mathcal{D}}(\boldsymbol{\theta}_{S,\lambda}) &= R_{\mathcal{D}}(\tilde{\boldsymbol{\theta}}) + [R_{\mathcal{D}}(\boldsymbol{\theta}_{S,\lambda}) - J_{S,\lambda}(\boldsymbol{\theta}_{S,\lambda})] + [J_{S,\lambda}(\boldsymbol{\theta}_{S,\lambda}) - J_{S,\lambda}(\tilde{\boldsymbol{\theta}})] \\
&\quad + [J_{S,\lambda}(\tilde{\boldsymbol{\theta}}) - R_{\mathcal{D}}(\tilde{\boldsymbol{\theta}})].
\end{aligned}$$

By definition, $J_{S,\lambda}(\boldsymbol{\theta}_{S,\lambda}) - J_{S,\lambda}(\tilde{\boldsymbol{\theta}}) \leq 0$. By Proposition 5.1, there exists $\tilde{\boldsymbol{\theta}}$ such that $R_{\mathcal{D}}(\tilde{\boldsymbol{\theta}}) \leq \frac{6M^2\|f\|_{\mathcal{B}}^2}{m}$. Therefore,

$$R_{\mathcal{D}}(\boldsymbol{\theta}_{S,\lambda}) \leq \frac{6M^2\|f\|_{\mathcal{B}}^2}{m} + [R_{\mathcal{D}}(\boldsymbol{\theta}_{S,\lambda}) - J_{S,\lambda}(\boldsymbol{\theta}_{S,\lambda})] + [J_{S,\lambda}(\tilde{\boldsymbol{\theta}}) - R_{\mathcal{D}}(\tilde{\boldsymbol{\theta}})]. \quad (5.7)$$

By Theorem 3.2, we have with probability at least $1-\delta/2$,

$$
\begin{aligned}
&R_{\mathcal{D}}(\boldsymbol{\theta}_{S,\lambda}) - J_{S,\lambda}(\boldsymbol{\theta}_{S,\lambda}) \\
&= R_{\mathcal{D}}(\boldsymbol{\theta}_{S,\lambda}) - R_S(\boldsymbol{\theta}_{S,\lambda}) - \frac{\lambda}{\sqrt{n}}\|\boldsymbol{\theta}_{S,\lambda}\|_{\mathcal{P}}^2 \log[\pi(\|\boldsymbol{\theta}_{S,\lambda}\|_{\mathcal{P}}+1)] \\
&\le \frac{1}{\sqrt{n}} 2M^2(\|\boldsymbol{\theta}_{S,\lambda}\|_{\mathcal{P}}+1)^2\{\log[\pi(\|\boldsymbol{\theta}_{S,\lambda}\|_{\mathcal{P}}+1)] + 14d^2\sqrt{2\log(2d)} \\
&\quad + \sqrt{2\log(2/3\delta)}\} - \frac{\lambda}{\sqrt{n}}\|\boldsymbol{\theta}_{S,\lambda}\|_{\mathcal{P}}^2 \log[\pi(\|\boldsymbol{\theta}_{S,\lambda}\|_{\mathcal{P}}+1)] \\
&\le \frac{1}{\sqrt{n}} 4M^2(\|\boldsymbol{\theta}_{S,\lambda}\|_{\mathcal{P}}^2+1)\{\log[\pi(\|\boldsymbol{\theta}_{S,\lambda}\|_{\mathcal{P}}+1)] + 14d^2\sqrt{2\log(2d)} \\
&\quad + \sqrt{2\log(2/3\delta)}\} - \frac{\lambda}{\sqrt{n}}\|\boldsymbol{\theta}_{S,\lambda}\|_{\mathcal{P}}^2 \log[\pi(\|\boldsymbol{\theta}_{S,\lambda}\|_{\mathcal{P}}+1)] \\
&\le \frac{1}{\sqrt{n}}\|\boldsymbol{\theta}_{S,\lambda}\|_{\mathcal{P}}^2 \log[\pi(\|\boldsymbol{\theta}_{S,\lambda}\|_{\mathcal{P}}+1)] \\
&\qquad \times \left\{4M^2[1 + 14d^2\sqrt{2\log(2d)} + \sqrt{2\log(2/3\delta)}] - \lambda\right\} \\
&\quad + \frac{4M^2}{\sqrt{n}} \log[\pi(\|\boldsymbol{\theta}_{S,\lambda}\|_{\mathcal{P}}+1)] + \frac{1}{\sqrt{n}} 4M^2(14d^2\sqrt{2\log(2d)} + \sqrt{2\log(2/3\delta)}) \\
&\le \frac{1}{\sqrt{n}}\|\boldsymbol{\theta}_{S,\lambda}\|_{\mathcal{P}}^2 \log[\pi(\|\boldsymbol{\theta}_{S,\lambda}\|_{\mathcal{P}}+1)] \\
&\qquad \times \left\{4M^2[2 + 14d^2\sqrt{2\log(2d)} + \sqrt{2\log(2/3\delta)}] - \lambda\right\} \\
&\quad + \frac{1}{\sqrt{n}} 4M^2\left[\log(2\pi) + 14d^2\sqrt{2\log(2d)} + \sqrt{2\log(2/3\delta)}\right] \\
&\le \frac{1}{\sqrt{n}} 4M^2\left[\log(2\pi) + 14d^2\sqrt{2\log(2d)} + \sqrt{2\log(2/3\delta)}\right],
\end{aligned} \tag{5.8}
$$

where we have used the facts that $(a+b)^2 \le 2a^2+2b^2$ for all $a, b \ge 0$ and that $\lambda \ge 4M^2[2 + 14d^2\sqrt{2\log(2d)} + \sqrt{2\log(2/3\delta)}]$ in the second and last inequalities, respectively. By Theorem 3.2 again, with probability at least $1-\delta/2$, we have

$$
\begin{aligned}
J_{S,\lambda}(\tilde{\boldsymbol{\theta}}) - R_{\mathcal{D}}(\tilde{\boldsymbol{\theta}}) &\le \frac{1}{\sqrt{n}} 2M^2(\|\tilde{\boldsymbol{\theta}}\|_{\mathcal{P}}+1)^2\{\log[\pi(\|\tilde{\boldsymbol{\theta}}\|_{\mathcal{P}}+1)] \\
&\quad + 14d^2\sqrt{2\log(2d)} + \sqrt{2\log(2/3\delta)}\} \\
&\quad + \frac{\lambda}{\sqrt{n}}\|\tilde{\boldsymbol{\theta}}\|_{\mathcal{P}}^2 \log[\pi(\|\tilde{\boldsymbol{\theta}}\|_{\mathcal{P}}+1)] \\
&\le \frac{1}{\sqrt{n}} 4M^2(\|\tilde{\boldsymbol{\theta}}\|_{\mathcal{P}}^2+1)\{\log[\pi(\|\tilde{\boldsymbol{\theta}}\|_{\mathcal{P}}+1)] \\
&\quad + 14d^2\sqrt{2\log(2d)} + \sqrt{2\log(2/3\delta)}\}
\end{aligned}
$$

$$+\frac{\lambda}{\sqrt{n}}\|\tilde{\boldsymbol{\theta}}\|_{\mathcal{P}}^2 \log[\pi(\|\tilde{\boldsymbol{\theta}}\|_{\mathcal{P}}+1)]. \tag{5.9}$$

Note that, by Proposition 5.1, we have $\|\tilde{\boldsymbol{\theta}}\|_{\mathcal{P}} \leq 2\|f\|_{\mathcal{B}}$. Hence, the inequality (5.9) becomes

$$\begin{aligned} J_{S,\lambda}(\tilde{\boldsymbol{\theta}}) - R_{\mathcal{D}}(\tilde{\boldsymbol{\theta}}) \leq & \frac{1}{\sqrt{n}} 4M^2(4\|f\|_{\mathcal{B}}^2+1)\{\log[\pi(2\|f\|_{\mathcal{B}}+1)] \\ & + 14d^2\sqrt{2\log(2d)} + \sqrt{2\log(2/3\delta)}\} \\ & + \frac{4\lambda}{\sqrt{n}}\|f\|_{\mathcal{B}}^2 \log[\pi(2\|f\|_{\mathcal{B}}+1)]. \end{aligned} \tag{5.10}$$

Adding the estimates in (5.7), (5.8), and (5.9) together completes the proof. □

6 Conclusion

In this paper, we theoretically analyzed the optimization problem arising in deep learning-based PDE solvers for second-order linear PDEs and two-layer neural networks under the assumption of overparametrization (i.e., the network width is sufficiently large). In particular, we show that gradient descent can identify a global minimizer of the least-squares optimization problem for solving second-order linear PDEs. Note that we have fixed the samples in the least-squares optimization, while practical algorithms would randomly sample the PDE domain and its boundaries in every iteration of gradient descent. Hence, there is still a gap between the optimization problem analyzed in this paper and the practical algorithm. This gap can be filled by studying the convergence behavior of stochastic gradient descent, which will be left as future work.

We have also analyzed the generalization error of deep learning-based PDE solvers for second-order linear PDEs and two-layer neural networks, when the right-hand-side function of the PDE is in a Barron-type space and the least-squares optimization is regularized with a Barron-type norm, without the overparametrization assumption. The Barron-type space and norm are adaptive to PDE problems and are different from those for regression problems. The global minimizer of the regularized least-squares problem can generalize well with a scaling of order $\frac{1}{m}+\frac{1}{\sqrt{n}}$, where m is the number of neurons and n is the number of data samples. Note that whether gradient descent methods can identify a global minimizer of the regularized least-squares problem is still unknown. This is left as interesting future work.

Acknowledgments

H. Y. was partially supported by the US National Science Foundation under awards DMS-2244988, DMS-2206333, and the Office of Naval Research Award N00014-23-1-2007.

References

Barron, A.R., 1993. Universal approximation bounds for superpositions of a sigmoidal function. IEEE Trans. Inf. Theory 39 (3), 930–945.

Berg, J., Nyström, K., 2018. A unified deep artificial neural network approach to partial differential equations in complex geometries. Neurocomputing 317, 28–41.

Berner, Julius, Grohs, Philipp, Jentzen, Arnulf, 2018. Analysis of the generalization error: empirical risk minimization over deep artificial neural networks overcomes the curse of dimensionality in the numerical approximation of Black-Scholes partial differential equations. CoRR. arXiv: 1809.03062.

Cao, Yuan, Gu, Quanquan, 2019. Generalization bounds of stochastic gradient descent for wide and deep neural networks. CoRR. arXiv:1905.13210.

Carleo, G., Troyer, M., 2017. Solving the quantum many-body problem with artificial neural networks. Science 355, 602–606.

Chen, Liang, Wu, Congwei, 2019. A note on the expressive power of deep rectified linear unit networks in high-dimensional spaces. Math. Methods Appl. Sci. 42 (9), 3400–3404.

Chen, Zixiang, Cao, Yuan, Zou, Difan, Gu, Quanquan, 2019. How much over-parameterization is sufficient to learn deep relu networks? CoRR. arXiv:1911.12360.

Dissanayake, M.W.M.G., Phan-Thien, N., 1994. Neural-network-based approximations for solving partial differential equations. Commun. Numer. Methods Eng. 10, 195–201.

Du, Simon S., Lee, Jason D., Li, Haochuan, Wang, Liwei, Zhai, Xiyu, 2018. Gradient descent finds global minima of deep neural networks. CoRR. arXiv:1811.03804.

Du, Simon S., Zhai, Xiyu, Poczos, Barnabas, Singh, Aarti, 2019. Gradient descent provably optimizes over-parameterized neural networks. In: International Conference on Learning Representations.

E, W., Yu, B., 2018. The deep Ritz method: a deep learning-based numerical algorithm for solving variational problems. Commun. Math. Stat. 6, 1–12.

E, Weinan, Han, Jiequn, Jentzen, Arnulf, 2017. Deep learning-based numerical methods for high-dimensional parabolic partial differential equations and backward stochastic differential equations. Commun. Math. Stat. 5 (4), 349–380.

E, Weinan, Ma, Chao, Wang, Qingcan, 2019a. A priori estimates of the population risk for residual networks.

E, Weinan, Ma, Chao, Wu, Lei, 2019b. A priori estimates of the population risk for two-layer neural networks. Commun. Math. Sci. 17 (5), 1407–1425.

E, Weinan, Ma, Chao, Wu, Lei, 2019c. Barron spaces and the compositional function spaces for neural network models. ArXiv. arXiv:1906.08039 [abs].

E, Weinan, Ma, Chao, Wu, Lei, 2020. A comparative analysis of optimization and generalization properties of two-layer neural network and random feature models under gradient descent dynamics. Sci. China Math. 63 (7), 1235–1258.

E, Weinan, Wang, Qingcan, 2018. Exponential convergence of the deep neural network approximation for analytic functions. CoRR. arXiv:1807.00297.

Goodfellow, I., Bengio, Y., Courville, A., 2016. Deep Learning. MIT Press, Cambridge.

Gu, Yiqi, Wang, Chunmei, Yang, Haizhao, 2020a. Structure probing neural network deflation. CoRR.

Gu, Yiqi, Yang, Haizhao, Zhou, Chao, 2020b. Selectnet: self-paced learning for high-dimensional partial differential equations. CoRR. arXiv:2001.04860.

Han, J., Jentzen, A., E, W., 2018. Solving high-dimensional partial differential equations using deep learning. Proc. Natl. Acad. Sci. USA 115, 8505–8510.

Han, Jiequn, Long, Jihao, 2018. Convergence of the deep bsde method for coupled fbsdes. ArXiv. arXiv:1811.01165 [abs].

He, Kaiming, Zhang, Xiangyu, Ren, Shaoqing, Sun, Jian, 2015. Deep residual learning for image recognition. CoRR. arXiv:1512.03385.

Huang, Gao, Liu, Zhuang, Weinberger, Kilian Q., 2016. Densely connected convolutional networks. CoRR. arXiv:1608.06993.

Huang, Jianguo, Wang, Haoqin, Yang, Haizhao, 2020. Int-deep: a deep learning initialized iterative method for nonlinear problems. J. Comput. Phys., 109675.

Hutzenthaler, M., Jentzen, A., Kruse, Th., Nguyen, T.A., 2019. A proof that rectified deep neural networks overcome the curse of dimensionality in the numerical approximation of semilinear heat equations. Technical Report 2019-10, Seminar for Applied Mathematics. ETH Zürich, Switzerland.

Jacot, Arthur, Gabriel, Franck, Hongler, Clément, 2018. Neural tangent kernel: convergence and generalization in neural networks. CoRR. arXiv:1806.07572.

Khoo, Y., Lu, J., Ying, L., 2019. Solving for high-dimensional committor functions using artificial neural networks. Res. Math. Sci. 6, 1–13.

Lagaris, I.E., Likas, A., Fotiadis, D.I., 1998. Artificial neural networks for solving ordinary and partial differential equations. IEEE Trans. Neural Netw. 9, 987–1000.

Liang, Shiyu, Srikant, R., 2016. Why deep neural networks? CoRR. arXiv:1610.04161.

Liao, Y., Ming, P., 2019. Deep Nitsche method: deep Ritz method with essential boundary conditions. arXiv e-prints. arXiv:1912.01309.

Lu, Jianfeng, Shen, Zuowei, Yang, Haizhao, Zhang, Shijun, 2020a. Deep network approximation for smooth functions. arXiv e-prints. arXiv:2001.03040.

Lu, Yiping, Ma, Chao, Lu, Yulong, Lu, Jianfeng, Ying, Lexing, 2020b. A mean-field analysis of deep resnet and beyond: towards provable optimization via overparameterization from depth. CoRR. arXiv:2003.05508.

Mei, Song, Montanari, Andrea, Nguyen, Phan-Minh, 2018. A mean field view of the landscape of two-layer neural networks. Proc. Natl. Acad. Sci. 115 (33), E7665–E7671.

Montanelli, Hadrien, Du, Qiang, 2019. New error bounds for deep relu networks using sparse grids. SIAM J. Math. Data Sci. 1 (1), 78–92.

Montanelli, Hadrien, Yang, Haizhao, 2020. Error bounds for deep relu networks using the Kolmogorov–Arnold superposition theorem. Neural Netw. 129, 1–6.

Montanelli, Hadrien, Yang, Haizhao, Du, Qiang, 2019. Deep ReLU networks overcome the curse of dimensionality for bandlimited functions.

Opschoor, Joost A.A., Schwab, Christoph, Zech, Jakob, 2019. Exponential relu dnn expression of holomorphic maps in high dimension. Technical report, Zurich.

Poggio, T., Mhaskar, H.N., Rosasco, L., Miranda, B., Liao, Q., 2017. Why and when can deep—but not shallow—networks avoid the curse of dimensionality: a review. Int. J. Autom. Comput. 14, 503–519.

Raissi, M., Perdikaris, P., Karniadakis, G.E., 2019. Physics-informed neural networks: a deep learning framework for solving forward and inverse problems involving nonlinear partial differential equations. J. Comput. Phys. 378, 686–707.

Rudd, K., Ferrari, S., 2015. A constrained integration (CINT) approach to solving partial differential equations using artificial neural networks. Neurocomputing 155, 277–285.

Shalev-Shwartz, S., Ben-David, S., 2014. Understanding Machine Learning: From Theory to Algorithms. Cambridge University Press.

Shen, Zuowei, Yang, Haizhao, Zhang, Shijun, 2019. Nonlinear approximation via compositions. Neural Netw. 119, 74–84.

Shen, Zuowei, Yang, Haizhao, Zhang, Shijun, 2020. Deep network approximation characterized by number of neurons. Commun. Comput. Phys. 28 (5), 1768–1811.

Shen, Zuowei, Yang, Haizhao, Zhang, Shijun, 2021a. Neural network approximation: three hidden layers are enough. Neural Netw. 141, 160–173.

Shen, Zuowei, Yang, Haizhao, Zhang, Shijun, 2021b. Deep network approximation with discrepancy being reciprocal of width to power of depth. Neural Comput. 33 (5), 1005–1036.

Shin, Yeonjong, Darbon, Jerome, Karniadakis, George Em, 2020. On the convergence of physics informed neural networks for linear second-order elliptic and parabolic type pdes.

Sirignano, J., Spiliopoulos, K., 2018. DGM: a deep learning algorithm for solving partial differential equations. J. Comput. Phys. 375, 1339–1364.

Yang, Yunfei, Wang, Yang, 2020. Approximation in shift-invariant spaces with deep ReLU neural networks. arXiv e-prints. arXiv:2005.11949.
Yarotsky, Dmitry, 2017. Error bounds for approximations with deep ReLU networks. Neural Netw. 94, 103–114.
Yarotsky, Dmitry, Zhevnerchuk, Anton, 2019. The phase diagram of approximation rates for deep neural networks. arXiv e-prints. arXiv:1906.09477.
Zhang, Y., Xu, Z.-Q.J., Luo, T., Ma, Z., 2019. A type of generalization error induced by initialization in deep neural networks. arXiv e-prints. arXiv:1905.07777.

Index

D

E

F

O

P

Q

R

T

U

V

W

9780443239847